# THE CAMBRIDGE HISTORY OF WAR

Volume III of *The Cambridge History of War* covers the early modern world, offering a 400-year perspective from the last Eurasian nomadic empires to the advent of ironclad, steam-driven warships in the mid-nineteenth century. Together, the chapters cover the rise of professional armies and purpose-built warships in Europe; the evolution of military societies in the great Islamic empires; the vicissitudes of Ming and Qing military organization and that of their Asian neighbors; and the raising and maintaining of armies in Africa and the Americas. Numerous processes of imperial expansion, both on land and sea, are examined, as are the processes of global confrontation and interchange across different military systems. Technology, organization, finance, and military cultures are each explored from a broad perspective. Bringing together an impressive team of experts in their fields, the volume provides a comprehensive and accessible history of war from 1450 to 1850.

David Parrott is Professor of Early Modern European History, and Fellow and Tutor at New College, University of Oxford. His books include *Richelieu's Army: War, Government and Society in France, 1624–42* (2001), *The Business of War: Military Enterprise and Military Revolution in Early Modern Europe* (2012), and *1652: The Cardinal, the Prince and the Crisis of the Fronde* (2020).

Gábor Ágoston is Professor of History at Georgetown University. His works include the *Encyclopedia of the Ottoman Empire* (2009), *Guns for the Sultan: Military Power and the Weapons Industry in the Ottoman Empire* (2005), and *The Last Muslim Conquest: The Ottoman Empire and Its Wars in Europe* (2021).

# THE CAMBRIDGE HISTORY OF WAR

★

VOLUME III

# War and the Early Modern World

★

*Edited by*

DAVID PARROTT

*and*

GÁBOR ÁGOSTON

CAMBRIDGE
UNIVERSITY PRESS

CAMBRIDGE
UNIVERSITY PRESS

Shaftesbury Road, Cambridge CB2 8EA, United Kingdom

One Liberty Plaza, 20th Floor, New York, NY 10006, USA

477 Williamstown Road, Port Melbourne, VIC 3207, Australia

314–321, 3rd Floor, Plot 3, Splendor Forum, Jasola District Centre, New Delhi – 110025, India

103 Penang Road, #05–06/07, Visioncrest Commercial, Singapore 238467

Cambridge University Press is part of Cambridge University Press & Assessment, a department of the University of Cambridge.

We share the University's mission to contribute to society through the pursuit of education, learning and research at the highest international levels of excellence.

www.cambridge.org
Information on this title: www.cambridge.org/9780521874281

DOI: 10.1017/9780511843136



When citing this work, please include a reference to the DOI 10.1017/9780511843136

First published 2025

Printed in the United Kingdom by CPI Group Ltd, Croydon CR0 4YY

*A catalogue record for this publication is available from the British Library*

*A Cataloging-in-Publication data record for this book is available from the Library of Congress.*

ISBN 978-0-521-87428-1 Hardback

# Contents

*List of illustrations* *page viii*
*List of maps* *x*
*Notes on contributors* *xii*
*Acknowledgments* *xvi*

Introduction to volume III *1*
DAVID PARROTT

1. The legacy of the nomadic empires *53*
BEATRICE F. MANZ

2. Ming conflicts with its neighbors, 1368–1644 *71*
MORRIS ROSSABI

3. Ottoman wars and military transformation, 1453–1826 *98*
GÁBOR ÁGOSTON

4. Spain's wars and the rise of the Hispanic world empires, 1516–1828 *122*
FELIPE FERNÁNDEZ-ARMESTO

5. Wars of the European continent, 1500–1650 *145*
JAMES D. TRACY

6. The evolution of naval warfare, 1450–1650 *170*
NICCOLÒ CAPPONI

7. Military finance, organisation, and state development, 1450–1650 *191*
PETER H. WILSON

8. Fortification and siege warfare *205*
SIMON PEPPER

9. European expansion in the Indian Ocean and Pacific, 1450–1850 *231*
LOUIS SICKING

10. Conflict in Northeastern Europe and the emergence of Russia to 1721 *255*
CAROL B. STEVENS

11. The military in Safavid Iran, 1501–1736 *283*
RUDOLPH MATTHEE

12. Wars of dynasties, wars of empires: The nature of European conflicts, 1648–1792 *309*
HAMISH SCOTT

13. The standing army: France and Brandenburg-Prussia – a comparison *342*
BERNHARD R. KROENER

14. Navies, technology, and organisation, 1648–1815 *366*
ROBIN BRIGGS

15. Warfare in early modern Africa, *c.* 1450–*c.* 1850 *387*
RICHARD REID

16. Warfare in India, 1450–1850 *409*
KAUSHIK ROY

17. Japan and Korea, 1450–1850 *430*
EUGENE Y. PARK

18. Warfare in South-East Asia, 1448–1851 *449*
MICHAEL CHARNEY

19. War in the era of Qing imperial consolidation and expansion, 1587–1804 *466*
PAMELA CROSSLEY

20. The seventh continent: Russian territorial expansion, 1450–1850 487
DAVID SCHIMMELPENNINCK VAN DER OYE

21. The maintenance of Hispanic dominance in Latin America 513
GABRIEL PAQUETTE

22. The United States from Independence to the war with Mexico, 1776–1848 530
KEVIN D. MCCRANIE

23. The wars of the French Republic and the Napoleonic Wars, 1789–1815 552
ALAN FORREST

24. European sea power and the origins of European domination, 1800–1850 573
ANDREW LAMBERT

*Select bibliography* 602
*Index* 633

# *Illustrations*


1. Mounted standard bearers and archers. Ex coll.: C. C. Wang Family, Gift of The Dillon Fund, 1973. Retrieved from The Metropolitan Museum of Art, Washington. *page* 66
2. Mounted archer. Dietz Album, A, f. 72, p. 13. Staatsbibliothek zu Berlin – Preußischer Kulturbesitz, Berlin, Germany. 69
3. Fort of Los Adaes. Courtesy of the British Library. 125
4. Lienzo de Tlaxcala. Digital image © 2012 Fray Angélico Chávez History Library, New Mexico Digital Collections. 129
5. Parlamento of Negrete of 1793. Published in *Atlas of the Physical and Political History of Chile* by Claudio Gay. 1854. Retrieved from Memoria Chilena. 142
6. Siege of Vienna, 1529. Getty Images / mikroman6. 151
7. Battle of Pavia, 1525. British Museum: 1895,0122.79. 158
8. Idealised Mediterranean port with ships. Permission granted by Capponi family for reproduction. 185
9. Battle of the Sound fought between Swedish and Dutch fleets, 1658. Rijksmuseum, Amsterdam. 188
10. Medieval and early modern fortifications compared 210
11. Fortifications in seventeenth-century Netherlands. Rijksmuseum, Amsterdam. Public domain. 211
12. Siege of Ostend, 1601–4 219
13. The perfect siege of Ath, 1697 222
14. Sea battle between Dutch warships and Spanish galleys. SK-A-4250, Rijksmuseum Amsterdam. 239
15. A Dutch squadron commanded by Cornelis Simonsz van der Veer carries out a surprise attack on three Portuguese galleons in the Bay of Goa, 30 September 1639. SK-A-2126, Rijksmuseum Amsterdam. 241
16. Battle of Poltava, 1709. Courtesy of Museums of Russia. 278

17. Army battle standard – Safavid era. TKS.1664, Topkapi Museum, Istanbul. 287
18. Helmet – Safavid era. TPM 1–348, Topkapi Museum, Istanbul. 288
19. Model of the French eighty-gun two-deck warship *Foudroyont*. Musée d'histoire de Nantes, 294. 373
20. Opening stage of the battle of the Nile, 1798. National Maritime Museum, Greenwich. 383
21. Contrasting depictions of Korean military men. Sources: Jinju National Museum, blog, last modified 23 August 2017, https://blog.daum.net/navykanggu/2516 and Korean History and Culture News, 22 February 2019, www.ns-times.com/news/view.php?bIdx=1898. 444
22. Contrasting depictions of Japanese military men. Sources: Edo Guide, https://edo-g.com/images/view/2 and History Guide, https://history-g.com/eng/images/view/59. 446
23. Naval cannon. Source: author's photograph. 457
24. Muscovite cavalry, *c.* 1550. Source: Getty Images. 493
25. Preobrazhenskii Life-Guard Regiment. From *Historical Description of the Clothing and Armament of the Russian Troops, Compiled by the Highest Command, with Drawings* [in Russian], ed. A. V. Viskovatov (1841–62). 508
26. HM steam sloops *Rattler* and *Alecto*. National Maritime Museum, Greenwich. 586
27. Bombardment of St. Jean d'Acre, 1840. Engraved by Charles Frederick de Brocktorff, from a painting by J. K. Wilson, 1841. National Maritime Museum, Greenwich. 591
28. Signing of the Treaty of Nanking, 29 August 1842. Painting by Capt. John Platt, 1846. Anne S. K. Brown Military Collection, Brown University. 592

# *Maps*


1. The legacy of the nomadic empires *page* 55
2. China under the Ming dynasty, 1368–1644. Created by Béla Nagy. 73
3. The defeat and capture of the Zhengtong Emperor, 1449 85
4. The growth of the Ottoman Empire, sixteenth and seventeenth centuries. Created by Béla Nagy. 100
5. Empire of Charles V (r. 1519–56). Created by Béla Nagy. 146
6. European exploration and empires, 1492–1700. Created by Béla Nagy. 233
7. The Great Northern War, 1700–21. Created by Béla Nagy. 274
8. The Safavid Empire 285
9. Europe after the treaties of Utrecht and Rastatt (1713, 1714). Created by Béla Nagy. 317
10. The partitions of Poland, 1772–5. Created by Béla Nagy. 340
11. Warfare in early modern Africa, *c.* 1450–*c.* 1850. Source: Richard Reid, *Warfare in African History* (Cambridge, 2012). Reproduced with permission. 389
12. Map of India 411
13. Map of Japan and Korea 431
14. Mainland South-East Asia, *c.* 1340. Source: Victor Lieberman, *Strange Parallels: Southeast Asia in Global Context, c. 800–1830* (Cambridge, 2003), p. 13. Reproduced with permission. 450
15. Mainland South-East Asia in 1824. Retrieved from Victor Lieberman, *Strange Parallels: Southeast Asia in Global Context, c. 800–1830* (Cambridge, 2003), p. 32. Reproduced with permission. 461
16. China under the Manchu (Qing) Empire, 1644–1912. Created by Béla Nagy. 469
17. The expansion of Russia in Europe and Asia, 1533–1900. Created by Béla Nagy. 490

18. Spanish Central and South America 520
19. European Colonies in North America, *c.* 1750. Created by Béla Nagy. 531
20. Wars in Europe between 1804 and 1815. Created by Béla Nagy. 571

# *Notes on contributors*

Gábor Ágoston is Professor of History at Georgetown University. He is the author of *The Last Muslim Conquest: The Ottoman Empire and Its Wars in Europe* (2021), *Guns for the Sultan: Military Power and the Weapons Industry in the Ottoman Empire* (2005), and several other monographs, and also co-author and co-editor of the first English-language *Encyclopedia of the Ottoman Empire* (2009).

Robin Briggs is Emeritus Senior Research Fellow, All Souls College, Oxford, and a Fellow of the British Academy. He is the author of *Early Modern France* (1977 and 1998), *Communities of Belief: Cultural and Social Tensions in Early Modern France* (1989), *Witches and Neighbours: The Social and Cultural Context of European Witchcraft* (1996 and 2002), and *The Witches of Lorraine* (2007).

Niccolò Capponi, independent scholar, writer, and vice-president of the Centro Studi Famiglia Capponi, Florence. He is the author of *Victory of the West: The Story of the Battle of Lepanto* (2006), *Al traditor s'uccida. La congiura de' Pazzi, un dramma italiano* (2014), plus a number of other monographs and essays.

Michael Charney is Professor of Asian and Military History at SOAS, University of London, where he teaches the history of violence and cities in Africa and Asia, and the emergence of the contemporary world. He has published numerous books and articles on Southeast Asian warfare, including his most recent book, *Imperial Military Transportation in British Asia: Burma, 1941–42* (2019). His current research focuses on warfare and mobility in West Africa and Southeast Asia.

Pamela Crossley is Collis Professor of History, Emerita, at Dartmouth College (USA). Her most recent book is *Hammer & Anvil: Nomad Rulers at the Forge of the Modern World* (2019) and her forthcoming book is *China's Global Empire: Qing, 1636–1912*.

Felipe Fernández-Armesto, who occupies the William P. Reynolds Chair for Mission in Arts and Letters at the University of Notre Dame, edited *The Oxford History of the World*, and, with his graduate student class as co-editors, *Approaches to Global History: To See the World Whole*. Recent books include *Straits: Beyond the Myth of Magellan* and, with Manuel Lucena Giraldo, *Un imperio de ingenieros*.

ALAN FORREST is Emeritus Professor of History at the University of York, where he taught from 1989 until 2012. His works include *Napoleon's Men* (2002), *The Legacy of the French Revolutionary Wars* (2009), *Waterloo* (2015), and *The Death of the French Atlantic: Trade, War and Slavery in the Age of Revolution* (2020). He is also general editor of the three-volume *Cambridge History of the Napoleonic Wars* (2022–23).

BERNHARD R. KROENER is Emeritus Professor of Military History at the University of Potsdam. His publications include *Militär, Staat und Gesellschaft im 20. Jahrhundert (1890–1990)* (2011), *Kriegswesen, Herrschaft und Gesellschaft. 1300–1800* (2013), and he is the editor of *Europa im Zeitalter Friedrichs des Großen. Wirtschaft, Gesellschaft, Kriege* (1989), *Krieg und Frieden. Militär und Gesellschaft in der frühen Neuzeit* (1996), and *Die preußische Armee zwischen Ancién Regime und Reichsgründung* (2008).

ANDREW LAMBERT FKC is the Laughton Professor of Naval History at Kings College London. His work focuses on sea power, British strategy, and naval historiography. He has published widely: his most recent books are: *Seapower States* (2018), which won the 2018 Gilder Lehrman Book Prize in Military History, and *The British Way of War: Julian Corbett and the Battle for a National Strategy* (2021). His current research focuses on the reconstruction of British strategy after 1915.

KEVIN D. MCCRANIE is the Philip A. Crowl Professor of Comparative Strategy at the US Naval War College. He is the author of *Admiral Lord Keith and the Naval War against Napoleon* (2006), *Utmost Gallantry: The U.S. and Royal Navies at Sea in the War of 1812* (2011), and *Mahan, Corbett, and the Foundations of Naval Strategic Thought* (2021).

BEATRICE F. MANZ is Professor Emerita of Tufts University. She is the author of *The Rise and Rule of Tamerlane* (1989), *Power, Politics and Religion in Timurid Iran* (2007), and *Nomads in the Middle East* (2022).

RUDOLPH MATTHEE is the John and Dorothy Munroe Distinguished Professor of History at the University of Delaware. He is the author of four prize-winning monographs on early modern Iranian history. His most recent book is *Angels Tapping at the Wine-Shop's Door: A History of Alcohol in the Islamic World* (2023).

EUGENE Y. PARK is Professor of History at the University of Nevada, Reno. He is the author of *Between Dreams and Reality: The Military Examination in Late Chosŏn Korea, 1600–1894* (2007), *A Family of No Prominence: The Descendants of Pak Tŏkhwa and the Birth of Modern Korea* (2014), *A Genealogy of Dissent: The Progeny of Fallen Royals in Chosŏn Korea* (2018), and *Korea: A History* (2022). He is also co-editor of *Peace in the East: An Chunggŭn's Vision for Asia in the Age of Japanese Imperialism* (2017).

GABRIEL PAQUETTE is Professor of History at the University of Maine. He previously held research and teaching positions at Oregon, Johns Hopkins, Harvard, Cambridge, and Wesleyan Universities. His publications include *The European Seaborne Empires* (2019) and (as co-editor) *Spain and the American Revolution* (2020).

DAVID PARROTT is Professor of Early Modern European History, and Fellow and Tutor at New College, University of Oxford. His books include *Richelieu's Army: War, Government and Society in France, 1624–42* (2001), *The Business of War: Military Enterprise and Military Revolution in Early Modern Europe* (2012), and *1652: The Cardinal, the Prince and the Crisis of the Fronde* (2020).

SIMON PEPPER is Emeritus Professor of Architecture, University of Liverpool. He has published (with Nicholas Adams), *Firearms and Fortifications: Military Architecture and Siege Warfare in Sixteenth-Century Siena* (1986) as well as books on social housing, free libraries, and urban planning.

RICHARD REID is Professor of African History in the Faculty of History and a Fellow of St. Cross College at the University of Oxford. He is the author of *Frontiers of Violence in Northeast Africa: Genealogies of Conflict since c. 1800* (2011), *Warfare in African History* (2012), *A History of Modern Uganda* (2017), and *Shallow Graves: A Memoir of the Ethiopia–Eritrea War* (2020).

MORRIS ROSSABI is Distinguished Professor of History at City University of New York and Adjunct Professor of History at Columbia University. He is the author of *China and Inner Asia* (1975), *Khubilai Khan* (1988), *Modern Mongolia* (2005), *From Yuan to Modern China and Mongolia* (2014), and *A History of China* (2nd ed., 2021). In 2009, the National University of Mongolia awarded him an honorary doctorate.

KAUSHIK ROY is Guru Nanak Chair Professor in the Department of History, Jadavpur University, Kolkata, India. He was associated with PRIO's 'Warring with Machines' project until 2023. His latest publication is *A Global History of Warfare and Technology* (2022).

DAVID SCHIMMELPENNINCK VAN DER OYE (1957–2022) was Professor of Russian history at Brock University in St. Catharine's, Canada, and was a Fellow of the Royal Society of Canada. His publications included *Toward the Rising Sun: Russian Ideologies of Empire and the Path to War with Japan* (2001), *Reforming the Tsar's Army: Military Innovation in Imperial Russia from Peter the Great to the Revolution* (2004), and *Russian Orientalism: Asia in the Russian Mind from Peter the Great to the Emigration (2010)*. He was joint editor of the two-volume collection *Rethinking the Russo-Japanese War, 1904–05* (2007).

HAMISH SCOTT (1946–2022) was Professor of International History (2000 to 2006) then Wardlaw Professor of International History (2006 to 2009) at St. Andrew's University, and subsequently a Research Fellow at Jesus College, Oxford. In 2006 he was elected a Fellow of the British Academy. His publications include *The Emergence of the Eastern Powers, 1756–1775* (2001), *The Birth of a Great Power System, 1740–1815* (2006), and his edited two-volume *Oxford Handbook of Early Modern European History, 1350–1750* (2015). His last major work, *Forming Aristocracy: The Reconfiguration of Europe's Nobilities, c. 1300–1750*, is in the process of publication.

LOUIS SICKING is Aemilius Papinianus Professor of History of Public International Law at Vrije Universiteit Amsterdam and lecturer in medieval and early modern history at the

University of Leiden. He is the author of *Neptune and the Netherlands: State, Economy, and War at Sea in the Renaissance* (2004), *Colonial Borderlands: France and the Netherlands in the Atlantic in the 19th Century* (2008), *La Naissance d'une thalassocratie. Les Pays-Bas et la mer à l'aube du Siècle d'or* (2015), and co-editor of *Conflict Management in the Mediterranean and the Atlantic, 1000–1800* (2020)

CAROL B. STEVENS is Emerita Professor of History and of Russian and Eurasian Studies at Colgate University, Hamilton NY, USA. She is the author of *Russia's Wars of Emergence* (also in Polish and Russian: 2007–23) and *Soldiers on the Steppe Frontier* (1995), and currently editor of the journal *Canadian–American Slavic Studies*.

JAMES D. TRACY is Emeritus Professor of History at the University of Minnesota, where he taught from 1966 through 2009. He is the author of *Emperor Charles V, Impresario of War: Campaign Strategy, International Finance and Domestic Politics* (2002) and *Balkan Wars: Habsburg Croatia, Ottoman Bosnia, and Venetian Dalmatia, 1497–1617* (2016).

PETER H. WILSON is the Chichele Professor of the History of War at the University of Oxford, and a Fellow of All Souls College. His books have been translated into Chinese, German, Italian, Japanese, Macedonian, Polish, and Spanish, and include *Iron and Blood: A Military History of the German-Speaking Peoples since 1500* (2022), *Lützen* (2018), *The Holy Roman Empire: A Thousand Years of Europe's History* (2016), and *Europe's Tragedy: A History of the Thirty Years War* (2009).

# *Acknowledgments*

This volume has been long in gestation and the two present editors wish to thank their predecessors, Professor John Childs and Professor Arthur Waldron, for their initial work in conceptualizing the project, and commissioning many of the original authors of the chapters. We hope that they see the finished volume as a worthy realization of their original project.

In the course of commissioning and receiving the chapters and preparing the volume, the editors record with great sadness two untimely deaths amongst the contributors. David Schimmelpenninck van der Oye (1957–2022), Professor at Brock University Canada, a leading expert on Russian imperial history, has distilled much of his accumulated knowledge of Russia's military expansion into the chapter in this volume. Hamish Scott (1946–2022), FBA, Professor at St. Andrew's and Senior Research Fellow at Oxford University, was the leading British scholar of early modern international relations, author of numerous books on eighteenth-century European politics and the emergence of the great power system. His chapter in the current volume is also influenced by his other substantial research interest, the continued power and influence of the European aristocracy, not least in the waging of warfare, a subject which will be developed in a substantial, posthumously published study.

The editors would like to express their warmest thanks to Eugene Ludwig, long-standing supporter of the Humanities in New College, Oxford, the support of whose Fund allowed us to pay for editorial assistance and to cover the costs of drafting the maps for the volume. We are most grateful as well for the work of our editorial assistant, Basil Bowdler, in preparing the chapters for production, assembling the bibliography, and numerous other tasks, all accomplished with efficiency and enthusiasm. We would also like to thank the Cambridge University Press editorial team, Michael Watson, Elizabeth Hanlon, and Victoria Phillips, for their patience, support, and helpfulness in seeing this project through to production.

David Parrott
Gábor Ágoston

# Introduction to volume III

DAVID PARROTT

The four volumes of the *Cambridge History of War* were conceived in global terms. The aim was to go beyond a history centred on warfare in Europe, in which the global context emerged solely through the eyes of European exploration, trading, and colonisation. Instead, the volumes would seek to provide the reader with a broader approach to warfare across the world, in which the experiences and trajectories of states and their military systems could be examined and compared. Europe and Europe's military engagement with the wider world would have a place, but would not be the single point of reference from which global warfare would be seen. This aim was the starting point for Volume III, both as initially conceived by the first editors, John Childs and Arthur Waldron, and then by the current editorial team.

The challenges in achieving this goal are of course considerable. At a practical level, the volume is made up of a series of interpretative essays written by historians with established expertise in particular areas or states. Any pretension to complete global coverage would be futile, and our intention had always been to draw out common themes, comparisons, and divergence rather than to aim for comprehensiveness for its own sake. We set a challenging agenda for our authors, many of whom were invited to examine 400 years of the military experience of a state, an empire, or even a continent, in a few thousand words. The success of their work should be seen not so much in the accumulation of historical detail, but in the establishment and exploration of broader, shared, or comparative themes such as the role of environment, use of weaponry, equipment, and tactics, the development of military organisation and command, and the relationship of war to the social or political structures which sustained it.

The challenges of a global approach to a volume examining the period from *c.* 1450 to *c.* 1850 are also present at a conceptual level. The subtitle of convenience 'War in the Early Modern World' is a well-established means to divide 'medieval' from 'modern' Europe, but it carries considerable baggage. Most obviously it is linked to a European perspective on historical development which identifies key shifts in the period after 1450 – phenomena such as printing, changing conceptions of education, religious reformations and the fracturing of Western Christendom, economic diversification and expansion – as a decisive watershed in the move towards modernity, a departure from some holistically conceived notion of the Middle Ages. Even applied to European history, 'early modernity' can be problematic in its assumptions of continent-wide shared chronology and experience. And such periodisation makes no sense at all in global terms: labelling a period of Chinese or Persian history 'medieval' and speaking of a transition to 'early modernity' is meaningless within a European chronology.[1]

The problem is intensified by another consequence of the long period covered by this volume. If the European 'early modern' period might be defined as roughly the three centuries between 1450 and 1750, the remaining century incorporated into this volume is widely understood to bring transformative change to Europe: a first century of modernity shaped by revolutions in politics, agriculture, industry, and communications. The impact of earlier developments in early modern Europe on the rest of the world can be debated and set against alternative models of power-projection and social organisation. But it is widely assumed that advances in technology, resource-mobilisation, and administrative effectiveness after 1750, above all in the four decades after 1810, propelled Europe into a dominant position across the world stage.

This periodisation has considerable significance for any discussion of military affairs and warfare. Conventionally the history of war throughout this entire 400-year period 1450–1850 has been interpreted as one of European military dynamism contrasted with stasis or inflexibility on the part of the non-European world. The fate of Aztec and Inca empires in Central and South America in the first part of the sixteenth century is treated as paradigmatic of how European combinations of weaponry and tactics, combined with superior strategy and organisational efficiency,

1 Jack A. Goldstone, 'The Problem of the "Early Modern World"', *Journal of the Economic and Social History of the Orient* 41 (1998), 249–84; Alan Strathern, 'Global Early Modernity and the Problem of What Came Before', *The Global Middle Ages: Past and Present Supplement* 238 (2018), 317–44; David Porter, *Comparative Early Modernities 1100–1800* (Basingstoke, 2012); Charles Parker, *Global Interactions in the Early Modern Age, 1400–1800* (Cambridge, 2010).

could defeat vastly larger and better-resourced military and political structures. Thereafter European warfare, honed through continual inter-state competition, proved itself continually superior to the military capacities of the extra-European world. Though states beyond Europe might adopt the technologies and organisation of European armies and navies, the capacity of European states to develop and innovate in military affairs maintained a consistent advantage. From this perspective there was nothing startling about the dramatic intensification of the encounters between Europe, the United States, and the rest of the globe in the first half of the nineteenth century; it was simply the logical outcome of four centuries of military advantage brought about by states with a particular capacity to prioritise and facilitate war-making through political, social, and economic institutions.

A central aim of the contributions to this volume is to challenge this familiar single-track narrative of the history of warfare based on the centrality of European military developments and their export or adoption, usually with limited success, in a wider global context. Yet at the same time it is impossible to ignore, and pointless to marginalise, the role of European states in a global history of warfare beginning in 1450. The course charted here, between a global approach to the subject facilitating the exploration and evaluation of military systems and traditions that developed independently of any European engagement, while nonetheless recognising the significance of European states in the larger context, is seen most clearly in the volume's emphasis on navies and naval development.

The history of war has frequently been taken as synonymous with the history of armies and land warfare, and it was decided from the outset that war at sea should receive appropriate attention across these centuries. This justified both the decision to commission four chapters from specialists in the history of navies – Niccolò Capponi, Louis Sicking, Robin Briggs, and Andrew Lambert – and to encourage other authors to devote attention where appropriate to naval power and maritime conflict. This attention to the maritime dimension of warfare necessarily entails a strongly European emphasis. As Louis Sicking, Morris Rossabi, and Michael Charney all point out, the great Chinese fleets, commanded by Zheng He and despatched on multiple expeditions across the Indian ocean as far as the East African coast, were abandoned after 1433; with them seems to have disappeared any ambition by Asian empires or states to gain deep-water naval hegemony. Korean naval successes against the Japanese with armoured 'turtle ships' are specifically discussed in Morris Rossabi's chapter. The Ottoman naval commitment to both the Mediterranean and into the Red Sea and beyond receives attention in the chapters by Gábor Ágoston and Louis Sicking. But the full development of the oceanic sailing warship, naval firepower, and maritime tactics and strategy, and, at the very end of our period, the emergence of the ironclad steamship, are unarguably the product of a European military trajectory of which other nations in the centuries before 1850 were spectators or modest imitators.

European militaries and war-making thus constitute a large presence in the volume. It should be stressed moreover that discussions of 'European' warfare cover an immense variety of military experiences across the continent and across time: there is no single European 'way of war' emerging through some composite evolutionary process. We will return to this in the discussion of 'military revolution' below. Nor are the chapters conceived as a narrow history of battles and campaigns, and still less is the volume intended to produce a global scorecard of military winners, measured in terms of battles won and defeats avoided. Looking at why military systems failed to deliver; how they stagnated and decayed; the tensions between social status, military hierarchy, and discipline; the slippery notions of professionalism and experience: these are no less important themes explored in the volume, whether in Bernhard Kroener's account of decaying professionalism in the eighteenth-century officer corps of France and Prussia, or Michael Charney's and Pamela Crossley's chapters on problems of command and control, and failure to adapt to new threats, in the armies of South-East Asia and Qing China.

What does the volume contribute to a broader understanding of warfare over these four centuries? Certain overarching themes emerge from the chapters and reach across both geographical and chronological divisions.

## Environment

For the greatest part of this period the natural environment determined most human activities, including warfare. The volume opens with an exploration of one of the most enduring, pervasive, and influential military phenomena of the period: nomadic warfare shaped by human adaptation to the open expanses of the Asian steppes. Beatrice Manz's chapter examines the military organisation of societies based on horse-breeding and herding of livestock, from the eruption of Mongol imperial expansion under Chinggis Khan in the early thirteenth century through to their final demise in the nineteenth century, squeezed into submission or extinction by Russian and Chinese empires and their garrison-based militaries. Between these dates, Manz demonstrates the formidable power and resilience of a military system whose essential elements – horsemanship and archery – had evolved in response to the challenges of subsistence through hunting and herding. The vast empire of Chinggis Khan and his heirs and successors, and the imposing conquests of Tamerlane (r. 1370–1405), point to what would remain an enduring factor in global warfare throughout most of our period: the military balance between nomadic and settled populations. As Rudolph Matthee comments in his chapter on the rise of the Safavid Empire, the entire history of the further Middle East after the Seljuq and Mongol conquests may be seen as a quest for ways to harmonise sedentary–nomadic relations. By no means did the advantage always lie with settled peoples in this struggle. As late as the mid-eighteenth century the nomad warlord Nadir Shah Afshar (r. 1736–47) built an army dominated by cavalry contingents raised and commanded by tribal rulers, and used it to reunite much of the Safavid Empire.

The formidable success of the nomadic empires demonstrated the military effectiveness of communities where the entire male population were trained as horsemen. Empowered by skilful tactics, sophisticated strategic thinking, and the ruthless exploitation of subject territory and populations, the military system of the Mongols and their successors was a formidable engine of military conquest. Its organisation and traditions were subsequently integrated as foundational features of the military systems of the Middle Eastern Islamic empires, and carried into the Russian and Chinese imperial states. The Jurchen heirs of the Mongols ruled China as the Qing dynasty from 1644, having deployed traditional styles of nomadic warfare to weaken and break Ming rule in China.

Similar conflicts between the military systems of nomadic and settled societies play an enduring role in warfare in Africa. Richard Reid's chapter examining warfare across the continent shows how structures and styles of warfare were heavily determined by the physical environment. On the dry savannah across North Africa local communities lived largely from cattle herding, and young men combined the roles of herdsmen and warriors fighting on behalf of the rest of the community. Local wars were waged for control of livestock and grazing land by mobile and autonomous forces. More fertile areas, such as the Ethiopian highlands or the Great Lakes in the east, permitted the growth of denser populations drawing on a strong and productive agrarian base, and very different types of military organisation grew up, characterised by larger, better-integrated armies, and a much greater willingness to experiment with new weaponry, especially firearms.

Nomadic cultures may appear to have enjoyed impressive tactical and organisational military strengths: mobility and surprise, well-integrated weaponry – above all the quick-fire composite bow, used on horseback – high levels of military participation, and home territory that was apparently less vulnerable to counter-attack. But settled populations gradually began to gain the upper hand in clashes with nomadic warrior-states, though this process took centuries and, as the example of Nadir Shah demonstrates, the outcome was by no means predetermined. More efficient resource mobilisation allowed the creation of larger armies, permitted extensive – and often expensive – garrisoning and fortification, and facilitated initiatives in developing artillery and infantry firearms, which in turn required sophisticated and expensive logistical organisation to ensure that troops could be supplied with powder, match, and shot. The Qing dynasty's remarkable series of campaigns through the eighteenth century, described in the chapter by Pamela Crossley, deployed the resources and standing armies of Imperial China to defeat and subdue all the nomadic peoples to the west, bringing them under administrative and military control, thus concluding what the Chinese had perceived as a struggle lasting over a millennium.[2]

Though it might be generalised that nomadic societies reflected an earlier stage of human development compared to the social, and often structural, organisation required to maintain settled agriculture, the extent to which human intervention could modify existing ecologies should not be ignored. Felipe Fernández-Armesto's chapter considers how the Spaniards' introduction of horses into the Americas transformed native communities – including those outside of Spanish control. The 'horse revolution' brought a decisive shift back towards hunting and herding, dramatically changing the social and military environment of native North American peoples such as the Apaches and Comanches, and in Chile, the Mapuche. In most of these cases, the new possibilities for hunting bison and herding cattle from horseback were accompanied by widespread raiding and intimidation of neighbouring peoples for tribute.

Yet settled, agrarian communities remained the essential, stable base on which states could be formed and specialised armies created and maintained. The shift, described for Europe in Peter Wilson's chapter, from a domain-state

2 More detail in Peter C. Perdue, *China Marches West: The Qing Conquest of Central Eurasia* (Cambridge, MA, 2005).

dominated by kinship relationships and obligations towards a tax-state based on formalised fiscal obligations, was hard to achieve in a nomadic society in which warfare was a collective activity undertaken by almost all adult males, organised tribally or through small, semi-autonomous communities. But the economic strength of settled agrarian communities, for most of these four centuries and across the globe, was anything but assured. Agriculture based on cultivating grain crops was both land- and labour-intensive, and even on good soils generated limited yields; many communities achieved bare levels of subsistence, and surpluses over and above basic needs were occasional and unreliable. Without the presence of waterways or seas – the only commercially viable means to transport bulk goods – markets for grain were local and the redistribution of surpluses heavily restricted. The northern hemisphere in particular was affected by the 'Little Ice Age', a sustained period of overall cooling that ran from the sixteenth to the mid-nineteenth century, and led to a significant decline in agricultural productivity as growing seasons were shortened and extreme weather became more common; higher altitudes became unsuitable for grain farming, while glaciers rolled over pasturage.[3] European populations, which had achieved a dramatic recovery during the later fifteenth and sixteenth centuries after the famine and plague-induced collapse of the fourteenth, stalled in the seventeenth, and did not begin rapid growth again until well into the eighteenth century.

Environmental factors other than agrarian versus pastoral land-use could also shape military systems throughout this period. One of the great divides in African warfare, as Reid points out, was a line south of the great savannah, below which the tsetse fly was endemic. The fly spread sleeping sickness, making it impossible to keep horses, and therefore ensuring that the cavalry-centred pattern of warfare that characterised North Africa was not repeated in the south of the continent. Several contributors also stress the prevalence of the malarial mosquito in equatorial Africa, and its deadly impact on European forces moving outside a narrow coastal strip into the African heartlands. Only at the very end of the period under discussion did armies start to escape from the grip of contagious diseases spread widely over the globe which had hitherto been the greatest cause of mortality amongst soldiers. For army commanders from the fifteenth to the eighteenth century, plague, dysentery, typhus, and malarial and other fevers were far graver threats to the military effectiveness and strength of their forces than enemy

3 Geoffrey Parker, *Global Crisis: War, Climate Change and Catastrophe in the Seventeenth Century* (New Haven and London, 2013); Jean Grove, *The Little Ice Age* (London, 2001); Emmanuel Le Roy Ladurie, *Histoire humaine et comparée du climat*, 3 vols. (Paris, 2004–9).

action. Gabriel Paquette's remark that 'Yellow Fever proved a crucial, and reliable, ally in maintaining Spanish dominance in the Americas', has plenty of wider echoes.[4] It was perhaps consciousness of the particular deadliness of disease in the confined and crowded spaces of warships that led naval authorities to embrace more rapidly and more effectively concerns with hygiene and diet, as Robin Briggs discusses in his chapter on seventeenth- and eighteenth-century navies. Yet all of the chapters dealing with warfare at the end of the period point to a continuing struggle with infectious disease which armies and navies might have tamed to some extent, but had certainly not beaten. Michael Charney points out that in the first Anglo-Burmese war of 1824–6, the forces of the East India Company suffered extensive casualties that were in part due to Burmese weapons, but much more the consequence of tropical disease.

## Processes of military change

The dominant paradigm through which military change has been viewed across these centuries is that of 'military revolution'. Launched by Michael Roberts's eponymous lecture/article of 1955, its explanatory attractiveness is undeniable.[5] Roberts's thesis links together military technology, tactical evolution, strategy, and army growth in the period between 1560 and 1660, showing how these were financed, controlled, and administered through a consequent growth of centralising state power and bureaucracy. Driven by the growing demands of waging war, more powerful agencies of resource extraction and control emerged and asserted themselves across every aspect of state and society. Military change, initially seen as an organisational response to the development of viable infantry firearms, was the principal force creating the modern state. Though influential, especially among non-military historians, the thesis has always had its critics. Early challenges drew attention to its narrow and unconvincing chronology, and its excessive focus on Protestant Northern Europe – the Dutch and the Swedes – as the instruments of revolution. Adopted and partially reconstituted by Geoffrey Parker in both an article from 1977 and his 1988/96 book, 'military revolution' was given a broader and firmer

4 Ch. 21 below, p. 520.
5 Michael Roberts, *The Military Revolution, 1560–1660* (Belfast, 1956); reprinted *inter alia* in Clifford J. Rogers (ed.), *The Military Revolution Debate: Readings in the Transformation of Early Modern Europe* (Boulder, 1995), pp. 13–35.

grounding, and much wider range.[6] Parker agreed that the spread of firearms was crucial to the development of warfare, but the process was experienced more broadly and incrementally: Spanish armies of the early sixteenth century were already evolving tactics and training to deploy infantry firepower more effectively.[7] More importantly, Parker's modified thesis gave a distinct role to artillery, linking it to siege warfare and the protracted struggle that gunpowder opened up between offensive and defensive technologies. If cannon could easily destroy the traditional high, narrow curtain walls of medieval fortifications, they might be countered by new styles of low-lying fortification: immensely thick, earth-packed walls, protected by the crossfire of projecting triangular bastions, and by increasingly elaborate outworks that made it difficult for the cannon of the besieging army to get near enough to damage the main lines of defences. The financial and organisational demands of building these new fortifications strengthened the power of the state, while their scale pushed the size of armies upwards in order to conduct effective sieges. Emphasis on the importance of artillery, rather than the infantry handguns of Roberts's original thesis, also enabled Parker to add an area that had been neglected hitherto, arguing for a transformative revolution in naval warfare brought about primarily by the development of shipboard gunnery.

The subsequent extended and lively historiographical debate has engaged with numerous aspects of the thesis, from the technology of muskets through to discussion about the appropriateness of the word 'revolution' to describe an extended historical process: one or multiple military revolutions?[8] Fully fledged military revolution or the more limited concept of a 'revolution in military affairs' (RMA)?[9] The essential points over which controversy has raged can, however, be reduced to three areas. How fundamental was

6 Geoffrey Parker, 'The Military Revolution, 1560–1660: A Myth?', *Journal of Modern History* 48 (1976), 195–214; Geoffrey Parker, *The Military Revolution, 1500–1800: Military Innovation and the Rise of the West*, 2nd ed. (Cambridge, 1996).

7 Further developed in Fernando Gonzalez de Léon, '"Doctors of the Military Discipline": Technical Expertise and the Paradigm of the Spanish Soldier in the Early Modern Period', *Sixteenth Century Journal* 27 (1996), 61–85; René Quatrefages, *La Revolución Militar Moderna. El Crisol Espanol* (Madrid, 1996).

8 Clifford J. Rogers, 'The Military Revolutions of the Hundred Years' War', in Rogers (ed.), *Military Revolution Debate*, pp. 55–93; Jeremy Black, *European Warfare 1494–1660* (London, 2002), pp. 32–54; Jeremy Black, *Beyond the Military Revolution: War in the Seventeenth Century* (Basingstoke, 2011); John Childs, *Warfare in the Seventeenth Century* (London, 2001); Andrew Ayton and Leslie Price (eds.), *The Medieval Military Revolution: State, Society and Military Change in Medieval and Early Modern Europe* (London, 1998); David Eltis, *The Military Revolution in Sixteenth-Century Europe* (London, 1998).

9 Jeremy Black, *War and Technology* (Bloomington, 2013); Colin Gray, *Strategy for Chaos: Revolutions in Military Affairs and the Evidence of History* (London, 2004).

military technology and technological change as a transformative force? Was warfare the revolutionary catalyst of wider change in states and societies, or were these structural changes in fact a necessary *precondition* for extensive developments in resourcing and waging war? Was military revolution an essentially European phenomenon or could it be understood as a global transformation, and to what extent did Europe 'export' revolution to the rest of the world?

No chapter in this volume specifically addresses the 'military revolution', and this was a decision by the editors. However four chapters consider central aspects of the thesis and the surrounding debates. Peter Wilson's chapter takes a wide-ranging view of the relationship between finance, evolving military systems, and state-formation in Europe; Simon Pepper gives attention to the evolution of siege warfare and fortifications and its wider impact on war and resource-mobilisation; Niccolò Capponi's and Robin Briggs's chapters sequentially consider the nature of change in European naval warfare from the early sixteenth century to the end of the eighteenth century. Meanwhile the three controversial issues at the heart of the revolution thesis inform the discussions of many of the other contributions to the volume.

### *Military technology*

How far were the weapons of war transformed in the course of these 400 years, and did changing technologies have a decisive effect on the shape, duration, and character of warfare? Running through this entire period is the assumption that 'early modernity' in warfare was, above all, defined by the evolving and systematic use of gunpowder as a missile propellant. Whether linked to the so-called Islamic 'gunpowder empires' of the Ottomans, Mughals, and Safavids, the struggles for political hegemony in China, Japan, and South-East Asia, or the shifting patterns of European warfare, the development and proliferation of artillery and infantry firearms has been treated as the decisive factor in successful warfare. Much material in the chapters underpins the importance of the evolution of artillery and handguns, and does so on a global scale. But the global dimension is no less problematic and challenging when it comes to making broad assumptions about the central significance of gunpowder weaponry. The adoption of firearms and artillery was far from universal, even in Europe; almost to the end of our period there were military contexts in which firearms remained of marginal importance. The chapters offer plenty of evidence of the limitations, as well as the strengths, of gunpowder weaponry – both in terms of technology and in terms of the environments in which warfare was waged. Moreover, the difficulties of assimilating firepower into existing military structures and organisations emerge in many contexts: artillery, or infantry weighed down with muskets and their accoutrements, sacrificed speed and flexibility; the advantages of shooting arrows from horseback, rather than firing a powder weapon, were accepted up to the end of the eighteenth century. Where heavy armour had never been adopted by rival military elites the advantage of the musket – its impressive penetrative power – was largely irrelevant.[10]

The Ottoman siege of Constantinople in 1453 is frequently cited as a decisive moment in the rise of gunpowder warfare. Yet Gábor Ágoston cautions against the well-worn assumption that Ottoman success was owed to giant bombards operated by renegade Christian gunners, stressing that 'there were a number of other factors that explain Ottoman success. Of these, overwhelming numerical superiority (70,000 besiegers versus 10,000 defenders), abundant supplies of food, weapons, shot, and powder in the Ottoman army all demonstrated the Ottomans' organizational and logistical capabilities.'[11] Over a much longer timespan, Simon Pepper's chapter also cautions against fixating on artillery as the single determinant of success and failure in siege warfare. Undermining walls, with or without gunpowder, was often seen as a more effective way of bringing besieged garrisons to terms. The science of siege warfare, reaching its apogee in the careful placing of gun batteries and the methodical advance of lines of trenches associated with the French marshal Sébastien Le Prestre de Vauban, was often sidelined in favour of less time-consuming assaults or surprise attacks, aiming to take enough of the fortifications to force the besieged to terms.[12] In some cases success in siege warfare was indeed a question of who could muster the most artillery in attack or defence, but outcomes could depend on many other factors.

The adoption of firearms and cannon did not proceed in a universal and uncritical global fashion. Gunpowder has been used in Asia for centuries in catapult, throwing, or incendiary weapons – grenades, primitive flame-throwers, and firebombs – rather than for its ability to propel stone or

10 On the penetrative power of infantry firearms, see Burt Hall, *Weapons and Warfare in Renaissance Europe* (Baltimore and London, 1997); Clifford Rogers, 'Tactics and the Face of Battle', in Frank Tallett and David Trim (eds.), *European Warfare, 1350–1750* (Cambridge, 2012), pp. 203–35.

11 Ch. 3 below, pp. 103–4.

12 See also Jamel Ostwald, *Vauban under Siege: Engineering Efficiency and Martial Vigor in the War of the Spanish Succession* (Leiden, 2007).

metal shot.[13] By the sixteenth century, as Morris Rossabi points out, the Ming armies were copying Western cannons and firearms, but subjected them to critical evaluation and selective deployment. The Ming and their Qing successors recognised that for much of the warfare in which they found themselves engaged, particularly struggles against nomadic peoples to the west and north, firearms were of limited utility. In a struggle against mobile, cavalry-based forces, the best military response remained horsemen with composite bows – arguably the most widespread and enduring piece of military technology in human history. Against nomads, artillery and handguns might be more usefully deployed in protected defensive positions and by established garrisons – along the Great Wall, for example. As Pamela Crossley shows in her chapter, all these issues were the subject of lively discussion, theorising, and practical experiments with different types of gunpowder weaponry throughout the late Ming and early Qing period.

In contrast, Kaushik Roy is positive about the developing role of firearms in India both before and after the establishment of the Mughal Empire. By the mid-sixteenth century he argues that warfare in the Deccan was being shaped by large infantry forces making disciplined use of muskets and supported by iron cannon. Yet the subsequent military success of the Mughal rulers rested on the combination of musketeer infantry supported by mass forces of cavalry bowmen utilising the traditional nomadic tactics of rapid-fire archery, outflanking, and envelopment. The later Maratha struggle against the Mughals was another war dominated by light cavalry, once again resorting to the durable tactics of nomad warfare. Similarly, Eugene Park notes the initial enthusiasm with which both Japanese and Korean armies adopted and improved imported European muskets and developed distinctive tactics attuned to their cumbersome deployment. This was especially evident in Odo Nobunaga's division of his musketeer infantry into loaders and shooters at the battle of Nagashino (1575), with three times the number of loaders feeding primed muskets to the shooters. The deployment of musketeers was repeated with equal success at Sekigahara (1600) and by Japanese forces in their invasions of Korea in the 1590s. Yet in the ensuing centuries neither power sought to integrate evolving gunpowder arms into military forces that were secluded from most forms of warfare – in Japan as a result of the deliberate imposition of *sakoku* or the near-total exclusion of outsiders.

13 Tonio Andrade, *The Gunpowder Age: China, Military Innovation and the Rise of the West in World History* (Princeton, 2016); Peter Lorge, *The Asian Military Revolution: From Gunpowder to the Bomb* (Cambridge, 2008); Kenneth Chase, *Firearms: A Global History* (Cambridge, 2003).

Weaponry reflected a blend of social signifiers, established tradition, and vested interests. In Korea, the 'Three Skills Army' (Samsubyŏng) in the late sixteenth century comprised musketeers, archers, and close-combat 'killers' (*salsu*, comprising swordsmen, pikemen, and spearmen). But, as in Japan, the development of musketry and associated tactics atrophied after the defeat of Korea by the Qing in 1637, and the long subsequent period of peace as a tributary of Qing China.

As Richard Reid's chapter demonstrates, Africa offers an immense variety of military systems shaped by politics, environment, and available resources. However, the relative slowness and reluctance of so many states and military regimes to adopt firearms was notable. The Ethiopian army had adopted handguns to some extent by the end of the sixteenth century, yet they remained secondary to infantry armed with spears and swords. Likewise, even following the collapse of the great cavalry empires of the western savannah by the late sixteenth century, there was no immediate adoption of firearms as an alternative. It was not until the nineteenth century that guns had a significant impact on savannah warfare.[14] David Schimmelpenninck's account of Muscovite warfare also marginalises firearms: the army was preponderantly composed of light cavalry, and its tactics and weapons imitated those of the Tatar khanates on the eastern and southern frontier. 'The descriptions of Muscovite warfare by the sixteenth-century Habsburg diplomat Baron Sigismund von Herberstein resemble accounts of the Scythians by Herodotus two millennia earlier.' This was not a reflection of military backwardness, but rather a recognition that environment and enemy military organisation dictated a style of warfare in which cumbersome firearms and slow-moving artillery had little place. Artillery, as Rudolph Matthee observes in his chapter on the Safavid empires, is the best example of 'the dictum that the adoption of new technology is of limited value unless accompanied by cultural changes'.[15] Cannon, hard to cast and difficult to move over land, remained even more marginal to an army that remained cavalry-focused and less intent on conquering land and urban hubs than on intimidating enemies for tribute.

Practical and cultural factors could therefore constrain the use of firearms and encourage reliance on traditional weaponry. It is also important to recognise the limitations of the technology itself. At no point from the advent of early firearms through to the end of the period covered by this volume was there a single development in guns and gunpowder technology

14 Ch. 15 below, p. 395. 15 Ch. 20 below, p. 494; Ch. 11 below, p. 300.

so significant as to count as 'revolutionary': that is to say, an improvement in speed of fire, range, accuracy, facility of use, so dramatic that it immediately rendered previous weaponry or tactics obsolete. Such revolutions in military technology have occurred: the cavalry stirrup, the breech-loading rifle, and the machine gun are transformative in exactly this way. But no such revolution took place between 1450 and 1850. Advancing such an unsustainable claim for the matchlock infantry musket of the late sixteenth century, combined with a few supposed innovations to enhance its effectiveness like the counter-march and volley fire, was the original fault of Michael Roberts's 1955 thesis, from which many misconceptions about military revolution continue to flow.[16] Instead, these centuries witnessed small and incremental change: the shift after the 1680s from matchlock to flintlock firing mechanism for standard infantry muskets improved their speed of fire and allowed units of infantry to mass in close order, impossible when each man needed to hold a length of burning match in conjunction with musket and fork-rest.[17] Rifling of barrels and greater uniformity of production finally gave the standard musket a degree of accuracy that it had entirely lacked before. In the South-East Asian context, Michael Charney comments of infantry firearms that: 'they became sufficiently accurate and powerful by the early eighteenth century – as South-East Asian armies shifted from the matchlock to the flintlock – to end the role of the elephant as a tactical combat animal'.[18] And insofar as effectiveness was more a matter of production quality than innovative technology, it is worth noting Kaushik Roy's comment that: 'Qualitatively, the muskets manufactured in Punjab were on a par with, if not better than, the Brown Bess used by the EIC personnel.'[19] The point is echoed by other contributors, who argue that in terms of sophisticated metallurgy and craftsmanship Middle Eastern or Asian gun-founders were capable of equalling or surpassing muskets and cannons produced in Europe.[20]

The same incremental picture holds for artillery: the basic technology of the smooth-bore, muzzle-loaded cannon, firing solid shot or a variety of anti-personnel projectiles, hardly changed across the period. Developments in metallurgy, reductions in weight, and improvements

16 Seen in its most extreme form in Martin D. Feld, 'Middle-Class Society and the Rise of Military Professionalism: The Dutch Army, 1589–1609', *Armed Forces and Society* 1 (1975), 419–42.

17 Jeremy Black, *European Warfare, 1660–1815* (London, 1994), pp. 38–66.

18 Ch. 18 below, p. 457.

19 Ch. 16 below, p. 427.

20 See also Andrade, *The Gunpowder Age*.

in the design of gun carriages were progressively and widely adopted.[21] By the mid-eighteenth century European field and siege artillery was lighter and more manoeuvrable, and in general fired shot of lesser weight but with more velocity and range. As Alan Forrest notes: 'If the Revolutionary and Napoleonic Wars were notable for the deployment of larger massed armies in the field, it is more difficult to point to any technical or strategic innovations . . . Artillery pieces were lighter and more manoeuvrable than in previous wars, but they owed the innovation to Gribeauval and the Bourbons.'[22] By far the most significant change was quantitative: a lot more artillery was deployed on battlefields and sieges during the major European wars of the eighteenth century, and this was largely responsible for the steady increase in battlefield casualties.

Insofar as gunpowder weaponry was the catalyst of fundamental military change it was in conjunction with other technologies. As we have seen, the response to armies equipped with siege artillery was the adoption of a completely new style of fortification both for city defences and in the form of free-standing 'star-forts'. These thick-walled, multi-bastioned structures, provisioned with plentiful artillery, became the key to defensive strategies along contested frontiers, notably, as James Tracy shows in his chapter, in the Hungarian and Balkan borderlands between the Austrian Habsburgs and the Ottoman Empire.[23] The most familiar role of such artillery forts was as a means of defence and power-projection by European states and trading companies as they established their coastal toeholds in Africa and the Far East. Louis Sicking's chapter points to the progressive elaboration and expense of these fortification projects as one factor in the gradual decline of the Dutch East India company in the eighteenth century. New-style fortifications were hugely expensive to build and equip, and when constructed for urban defence, extremely disruptive to existing buildings and town layouts. As such, the decision whether or not to fortify could be hotly debated by states and communities in Europe; projects for the reworking or supplementing of existing 'medieval' walls and towers as an alternative to

21 Gábor Ágoston, *Guns for the Sultan: Military Power and the Weapons Industry in the Ottoman Empire* (Cambridge, 2009), for parallel evidence of Ottoman innovation and development of artillery.

22 Ch. 23 below, p. 566; Gunther Rothenberg, *The Art of Warfare in the Age of Napoleon* (London, 1977).

23 See also James Tracy, *Balkan Wars: Habsburg Croatia, Ottoman Bosnia and Venetian Dalmatia, 1499–1617* (Lanham and Boulder, 2016); Géza Pálffy, *The Kingdom of Hungary and the Habsburg Monarchy in the Sixteenth Century* (New York, 2009); A. Peacock (ed.), *Ottoman Frontiers* (Oxford, 2009).

complete rebuilding, as Simon Pepper shows, were widely adopted.[24] Pepper's chapter also points to the massive proliferation of improvised fortifications in military campaigning, mostly earthworks.[25] This was a logical extension of the ever-more elaborate siege-works constructed by besieging armies from the sixteenth century to protect against the artillery fire and sorties of the besieged and from relief forces. In 1810 the lines of Torres Vedras protected Wellington's retreating army so effectively that the French declined to attack the 29-mile-long double chain of forts, redoubts, artillery batteries, and obstacles. Such elaborate but temporary military defences could serve their purpose as effectively as monumental stone fortification projects such as Vauban's *Pre carré*, the double line of fortresses that guarded France's eastern and north-eastern frontier.

Artillery and fortification came together in a gradual transformation of the military landscape of Europe and those parts of the world that were part of European colonial or trading networks. The slowness of Asian powers to adopt European fortification techniques was certainly in part a consequence of their own lengthy tradition of building massive, stone and packed-earth city walls both for defence and for prestige: Chinese cities already possessed what were effectively artillery-resistant fortifications.[26] The eclectic adoption of European frontier-fort design and field fortifications nonetheless became a much more widespread phenomenon as the period continued. Schimmelpenninck details the two Chinese sieges of Albazin, an outpost of the Russian expansion into Siberia. The Chinese took Albazin in 1685 after a short siege, but the Russians returned later that year, rebuilt the citadel under the direction of a Prussian military expert, and the next year held out during a months-long siege until the Chinese finally withdrew.[27]

The other major application of gunpowder and its technologies, in the reshaping of naval warfare, was more clearly decisive, and perhaps the nearest thing to a genuine military revolution, or at least a 'revolution in

24 John Hale, 'To Fortify or Not to Fortify? Machiavelli's Contribution to a Renaissance Debate', in *Renaissance War Studies* (London, 1983), pp. 189–209; Richard Tuttle, 'Against Fortifications: The Defence of Renaissance Bologna', *Journal of the Society of Architectural Historians* 41 (1982), 189–201; Martha Pollack, *Cities at War in Early Modern Europe* (Cambridge, 2010); James Tracy (ed.), *City Walls: The Urban Enceinte in Global Perspective* (Cambridge, 2009).

25 See also James Tracy's account of the siege of Metz in 1552, Ch. 5 below, pp. 148, 160; and James Tracy, *Charles V: Impresario of War* (Cambridge, 2010).

26 Andrade, *The Gunpowder Age*, pp. 96–102; Edward Farmer, 'The Hierarchy of Ming City Walls', in James Tracy (ed.), *City Walls: The Urban Enceinte in Global Perspective* (Cambridge, 2009), pp. 461–87.

27 Ch. 20 below, p. 499.

military affairs', seen in this period. Decisive changes in the technology of warfare at sea took place in three distinct stages, sequentially examined through the chapters on navies by Niccolò Capponi, Robin Briggs, and Andrew Lambert. These technological shifts are marked first by the emergence of the stand-off naval engagement, deploying firepower rather than the 'ram and board' tactics dominant hitherto; the next phase, in the mid-seventeenth century, saw the development of the line-astern tactic and led to the rapid predominance of the purpose-build warship; the final phase in the first half of the nineteenth century saw the utilisation of steam power and the shift from wood to iron as the basis of warship construction.

Capponi's chapter traces the shift in naval technology through armed galleys and the experiments with the heavy-armed galleass, towards the broadside-mounted gun battery deployed on galleons and other square-rigged ocean-going sailing vessels. All of them marked a decisive shift away from naval tactics based on close-quarter engagement, boarding, and capture. The galleass, used with impressive effect against the Ottoman galleys at Lepanto (1571), was, Capponi argues, a striking example of a hybrid 'wonder weapon' that rapidly lost any decisive impact and was soon abandoned by its Venetian inventors. Instead, an evolution in sailing-warship design culminated, as both Capponi and Briggs agree, in the mid-seventeenth century.[28] The shift towards firepower, concentrated on two or three decks of broadside-mounted cannon, led to the adoption of tactics that moved away from individual fire-fights between combatant vessels in favour of squadrons arranged in line astern and raking each other with the full weight of their gun batteries.[29] During the couple of decades surrounding the Anglo-Dutch Wars (1650s–60s), the century-old deployment of fleets composed largely of co-opted merchant vessels temporarily fitted out with extra cannon became obsolete. Instead, vessels specifically designed to engage in this new type of artillery combat were built to new and massively stronger specifications, designed to carry batteries of heavy cannon on the lower decks, to accommodate the very large crews require to service anything up to 80 or 100 guns,

28 Geoffrey Parker, *The Military Revolution: Military Innovation and the Rise of the West, 1500–1800*, 2nd ed. (Cambridge, 1998), ch. 3; Geoffrey Parker, 'The *Dreadnought* Revolution of Tudor England', *Mariner's Mirror* 82 (1996), 269–300; Nicholas Rodger, 'The Development of Broadside Gunnery, 1450–1650', *Mariner's Mirror* 82 (1996), 301–24.

29 M. A. J. Palmer, 'The "Military Revolution" Afloat: The Era of the Anglo-Dutch Wars and the Transition to Modern Warfare at Sea', *War in History* 4 (1997), 123–49; Jan Glete, *Navies and Nations: Warships, Navies and State-Building in Europe and America, 1500–1860*, 2 vols. (Stockholm, 1993).

and to survive both the impact of enemy shot and the stresses of cannon battery discharges within the hull. Massive investment in a purpose-built weapon system was now the rule, and states not prepared, or able, to invest in both fleets and the supporting infrastructure of dockyards, construction, and maintenance were at a clear and permanent military-technological disadvantage.[30]

Yet after this initial and irrevocable leap from hybrid merchant / warship to the expensively purpose-built vessel rated by its number of decks and guns, there was a long period of relative stasis. As Briggs's chapter shows, most subsequent developments in the sailing warship were not the product of any radical technological transformation but the incremental effect of numerous smaller, and in many cases experience-related, innovations and modifications. To an uninformed observer the warships that fought at Trafalgar do not appear significantly different from those being built 150 years earlier. Not so the transformative leap forward described by Lambert over the next three or four decades, as the screw-propeller-driven steamship emerged from the confusing mass of rapid experiments, trials, and dead-ends following on from the development of steam power and the industrialised potential of iron and steel. As in the mid-seventeenth century, this was irrevocably to transform naval warfare and maritime power-projection across the globe. Initially steam-powered wooden battleships with familiar broadside-mounted gun batteries were developed in parallel with iron-hulled prototype warships, but by 1850 the supplanting of wood by iron was gaining momentum and pointed the way to another major technological transformation.

### *Military organisation and military culture*

Revolutionary or not, technological innovation, if it is to be effective, must be integrated into larger systems of military organisation: command and control, the management of logistical systems to ensure supply of food and munitions, and the overall strategic vision – the sense of what can realistically be achieved by the resources at the commander's disposal.

In all these areas comparisons and contrasts emerge amongst the chapters in this volume. An essential tenet of the original European 'military revolution' thesis was that changes in weaponry and tactics were managed by a far more professionalised officer corps. If the social background of officers remained largely that of the titled and landed elites, it is widely assumed

30 Nicholas Rodger, 'From the "Military Revolution" to the "Fiscal-Naval State"', *Journal for Maritime Research* 13 (2011), 119–28.

that their standards of military knowledge and institutionalised self-discipline and respect for military hierarchy were transformed over these centuries. Much has been made of the revival of classical military texts and their use by a generation of European military writers and staff officers in the late sixteenth century in theorising a new 'art of war' which consciously rejected what was assumed to be the individualist anarchy of feudal and knightly warfare.[31] Hitherto, it is argued, elite warriors had been more concerned to seek single combat and individual glory than to participate in the collective exercise of disciplined combat; their preparation for war was based on the cultivation of personal martial skills and an early apprenticeship in the military entourage of rulers or grandee commanders.[32] This 'before and after' story is exaggerated even in a European context: treatises on the art of war which drew on models of Roman and Greek military organisation and discipline and sought to accommodate these to contemporary weapon-systems had proliferated, above all in Italy, Spain, and France, for well over a century before the supposed 'rediscovery' of military antiquity by those indefatigable self-publicists, the Orange-Nassau clan.[33] And what is presented as the decisive rise of a literate and educated officer class, ready to subordinate instinct and impulsive leadership to a reasoned grasp of 'military science', looks altogether less distinctive from a global perspective. One section of Morris Rossabi's chapter focuses on the extraordinary quantity and detail of printed texts in Ming China concerning every aspect of the art of war, reflecting developments in weaponry, military deployment, and strategic thinking. Ming officers could draw upon massive resources of accumulated written knowledge, ideas about discipline, tactics, and military organisation.[34] Examinations for entry into, or promotion within, the officer corps took place in China, but also in Korea – admittedly a satellite state of the Qing after 1637, but where a Directorate General for military training had existed since the later sixteenth century. Indeed, in matters of acquiring regular skills of disciplining, controlling, feeding, and supplying armies effectively, European

31 M. van der Hoeven (ed.), *Exercise of Arms: Warfare in the Netherlands (1568–1648)* (Leiden, 1998); H. Hahlweg, *Die Heeresreform der Oranier und die Antike* (Berlin, 1941).

32 For a bracing rejection of these assumptions about medieval military organisation, see Christopher Tyerman, *How to Plan a Crusade: Religious War in the High Middle Ages* (London, 2015).

33 Marcello Fantoni (ed.), *Il 'Perfetto Capitano'. Immagini e realtà (secoli XV–XVII)* (Rome, 2001); Gonzalez de Léon, '"Doctors of the Military Discipline"'.

34 Equally the case under the Qing: S. R. Gilbert, 'Mengzi's Art of War: The Kangxi Emperor Reforms the Qing Military Examinations', in Nicola di Cosmo (ed.), *Military Culture in Imperial China* (Cambridge, MA, 2009), pp. 243–56.

officer corps only caught up with the more precocious examples of Middle Eastern and Asian military organisation in the later seventeenth century.[35]

The scale of military forces raised and maintained by powers from the Ottoman Empire eastwards as far as Japan required early engagement with the issues of command and control and logistical support systems. All the lastingly successful powers which emerged from the nomadic tradition of warfare had managed to find means to subordinate the localist, tribal-dominated recruitment of troops into some form of overarching command structure. Manz's chapter points to the original achievement of Chinggis Khan in creating an organisational and disciplinary structure that could counter the friability and clan rivalries of a force drawn from numerous extended tribes with their conflicting loyalties and antagonisms:

> Like earlier steppe armies, it was organised in decimal units from 10 to 10,000, but unlike most others, it was not organised tribally. He divided it into three sections, right wing, left wing, and centre, with commanders and troops assigned permanently to each. Chinggis also created an imperial bodyguard of 10,000 men, known as the *kesig*, which served as a training ground for military commanders. In both army and *kesig*, the central personnel were recruited from Chinggis's personal following, made up of people who left their own tribes to serve him. Tribes continued to exist but were not central to either military or administrative structure in the early Mongol empire.[36]

This model of internal restructuring to define command structures was a common legacy of the Mongols' heirs. The Manchu armies that were ultimately to conquer Ming China had grasped this nettle with the creation of the Eight Banners by the khan/military leader Hung Taiji – incorporating the tribal recruits from across Manchuria into a military structure under the overall command of members of the ruling family.[37] In some cases this aim to break down tribal substructures in the armies required more drastic measures. The initial achievements of the Safavid rulers had depended on the military skills and commitment of the Qizilbash, tribal contingents of horsemen serving for booty rather than pay, whose subordination to Safavid authority and military discipline was unreliable. As the shahs grew more successful and wealthier, the response to this conditional obedience was to create a parallel force of *gholams* – slave soldiers drawn from neighbouring territories such as Georgia and Armenia and paid directly from the ruler's

35 The study by Erik A. Lund, *War for the Every Day: Generals, Knowledge and Warfare in Early Modern Europe, 1680–1740* (Westport and London, 1999), points to a more significant transformation in the technical skills of officer corps.

36 Ch. 1 below, p. 54. 37 Ch. 19 below, pp. 471–2

revenues. These troops and their officers were the object of sustained training, promotion through merit, and tactical experimentation. The model was partly owed to the far better-known system of highly trained slave soldiers developed by the Ottoman sultans – the corps of Janissaries, recruited through the *devshirme* – levies of boys taken from Christian families in the Balkans. As Ágoston points out, the trajectory of Ottoman armies throughout this period was towards a larger and larger proportion of permanently salaried and directly appointed troops within the army, principally the Janissary corps. The Mughal Empire faced the same challenge of how to mould an army composed of huge tribal contingents of cavalry archers into a disciplined, cohesive force able to combine effectively with infantry and artillery units. As initial military success allowed the Mughals to consolidate their conquests on the subcontinent, and to recruit heavily from indigenous populations, the *mansabdari* system was introduced. Establishing a highly structured system of thirty-three military ranks through which career officers could be promoted and demoted, loyalty could be focused on the Mughal sultans, and a military structure created that overrode tribal groupings: officers raised troops from land grants conditionally made to them by the sultan, and were moved around the empire in fulfilment of military duties.

In all these cases we see what were at the outset steps to escape the fractious loyalties of armies composed of constituent tribal or clan groupings; these subsequently developed into a more systematic attempt to create a professionalised, formally trained officer corps, whether this was for the whole, or just an elite cutting edge for the army. Equally apparent from the discussions in the volume is the difficulty that most regimes encountered in maintaining military professionalism over any length of time. Both political expediency and social pressures weighed on professionalised officer corps, cankering promotional structures and encouraging sinecures and token appointments. Bernhard Kroener's study of the officer corps in eighteenth-century France and Prussia is an excellent example of this process in action. Of France, Kroener writes: 'Admission into the highest military ranks remained first and foremost a process based on social exclusion and social connections. This defeated all those initiatives and efforts since the reign of Louis XIV that had aimed to establish the counter-principle of military professionalism based on seniority and performance.'[38] A similar picture of mid-eighteenth-century military decline is offered by Michael Charney in his chapter on South-East Asia, where again court and factional groupings

38 Ch. 13 below, p. 355.

increasingly penetrated the officer corps, and elite status dominated promotion. 'Consequently, lethargy, indifference, and even resistance to change, as well as incompetence, slowed and confused the trajectory of martial developments in the region.'[39] Other cases of stagnation and regression are numerous: a burgeoning, innovative literature concerned with military theory and practice in Ming China could not prevent a military collapse that reflected economic weakness, political division, and demoralisation amongst an officer corps that had become in practice hereditary. Similar problems afflicted Safavid and Ottoman command and control as initially robust and professional structures started to lose out to placemanship, favour, and the lapsing of hitherto tight hierarchical control over the officer corps.

No less important in determining military success throughout this period was a grip on the logistical requirements of armies and navies. Forces of nomadic cavalry could support themselves through requisitioning and plunder in many operational circumstances, at least in the short term, but even these armies needed to be deployed with a strategic eye on the availability of forage for the horses.[40] Agriculture may have served as a more reliable and functional basis for military organisation in the long haul, but the problems of raising and deploying armies on the basis of primitive methods of cultivation and low productivity should not be underestimated. Commanders who assumed that they could maintain their armies on campaign through *ad hoc* requisitioning and pillage of supplies regularly discovered that an army of even 20,000–30,000 troops – the population of a medium-to-large city – imposed demands for food far beyond the capacity of local communities to produce or provide. The movement of immense armies dominated by infantry, probably with an artillery train and extensive supporting resources, allowed no such flexibility. The Ottoman armies that were brought together from military fiefs (*timar*) and standing troops, and assembled around Istanbul at the beginning of the campaign season, presented a massive logistical challenge. The ability to supply Ottoman armies on campaign, drawing heavily on the disposition of maritime and riverine transport to maintain supplies of food and munitions, was

39 Ch. 18 below, p. 459.

40 And as Rossabi points out, major campaigns by nomadic armies were based on sophisticated logistics: Tamerlane set up extensive supply lines running across the inhospitable territories of Central Asia in preparation for his massive attack on Ming China, which was forestalled by Tamerlane's own death: Ch. 2 below, p. 87; see also Beatrice Manz, *The Rise and Rule of Tamerlane* (Cambridge, 1989).

universally admired and respected.[41] Yet even the Ottoman troops greatly preferred campaigning in the Balkans, where supply mechanisms were more assured, to the harsh and frugal frontier wars with the Safavid Empire.

Effective logistical support was, first of all, a question of adequate financial support for operations; but it also invariably drew on managerial skill sets that could link the projected needs of military operations with private contractors and suppliers who could most effectively tap both local and far-flung markets in foodstuffs and munitions, and organise the transport, storage, and distribution of goods. The most effective examples of this sort of cooperation between military officials and the private sphere are seen in navies, where the negotiation of supply contracts ran parallel to a mass of carefully managed contracts for shipbuilding, maintenance, and the stockpiling of naval stores. Robin Briggs's chapter emphasises the central importance of this managerial culture to the outstanding success of the Royal Navy during the eighteenth century, increasing operational effectiveness by allowing ships to remain at sea for longer, and reducing losses of sailors to disease and poor diet.[42]

Reliable provisioning of armies remained the greatest single challenge to military operations, and the crucial question for much land warfare was how far commanders were prepared to allow their military operations to be dominated – or constrained – by the needs of amassing, transporting, stockpiling, and distributing supplies to maintain their forces with food and munitions. These chapters provide many examples of the devastating impact on the survival of armies of neglecting logistical needs. The price of avoiding this might be the adoption of a style of positional warfare constrained by magazines and supply convoys, focused on sieges of cities chosen more for their accessibility than their strategic significance.[43] But though this *ancien régime* military culture of slow-moving armies, with set-piece sieges and battles, undoubtedly existed, and not only in Europe, it was by no means the only line of development.

41 Ch. 3 below, pp. 105, 111; Caroline Finkel, *The Administration of Warfare: The Ottoman Military Campaigns in Hungary, 1593–1606* (Vienna, 1988); Rhoads Murphey, *Ottoman Warfare, 1500–1700* (London, 1999).

42 For more detail see Nicholas Rodger, *The Wooden World: An Anatomy of the Georgian Navy* (London, 1986).

43 John A. Lynn, 'The Evolution of Army Style in the Modern West, 800–2000', *International History Review* 18 (1996), 505–45; John A. Lynn, 'Linear Warfare: Images and Ideals of Combat in the Age of Enlightenment', in *Battle: A History of Combat and Culture* (Westview, 2004), pp. 111–44.

The latter half of the Thirty Years' War was fought by commanders who understood that military success was not necessarily about pitched battles, still less large-scale sieges, but an understanding of campaigning in which mobility, surprise, and flexibility would allow small, cavalry-heavy armies to notch up cumulative advantages against an enemy and his territory.[44] The legacy of this could be seen, as Carol Stevens points out, in the army composition and strategic approach of the Swedes: '[Charles XII's] army was . . . by comparison with other armies of the era, particularly mobile and aggressive in attack. It boasted effective heavy cavalry, which built upon the model of the Polish hussars . . . [Charles XII] famously distrusted firepower; he may also have minimised his use of artillery during the Great Northern War because of lengthy supply lines.'[45]

One enduring manifestation of this military culture, primarily adapted to the limitations of logistical support, was the style of irregular 'small war' (*kleiner Krieg*) which gained a lot of theoretical attention in the following century as well as some distinguished practitioners.[46] Small war became almost synonymous with European colonial warfare wherever large-scale local soldier-markets could not easily be tapped – as they could, for example, in India. Colonial warfare in the Americas, as waged by the Spanish, Dutch, English, and French was a constant struggle to make effective use of small – and often, thanks to disease, rapidly diminishing – forces in the face of what were often near-insuperable logistical challenges. As Kevin McCranie shows in his chapter on warfare in America, the British held a major advantage over their colonial opponents in the Revolutionary War so long as they could supply and move their armies by sea. As soon as warfare moved further inland, conventional forces became acutely vulnerable to supply failure, shown, for example, in General John Burgoyne's fatal advance down from Canada towards Saratoga. When the Royal Navy lost control of the Atlantic to the European fleets allied with the American colonists, the whole logistical base on which Britain could sustain its military operations in America began to crumble. This was nothing new: back in the 1590s the huge Japanese

44 David Parrott, *The Business of War: Military Enterprise and Military Revolution in Early Modern Europe* (Cambridge, 2012), pp. 139–95; Peter H. Wilson, *Europe's Tragedy: A History of the Thirty Years War* (London, 2009); P. Sörensson, 'Das Kriegswesen während der letzten Periode des Dreißigjährigen Krieges', in Hans Ulrich Rudolf (ed.), *Der Dreißigjährige Krieg. Perspektiven und Strukturen* (Darmstadt, 1977), pp. 431–57.

45 Ch. 10 below, p. 276.

46 George Satterfield, *Princes, Posts and Partisans: The Army of Louis XIV and Partisan Warfare in the Netherlands (1673–1678)* (Leiden, 2003); Johann Kunisch, *Der kleine Krieg. Studien zum Heerwesen des Absolutismus* (Wiesbaden, 1973); M. Ultee (ed.), *Adapting to Conditions: War and Society in the Eighteenth Century* (Alabama, 1986).

invasion of Korea collapsed when the Japanese could no longer supply their troops by sea, having lost naval command of the straits to the Koreans. Felipe Fernández-Armesto stresses another response to logistical and costing constraints for European forces waging war outside Europe: a massive reliance on co-opting indigenous peoples as fighting forces. The Spanish conquests and the maintenance of their empire in the New World rested heavily on the integration of native elites and their military resources. Extremely small forces of hardened Spanish troops, habituated to local diets and diseases, coordinated the military activity of indigenous warriors whose local survival skills and tolerance of inadequate support systems were far greater.

## State capacity and war: Military-fiscal state or contractor state?

### *The military-fiscal state*

The 'military revolution' has provided one lens through which warfare over these centuries has repeatedly been viewed, but as we have seen it is one that always threatens to view global military developments through a narrow focus on a European model of military evolution which does not even fit all European contexts. Gábor Ágoston encapsulates the issue in relation to the Ottoman Empire: 'For too long, Ottoman military and naval developments have been discussed in contrast to developments in a handful of paradigmatic West European polities, against whose armies and navies the Ottomans waged no war in the early modern era.'[47] Yet the fully fledged 'military revolution' thesis extends beyond changes in military technology, tactics, organisation, and strategy. In the usual iteration, these changes force an even more extensive transformation of state capacity through the expansion of centralised and bureaucratic governments, able to meet the challenges of maintaining larger, more sophisticated, and more expensive military systems. And this transformation in government has traditionally been linked to the rise of autocratic regimes, and in particular the establishment of 'absolute' monarchies.[48]

47 Ch. 3 below, p. 101.

48 C. Tilly, *Coercion, Capital and European States,* AD *990–1992* (Oxford, 1992); William H. McNeill, *The Pursuit of Power: Technology, Armed Force and Society since* AD *1000* (Chicago, 1984). Other examples include: R. Bean, 'War and the Birth of the Nation State', *Journal of Economic History* 33 (1973), 203–21; Bruce Porter, *War and the Rise of the State* (New York, 1994); Martin Van Creveld, *The Rise and Decline of the State* (Cambridge, 1999).

This direct link between military change and state formation has aroused its fair share of scepticism and criticism even within a Europe-focused historiography. Jeremy Black, in a prescient study from 1991, questioned the military significance of developments in the period from 1560–1660, and argued that military changes in the period from 1660–1760 had a better claim to be considered transformative, not least citing the exponential increase in the size of European armies after 1660. Linked to these assertions was the larger case that this expansion of military capacity and resources should be seen not as the *cause* of subsequent developments in administration and government, but as the *consequence* of such changes. It was the greater capacity of governments to raise taxes, quell opposition, and manage their resources that made possible the transformation of warfare, whether in the form of armies running to hundreds of thousands of troops or navies composed of purpose-built rated warships.[49]

This challenge to an assumed relationship between war and state building was taken in another direction by what has been an extended debate about whether, in Peter Wilson's words, 'European states either adopted an authoritarian "coercion-intensive" solution to the pressures of war, or followed a constitutional "capital-intensive" route'.[50] Against the model of the absolutist states of France, Spain, or those of East-Central Europe, with their ability to coerce the payment of taxes and impose service obligations, could be set the constitutional regimes of the Dutch Republic and Britain. The traditional assumption that powerful, centralised monarchies were the model for the effective mobilisation of resources and waging of war could be challenged by the evidence that constitutional regimes were more efficient in collecting taxes – largely because they had more ability to extract resources from the wealthier classes within their societies.[51] Still more important, these states had the capacity to finance war by borrowing money, if not in greater

49 Jeremy Black, *A Military Revolution? Military Change and European Society, 1550–1800* (London, 1991); also Jeremy Black, *War and the World: Military Power and the Fate of Continents, 1450–2000* (New Haven, 1998).

50 Ch. 7 below, p. 193.

51 Brian Downing, *The Military Revolution and Political Change: Origins of Democracy and Autocracy in Early Modern Europe* (Princeton, 1992); Thomas Ertman, *Birth of the Leviathan: Building States and Regimes in Medieval and Early Modern Europe* (Cambridge, 1997). More specifically related to the case of Britain: Michael Braddick, *The Nerves of State: Taxation and the Financing of the English State, 1588–1714* (Manchester, 1996); John Brewer, *The Sinews of Power: War, Money and the English State, 1688–1783* (London, 1989) – but see also Lawrence Stone, 'Introduction', in Lawrence Stone (ed.), *An Imperial State at War: Britain from 1689–1815* (London, 1994), pp. 9–21.

quantities than absolute monarchies, then certainly at lower rates of interest and on more sustainable terms.[52]

This debate has been further nuanced by several decades of research and reconsideration of 'absolutism' and the basis on which state authority was exercised. Starting with that traditional paradigm of absolute monarchy, the France of Louis XIV, historians have argued that a rhetoric of absolutist assertion and representation, combined with such monarchs' undisputed legal authority, has concealed the reality of a state in which effective power depended on compromise and collaboration with the socio-political elites.[53] An alternative pattern has been depicted in which monarchical authority is sustained through policies which shared the priorities and interests of provincial or institutional elites, or which respected the privileges and resources of these groups. This had clear implications for military organisation. In Russia, Stevens shows how the creation of a vast, permanent army from the late seventeenth century, based on military conscription, high taxation on the unprivileged, and enforced noble military service in the officer corps, was bought through the permanent concession of peasant enserfment to the landowning class, the granting of massive power and influence to the court aristocracy, and tacit acceptance of an officer corps dominated by noble privilege and interests.

One of the most characteristic features of states in which authority rested on collaboration between ruler and elites was purchase and proprietorship of military office. This is the central theme of Bernhard Kroener's chapter on France and Prussia, and in both cases he demonstrates how military venality was a concession to the social aspirations and established influence of the nobilities. Of course the 'public–private partnership' implied by purchased military office brought financial benefits to the ruler, as wealthy proprietors invested in their regiments for reasons of personal prestige or against the office's resale value.[54] But at the same time proprietorship embedded

52 Richard Bonney (ed.), *Economic Systems and State Finance* (Oxford, 1995); Guy Rowlands, *The Financial Decline of a Great Power: War, Influence and Money in Louis XIV's France* (Oxford, 2012); Guy Rowlands, *Dangerous and Dishonest Men: The International Bankers of Louis XIV's France* (Basingstoke, 2014).

53 William Beik, *Absolutism and Society in Seventeenth-Century France: State Power and Provincial Aristocracy in Languedoc* (Cambridge, 1985); William Beik, 'Absolutism as Social Collaboration in Lous XIV's France', *Past and Present* 188 (2005), 195–224; Darryl Dee, *Expansion and Crisis in Louis XIV's France: Franche-Comté and Absolute Monarchy, 1674–1715* (Rochester, 2009); James B. Collins, *The State in Early Modern France* (Cambridge, 2012).

54 Fritz Redlich, *The German Military Enterpriser and His Work Force, 14th to 18th Centuries*, 2 vols. (Wiesbaden, 1964); Guy Rowlands, *The Dynastic State and the Army under Louis XIV: Royal Service and Private Interest, 1661–1701* (Cambridge, 2002); Hervé Drévillon,

networks of patronage and influence, limited direct administrative control, created conflicting interests through the proprietors' wish to recoup money through regimental management, and led to the tightening of social and financial exclusivity in awards of commissions and promotion.[55]

All of this has an impact on the way that European state-formation and the impact of war is viewed, and challenges 'preconceptions about functioning hierarchies and chains of command, an increasingly effective military administration, rigid discipline, and corresponding efficiency in the waging of warfare'.[56]

One way of approaching the relationship is to abandon assumptions that war-waging should be tied to a broad notion of state development. Recent focus on what have been termed 'military-fiscal states' has been less interested in categorising states in terms of the general types of government they possess: authoritarian versus constitutional, centralised or decentralised, coercive or collaborative. Consideration has started instead from the premise that the early modern state is fundamentally geared to the waging of war, whether in pursuit of the dynastic objectives and reputation of its rulers, the commercial and economic ambitions of a ruling oligarchy, the interests of a militarised ruling elite, or simply the defence of territorial or confessional integrity.[57] This requires jettisoning the assumption that such states are driven by a wide range of policy objectives and ambitions: economic planning and development, social and confessional regulation, the development of central authority at the expense of localities and institutions. If these are pursued at all, they will be of secondary and negotiable importance to the waging of war. So rather than thinking holistically about the state and its purposes, military-fiscal capacity focuses on its central priority, and ultimately its *raison d'être*, to which both ideology and practice were subordinated.

*L'Impôt de sang. Le métier des armes sous Louis XIV* (Paris, 2005); Alan Guy, *Oeconomy and Discipline: Officership and Administration in the British Army, 1714–63* (Manchester, 1985); A. P. Bruce, *The Purchase System in the British Army, 1660–1871* (London, 1980).

55 Ch. 13 below, p. 348; Hamish Scott and Christopher Storrs, 'The Military Revolution and the European Nobility, *c.* 1600–1800', *War in History* 3 (1996), 1–41; E. Léonard, *L'Armée et ses problèmes au XVIIIe siècle* (Paris, 1958); Bernhard Kroener, 'Militärischer Professionalismus und soziale Karriere. Der französische Adel in den europäischen Kriegen, 1740–1763', in Bernhard Kroener (ed.), *Europa in Zeitalter Friedrichs des Großen. Wirtschaft, Gesellschaft, Kriege* (Munich, 1989), pp. 99–132. More broadly, see William Doyle, *Venality: The Sale of Offices in Eighteenth-Century France* (Oxford, 1996).

56 Ch. 13 below, p. 342.

57 Christopher Storrs (ed.), *The Fiscal-Military State in Eighteenth-Century Europe* (Farnham, 2009); Bernhard Kroener, '"Das Schwungrad an der Staatsmaschine?" Die Bedeutung der bewaffneten Macht in der europäischen Geschichte der Frühen Neuzeit', in Bernhard Kroener and Ralf Pröve (eds.), *Krieg und Frieden. Militär und Gesellschaft in der Frühen Neuzeit* (Paderborn, 1996), pp. 1–23.

Given this priority, with its consequent imperative to generate resources to wage war, the comparative questions to be asked about different states relate directly to the means and the efficiency with which they gather resources – especially finance – and the ways in which such resources are used to wage war. In raising money to meet military expenditure we should not seek political consistency in policy-making and practice: a state like Louis XIV's France was readily prepared to surrender direct royal authority or to confirm privileges in return for injections of cash, or even rely upon financial intermediation by the crown's privileged subjects to allow it to raise more money at lower rates of interest.[58] Military-fiscal states should also be examined in terms of how they waged war, measured in terms of sustainable military capacity, the overall size of armies and navies relative to population, and the ways in which these forces were managed and controlled. As Carol Stevens points out, Russia possessed a huge resource superiority over the Scandinavian states, and from the early eighteenth century developed a centralised, coercive mechanism to raise and maintain its army, but one which accepted high levels of corruption and inefficiency in military management, was inherently expensive to run, and was reliant on draconian policing. It was not, Stevens adds, a model that would be widely imitated.

Does the concept of the military-fiscal state, especially given its very particular origins in debates about European state formation, have relevance to the wider global world? A major attraction of the concept to historians of Europe has been its potential to explain the dramatic increases in the size and permanence of European armies and navies, whether the first round of sustained increases in the period from roughly 1500 to 1560, or the much more significant and permanent growth from the later seventeenth century onwards. The global challenge here is that almost all European accounts of army size and troop mobilisation before the 1680s look puny in comparison with the resources raised by the great empires of the Far East and the Islamic empires, especially the Ottomans and the Mughals. Roy suggests that at the height of its strength, and drawing on a population of about 150 million, the Mughal army had on its payroll about 600,000 cavalry and 500,000 infantry. Meanwhile, Ming campaign armies raised against the Mongols under the Yongle Emperor in the first half of the fifteenth century are reported – in not entirely reliable sources – to have been of between 250,000 and 500,000

58 Rowlands, *The Financial Decline of a Great Power*; Mark Potter, *Corps and Clienteles: Public Finance and Political Change in France, 1688–1715* (Aldershot, 2003); Julian Swann, *Provincial Society and Absolute Monarchy: The Estates General of Burgundy 1661–1790* (Cambridge, 2003).

troops. At the same time the first of the great Treasure Fleets despatched into the Indian Ocean in 1406 consisted of 317 vessels and around 27,000 men. The initial Japanese invasion of Korea involved an army of 160,000 men, and was opposed by an alliance of up to 60,000 Chinese and 84,500 Korean troops.[59] Ágoston draws attention to impressive levels of military mobilisation by the Ottomans, pointing in particular to the military effort during Istanbul's 'Thirty Years' War' (1578–1611), when Ottoman troops fought, often simultaneously, against the Safavids in the east (1578–90, 1603–11), the Habsburgs in the north (1593–1606), and the Celali rebels in eastern Anatolia and northern Syria (*c.* 1595–1610). Even the Safavid Empire, based on a much more limited total population, could still draw on large-scale cavalry forces drawn from tribal communities to raise up to 100,000 troops.[60]

In contrast, it was not until the later seventeenth century that the military establishments of major European states permanently crossed a threshold of 100,000 troops, though from the 1680s the upward push was dramatic. During the War of the League of Augsburg (1688–97) the French crown achieved the remarkable feat of maintaining a military establishment of up to 340,000 troops on land, and 50,000–70,000 men in its navy. This was an entirely new development; despite figures regularly cited to suggest that armies of 100,000 were a feature of the Thirty Years' War, these were never more than the shortest-term 'spikes'. Overall military establishments were maintained at a typical strength of 50,000–60,000, with France and Spain as outliers whose total military strength might edge closer to 70,000–80,000.[61] European army size essentially flatlined between the mid-sixteenth and the mid-seventeenth century, and fell considerably short of the levels of military mobilisation that had been achieved elsewhere in the world centuries earlier.

How should this divergence be understood in terms of military-fiscal systems? For some it is an obvious means to challenge a European perspective which, as we have seen, is deeply rooted in interpretations of the history of early modern war. 'To a historian specializing in the non-European world there is something puzzling about the excitement with which European historians hail the arrival of . . . standing armies, legal codes, bureaucracies,

59 Ch. 17 below, p. 434; see more generally Kenneth Swope, *A Dragon's Head and a Serpent's Tail: Ming China and the First Great East Asian War, 1592–98* (Norman, 2016).

60 Ch. 3 below, p. 115; Ch. 11 below, p. 293.

61 Wilson, *Europe's Tragedy*; Jürgen Pohl, *Die Profiantirung der keyserlichen Armaden ahnbelangendt. Studien zur Versorgung der Kaiserlichen Armee 1634–35* (Vienna, 1994); David Parrott, *Richelieu's Army: War, Government and Society in France, 1624–42* (Cambridge, 2001); Geoffrey Parker, *The Army of Flanders and the Spanish Road, 1567–1659* (Cambridge, 1994).

absolutist rulers ... as if they were unique to Europe and self-evident stepping stones to modernity.'[62] Certainly the logistical and organisational challenges of mustering, feeding, equipping, and controlling armies of over 100,000 men were daunting: the seventeenth-century Italian commander and theorist Raimondo Montecuccoli wrote that no commander could realistically manage a campaign army of more than 30,000 men.[63] In contrast, Ottoman, Chinese, Japanese, and other non-European governments had achieved impressive levels of effectiveness in managing supply systems, communication networks, garrisoning, and weapons procurement early in their respective histories. In many of these the acquired management skills for these enterprises were passed on thanks to the records and writings of specialised administrators working in structured bureaucracies.

One global distinction seems essentially cultural. A shared characteristic of European military-fiscal states was that, by whatever means – coercive, consensual, or concessionary – the ruler and government sought to maximise the financial resources made available for the support of their military. Even if this weakened the ruler's direct authority, undermined the principle of a 'monopoly of violence', or embedded medium- and long-term weaknesses into the fiscal system, the priority of maximising revenue extraction prevailed. This priority, as Roy indicates, could also be seen in the Delhi Sultanate and the policies of successor Mughal rulers. The Mughal state at the height of its power was extracting up to 50 per cent of the gross agricultural product of the territories over which it ruled, while about 80 per cent of the Mughal budget went to meet military expenditure. Similar initiatives can be seen in the increased demands for revenues and in new fiscal expedients adopted to support the growing proportion of permanent troops in the Ottoman armies. No less typical was the considerable provincial resistance evident by the late sixteenth century to this tightening of the Ottoman fiscal vice.

Yet in contrast, in both Ming and Qing China there was considerable, deep-seated opposition to the assumption that the purpose of government was to maximise resources that could be devoted to the waging of war. Historically China, and various of the political and cultural 'satellite states' in South-East Asia, had imposed comparatively low levels of taxes. Political virtue was strongly associated with fiscal restraint and concern for the welfare of subjects. This could certainly clash with military demands: even in peacetime, expenditure on the

62 Patricia Crone, *Pre-Industrial Societies* (Oxford, 1989), p. 48.
63 Thomas M. Barker, *The Military Intellectual and Battle: Raimondo Montecuccoli and the Thirty Years War* (Albany, 1975), p. 60.

two types of Qing troops (Eight Banners and Green Standard) ran at around 55 per cent of the imperial budget during the mid-eighteenth century. Rossabi shows how, during the reign of the Yongle Emperor in the early fifteenth century, extended and large-scale military campaigns in Annam and Mongolia, combined with the great naval expeditions of Zheng He, involved levels of expenditure which horrified the court officials, and led to the abandonment of all these military policies within three years of the emperor's death.

For less strongly ideological reasons, both Japan, from the mid-seventeenth century and the beginning of its period of *sakoku* (isolation), and Korea, effectively a Manchu protectorate, enjoyed a lengthy period of low-intensity military activity; both had military establishments, but these were embedded within government, court, and wider society. The huge costs of earlier military operations – the Japanese invasions of Korea in the 1590s, for example – were not to be incurred subsequently. The frantic fiscal extraction and innovation seen elsewhere as states grappled to fund upward spiralling military budgets could be sidelined: the fiscal regimes were by no means economically enlightened, but financing war was not the dominant fact of the state's existence.

## *The contractor state*

Awareness of how states created mechanisms to raise and borrow money and to expand their capacity to meet the costs of warfare has been matched in recent years by attention devoted to how the money was spent. Are there common patterns to the ways in which states set about funding and administering warfare? One important observation in an overview of the 'spending of states', especially in the period of this volume, is that a large, and often growing, part of the management of war was outsourced into the hands of private agents. How far in military terms was this the age of the 'contractor state'?[64] Rather than meeting the challenges of waging large-scale and protracted warfare through an expansion of centralisation and bureaucracy, the response of states was to make use of a huge variety of private agents to

64 Recent writing on the 'contractor state includes Roger Knight and Martin Wilcox, *Sustaining the Fleet, 1793–1815: War, the British Navy and the Contractor State* (Woodbridge, 2010); Gordon Bannerman, *Merchants and the Military in Eighteenth-Century Britain: British Army Contracts and Domestic Supply, 1739–1763* (London, 2008); Rafael Torres Sánchez (ed.), *War, State and Development: Fiscal-Military States in the Eighteenth Century* (Pamplona, 2007); Stephen Conway and Rafael Torres Sánchez (eds.), *The Spending of States: Military Expenditure during the Long Eighteenth Century: Patterns, Organization and Consequences, 1650–1815* (Saarbrücken, 2011); Rafael Torres Sánchez, *Military Entrepreneurs and the Spanish Contractor State in the Eighteenth Century* (Oxford, 2016).

manage and support most aspects of their war effort. These contractors – merchants, financiers, manufacturers, transport operatives – provided specialist expertise, interconnecting networks of contacts and agents, more efficient and reliable access to credit, and an extended reach when it came to procurement of goods and skills. Recognition that private networks of merchants or financiers had an international, sometimes even a global, reach, which extended well beyond anything that could be established by state agencies, was a major factor in determining heavy reliance on the private sector. No less important was the awareness that more capital could be raised more quickly and at better rates of interest through the agency of private contractors than via the state and its direct representatives. Adopting James Tracy's memorable phrase: 'In fiscal terms, the putative age of centralization is better called an age of intermediation.'[65]

How far did this willingness to depend on military contracting extend? Amongst European states, the most extensive period of military outsourcing occurs, with some irony, in exactly the century – 1560–1660 – that is traditionally identified with the chronology for 'military revolution'. For it was in these decades, culminating in the Thirty Years' War (1618–48), that most European states did not just outsource the provisioning, equipping, and transportation of their armed forces, but contracted out the direct recruitment, maintenance, and control of their armies and navies as well. The rise of the 'military contractor' is described in Peter Wilson's chapter and is a distinct phenomenon standing between the almost universal use of hired mercenaries, and the more 'managed' systems of military proprietorship – regimental and company ownership – which emerged after 1660, and whose benefits and problems are described in Bernhard Kroener's chapter. The state's primary concern in making a contract with a military enterpriser, who would take charge of the recruitment, appointments, ongoing costs, and discipline of one or more military units, was financial. Rulers grasped the opportunity to intermediate, to offload the heavy costs of raising and operating all, or large parts, of their armies and navies onto the shoulders of their senior officers, most of whom raised the money for this through their personal connections or their access to bankers and other credit sources. Reimbursement for the financial commitment of the

65 Ch. 5 below, p 148; James Tracy, 'The Age of Fiscal Intermediation in Europe, 1300–1800', in Jaco Zuijderduijn and Dries Raeymaekers (eds.), *Publieke financiën in de Lage Landen (1300–1800)* (Maastricht, 2015), pp. 13–26; James Tracy, 'Taxation and State Debt', in Hamish Scott (ed.), *The Oxford Handbook of Early Modern European History, 1350–1750*, 2 vols. (Oxford, 2018), vol. 2, pp. 512–37.

enterprisers was most often provided through further outsourcing, this time to delegate the right to collect taxes. Military commanders sought to make their profits from this fiscal decentralisation, through which they could collect military taxes or 'contributions' over designated areas, imposing greatly increased financial burdens on communities under the direct threat of military coercion.

The most notorious military enterprisers were the so-called 'general contractors', principally associated with the Thirty Years' War, who were prepared to raise entire armies under their own authority and credit. Sometimes, as with the best known of these, Albrecht von Wallenstein, duke of Friedland (1583–1634), this was in response to an invitation from the contractor's sovereign, in this case the Holy Roman Emperor. In other cases, like those of Bernhard von Sachsen-Weimar (1604–39), or Duke Charles IV of Lorraine (1604–75), the armies were essentially freebooting forces hired out to the highest bidder.[66] In all these cases money raised or borrowed by the commander was combined with resources advanced by regimental officers, acting as subcontractors to raise and maintain the army, extending the network of credit and spreading the risk.[67]

Given that it so directly challenges the assumption that 'modernising' governments and states were working towards a monopoly of violence, it is unsurprising that private enterprise at the immediate level of raising, maintaining, and directly controlling sizable portions of the state's armed forces has attracted scrutiny. Too often, though, it has been marginalised as a failed experiment, associated principally with the Thirty Years' War, and abandoned by rulers and their states once circumstances permitted. In fact, military outsourcing was a logical development in the context of military-fiscal states whose *raison d'être* was the raising and maintenance of internationally competitive armies and navies as cost-efficiently as possible. As Wilson stresses when examining the military policies of early modern rulers: 'Centralisation and consolidation were rarely desired and certainly not goals in themselves.'[68] Moreover, in the haste to write off military contracting as a failed and rejected experiment, insufficient attention has been given to the ways in which outright military enterprise evolved into much longer-term

66 Golo Mann, *Wallenstein: His Life Narrated* (London, 1976); Geoffrey Mortimer, *Wallenstein: The Enigma of the Thirty Years War* (Basingstoke, 2010).

67 Redlich, *The German Military Enterpriser*; Parrott, *The Business of War*.

68 Ch. 7 below, p. 198; these issues are examined in S. Förster, C. Janesen, and G. Kronenbitter (eds.), *Rückkehr der Condottieri. Krieg und Militär zwischen staatlichem Monopol und Priviatisierung. Von der Antike bis zur Gegenwart* (Paderborn, 2010).

systems of military venality and proprietorship.[69] These systems retained many of the features of a public–private partnership: the individual, family, and corporate interests of the office-holders remained, as Kroener shows, embedded in the military systems of *ancien régime* states. They also provided a substantial source of informal financial support for the levy and maintenance of cavalry and infantry units, and cemented the political ties between rulers and social elites.[70]

Less controversial, but no less noteworthy in challenging assumptions that the 'rationalising' pressure of managing and administering warfare always tended towards greater direct administration and centralised control, is the massive and continuous presence of private contracting in every aspect of military supply, maintenance, and support. From the individual sutlers, who were very frequently women – *vivandières* – who sold supplementary food, drink, and other goods to soldiers in encampments or on the march, up to the private owners of shipyards where warships were built and repaired under contract, the entire military support system remained dependent on the functions, skills, and access to resources of private contractors. Far from diminishing in importance, this private component of military and naval infrastructure expanded in the eighteenth century. The main development in this century came in the closer and more effective supervision of the 'contractor state' by government agents both at the centre and on the ground, whether in shipyards, magazines, supply depots, or abattoirs.[71]

Better oversight was also a factor in another area of continuity, the direct hiring of mercenary regiments, naval squadrons, or other military units by European states. Military enterprise should be seen as an activity that temporarily overshadowed, but never replaced, the time-honoured practice of hiring foreign mercenaries to serve alongside and supplement forces that were raised, by whatever means, within the state. Whether because of their specialist skills – the Swiss and *Landsknecht* pike-squares; German pistol-armed cavalry or *Reiters*; East European or Balkan light cavalry often known as Cossacks or Croats – or the flexibility of 'hiring and firing' such units, the attraction of recruiting mercenary contingents was a constant through to the end of the eighteenth century, and in some cases beyond.

69 An eclectic survey is provided in Erik-Jan Zürcher (ed.), *Fighting for a Living: A Comparative History of Military Labour 1500–2000* (Amsterdam, 2013).

70 Rowlands, *Dynastic State and the Army*; Christopher Duffy, *The Army of Maria Theresa* (London, 1977).

71 Roger Knight, *Britain Against Napoleon: The Organization of Victory, 1793–1815* (London, 2014); Sánchez, *Military Entrepreneurs*; Bannerman, *Merchants and the Military*.

The obvious distinction lay in the relative balance of control between mercenary and contracting state. Swiss mercenaries or German *Reiters* in the sixteenth century occupied a sellers' market in which the heavy expense of hiring their services was combined with *de facto* autonomy of disciplinary and judicial status, and a great deal of unpredictable small print about what services their contract might or might not bind them to offer. In contrast, by the eighteenth century the balance had decisively shifted: Hessians hired for the British army, Swiss regiments in French service, and Germans in Danish or Russian service were all expected to conform to the disciplinary authority and military commands of the host army, and often served under lengthy contracts which assimilated them more effectively into the command structure and military culture of the national force.[72]

Military contracting has been discussed so far in a European context, but how far is that to provincialise a wider, global phenomenon? Gábor Ágoston tersely notes the double standard at work when discussing the outsourcing of military functions. 'For too long, historians have viewed fiscal decentralization and military devolution as signs of "Ottoman decline", responsible for Ottoman military defeats in the seventeenth and eighteenth centuries. However, military devolution and governments' reliance on local elites and military entrepreneurs in war-making had been the norm in most European states until about the mid-eighteenth century.'[73] In reality, it was precisely their increasing reliance on the outsourcing of military functions that provided the Ottomans with the means to sustain their ever-growing armies and gave them vital flexibility in financing and supplying these forces. The Ottomans had drawn on the services of *martoloses* and *voynuks*, Christian military forces hired from the Balkans and Greece, and they had consistently used the support of the Barbary corsairs as contractors raising auxiliary naval forces to serve with or in place of the Ottoman fleets.[74] The same is true of India, where armies, from the Rajput rulers through the Delhi Sultanate and the Mughals, could be built on a vast pool of potential military labour: peasant-soldiers, trained in weapons use, and easily drawn into military service. Faced with Mughal invasion, the Lodhi Sultan, for example, successfully competed in what were traditional Mughal recruiting grounds, notably Afghanistan, to raise the army that was defeated at Panipat in 1526. Kaushik

72 Peter H. Wilson, 'The German "Soldier Trade" of the Seventeenth and Eighteenth Centuries: A Reassessment', *International History Review* 18 (1996), 757–92.

73 Ch. 3 below, p. 113.

74 Gábor Ágoston, *The Last Muslim Conquest: The Ottoman Empire and Its Wars in Europe* (Princeton, 2021), p. 51; Ch. 3 below, p. 106.

Roy's chapter raises many important points about how the predecessor regimes to the British East India Company (EIC) managed the hiring, maintenance, and control of their armies. He has little time for the argument that the EIC had the administrative and cultural mindset to take control of contracted soldiers in a way that had not previously been achieved in India, and makes a strong case for the ability of the Mughal emperors to tap and control the military labour market via the *mansabdari* system, closing off large sections of the market from their potential opponents.

The hiring of foreign mercenaries was a global phenomenon, and the volume provides numerous examples. Felipe Fernández-Armesto points to the heavy use of hired soldiers in the Spanish Empire, no less than in the armies which asserted Spanish power in Europe. Native auxiliaries, serving with a strong sense of their local identities – and local rivalries – were a mainstay of conquest and colonisation in the New World; they formed the great bulk of the forces whose leadership and 'cutting edge' were the small private forces of self-financing European adventurers. Charney points to Christians, driven out of Japan to escape persecution in the seventeenth century, who joined the courts of Ayudhya, Arakan, and even the Spanish Philippines to serve as mercenary bands. Even the general contractor raising entire private armies was by no means unique to Europe. Richard Reid discusses territories in the African interior where the trend was towards private armies raised and maintained by entrepreneur warlords. Powerful leaders across East Africa built and retained – usually through payments in cattle – networks of military clients that could be mobilised for military service. In West Africa, fortified city-states, the most powerful of which was Ibadan, were dominated by armed entrepreneurs who commanded professional armies of young men. Drilled, disciplined, and trained in battle-field manoeuvre and in the use of firearms, these forces hired themselves out to the highest bidder.

Sea power was often placed into the hands of private contractors, especially when the size and design of vessels made them more accessible to individual financial initiatives. One consequence was that the line between a state's naval policy and self-interested privateering might well be blurred. Endemic Japanese piracy in the China Sea did not serve any discernible military interest of the Tokugawa Shogunate. But the successful Maratha navy, which was privately financed and run by local chiefs, was as much an agent of state power as any privately funded galley squadron in the Mediterranean. This fleet, which emerged in the later seventeenth century, was composed of small, lightly armed ships able to operate in shallow waters out of reach of European warships, and buttressed state opposition to the encroachment of the EIC by inflicting heavy losses on British merchant vessels.[75]

Only in the Far East, and in those areas where the nomadic warrior tradition of universal military skills and service prevailed, was the use of mercenaries and contracted expertise less evident. Though even in Qing China, where military activity was undertaken by a semi-hereditary soldier-population under – theoretically – tight bureaucratic control, the supply of foodstuffs to armies was regularly outsourced to merchant contractors.[76] The imperial government devised the so-called *kaizhong* system, by which it commissioned merchants to transport supplies, and provided traders with licences to purchase salt and tea from the state monopolies of those products. Russia presents two distinctive extremes. Carol Stevens describes a Russian army operating on the Baltic and in Eastern Europe that increasingly rejected reliance on foreign mercenary units (though not foreign military expertise) and developed into a state-run institution under the autocratic initiatives and centralising tendencies of Peter I and his successors. In contrast, David Schimmelpenninck's account of the push into Siberia from the later sixteenth century provides an account of private war being waged, on behalf of the tsar, by the hired Cossacks of the Stroganov family against the Tatar khanates.

## War and empire

Military revolution, the fiscal-military state, and the contractor state have all been organising concepts through which warfare has been interpreted across this period. A slightly different focus has been the concept of empire and empire-building. By this is usually meant the construction of global empires by the European powers, most obviously the European maritime powers. The period is interpreted through the various ways in which these powers developed their military technology and military organisation in order to facilitate a steady process of trading and territorial encroachment on the rest of the world, culminating in the great scramble for territorial possession in the nineteenth century. But especially if the period under scrutiny is that of 1450–1750, very different conclusions about global empire and military force can be drawn from the evidence. From the fall of Constantinople in 1453 to

75 Ch. 16 below, pp. 423–4.
76 Further details in Yingcong Dai, 'Military Finance of the High Qing Period: An Overview', in Nicola di Cosmo (ed.), *Military Culture in Imperial China* (Cambridge, MA, 2009), pp. 296–316.

the Qing conquests of Tibet and much of Mongolia, this was the great expansionist age of Asian empires. Only in the last century – perhaps even the last half-century – of the period did a combination of European technological and resource advantage and a shift in ideas of empire lead to a dramatic readjustment towards European global predominance.

Asian and Near-Eastern empire-building raises important questions about the nature of military organisation and success, addressed by the authors in this volume in their discussions of China, Japan, maritime South-East Asia, Russia, and the great Islamic empires of the Ottomans, Mughals, and Safavids. It was no less the case, as Richard Reid points out, that some of the most successful and aggressively expansionist of the sub-Saharan African states proved highly effective empire-builders. What explains the military successes of these states – their capacity to both conquer and then exercise cost-efficient control over land and peoples, and to hold together contiguous territories from an imperial centre? The shared legacy of many of the Asian imperial projects was Chinggis Khan and his successors, the Mongol empires that at their height stretched from Poland to Japan.

The greatest single strength of the nomadic legacy empires was the ability to mobilise virtually the entire adult male population for military operations. They could wage war with daunting numbers, far beyond the capacity of sedentary societies, who either came to regard warfare as a specialist skill or could only mobilise large proportions of the populations for short periods in between agricultural activities. Moreover, the aims of nomadic empire-building were straightforward: unlimited outwards expansion, crushing the military opposition of adjoining states, and profiting first by plunder, then by the imposition of regular tribute payments on conquered territory. Where the conquered peoples could match or, more often, complement the military systems of the nomads, they were offered the opportunity to be incorporated into the military forces. The primary weakness of the nomadic imperial system was its tendency to fragment; as we have seen, the essential building blocks of the military system were tribal. The success of Tamerlane and his successors of the Timurid dynasty lay in subordinating the tribal loyalties and factional instincts of their supporters to overarching imperial organisation, ideologies, and cultures. In Tamerlane's case this involved the construction of an elaborate city and court at Samarqand. But lasting stability proved impossible to achieve, and it was to be the heirs of these nomad empires who managed to create the durable empires of the Middle and Far East.[77]

The obvious contrast with European powers for much of this period, even allowing for a large degree of exaggeration in surviving accounts, is the scale of military mobilisation and permanent forces achieved by these heirs to the Mongols. And this was not merely the ability to mobilise the adult male population for occasional military service. After the Manchus' conquest of China, the Qing dynasty retained the Eight Banner army with its potential to mobilise up to a million soldiers. Crossley comments that the banner system, rather than being based on a population of farmers who sometimes went to war, was a vast, funded caste of warriors who sometimes performed other activities.[78]

These imperial 'heirs of Tamerlane' also possessed ideologies of imperial expansion which were both robust and in many cases pragmatic – in sharp contrast to Europe. For Europeans from the fifteenth to the seventeenth century there was only one empire, the Holy Roman Empire, heir both to an ideal of a united Christendom and to classical Rome. The diverse jigsaw puzzle of states, cities, lordships, and institutions stretching across Germany and down into Italy, subject to the ultimate juridical sovereignty of an elected emperor, provided no ideological model for imperial expansion; quite the contrary, it acted to preserve territorial fragmentation.[79] The erosion of this notion of empire as the expression of an ideal of Christendom united in its territorial diversity was a much lengthier process than is often supposed. The determination of generations of historians to see its death knell either in the consolidation of Protestantism or in the peace treaties of Westphalia (1648) is misplaced.[80] Its resilience was closely linked to a system of international politics dominated by the aspirations of rulers and their families rather than by the priorities and ambitions of states seeking to draw legitimacy from ideas of national identity and expansionary national interest. For, as both James Tracy and Hamish Scott demonstrate in their chapters, the dominant force behind war and policy-making within Europe until the mid-eighteenth century remained dynasticism.

77 John Darwin, *After Tamerlane: The Rise and Fall of Global Empires* (London, 2008).

78 Ch. 19 below, p. 472.

79 Peter H. Wilson, *The Holy Roman Empire: A Thousand Years of Europe's History* (London, 2016).

80 Andreas Osiander, 'Sovereignty, International Relations and the Westphalian Myth', *International Organization* 55 (2001), 251–88; Axel Gotthard, 'The Peace of Westphalia: A European Peace', in O. Asbach and P. Schröder (eds.), *The Ashgate Research Companion to the Thirty Years' War* (Farnham, 2014),

The great territorial agglomeration inherited by Charles of Habsburg at the end of the 1510s was explicitly not an empire, but a group of territories aligned under the same ruler, not subordinated to any one of those territories. Charles V's grandiose example of what historians have come to refer to as 'composite monarchy' in fact epitomises the dynastic principle underpinning international relations and conflict.[81] As some of the territories inherited by Charles, notably Naples and Flanders, were subject to family counterclaims by his arch-rivals, the Valois kings of France, decades of warfare ensued.[82] Tracy's chapter focuses attention on that dynastic rivalry between the Habsburgs and the French monarchs of the Valois and Bourbon dynasties as the enduring source of conflict in Europe down to 1648. It is a central theme picked up by Scott in explaining European bellicosity from the Peace of Westphalia through to the Revolutionary Wars of the late eighteenth century: the age of the 'Wars of Succession', denoting the continued importance of dynastic claims and rivalries in driving diplomacy and warfare.

Both Tracy and Scott also, and justifiably, devote attention to the self-lacerating religious warfare that from the Reformation onwards tore apart whole areas of Europe. Scott makes a strong case that religiously motivated conflict has too readily been treated as a burnt-out *casus belli* after the Peace of Westphalia, when it was supposedly replaced by an international order whose political motivations were based on rational calculations of *Realpolitik*. In fact, Scott argues, while all-out religious civil war had mostly – but not entirely – disappeared from Europe after 1648, conflicts over enforced conformity, suspicion of religious policies, and alliances bound together by religious ideals were part of the fabric of international conflict well into the eighteenth century.

And in one key respect religious war remained central both to motives for war and in the European political imagination throughout the period. Ottoman expansion both in South-East Europe and in the Mediterranean down to the end of the sixteenth century reflected the efficiency of the empire's military and civil organisation and an ideology that combined religious aspirations with the territorial ambitions of a militarised ruling class. In attempting to resist this Ottoman imperialism, European authorities – most obviously the pope and the (Habsburg) Holy Roman Emperor –

81 John H. Elliott, 'A Europe of Composite Monarchies', *Past and Present* 137 (1992), 48–71; Geoffrey Parker, *Emperor: A New Life of Charles V* (New Haven and London, 2019).

82 Christine Shaw and Michael Mallett, *The Italian Wars 1494–1559: War, State and Society in Early Modern Europe* (Abingdon, 2018); David Potter, *Renaissance France at War: Armies, Culture and Society, c. 1480–1560* (Woodbridge, 2008).

could draw on a still lively, and occasionally effective, rhetoric calling for unifying warfare to push back the infidel and – ultimately – reconquer the Holy Land. Of course, the language of crusade was deployed with self-conscious cynicism by some European powers – Catholic opponents of the Habsburgs, or Protestant powers – who in reality found the Ottomans to be 'useful enemies'.[83] But, especially within the Habsburg lands, the ethos of crusade, shaped in Spain by the centuries-long *reconquista* of the Muslim-ruled south, was a powerful ideological force. It was a strong element in an expansionist drive that led to the creation of the greatest of the early modern European overseas empires, Spain's dominions in the Americas.

Yet for the most part the first centuries of European empire-building lack anything of the epic scale and massive resources drawn into the expansion of the Asian imperial powers. Empires in which the metropole is linked by sea with its various colonial or tributary territories are not necessarily weaker than empires where authority is exercised over a contiguous landmass. Indeed, water may represent a more effective, in many cases a faster, means to project power than land-based communications, reliant, in most places and for most of this period, on primitive or decaying systems of roads, and unreliable transportation for bulk goods or heavy weaponry. But the reality was that the scale of trans-maritime military activity was considerably smaller than most operations undertaken overland. State-sponsored and state-financed naval expeditions such as those of the Chinese in the first half of the fifteenth century, displays of imperial prestige and power intended to overawe states around the Indian Ocean, were rarities – and entirely abandoned by the Chinese government after 1433.[84] In Europe a few great sea-battles like Lepanto (1571), and amphibious expeditions like the Ottoman siege of Malta (1565) or William III's invasion of England in 1688, were again exceptions which underlined the generally smaller scale of naval operations, at least in terms of manpower.

Certainly, European global expansion from the Portuguese and Spanish, through to the Dutch, French, and English, was modest in the scale of its resources, and, with the exception of the Portuguese, almost entirely the result of private initiatives. As Louis Sicking demonstrates in his survey of European

83 Noel Malcolm, *Useful Enemies: Islam and the Ottoman Empire in Western Political Thought, 1450–1750* (Oxford, 2019).

84 Louise Levathes, *When China Ruled the Seas: The Treasure Fleets of the Dragon Throne, 1405–1433* (Oxford, 1997); Charney adds that the huge battle fleets that had besieged Melaka in the sixteenth and early seventeenth centuries largely disappeared after the sinking of Iskander Mudah of Aceh's (r. 1607–36) huge fleet of several hundred ships in 1629.

expansion into the Indian and Pacific Oceans, the primary motivation for voyages was the opening up of trade routes to bring Asian goods back to European markets, followed by the realisation that equal or greater profits could be made by acting as carriers and traders operating within Asian networks. The small scale of a few squadrons of galleons or armed merchantmen posed no significant threat to any except the weakest or most politically fragmented local territories. The establishment of what were more than trading toeholds – the expansion of the Portuguese *Estado da India* around Goa, the Dutch seizure of the Moluccas – depended on the exploitation of particular, local political circumstances and weaknesses. The few attempts to make inroads into the territory of major Asian powers were abruptly terminated: Michael Charney describes the failure of Portuguese adventurers to carve out territory in Burma in the early sixteenth century; Sicking cites the failure of the Dutch to hold Taiwan against rebel Ming forces in the 1660s. Expanding beyond the territorial constraints that the major Asian powers were prepared to tolerate was not a serious option before the later eighteenth century.

The one assured military asset the Europeans possessed was the quality of their shipbuilding, superior to Asian techniques in its robustness of construction and gun-carrying capacity. But this was a distinction of quality not quantity – it gave the Europeans some advantages in asserting control of sea routes, and in commanding fees for maritime 'protection' of goods moving between Asian ports. But this naval advantage was not enough to tip the balance of advantage towards extensive occupation or conquest. Moreover, it remains an open question why Asian powers, who were more than capable of adopting and improving on European gunpowder weaponry, seem to have been reluctant to copy European shipbuilding.[85]

Even in the eighteenth century, notably in the case of the British EIC, early expansion onto the Indian subcontinent was opportunist and contingent, rather than the product of any vision of empire. It was only with Britain's extensive gains overseas during the third quarter of the eighteenth century, especially following the Seven Years' War, that there was a significant adjustment of official attitudes, which, as Scott argues, became more consciously imperialist. Yet this did not change the basic calculations across most of Asia: accepting that Europeans would for the most part work within the political framework of the major Asian states and empires brought willingness to tolerate a status quo that offered attractive trading benefits. This acquiescence meant that the heaviest cost incurred by the

85 Andrade, *The Gunpowder Age*, pp. 196–210.

shareholder-investors in European trading enterprises was confronting the military threat posed by other Europeans. Sicking's chapter shows how the Dutch built their Far Eastern and Brazilian commercial networks at the expense of bases and routes held by the Portuguese, but the Dutch were in turn preyed upon by the British in the following century. The profits from Asian trade with Europe and from inter-Asian trade – which remained a small element of the early modern European economy – were substantially reduced throughout these centuries by protection costs incurred against other European powers. Sicking summarises the situation: '(The) European presence in Asia and East Africa involved small numbers of people, based in trading posts and forts whose purposes were overwhelmingly commercial . . . Contact with local rulers was limited to the coast or along the fringes of major rivers, leaving great empires in the interior untouched.'[86]

The striking exception to this process of trading networks, territorial toeholds, and inter-European global conflict was Spain's empire in the New World. The chapters by Felipe Fernández-Armesto and Gabriel Paquette examine the fortunes of what was a wholly different European early modern imperial undertaking. Yet in its origins, as Fernández-Armesto points out, Spanish expansion westwards had much in common with the activities of other global Europeans. Bands of self-financing adventurers, fuelled by hopes of profit from trade or plunder, successively took over the Caribbean islands and expanded westwards to the Central American coast, the Panama Isthmus, and then down into South America.[87] These were paralleled by similar, albeit more limited Portuguese initiatives in coastal Brazil – along lines of demarcation that it suited both powers to accept. Initially confronting no serious opposition on the Caribbean islands of Hispaniola and Cuba, the first generation of *conquistadors* moved from trade/plunder to territorial occupation and the effective enslavement of the local population. The next stage – the destruction of the Aztec state and is tributaries, and the even more remarkable conquest of the Inca Empire – brought together a combination of European military technology and tactics, and the leveraging of indigenous hatred for the existing hegemonic regimes, together with deep-seated local rivalries and antipathies. A combination of good fortune, courage, and brutality gave the Spanish war-bands a temporary advantage in destabilising their opponents and

86 Ch. 9 below, p. 252.

87 Hugh Thomas, *Rivers of Gold: The Rise of the Spanish Empire, from Columbus to Magellan* (London, 2013).

allowed them to lead coalitions of native allies into completing the destruction.[88]

Fernández-Armesto reminds us that through initial conquests, subsequent expansion, and the maintenance of control, the principal means of the Spanish lay in co-opting and drawing upon the support of indigenous elites and their political and military resources. As with the great empires of the Middle and Far East, the ability of the conquerors to assimilate and reward previous rulers was central to the cost-efficient management and exploitation of the new territories. Despite the terrifying demographic collapse of the native populations of Central and South America following the influx of European diseases, a system of imperial rule based on co-opting native elites proved enduring. These elites were themselves more likely to form part of a mixed – *mestizo* – population through intermarriage and cultural assimilation. Paquette's chapter stresses that even after 1700, there were only 6,000 Spanish regular troops in all of Spanish America and the figure never exceeded 14,000 before 1750. The suppression of the massive Andean revolt of 1781–83 could not have been achieved without the active collaboration of the native elites.

The Spanish territories in the New World raise some key questions, explored in both the relevant chapters, about metropolitan Spain's attitude to its colonies. Initially modest revenues to be gained from the colonists were set against the potentially massive cost of defending the colonies against other European powers and their merchants and adventurers. The balance sheet for imperial expansion in the Americas would have looked overwhelmingly unfavourable to the Spanish crown had it not been for the discovery and exploitation of silver from the mines of northern Mexico and Bolivia. Both the crown's tax on silver production and, more importantly, the massive borrowing that it facilitated financed the Spanish war effort in Europe. But protecting the colonial geese laying the silver eggs also became a military priority for the crown, both in the maintenance of the galleon fleets to provide escorts for the flow of silver and high-value goods from and to the New World, and in massive, state-of-the-art fortifications for ports and coastal cities in the Spanish possessions. With that need for military protection came an increasing administrative and judicial assertiveness. The crown continued to appoint senior administrators – viceroys, judges of the major courts, bishops – from metropolitan Spaniards, *peninsulares* as they were

88 Matthew Restall, *When Montezuma Met Cortés: The True Story of the Meeting That Changed History* (New York, 2019); Ross Hassig, *Mexico and the Spanish Conquest* (London, 1994).

termed by the resentful elites of colonial-born Spaniards.[89] As we have seen, the presence of regular troops was severely limited, but despite the losses sustained through disease and acclimatisation the crown insisted far into the eighteenth century on drawing these forces from regular officers and troops raised in Spain. It was not until after 1763, Paquette reminds us, that the decision was taken to create a genuinely colonial army.

On these seemingly slender imperial resources, and a lasting degree of self-interested cooperation between crown and colonial elites, the Spanish Empire did not merely hold its own against the attacks of outsiders, but vastly expanded its territorial extent during the seventeenth and eighteenth centuries, stretching from the River Plate northwards as far as Arizona and Louisiana. Like the great Asian empires, it was the new challenges emerging in the first decades of the nineteenth century that undermined an imperial system that until then had shown not just resilience, but an impressive capacity for expansion.

## Military transformations and 'total war' (1750–1850)

The final century of the 400-year period considered in this volume has been identified as one in which the military dynamic undergoes a dramatic change, heralding the beginning of modern war. Over and above the facts of an impressive expansion in the scale of conflict, it has been argued that from the 1790s the Revolutionary and Napoleonic conflicts in Europe assume a quality of 'total war', a notion which is charged with significance for later, global conflict.[90]

In his introduction to the first section of the succeeding and final volume of the *Cambridge History of War*, Hans van de Ven identifies the distinctiveness of warfare from 1850 to 1914 as a product of industrialisation, the development of strong European nationalism, and the huge increase in bureaucratic capacity with its ability to increase both the reach of the state and its extractive power. All of these come together to bring about the mass participation in warfare that reached its apocalyptic climax in the twentieth century. To what extent, however, were these three factors prefigured in

89 David Brading, *The First America: The Spanish Monarchy, Creole Patriots and the Liberal State, 1492–1867* (Cambridge, 1993); John H. Elliott, *Empires of the Atlantic World: Britain and Spain in America, 1492–1830* (New Haven and London, 2020).
90 David Bell, *The First Total War: Napoleon's Europe and the Birth of Warfare as We Know It* (Boston and London, 2007); Jean-Yves Guiomar, *L'Invention de la guerre totale* (Paris, 2004).

late eighteenth- and earlier nineteenth-century developments? Had a break with the entire pre-modern period already emerged by 1800? Were the wars that ravaged Europe from the 1790s fundamentally different from those of the *ancien régime*? Was there, as historians such as David Bell and Jean-Yves Guiomar argue, a profound shift in attitudes to war in the decades after 1790? Was war turned from an extension of normal political and diplomatic behaviour, sanctioned by a social elite for whom martial values were almost universal in defining status, into something seen as quite outside normal social and cultural experience?

An obvious starting point is military technology. As was noted earlier, essential weaponry did not change during the period of the Revolutionary and Napoleonic Wars: in contrast to the industrialised societies of the later nineteenth and twentieth centuries, large-scale, intense, and continuous war did not drive rapid technological innovation. Change was incremental and often piecemeal rather than radical and transformative. The successive technical improvements that Robin Briggs describes for developments in eighteenth-century navies had enhanced the firepower, operational range, and seaworthiness of warships without bringing any fundamental change to their design or tactical conception. On land, both musket and cannon continued to be based on muzzle-loaded designs which constrained the rate of fire and determined tactics. If there was a 'revolution' in artillery it was one of quantity: the advent of manoeuvrable middle-weight field guns, deployed in their hundreds, turned the battles of the Seven Years' War into killing fields far bloodier than their predecessors. This was a process that was simply carried further by Napoleon.

For dramatic technological change with transformative impact, we must turn to the naval innovations described by Andrew Lambert, developed during the period of European peace in the decades from 1820 to 1850, but which were also, unmistakably, a product of states whose industrial capacity had taken a massive leap forward.[91] Industrialisation did not simply speed the sequence from development to manufacturing to distribution, but introduced precision machining into the production process: building an iron warship was about engineering by machine tools rather than the human skills exercised in craft workshops. Non-European powers without a comparable industrial base could no longer simply take European workshop-produced weaponry and in many cases produce superior versions through their own craftsmen. The lead in naval technology was growing

91 See Ch. 24 below, and above, p. 18.

unbridgeable in the decades down to 1850, and in subsequent decades European states would achieve the same industrialised advantages in the production of breech-loading rifles, artillery, and machine guns.[92]

More sophisticated and extensive administration, better communications, and the growth of the bureaucratic state can also be seen as part of a lengthier process of development, stretching back to the seventeenth century. The organisational skills and capacity to raise, clothe, arm, and feed an army of over 300,000 men in 1690s France presented an administrative challenge that was a whole order of magnitude greater than any previous military establishment, and connects more obviously with the military world of the mid-eighteenth century, when Russia, Prussia, and Austria were seeking to achieve the same levels of resource mobilisation. Equally the immense dockyards and the naval administrations that from the later seventeenth century were required to build and maintain fleets of hundreds of specialised, rated warships could be seen as another watershed: subsequent developments more obviously build from this innovatory infrastructure than from the more haphazard organisation before that time.

Yet, making full allowance for an imposing increase in the capacity of the major European powers to administer, resource, and finance their military establishments between the 1690s and 1780s, it could nonetheless be argued that there was a further decisive leap upwards from the 1790s in a way that set the scene for patterns of 'modern' resource mobilisation. The Jacobin Republic could put three-quarters of a million men under arms in the 1790s; in 1812 Napoleon could invade Russia with 650,000 troops, see that army destroyed, and raise a similar-sized army for the campaign of 1813.

Various *ancien régime* states had made use of conscription to raise their armies, though usually as a means to circumvent the costs of trying to recruit in a labour market where other forms of employment would be more attractive. Since the sixteenth century, Sweden had tried to compensate for a sparse population and a weak financial base with various forms of compulsory military service, above all seen in the introduction of the *indelningsverk* in the 1680s. This became the model for the Prussian cantonal system in its evolving forms described by Bernhard Kroener in his chapter. The Russian state under Peter the Great meanwhile enforced the lifetime conscription of a proportion of the peasantry. Such lifetime service, or at least regular periods under arms, also acknowledged the

92 Andrade, *The Gunpowder Age*, pp. 244–56.

challenge of *ancien régime* battlefield tactics which placed a premium on extensive training in drill and discipline.[93]

As Alan Forrest notes, the initial attempts by the French Revolutionaries to introduce a form of conscription were born of desperation. The crumbling of the *ancien régime* French army, which had largely relied on voluntary recruitment and some foreign mercenary units, left the Revolution vulnerable in the face of external military pressure and major uprisings within France. Stirring up patriotic fervour to defend the nation, and urging the responsibilities of civic duty on the lukewarm, seemed the last resort in trying to raise an army large enough that numbers and *élan* might prevail against the disciplined ranks of traditional armies or mass risings in the provinces. By 1793 manpower was desperately needed, and the Revolutionaries turned first to a system of local levies and then to the *levée en masse*, and the principle that the whole of French society was requisitioned for some form of war service – the nation in arms.[94] This was not the first time that patriotic fervour towards the nation, rather than professionalised military service undertaken in the name of the ruler, had been invoked. Kevin McCranie points out that the American colonists who volunteered to fight against British rule in 1775–6 were ill-equipped and haphazardly organised, but driven by a strong sense of collective grievance. Numbers allowed them to prevail against British regulars, who were forced to evacuate Boston. But fervour and enthusiasm for the cause of independence proved less militarily effective from later 1776 as the British applied strategies that played to maritime strength and the operational flexibility this brought. Prefiguring the way in which French recruits' 'service to the nation' was turned under Napoleon into armies that were disciplined and drilled to levels comparable to *ancien régime* professionalism, George Washington in 1777–8 sought to convert the volunteers to the Patriot cause into troops more capable of engaging in open campaign with the British.

The language of citizenship and patriotic obligation, which in France from the late 1790s was to fuel ever-more demanding and extensive military conscription of the male population, had another effect often associated with modernity. The assertion that military service was a duty owed by citizens to the nation appeared incompatible with military operations that were underpinned by the credit and profit-opportunities of private contractors. If war was

93 John Keep, *Soldiers of the Tsar: Army and Society in Russia, 1462–1874* (Oxford, 1985).

94 John A. Lynn, *The Bayonets of the Republic: Motivation and Tactics in the Army of Revolutionary France, 1791–94* (Boulder, 1996); Alan Forrest, *The Soldiers of the French Revolution* (Durham, NC, 2003); Thomas Hippler, *Citizens, Soldiers and National Armies: Military Service in France and Germany, 1789–1830* (London, 2008).

being presented as a terrible necessity, undertaken in defence of the nation and requiring ultimate sacrifices from its citizens, then any suggestion that parts of the military machine could be run for private profit was anathema. The proven efficiency and flexibility of well-managed contracts in equipping, supplying, and maintaining armies and fleets was outweighed by the ideological concern to ensure that all aspects of national war-making were controlled by and on behalf of the nation and not the subject of mercenary profiteering. The Revolutionaries' attitude to supply contractors and those involved in military financing, denounced as war-profiteers and scourges of the nation, was an inevitable corollary of the language of the *levée en masse*. And so long as French armies could live by pillage and requisitioning across the frontiers, the notion of the military as an enterprise of the state could be maintained.

As Napoleon restructured and institutionalised the army, the need for more orderly support mechanisms grew, and with it the benefits of competitive contracting for supplies of uniforms, weaponry, munitions, transport, and food stockpiling.[95] Yet, despite this return to pragmatism, a real shift both of ideology and of structure had taken place: the overt and direct dependence on the private sector, the dominance of the 'contactor state' in European warfare and its overseas contexts, entered into a 150-year eclipse. A huge growth in military bureaucracy with direct responsibility for a large number of military functions; a determination to keep private suppliers at the far end of chains of procurement and supply: both were modern phenomena that could be traced back to the advent of conscription sustained by an ideology of service to the nation. Related to this was the way that national conscription created large pools of cheap military labour, so that ancillary services that might previously have been performed by contractors, whether related to food supply, maintenance, construction, or distribution, could now be undertaken by soldiers in uniform. This comprehensively direct military control of support functions was one of the most distinctive aspects of 'modern' military forces. It was a process that began with the ideologies of revolutionary citizenship and only lost its grip in the later twentieth century, when a new age of military outsourcing was inaugurated following from the disappearance of compulsory military service and growing levels of expensive military specialisation.

Did the character of war change? Did conflict become, in a Clausewitzian sense, more absolute – or total – from the Revolutionary and Napoleonic

95 Rafe Blaufarb, *The French Army, 1750–1820* (Manchester, 2017); Rothenberg, *The Art of Warfare*.

Wars onwards? The global perspective of this volume certainly enjoins caution. Forrest speaks of the period of war between the 1790s and 1815 resulting in between 3.5 and 5 million deaths across Europe, but both in Europe and globally these figures were not dramatically higher than for any previous conflict, above all if deaths from disease and war-related malnutrition are included. The European Thirty Years' War, or the Manchu conquest of China in the 1640s and 1650s, brought mortality levels and destruction considerably greater than the Napoleonic Wars. World War I and, above all, World War II still cross a threshold in violent deaths of soldiers and civilians which no previous conflict can rival. Intensity of ideologically driven conflict could be seen as another aspect of modern/total war – not least in the paradox explored by David Bell that Enlightenment hostility to war as an abnormal and abhorrent rejection of civilised values meant that war itself, when it was required, should be pursued with all the more intensity and extremism in order, supposedly, to prevent its future recurrence.

The idea of 'war to end all wars' had its roots in an Enlightenment rejection of the idea that war was a normal and unavoidable aspect of politics and international relations; once launched in the aftermath of the French Revolution, it was to have a distinct and grim afterlife far into the twentieth century. That it could incite exterminatory violence is certainly attested, whether in the campaigns against anti-revolutionary forces within France or in the conduct of campaigns during the Peninsular War or in Eastern Europe. Yet, if Forrest reminds us of the impact of 'total war' in this European context in intensifying and systematising extreme violence in pursuit of a changed conception of war, Pamela Crossley, to take one example from the volume, points to similar exterminatory violence as a calculated policy of the Qing government. In the wake of the conquest and occupation of the Xinjiang territory from the 1770s, the Qing army set about a systematic policy of pacification via the extermination of the Zunghar population, killing all males and enslaving and distributing the women to Qing soldiers. This was certainly an experience of 'total' war, but the product of calculated government policy rather than ideological fervour. And perhaps that fundamental dynamic of an interplay between the two was ultimately characteristic of the entire period. Civil and religious strife, together with the ideologies that drove much imperial expansion, have been fuelled by mental constructs of enemies or the subjugated that, time and again, fed extreme violence in pursuit of ethnic or confessional 'cleansing', or the 'pacification' of societies or races deemed inferior. Yet throughout these 400 years at least as much warfare had been driven by brutal types of military logic in which extreme violence against both soldiers and civilians has been justified and practised not with ideological fervour but in pursuit of what were deemed vital objectives. Systemic, unrestrained violence in pursuit of military and political objectives did not emerge from the 1790s; from a global perspective it had never gone away.

I

# The legacy of the nomadic empires

BEATRICE F. MANZ*

For almost three millennia the pastoral nomads of the Eurasian steppe formed a great reserve of mounted cavalry, threatening their settled neighbours while offering them goods and services of great value – in particular horses and skilled soldiers for their armies. The Eurasian nomads were also empire-builders, creators of imperial ideology and administrative structures that were passed down through generations of successor states. Their imperial centre in Mongolia was home to two related peoples – the Turks and the Mongols – each defined by the powerful empires they erected. The Türk Empire, which flourished from the mid-sixth to mid-eighth centuries, was the first of these and it controlled the steppe from Mongolia to the Volga river, fighting and trading with China, the empires of the Middle East, and Byzantium. The second great state was the Mongol Empire, founded by Chinggis Khan in 1206. The Mongols extended their power yet further than the Turks, conquering much of Eurasia.

The extraordinary success of the Türk and Mongol empires gave them prestige well beyond the period of their actual rule. Connection to the former Türk Empire remained important long after its fall, and for Eurasian nomads, the descendants of Chinggis Khan were for centuries the only dynasty with the right to claim the sovereign titles *khan* or *khaghan*. It was not only the nomads who honoured steppe imperial traditions. During periods of internal strife in the Türk Empire, part of its population had migrated west and formed states both within the central Islamic region and on its frontiers. Even after the Mongol invasion, some Türk traditions survived in the Middle East. The Mongol heritage proved even more enduring. Both the Muscovite state and the Qing dynasty bore the stamp of Mongol tradition, and in the central Islamic lands the Mongol legacy remained active for several centuries.

* I would like to thank Nicola Di Cosmo for his help with this chapter.

Thus, for most of the period covered in this volume, the imperial traditions of the steppe remained an important part of political life across a wide expanse of Eurasia.

## The imperial legacies

Many of the institutions of the Türk Empire were still alive in Mongolia during the life of Chinggis Khan. When Chinggis Khan became ruler in 1206, he centred his regime in the sacred region of the earlier empire, adopted many of the offices and titles of the Türks, and adapted their alphabet for Mongolian. His army combined established steppe traditions with an unusually centralised political structure. Like earlier steppe armies, it was organised in decimal units from 10 to 10,000, but unlike most others, it was not organised tribally. Chinggis divided his forces into three sections: the right wing, left wing, and centre, with commanders and troops assigned permanently to each. Chinggis also created an imperial bodyguard of 10,000 men, known as the *keshig*, which served as a training ground for military commanders.[1] In both army and *keshig*, the central personnel were recruited from Chinggis's personal following, those who had left their tribes to serve him. Tribes continued to exist but were not central to either the military or administrative structures of the early Mongol Empire. From the beginning of his conquest of settled territory, Chinggis incorporated local troops; soldiers from conquered populations made up a significant proportion of the army on many campaigns.[2]

After Chinggis Khan's death his realm developed into four separate sections, known as *uluses*, under the lines of his sons Jochi, Chaghatai, Ögedei, and Tolui, with Ögedei recognised as supreme khan, or *khaghan*. However, Mongol unity collapsed in 1259, and from then on the sections of the empire were often at war with each other. Nonetheless, the Mongol realm continued to grow, retaining inherited institutions and a strong sense of common identity. The figure of Chinggis Khan embodied the power of world empire and remained a key factor in dynastic legitimation. Chinggis's practice and decrees were enshrined in the *jasaq* or *yasa*, a shadowy entity – maybe

1 Thomas Allsen, 'The Rise of the Mongolian Empire and Mongolian Rule in North China', in Herbert Franke and Denis Twitchett (eds.), *Cambridge History of China*, vol. 6: *Alien Regimes and Border States, 907–1368* (Cambridge, 1994), pp. 343–8.

2 Beatrice Manz, 'Nomads and Regional Armies in the Middle East', in Kurt Franz and Wolfgang Holzwarth (eds.), *Nomad Military Power: Iran and Adjacent Areas in the Islamic Period* (Wiesbaden, 2015), pp. 11–14.

a collection of precepts, maybe law; maybe written, maybe not – but a useful symbol of Mongol tradition.[3] Thus from the thirteenth through much of the fourteenth century most of Eurasia was ruled by an elite loyal to the traditions of the Mongol Empire.

In the later thirteenth century, Khubilai Khan, ruling in China and Mongolia, began to patronise Buddhism and was declared *Chakravartin* (universal king). In the Eastern Mongol realms, Mongolian remained the predominant language and Buddhism became the dominant religion. To the west – in the Middle East, Central Asia, and the Russian steppes – most khans converted to Islam in the early fourteenth century and the Mongols of that region began to speak Turkic, the language of most of their nomadic followers. They also adopted much of the bureaucratic and cultural tradition of their Perso-Islamic subjects, which spread into the Golden Horde and eastern Turkestan. This new elite combining Mongol, Turkic, and Islamic culture has come to be known as Turco-Mongolian. (See Map 1.)

The Turco-Mongolian synthesis was codified through the career of the warlord Tamerlane (r. 1370–1405), founder of the Timurid dynasty. Tamerlane was of Mongol but not Chinggisid descent and ruled through

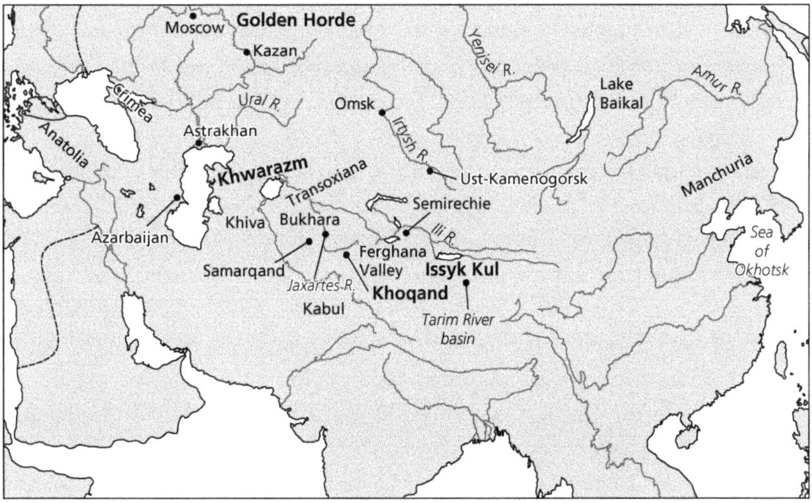


MAP 1. The legacy of the nomadic empires

3 David Morgan, 'The "Great *Yasa* of Chinggis Khan" Revisited', in Reuven Amitai and Michal Biran (eds.), *Mongols, Turks and Others: Eurasian Nomads and the Sedentary World* (Leiden and Boston, 2005), pp. 291–308; Denise Aigle, 'Loi mongole vs loi islamique. Entre mythe et réalité', *Annales IISS* 5–6 (2004), 974–82.

a Chinggisid puppet khan. From his centre in Samarqand, he set out to symbolically recreate the Mongol Empire and to assert pre-eminence within the Islamic and Mongol worlds. Like Chinggis Khan, he formed a centralised army of conquest in which tribes were divided among decimal units and settled soldiers played a significant part.[4] His descendants presided over a brilliant court culture combining Iranian, steppe, and Chinese elements. The combination of military and cultural achievement gave the Timurid dynasty lasting prestige and helped to incorporate steppe traditions into the central Islamic lands.

## The rise of new powers in the fifteenth century

By the fifteenth century, the Chinggisid dynasty had lost its dominance over Eurasia; China had expelled the Mongols in 1368, and the former Mongol territories in the Middle East were controlled by the Timurids and other dynasties of tribal origin. In eastern Turkestan and the steppe, Chinggisids continued to rule but were challenged by tribal powers. Nonetheless, steppe imperial tradition remained important. In the central Islamic lands, Turkmen – western Turks mostly descended from the Oghuz confederation of the Türk Empire – had formed states and tribal confederations dominating Anatolia and parts of Azarbaijan. Ultimately the most successful of these was the Ottomans, but at this time two tribal confederations, the Qaraqoyunlu and the Aqqoyunlu, held decisive power in the eastern regions. To counter the pretentions of the Timurids who linked themselves to the heritage of the Mongol Empire, the Anatolian powers based their legitimacy on a mythical ancestor, Oghuz Khan, connected to the Türk Empire.[5] In the later fifteenth century, the Anatolian confederations took over western Iran, bringing their Turkmen followers with them.

The steppe stretching from the Crimea to the Issyk Kul – the *ulus* of Chinggis Khan's son Jochi – remained largely under Turco-Mongolian rule but broke into smaller entities. In the western regions, the Golden Horde (now the 'Great Horde') declined as smaller Chinggisid states arose. From about 1420–40 the Chinggisid prince Hajji Giray established an independent khanate in the Crimea.[6] In 1447–8 a khanate arose in Kazan, competing with Muscovy to the south and with yet another Chinggisid state, the

4 Beatrice Manz, *The Rise and Rule of Tamerlane* (Cambridge, 1989), pp. 66–7.

5 John Woods, *The Aqquyunlu: Clan, Confederation, Empire* (Minneapolis and Chicago, 1976), pp. 4–5.

6 Dariusz Kołodzieczyk, 'Girāy Khāns', in Kate Fleet, Gudrun Krämer, Denis Matringe, John Nawas, and Everett Rowson (eds.), *The Encyclopedia of Islam*, 3rd ed. (Leiden, 2018).

khanate of Sibir, to the north. Muscovy should also be counted among the successor states to the Golden Horde. Its dukes had begun their rise under the aegis of the Golden Horde, for whom they collected dues from the other Russian principalities. By the fifteenth century they were largely independent and took their place among the powers fighting over the remains of the Golden Horde. In state administration, military organisation, and imperial ideology Muscovy owed a great deal to Mongol institutions. The Muscovites used the Mongol compound bow, adopted Mongol techniques in riding, and took over many of the insignia and symbols of the empire.[7] The Muscovite army included numerous Mongol soldiers – known in this region as Tatars – some of whom became full members of the Russian landed service elite.[8]

The western steppe was a constantly boiling cauldron of shifting alliances. Muscovy and Lithuania were rivals for power and both paid off the Crimean khans in the hope of curtailing their raids. Muscovy often allied with the Crimeans against the khanate of Kazan, and also gave refuge to unsuccessful claimants to the Kazan throne. In 1456 its rulers created a Mongol vassal state, the khanate of Kasimov, under the fugitive son of the Kazan khan.[9] The area north of the Volga came into the hands of the Manghit tribe, known more widely as the Noghay. Its rulers were not Chinggisids and had no imperial pretensions, but the confederation was an important player in steppe politics. In 1428–9, the Noghay were among the powers who elected a new Chinggisid ruler, Abu'l Khayr Khan (r. 1428/9–68), over the Turco-Mongolian tribes north of the Aral Sea, known as Uzbeks. Over the next forty years, Abu'l Khayr united a large region, including many of the cities on the lower Jaxartes river.[10] His career laid the foundation for the later Uzbek khanates.

The eastern steppe – the original centre of the Mongolian Empire – was the scene of constant battles, as various tribal leaders vied for pre-eminence. There were two main groups of tribes. One was the Eastern Mongols, a number of whose leaders claimed descent from Chinggis Khan, and who

7 Donald Ostrowski, *Muscovy and the Mongols: Cross-Cultural Influences on the Steppe Frontier, 1304–1589* (Cambridge, 1998), pp. 50–60, 180–4; Michael Cherniavsky, 'Russia', in Orest Ranum (ed.), *National Consciousness, History and Political Culture in Early-Modern Europe* (Baltimore and London, 1975), pp. 121–6.

8 Janet Martin, 'Multiethnicity in Muscovy: A Consideration of Christian and Muslim Tatars in the 1550s–1580s', *Journal of Early Modern History* 5 (2001), 1–23.

9 Kołodzieczyk, 'Girāy Khāns'; V. Bartol'd and A. Bennigsen, 'Ḳāzān', in H. A. R. Gibb (ed.), *The Encyclopaedia of Islam*, 2nd ed. (Leiden, 1960–2006).

10 Yuri Bregel, *An Historical Atlas of Central Asia* (Leiden and Boston, 2003), pp. 44–5.

had been part of the Mongol Yüan dynasty of China. To the west, from the upper Yenisei to the Tarim basin, were the Western Mongols or Oirat, most of whom traced their origins back to the Türkic tribes predominant before the rise of the Mongols. They had also formed part of the Mongol Empire but had no genealogical ties to the Chinggisids.

The rise of the Oirat commenced at the dawn of the fifteenth century, and their leader Esen (r. 1439–55) succeeded in uniting most of the Oirat and Eastern Mongols. At first, Esen ruled through a Chinggisid puppet khan, as his father had done. However, in 1452 he killed his khan and took the title for himself. This was not acceptable to his followers; he was overthrown and killed in 1455. A Mongol chronicle records this as heavenly punishment for daring to claim the sovereignty reserved for the descendants of Chinggis.[11] The Oirat also looked west, and shortly after Esen's fall, they inflicted a devastating defeat on Abu'l Khayr Khan Uzbek, dealing a severe blow to his prestige. However, they did not attempt to take the whole of his territory.

## State building in the sixteenth century

The sixteenth century marked the establishment of the great regional states which formed modern Eurasia, almost all of which showed the influence of steppe traditions. The Middle East was divided among four empires, the Ottomans, Safavids, Uzbeks, and Mughals, all connected with some part of the Turco-Mongolian heritage. Muscovy moved from principality to empire, swallowing the remains of the Golden Horde. The Uzbek and Kazakh khanates emerged in the eastern Dasht-i Qipchaq or Qipchaq Steppe, while in Manchuria the Jurchen state, later to become the Qing dynasty, began its rise. At this point, nomad rule was still accepted in much of Eurasia.

Two new powers emerged from the confederation of Abu'l Khayr Khan. After Abu'l Khayr's defeat by the Oirats, the Chinggisid princes Giray and Janibeg split off from the Uzbek confederation with a number of tribes and moved into the Semirechie as *qazaqs*, free adventurers. This is the origin of the name Kazakh. After Abu'l Khayr's death and the collapse of his confederation in 1468–9 these men returned and gained the upper hand within his

11 Veronika Veit, 'The Eastern Steppe: Mongol Regimes after the Yuan (1368–1636)', in Nicola Di Cosmo, Allen J. Frank, and Peter B. Golden (eds.), *The Cambridge History of Inner Asia: The Chinggisid Age* (Cambridge, 2009), pp. 163–5; Junko Miyakawi, 'The Legitimacy of Khanship among the Oyirad (Kalmyk) Tribes in Relation to the Chinggisid Principle', in Reuven Amitai-Preiss and David Morgan (eds.), *The Mongol Empire and Its Legacy* (Boston and Leiden, 1999), p. 324.

territories.[12] At the end of the fifteenth century Abu'l Khayr's grandson Muhammad Shibani (d. 1510) regathered the Uzbeks, then in 1508 overthrew the Timurids and moved into Transoxiana with a large number of Uzbek followers. As a descendant of Chinggis replacing the non-Chinggisid Timurids, Muhammad Shibani presented his rule as a restoration of Mongol tradition. The state he established, however, was much less centralised than that of Chinggis Khan. On Shibani Khan's death in 1510 rule was apportioned among his relatives, with the senior member of the dynasty presiding as khan. The dynastic apanages became hereditary, and the senior khan ruled as the first among equals.[13]

While the Uzbek khanate consolidated itself in Transoxiana, the Kazakhs achieved predominance north of the Jaxartes under their khans. The term Uzbek now applied to the population which had migrated into Transoxiana, and the erstwhile followers of Giray and Janibeg were called Kazakhs. Over time, cultural and linguistic differences emerged between the two groups, as the Uzbeks took over much of the Timurid cultural heritage, while the Kazakhs remained self-consciously nomads of the steppe.

Uzbek expansion to the south was blocked by a new power in Iran, the Safavids. This dynasty was probably of Kurdish origin but most of its manpower came from the Turkmen nomads of Anatolia and northern Syria, some of whom had moved into Iran with the conquests of the Qaraqoyunlu and Aqqoyunlu. Despite later Safavid moves towards centralisation the nomadic tribes remained a powerful force in Iran, providing a large proportion of the military up to the twentieth century. As the former Timurid realm was divided between the Uzbeks and the Safavids, the Timurid prince Babur retreated south to Kabul and then moved on to conquer northern India in 1526. On his father's side, Babur was descended from Tamerlane and on his mother's from Chinggis Khan; the state he founded in India is known both as the Later Timurid and as the Mughal (Mongol) Empire. The historiography and imperial legitimation of the Mughals retained a large element of Mongol and Timurid traditions.

While the Turks were expanding in the Middle East, in the western steppe Russia emerged as the pre-eminent power. In 1502 the Great Horde was replaced by a more modest khanate centred on Astrakhan. Both the Crimean

12 S. G. Kljaštornyj and T. I. Sultanov, *Staaten und Völker in den Steppen Eurasiens. Altertum und Mittelalter*, trans. Reinhold Schletzer (Berlin, 2006), pp. 269–77.

13 Robert D. McChesney, 'The Chinggisid Restoration in Central Asia: 1500–1785', in Cosmo, Frank, and Golden (eds.), *Cambridge History of Inner Asia*, pp. 291–4.

khans and the Noghays interfered in the affairs of Kazan and Astrakhan, but found themselves increasingly outflanked by the Russians. In 1547 Ivan IV was crowned as tsar, a term formerly applied to the Byzantine Basileus and the Khan of the Golden Horde. In 1552 Ivan destroyed and incorporated the khanate of Kazan, an event celebrated with great pomp and described as the true beginning of the Tsardom of Muscovy.[14] Astrakhan fell to his armies in 1556. The Crimean khanate was now a vassal of the Ottomans, but in the steppe its khans remained autonomous, playing a significant part in Eastern European politics.

Events in the eastern steppe were less striking, but here too a new political order was being ushered in. In the 1480s, the Mongols had achieved partial unity under a series of Chinggisid khans, beginning with Dayan Khan. Towards the end of the century, they turned to Tibetan Buddhism to strengthen their authority, a move that had far-reaching consequences. In 1578, Dayan's grandson Altan Khan granted the title of Dalai Lama to a Tibetan lama and in return the new Dalai Lama recognised him as the reincarnation of Khubilai Khan, who had held the title of Chakravartin. Tibetan Buddhism now incorporated Chinggis Khan as a sacral figure; he became the incarnation both of the Chakravartin – the universal ruler turning the wheel of Buddhist law – and of the *bodhisattva* Vajrabani, a martial figure deriving from the Indian god of thunder. He was also associated genealogically with the ancient kings of India and Tibet, and later even with the Buddha. From this time on Mongolian and Tibetan politics remained closely intertwined. Chinggis's inclusion in the Buddhist tradition had mixed results. Religious legitimation added to the stature of the Mongol khans and thus might promote unification. On the other hand, Tibetan Buddhists were divided among rival sects and sought allies among the disunited Mongols, making both khans and lamas increasingly numerous.[15]

## The seventeenth century

In the seventeenth century, the regional states of the Middle East and Central Asia remained relatively stable, but this century brought a decisive change to the Eurasian steppe with the Russian penetration of Siberia and the rise of the Jurchen Qing dynasty in Manchuria, Mongolia, and China. The Russians

14 Ostrowski, *Muscovy and the Mongols*, pp. 180–2; Cherniavsky, 'Russia', p. 125.
15 Veit, 'The Eastern Steppe', pp. 170–2; Peter C. Perdue, *China Marches West: The Qing Conquest of Central Eurasia* (Cambridge, MA, 2005), p. 66; Michal Biran, *Chinggis Khan* (Oxford, 2007), pp. 139–40.

began to advance into Siberia towards the end of the sixteenth century and by 1648 they had reached the Pacific. As they moved east they erected a series of forts, which also served as trading posts, along the Siberian river system from Omsk on the Ural river to the Sea of Okhotsk. By the end of the century, the Russians and Manchus had reached a common border in the Baikal and Amur regions and developed a system of treaties culminating in the Treaty of Nerchinsk in 1689. This agreement defined the steppe peoples as subjects of either Russia or China and stipulated that neither power should ally with rulers within the other's sphere.[16] Little room for manoeuvre was left to the peoples of the steppe.

In the eastern steppes, three groups attempted to form new states modelled on Chinggisid traditions: the Oirats, the Eastern Mongols, and a new, ultimately successful power, the Jurchen. From the 1580s onwards, a new Jurchen leader, Nurhaci, rose to power in Manchuria. He developed a written language which used the Mongolian alphabet – the third incarnation of the alphabet of the steppe empires. Nurhaci soon began to attract his Mongol neighbours, strengthening alliances through dynastic marriage ties, a policy which the Qing practised throughout most of their history.[17] In 1616 he founded a new dynasty, the Aisin, later renamed Qing, and began to compete with the Chinggisid Dayan Khan's descendant Ligdan for the loyalty of the Eastern Mongols.

In 1632 Nurhaci's son and successor, Hung Taiji, used his Mongol allies to defeat Ligdan and adapted Mongol traditions to the new state. He incorporated most of the eastern and southern Mongol tribes and organised Mongols, Manchus, and Chinese into banners – military and administrative divisions deriving from the Mongol system of decimally organised *tümens*. Banners were headed by government officials; in the southern Mongolian regions, these were local Chinggisids, and their position was hereditary.[18] A major stroke of fortune was the recovery of the so-called seal of the Yüan dynasty, announced in a lavish public ceremony in 1635. The Qing emperors now presented themselves as heirs to the Mongol Empire. The partnership with the Dalai Lama initiated by Altan Khan provided another useful tool, and the Qing rulers began to develop ties with Tibetan lamas.[19]

16 Perdue, *China Marches West*, pp. 161–3.
17 Ibid., pp. 122, 124, 126, 135; Nicola Di Cosmo, 'The Qing and Inner Asia: 1636–1800', in Cosmo, Frank, and Golden (eds.), *Cambridge History of Inner Asia*, p. 340.
18 Veit, 'The Eastern Steppe', pp. 177–81; Di Cosmo, 'The Qing and Inner Asia', pp. 337–42.
19 Di Cosmo, 'The Qing and Inner Asia', p. 345; Perdue, *China Marches West*, pp. 122–6.

While the Manchus were consolidating their state, the Western Mongols formed themselves into the last of the major steppe powers, the Zunghar confederation. Unification was facilitated by the departure of several independent-minded Oirat groups, notably the western Oirats (known to their western steppe neighbours by their Türkic name, Kalmyk), who migrated to the Volga, where they formed a vassal state to the Russians.[20] In 1640 the chief of the Zunghar confederation, Batur Khong-Taiji, convened a *quriltay* of the major Mongol tribes not under Qing control. The congress declared Tibetan Buddhism as the religion of the Mongols and promulgated a Mongol law code.[21] In the 1670s a new leader, Galdan (1644–97), formed a state stretching from eastern Turkestan into the Kazakh regions of the Ili and forged ties with the Dalai Lama. Over the next eighty years, the Zunghars and Qing competed for the control of modern-day Xinjiang and the remaining independent Mongolian tribes. Unable to advance east against the Qing, the Zunghars expanded west at the expense of the Kazakhs. Galdan's nephew, Tsewan Rabdan (r. 1697–1727), brought the Zunghars to the height of their power, maintaining autonomy from both Russia and China and competing with the Russians for influence over the fur-trapping tribes of Siberia. The struggle with the Zunghars brought the Qing into the politics of ever-more distant regions. They attempted to disrupt Zunghar ties with Tibet and the Dalai Lama, and by the early eighteenth century had begun to involve themselves in Tibetan politics. Now they saw themselves as the rightful rulers over Chinese, Manchu, Mongol, and even Muslim peoples.[22] A succession struggle within the Zunghar confederation following the death of Tsewan Rabdan allowed the Qing to occupy the Ili region in 1755 and destroy the Zunghar state.

While the Chinese were conquering Xinjiang, the Russians were moving into the nomad steppe beyond the Volga. From 1715 to 1743 they constructed several lines of fortifications stretching from the mouth of the Ural (Yayiq) river to Omsk and along the Irtysh to Ust-Kamenogorsk.[23] As the Russians advanced, they encouraged colonisation in nomad territories, offering land to foreign populations.[24] Thus the Kazakhs were hemmed in by the Russians in

20 Perdue, *China Marches West*, pp. 100–3.

21 James A. Millward, 'The Qing Formation, the Mongol Legacy, and the "End of History" in Early Modern Central Eurasia', in Lynn A. Struve (ed.), *The Qing Formation in World-Historical Time* (Cambridge, MA, 2004), p. 107.

22 Perdue, *China Marches West*, pp. 174–235.

23 Bregel, *An Historical Atlas of Central Asia*, pp. 58–9.

24 Andreas Kappeler, *The Russian Empire*, trans. Alfred Clayton (Edinburgh, 2001), pp. 41, 50; Carol B. Stevens, *Russia's Wars of Emergence, 1460–1730* (Harlow, 2007), pp. 270–1.

the west and in the east by the Zunghars. By the early seventeenth century, the Kazakhs had split into three hordes, or *zhuz*, and they were further divided by rivalry among the Chinggisids, making it almost impossible for their khans to achieve sustained power. A disastrous Zunghar invasion in 1723 pushed the Kazakh khans to turn to the Russian Empire for protection; the Junior Zhuz in 1730, the Senior Zhuz in 1738, and the Middle Zhuz in 1740.[25] Thus by the mid-eighteenth century Russia and China, which had formed the western and eastern edges of the Mongol Empire, had taken control over almost all of its northern territory.

## The rise of tribalism in the Islamic lands

While the northern Turco-Mongolian powers lost out to the great land empires, the political order in the south developed in the opposite direction: increasing decentralisation and a rise in tribalism. Rulers began to move away from the Chinggisid order, turning instead to the mixed heritage of the non-Chinggisid Tamerlane. The Uzbek khans had always suffered from disunity, and towards the end of the seventeenth century, the tribes began to interfere in dynastic politics to enthrone their Chinggisid candidates or take power themselves. In 1747, the Manghit tribe unseated the khan of Bukhara and founded a new dynasty under its own name. The Manghits used the figure of Tamerlane for legitimation, and a new popular literature in Central Asia enshrined his memory. On the death of the Khivan khan in 1764 power in Khwarazm devolved to tribal emirs, who invited in Kazakh khans only to depose them, and in 1798 the Ming tribe founded a dynasty in the fertile Ferghana valley, using an invented genealogy leading back to the Timurids.[26]

In Iran and Afghanistan, the eighteenth century witnessed an upsurge in tribal power and the rise of the nomad warlord Nadir Shah Afshar (r. 1736–47) from a Turkmen tribe which had been prominent under the Safavid dynasty. Nadir Shah created an army that combined infantry and artillery with a cavalry composed of nomad contingents under their own leaders. By 1736 he had reunited much of the Safavid realm and could claim the Iranian

25 Allen J. Frank, 'The Qazaqs and Russia', in Cosmo, Frank, and Golden (eds.), *Cambridge History of Inner Asia*, pp. 369–71; K. M. Baipakov and B. E. Kumekov, 'The Kazakhs', in Chahryar Adle and Irfan Habib (eds.), *History of Civilizations of Central Asia*, vol. 5: *Development in Contrast: From the Sixteenth to the Mid-Nineteenth Century* (Paris, 2003), pp. 97–9.

26 McChesney, 'The Chinggisid Restoration in Central Asia', p. 289; Yuri Bregel, 'The New Uzbek States: Bukhara, Khiva and Khoqand: *c.* 1750–1886', in Cosmo, Frank, and Golden (eds.), *Cambridge History of Inner Asia*, pp. 392–6, 400–1.

sovereign title of Shah. For further legitimation, he also forged a connection to Tamerlane. Having secured Iran, he conquered Delhi in 1739 and both Bukhara and Khiva in 1740, but did not retain power in either. Nadir Shah's army was extremely expensive, and his extortionate taxation eventually led to rebellion. On his death in 1747 his army and most of his realm collapsed.[27] For the next 150 years Iran was ruled by dynasties of Turkic tribal provenance, relying on armies made up largely of light cavalry from regional nomad tribes.

## The military balance between nomad and settled

The soldiers of the steppe were rightly famed as warriors; they inflicted damage on Russian and Chinese armies up into the eighteenth century and remained prized troops in both empires. Their mobility, their hardiness, and their skill in horsemanship and archery remained unsurpassed. However, although these attributes made them formidable soldiers, other factors were also crucial to military success and here many nomad societies had less of an advantage. In terms of command structures, discipline, and the logistics of supplying troops and livestock, settled states often had the edge over their nomadic rivals and over time these advantages enabled the Ottoman, Russian, and Chinese empires to achieve a decisive superiority on the steppe. Scholarship on nomad–sedentary relations usually emphasises the difficulties faced by settled armies against nomads. Nomad armies were famously averse to open battle, preferring to retreat and lead their pursuers into hostile territory, sometimes directly into an ambush. They had relatively few crops and could therefore allow a foreign army to cross their land. Nomad armies had limitations of their own, however. While they needed fewer food supplies than settled armies, they required more pasture. Even on the steppe, this was a dependence which left nomadic forces vulnerable. Settled commanders achieved notable successes by attacking winter pastures, blocking seasonal migrations, or mounting surprise attacks on the central camp of nomad armies on campaign.[28] While nomads could retreat across their own territory, they could not well afford to give up land

27 Peter Avery, 'Nādir Shāh and the Afsharid Legacy', in Peter Avery, Gavin Hambly, and Charles Melville (eds.), *Cambridge History of Iran*, vol. 7: *From Nadir Shah to the Islamic Republic* (Cambridge, 1991), pp. 35–59.

28 Nicola Di Cosmo, 'Introduction: Inner Asian Ways of Warfare in Historical Perspective', in Nicola Di Cosmo (ed.), *Warfare in Inner Asian History* (Leiden and Boston, 2002, p. 5; Michael Khodarkovsky, *Russia's Steppe Frontier: The Making of a Colonial Empire, 1500–1800* (Bloomington, 2002), pp. 18–19; L. J. D. Collins, 'The Military Organization and Tactics of the Crimean Tatars, 16th–17th Centuries', in Vernon Parry and Malcolm Yapp (ed.), *War, Technology and Society in the Middle East* (London, 1975), p. 272.

permanently. The loss of good pasture was a serious blow. Furthermore, if they were to avoid giving battle before invading armies, nomad populations needed an interior space into which to retreat. For these reasons the encroachment of the Russians and the Qing into the steppe weakened the nomad powers both economically and militarily.

What set nomad societies apart was the universal training in horsemanship and archery that their lifestyle provided. Many pastoralists also practised organised communal hunts which provided an element of paramilitary training.[29] A steppe ruler, moreover, could assemble an army with little direct investment. When called to arms, most soldiers brought their own mounts and equipment; if they needed outside help to acquire these, the burden was usually placed directly on those not joining the expedition.[30] Despite their common lifestyle and training, not all nomad soldiers were equally skilled. Almost all nomad forces were drawn up from two sources. One was the personal following or military warband collected around a ruler or a chief. Its members were essentially professional soldiers whose primary allegiance was to their leader and they were ready for service at any time. Then there was the general army, which was drawn from contingents provided and led by tribal leaders. These troops were called up for specific campaigns and returned to their usual occupation when not in service. While they had military skills, they were less trained than the members of the tribal aristocracy and the warbands.[31] (See Figure 1.)

A tribally organised army could often be mobilised with remarkable speed to undertake sophisticated campaigns. Troops were usually decimally organised in contingents with designated banners and passwords, and their campaigns – including raids – often involved careful planning, with scouts sent ahead, and the army deployed in separate, coordinated movements.[32] Under strong commanders, such armies could enjoy notable successes, as shown by Shibani Khan's conquest of Transoxiana with a largely tribal Uzbek army. The weakness of tribal armies lay in command structure and discipline. The primary allegiance of tribesmen was to their tribal leader, and even these men

29 Timothy May, 'The Training of an Inner Asian Nomad Army in the Pre-Modern Period', *Journal of Military History* 70:3 (2006), 617–35, 620–2, 630–1.

30 Jürgen Paul, 'Perspectives nomades: État et structures militaires', *Annales: Histoire, Sciences Sociales* 59:5–6 (2004), 1069–93, 1074; Khodarkovsky, *Russia's Steppe Frontier*, p. 17; Collins, 'The Military Organization and Tactics of the Crimean Tatars', p. 259.

31 Paul, 'Perspectives nomades', pp. 1084ff.; Collins, 'The Military Organization and Tactics of the Crimean Tatars', p. 259.

32 See, for example, Collins, 'The Military Organization and Tactics of the Crimean Tatars', pp. 265–8; Kljaštornyj and Sultanov, *Staaten und Völker*, pp. 297–300.

FIGURE 1. Mounted standard bearers and archers. *Eighteen Songs of a Nomad Flute: The Story of Lady Wenji*. The image dates from the fourteenth century but is a copy of a twelfth-century scroll which has not fully survived. It is thought that it represents the material culture of the Liao Khitan of Manchuria and northern China in the tenth and eleventh centuries. However, it appears much like later steppe nomad material culture. Unidentified artist, early fifteenth century. Ex coll.: C. C. Wang Family, Gift of The Dillon Fund, 1973. Retrieved from The Metropolitan Museum of Art, Washington.

could struggle to control them. Because tribal troops were usually unsalaried and rewarded through booty, they were often distracted by the allure of plunder, allowing a defeated enemy to regroup.

It was not always easy for a steppe ruler to assemble all of the troops that were theoretically available to him. A dissatisfied group could break off or ignore a call to arms. Tribal armies were also prone to break up when they suffered reverses; we have already encountered the desertion of the Jochid leaders Janibeg and Giray from Abu'l Khayr Khan after his defeat by the Oirats. Furthermore, Turkic and Mongolian rulers often had to cope with numerous rivals and subordinate dynastic leaders, which made it difficult for a single ruler to monopolise power. This was a problem for Uzbek and Kazakh khans, for instance, who were frequently unable to muster forces sufficient for their needs.[33] However, a strong central government was not necessary to conduct a successful raid. The centrifugal Crimean and Mongol tribes effectively raided their settled, more centralised, neighbours to extract concessions from them. Raids did not require a large army to succeed, and even those undertaken by individual tribal leaders could serve the interests of the state. Indeed, disunity may sometimes have been exaggerated as a way for the central ruler to avoid taking responsibility for raids.[34]

Although some tribal armies did achieve notable conquests, the greatest armies of the steppe powers were those that transcended tribal structures. Chinggis Khan, and later Tamerlane, used the decimal system against the tribal structure, forming *tümens* (divisions of 10,000) out of sections of several different tribes and placing personal followers in command of all large contingents. Nurhaci and Hung Taiji created banners modelled on the Mongolian system, which retained some of the ethnic and tribal structure but divided up larger groups.[35] All of these leaders moreover made significant use of settled troops almost from the beginning of their careers. These large mixed armies combined the advantages of nomad skill and mobility with a more centralised command and discipline. To build such an army, however, was a major feat, and a rare one.

33 Cosmo, 'Introduction', in Cosmo (ed.), *Warfare in Inner Asian History*, p. 7; Baipakov and Kumekov, 'The Kazakhs', p. 202; Kljaštornyj and Sultanov, *Staaten und Völker*, pp. 295ff.; Collins, 'The Military Organization and Tactics of the Crimean Tatars', pp. 258, 270.

34 Collins, 'The Military Organization and Tactics of the Crimean Tatars', pp. 258–9.

35 Di Cosmo, 'The Qing and Inner Asia', pp. 340–2.

## The role of firearms

One major military legacy of Mongol rule was the spread of gunpowder – and perhaps of cannon – from China to the rest of Eurasia.[36] By the fifteenth century firearms had become a major element in warfare over much of Eurasia, but in the Eurasian steppe mounted archers retained their ascendancy throughout most of the period under review. Though firearms played some part in shifting the balance between nomad and settled powers, they were not the decisive factor. Up to the mid-eighteenth century, bows – particularly the Mongol compound bow, which had great force for its size – had several advantages over guns. Archers could shoot on horseback, and they could discharge six or more arrows a minute, with greater range and accuracy than early guns (see Figure 2). Until the nineteenth century guns remained slow to reload and were difficult to discharge from horseback, so that many horsemen dismounted to shoot.[37] In field battles, light cavalry could decimate unprotected gunmen and artillery after the first barrage was passed. In the fifteenth century, the Ottomans developed light firearms mounted on wagons to use against nomads; when the arms were deployed, the wagons were ranged in a circle to provide protection. The tactic was copied by the Chinese and several other powers, while the Russians developed moveable wooden fortresses to protect gunmen. With this equipment settled powers scored some notable victories over nomadic adversaries, but at considerable economic cost, and at the sacrifice of mobility.[38]

Firearms were most useful in defending border fortresses. Starting in the fifteenth century, the Chinese developed a fortified line of defence on the steppe frontier – the beginnings of the Great Wall – and in the sixteenth century they began to deploy firepower along it.[39] Muscovy started building lines of fortresses in the early sixteenth century, which expanded ever further into the steppe over the next two centuries. These were manned by gunpowder-armed garrisons. When protected by fortress walls, artillery was effective against nomads, but it remained a supplement to cavalry rather than a central component of the

36 Thomas Allsen, 'The Circulation of Military Technology in the Mongolian Empire', in Cosmo (ed.), *Warfare in Inner Asian History*, pp. 273–83; Kenneth Chase, *Firearms: A Global History to 1700* (Cambridge, 2003), pp. 1–2.

37 Timothy May, *The Mongol Conquests in World History* (London, 2012), p. 130; Chase, *Firearms*, pp. 23–6, 73–4.

38 Chase, *Firearms*, pp. 162–7; Nicola Di Cosmo, 'Did Guns Matter? Firearms and the Qing Formation', in Struve (ed.), *The Qing Formation*, pp. 132, 135–6; Stevens, *Russia's Wars* of Emergence, p. 46; Perdue, *China Marches West*, pp. 178–98.

39 Perdue, *China Marches West*, pp. 60–1; Di Cosmo, 'Did Guns Matter', pp. 132–4, 144–5.

FIGURE 2. Mounted archer. Dietz Album, A, f. 72, p. 13. The five Diez albums in Berlin, acquired by Heinrich Friedrich von Diez in Constantinople around 1789, contain more than 400 figurative paintings, drawings, fragments, and calligraphic works. Staatsbibliothek zu Berlin – Preußischer Kulturbesitz, Berlin, Germany.

military.[40] In addition, the Chinese and Russians developed fortified supply depots in the steppe frontier which gave them greater reach into the steppe, just as the nomads were losing space into which to retreat.[41] By 1750, when guns began to become effective against mounted archers, the steppe powers had already been weakened beyond recovery.

## The end of the imperial nomad order

By the end of the nineteenth century, almost all the successor states of the nomad empires had lost power. In the steppe, the Russians and Chinese had left little space to the nomads, while in the south colonial powers exerted

40 Stevens, *Russia's Wars of Emergence*, pp. 46–55, 93–4, 133–4.
41 Peter Purdue, 'Military Mobilization in Seventeenth and Eighteenth-Century China, Russia and Mongolia', *Modern Asian Studies* 30:4 (1996), 757–93, 775–6, 780.

increasing control. In 1783, the Crimean khanate was annexed by the Russians. From this period the Kazakh khanates, already protectorates, were increasingly integrated into the Russian Empire; by 1860 Kazakhstan had been brought into the Russian administrative system, though some Chinggisids were still used as administrators.[42] The Kazakh khans of the Senior Horde lost their throne in 1856, and the Mughal dynasty of India ended in 1858. Two states lasted into the twentieth century: the Ottomans sultans, descended from western Turks, ruled to 1923 and the last Chinggisid ruler of Mongolia died in 1924.

42 Frank, 'The Qazaqs and Russia', pp. 369, 371–9; Kljaštornyj and Sultanov, *Staaten und Völker*, p. 330.

2

# Ming conflicts with its neighbors, 1368–1644

MORRIS ROSSABI

## Introduction

In her book *A Translucent Mirror*,[1] Pamela Crossley wrote that "The Ming empire (1368–1644) was perpetually engaged in a struggle against various peoples along its northern borders." This bald statement contradicts conventional wisdom that the Ming dynasty, having endured Mongol rule for about a century, attempted to limit relations with foreign lands and developed a policy of isolation. In fact, the Ming empire became embroiled in numerous conflicts along its frontiers.[2] Some of these clashes were defensive, but others were attempts to annex additional territory. The court resolved a few of these conflicts through diplomacy or withdrawal from alien lands, yet others festered throughout the dynasty. It scarcely enjoyed clear-cut victories. Nonetheless, such repeated battles necessitated improvements in military strategy and technology, and it is no accident that a spate of texts on the military appeared during this time. The first Ming emperor himself had to master principles of land and naval warfare in the course of defeating other rebel groups and in seeking to gain the throne.

The Ming perceived that the Mongols posed the greatest threat. The Mongols, under Chinggis Khan's son Ögödei, had seized north China in 1234, and, under Ögödei's nephew Khubilai Khan, had conquered south

1 Pamela Crossley, *A Translucent Mirror: History and Identity in Qing Imperial Ideology* (Berkeley, 1999), p. 57. James Tong, in his *Disorder Under Heaven: Collective Violence in the Ming Dynasty* (Stanford, 1992), cites 630 incidents of collective violence during the Ming, 80% of which derived from the second half of the dynasty. See also David Robinson, "Why Military Institutions Matter for Ming History," *Journal of Chinese History* 1 (2017), 235–42 and 297–327. For an important translation, see Foon Ming Liew, *The Treatise on Military Affairs of the Ming Dynasty (1368–1644): An Annotated Translation* (Hamburg, 1998), chs. 89 and 90.

2 In *The Making of the Chinese State: Ethnicity and Expansion in the Ming Borderlands* (Cambridge, 2006), p. 112, Leo Shin echoes Crossley's view, noting that "By all accounts, the southern border zone remained during the Ming a violent place."

China in 1279. They had withdrawn from China in 1368 but had not been totally pacified. Nor could the Ming point to a specific battle that crushed the Mongols and prompted them to abandon rule over China. Ming rulers and officials also had vivid memories of the Jurchen peoples of Manchuria occupying north China from 1127 to 1234. Freed from Mongol rule in 1368, the Jurchens reassembled into various clan and tribal groups in Manchuria and were potentially troublesome to the Ming.[3] To the northwest of China, Türkic and Mongol peoples in the oases and steppes along the traditional Silk Roads, which the Yuan or Mongol dynasty had controlled, were now independent. Thus, the Ming had legitimate concerns about China's northern borders. (See Map 2.)

Japanese pirates, known as *Wakō* (or *Wokou*, in Chinese), preyed upon settlements along China's east coast, as well as ships in that region.[4] Paradoxically, the abortive Mongol invasions of Japan in 1274 and 1281 had renewed Japanese interest in Chinese products, which they had observed or found in Yuan dynasty ships.[5] A demand for these goods developed. Early Ming restrictions on trade prompted the *Wakō* to ravage the Chinese coast to meet the needs of the Japanese market. The court had to devise a solution in order to prevent depredations on its prosperous east coast.[6]

Early in the fifteenth century, it ensured relative peace for about 150 years by allowing the Japanese to offer tribute, which really amounted to trade.[7] The *Wakō* threat dissipated, as Japanese merchants and monks / traders could secure Chinese goods without raiding coastal towns or attacking ships. In 1549, turbulence in Japan, which led to battles between warring Japanese families on Chinese soil, prompted the Ming to suspend tribute and trade.

3 On these developments, see Morris Rossabi, *The Jurchens in the Yuan and Ming* (Ithaca, 1982).

4 On *Wakō* activities in Korea and China, see Benjamin Hazard, "Japanese Marauding in Medieval Korea: The Wakō in Late Koryō," PhD diss. (University of California at Berkeley, 1968).

5 See Thomas Conlan, *In Little Need of Divine Intervention: Takuzaki Suenaga's Scroll of the Mongol Invasions of Japan* (Ithaca, 2001), for a Japanese view of these invasions. In *Khubilai Khan's Lost Fleet: In Search of a Legendary Armada* (Berkeley, 2008), James Delgado describes some of the objects discovered in underwater exploration and salvaging of the ships sunk during these expeditions. Although his historical account is not entirely reliable, his analysis of the nautical aspects of the abortive invasions is valuable.

6 See Wang Yi-t'ung, *Official Relations between China and Japan, 1368–1549* (Cambridge, MA, 1953), for the initial resolution of this conflict.

7 Yet maritime prohibitions on trade persisted. As Li Kangying notes in his *The Ming Maritime Trade Policy in Transition, 1368 to 1567* (Wiesbaden, 2010), p. 33: "Whereas the early Ming was a paradise for petty landholding peasants, it was a hell for merchants. Commerce was viewed as a damaging factor, something that threatened the stability-orientated social and economic landscape."

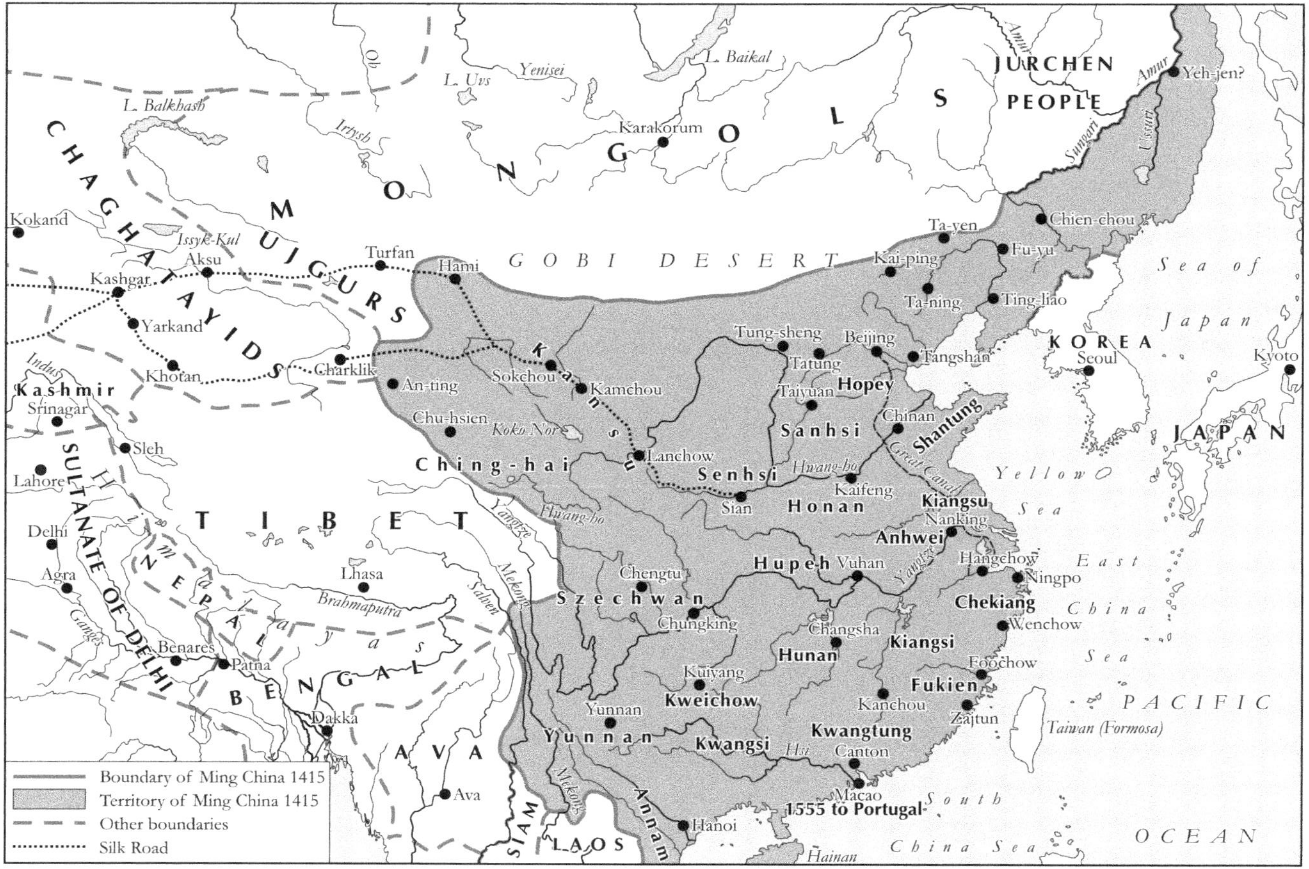


MAP 2. China under the Ming dynasty, 1368–1644. Created by Béla Nagy.

The *Wakō* responded with incursions and attacks along the southeast coast, and the unpaid and demoralized Ming soldiers and the court's navy, lacking sufficient ships, could not cope with the Japanese pirates. However, within a decade, the court shored up its defenses and assigned more soldiers and provided more ships and weapons to its navy. By the mid-1560s, the Ming armies and navy had kept the *Wakō* at bay.[8]

The south offered different challenges. Chinese dynasties had repeatedly sought to expand into modern Vietnam. The Han and Tang dynasties had failed. The Yongle Emperor (r. 1403–24), who had usurped the throne and ousted his young nephew, wished to succeed where other rulers had failed, partly to burnish his credentials and thus to bolster his legitimacy as the emperor. Other areas in mainland and overseas Southeast Asia hardly posed a threat. Yet the court needed at least a minimal naval force to protect coastal commerce.

The most formidable foreign threat thus appeared to be the Mongol armies, who a mere century earlier had created the largest contiguous land empire in world history. Their peerless cavalry, their proficiency with bow and arrow, and their disciplined forces concerned the Ming emperors, who attempted to bring them under control. They were perceived as a genuine threat to the Ming. However, the court did not realize that divisions among the Mongols precluded united action, which would indeed have posed a serious challenge. The Ming set up defenses based on the mistaken assumption that the Mongols would remain the most dangerous of its neighbors.

## Agencies of foreign relations

Diplomacy was always the first tool the Ming deployed in its defensive arsenal, partly because the court, at least in theory, disdained dependence on military officials, who were less esteemed than civil officials. In fact, recent studies have shown that Ming military officials were highly sophisticated. One scholar has asserted that "members of the hereditary military officialdom were active in writing poetry; receiving, commissioning, and collecting works of art; and even practising painting."[9] Nonetheless, it was the creation of trading permissions which became the first step in establishing diplomatic relations with the Ming

8 For these relations, see Wang Yi-t'ung, *Official Relations between China and Japan, 1368–1549* (Cambridge, MA, 1953), and So Kwan-wai, *Japanese Piracy in Ming China during the Sixteenth Century* (East Lansing, 1975).

9 Kathleen Ryor, "*Wen* and *Wu* in Elite Cultural Practices during the Late Ming," in Nicola Di Cosmo (ed.), *Military Culture in Imperial China* (Cambridge, MA, 2009), p. 241.

state. The court sporadically allowed the various Mongol groups to send tribute embassies, which were often disguised trade missions. The Mongols tried to send as many embassies as possible to the court to obtain both essential goods and luxuries, such as silk. The court also established horse fairs along its border, where Mongols could trade their horses for Chinese goods, reducing expenses for both sides in traveling to the Ming capital, which was originally Nanjing but became Beijing by the early fifteenth century.[10] Recognizing that the Mongols needed Chinese goods, the court attempted to avert conflict by using commerce. It would allow the Mongols to trade for essentials if they refrained from belligerent actions. However, if they raided or made incursions on Chinese territory, the court would suspend their rights to offer tribute and to trade with Chinese merchants. Nonetheless, the Ming, on occasion even without provocation, reduced the number of Mongol trade missions, compelling them to attack border towns to secure the products they required.

Part of China's diplomacy entailed the establishment of organizations specifically to facilitate foreign relations. The *Huitong guan* (Interpreters' Institute), which served as the hostelry for envoys, also had *tongshi* (interpreters) who provided valuable services.[11] In 1469, the Interpreters' Institute consisted of sixty interpreters, with nine for Central Asian languages and Mongolian and two for Uyghur. However, because the court granted only low status and poor pay to them, the interpreters were not always competent. The *Siyi guan* (Translators' Institute) consisted of eight Bureaus, including Mongolian, Jurchen, and Uyghur, but faced problems similar to those of the Interpreters' Institute. Nonetheless, it still was sufficiently competent to provide adequate translations.[12]

## Proliferation of military texts

The Ming's defenses also entailed preparation for war. It is no accident that numerous compilations on military matters appeared during the Ming. In fact, the dynasty produced 33.3 percent of all written works on the military,

10 See Henry Serruys, *Sino-Mongol Relations during the Ming*, vol. 3: *Trade Relations: The Horse Fairs* (Brussels, 1975).

11 Translations of the titles of these offices are based on Charles Hucker, *A Dictionary of Official Titles in Imperial China* (Stanford, 1985).

12 See Pamela Crossley, "Structure and Symbol in the Role of Ming–Ching Foreign Translation Bureaus (*siyiguan*)," *Central and Inner Asian Studies* 5 (1991), 38–70; and Morris Rossabi, "Ming China's Relations with Hami and Central Asia, 1404–1513: A Reexamination of Traditional Chinese Foreign Policy," PhD diss. (Columbia University, 1970), pp. 228–37, for additional information on these agencies.

from the Zhou through the Qing or over about 2,500 years.[13] Although most would ultimately be edited and printed in the dynasty's later years, they reflect developments, information, and innovations which stemmed from an earlier period. The available texts offer guides for various aspects of warfare. In his *Dengtan bijiu* (*Necessary Investigations at Mounting the Altar*) of *c.* 1600, Wang Minghao, who had passed the highest-level military examination (*jinshi*), provided a summary of knowledge on tactics, military organization, and armaments, among other subjects.[14] Qi Jiguang (1528–88), a hereditary military official who fought against both the Mongols and Japanese pirates, wrote *Lianbing shiji* (*Practical Arrangement of Military Training*) and *Jixiao xinshu* (*New Book on Effectful Discipline*), both treatises on military training and organization.[15]

The most important compendium was Mao Yuanyi's *Wubeizhi* (*On Military Preparedness*), a massive work of 240 chapters (*juan*) and 200,000 Chinese characters completed in 1594, which was the culmination of the need for reliable guides in this period of violence.[16] Several commanders recruited Mao (1594–1641) as an adviser, especially in Ming engagements against the Manchus, but he was repeatedly challenged, demoted, or dismissed. He ended his days distressed and embittered, but his work became a Chinese military classic. His encyclopedic text attests to the profound concern and interest in military affairs. It encompasses an extraordinary array of subjects, ranging from early military theories to explications of major battles in Chinese history and consideration of weather and topography in armed conflicts. Mao offers practical advice on the training of infantry and cavalry and marching and arraying of troops. He also provides details on attacking and defending cities, medical care, transport, and provision of food and other supplies. Perhaps the most critical sections dealt with maps, weapons, and border defenses, which were all enriched by charts and illustrations. In addition, the maps of China's east coast and of sailing charts were among the valuable parts of the work.

The chapters on armaments range from descriptions of the earliest and most rudimentary weapons to the latest arms, including those

13 Joseph Needham and Robin D. S. Yates, *Science and Civilisation in China*, vol. 5 pt. 6: *Chemistry and Chemical Technology: Military Technology: Missiles and Sieges* (Cambridge, 1994), pp. 29–31.

14 On this text, see Wolfgang Franke, *An Introduction to the Sources of Ming History* (Kuala Lumpur, 1968), p. 208.

15 L. C. Goodrich and Chao-ying Fang (eds.), *A Dictionary of Ming Biography* (New York, 1976), pp. 220–24 for a short biography of Qi; and James Millinger, "Ch'i Chikuang: Chinese Military Official," PhD diss. (Yale University, 1968), for a fuller consideration.

16 On Mao, see Goodrich and Fang, *A Dictionary of Ming Biography*, pp. 1053–54.

Westerners had developed and introduced. Mao described bows and arrows, which Song Yongxing, a seventeenth-century writer on technology, also pictured in his *Tiangong kaiwu* (*The Creations of Nature and Man*) of 1637.[17] Both Mao and Song devoted attention to different kinds of bows and arrows, including the multiple-arrows crossbow and the rocket arrow launcher.[18] They described the making of the bow, emphasizing the materials and glues, and the iron arrowheads. They then moved on to gunpowder and incendiary weapons, pointing out that "the explosive property of gunpowder is the result of the mixing of sulphur with saltpeter."[19] The armaments also included a wide variety of cannons, including such picturesquely named but deadly ones as the "red barbarian" or the "divine frightening." Other uses of gunpowder were for muskets, bronze guns or bombards, field guns, land mines, and underwater or submarine mines. Siege weapons for attacks on cities were also described and illustrated.[20] A Muslim merchant who visited China dates the use of cannons and cannonballs as early as the late fifteenth century or the first decade of the sixteenth century. He also comments that "700,000 soldiers have 700,000 guns, and the entire army is thus equipped with all the implements of war."[21] His overly sanguine view of the Ming army may have been based on the troops he saw in the capital, where one-sixth of the army was stationed, and the Northern Metropolitan Area, the region around Beijing, where another one-tenth of the total military forces was quartered.[22] As noted later, starting in the middle of the fifteenth century, the Ming troops along the frontiers began to weaken.[23]

17 See E-tu Zen Sun and Shiou-chuan Sun (trans.), *Tien-kung k'ai-wu: Chinese Technology in the Seventeenth Century by Sung Ying-hsing* (Pennsylvania, 1966).
18 Needham and Yates, *Science and Civilisation in China*, vol. 5 pt. 6, p. 189.
19 E-tu Zen Sun and Shiou-chuan Sun (trans.), *Tien-kung k'ai-wu: Chinese Technology*, p. 271.
20 Mao Yuanyi, *Wubei zhi* (Qing, 1912), 113/8a and 113/19b; see also Joseph Needham, *Science and Civilisation in China*, vol. 5 pt. 7: *Chemical Technology: Military Technology: The Gunpowder Era* (Cambridge, 1986), pp. 234, 297, 462–63, 490.
21 Kaveh Hemmat, "Itineraries of Political Information in the Khataynamah Critique of Ottoman Governance in a Transoxian Merchant's Description of China," unpublished paper. I am grateful to Kaveh Hemmat for allowing me to read this paper before publication.
22 David Robinson, *Bandits, Eunuchs, and the Son of Heaven: Rebellion and the Economy of Violence in Mid-Ming China* (Honolulu, 2001), p. 38.
23 Kenneth Swope, "The Beating of Drums and Clashing of Symbols: Music in Ming Dynasty Military Operations," *Chinese Historical Review* 16:2 (2009), 147–77, 163, challenges this view, asserting that the "much maligned later Ming dynasty" was "a model of competence."

The Ming military developed battle tactics which ironically it may have borrowed from the Mongols. It used flag and smoke signals and drums to coordinate action and to bolster morale and rang bells to transmit orders. A military historian has written that war began to resemble an orchestra and that the Ming employed signals and noise as psychological weapons to intimidate foes. Victory parades and imperial processions, as well as martial arts demonstrations, served the same objective.[24]

In short, war and violence were not far from the minds of many Chinese policy makers. Naturally, they first exhausted all avenues for diplomacy, but if negotiations proved fruitless, they frequently resorted to violence. Even in its heyday, the Ming state domestically was less stable and orderly than has commonly been accepted. As a historian of the dynasty has stated, "violence was fully integrated into Ming life, during good times and bad."[25] He has uncovered a level of violence which challenges the view of a China of "order, controls, and triumphant virtue."[26] Banditry and rebellion erupted in response to economic discrimination or exploitation, and the hotbeds for such violence were places far from Beijing such as Guangdong and Fujian provinces, which accounted for more than one-half of the total number of incidents.[27] Even in Beijing, there were numerous examples of violence. The extraordinary wealth of corrupt officials, rapacious eunuchs, and other elites in the capital tempted urban gangs, while pirates and highwaymen preyed upon the caravans and ships attempting to supply the urban centers. Texts on defense and warfare were therefore essential for the government.

Nearly all the armaments depicted in the *Wubeizhi* were used in land warfare, not in naval conflicts.[28] By the early seventeenth century, when the text was compiled, Chinese seafaring had declined. During the Song, Yuan, and early Ming periods, Chinese ships traveled to South and Southeast Asia. The seven renowned Zheng He expeditions had reached West Asia and the east coast of Africa, the first direct contacts China had with that continent.[29] However, once the Yongle Emperor (r. 1403–24)

24 Ibid., pp. 152–54, 166. 25 Robinson, *Bandits*, p. 5. 26 Ibid., p. 172.

27 Tong, *Disorder Under Heaven*, p. 26.

28 For a recent study of military expenditures and armaments in the Ming period, see Kai Filipiak, *Krieg, Staat und Militär in der Ming-Zeit (1368–1644)* (Wiesbaden, 2008), pp. 156–72.

29 On the Zheng He expeditions, see J. J. L. Duyvendak, "The True Dates of the Chinese Maritime Expeditions in the Early 15th Century," *T'oung Pao* 34 (1939), 341–412; J. J. L. Duyvendak, *China's Discovery of Africa* (London, 1949); Paul Pelliot, "Les Grands Voyages maritimes chinois au début du XVe siècle," *T'oung Pao* 30 (1933), 237–452; Paul Pelliot, "Notes additionelles sur Tcheng Houo et sur ses voyages," *T'oung Pao* 31 (1935), 274–314; Paul Pelliot, "Encore à propos des voyages de Tcheng Houo," *T'oung Pao* 32 (1936), 210–22. J. V. G. Mills provides a translation of an important primary source on

had died, influential court officials turned against such costly expeditions, and within a few years ended them.[30] Eunuch involvement and support for these voyages – Zheng He was himself a eunuch – may also have prompted official calls to terminate them.[31] The simultaneous withdrawal from and abandonment of efforts to impose Chinese rule on Vietnam led to a de-emphasis on sea power.[32] By the middle of the fifteenth century, the Ming navy had declined, and the Chinese would not develop new naval armaments. Most of the violence afflicting the Ming would derive from the northern frontiers and would involve battles on land.

## Military organization

Like the preceding Yuan dynasty, the Ming established a hereditary and registered military officer class, with considerable power and a clear distinction between it and ordinary soldiers.[33] The court started with a new premise: "The deliberate creation of a military elite that was set above the civil service exams was a new departure for a dynasty of Chinese origin."[34] The Ming restored the civil service examinations only in 1384, offering the military nobility sixteen years from the dynasty's founding to enhance its own power and authority. The military nobility capitalized on this head start to assume the highest policy-making

the expeditions in Ma Huan, *Ying-yai Sheng-lan, "The Overall Survey of the Ocean's Shores" [1433]* (Cambridge, 1970). A recent study is Edward Dreyer, *Zheng He: China and the Oceans in the Early Ming Dynasty* (New York, 2005).

30 See Lo Jung-pang, "The Decline of the Early Ming Navy," *Oriens Extremus* (Dec. 1958), 149–68.

31 Shih-san Henry Tsai, *The Eunuchs in the Ming Dynasty* (Albany, 1996), p. 163; and Louise Levathes, *When China Ruled the Seas: The Treasure Fleet of the Dragon Throne, 1405–1433* (New York, 1994), pp. 175–77.

32 In "Intervention in Vietnam: A Case Study of the Foreign Policy of the Early Ming Government," *The Tsing-hua Journal of Chinese Studies* 8:1–2 (1970), 154–85, 177, Lo Jung-pang perhaps goes too far in writing that "the abandonment of Annam [Vietnam], along with the suspension of the naval expeditions, and the assumption of a defensive posture against the Mongols in the north, were symptoms of the recession of China's military power. China lost her place of international leadership and the respect and fear of her neighbors." Actually, in 1449, the Ming would launch a major offensive against the Mongols.

33 Hidehiro Okada, "China as a Successor State to the Mongol Empire," in Reuven Amitai-Preiss and David Morgan (eds.), *The Mongol Empire and Its Legacy* (Leiden, 1999), pp. 266–68; Romeyn Taylor, "Yuan Origins of the Wei-so System," in Charles Hucker (ed.), *Chinese Government in Ming Times: Seven Studies* (New York, 1969), pp. 23–27, 32–35.

34 Edward Dreyer, *Early Ming China: A Political History, 1355–1435* (Stanford, 1982), p. 5.

positions, the highest ranks, and the greatest pay, at least until 1435.[35] Each generation of a household provided one of its members for such a post, but by the early sixteenth century this military nobility had, due to corruption and mismanagement, considerably deteriorated.

In seeking to protect its northern borders and yet to reduce expenditures, the Ming restored a system of colonies which had a venerable history of many centuries but reached its height during this dynasty. These so-called *tuntian* consisted of colonies of civilians and convicts, although the military ones were the most pervasive. Soldiers along the frontiers farmed the land and were expected to be self-supporting and even to provide their own military supplies. However, it turned out that the soldiers could not earn sufficient income to furnish provisions for themselves.[36] In response, the government devised the so-called *kaizhong* system, by which it commissioned merchants to transport the needed supplies and then to provide the traders with licenses to purchase salt and tea from the state monopolies of those products.[37] The system never worked properly. Peasants who supplied the grain were overtaxed by corrupt officials who then siphoned off some of the income, which translated into fewer and fewer deliveries to the frontiers. From the mid-to-late fifteenth century, the government demanded tax payments in silver instead of grain. Merchants would collect the ever-increasing amount of silver and would turn it over to officials to secure their licenses. The result was that "the merchants could draw the salt immediately after submitting the appropriate value of silver, [and they] no longer had to deliver grain to frontier military granaries." In turn, officials used the silver for other purposes, including "to pay the official salaries of imperial princes . . . The silver was spent but no horses were purchased."[38] Frontier soldiers who did not receive salaries and were not provided with the supplies they required would often simply desert, making some border regions vulnerable.[39]

35 For a detailed description of the military organization, see Foon Ming Liew, *The Treatises on Military Affairs of the Ming Dynastic History (1368–1644)* (Hamburg, 1998).

36 See, for example, Shin, *Making of the Chinese State*, pp. 36–37. He also cites the devastating war in Vietnam and the tropical climate to which Chinese soldiers were not accustomed as other factors in the decline of the *tuntian*. Thus, the Ming faced great difficulties in the present province of Guangxi.

37 Foon Ming Liew, *Tuntian Farming of the Ming Dynasty (1368–1644)* (Hamburg, 1984), pp. 15–35, 97. On the use of tea licenses, see Morris Rossabi, "The Tea and Horse Trade with Inner Asia during the Ming," *Journal of Asian History* 4:2 (1970), 136–68, 152.

38 Liew, *Tuntian Farming*, pp. 30, 32.

39 On the poor pay for soldiers, see Liew, *The Treatises*, p. 75.

## Mongol campaigns

Facing such organizational difficulties, the Ming naturally feared the Mongols and devoted most of their strategic planning to defense against their former overlords. The majority of Ming wars were thus directed against the Mongols. The two sides did not achieve any lasting rapprochement, and sporadic armed conflict characterized their relationship. Ming emperors and statesmen concentrated so much on the Mongols that they often ignored, to their detriment, other potent threats. Conflicts repeatedly flared up along the frontiers but also in the steppes in one direction and on the fringes of Beijing in the other.

Knowledge of the ensuing campaigns is skewed because the Chinese sources are almost exclusively the only accounts of these engagements. Mongols did not possess the lengthy historical and literary traditions in which the Chinese were preeminent. Nearly all the information on these struggles derives from Chinese sources, with all the attendant biases. What remains is one viewpoint, which needs repeatedly to be evaluated. Statistics are especially suspect. Chinese sources often overstate the number of troops on specific campaigns and tend to exaggerate the damage and loss of life they inflicted on their Mongol enemies. Some battles, which they portray as spectacular victories during which the Mongols suffered the loss of impossible numbers of men and animals, on occasion did not take place at all, or at least not on the scale the sources depict them. The Mongols repeatedly tried to avoid full-scale battles, opting instead for hit-and-run attacks or guerrilla raids. They would scatter far and wide to avoid confrontation with large enemy forces, opting instead to flee and elude their Chinese pursuers.

The Yongle Emperor, the third Ming ruler (r. 1402–24), designed policies to pacify the Mongols. His father, the Hongwu Emperor, had succeeded in defeating Töghus Temür, the last influential Mongol Khan, but other leaders arose to command various Mongol groups. Yongle sought to use the tried and trusted tactic of divide and rule to deflect the threat posed by the Oyirad and the Eastern Mongols, the two major contingents among these steppe peoples. Even without Yongle's deliberate attempt to cause divisions, the Oyirad and the Eastern Mongols would have faced formidable obstacles in unifying. Mongol pastoral nomads rarely joined together into large confederations because the dominant and generally optimal unit of organization was the tribe. They only infrequently transferred their loyalty and allegiance from a tribal leader to a khan. Yongle probably did not recognize these inherent fissures, but he also did not acknowledge the possibility of peace if

China had engaged with the Mongols' desperate need for commerce. If the Ming had permitted considerably more opportunities for the Mongols to trade, it would probably have reduced conflicts along the northern borders.

Instead Yongle adopted a belligerent policy. By favoring the Oyirad, he earned the Eastern Mongols' wrath. Nonetheless, his support for the Oyirad helped them to defeat Arughtai, the Eastern Mongols' leader. Judging from later events, however, the Oyirad initial victory was hardly decisive. Misinformed about the scope of Arughtai's defeat, Yongle dispatched a commander named Qiu Fu, allegedly with 100,000 troops, once and for all to crush the Eastern Mongols. It is clear the Ming dynastic history exaggerated the scale of this force since supplies for that many men and animals could not have been prepared so rapidly. Confronted by Qiu, Arughtai used the typical Mongol tactic of a feigned retreat to lure the Chinese force into a deadly trap. Qiu pursued the elusive Arughtai and his relatively small force farther and farther into the steppes until the remainder of the Mongol leader's larger armies suddenly appeared and overwhelmed Qiu's 1,000 cavalry – a more credible number than the 100,000-man army originally cited in the Ming sources – and killed Qiu himself.

This humiliating defeat prompted Yongle personally to lead five campaigns against the Mongols.[40] The Chinese sources described these expeditions in almost the same fashion. Yongle would organize a sizable force, allegedly half a million in one campaign and 230,000 in another, and set forth for the steppes. In several cases, the Ming reputedly intimidated the Mongols, who fled, and the Chinese forces thus did not find substantial enemy troops. When they defeated the Mongols, they encountered small detachments, not a major force. The only real battle was a confrontation in 1415 with the Oyirad who, perhaps buoyed by their highly successful hit-and-run policies, decided to challenge Yongle's army in open combat. Both sides used cavalry, archers, and swordsmen, but the ensuing battle did not lead to a clear-cut victory. The Ming advantage in guns and cannons was scarcely useful in steppe warfare.[41]

40 On these campaigns, see Wolfgang Franke, "Yung-lo's Mongolei-Feldzüge," *Sinologische Arbeiten* 3 (1945), 1–54; Wolfgang Franke, "Chinesische Feldzüge durch die Mongolei im fruhen 15. Jahrhundert," *Sinologica* 3 (1951–53), 81–88; V. M. Kasakevich, "Sources to the History of the Chinese Military Expeditions into Mongolia," *Monumenta Serica* 8 (1943), 328–35; Dmitrii Pokotilov, *History of the Eastern Mongols during the Ming Dynasty from 1368 to 1644*, trans. Rudolph Löwenthal (Chengtu, 1947–49); and Morris Rossabi, "The Ming and Inner Asia," in Denis Twitchett and Frederick Mote (eds.), *The Cambridge History of China*, vol. 8 pt. 2: *The Ming Dynasty, 1368–1644* (Cambridge, 1998), pp. 228–31.

41 Peter Lorge, *The Asian Military Tradition: From Gunpowder to the Bomb* (Cambridge, 2008), p. 71.

In any event, both the Ming and the enemy suffered serious losses, and the Oyirad became especially vulnerable. Arughtai capitalized on the Oyirad's weakness to attack and kill their leader. The death of the Oyirad ruler was to be the only important consequence of Yongle's campaigns. The Ming emperor was unable to pacify the various Mongol groups, yet he devoted considerable financial resources to these expeditions. As one prominent historian wrote, "The northern frontier became the fixation, the virtual obsession of many Chinese statesmen throughout the mid and late Ming times."[42]

Some court officials, however, perceived these campaigns as wasteful and persuaded Yongle's immediate successors to abandon such expeditions. Aware that the Ming had begun to face financial difficulties, they resented the squandering of resources which these inconclusive, if not failed, campaigns represented. In the light of later events, they were right. However, they neither recognized nor acknowledged that permission for the Mongols to trade might have averted border disturbances. Disputes over trade and tribute continued to bedevil relations, leading to minor military engagements and hit-and-run raids. The court could not prevent these raids, partly because of the inadequacy of its cavalry. As the astute Jesuit observer Matteo Ricci explained, "the Chinese know little about the taming or training of horses . . . They have countless horses in the service of the army, but these are so degenerate . . . that they are put to rout even by the neighing of the Tartars' steeds."[43] This deficiency placed the Ming at a disadvantage in dealings and conflicts with the Mongols, consummate horsemen.

The court ultimately decided on an aggressive campaign rather than seeking peace through increasing trade to fulfill Mongol needs for Chinese products. This bellicose policy of initiating an expedition into the steppes appeared to be no different from the earlier Yongle campaigns, which had been inaccurately portrayed as smashing victories. When the eunuch Wang Zhen proposed such an offensive to crush the Oyirad ruler Esen Taishi, who had led incursions into China, the Zhengtong Emperor (r. 1436–39) saw this as a means to imitate his illustrious predecessor and readily agreed. The expedition was star-crossed from the outset. Inadequate preparations, as well as lack of water and fodder in the area, did not deter the Ming. Esen himself had divided his forces into three divisions and had prepared for attacks as far east as Liaodong (in modern Manchuria) and as far west as Datong. In early

42 Rossabi, "The Ming and Inner Asia," 271.
43 Matthew Ricci, *China in the Sixteenth Century: The Journals of Matthew Ricci, 1583–1610*, trans. Louis Gallagher (New York, 1953), p. 13.

August of 1449, Wang Zhen and the emperor led an army supposedly composed of half a million men, without doubt an exaggerated number. They ignored the warnings of the Minister of War and other officials not to cross the walls along the northern frontiers. Marching into unfamiliar territory in a driving rain, they did not realize that Esen's army waited in ambush. Once they learned about the danger, Wang Zhen ordered a retreat. It was too late as the Mongol pursuers isolated and crushed the rearguard. Instead of hurrying back to the safety of a walled town near the border, the Ming forces stopped at Tumu to await the arrival of the emperor's and Wang's treasure trove of goods. Esen's forces attacked them out in the open rather than having to lay siege to a walled town. They routed the Chinese army, killed Wang, and captured the emperor. (See Map 3.) Esen then erred by not following up his victory with an immediate assault on Beijing, permitting the Ming to set up its defenses and to appoint the Zhengtong Emperor's brother as his successor.[44] Esen himself was assassinated when he antagonized many Mongols by disobeying tradition and adopting the title of "khan," which was reserved for descendants of Chinggis Khan. The Ming survived this debacle but in a weakened state and with a deteriorating military force.

For the next century, relations between the Ming and the Mongols alternated between peace and violence. Yet the Ming no longer ventured into the steppes. Instead the Chinese began to construct walls along the border to deter attacks, a project which continued well into the sixteenth century. Although the walls provided a modicum of protection, they failed to prevent incursions. As one historian observed, "Like the Maginot Line, [wall-building] was brilliantly conceived and executed . . . like it as well, it was based on strategic miscalculations: it ignored, for instance, the ease with which the Mongols could find ways through or around it."[45] Batu Möngke (*c.* 1464–*c.* 1532), the so-called Dayan Khan, tried to unify the Mongols and engaged in repeated raids on China, reaching close to Beijing. Altan Khan (r. 1571–82), his grandson, continued these belligerent policies. When the Ming curbed his tribute and trade missions, he retaliated

44 On these campaigns, see Frederick Mote, "The T'u-mu Incident of 1449," in Frank A. Kierman and John K. Fairbank (eds.), *Chinese Ways in Warfare* (Cambridge, MA, 1974), pp. 243–72; Morris Rossabi, "Notes on Esen's Pride and Ming China's Prejudice," *The Mongolia Society Bulletin* 9:2 (1970), 31–39. Li Shih, *Pei shih lu* (reprinted in Shen Chieh-fu, *Chi-lu hui-pien* [Taipei, 1960], ch. 17) and Yang Ming, *Cheng-t'ung lin jung lu* (reprinted in Shen Chieh-fu, *Chi-lu hui-pien* [Taipei, 1969]) are primary sources but with definite biases.

45 Arthur Waldron, *The Great Wall of China: From History to Myth* (Cambridge, 1990), p. 164.

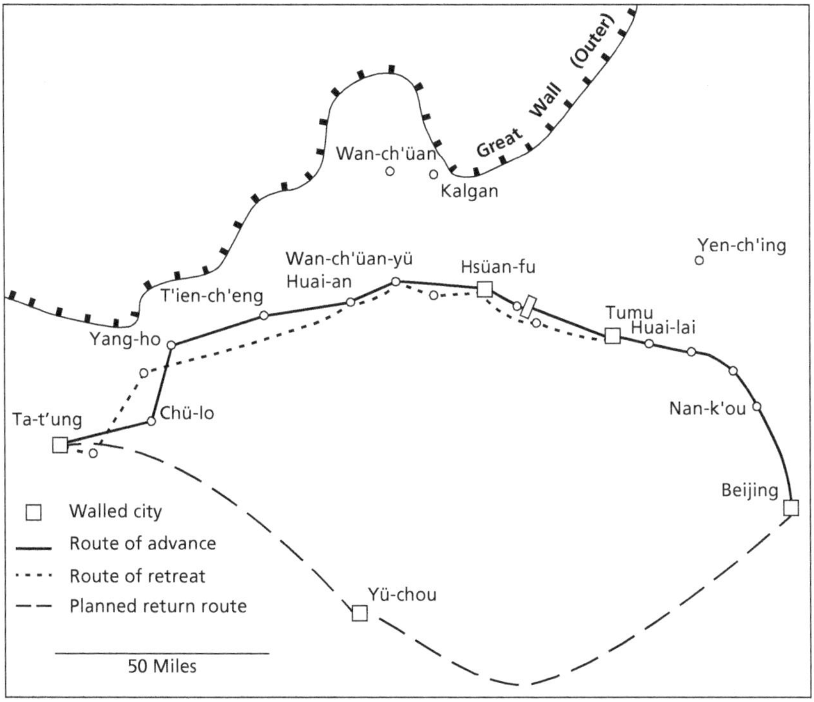


MAP 3. The defeat and capture of the Zhengtong Emperor, 1449

with attacks in 1570 and again in 1571; his troops approached Beijing, compelling the court to permit greater opportunities for trade. Having gained the commercial concessions he sought from the Ming, Altan Khan turned his attention to unification of the Mongols. Believing that religious unity would translate into political centralization, he attempted to convert the population to the Yellow Sect of Tibetan Buddhism. His assumption about religious unity leading to political unity proved to be incorrect, but he and his descendants' emphases on domestic affairs diverted them from armed conflicts with the Ming.

One tactic that developed in response to periodic raids along the Chinese borders has been labeled "clearing the fields and strengthening the walls."[46]

46 Kenneth Swope, "Clearing the Fields and Strengthening the Walls: Defending Small Cities in Late Ming China," in Kenneth Hall (ed.), *Secondary Cities and Urban Networks in the Indian Ocean Realm* (Boulder, 2008), pp. 123–54, provides a description of this tactic.

Aware of an impending attack, often via warnings from watchtowers, Chinese settlers would retreat into small walled towns and would defend themselves with cannons from the wall tops. Their pastoral enemies would lay siege to the town, but the inhabitants had supplies and firepower, which gave them the advantage, at least in theory.[47] The Chinese could afford to wait out the enemy because the latter would eventually have to move to find grass for their horses and other animals.

Ironically, despite repeated struggles with the Mongols, the Ming court adopted much from its steppe predecessors. Some of these borrowings were due to the first emperor's decision to permit Mongols to remain in China. In addition, conflicts among the Mongols, famines or at least lack of assurance of basic necessities, and the seductiveness of a more abundant economy combined to induce some Mongols to flee to China. A few performed valuable services as translators or interpreters and as members of the Imperial Guard (*Jinyiwei*), a special elite force that protected the emperor and probably derived from the Mongol institution known as the *keshig*.[48] The court provided some with a rice allowance and paper money and others with pastureland and hay for their horses.[49] The Ming also adopted features of Mongol military organization. Like the Yuan dynasty, the Ming adopted a structure based on a decimal system, with a chiliarchy (*qianhuso*) and five chiliarchs comprising a Guard (*wei*).[50]

## Northwestern frontier

The Ming's relations with other northern neighbors appeared, at first, to be more peaceful. The only major untoward incident was Tamerlane's projected invasion early in the dynasty. Tamerlane (or Temür), an ardent believer in Islam, heard rumors that the Ming state was persecuting Muslims and decided to save his co-religionists. Like Chinggis Khan, his

47 Lorge, *The Asian Military Tradition*, p. 75.

48 On the *keshig*, see C. C. Hsiao, *The Military Establishment of the Yuan Dynasty* (Cambridge, MA, 1978), pp. 33–43; and Christopher Atwood, "*Ulus* Emirs, *Keshig* Elders, Signatures, and Marriage Partners: The Evolution of a Classic Mongol Institution," in David Sneath (ed.), *Imperial Statecraft: Political Forms and Techniques of Governance in Inner Asia, Sixth–Twentieth Centuries* (Bellingham, 2006), pp. 141–73.

49 Henry Serruys, "Foreigners in the Metropolitan Police during the 15th Century," *Oriens Extremus* 8:1 (1961), 59–83, 62–63; and Henry Serruys, "Landgrants to the Mongols in China, 1400–1460," *Monumenta Serica* 25 (1966), 394–405.

50 On the Mongol influences on the Ming, see David Robinson, "The Ming Court and the Legacy of the Yuan Mongols," in David Robinson (ed.), *Culture, Courtiers, Competition: The Ming Court (1368–1644)* (Cambridge, MA, 2008), pp. 365–411; and Morris Rossabi, "Notes on Mongol Influences on the Ming Dynasty," in Morris Rossabi (ed.), *Eurasian Influences on Yuan China* (Singapore, 2013), pp. 200–23.

illustrious predecessor, he painstakingly set up supply lines through the daunting mountain and desert terrain of Central Asia. Chinggis had traveled east to west from Mongolia to Central Asia with a 200,000-man army, while Temür marched eastward from Central Asia to China. Fortunately for the Ming, who knew nothing of his plans and had not prepared to counter his assault, Temür died early in 1405 while *en route* to China. His son and successor Shāhrukh achieved a diplomatic settlement with the Ming, facilitating trade and ending Central Asian military threats to China.

The oases and towns closer to the northwestern border maintained generally peaceful relations with the Ming until the late fifteenth century. Trade and tribute missions arrived in Beijing from Mongolian and Türkic peoples in Hami, Turfan, and Besh Baliq. The Chinese sources referred to Hami as the gateway to the so-called Western Regions, and the court assertively attempted to prevent enemies from controlling the oasis. It succeeded for a century, but, as it declined, a hostile state occupied Hami in 1513. Without a strong wall to protect them, Hami's inhabitants were vulnerable, and by this stage the Ming army was unable to defend them. The loss of Hami contributed to the undermining of the Ming's defense of the northwestern border.[51]

## Southern attack

The court's policy in Annam (present-day North Vietnam) also diminished its military stature. The founder of the Ming dynasty had adopted a cautious policy, warning "against indulging in military adventurism" and "indicating [his] intention to maintain peace with the maritime countries of the south and east while [he] focused . . . attention on the land frontier of the north and west."[52] He urged his successors to focus on threats from the north and west, rather than pursuing expansionist policies in the south and east. Even when a usurper challenged and overthrew the Trān dynasty (1225–1400) in Annam, he refrained from sending troops to crush the rebels.

Yet again, the Yongle Emperor, his son and successor, diverged from his father's policies and adopted a more bellicose strategy. He was provoked when the usurper acquiesced to a Ming demand to restore a Trān descendant as the king but then ambushed and killed the hapless monarch. Yongle felt compelled to avenge this betrayal by the rebels. In 1406, he dispatched two expeditionary

51 See Morris Rossabi, "Ming Foreign Policy: The Case of Hami," in S. Dabringshaus and R. Ptak (eds.), *China and Her Neighbours* (Wiesbaden, 1997), pp. 79–97.
52 Lo, "Intervention in Vietnam," p. 157.

armies and one naval force, composed of more than 200,000 men. The so-called Eastern Army attacked via the province of Guangxi while the Western units crossed over from Yunnan province into Vietnam. This two-pronged assault moved with considerable alacrity, and Ming forces occupied the Annamese capitals. The naval contingent captured the rebel leader when he tried to escape by sea. By early 1407, the Ming seemed to have been singularly successful, but appearances were deceptive.[53]

Attempting to govern the newly subjugated territory as part of a newly founded province, Jiaozhi, within China, Ming forces faced unusual and eventually devastating challenges. Although they seized the capitals, the vast majority of the Annamese population lived in the countryside, and native leaders capitalized on familiarity with the terrain in resisting the Ming. Their Annamese opponents generally did not engage in full-scale battles. Instead they developed what would later be termed guerrilla warfare. Ambushes, night-time raids on Ming forces, and small-scale attacks on Chinese troops who were separated from their armies were the principal Annamese tactics. All these engagements required the support of the native inhabitants, the key to the guerrillas' successes. Native leaders received intelligence information, sanctuary, and food and supplies from the local populations who resented the Chinese occupation and the taxes imposed upon them by Ming officials. This was classic guerrilla warfare, in which weapons and seemingly overwhelming force did not guarantee success. The objective was to wear down the enemy and generate demoralization of his troops. Goodwill and assistance from the Annamese population were more critical than manpower or firepower. In addition, Ming troops were unaccustomed to Annam's tropical and semitropical heat, and many died due to the attendant parasitic and infectious diseases.

The Yongle Emperor also misjudged the closeness of Annamese and Chinese civilizations. Although the Annamese had borrowed the use of the Chinese written language and had adopted features of Confucian civilization, partly under their late fourteenth- and early fifteenth-century leader Hô Quý Ly,[54] they had their own strong identity and did not wish to be subsumed by the Ming.[55] The Chinese were also distressed by the Annamese Confucians' incorporation of the ideas of local cults.

53 Alexander Woodside, "Early Ming Expansionism (1406–1427): China's Abortive Conquest of Vietnam," *Harvard University Papers on China* (1963), 1–37, is still useful on this invasion.

54 John Whitmore, *Vietnam, Hô Quý Ly, and the Ming (1371–1421)* (New Haven, 1985), p. 59, on Hô Quý Ly and his attempts to foster Confucianism in Annam.

55 Wang Gungwu, "China and Southeast Asia," in Jerome Ch'en and Nicholas Tarling (eds.), *Studies in the Social History of China and Southeast Asia: Essays in Memory of Victor Purcell* (Cambridge, 1970), pp. 381–85.

Like many campaigns waged against guerrilla resistance, the Ming invasion had appeared at first to be successful and able to sustain the establishment in July 1407 of Jiaozhi as a Chinese province. In successive later expeditions, Ming troops defeated the Annamese forces. In 1408–9 they overwhelmed a Trān descendant, and in 1411–14 they crushed yet another Trān leader. Yet in 1417–18, another outbreak of resistance against Chinese rule erupted. On this occasion, a vibrant military leader named Lê Loi (*c.* 1385–1433) emerged and galvanized the population. He lost to the Ming forces on the battlefield several times, but he then reverted to the policy of guerrilla warfare and continued to wear down the Chinese invaders. The Chinese advantage in weaponry and firepower could not match knowledge of the terrain, support from the local populace, ambushes, and unorthodox tactics of meeting the enemy.

The war of attrition ground down the Ming forces. The remarkable Ming innovations in weaponry scarcely made a difference in small-scale combat. Even the new tactics the Chinese armies had developed, including better ways to undertake sieges, were irrelevant in this kind of warfare. Huge expenses added to Chinese difficulties. Court officials, horrified by the costs of the expeditions in Annam and Mongolia, not to mention the voyages of Zheng He, wanted to abandon these aggressive policies. Within three years of the Yongle Emperor's death, the court ordered the withdrawal of Ming troops. Proper analysis and implementation of tactics and strategy proved to be more important than weapons and more numerous and allegedly more powerful forces.

Although periods of peace prevailed across much of the rest of the dynasty's territories, the Ming's southern border continued to be troublesome. In the middle of the sixteenth century, Mac Dang-dung, a usurper, overthrew the Lê dynasty in north Vietnam, and the Ming responded by threatening an invasion. Fearing an attack, Mac Dang-dung submitted and accepted his status as leader of a chieftaincy (*tusi*) within China rather than ruler of an independent country. Earlier, the southwestern kingdom of Luchuan, adjacent to the border with Burma, had challenged Ming rule over the province of Yunnan, which had come under China's control only under the Yuan (1271–1368) dynasty. In 1442, a Ming force attacked and destroyed Luchuan. Yet other chieftains arose throughout the late fifteenth and sixteenth centuries to undermine Ming authority. Violence persisted, culminating in the She-An rebellion of 1621, led by a people named the Lolo, in the provinces of Sichuan and Guizhou. Eight years elapsed before Ming troops could suppress them.

## War against Japan in Korea

Tactics and strategy also trumped a well-trained and professional army in a late Ming conflict. By the last quarter of the sixteenth century, the Ming faced financial shortages, corruption, demoralization, and a declining military. By contrast, Japan was unified for the first time in more than a century. Its samurai troops had been instrumental in pacifying the country and appeared to have a fine esprit de corps. Its leader Toyotomi Hideyoshi (1536–98), who took power after the death of Oda Nobunaga (1532–82), the first of the triumvirate to unify Japan, issued edicts which stabilized the country and ensured samurai monopoly of weapons. Facing disunited enemies and no other charismatic figure, Hideyoshi eventually controlled nearly all of Japan by 1590. Aspiring to even greater recognition of his stature, he dispatched an embassy to China, seeking diplomatic parity with the Ming emperor. The Chinese court rebuffed him and instead referred to him as the vassal king of Japan.[56] Hideyoshi wished to avenge this insult and immediately approached the king of Korea to form an alliance to attack China. The Korean king rejected this proposal, prompting Hideyoshi to initiate an invasion of Korea. It appears likely that Hideyoshi intended to launch an attack on Korea and possibly China even before he sent envoys to both countries. The two courts' allegedly insulting responses served as pretexts for decisions that he had already made.[57]

The ensuing conflict among Japan, Korea, and China yields insights into East Asian warfare. Because of its great interest in military matters, as evidenced by the extraordinary number of texts on armaments and strategy, and in innovations in weaponry, the Ming had valuable technology, despite the Chinese Confucians' often-mentioned disparagement of the military.[58] It had fairly reliable cannons, based on Portuguese models, muskets, and fire arrows and could use bombards at sea, which frequently offered advantages in naval combat.[59] The bronze bombards, which weighed about 600 pounds,

56 Kenneth Swope, "Deceit, Disguise, and Dependence: China, Japan, and the Future of the Tributary Empire, 1592–96," *International History Review* 24:4 (2002), 757–82, 775.

57 Hideyoshi stated that "Even China will enter my grip. I will command it during my lifetime." After initial military successes, he proclaimed that he would soon move to the Ming capital. See Mary Elizabeth Berry, *Hideyoshi* (Cambridge, MA, 1982), pp. 207, 212.

58 Kenneth Swope, "A Few Good Men: The Li Family and China's Northern Frontier in the Late Ming," *Ming Studies* 49 (2004), 34–81, 35.

59 Kenneth Swope, "Crouching Tigers, Secret Weapons: Military Technology Employed during the Sino-Japanese-Korean War, 1592–1598," *Journal of Military History* 69 (2005), 22–27.

could fire 60-pound lead balls. On the other hand, the Japanese possessed muskets and small-bore guns, as well as spears and halberds, which were useful in the close combat at which they excelled. In light of the sharp decline and demoralization of the Ming military, the Japanese appeared to have a distinct advantage in the quality of its forces. The Chinese depended upon a few hereditary military families which had retained their fighting spirit and skills.[60] The Li family, of Korean heritage, was an example of these loyal troops who fought against the Mongols and now in opposition to Japan. Yet the Japanese seemed to be more than a match for a Sino-Korean force.[61]

In May of 1592, the Japanese campaign started auspiciously. One hundred and fifty thousand troops landed and overwhelmed Korean resistance in the port of Pusan. They then divided into three groups and headed directly for Seoul, compelling opposition forces, including the Korean king, to seek refuge north in P'yŏngyang. Japanese troops pursued them and occupied P'yŏngyang within a month. Yet again, the Korean monarch fled northward to Ŭiju on the Yalu river. The Koreans had earlier requested assistance from the Ming court, which had equivocated. The rapid Japanese thrust northwards did now provoke a response, but it proved to be tepid. In August, the court sent a Ming force of 3,000 troops to relieve P'yŏngyang, which the larger Japanese army easily crushed. Aware now of the threat posed by the Japanese forces, the Ming eventually dispatched a 36,000-man army to help the Koreans regain P'yŏngyang. Joined by Koreans, the Ming army besieged P'yŏngyang, trying to starve and force out the Japanese troops. Koreans also burned crops in order to prevent the Japanese from obtaining food.[62] Lacking reinforcements and provisions, the Japanese withdrew.

The Sino-Korean army succeeded, in large part, because it denied supplies to the Japanese. Command of the sea proved to be crucial in this endeavor. Korean sources attribute the naval victories to Yi Sun-sin, although Chinese vessels supplemented the Korean ships. Yi, however, possessed a tremendous

60 Kenneth Swope, "Restoring the Double-Edged Sword: Wanli as Supreme Military Commander," in David Robinson (ed.), *Culture, Courtiers, and Competition: The Ming Court (1368–1644)* (Cambridge, MA, 2008), asserts that, despite the corruption, mismanagement, and demoralization at court, the Wanli Emperor participated in the military decision-making and appeared to welcome the conflict. For additional details, see Kenneth Swope, *A Dragon's Head and a Serpent's Tail: Ming China and the First Great East Asian War, 1592–98* (Norman, 2009).

61 Professors Kenneth Swope and JaHyun Kim Haboush offer invaluable bibliographic advice on these wars. Two sources, in addition to the ones cited, are Yoshi Kuno, *Japanese Expansion on the Asiatic Continent*, 2 vols. (Berkeley, 1937/1940); and Samuel Hawley, *The Imjin War* (Berkeley, 2005).

62 Berry, *Hideyoshi*, p. 214.

advantage over the Japanese thanks to Korean deployment of the so-called turtle ships. Larger than the fishing and cargo vessels which were used to supply the Japanese troops, the Korean ships had metal roofs and were reinforced with iron plates and spikes, thus resembling turtles. Also, because they were speedier than the Japanese vessels, Yi's turtle boats prevented most resupplying of the Japanese in Korea. Thus, after considerable loss of life and destruction, Hideyoshi temporarily abandoned his dream of conquering Korea and possibly China and becoming the dominant ruler in East Asia. During their retreat, however, Japanese forces defeated a Sino-Korean army at Pyŏchegwan and inflicted considerable casualties, but they withdrew from Seoul. They fled to the south, persisting in major campaigns against Korean troops. They occasionally received supplies from their homeland but also secured provisions from the areas they occupied. Nonetheless, they did not initiate an offensive.

The lull in all-out conflict ended when Hideyoshi initiated a second invasion in 1597. He dispatched an army, allegedly composed of 140,000 troops, to join the existing forces in southern Korea. Unlike the first invasion, the second attack did not succeed in reaching Seoul. Yet Japanese troops either won or achieved stalemates in the battles. Even though they were often on the defensive, they had a more sophisticated grasp of tactics. For instance, when Chinese and Korean forces surrounded the outnumbered and poorly supplied Japanese forces in their base at Ulsan, their position appeared precarious. But by building a fort on high ground they virtually neutralized the Sino-Korean army's comparative advantage in weaponry. The Chinese could not transport their powerful cannons along the steep and narrow paths leading up to the enemy fort. Instead they were forced to rely on unreliable rockets and grenades, and on fire arrows, but neither these weapons, nor a blockade, managed to dislodge the Japanese forces. A group of Japanese, arriving by sea, managed to evade the blockade and joined their countrymen in Ulsan. The arrival of reinforcements panicked the Chinese commander, who lifted the siege and fled. The clear lesson of Ulsan was that, for all their advanced weaponry, the Chinese could not expel the Japanese from their bases in southern Korea.[63] Despite a Sino-Korean naval victory at Sunchōn, in which the great Korean admiral Yi was killed, stalemate ensued. The war only concluded with Hideyoshi's illness and death in 1598 and his successors' decision to abandon the costly struggle with Korea.

63 See Kenneth Swope, "Beyond Turtleboats: Siege Accounts from Hideyoshi's Second Invasion of Korea, 1597–1598," *Sungkyun Journal of East Asian Studies* 6:2 (2006), 177–206, for a fine account of the siege at Ulsan.

The conflict proved that more sophisticated weaponry did not inexorably lead to victory. Like the Ming invasion of Annam, where guerillas with rudimentary weapons eventually outlasted the much more powerful Chinese enemy, in Korea, access to provisions and opportunities to resupply, which were ordinary logistical concerns, proved as critical as weaponry to success and failure. Once again, Korea demonstrated that intelligent tactics could overcome greater firepower.

## Ming decline and fall

The performance of Ming forces in their final, decisive series of wars offered further proof that they had still not learnt the lessons of conflict from the late fourteenth to the late sixteenth century. As we have seen, the Ming had long assumed that their principal threat came from the Mongols and had focused their defenses accordingly. They continued to mass most of their military resources along the borders with the Mongols, and several of their military campaigns ventured into the Mongol steppes. Such policies had been justifiable in the pre-1368 period and in its immediate aftermath, but by the fifteenth century divisions among the Mongols should have indicated that they no longer posed a serious threat to the Ming.

Misperceptions in this case would have a greater impact on Ming defenses than actual weaponry. While policy makers concentrated on the Mongols and built walls to prevent Mongol incursions, they neglected to pay as much attention to other threats. Better gathering of intelligence information and more prescient strategic prioritization might have assisted the Ming in coping with the Manchu threat from across their northeastern borders. Such changes in outlook would probably not have altered the outcome, but would at least have highlighted the real perils the Chinese faced. Yet again factors other than weaponry played a crucial role in a Chinese military encounter with an enemy. Despite the proliferation of Ming military texts and remarkable advances in weaponry in the sixteenth century, the court's political and economic failures and its policy makers' myopic concentration on the Mongols left it vulnerable to a new and more powerful confederation across the Ming's northeastern frontiers.

The Jurchens had inhabited modern-day Manchuria for centuries, even ruling north China as the Jin dynasty (1127–1234) before they were ejected by the Mongols, and their relations with the early Ming had generally been harmonious. So long as they remained divided into numerous confederations and tribes, the Manchus could not pose a substantial threat to

China.[64] The early Ming emperors sought to exacerbate the disunity by dealing with them as separate Guards units (*wei*), an institution they borrowed from the Yuan dynasty military structure. The court trusted them sufficiently to recruit them as envoys, translators, spies, and even soldiers. The Yongle Emperor, in particular, developed equable relations with the Jurchens. The Chinese perceived the Jurchens as vassals, and in turn the Jurchens acquiesced to this relationship in order to secure trade with China. Yet conflict arose between the two peoples following Yongle's death as disputes broke out over the quality of goods offered to each other, the smuggling of Chinese and Jurchen goods, and the frequency of Jurchen missions to China. The court's inability to attract qualified interpreters, incorruptible officials, and capable military commanders contributed to tensions and conflicts. The Jurchens responded by appearing to support the Oyirad leader Esen, who captured the Ming emperor in 1449, and this ensured a tenuous and fraught relationship with the court from the late fifteenth to the middle of the sixteenth century. These tensions were exacerbated by internal Ming failures and weaknesses, including tax evasion, corrupt officials – including the excesses of such powerful eunuchs as Wei Zhongxian – and vast expenditure on the campaigns in Korea and for the suppression of domestic insurrections and rebellions.[65]

Meanwhile, Jurchens directly across the border in southern Manchuria were shifting toward a sedentary lifestyle. As they moved in that direction, they developed skills which would permit them to become a more potent threat to the Ming. However, they required unity and a charismatic and well-organized leader. Nurhaci was just such a leader, whose grandfather and father had been prominent among the Jurchen hierarchy. Both had died while collaborating with a Ming general to crush other, more recalcitrant Jurchens who were at war with China. Pointing to the assistance his family had provided, Nurhaci demanded and received the Ming court's designation as ruler of the Jianzhou Jurchens.[66]

64 See Henry Serruys, *Sino-Jurchen Relations during the Yung-lo Period, 1403–1424* (Wiesbaden, 1955) and Rossabi, "The Ming and Inner Asia," pp. 261–64, for a description of their early relations.

65 On him, see the brief summary of his influence in William Atwell, "The T'ai-ch'ang, T'ien-ch'i, and Ch'ung-chen Reigns, 1620–1644," in Frederick Mote and Denis Twitchett (eds.), *The Cambridge History of China*, vol. 7 pt. 1: *The Ming Dynasty, 1368–1644* (Cambridge, 1988), pp. 596–614.

66 On so-called myths surrounding Nurhaci's rise to power, see Pamela Crossley, *The Manchus* (Oxford, 1997), pp. 48–71.

Nurhaci's power and eventual military success derived not solely from weaponry or other military assets, but from his administrative skills. Recognizing that he needed revenue to build a state, he first monopolized ginseng and furs, for both of which there was considerable Chinese demand, and hastened the development of an iron industry, which provided tools and weapons and lessened dependence on China and Korea. Increased revenues permitted him to lure Chinese defectors to assist in devising institutions for census taking, tax collection, budgeting, and organization of government agencies. Simultaneously, he assembled, through conflicts, peaceful submissions, and marital alliances, his own people and followers into banners (*gusa*), a military and political organization, each of which was commanded by relatives or loyal retainers.[67] As he incorporated more non-Jurchen lands and men, including Mongols, a new identity began to emerge, which came to be identified as Manchu and led eventually to the creation of a new written language, based on the Mongols' Uyghur script. The new structure Nurhaci established, consisting of a political and military organization, substantial sources of revenue, a new identity, and Chinese advisers, made him an important presence in Northeast Asia by the late sixteenth century. He desired to be a khan and did not, at this stage, aspire to be emperor of China. Although he knew of Ming China's weakness, he scarcely conceived of a conquest of all of China.[68] Only in 1616 did he truly challenge the Ming by designating himself as Emperor of the Later Jin, a reference to the Jurchen dynasty that ruled north China from 1127 to 1234. Even then, he may not have known how much of China he could actually govern.

Meanwhile the Ming court continued to face the internal disturbances and insurrections which would eventually permit Nurhaci's son and more ambitious successor Hung Taiji to build an empire with China as its centerpiece.[69] Weavers, potters, and other craftsmen rioted in urban areas.[70] A White Lotus rebellion erupted in Shandong from 1622 to 1624. So-called aborigines in

67 On the banner system, see Mark Elliott, *The Manchu Way: The Eight Banners and Ethnic Identity in Late Imperial China* (Stanford, 2001), pp. 56–63.

68 On his limited aspirations, see Gertraude Roth, "The Manchu–Chinese Relationship," in Jonathan Spence and John Wills (eds.), *From Ming to Ch'ing: Conquest, Region, and Continuity in Seventeenth-Century China* (New Haven, 1979), pp. 3–6 and Crossley, *The Manchus*, p. 65.

69 In *The Manchus*, Pamela Crossley, writes "All his life Hung Taiji had excited the fears of his father, brothers and cousins who were disturbed by his combination of deep talent and unfettered ambition . . . In the years 1635 and 1636 Hung Taiji publicly transformed the khanate into an empire." (pp. 75, 78)

70 On such disturbances, see Tsing Yuan, "Urban Riots and Disturbances," in Spence and Wills (eds.), *From Ming to Ch'ing*, pp. 277–320.

Guizhou and Sichuan repeatedly challenged local authorities, and a major Shaanxi rebellion, which the dynasty could not crush, broke out in 1628.

In 1618, capitalizing on these disturbances, Nurhaci had initiated a campaign against Liaodong. In a standard steppe maneuver of feigned retreat, he trapped and defeated Ming forces and killed the Liaodong commander. Having decided that Nurhaci intended to overthrow the Ming, the court ordered commander Yang Hao to undertake a campaign to oust the Manchus from Liaodong. Again, morale, logistics, and tactics outweighed numbers and weaponry. The Chinese had the advantage in manpower and possibly weapons, but they lacked esprit de corps, and their battle plans went awry. In a tactical mistake Yang dispersed his forces into several units in the attack against Liaodong, making each vulnerable as Nurhaci kept his troops together and had a numerical advantage in battles. Cutting them off from their main forces, Nurhaci vanquished one after another of Yang's smaller divisions, culminating in a final assault at Niu-mao zhai. Nurhaci's son Hung Taiji led a cavalry detachment to take command of the heights surrounding the battle site.[71] His troops then stormed down on and defeated the Ming forces. His Chinese enemies often broke ranks, and numerous soldiers defected. Having not counted on such smashing victories, Nurhaci had not prepared the logistical support, the supply lines, and the administrative system to encompass these new territories. Almost a decade elapsed before his son Hung Taiji finally capitalized on these successes to overwhelm Ming forces in 1629–30.

The real military story and the basis for Manchu triumphs was the Ming's political deterioration. Politics trumped military concerns or advantages. The court was unstable, with the eunuch Wei Zhongxian dominant through part of the 1620s, and factionalism rampant among officials; epidemics and natural catastrophes, including earthquakes and droughts, struck the country; rebellions erupted in various parts of the realm; and pirates attacked shipping along the southeast coast. Many Chinese were appalled by court corruption and maladministration.

Perhaps most debilitating to the Ming cause were the internal insurrections. By the late 1630s, Li Zicheng, a rebel who may have had Muslim ancestors, was advancing from Shaanxi province toward the Ming capital, and Zhang Xianzhong gathered forces from Sichuan province to challenge the dynasty.[72]

71 Ray Huang, "The Lung-ch'ing and Wan-li Reigns, 1567–1620," in Mote and Twitchett (eds.), *The Cambridge History of China*, vol. 7 pt. 1, p. 583. For additional details, see Kenneth Swope, *The Military Collapse of China's Ming Dynasty* (Abingdon, 2013).

72 On these disturbances, see the early work by James Parsons, *The Peasant Rebellions of the Late Ming Dynasty* (Tucson, 1970); and for a comprehensive study of the Manchus'

Both made steady progress and encountered scant resistance, as Ming troops submitted, fled, or defected to the rebels. In April of 1644, Li entered the capital, and the last Ming emperor hanged himself. The Manchus, with the aid of Chinese defectors, capitalized on these internal struggles to move into China proper, crush the rebels, and establish the Qing dynasty.

In sum, no matter how superior the Ming's weaponry and manpower, corruption, demoralization, strategic ineptitude, and economic problems outweighed their other assets.[73] The profusion of military texts indicated that the court had considerable knowledge about war, weapons, tactics, and considerations such as treatment for the injured and wounded; but lack of political and economic stability proved to be more important in Ming China's last years.

success, see Frederic Wakeman, *The Great Enterprise: The Manchu Reconstruction of the Imperial Order in Seventeenth-Century China*, 2 vols. (Berkeley, 1985).

73 As one military historian notes, "larger issues of morale, training, command, and supply far exceeded the importance of guns by themselves." Lorge, *The Asian Military Tradition*, p. 69.

3

# Ottoman wars and military transformation, 1453–1826

GÁBOR ÁGOSTON

## Introduction

Ottoman wars from the mid-fifteenth century through the end of the eighteenth century fundamentally changed the geopolitics of Southeast Europe, West Asia, and the Middle East. The Ottomans emerged in West Asia Minor toward the end of the thirteenth century. Known as "Turks" in contemporaneous Europe, the Ottoman ruling elite incorporated people of a wide variety of ethnic origins who considered themselves the followers and descendants of Osman (d. 1324?), the dynasty's founder. Those loyal to the dynasty of Osman used the Turkish designation of "Osmanlı" (Turkish, "of Osman"), which over time came to be rendered in English as "Ottoman." Within three generations after Osman's death, the Ottomans had either conquered or subjugated into vassalage most of the preexisting polities and rival dynasties in the Balkans and Asia Minor. Timur Lenk's victory over the Ottomans near Ankara in 1402 temporarily checked Ottoman expansion. Still, the dynasty recovered, and by 1453 Sultan Mehmed II (r. 1444–46, 1451–81) sealed the Ottomans' status as a formidable military power by conquering Constantinople, the capital of the thousand-year-old Byzantine or Eastern Roman Empire. Expansion through conquest continued well into the sixteenth century. Ottoman battlefield victories under Sultan Selim I (r. 1512–20) against the Safavids of Persia (Chaldiran in 1514) and the Mamluks of Syria and Egypt (Marj Dabiq in 1516 and Raydaniyya in 1517) and under Sultan Süleyman (r. 1520–66) against the Hungarians (Mohács in 1526) resulted in spectacular Ottoman territorial gains in eastern Anatolia, Azerbaijan, Syria, Egypt, and Hungary.

Ottoman expansion, a consequence of imperial ambitions and Ottoman responses to geopolitical challenges, led to an imperial rivalry with Habsburg Spain and Austria (in the Mediterranean and Hungary) and with the Safavids of Persia (in Azerbaijan and Iraq.) By the middle of the sixteenth century, the Ottomans consolidated their conquests against both rivals, and the empire's borders in Hungary saw only minor adjustments until the end of the seventeenth century. The eastern borders, established in 1555 and modified in 1639 in peace treaties with the Safavids, also stabilized, changing little until World War I.

In the latter part of the sixteenth century, the pace of Ottoman expansion slowed, and attention shifted toward defending the empire's frontiers. Two long wars at opposite ends of the empire, one against Safavid Persia (1578–90) and the other against the Habsburgs in Hungary (1593–1606), exhausted the empire and necessitated changes in military recruitment, financing, and provisioning. These transformations, long viewed as evidence of "Ottoman decline," fundamentally reshaped the power relations within the Ottoman ruling elite. Militarily, the Ottomans remained a force to be reckoned with, as demonstrated by the conquests achieved under the able leadership of the Köprülü grand viziers – including the conquests of the fortresses of Várad in Transylvania (1660, Oradea in modern Romania), Érsekújvár in royal Hungary (1663, Nové Zámky in modern Slovakia), Candia in Crete (1669), and Kamieniec in the Polish-Lithuanian Commonwealth (1672, Kamianets Podilskyi in modern Ukraine). With these victories, the Ottoman Empire reached its greatest territorial extent. (See Map 4.)

Following their unsuccessful siege of Vienna (1683), the Ottomans suffered a series of defeats at the hands of the forces of the anti-Ottoman coalition of the Habsburgs, Venice, Poland-Lithuania, and Russia, known as the Holy League. In the peace treaties of Karlowitz (1699) and Istanbul (1700), the Ottomans ceded Hungary to the Habsburgs, the Morea (Peloponnese) and most of Dalmatia to the Venetians, Podolia to the Polish-Lithuanian Commonwealth, and Azov to Russia.

Even in this period, the Ottomans had victories. Russian expansion was checked in 1711 by a resounding Ottoman victory against Peter the Great at the Prut river, just two years after Peter's victory over Sweden. In the 1715–18 wars against Venice and Austria, the Ottomans retook the Morea from Venice even as they were defeated in Hungary and the Balkans by Eugene of Savoy and (temporarily) lost Belgrade. Ottoman resurgence in the 1736–39 war against the Austrian Habsburgs and Russians was noticeable. However, the Ottomans' recapture of Belgrade owed as much to Austrian military and diplomatic inability and French diplomatic mediation as to Ottoman military recovery. Ottoman victories against Russia at Özi (Ochakiv, in modern southern Ukraine) in 1737 and Austria at Belgrade in 1739, and the long peace along the northern frontier after the Treaty of Belgrade (1739) – a result of a conscious policy of successive grand viziers to keep the empire

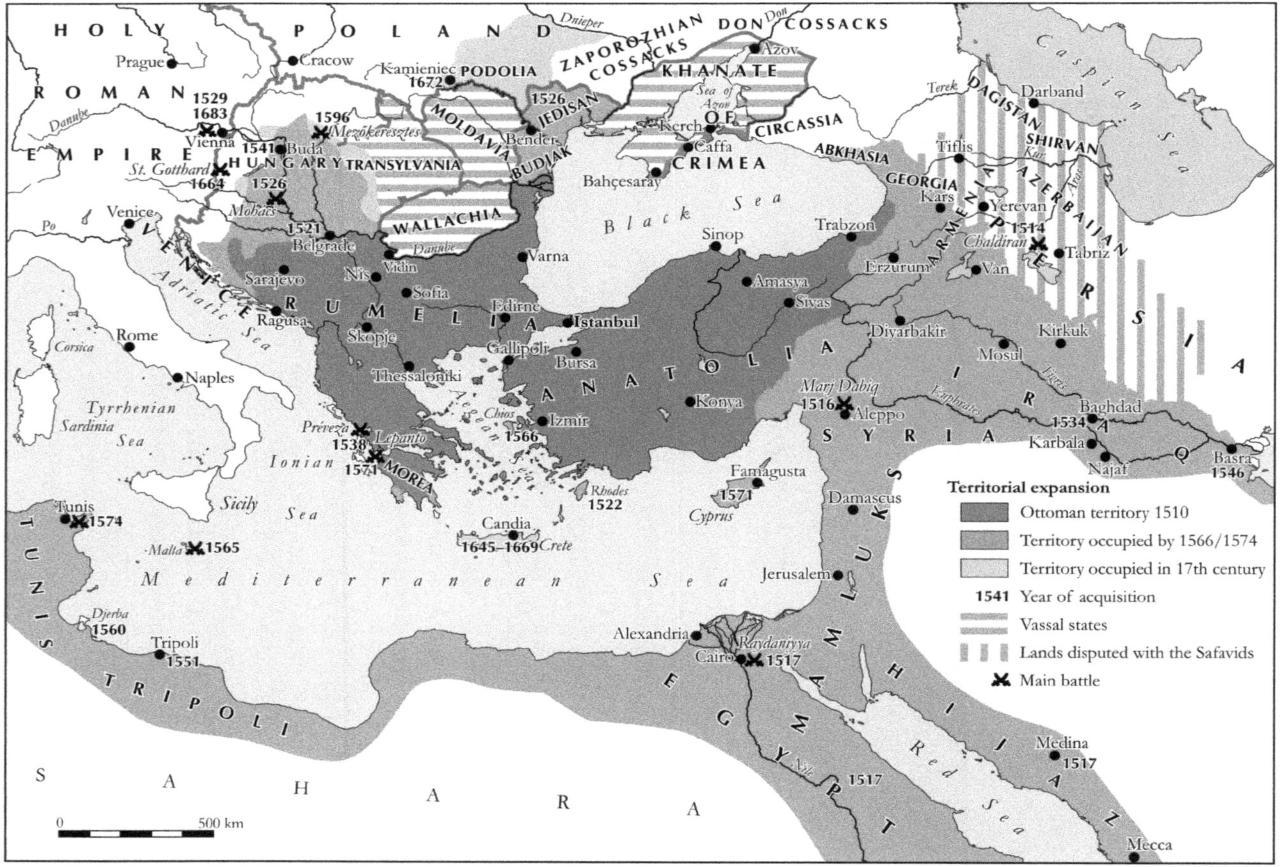


MAP 4. The growth of the Ottoman Empire, sixteenth and seventeenth centuries. Created by Béla Nagy.

off the battlefield – delayed much-needed reforms in the Ottoman military, administration, and finances. Consequently, when war broke out with Russia again in 1768, the Ottoman government and society were unprepared.

By 1768, Russia had emerged as a significant military power, with a large conscript army led by able generals, who – along with their foreign officers – had been schooled in the Seven Years' War (1756–63). In the 1768–74 Russo-Ottoman war, Russian forces humiliated the Ottoman military, exposed its weaknesses, and achieved deep strategic penetration into Ottoman territory. In 1783 Russia annexed the Crimean peninsula, home of the Muslim Turkic Crimean Tatars and an important Ottoman satellite and buffer on the empire's Danube–Black Sea frontier. The failures of the Russo-Ottoman war of 1768–74 and the wars in 1787–92 against Austria and Russia forced Sultan Selim III (r. 1789–1807) to reorganize the empire's military. However, his "New Order" (*Nizam-i Cedid*) army was disbanded in 1807, during the Russo-Ottoman war of 1806–12 after a popular uprising, led by the Janissaries, who had vested interest in maintaining the "Old Order," dethroned and killed the sultan. Reforms resumed after Sultan Mahmud II (r. 1808–39) abolished the Janissary corps (1826). The sultan's new conscript army, called the "Victorious Soldiers of Muhammad" (*Asakir-i Mansure-i Muhammediye*), again proved inferior both against the Russians (1828–29) and against Mehmed Ali's (r. 1805–49) modern Egyptian army, which invaded Anatolia in 1832. It was only during the Tanzimat reforms (1839–76) that the Ottoman government succeeded in establishing a modern conscript army.

Until recently, historians have demonstrated little interest in the history of Ottoman warfare, despite its great significance. With the opening of the Ottoman archives in the 1990s and the growing number of Ottomanists now writing about military history, this picture is rapidly changing.[1] For too long, Ottoman military and naval developments have been discussed in contrast to developments in a handful of paradigmatic West European polities, against whose armies and navies the Ottomans waged no war in the early modern era. This chapter has a different approach and examines the evolution of the Ottoman military within the Ottomans' changing geopolitical context. Viewed from this angle, the Ottomans' military effectiveness and staying

1 See, for example, Rhoads Murphey, *Ottoman Warfare, 1500–1700* (New Brunswick, 1999); Virginia H. Aksan, *Ottoman Wars, 1700–1870: An Empire Besieged* (Harlow, 2007), and its new, revised edition: *The Ottomans, 1700–1923: An Empire Besieged* (London and New York, 2022); Mesut Uyar and Edward J. Erickson, *A Military History of the Ottomans from Osman to Atatürk* (Westport, 2009); Gábor Ágoston, *The Last Muslim Conquest: The Ottoman Empire and Its Wars in Europe* (Princeton, 2021), material from which is used in this chapter.

power are remarkable, for it was achieved in different military environments against various enemies, whose military cultures, strategies, armies, fighting techniques, and tactics were very dissimilar. These enemies and rivals included Turkic principalities and Turcoman confederations of Asia Minor and Iran; Byzantines and their Genoese and Catalan mercenaries; European crusaders; the Safavids of Persia and their partisans in Anatolia; the Mamluks of Syria and Egypt and their allied Arab tribesmen; Georgians of the southern Caucasus; the medieval Balkan states (Bulgaria, Serbia, Bosnia, Albania, Wallachia, Moldavia); Venice; Hungary; Poland-Lithuania; Zaporozhian Cossacks; Spanish and Austrian Habsburgs and the latter's Italian, Spanish, German, Walloon, and French mercenaries; as well as Muscovy / Russia. To effectively engage so many and so diverse an array of adversaries required constant adjustment and adaptation, but also offered opportunities for military acculturation. It also influenced the composition of Ottoman military manpower and ruling elites, and their ideology, strategy, and tactics.

These multiple challenges played a vital role in the evolution of Ottoman military capabilities. For instance, in addition to the persistence of military cultures and systems of recruitment and remuneration, the high proportion of cavalry in the Ottoman military until the mid / late seventeenth century can also be explained by the continued effectiveness of light horsemen in the Iranian, Hungarian, and Pontic theaters of war against the Safavid light cavalry, Hungarian and Polish Hussars, and Cossack horsemen. On the other hand, to successfully besiege Byzantine and Balkan medieval castles, and later artillery-resistant fortresses in Hungary, and Venetian-held Cyprus and Crete required familiarity with the evolving art of siege warfare as well as the creation of adequate weapons manufacturing capabilities. Wars and battles with the Hungarians in the 1440s acquainted the Ottomans with field artillery and acted as a catalyst for integrating firearms into the Ottoman standing forces, and further developing the old Turkic *tabur* in response to the Hussite / Hungarian *Wagenburg* ("wagon fortress"). Similarly, the combined use of *tabur* and cavalry on the wings provided a means of protection against enemy cavalry charges, which explains the lack of pikemen and the bayonet in the Ottoman army, despite the Ottomans' exposure to Habsburg infantry using pikes and bayonets. While this chapter synthesizes recent research from the above perspective, it focuses on the evolution of Ottoman military capabilities from the mid-fifteenth through the late eighteenth centuries.

## Ottoman expansion and military superiority

Ottoman expansion is too often considered only in terms of military conquest, but it also entailed the Ottomans' skillful use of politics, as well as the organization, pacification, and integration of conquered lands. A seldom noticed feature of the Ottoman expansion is the Ottomans' clever exploitation of succession quarrels and civil wars in lands adjacent to the nascent Ottoman state. The Ottoman conquests of Serbia, the Morea, the Emirate of Karaman, Herzegovina, Wallachia, and the Crimea were all preceded by such internal rivalries. In these states, the Ottomans supported whichever faction was most pro-Ottoman. For instance, in Serbia, they backed the pro-Ottoman elite led by Michael Angelović, joint regent of Serbia and the brother of Mehmed II's grand vizier or first minister, Mahmud Pasha Angelović.[2]

The various methods that the Ottomans employed to incorporate the conquered peoples into their growing empire in the pre-1450s included inter-dynastic marriages, forced resettlement, integration of the conquered nobility into the Ottoman military and administrative elite, and flexible tax policies.[3] Inter-dynastic marriages, important in the early phase of Ottoman emergence, had waned by the 1450s. From the sixteenth century on, resettlement was employed sparingly and served primarily economic rather than military purposes. Flexibility and pragmatism, however, remained the true hallmark of Ottoman provincial administration and taxation after the 1450s, especially in frontier provinces. This fact helps to explain the Ottomans' consolidation of power over the conquered lands.[4]

### *The conquest of Constantinople and its impact*

Older military accounts of the conquest of Constantinople have put undue emphasis on Mehmed II's giant bombards and Christian gunners. Such

2 Donald Edgar Pitcher, *An Historical Geography of the Ottoman Empire from Earliest Times to the End of the Sixteenth Century* (Leiden, 1973), p. 76. Related to the Byzantine nobility, the Angelović family belonged to the ruling elite of Serbia. The future Mahmud was captured in an Ottoman raid in 1427, educated as a Muslim in the Ottoman palace school, and made grand vizier in 1456. His brother became joint regent of Serbia in 1458. See Theoharis Stavrides, *The Sultan of Vezirs: The Life and Times of the Ottoman Grand Vezir Mahmud Pasha Angelović (1453–1474)* (Leiden, 2001).

3 Halil İnalcık, "Ottoman Methods of Conquest," *Studia Islamica* 2 (1954), 104–29, and Gábor Ágoston, "The Ottomans: From Frontier Principality to Empire," in John Andreas Olsen and Colin S. Gray (eds.), *The Practice of Strategy: From Alexander the Great to the Present* (Oxford, 2011), pp. 114–17.

4 Gábor Ágoston, "A Flexible Empire: Authority and Its Limits on the Ottoman Frontiers," *International Journal of Turkish Studies* 9:1–2 (2003), 15–31.

presentations celebrated the Ottomans as one of the new "gunpowder empires," whose military might was based on the newly harnessed gunpowder technology. While effective cannon fire and mining played an essential role in breaching Constantinople's walls, once thought impenetrable, several other factors explain Ottoman success. Of these, overwhelming numerical superiority (70,000 besiegers versus 10,000 defenders) and abundant supplies of food, weapons, shot, and powder in the Ottoman army all demonstrated the Ottomans' organizational and logistical capabilities. At the siege, the Ottomans exhibited all the hallmarks of their military superiority, which would characterize their art of war until the late sixteenth century: careful planning, outstanding and resourceful leadership, a large and disciplined professional army, cutting-edge military technology, prowess in siege warfare, and superior supply and logistics.

The conquest of Constantinople united the Asian and European provinces of the sultan and afforded the Ottomans a commanding position over military and trade routes between Asia and Europe. Increased revenues from trade and territorial expansion enabled the sultan to enlarge his salaried standing forces, known as "servitors of (his household's) gate" (*kapukulu*), as well as the provincial light cavalry (timariot *sipahi*), who were paid through military fiefs (*timar*). Of the standing troops, Mehmed II is said to have raised the number of Janissary infantrymen from 3,000 to about 10,000 men by the end of his reign. However, shortly after his death, an official pay register recorded fewer than 8,000 Janissaries. The salaried standing troops (Janissaries, artillery gunners, and palace cavalry) together numbered some 15,000 men and constituted about one-fourth of the professional combat forces. With the salaried and timariot forces, the sultan was able to counterbalance the private cavalry forces and raiders (*akıncı*) under the command of the Turkish frontier lords.[5]

Constantinople was an ideal logistical center for further conquests in the Balkans and Asia Minor. Serbia fell in 1459, Bosnia in 1463, parts of Herzegovina in 1467, Albania in 1468, the Byzantine Despotate of Morea in 1460, and Venetian-held Negroponte in 1470. Wallachia was made a vassal in 1473. In Asia, the Ottomans were victorious over the Karamans (1468) and the Aqqoyunlu ("White Sheep") Turcoman confederation (1473), which controlled eastern Anatolia, Azerbaijan, Iraq, and western Iran from its capital Tabriz. Ottoman victory demonstrated the superiority of Ottoman

5 Gyula Káldy-Nagy, "The First Centuries of the Ottoman Military Organization," *Acta Orientalia Academiae Scientiarum Hungaricae* (henceforth *AOH*) 31 (1977), 147–62.

standing forces over the Aqqoyunlus' traditional Turcoman tribal military organization.

Mehmed II's victory over the "Empire" of Trebizond (1461) – the last vestige of Byzantium in Anatolia – secured Ottoman rule along the southern shore of the Black Sea. The conquest of the Genoese colonies of Caffa and Sudak in the Crimea, and the Venetian port of Tana (Azov) at the mouth of the Don river in 1475, as well as the subjugation of the Crimean Tatars into vassalage in 1478, were important steps in integrating the Black Sea littoral into the Ottoman economy.

### *A naval power*

Ottoman military history is too often equated with land warfare, and the importance of the fleet in Ottoman power projection is little appreciated. The Ottoman navy was instrumental in protecting the new imperial capital, the Levantine trade routes, and the maritime lanes of communication that connected Constantinople with Egypt and the Black Sea littoral – vital links for the supply of the capital, and for the Ottoman economy in general. The Mediterranean and Black Sea fleets, in cooperation with the riverine flotillas on the Danube, Euphrates, and Tigris rivers and their tributaries, also played a crucial role in campaign logistics, by transporting troops and hundreds of tons of food, cannon, shot, and gunpowder to the theaters of war in Hungary, Eastern Europe, North Africa, the Mediterranean islands of Rhodes, Cyprus, and Crete, and the Caucasus and Iraq. These tasks required certain types of ships and crews, all shared by the protagonists in the two seas.

Mehmed II's conquest of the southern and northern shores of the Black Sea was completed in 1484 under Bayezid II (r. 1481–1512), by the latter's conquests of Kilia and Akkerman in the mouth of the Danube and Dniester rivers. Ottoman chroniclers recognized the geostrategic importance of these towns by calling them "keys to those realms," whose conquest "opened" Moldavia to the Ottomans. With their conquests, the Ottomans also denied Poland-Lithuania, the strongest Christian kingdom in the north, access to the Black Sea.[6] Moldavia was soon turned into an Ottoman vassal (1486). However, it was not until Süleyman I's 1538 campaign that Bessarabia and the coastline between the Dniester and Dnieper rivers were secured. It is often claimed that due to these conquests, the Black Sea became an "Ottoman lake." Nevertheless, Ottoman authority waned in the "Wild Fields," where the

6 Reha Bilge, *II. Bayezid. Deniz Savaşları ve Büyük Strateji* (Istanbul, 2012), pp. 59–69.

nomadic Little Nogay Horde and the sedentary Zaporozhian Cossacks challenged Ottoman and Polish-Lithuanian authority.[7]

Coordinated land and naval operations, an often overlooked Ottoman military capability, accomplished the conquest of Kilia and Akkerman. During the Venetian–Ottoman war of 1499–1503, similar synchronized land and naval campaigns resulted in the Ottoman conquest of Venice's naval bases of Lepanto, Navarino, Modon, and Koron. In the possession of Koron and Modon, which the Turkish *ghazi*-corsair Kemal Reis (d. 1511) called the "two eyes of Venice," the Ottomans could now monitor the Venetian navy as it passed from the Adriatic into the Mediterranean.

With the employment of Kemal Reis and his fellow Turkish *ghazi*-corsairs in official Ottoman service, Sultan Bayezid established a precedent for a mutually beneficial relationship between the Ottoman dynasty and the Turkish (and later Barbary) maritime corsairs. These enriched the Ottomans with precious naval knowledge and personnel, such as Kemal's nephew, Piri Reis (d. 1553), the author of the "Book of Navigation" and two world maps. Under Sultans Selim I and Süleyman I, the famed corsair Hayreddin Barbarossa (d. 1546) of Algiers and his seamen extended the Ottomans' naval reach to North Africa.[8] The fleet was also instrumental in the Ottoman conquest of Mamluk Egypt in 1516–17, and in halting Portuguese expansion in the Red Sea and the Persian Gulf, and thus protecting Islam's holy places, Mecca and Medina.[9]

The size of the Ottoman navy was already impressive in the late fifteenth century. Mehmed II mobilized 380 galleys in his Caffa campaign (1475), whereas Bayezid II ordered the construction of no fewer than 250 galleys in 1500, during the 1499–1503 Ottoman–Venetian war. In the late sixteenth and seventeenth centuries, Ottoman galleys were about 42 meters long, with twenty-five banks and forty-nine oars – one oar per bank on both sides, except the last bank, where one side was reserved for the vessel's kitchen. These galleys had 196 or 245 oarsmen, four or five oarsmen to a bench. Oarsmen for the Ottoman galleys came as forced levies from western Anatolian and Balkan households in varying ratios according to demand. Azab infantrymen, the combatants of the galley crews, were recruited via

7 Victor Ostapchuk and Svitlana Bilyayeva, "The Ottoman Northern Black Sea Frontier at Akkerman Fortress: The View from a Historical and Archeological Project," in A. C. S. Peacock, *The Frontiers of the Ottoman World* (Oxford, 2009), pp. 137–70.

8 Svat Soucek, *Studies in Ottoman Naval History and Maritime Geography* (Istanbul, 2009).

9 Palmira Brummett, *Ottoman Seapower and Levantine Diplomacy in the Age of Discovery* (Albany, 1994); Salih Özbaran, *Ottoman Expansion toward the Indian Ocean in the 16th Century* (Istanbul, 2009).

similar levies. These were supplemented by Janissaries and timariot cavalrymen, resulting in about 100, 120, or 150 fighting troops per galley, two-thirds of whom fought with guns, while the rest were archers.[10] However, gathering timariot troops was not easy and often delayed naval operations. For instance, in June 1570, the navy that sailed against Venetian-held Cyprus waited for the Anatolian provincial troops twenty days in Finike, west of Antalya on the Mediterranean coast.[11]

After Lepanto, the Ottomans rebuilt their destroyed navy with surprising speed, proof of their enormous resources and organizational skills. In 1574, in a bold strategic move and naval gamble that stretched the logistical and temporal limits – from March to late October, known in Ottoman parlance as "sea season" (*derya mevsimi*) – of Mediterranean galley warfare, the Ottomans retook Tunis from Habsburg Spain, with a naval force more massive than either party had at Lepanto. The victory, off the coast of Spanish Habsburg Sicily – at a great distance from Istanbul, the logistical center of Ottoman operations – demonstrated Ottoman naval resurgence. However, after the Spanish–Ottoman truce of 1581 the two empires disengaged. Ottoman naval capabilities decayed in the following decades as the Ottomans concentrated on their long wars on land against the Safavids and Habsburgs.[12]

### *Land forces and military strength*

Steady treasury surplus until the 1590s enabled the sultans to pay the standing troops regularly, thus enhancing their morale and combat capabilities. In the early 1500s, the standing army numbered 12,000 to 17,000 men. (See Table 1.)

Provincial timariot horsemen formed the bulk of the army until about the mid-1700s. The *timar* land tenure system enabled the Ottomans to maintain a sizable semiprofessional army whose members were stationed in the provinces, collected revenues, and upheld law and order in the countryside. According to a unique treasury balance sheet from 1527–28, which recorded revenues and expenditures both in cash and *timar*, the potential strength of Sultan Süleyman's salaried and fief-based

10 İdris Bostan, *Kürekli ve Yelkenli Osmanlı Gemileri* (Istanbul, 2006), pp. 198–202; Colin Imber, *The Ottoman Empire, 1300–1650: The Structure of Power*, 3rd ed. (London, 2019), pp. 306–10.

11 İdris Bostan, *Beylikten İmparatorluğa Osmanlı Denizciliği* (Istanbul, 2006), p. 94.

12 Andrew C. Hess, *The Forgotten Frontier: A History of the Sixteenth Century Ibero-African Frontier* (Chicago, 1978), pp. 94–99.

Table 1 Paper strength of the salaried standing troops, 1484–1530

| Date | Janissary | Artillery | Palace Cavalry | **Total** |
|---|---|---|---|---|
| 1484 | 7,841 | – | 4,306 | **12,147** |
| 1512 | 8,164 | 1,078 | 3,896 | **13,138** |
| 1514 | 10,065 | 1,182 | 6,202 | **17,449** |
| 1520/Selim | 7,780 | 1,217 | 5,667 | **14.664** |
| 1520/Süleyman | 8,361 | 1,226 | 6,380 | **15,967** |
| 1521 | 8,349 | 1,608 | 6,192 | **16,149** |
| 1522–23 | 7,010 | 1,715 | 6,022 | **14,747** |
| 1523 | 7,164 | 1,659 | 6,118 | **14,941** |
| 1523–24 | 8,641 | 1,705 | 5,882 | **16,228** |
| 1524–5 | 9,390 | 1,676 | 5,997 | **17,063** |
| 1527–28 | 7,886 | 2,162 | 5,088 | **15,136** |
| 1530 | 8,407 | 2,383 | 4,906 | **15,696** |

*Source:* Istanbul, Başbakanlık Osmanlı Arşivi (BOA), Maliyeden Müdevver Defterleri (MAD.d.) 23; Gábor Ágoston, *The Last Muslim Conquest: The Ottoman Empire and Its Wars in Europe* (Princeton, 2021), p. 278.

Table 2 Potential strength of the Ottoman salaried and fief-based forces in 1527–28

| | |
|---|---|
| Provincial timariot light cavalry | 27,888 |
| *Cebelü* retainers (estimated) | 41,832 |
| Garrison soldiers paid via *timars* | 9,653 |
| Central salaried (*kapukulu*) troops | 15,136 |
| Salaried garrison troops | 27,617 |
| Salaried troops in Egypt | 3,761 |
| **Total timariot and salaried troops** | **125,887** |

army along with garrison forces can be estimated at about 125,000 men.[13] (See Table 2.)

Alongside these professional troops, the Ottomans also used peasant militia footmen (*azab*) and irregular frontier raiders (*akıncı*), numbering together perhaps 10,000 to 20,000 men per campaign in the first decades of the sixteenth century.[14] The Tatar horsemen of the vassal Crimean khanate

13 Ömer Lütfi Barkan, *Osmanlı Devletinin Sosyal ve Ekonomik Tarihi: Tetkikler-Makaleler*, ed. Hüseyin Özdeğer, 2 vols. (Istanbul, 2000), vol. 1, pp. 649–702; Ágoston, *The Last Muslim Conquest*, 282.

14 Káldy-Nagy, "First Centuries," pp. 163, 170.

Table 3 Mobilized salaried troops, 1514–43

| Date | Janissary | Artillery | Palace Cavalry | **total** |
|---|---|---|---|---|
| 1514 | 10,065 | 1,005 | 5,262 | **16,332** |
| 1541 | 6,362 | 1,667 | 3,670 | **11,699** |
| 1543a | 8,166 | 4,167 | 2,624 | **14,957** |
| 1543b | 7,365 | 3,959 | 2,502 | **13,826** |

*Source:* 1541: Istanbul, Topkapı Sarayı Müzesi Arşivi Defterleri (TSMA. D) 9619 and Rhoads Murphey, *Ottoman Warfare, 1500–1700* (London, 1999), p. 46; 1543: Istanbul, BOA, Kamil Kepeci Defteri (KK.d.) 1765, 10v and Mehmet İpçioğlu, "Kanuni Süleyman'ın Estergon (Esztergom) Seferi 1543," *Osmanlı Araştırmaları* 10 (1990), 137–61, 140.

might have added another 10,000 to 20,000 men to the Ottoman expeditionary armies in the late sixteenth century.[15] In short, Süleyman's army consisted of 70,000–80,000 provincial cavalrymen remunerated with military fiefs; a much smaller standing army, numbering about 15,000 men (Janissary infantry, palace cavalry, and artillery); and 42,000 garrison soldiers who served in the empire's 292 fortresses and forts. With 35,000–40,000 frontier raiders and peasant militiamen, Süleyman's military potential reached 160,000–170,000 men.

While none of the Ottomans' neighbors could match this military potential in the early sixteenth century, it was only about 1 percent of the empire's estimated 13 million population. This is hardly a sign of "a near-perfect military society," as once suggested.[16] Not counting the auxiliaries, Süleyman regularly mobilized 60,000 men for his campaigns, of which 45,000 were provincial timariot cavalry and 12,000 to 16,000 salaried foot and horsemen.

The relatively high number of Janissaries and palace cavalry in 1514 and that of the artillerymen in 1543 (see Table 3) reflect force structures adjusted according to tasks (field battles and sieges) and enemies (Safavid cavalry and Hungarian garrisons and defense artillery, respectively). The higher numbers in 1543 show troop strength at the beginning of the campaign in southwestern Hungary, whereas the lower numbers are of those present in the roll call after the last siege of the campaign. The difference suggests a general casualty rate of 7.5 percent in a campaign which saw a handful of minor sieges, but no field

15 Mária Ivanics, *A Krími Kánság a tizenöt éves háborúban* (Budapest, 1994), p. 177.
16 Peter Sugar, "A Near-Perfect Military Society: The Ottoman Empire," in L. L. Farrar (ed.), *War: A Historical, Political and Social Study* (Santa Barbara, 1978), pp. 95–104.

battles. The 10 percent casualty rate among the Janissaries, who stormed the walls, was double that of the artillery and cavalry.

### *Weapons, supply, and logistics*

The Ottomans adopted firearms at the end of the fourteenth century, some two generations after the weapons' appearance in Western Europe. They established a separate artillery corps as part of the sultans' standing army in the early fifteenth century, well before their European opponents. Before firearms, the Janissaries fought with their formidable recurve bow, saber, shield, and light coat of mail, whereas other infantry units used crossbows, javelins, and war axes. Ottoman horsemen used swords and bows. Under Murad II (r. 1421–44, 1446–51) the Janissaries began to use matchlock arquebuses, and by the latter part of the sixteenth century, most of them carried firearms. Murad III (r. 1574–95) equipped the Janissaries with heavier matchlock muskets, although flintlock muskets with the Spanish *miquelet*-lock were also manufactured in the empire from the late sixteenth century.[17]

The Ottomans were self-sufficient in producing weapons and ammunition well into the eighteenth century. Major cannon foundries operated throughout the empire: in the Adriatic, Hungary, the Balkans, Anatolia, Iraq, and Egypt. The center of cannon casting was the Imperial Cannon Foundry in Istanbul, which Mehmed II established after the city's capture. The Istanbul foundry could quickly multiply its capacity during major wars, casting several hundreds of cannons before the campaign season. In addition to the Istanbul gunpowder works, the Ottomans produced gunpowder in Cairo, Baghdad, Aleppo, and Yemen in the Arab provinces; in Buda, Esztergom, Pécs, Temesvár, Belgrade, Salonica, and Gallipoli in the European provinces; and in İzmir, Bor, Erzurum, Diyarbakır, Oltu, and Van in Asia Minor. These works met the demand of the army, navy, and garrisons until about the 1770s, when diminishing production forced Istanbul to import substantial quantities of gunpowder from Europe. At the end of the eighteenth century the new Azadlı gunpowder factory in Istanbul, modernized with French assistance, was again able to manufacture sufficient quantities of gunpowder of good quality. The Ottomans also managed to keep pace with Europe regarding weapons technology. More importantly, their military-industrial complex in the capital, supplemented by smaller provincial

17 Gábor Ágoston, *Guns for the Sultan: Military Power and the Weapons Industry in the Ottoman Empire* (Cambridge, 2005), pp. 15–60, 88–93; and Gábor Ágoston, "Firearms and Military Adaptation: The Ottomans and the European Military Revolution, 1450–1800," *Journal of World History* 25:1 (2014), 85–124, 88–98, 105–7.

cannon foundries and gunpowder workshops, enabled the Ottomans to establish a long-lasting firepower superiority over their European and Asian rivals.[18]

The Ottomans also had a well-oiled financial and bureaucratic apparatus and advanced provisioning, supply, and logistical systems. They had a sophisticated road network, partly inherited from Roman and Byzantine times, and elaborate and well-functioning courier and relay system, the stations of which were also used as grain-storage depots.[19] Roads, mountain passes, and bridges were repaired before the campaigns, and substantial quantities of wheat, barley, flour, and biscuit were stored in the depots along the campaign routes. The mobilization, storage, and distribution of food supplies to the fighting army remained the strength of the Ottomans until about the mid-eighteenth century. Thanks to their efficient supply system, Ottoman soldiers were usually better-fed than their opponents, which positively affected discipline and morale.[20] However, during the Russo-Ottoman wars of 1768–74, the Ottoman supply system collapsed, contributing to the Ottomans' disastrous defeat in that conflict.[21]

## Military devolution and the Ottomans' staying power

### *Geopolitics after Karlowitz (1699)*

As elsewhere in Europe, the seventeenth and eighteenth centuries brought longer wars, overburdened frontiers, military expansion, and socioeconomic crises in the Ottoman realms. The loss of Hungary, which the Ottomans accepted in the peace treaties of Karlowitz (1699) and Passarowitz (1718), called for the reorganization of the Danube frontier, challenging both central and provincial administration. New geopolitical realities fundamentally affected the frontiers' population and economy, as did the Ottomans' acceptance of international principles of respecting their neighbors' sovereignty and territorial integrity. Joint border commissions demarcated and international treaties guaranteed the empire's new borders. Of these, the borders with the Habsburg monarchy were effectively guarded by the Habsburg military administration. Combined, these

18 Ágoston, *Guns for the Sultan*, pp. 128–89; and Ágoston, "Firearms and Military Adaptation," pp. 99–108.

19 Colin Heywood, *Writing Ottoman History: Documents and Interpretations* (Aldershot, 2002), articles IX, X, XI (no continuous pagination)

20 Caroline Finkel, *The Administration of Warfare: The Ottoman Military Campaigns in Hungary, 1593–1606* (Vienna, 1988); Murphey, *Ottoman Warfare*, pp. 65–103.

21 Aksan, *Ottoman Wars*, pp. 67–75, 147–51.

mutually accepted principles and the new border surveillance regime, including the new institution of *cordon sanitaire* for enhanced mobility control, made cross-border raids – a vital revenue source for the Ottoman border guards and societies for centuries – impossible. The reconquest of the Morea from the Venetians in the 1715–18 war consolidated Ottoman positions along the western (Mediterranean) frontier, but the principles mentioned above affected the border's economy and society there, too.

The rise of Russia as a significant military power with interests in and grand political designs on the Black Sea region and the Balkans was the single most important geopolitical development in the eighteenth century. The defense of Ottoman positions along the Danube, Black Sea, and Caucasus frontiers against Russian encroachments had become the primary policy concern for the Ottoman ruling elite.

The collapse of the Safavids in Persia in the early 1720s, the rise of Nadir Shah (r. 1736–47), and his siege of Mosul in 1743 temporarily upset the balance of power along the empire's eastern frontiers. However, Mosul was recaptured, and the peace treaty of Kurdan (1746) restored the borders according to the 1639 Ottoman–Safavid treaty. Ottoman military success and the consolidation of Ottoman hold over the region were mostly due to the pro-Ottoman Jalili household, which monopolized Mosul's governorship by the beginning of the eighteenth century. The Jalilis' rise, in turn, was closely related to their defending and administering the region on Istanbul's behalf. This was a clear example of a new era, sometimes labeled as the "age of the *ayan*"; that is, of local notables and gentry, with whom the Istanbul government shared revenues and administrative and military tasks.[22]

To fight against multiple enemies and defend the empire's frontiers in the north, west, and east required the expansion of the military. Unlike in the sixteenth century when the timariot cavalry comprised the majority of combat forces and local (*yerlü*) troops financed by provincial treasuries provided the bulk of garrison troops, in the seventeenth century the imperial treasury increasingly paid both the combat forces and frontier garrisons. To alleviate the resulting fiscal pressure, the government relied

22 Dina Rizk Khoury, *State and Provincial Society in the Ottoman Empire: Mosul, 1540–1834* (Cambridge, 1997), pp. 120–33. On the *ayan* see Ali Yaycioglu, *Partners of the Empire: The Crisis of the Ottoman Order in the Age of Revolution* (Stanford, 2016); and the chapters by Dina Rizk Khoury, Fikret Adanir, and Bruce Masters in Suraiya Faroqhi (ed.), *The Cambridge History of Turkey*, vol. 3: *The Later Ottoman Empire, 1603–1839* (Cambridge, 2006).

more and more on the household armies of provincial governors and local grandees, which changed the composition of the Ottoman military. In return for their contribution to the Ottoman war effort, provincial officials and notables administered and controlled an ever-growing share of state revenues. This significantly affected the collection and management of revenues and distribution of wealth, as well as power relations between the imperial household and the provincial grandees.

For too long, historians have viewed fiscal decentralization and military devolution as signs of "Ottoman decline," responsible for Ottoman military defeats in the seventeenth and eighteenth centuries. However, military devolution and governments' reliance on local elites and military entrepreneurs in war-making had been the norm in most European states until about the mid-eighteenth century, because monarchs committed more human and economic resources to warfare than they had direct access to. Despite fiscal expansion, state-funded and state-administered standing armies had emerged only by the 1760s in most of Europe.[23] The Ottomans both preceded and lagged behind this European trend. Between the 1450s and 1570s, with their sizable standing army and efficient supply and logistical systems, the Ottomans were an exception to the European trend, and achieved military superiority over their rivals. From the 1580s onward, the Ottomans too followed a pattern of military devolution, for the very same reasons other states had done so: insufficient direct revenues relative to military demands.[24] However, by the second part of the eighteenth century both Austria and Russia had emerged as strong military-fiscal states with large standing armies, supported by centralized bureaucracies and finances.[25] It was not until the 1830s that the Ottomans established their own state-funded conscript army along European standards.

23 David Parrott, *The Business of War: Military Enterprise and Military Revolution in Early Modern Europe* (Cambridge, 2012).

24 Some of the material used in this section has been developed in Gábor Ágoston, "Military Transformation in the Ottoman Empire and Russia, 1500–1800," *Kritika: Explorations in Russian and Eurasian History* 12:2 (2011), 281–319; Gábor Ágoston, "Empires and Warfare in East-Central Europe, 1550–1750: The Ottoman–Habsburg Rivalry and Military Transformation," in Frank Tallett and D. J. B. Trim (eds.), *European Warfare, 1350–1750* (Cambridge, 2010), pp. 110–34; and Ágoston, *The Last Muslim Conquest*, pp. 315–28.

25 See the chapters by Hamish Scott, Michael Hochedlinger, and Janet Hartley in Christopher Storrs (ed.), *The Fiscal-Military State in Eighteenth-Century Europe: Essays in Honour of P.G.M. Dickson* (Farnham, 2009). Hamish Scott has argued that Russia was a military-fiscal state.

Table 4 Paper strength of the salaried troops, 1567–1711

| Date | Janissary | Artillery | Palace Cavalry | **Total** |
|---|---|---|---|---|
| 1530 | 8,407 | 2,383 | 4,906 | **15,696** |
| 1567–68 | 12,798 | 2,671 | 11,044 | **26,513** |
| 1568–69 | 11,535 | 2,460 | 5,395 | **19,390** |
| 1574 | 13,599 | 2,124 | 5,957 | **21,680** |
| 1582 | 16,905 | 3,736 | 8,346 | **28,987** |
| 1597 | 35,000 | 6,527 | 17,000 | **58,527** |
| 1609 | 37,627 | 7,966 | 20,896 | **66,489** |
| 1652–53 | 55,151 | 7,246 | 20,479 | **82,876** |
| 1654 | 51,047 | 6,905 | 19,844 | **77,796** |
| 1661–62 | 54,222 | 6,497 | 15,248 | **75,967** |
| 1666–67 | 47,233 | 6,193 | 13,267 | **66,693** |
| 1669–70 | 53,849 | 8,014 | 14,070 | **75,933** |
| 1687–88 | 62,826 | 17,995 | 19,800 | **100,621** |
| 1691–92 | 35,839 | 8,309 | 10,807 | **54,955** |
| 1694–95 | 78,798 | 21,824 | 13,395 | **114,017** |
| 1696–97 | 69,620 | 14,726 | 15,212 | **99,563** |
| 1699 | 67,729 | 15,470 | 13,447 | **96,646** |
| 1700 | 50,102 | 11,934 | 12,992 | **75,028** |
| 1701–2 | 39,925 | 10,893 | 12,999 | **63,817** |
| 1702–3 | 40,139 | 10,010 | 12,976 | **63,125** |
| 1704–5 | 52,642 | 11,851 | 17,133 | **81,626** |
| 1709–10 | 16,609 | 3,265 | 14,101 | **33,975** |
| 1710–11 | 43,562 | 8,775 | 15,625 | **67,962** |

*Source:* Gábor Ágoston, "Firearms and Military Adaptation: The Ottomans and the European Military Revolution, 1450–1800," *Journal of World History* 25:1 (2014), 85–124, 113, Table 7; and Gábor Ágoston, *The Last Muslim Conquest: The Ottoman Empire and Its Wars in Europe* (Princeton, 2021), p. 317, Table 9-2.

## *Expansion and metamorphosis of the salaried army*

Earlier studies dated the expansion of the Ottoman corps of salaried troops to the Long Habsburg–Ottoman War of 1593–1606 in Hungary and explained it by invoking the Ottomans' need to counter Habsburg infantry firepower by increasing the number of the Janissary footmen armed with hand firearms.[26] In light of new research, this suggestion has to be revised. Table 4 demonstrates

26 Halil İnalcık, "Military and Fiscal Transformation in the Ottoman Empire, 1600–1700," *Archivum Ottomanicum* 6 (1980), 283–337, 288–89; Gábor Ágoston, "Ottoman Warfare in Europe, 1453–1826," in Jeremy Black (ed.), *European Warfare, 1453–1815* (New York, 1999), pp. 118–44, 134–36.

that the expansion of all branches of the salaried army started during the reign of Sultan Süleyman and that the process accelerated during the wars against the Safavids (1578–90) and the Habsburgs (1593–1606). The number of the Porte's salaried standing troops rose from 15,696 in 1530 to 19,390 in 1569, 28,987 in 1582, and 58,527 in 1597. In the meantime, the number of Janissaries rose from 8,407 in 1530 to 12,131 in 1547 and 13,572 in 1563–64. It slightly dropped after the death of Süleyman but increased to over 14,000 men in the 1570s. In 1578, in the first year of the Ottoman–Safavid war, the number of Janissaries rose to 14,955, of which 5,017 were ordered to the Persian front by the government. In the last months of that war, the imperial treasury paid 22,760 Janissaries. In other words, during the Ottoman–Safavid war, the size of the Janissary corps increased by 52 percent. The corps expanded further during the wars of 1593–1606 in Hungary. In the fall of 1593, in the first campaign season of the war, the Porte paid 24,250 Janissaries, of whom 9,080 were ordered to the Hungarian front. The number rose to 42,632 men in 1606, the last year of the war, a 76 percent increase during the war.[27] The figures in Table 4 also show the general demand for, and availability of, military manpower during Istanbul's "Thirty Years' War" (1578–1611), when Ottoman troops fought, often simultaneously, against the Safavids in the east (1578–90, 1603–11), the Habsburgs in the north (1593–1606), and the Celali rebels in eastern Anatolia and northern Syria (*c.* 1595–1610). Other spikes in the numbers of the stipendiary army coincided with the wars against Venice over Crete (1645–69) and against the Holy League (1684–99). The data also reflect significant losses and policies of troop reduction, such as those after 1688 and 1699.

These paper numbers should be handled with caution, particularly regarding the Janissaries. Their metamorphosis is one of the most important and least understood processes of Ottoman social and military history in the seventeenth and eighteenth centuries. By the late seventeenth century, many Janissaries had become craftsmen, shop-owners, merchants, and tax-farmers, while ordinary tradesmen, merchants, and tax-farmers also bought their way into the corps using opportunities created by the exigencies of wartime economies. The Janissaries jealously guarded their privileges and fiercely opposed military reforms that could undermine their status. By the late

27 Istanbul, Başbakanlık Osmanlı Arşivi (BOA), Kamil Kepeci Defteri (KK.d.) 1866, p. 142 (1563–64); Maliyeden Müdevver Defterleri (MAD.d.) 7227, p. 794. (1578) KK.d. 1773, p. 243 (Nov. 10, 1589–Feb. 6, 1590). The parties ended the war by a treaty on March 21, 1590. For 1593 and 1606, see, BOA. KK.d 1775, p. 235 and MAD.d. 6913, p. 322. These and other data are collected in Ágoston, *The Last Muslim Conquest*, 318–19. See also Abdulkasim Gül, *Yeniçeriliğin Tarihi*, 2 vols. (Istanbul, 2022), vol. 1, p. 173.

seventeenth century, the Janissaries had evolved into a powerful social caste and political pressure group.[28]

In 1654 some 26% of the Janissaries were pensioners (*mütakaid*) and guards (*korocu*) not fit for combat. In the second half of the seventeenth century, between 25% and 60% of the total on payroll served in garrisons, and only somewhere between 17% and 37% of them participated in campaigns. By the early eighteenth century, many of those mobilized for campaigns were Janissaries only in name. The majority were new recruits, hired before and during each campaign. Before the 1711 campaign, the Ottoman government had fewer than 15,000 Janissaries on payroll, and thus planned to recruit almost 53,000 new ones. According to official mobilization lists, in the 1736–39 war, newly recruited Janissaries constituted between 72% and 80% of all mobilized Janissaries. The share of inexperienced recruits within the mobilized salaried army as a whole was between 54% and 64%.[29]

In addition to the corps of the salaried army (Janissary infantry, artillery, and cavalry), state-financed militiamen (*miri levend*) accounted for an ever-larger share of recruits.[30] In the 1736–39 Russo-Ottoman–Austrian War, for instance, the 13,000 to 16,000 *miri levend* troops comprised between 16% and 25% of the recruits and 10% and 15% of the deployed army. However, if one adds *levend* militiamen paid by the treasury and assigned to other fronts (Vidin, Bosnia, Bender), the number of these troops is much larger. Recruitment and pay lists mentioned 40,532 men in 1738, and between 24,050 and 28,132 men in 1739.[31] According to one estimate, by 1769, the number of mustered militiamen paid by the treasury could have risen to 100,000.[32]

28 Virginia H. Aksan, "Whatever Happened to the Janissaries? Mobilization for the 1768–1774 Russo-Ottoman War," in *Ottomans and Europeans: Contacts and Conflicts* (Istanbul, 2004), pp. 223–38; Cemal Kafadar, "Janissaries and Other Riffraff of Ottoman Istanbul: Rebels without a Cause?," in Baki Tezcan and Karl K. Barbir (eds.), *Identity and Identity Formation in the Ottoman World: A Volume of Essays in Honor of Norman Itzkowitz* (Madison, 2007), pp. 113–34; Baki Tezcan, *The Second Ottoman Empire* (Cambridge, 2010), pp. 175–90.

29 Ágoston, "Military Transformation," pp. 304–7. I am preparing an article about Ottoman troop mobilizations in the 1736–39 war.

30 İnalcık, "Military and Fiscal Transformation"; and Virginia Aksan, "Ottoman Military Recruitment Strategies in the late Eighteenth Century," in *Ottomans and Europeans*, pp. 191–207.

31 Istanbul, BOA, Baş Muhasebe Kalemi Defterleri (DBŞM.d.) 2286, pp. 2–3; DBŞM.d. 2390, pp. 2–3.

32 Aksan, *Ottoman Wars*, p. 57.

## *The rise of the household armies of governors and grandees*

As the military value of the small *timar*-holder horsemen declined in the seventeenth century, the importance of the household armies of provincial governors and *ayan* increased. By the seventeenth century, it was expected that provincial governors maintain large private armies, known as *kapu halkı*; that is, "troops of the gate" referring to the grandees' household. The ability to maintain such a force had become the condition for appointments, and many governors made use of the revenues of whole districts as "fodder money" (*arpalık*) in order to pay for their households and armies.[33]

In 1697, the grand vizier and the governor of Rumelia mobilized household troops numbering 2,600 and 2,400 men, respectively. The governors in Asia Minor and the Arab lands had smaller armies of 150 to 700 men each. The household troops of the pashas who held various court and provincial appointments often matched and surpassed these numbers. Ordinary district governors had much smaller forces, numbering as few as 25 men and as many as 150. Altogether these household troops comprised 18 percent of the mobilized army in 1697. These figures and similar mobilization calculations regarding the following year's campaign suggest the availability of large reserves of military manpower under the command of provincial governors.[34]

The household troops supplied by provincial grandees had become crucial by the 1768–74 Russo-Ottoman war, and remained so until the establishment of the reformed standing army in the 1830s. Between 1772 and 1774, the *ayan* and various Ottoman provincial officials were to supply more than 90,500 troops. Such an operation involved hundreds of provincial notables and tax collectors (often the same person). The number of soldiers each of them was to deploy varied between 50 men (one *bayrak* or company) and 1,000 men, depending on the wealth of the *ayan*.[35] Most of these troops were *levend* militiamen, known as *kapılı*, to distinguish them from their state-funded (*miri*) peers. In reality, the only difference between the two groups was that the former belonged to the grandees' households, whereas the imperial treasury paid the latter. In addition to their troops, the *ayan* also played an

33 İ. Metin Kunt, *The Sultan's Servants: The Transformation of Ottoman Provincial Government, 1550–1650* (New York, 1983), pp. 87–93.

34 Istanbul, BOA, MAD.d. 7483, pp. 2–7; for data regarding the 1698 campaign, see Géza Dávid and Pál Fodor, "Changes in the Structure and Strength of the Timariot Army from the Early Sixteenth to the End of the Seventeenth Century," *Eurasian Studies Yearbook* 4:2 (2005), 157–88, 178.

35 Based on Yuzo Nagata, *Muhsin-zâde Mehmed Paşa ve Âyânlık Müessesesi* (Tokyo, 1982), pp. 104–14. Later imperial decrees reduced this number by almost 11,000 troops, for several of the addressees were unable to supply the required number of soldiers.

important role in recruiting and mobilizing state-funded *levend* militiamen, as mentioned earlier.

### *Military and fiscal capabilities*

Historians have long noted the correlation between financial and military capabilities. A recent study thus attempted to explain the Ottomans' declining military performance vis-à-vis their Austrian and Romanov rivals by invoking the growing fiscal gap between the Ottomans and their rivals.[36] Generally, the correlation between fiscal and military capabilities is undeniable but not necessarily true for shorter periods. More importantly, dissimilar fiscal and military systems in the Habsburg, Romanov, and Ottoman empires make such comparisons misleading, if not futile. For instance, in the middle of the eighteenth century, Saint Petersburg's and Istanbul's revenues were comparable (239 tons of silver in Russia in 1751 versus 214 tons of silver in the Ottoman Empire in 1748). Yet, in the 1760s, the Russian government paid twice as many soldiers as its Ottoman rival (about 300,000 to 400,000 soldiers versus about 190,000 soldiers).[37]

In addition to the imperial treasury, whose account books are used for such comparisons, the Ottomans also had an inner treasury and some three dozen provincial treasuries. The inner treasury, which administered revenues and expenses related to the imperial household, functioned as a reserve treasury or "central bank," which could – and from the late sixteenth century onward increasingly did – lend money to the imperial treasury.[38] Between the middle of the sixteenth and the end of the seventeenth century, most frontier provinces had their local treasuries, from which local troops were paid who then manned the majority of the empire's frontier garrisons. Central treasury accounts do not reflect the revenues administered by, or the troops paid from, these provincial treasuries. This situation changed only toward the end of the seventeenth century, by which time the central treasury paid most of the local garrison troops in the Balkans and Anatolia.

Moreover, the budgets of the imperial treasury reflect only a portion of the empire's revenues and war-related expenditures. Except for a unique 1528

36 Kıvanç Karaman and Şevket Pamuk, "Ottoman State Finances in European Perspective, 1500–1914," *Journal of Economic History*, 73:10 (2010), 593–629.

37 Ágoston, "Military Transformation," pp. 299, 304, 310. Saint Petersburg's revenues rose from 925 tons of silver in 1786 to 1,342 tons of silver in 1796, and were about seven times more than Istanbul's in 1786, and nearly ten times more in 1796. The paper strength of the Russian army was about 500,000 men in 1782 and 1796.

38 Ahmet Tabakoğlu, "Osmanlı Devletinin İç Hazinesi," in Mehmet Genç and Erol Özvar (eds.), *Osmanlı Maliyesi. Kurumlar ve Bütçeler*, 2 vols. (Istanbul, 2006), vol. 1, pp. 51–56.

treasury account book, Ottoman budgets contained neither the *timar* revenues nor the number of timariot cavalrymen, which together made up the bulk of the army until the seventeenth century. Due to administrative and fiscal decentralization in the seventeenth and eighteenth centuries, the control of the central treasury over state revenues dropped from 58 percent in the 1520s to 25 percent by the 1660s.[39] While revenues that tax administrators and provincial grandees appropriated do not appear in the revenue section of the budgets, such revenues were not entirely lost. In return for their access to state revenues, provincial officials and local grandees played an increasingly important role in the Ottoman war effort, supplying troops, food, and draft animals, rendering logistical services, and maintaining order in the provinces. Alas, the monetary value of these military contributions and services is impossible to compute.

And finally, while European state "budgets" include not only the pay of the standing troops but also other war-related expenditures, such as weapons and ammunition purchases and costs associated with building and repairing fortresses, the Ottoman budgets list only a small portion of such expenses. The Ottoman government financed its cannon foundries, gunpowder factories, and saltpeter works from revenue sources which were specially designated for such purposes (*ocaklık*). The same applies for shipbuilding and the construction and repair of fortresses, roads, mountain passes, and bridges. Hundreds of service villages produced saltpeter, supplied raw materials for the various Ottoman military workshops, and performed logistical services in return for tax exemptions.[40]

This is not to say that war and state finances were not closely related. Fiscal pressure generated by wars prompted the Ottoman government to experiment with new fiscal policies, such as the introduction of the life-term tax farms (*malikane*) and the *esham* system – both instruments of domestic borrowing – in the wake of the wars of 1683–99 and 1768–74, respectively.[41] While these new fiscal policies increased Ottoman revenues only modestly compared with the Austrian and Russian fiscal expansion, fiscal decentralization and military devolution enabled the Ottomans to mobilize armies

39 Ahmet Tabakoğlu, "XVII ve XVIII yüzyıl Osmanlı bütçeleri," *İstanbul Üniversitesi İktisat Fakültesi Mecmuası* 41:1–4 (1985), 389–414, 402.

40 Ágoston, *Guns for the Sultan*, ch. 4; İdris Bostan, *Osmanlı Bahriye Teşkilâtı: XVII. Yüzyılda Tersâne-i Âmire* (Ankara, 1992) (for shipbuilding); Pál Fodor, "Bauarbeiten der Türken an den Burgen in Ungarn im 16–17 Jahrhundert," *AOH* 35 (1981), 55–88.

41 İnalcık, "Military and Fiscal Transformation"; Mehmet Genç, *Osmanlı İmparatorluğunda Devlet ve Ekonomi* (Istanbul, 2000), chs. 5 (on *malikane*), 7 (on *esham*), and 10 (on economy and warfare); Karaman and Pamuk, "Ottoman State Finances."

comparable both in size and in composition to those of their rivals, at least until the Russo-Ottoman war of 1768–74.

In March 1697, for instance, an official register listed 104,500 troops, including 10,000 men of the Danube flotilla, mobilized thus far for the Hungarian front. Several smaller armies, numbering between 17,000 and 5,000 men each, were sent to the Danube–Black Sea front and the Morea. With them, the number of mobilized troops in 1697 rose to almost 170,000 men. More importantly, the formerly cavalry-heavy Ottoman army now mirrored the infantry-to-cavalry ratio of its Habsburg rivals: in the late 1690s, infantrymen comprised 52 to 57 percent of the mobilized field armies.[42]

### *The Ottomans' staying power*

Fiscal decentralization and military devolution helped the Ottomans to maintain their ever-increasing armies. Ottoman and European military developments in the seventeenth century exhibit more similarities than are usually acknowledged. The divergent trajectories of Ottoman, Austrian, and Russian military evolution became pronounced only in the latter part of the eighteenth century. By the late 1700s, Ottoman troops were no match for the large standing armies of Russia and Austria. Before that, however, the Ottoman military system was still capable of challenging and even defeating the Habsburg, Venetian, and Russian armies.

It was with armies comprised of an expanded salaried corps and the household troops of provincial governors and grandees that the Ottomans nearly captured the capital of the Holy Roman Empire in 1683, defeated Peter the Great in 1711, defeated the Venetians and recaptured the Morea in 1715–17, and both retook Belgrade from Austria and scored victories against both Austria and Russia in the 1736–39 war. By the same token, European victories against the Ottomans prior to the Russo-Ottoman war of 1768–74 should not be overstated. Celebrated in European historiography as the "victory of the West," the battle of Lepanto (1571) only stopped Ottoman expansion in the western Mediterranean, while the eastern Mediterranean remained under Ottoman rule. Likewise, the Ottomans pressured the Habsburgs for a century and a half, nearly capturing their capital twice (1529, 1683), and not the other way around. The Habsburgs achieved their victories in the 1680s and 1690s only with the help of their allies, which forced the Ottomans to battle on several fronts simultaneously, an impossible undertaking with

42 Istanbul, BOA, MAD.d. 7483 and KK 39-Mükerrer (1697); Dávid and Fodor, "Changes in the Structure and Strength of the Timariot Army," p. 179 (1698); Ágoston, *The Last Muslim Conquest*, p. 328.

the military and fiscal means of the late seventeenth century. After their defeats, the Ottomans successfully halted a further Allied push to the Balkans and consolidated their rule along the Danube–Black Sea frontier. The Ottomans' staying power is remarkable in the face of multiple and diverse military challenges. This chapter has argued that this is mainly attributable to the adaptability of the Ottoman military, administrative, and fiscal systems. This topic deserves further research.

4

# Spain's wars and the rise of the Hispanic world empires, 1516–1828

FELIPE FERNÁNDEZ-ARMESTO

'Bella gerant alii.' In 1516, by means of traditional dynastic finagling, the house of Habsburg acquired the thrones of Castile and Aragon, or Spain for short: the most bellicose and spectacularly expanding state in Latin Christendom. Henceforth, it seemed, the Habsburgs would no longer be able to leave war to others.

Since concluding Castile's civil conflicts in the 1470s, the Spanish monarchs had, by force of arms, reconquered parts of French Catalonia, and added other acquisitions to their realms: southern Navarre, the western Canary Islands, Melilla, much of the Caribbean, and the kingdoms of Granada and Naples. For what came to be known as the Spanish monarchy it was the start of the most sustained period of success – measured by the crude, but decisive, standards of victory in the field and expansion on the frontiers – any Western European state had achieved since the Roman Empire. An almost uninterrupted record of victory continued until well into the seventeenth century, a period historians commonly refer to as that of Spanish hegemony or preponderance. Other empires of the early modern period were either seaborne affairs – dominating routes and linking ports – or vast products of traditional landward conquests, controlling production rather than trade. Spain's was the world's first great global empire of both land and sea, spanning the Atlantic and Pacific and uniting in a single allegiance huge territories in Europe, the Americas, and the Philippines. With some checks and losses at intervals in the seventeenth and eighteenth centuries, the monarchy continued to expand, even after the Habsburgs yielded to the Bourbons in a civil war of 1701–14, attaining its maximum extent in the 1790s.

Without exceptional prowess in and, perhaps, an exceptional vocation for war, such an achievement could hardly have been possible in a pre-industrial era of slow communications and rudimentary technology. Yet the extent to which war was responsible for the creation and maintenance of Spain's empire remains problematic: no early modern state had sufficient resources to keep conquered communities subject, and rivals abject, without deploying the arts of peace to advantage. 'War studies' in the current academic world encompasses what happens off, as well as on, the battlefield. The study of Spain's early modern wars demands a similarly balanced approach; an awareness that much of Spain's power was the product of negotiation, rather than of subjugation.

The question of how Spain became a great power and the heartland of a global empire is profound when one considers the dearth of natural advantages in a small country on the European periphery, with a modest population, not exceeding 6 million at the start of our period, and a land area of which a third is mountain and desert. Highly decentralised political traditions, moreover, were not conducive to military efficiency. The Spanish monarchy consisted of autonomous kingdoms and lordships, bound by no common institutions, only by allegiance – shaky in some places and at some times – to a common ruler, whose power to levy taxes and troops was in many places severely circumscribed by constitutional conventions. Castile, the biggest constituent state, which included most of the traditional polities of the Iberian peninsula, except for Portugal, Navarre, Aragon, Valencia, and Catalonia, was also the parent-monarchy on which colonies in the New World and, from 1572, the Philippines, depended. Subjects of other constituent parts of the monarchy were formally excluded, save on exceptional terms, from Castile's overseas dependencies, until 1778, heightening internecine resentments.

Though rebellions occurred, only the seven northernmost provinces of the Netherlands (part of the Habsburg inheritance) and Portugal (which joined the monarchy in 1580 as a result of dynastic accident, reinforced by intimidation) seceded, achieving independence during the convulsions induced by the Thirty Years' War. Even so, it was hard to engage Catalonia in full collaboration until 1716, when new statutes subjected that principality to the same fiscal rules as bound Castile. Before then, every attempt to get Catalans to make a proportionate contribution to defence had failed. Even thereafter, Navarre and the Lordship of Biscay (which was a privileged zone within Castile) retained traditional privileges.

Owing to the huge distances the monarchy encompassed, communications were haphazard beyond the Iberian peninsula (where, by the end of the seventeenth century, a semaphore network linked major towns). It took, for instance, between 59 and 163 days for a sailing ship to get from Spain to Mexico. Administrators in Madrid (where the court was located for most of the time from 1566 and definitively from 1606) had to reckon on six months to communicate with the viceroy of New Spain, or ten months for messages to reach Chile, and up to a year for them to get to Manila. Material resources, moreover, were never adequate to the upkeep of the remotest outposts on overextended and ultimately indefensible frontiers. In the 1580s the Portuguese commander of a fort on the coastal underbelly of the Safavid Empire, in the Persian Gulf, complained that he could not obtain sufficient arrows for his garrison of seven Portuguese and forty native mercenaries – the prospect of firearms did not even enter his calculations – nor could he garner enough opium to reconcile his men to a task too daunting to face without the aid of narcotics.[1] For want of pay, the Spanish Army of Flanders experienced more than forty serious mutinies in twenty years from 1590. According to a memoir presented to the Council of War – under the king, the body responsible for grand strategy – in 1602 the difficulties of recruitment derived partly from the loss of population to plague, partly from the high wages available to surviving labourers, partly from the bad reputation soldiers had with the administrators of justice, who inclined to harshness in consequence, and above all because 'they are so badly paid in their garrisons and in their destinations in the field, where they die of hunger and want of clothing'.[2] By and large, the disparity between commitments and resources grew throughout our period. In the 1760s, sixty-two soldiers manned the fort of Los Adaes, on the edge of the empire in Texas, but they had only one musket between them (see Figure 3). Spain's, in short, was an improbable empire. Like Dr Johnson's dog, one could hardly expect it to perform well. It was remarkable for performing at all.

Nevertheless, for some of the time, over some potential and actual opponents, Spain had advantages. In the sixteenth century, Spanish monarchs were relatively free to exploit opportunities for aggrandisement because of the temporary incapacity of rivals. France, the most formidable rival in terms of natural endowment, was rent by internal conflicts that dominated the second half of the century and recurred in the mid seventeenth. A contentious religious reformation helped to hobble England. Germany and Italy were 'geographical expressions' (and kings of Spain ruled much of them for much of the time). The Netherlands (also dominions of Spanish rulers for most of the century) and Sweden had not yet emerged as great

1 National Archives, State Papers 94/4; Felipe Fernández-Armesto, *Millennium* (London, 1995), p. 228.
2 Ian Thompson, 'El soldado del Imperio: Una aproximación al perfil del recluta español en el Siglo de Oro', *Manuscrits* 21 (2003), 17–38.

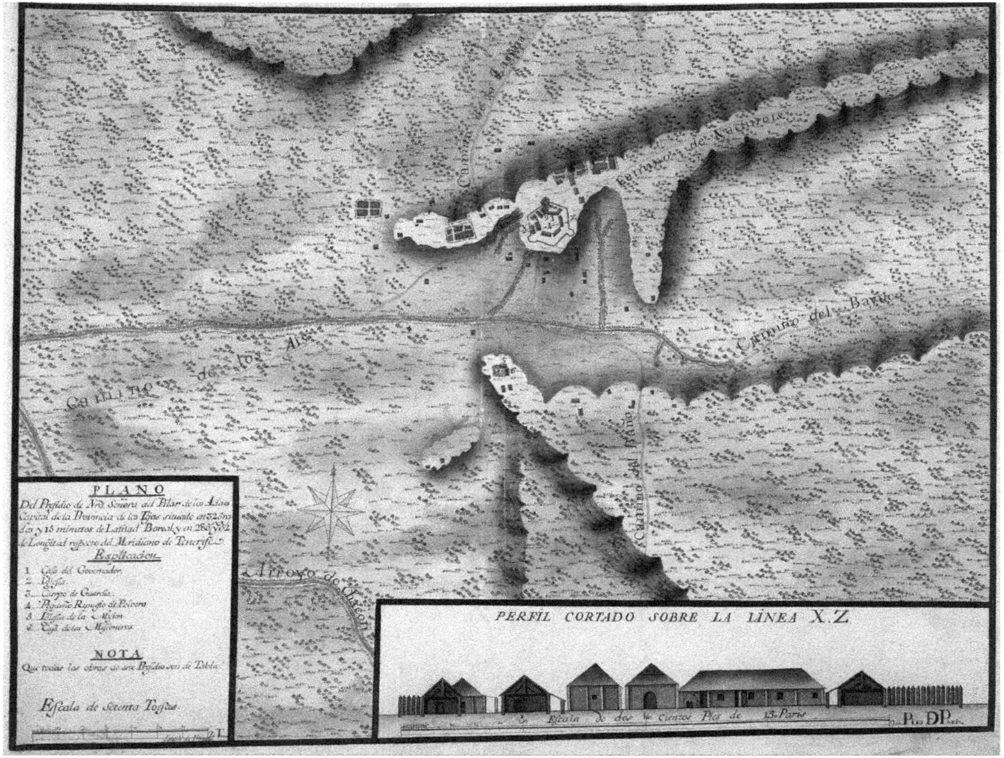


FIGURE 3. Fort of Los Adaes. The first capital of Spanish Texas was Los Adaes, located 33 miles east of the Sabine river, near Robeline, Louisiana. Joseph de Urrutia (1739–1803), a first lieutenant and cartographer in the Spanish army, drew this 1767 topographical rendering of Los Adaes as one of twenty-two plans of various presidios and towns he visited during the Marqués de Rubí's inspection tour across New Spain's northern frontier in 1766–8.
Courtesy of the British Library.

powers. Portugal was a potential rival, except during the Union of the Crowns from 1580 to 1640, but in the 1480s and 1490s a negotiated peace, dynastic alliance, and agreement to demarcate spheres of interest with Spain obviated trouble for most of the next century and a half. Spain's relative internal peace owed something to religious uniformity, which the Spanish Inquisition, founded in 1480, did much to enforce, while the troubles of England, France, and the Netherlands were compounded, if not caused, by *odium theologicum*. Wars, as long as they were successful, favoured Spain's internal peace by distracting or exporting troublemakers and creating common enemies.

The existence of a peculiar political ethos was probably decisive. In Castile and its dominions (albeit not in the Crown of Aragon to anything like the same extent) subjects identified service to the crown as the defining characteristic of nobility. When Don Quixote expressed it the tradition was dying, perhaps, but not yet dead:

> Nothing in the world confers so much honour and benefit as the service of God, first, and next of one's King and natural lord, especially in the exercise of arms, by means of which are acquired, if not riches, more honour, at least, than accrues to learning – as I have oftimes said.[3]

In partial consequence, uniquely in Western Europe, there were no aristocratic rebellions in Spain from 1506 to 1640. Great nobles and their resources were largely at the monarchs' disposal. Dukes and princes could receive commands without exciting fear of treachery. A sense of common interest bound crown and aristocracy tightly from 1520, when the noble estate collaborated with the king to suppress rebellious urban patriciates. Royal patronage – clout in the marriage market, the power to entail estates, the supply of rewards in the form of offices of profit, and the availability of many seigneurial rights for devolution back into noble hands (in what some historians call 'seigneurial reaction' in the late sixteenth and seventeenth centuries)[4] – helped to entrench aristocratic loyalty. Meanwhile the desire for ennoblement inspired many socially ambitious aspirants to serve the monarchs in war or in colonisation with little or no material reward. They filled the archives with *probanzas de méritos* – prolix collections of witness statements attesting to the value of the petitioners' deeds, the sacrifices made, and

3 *Don Quixote*, part II, ch. 24.

4 Fernand Braudel, *La Méditerranée et le monde méditerranéen à l'époque de Philippe II*, 3 vols., 2nd ed. (Paris, 1966), vol. 2, pp. 49–94; James Casey, *The Kingdom of Valencia in the Seventeenth Century* (Cambridge, 1979), pp. 101–26.

the wealth and blood expended.[5] Spain's wars overseas were startlingly cost-effective: they hardly made use of royal resources or officially recruited manpower. They were usually private enterprises under consortia of adventurers who rarely had any military experience, such as the bands of *conquistadores* who served with Cortés in Mexico and Pizarro in Peru. Only when the ethos of service eroded, as it did, gradually but inexorably, over the first half of the seventeenth century, did Spain's rate of success in war falter.

Spain was also blessed with large, relatively stable, exceptionally professional armed forces. On average, though all such calculations must be tentative, about 8,000 to 10,000 men were recruited a year from the Iberian peninsula and, in the seventeenth century, something approaching that annual figure in the New World. In the sixteenth and most of the seventeenth centuries, as in most European monarchies, recruiting was in the hands of intermediaries: contractors and captains. Warfare was a business farmed out to freebooters and mercenaries, available in conflict to either side, like Martin Schenck, who repeatedly switched between the service of king and rebels in the Netherlands 'until in 1589 he set up on his own', financing his force by extorting tolls, undertaking 'various exploits', as an observer remarked, 'to raise money or take a good prisoner' for ransom.[6] The Netherlands War led to the systematisation of exchanging prisoners, eroding freebooters' profits and stimulating a gradual shift to an army conscripted by representatives of the crown.

Foreign recruits were vital – companies raised by traditional methods in the various dependencies of the Spanish crown. Mercenaries were recruited from elsewhere, but the Spanish monarchy naturalised foreign recruits and focused on enlistment from its own lands, swapping forces as need arose, and maximising the usefulness of national contingents by transferring them far from home, rather as the Roman Empire had done in its day. Catholics gravitated to Spanish colours, especially from countries where they fled persecution. English Catholic garrisons in the Netherlands deserted en masse to the Spanish side – Guy Fawkes among them in 1587. By 1605 England's chief minister discerned a 'strong and visible torrent' of such defections. Even when Spain's cash and credit ran out, service to the Catholic monarch remained attractive, throughout the eighteenth century,

5 Murdo Macleod, 'Self-Promotion: The *Relaciones de Méritos y Servicios* and Their Historical and Political Interpretation', *Colonial Latin American Historical Review* 7 (1998), 25–42.

6 Geoffrey Parker, *The Army of Flanders and the Spanish Road*, 2nd ed. (Cambridge, 2008), p. 14.

to disaffected Irishmen and Jacobite exiles. Native auxiliaries, moreover, bore the brunt of fighting in the New World. In Mexico and Peru more natives fought with the Spaniards than against them. The Tlaxcalteca and Huexotzinca were the biggest contributors to the men and matériel that overthrew Aztec hegemony. Other communities joined the alliance Cortés – or rather, his native interpretrix, known to the Spaniards as Doña Marina – put together, sometimes to escape from the burdens of tribute to the predatory Aztecs, sometimes in anxiety to be on what came to look like the winning side. Maya chiefs, who hated their neighbours more than the Spaniards, rushed to proclaim themselves 'Maya conquistadores'.[7] Nahua warriors overwhelmingly composed the forces that followed up the conquest of Mexico in Honduras and Guatemala. (See Figure 4.)

The social composition of Spain's armies was also diverse. Even while the 'military revolution' replaced 'the plumed troop' and the panoply of chivalry with firepower, peasant levies, and professional soldiery, battle was still a setting for noble virtues and conduct and armigerous display, until well into the seventeenth century and, in lingering nostalgia, even beyond. There were still spurs to be won and accolades to be bestowed. In 1610 Gonzalo de Villagrá rosily recalled the gang of desperadoes who invaded New Mexico a dozen years before as if they were knightly companions. In the year of the publication of *Don Quixote*, Pedro Fernández de Quirós knighted all the members of his transpacific expedition, down to the black cooks. Velázquez captured the chivalric attitude to victory in 'The Surrender of Breda', his celebration, painted in the mid-1630s, of one of the spectacular Spanish military successes of the previous decade: the victorious commander courteously consoles the defeated defenders. Such images, however, gild a process of professionalisation, as firepower grew in importance and war ceased to be an aristocratic preserve. Don Quixote, who, as Marx said, 'protested with his lance against gunpowder', famously bemoaned the trend, calling down blessings on

> those centuries before the terrible ferocity of those diabolical machines of gunnery, whose inventor, I trust, is getting fully paid for his devilish invention in hell, because he provided the means whereby an infamous and cowardly hand can deal death to a valorous knight, who, without knowing how or why, in the midst of the courage and élan that ignite and sustain valiant hearts, is felled by a bullet discharged at random by someone who probably fled in terror from the flash of the flint when the accursed

7 Matthew Restall, *Maya Conquistador* (Boston, 1998).

FIGURE 4. Lienzo de Tlaxcala. Spanish–native military collaboration seen in an image from the Lienzo de Tlaxcala. Digital image.
© 2012 Fray Angélico Chávez History Library, New Mexico Digital Collections.

> contraption fired. Thus in an instant end the thoughts and life of one who should merit long ages of renown. And when I think of it I am tempted to aver that it pains me deep inside that I have embraced my vocation as a knight errant in an era so detestable as the present time in which we live; for no peril makes me afraid, and yet it vexes me to admit that I am anxious lest powder and shot deprive me of the chance of fame and of celebration for the strength of my arm and the edge of my sword across the whole wide world.[8]

Despite the Don's discomfort, which many field commanders and royal counsellors shared in their desire to preserve soldiering as a noble profession, Spain was fortunate in breeding, without aristocratic derogation, new

8 *Don Quixote*, part I, ch. 38.

generations of professionals and specialists. Technical specialists in support roles were abundant: gunsmiths, shipwrights, mapmakers, fortifications engineers. They were not (despite the persistence of a myth to the contrary among some modern historians) recruited from among genuinely noble and gentry ranks. Although gentlemen continued and, for a while, in the late sixteenth and early seventeenth centuries, increased in posts of command, and although there were many indigent hidalgos who served in the ranks, in the course of the seventeenth century nobles came increasingly to claim exemption from military service on grounds of status. Rather, soldiers below the higher levels of command came overwhelmingly from among the youth of the urban poor, often to the distress of their elders. As the author of a report of 1694 remarked,

> somewhere in Spain I saw a workman's boy volunteer as a soldier against his father's wishes, while his father went weeping through the streets declaring that his son would bring infamy on all their lineage. The people generally are so convinced that every soldier is an infamous fellow that there is no tailor or cobbler who will not think it very dishonourable to him if a son of his enlists.[9]

The pay, even when forthcoming, was little – perhaps, on average, never much more than a fifth of an urban journeyman's wages throughout our period – but it was desirable for the desperate; and the chances of booty, ransom, or reward animated soldiers' prospects. Luis de Miranda, a volunteer aboard the Armada against England in 1588, declared, 'We all thought we would be rich.' The much-quoted chorus Don Quixote heard on the road expressed the bottom line:

> It's poverty that makes me fight.
> I'd stay at home but for my plight.

Expectations of reward were rarely fulfilled. Pay, however, is not the only ingredient of morale. While all early modern soldiery was rude, licentious, and mutinous, there are reasons to think that superior medical services, which were normally attached to all units, and the ubiquity of the services of chaplains may have helped. The numbers of clergy assigned to the army in the sixteenth and seventeenth centuries, when every company was supposed to have a chaplain and a surgeon, were so great in theory as to defy belief: the typical tercio would contain twenty to thirty chaplains, and military service must have been one of the largest areas of clerical employment. Even if

9 Julián Juderías, *España en tiempo de Carlos II*, p. 301.

allowance is made for a shortfall in numbers in practice, the army was well served with pastoral care. In the eighteenth century one chaplain per battalion sufficed. Religion affected some men. Jerónimo de Pasamonte, who fought at Lepanto, chose soldiering in default of training for the priesthood. Alfonso de Contreras, a hard-living picaroon who seems to have been addicted to his own adrenalin, wrote extraordinary memoirs of his soldiering life, probably in the 1630s, in which he claims to have spent seven months as a hermit, never so happy as in his poor and lonely cave.[10]

Finance was critical, of course. Most warfare was positional and what counted was the money that kept armies in the field and fleets afloat. Spanish armies were superior when they stayed together longer than those of the enemy before mutinying or melting away. The money that supported military and naval expenditure came mainly from two sources: taxation in Castile and to a lesser extent in America; and credit – much of it foreign. The schoolroom perception that the monarchy was upheld by a silver bonanza from the New World is misleading. In the reign of Philip II, when American income was at its height, it rarely attained 20 per cent of the total.[11] Nevertheless, American silver consoled creditors. The creditworthiness of the crown depended on it. Repeated suspensions of payments to creditors – or 'bankruptcies' as historians sometimes misleadingly call episodes of mismatch between income and expenses – did not deter the bankers in Spain, Italy, Germany, and France who held the Spanish debt; on the contrary, only defaulters who are secure in the long term can get away with the practice. Supplementary sums could be raised from subsidies, 'crusading' indulgences the Church devolved to the crown, and sales of offices and honours. 'Counts are made in the counting-house', sang the poet Góngora in 1601, 'and ducats make dukes'. From 1603 currency revaluations became a frequent recourse.

Still, the financial basis of Spanish power was shaky. Honours and offices were wasting assets. Indulgences showed diminishing returns. The silver imposed costly defences in convoy against pirates – unsuccessfully in 1628 when the Dutch corsair Piet Heyn, according to a traditional Dutch rhyme, 'made the screams heard throughout Spain' as the bullion shipment for the year fell into his hands. Silver extraction depended on the availability of mercury with which to amalgamate the ore, and of which the supply was variable and critically low for much of the seventeenth century. The tax burden on Castile became insupportable. Philip III inherited debts of

10 Felipe Fernández-Armesto, *The Spanish Armada: The Experience of War in 1588* (Oxford, 1988), pp. 48–58.

11 Modesto Ulloa, *La hacienda real de Castilla en el reinado de Felipe II* (Madrid, 1977).

13 million ducats but could not conscientiously increase taxation. When the count-duke of Olivares proposed a 'Union of Arms' in 1624 in which other kingdoms would supplement the disproportionate taxation of Castile, he envisaged the Castilian input as providing 44,000 men out of a total of 140,000; Aragon was able to provide a subsidy sufficient to cover 2,000 men, rather than the 10,000 Olivares envisaged, and Valencia supplied 1,000 – half the number he had assigned to that kingdom. When he revived a version of the proposal in the next decade, the result was a rebellion that almost severed Catalonia from the monarchy.

Nonetheless, against this background – precarious yet less so, it appears, than in other monarchies – of recruitment, resources, sources of morale, and the favourable circumstances that deterred enemies in the Old World and drew allies in the New, the triumphs of the sixteenth century are intelligible. Success reached a moment of climax in about the 1590s. In the previous hundred years, Spain expelled French armies from Italy and came to control much of the peninsula. Footholds on the southern shore of the Mediterranean multiplied from the conquest of Melilla (1497) to the acquisition of Mámora (1614). Thanks to an alliance that, in the supreme moment of crisis in 1571, included the pope, Venice, and the Order of Malta, the Ottoman threat of conquests at Christian expense was more or less confined to the eastern Mediterranean. Amazingly, meanwhile, Spanish armies could operate effectively against Netherlandish rebels (a struggle which was unremitting from the 1570s), despite the lack of a viable harbour for reinforcement and supply. Laborious communications along the 'Spanish road' from the monarchy's dependency in Milan, across lands the Habsburgs never wholly controlled, helped keep the rebels at bay. By the 1590s, their suppression seemed inevitable, but for the temporary withdrawal of Spanish forces to invade France and impose Spanish policy there in one of the most successful military operations of the century. In 1588 Philip II had sent the first of a series of Armadas against England. The weather frustrated them, and storms turned the first expedition into a sort of Spanish Flodden or Spanish Somme, in which many of the most promising youth died in shipwrecks and their aftermath; but the fact that Spanish fleets could threaten a significant maritime power hundreds of miles from their nearest serviceable harbour served as a dramatic demonstration of the monarchy's reach and menace.

Meanwhile success in the New World dwarfed that in Europe. Historians have largely obscured the reasons: for war makes rattling good history and peace is poor reading. Yet, for every indigenous community that fell to

Spanish arms – or, more usually, to those of the Spaniards' native allies – scores or hundreds joined the monarchy after negotiation, enlivened in some cases by brief, token or ritual resistance. Spain was fortunate not only to pre-empt rival European powers in the most populous and economically rewarding parts of the hemisphere, but also to take advantage of the prevalence of cultures exceptionally partial to strangers. The 'stranger-effect' led indigenous peoples to accept rulers from outside their ranks and made them hospitable to visitors from afar, touched with the aura of a divine horizon, or suitable for exploitation as allies uncorrupted by existing links in the region, or as arbitrators who were not implicated in existing networks of kinship or obligation, or as marriage partners uncompromised by prejudicial ties, or as holy men impartial to incumbent gods.[12] Native communities hated one another with visceral enmity they could hardly feel, at least at first, towards new arrivals. Thus, in the valley of Mexico, conquistadores could play Tlaxcalteca and Huexotzinca, for instance, against detested hegemons in Cholula and Tenochtitlan. In Yucatán, the rivalries of Pech, Cocom, and Xiu outweighed any apprehension the newcomers aroused. In Peru, Huari and Cañari were eager for alliances that would enable them to avenge defeats and massacres at Inca hands. In the Huarochirí valley, the Checa explained that they abandoned their historic alliance with the Inca because the supreme Inca reneged on his annual obligation to dance at their principal shrine. Aided by such opportunities, Spaniards were able to succeed overthrown local elites. They gobbled almost at a gulp the hemisphere's most aggressive and dynamic empires, those of the Aztecs and Incas. By the end of the century they were dominant across the Caribbean and the Gulf of Mexico, from the Isthmus to the islands, and on the mainland of America from New Mexico to the Bío-bío, with outposts linking Peru to the Atlantic via Paraguay and the River Plate and Mexico to the extreme Orient via the Philippines.

The kinds of supremacy traditionally ascribed to Spaniards counted for little. Nautical proficiency was obviously important in getting them to the New World in the first place and perhaps in enhancing their and their allies' performance in riverine and lacustrine warfare. But other technologies – steel blades, firepower, horsemanship – were of marginal value, rapidly exhausted, and easily appropriated, as far as they were useful, by enemies. Nor is it credible that 'superstition' or corroded morale undermined native polities, many of which belonged to civilisations with impressive attainments and to cultures of abundant self-confidence. 'Culture shock' was at least as much of

12 Felipe Fernández-Armesto, *Our America* (New York, 2014), pp. 9–15.

a problem for Spaniards operating in an unfamiliar environment as it was for natives facing a perplexed, fearful, and generally underprepared foe, as anyone can attest who reads conquistadores' awestruck, dry-mouthed accounts of admired adversaries. Disease debilitated both sides impartially. To some extent, the Spaniards' allies were more exposed than their foes to the deadly infections Spaniards unwittingly introduced from the Old World among unimmunised populations.

Nor, once established, was the Spanish presence sustained by unaided arms. To keep collaboration going, the new order over which Spain presided had to reward the well disposed. Like every major pestilence, the demographic disaster that befell early colonial society benefited the survivors, not only by redistributing existing resources but also by encouraging the introduction of new economic expedients. Native chiefs could become ranchers. Indigenous communities could set up silk-weaving or leather-crafting or pig-farming enterprises. Peasants could farm cochineal on previously marginal or useless land or grow cacao for export. In many places, missions were successful examples of capitalist entrepreneurship, in which native residents, at the cost of enduring clerical paternalism or scowls from a divine termagant, could share economic benefits. None of these activities was possible until Spaniards introduced the relevant products or opened appropriate markets. All new chances of enrichment occurred in the context of the wider economic world to which Spain introduced America. While injected money lubricated exchange, inter-regional trade linked native civilisations, formerly unknown to one another, and reached across the Atlantic and Pacific Oceans. None of this was of much benefit to peasants whose labour was coerced, or peons who subsided into dependence on haciendas, or native aristocrats who failed to keep up with the pace of change. But a framework emerged in which prosperity crammed sufficient numbers of the elites, native and newcomer, into close relationships of mutual benefit.

When European rivals bestirred themselves, they failed to do more than peck at Spanish pickings, even though internecine struggles divided the defenders, who were few and largely unsupported by professional soldiery. Before the seventeenth century, rival powers managed only sporadic forays, largely confined to piratical exploits, flitting and stinging on the edge of the Spanish Main, like insects irritating the hide of some great beast. Elizabethan projectors raged at England's inability to challenge Spanish dominance. Not until 1607 would a lasting English presence in the New World begin to take shape, when a fragile colony settled at Jamestown in Virginia. Viable Dutch and French mainland colonies were even longer delayed and except on the Guiana coast were never securely established.

Events beyond Spanish control conspired, however, to interrupt success towards the middle of the seventeenth century. The Thirty Years' War, which strained or exhausted so many participants, was too severe a test. After the long record of victory, naval defeat began in earnest in 1639, with the battle of Downs, and defeat on land in 1643 with the battle of Rocroi. The former encounter arose because of a Spanish attempt to repeat a manoeuvre that had succeeded on previous occasions: using the North Sea port of Dunkirk to supply and reinforce troops in Flanders. In what proved to be an excess of confidence, a fleet of dozens of ships, carrying thousands of men, arrived in the English Channel in September. Disrupted by a Dutch attack, the fleet sheltered in the Downs, where, after a long blockade, the Dutch renewed their attack in force. After the loss of most of their ships and men, the Spanish were never again able to bid for command of the North Sea. The symbolic value of the defeat at Rocroi, when a French force relieved the beleaguered garrison by shattering the besieging army, was equally impactful. Though the Spanish recouped to some extent, the prospects of decisive victory receded and the treaties that ended the war in 1648 involved the loss of the northern Netherlands, and the effective end of the monarchy's chances of recovering Portugal. From her enemies' point of view, the Treaty was, at best, imperfectly effective in attempting to divide the Habsburg emperor from his Spanish cousins, and to exclude Spain from the key German mercenary market.

Though vexatiously debated among historians, the 'decline of Spain' is measurable by demographic collapse in the New World in the sixteenth century, and halting or decline in the peninsula for most of the seventeenth. Jittery morale coincided with this, as projectors and commentators lamented perceived decline and bemoaned the divine disfavour that seemed to condemn Spain's empire to a span depressingly short by the standards of the model to which everyone looked in early modern Europe: that of Rome. Relative to newly ascendant powers of the seventeenth century in France, England, Sweden, and the Netherlands, Spain's decline seems glaring. Meanwhile Europe's 'economic centre of gravity'[13] shifted northwards, away from the Mediterranean, implicating Venice, say, and the Ottomans, in a drift of which Spain's relatively diminished power seemed part. Artists marked the decline. Velázquez's Mars – in accord with his habit of derogating

13 John Elliott, *Europe Divided, 1559–98*, 2nd ed. (London, 2000).

ancient gods – is feeble and dejected. Juan Valdés Leal included among *vanitates* an engraving of the triumphal arch that commemorated Spain's greatest victory of the century at Nördlingen in 1634, when a joint Spanish–Austrian force had crushed the 'new model' army of imperial Sweden. After the Peace of Münster, which conceded independence to the Protestant-led rebel provinces of the Netherlands, Vermeer – the king's loyal Catholic subject – painted a sort of threnode, in which the trumpet of fame hangs limply from Clio's fingers, while the lights are extinguished in the chandelier, topped by the Habsburg eagle that crowns the composition.

Decline, as it turned out, was not definitive, but the period of climacteric continued. Renewed supplies of American silver helped stimulate partial recovery from the 1660s. In the Americas, expansion resumed in the late seventeenth century with the reoccupation of New Mexico – which the Spaniards conquered in 1598 but from which they withdrew in the 1680s in the face of indigenous disenchantment – and the submission of the last independent Maya kingdom in 1697. In 1700, however, King Carlos II died without an agreed heir and the War of the Spanish Succession – a brutal civil war exacerbated by the presence of foreign armies supposedly at the service of the rival factions – interrupted the revival of Spanish fortunes. When the war ended in 1714, the monarchy had been shorn of Menorca, Gibraltar, Sicily, the Southern Netherlands, and part of Milan. Instant reorganisation followed, with massive investment in the reconstruction of the fleet, and the recasting of the army in territorial regiments such as now constituted the armies of most other European powers. The outcome of the succession war was favourable in one respect. By installing a cadet branch of the Bourbons on the Spanish throne, it established the basis of the 'family compact' that kept peace with France for most of the eighteenth century.

Although in Europe the monarchy remained largely on the defensive, short-term gains in Sicily and Sardinia were made almost at once. Other Italian possessions returned during the War of the Austrian Succession in the 1740s and Menorca was retrieved in 1782. If such successes were surprising in view of the relative decline of Spain and the enormous material superiority of Britain, even more remarkable was the resumption of the Spanish advance in America. The Spanish conglomerate was weak, overshadowed by mightier rivals, surrounded by enemies, and condemned to engage on multiple fronts. Long, exposed frontiers and conflicting commitments made the empire effectively indefensible. Despite generous investment by the crown, the Spanish navy could not keep up with the growth of Britain's, although the security of transatlantic shipments was reorganised, with fast ships replacing convoys. Production of rigging and canvas increased in state-run factories, and the use of contractors was systematised, but the shortfall in resources compared with the Royal Navy grew ever wider. And defeats were frequent, so that Spain had to be in practice what Britain is proverbially supposed to be: a loser of battles but a winner of wars. The most remarkable case was that of the Seven Years' War, from which Spain emerged in 1763 with a huge accession of territory despite temporarily losing two of its most important posts – Havana and Manila – to British assailants.

In consequence, this empire was not only the biggest in the world of its day, encompassing the greatest diversity of environments and cultures, but also one of the most successful. It is worth recalling that the Spanish Empire in mainland America lasted until the 1820s, whereas the French lost most of their possessions in 1763 and the British in 1782. Spain's series of successes in the eighteenth century coincided with the rise of daunting competition. Previously, Spain's empire had been unique. Against it were ranged great territorial empires, such as China's, which controlled vast expanses to landward in order to levy tribute or monopolise production. Then there were maritime empires, such as those of other European powers, focused on coastal colonies, ports, and the policing of routes to control supply. Spain's global monarchy combined both characters from the second decade of the sixteenth century, linking an extensive land empire in America with a web of maritime routes, not only controlling the production of commodities as important as American silver, cochineal, cacao, and large amounts of sugar, but also channelling exports to much of the rest of the world. Towards the end of the seventeenth century the Portuguese in Brazil moved inland to settle Minas Gerais and from that moment, with increased intensity in the second half of the eighteenth century, other European empires began to acquire zones of production in hinterlands far from the ports: the Portuguese in Amazonia and in the Novas Conquistas in India, Dutch in Java, French in Louisiana, and British from 1750 in the Ohio valley beyond the Appalachians. Such events constituted a new threat to Spain. Meanwhile, growing British, French, and Russian interest in the Pacific threatened what had formerly been a 'Spanish lake', disdained by other powers.

Europeans were not Spain's only rivals. Imperialism is not merely a white vice, but a human one. In the second half of the eighteenth century Spain in America had to contend with indigenous imperialisms that resembled those of Huns or Mongols in the Eurasian steppes, or of Mali or Songhay in the African Sahel, or the phenomenon of the Oromo in East Africa, who, after acquiring horses in the fifteenth century, became the scourge of Ethiopia. Similar movements were impossible in America until Spaniards introduced horses. Thenceforth, unlike the pastoral peoples of the Old World, the grasslanders of the Americas lived chiefly as hunters, dependent on the exploitation of enormous herds of bison and guanacos. To everyone except these predators, New Mexico and the River Plate seemed like poor colonies. Yet Spanish settlements electrified the frontier folk, as surely as booty and payola from Rome transformed neighbouring barbarians, or the wealth and magnetism of China affected the Jurchen or Manchus. A horse-borne revolution changed ecologies, polities, and economies. Among the Apaches, who were hard to pin down and difficult to domesticate, white men's wealth, appropriated through raiding or ransom, broke down traditional links of kinship and hierarchy in favour of war leaders who arose to dispute the power of shamans and hereditary chiefs. The Comanche presented a further set of problems. Vastly outnumbering all neighbouring peoples put together, they reorganised their society for war, in a confederacy unattainable by the fissile Apaches. Like steppe-land or Sahelian imperialists in the Old World, they could control great swathes of territory and terrorise subject and victim peoples into becoming tributaries or slaves.[14] In the pampa, the Mapuche were hardly inferior to steppe-land or Sahelian imperialists in their capacity to exact tribute from neighbouring peoples. A fine example of the possibilities they realised is in the private museum of the owners of the Santa Rita vineyards in Chile, where galleries dedicated to the history of the Mapuche are full of stirrups, saddles, and bridles of silver, captured or cajoled from Spanish victims.

Yet none of this competition stopped Spain adding to her tally of territorial acquisitions, although she relied on local militias and rarely had more than 10,000 regular troops in play, except in moments of manifest crisis, throughout the hemisphere. Spanish policy hovered between contradictory strategies: playing 'barbarians' off against one another, like medieval Byzantines; or attracting them into peaceful submission or cooperation in the manner of Confucian mandarins with steppe-land nomads; intimidating them into quiescence; or forcing them into peace; or exterminating them in the style of English colonists with their unwanted Indians. Common to all these options were efforts to enhance frontier security by seeking allies to the north beyond Apache and Comanche territory. New settlements in New Mexico gave rise to new needs: opening communications towards the Pacific, the Gulf of Mexico, and the Mississippi valley, exploring the Great Plains and

14 Sara Ortelli, *Trama de una guerra conveniente: Nueva Vizcaya y la sombra de los apaches (1748–1790)* (Mexico City, 2007); Pekka Hämäläinen, *The Comanche Empire* (New Haven and London, 2009).

the Utah basin, excluding French interlopers, and pacifying the huge arc of frontier to the north of the new limits of the empire. Spain's empire therefore reached into Texas and Arizona, while expeditions emerged like tentacles to prod the plains. Paintings on hide by Pawnee artists, for instance, record an expedition of 1719 from New Mexico into Nebraska, with the aim of detaching the natives from their French paymasters; but the negotiations failed and the paintings show the misfortunes of the Spaniards, surrounded by hostile bowmen and French agents. With the Utes, on the other hand, Spanish efforts were successful and an alliance was formed in 1730.[15]

Arizona was attractive at first only as a route towards the Pacific. Natives, variously coy and hostile, protracted the agony of missionaries who formed the Spanish front line. From 1732, however, Spanish efforts were continuously sustained and by mid-century the territory was a reliable, albeit costly and unruly, outpost.[16] Texas, meanwhile, demanded attention to control Apaches and fend off the French. Neither enemy proved tractable, and missions were equivocal in effect, spreading deadly disease along with allegiance to Church and crown,[17] but in the second half of the century the policy of 'attraction' yielded positive results, which can be admired, for instance, in a plan of the settlement of San Juan Bautista, on the Rio Grande, of 1754, where between the arcaded and well-ordered streets processions of natives and Spaniards file with their armed escorts and attendant musicians to meet the mission folk in the main square and erect celebratory crosses at its corners.[18] In 1778 Bernardo de Gálvez, governor of Spanish Louisiana, claimed that peaceful attraction had done more to pacify natives at less cost than warfare, and that dependency was more easily created by gifts of foodstuffs and tools than enforced by violence.[19]

In 1786 Spain made peace with the Comanches, setting the frontier between the two empires at the Arkansas river. Common enmity against the Apaches was the decisive influence. The impact was invaluable: other communities joined the Spanish monarchy or withdrew from war. The relief made the Arizona frontier stronger. The population of New Mexico leapt from about 9,600 in 1769 to about 20,000 towards the century's end.

15 www.nmhistorymuseum.org/calendar.php?&id=37.
16 John Kessell, *Mission of Sorrows: Jesuit Guevavi and the Pimas, 1691–1767* (Tucson, 1970).
17 David Weber, *The Spanish Frontier in North America* (New Haven and London, 1992), p. 187.
18 Fernández-Armesto, *Millennium*, p. 192
19 Weber, *The Spanish Frontier in North America*, pp. 228–9.

Meanwhile, in 1763, despite the inglorious record of Spanish arms in the Seven Years' War, the treaty that ended hostilities with Britain enormously extended Spanish dominion by assigning most of formerly French Louisiana to Spain. The great province, which had struggled to attract settlers and show profits, benefited at once. Under Spanish rule, the population doubled to 40,000 and exports through New Orleans multiplied. In Florida, in the same period, the proximity of lands occupied by Britain exposed the Spanish frontier to constant threat of erosion or collapse. But the indigenous, despite British bribery and intimidation, remained surprisingly loyal to Spain.[20] The loss of the province in the treaty of 1763 was therefore only temporary and the Spaniards recovered it in 1784.

Florida was of value to Spain not for its people or products, but only because its harbours bordered the Gulf Stream route from the Caribbean back across the Atlantic. Similarly, on the Pacific coast, California was of interest only because it guarded the current-assisted route from the Philippines to New Spain. Control became critical in the second half of the eighteenth century when the Pacific became contested. In 1768 the crown's overseer in New Spain, José de Gálvez, decided that Spain had to incorporate the whole of California.[21] The project proved another great Spanish success. By 1780 a chain of missions reached as far as San Francisco. The colony wrenched natives out of nomadism and created a new economy, based on farming and ranching. Growth was spectacular. In 1783 grain production stood at 22,000 *toneladas*, in 1800 it reached 75,000. Over the same period, livestock quadrupled. Missions housed veritable industries, producing leather (exported to New England), wood products, building materials, farming implements, wagons, soap, and sails. Although Spain could never muster enough manpower to extend colonisation further north, the monarchy was fairly successful in keeping Russian and English interlopers at bay south of the forty-second parallel.

While the colonisation of California was under way, the American War of Independence – which pitched most of Britain's mainland colonies against their mother country – was an equivocal opportunity for a monarchy disinclined to validate rebellion and apprehensive of empowering an ambitious new neighbour. But, even before intervening directly, Spain resolved to do

20 Ibid., pp. 300–4.

21 Douglas Monroy, *Thrown Among Strangers: The Making of Mexican Culture in Frontier California* (Berkeley, 1993), p. 21.

'all we can to help the colonists'.[22] When the war ended, thanks to a string of victories over British forces, in the course of which Spanish expeditions penetrated as far north as Illinois and Michigan, Spain seemed to have attained all her objectives in America (as well as the reconquest of Menorca). The British menace had receded, in retreat between Canada and Belize. Florida was recovered. Louisiana and California seemed secure. The frontiers with native states were largely pacified.

Although most of the monarchy's growth was in North America, in the rest of the hemisphere Spain had achieved a comparable record of success, holding enemies at bay on most fronts, while continuing expansion into the cone of South America. The most critical victory was, perhaps, the defence of Cartagena, the acknowledged gateway to Spanish territories in South America, in 1741, against a British fleet of 186 ships and 23,600 troops and crew – the largest assault force Britain had ever assembled. The attackers withdrew after losing fifty ships and three-quarters of the men. Spanish victory was owed to the well-designed and equipped forts that guarded the harbour: although costly, fortification of the Spanish Main was a worthwhile investment, constantly renewed since the late sixteenth century. Land frontiers were harder to secure. Adjustments with Portugal on the Brazilian frontier involved some cessions – notably of thirty Jesuit reductions in a treaty of 1750, which generated the eleven-year-long Guaraní war, when mission *indios* resisted the Portuguese takeover. In exchange, Spain obtained a prize generally reckoned as more valuable: the Colonia del Sacramento in what is now Uruguay, reinforcing control of the River Plate to minimise contraband. In the southern cone the great 'parlamento' or parley that Ambrosio O'Higgins celebrated in 1793, reputedly with 261 Mapuche chiefs, extended the imperial frontier from the Bío-bío to the Beagle Channel.[23] (See Figure 5.)

The Spanish Empire in America reached its greatest extent in July 1796, when a Spanish agent, John Evans, who represented the Missouri Company, lowered the British flag and raised that of the king of Spain over a Mandan village in South Dakota, towards the edge of the Rocky Mountains.[24] Though a conscientious servant of Spanish interests, Evans was no typical subject of

22 Thomas Chavez, *Spain and the Independence of the United States: An Intrinsic Gift* (Albuquerque, 2002), p. 32.
23 Diego Barros Arana, *Historia general de Chile*, 15 vols. (Santiago, 2001), vol. 7, p. 51.
24 Glyn Williams, *Madoc: The Making of a Myth* (London, 1979), p. 146; Glyn Williams, 'John Evans' Strange Journey: Part I – the Welsh Indians', *American Historical Review* 54 (1979), 277–95.

FIGURE 5. Parlamento of Negrete of 1793. The 1793 Parlamento of Negrete was a diplomatic meeting between the Mapuches and Spanish authorities held in Negrete. Some 161 chiefs and 2,380 Mapuche warriors attended the meeting. Published in *Atlas of the Physical and Political History of Chile* by Claudio Gay. 1854. Retrieved from Memoria Chilena.

the Bourbons but a Welsh nationalist, born near Caernarfon and baptised a Methodist, who belonged to a conspiracy dedicated to separating Wales from England or, at least, founding a new Welsh colony beyond English influence in the New World. The inspiration for this romantic endeavour was the chivalresque legend of Prince Madoc, who supposedly conquered an empire in the ocean in the twelfth century. People were disposed to believe such effusions. In the reign of Elizabeth I, whose ancestry was Welsh, projectors of warfare against Spain invoked the Madoc myth to justify imperial ambitions.[25] In Evans's circle a further myth of Welsh-speaking Indians, commonly identified, for no good reason, with the Mandans, was treated as proof of Madoc's enterprise. Thus Evans, last of the great explorers

25 Felipe Fernández-Armesto, 'Inglaterra y el atlántico en la baja edad media', in Felipe Fernández-Armesto, Manuel Lobo Cabrera, and Antonio Rumeu de Armas, *Canarias e Inglaterra a través de la historia* (Las Palmas, 1994), pp. 11–28.

of America for Spain, upheld the tradition of forebears who sought such illusions as the terrestrial Paradise, the Fountain of Youth, El Dorado, the City of the Césares, the Land of Cinnamon, the cities of Cíbola, or the realms of Amazon queens. Like most other acquisitions, the adhesion of the Mandans was procured by negotiation, not conquest. 'Bella gerant alii' was hardly an adequate motto for a monarchy that was almost always at war on some front or other, but it came close to capturing the nature of ideal Spanish grand strategy in the Americas.

By then, the empire was so overextended that decision-makers realised that only peace could keep it going. The chances of that, however, were minimal. An expansionist United States menaced the Louisiana frontier; apart from native tribes, no reliable allies were available. The French Revolution had destroyed the family compact and left Spain teetering and tacking between equally unpalatable options. To fight France invited destruction. To ally with France against Britain prejudiced the empire. Neutrality exposed Spain to depredations by both sides. The ludicrous title Carlos IV conferred on his chief minister – 'Prince of the Peace' – reflected the relief felt throughout the monarchy at a mere postponement of reckoning in 1795. Peace was prolonged at high cost: the surrender of Louisiana in 1800–3, the loss of Spain's principal battle fleet at Trafalgar in 1805, and collaboration with Napoleon, who proved to be a treacherous ally. In 1807–8 he occupied Spain by subterfuge, dethroned and imprisoned the Bourbons, and imposed his brother on the throne. The crisis showed, however, that Spain's military potential was by no means exhausted. A British attempt to seize Buenos Aires had failed in 1807, as did an invasion of Venezuela by rebels with covert US support. Spanish armies, almost to a man, rallied against the French with British help and that of irregulars who mostly resembled social bandits and gave the term 'guerrilla' to the English language. Spanish forces scored a remarkable series of victories against a power that had carried all before it in conquering most of the rest of Western Europe and defeating Prussia and Austria. Napoleon's personal intervention helped restore French fortunes, but from 1810, after the emperor returned home, Spanish advances resumed. By 1814 they had driven the invaders away.

The effective interregnum, however, had strained political institutions to breaking point. In the absence of a central government able to command allegiance, local juntas ran the empire. In the New World, the result was, in effect, civil war among contenders for regional supremacy, in the course of which – as commonly happens in such circumstances – radical elements came to the fore, demanding autonomy or independence in varying degrees. The defeat of the French in 1814 released formidable numbers of Spanish veterans for the reconquest of rebel-held territory in America and, at first, the enterprise seemed successful. Spanish troops, however, fell in their thousands to an intractable enemy: mosquito-borne fevers that left the immunised rebel armies unscathed. Loyalist forces alone were insufficient to cope. In the course of the 1820s, Spanish ambitions collapsed throughout the mainland colonies. The persistence of success through climax, crisis, and climacteric, outlasting decline and defeat, was at an end. The ravages of the long and exhausting struggles to defend the global monarchy left neither Spain nor the newly independent states of formerly Spanish America with resources or infrastructure adequate to the demands of competition in an industrialising world. The chief legacy of Spain's early modern wars was a politicised army, glorified by its resolute defence of national independence against Napoleon, and elevated by *succès d'estime* above the apparently failed political institutions of the country.

5

# Wars of the European continent, 1500–1650

JAMES D. TRACY

Around 1900, scholars commonly marked modern history from the French invasion of Italy in 1494. The size of the army that crossed the Alps – about 30,000 men – and its use of field artillery to batter down the curtain walls of ancient towns was, supposedly, unprecedented. As France's claims in Italy were subsequently challenged by Spain and the Holy Roman Empire, the duchy of Milan and other Italian states collapsed, or changed hands, with astonishing abruptness. Today, it is no longer clear that the campaigns of the Wars of Italy (1494–1559) were so sharply differentiated from those of the last phase of the Hundred Years' War (1415–53). But the political cataclysms of our own time seem to confirm Niccolò Machiavelli's insights into the precariousness of power at the turn of the sixteenth century. No boundary was sacred, and no government lacked a portfolio of ideas for expansion, to be tested if circumstances seemed ripe. Since a power dominant in a given region often worked to keep things as they were, one might distinguish between ambitious governments eager for war and cautious governments concerned to preserve what they had. Any move by a hegemonic power was taken by its rivals as an attempt to reduce them to abject servitude.

In studies of early modern wars, the sovereign national states of later times have usually enjoyed the starring roles. Yet an observer standing on the threshold of the sixteenth century might well have seen the dynastic state, embracing multiple peoples, as the template for the future. The Spanish Habsburg monarchy included Naples (with Sicily), Milan, and the Netherlands (see Map 5). The core lands of the Austrian Habsburgs were mainly German-speaking, but after 1526/1527 their Bohemian lands incorporated a large Czech population, and Habsburg Hungary included what is now Slovakia and parts of Croatia. Denmark's kings ruled Norway and hoped to recover Sweden, lost to rebellion in 1523. The same hypothetical observer

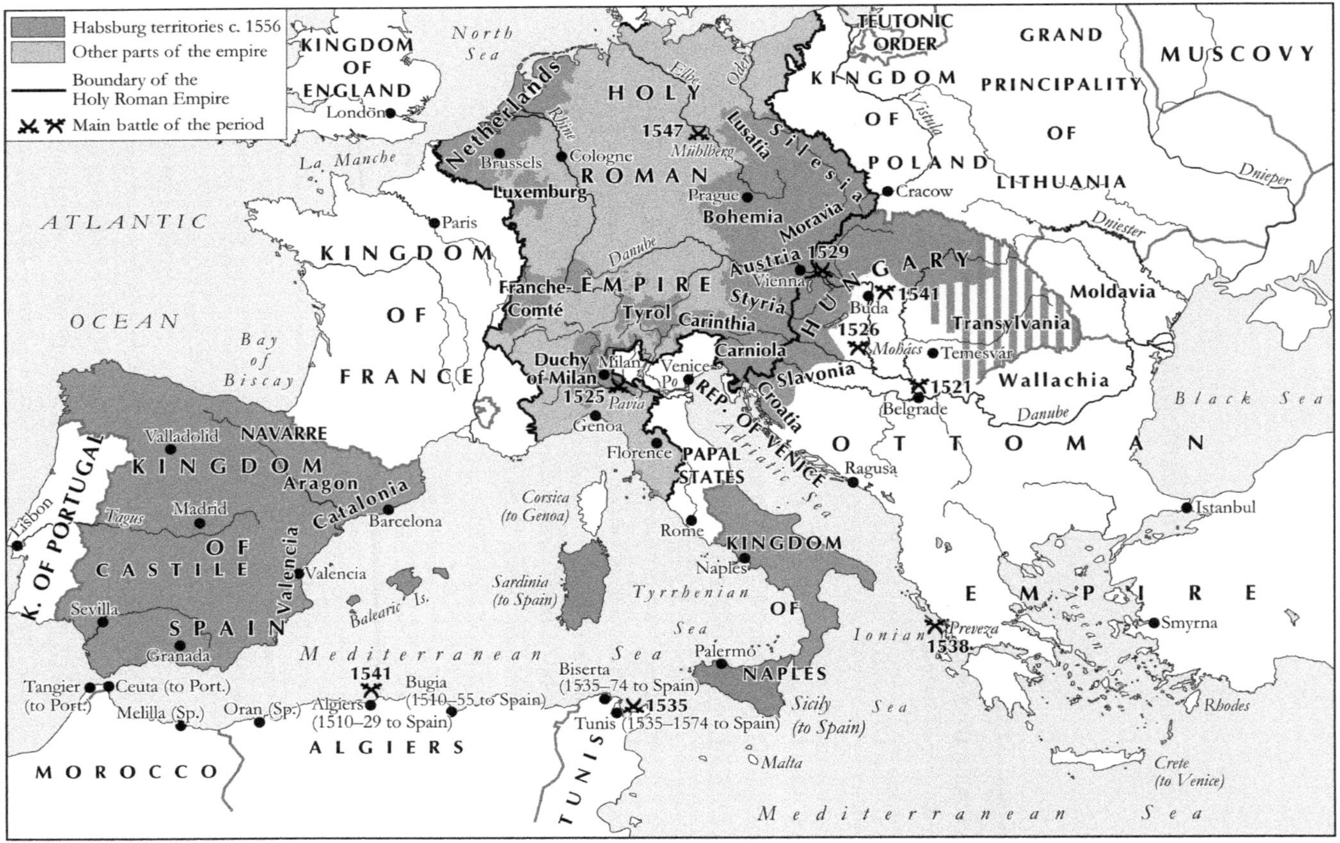


MAP 5. Empire of Charles V (r. 1519–56). Created by Béla Nagy.

might also have guessed wrongly about which princely territories were destined for greatness. In Germany, Bavaria and Electoral Saxony were of major importance in this period, whilst Brandenburg-Prussia was not. In Italy, Venice, not Savoy-Piedmont, was the major independent power.

It is also too easily assumed that modernizing states were the ones that came out ahead. Thus, if France emerged by about 1650 as Europe's dominant power, it was due to the efficiency of the king's bureaucrats. If the Dutch Republic became a major power, it was because bourgeois statesmen knew how to manage finance. If Sweden played an outsized role, it was because Gustavus Adolphus perfected infantry tactics pioneered by the Dutch. These arguments are based on verifiable processes of change. As states shifted from domain income to a reliance on tax revenue, tax-collection required professional administrators, and the slow march of bureaucratization does seem to have accelerated after about 1500.[1] Armies were, over time, better equipped, better trained, and better organized, even if the idea of a "Military Revolution" has its critics.[2] The Dutch and then the Swedes improved infantry tactics and drill, whilst *trace italienne* fortresses required a dramatic increase in the size of siege armies.

The problem lies in thinking that innovation had an immediate impact. In the sixteenth century, states did not need to access a certain kind of revenue to project power beyond their borders. Denmark, with its domain revenues, wielded a disproportionate influence in Northern Europe.[3] Moreover, bureaucratization did not necessarily entail the centralization often taken as synonymous with modernization: financial administration often developed more rapidly at intermediate levels of government, and lenders had a better chance of getting repaid when intermediate bodies interposed their credit. In Bavaria, taxes were managed by officials responsible to the estates or *Landschaft*, which regularly took over the duke's debts.[4] In the Dutch Republic, where provincial parliaments controlled

1 There were about 8,000 royal office-holders in France in 1515, and about 80,000 by 1660: Winfried Schulze, "The Emergence and Consolidation of the 'Tax-State,'" in Richard Bonney (ed.), *Economic Systems and State Finance* (Oxford, 1995), pp. 261–79, 268.

2 Michael Roberts, *The Military Revolution, 1560–1660* (Belfast, 1956); Geoffrey Parker, *The Military Revolution, 1500–1800: Military Innovation and the Rise of the West*, 2nd ed. (Cambridge, 1996); Jeremy Black, *A Military Revolution? Military Change and European Society, 1500–1800* (London, 1991); Clifford J. Rogers (ed.), *The Military Revolution Debate: Readings in the Transformation of Early Modern Europe* (Boulder, 1995).

3 Bjørn Poulsen, "Kingdoms on the Periphery," in Bonney (ed.), *Economic Systems and State Finance*, pp. 102–22.

4 Maximilian Lanzinner, *Fürst, Räte und Landstände. Die Entstehung der Zentralbehörden in Bayern, 1511–1598* (Göttingen, 1980).

revenues, the States of Holland carried in the early seventeenth century a funded debt ten times larger than the funded debt of the States General.[5] In the Holy Roman Empire, delays in the collection of taxes voted by the Diet were bridged by loans raised by the magistrates of Augsburg, Nuremberg, and other free imperial cities.[6] In France too, lenders dealt with corporate bodies – town governments, or corps of royal bureaucrats that raised loans for the king by subscription.[7] In fiscal terms, the putative age of centralization is better called an age of intermediation.

In warfare too innovation did not always bring immediate results. Sweden's crack infantry could be beaten by the cavalry of Poland-Lithuania, seen as backward by scholars.[8] A *trace italienne* fortress could change the strategic geography of a region, but victories against siege armies were often won by conventional methods, like the *retirata* that foiled Charles V at Metz in 1552, or the flooding that saved Leiden from the Spanish in 1575.[9] Governments tried to centralize the structure of command, but key decisions were still left to mercenary generals.

This chapter makes room for as many as possible of the state-actors who showed a capacity to project power. To keep things manageable, discussion is limited to inter-state wars. This means excluding wars fought within the borders of a single polity, with an exception for the Holy Roman Empire. Discussion is also restricted to wars fought on the Continent, thus excluding the British Isles and Scandinavia. The chapter will sketch patterns of conflict; that is, serial struggles involving many states, with wide repercussions. Three such patterns are selected: the Ottoman wars, the rivalry between France and the Habsburg dynasty, and battles between Catholic and Protestant powers for control of the Holy Roman Empire.

5 Marjolein 't Hart, Joost Jonker, and Jan Luiten van Zanden (eds.), *A Financial History of the Netherlands* (Cambridge, 1997), p. 20.

6 Marjolein 't Hart, "The Emergence and Consolidation of the Tax State, II: The 17th Century," in Bonney (ed.), *Economic Systems and State Finance*, pp. 281–93, 287.

7 Martin Wolfe, *The Fiscal System of Renaissance France* (New Haven, 1972), pp. 91–93; Mark Potter, *Corps and Clienteles: Public Finance and Political Change in France, 1688–1715* (Burlington, 2003).

8 Robert Frost, *The Northern Wars: State and Society in Northern Europe, 1558–1721* (Harlow, 2000), pp. 17–18, 55–64.

9 James Tracy, *Emperor Charles V, Impresario of War: Campaign Strategy, International Finance, and Domestic Politics* (Cambridge, 2002), pp. 239–40; Robert Fruin, *The Siege and Relief of Leiden*, trans. Elizabeth Trevelyan (The Hague, 1928). A *retirata* was an earthen rampart thrown up behind the point at which an old-style curtain wall had been breached by artillery.

## Ottoman wars, 1521–1617: A clash of civilizations

It has recently been suggested that sixteenth-century Ottoman history was part of European history.[10] To be sure, the Ottomans maintained a long-standing alliance with France, and peaceful relations with two Christian neighbors, Venice and Poland-Lithuania. They also granted trade concessions to the English and the Dutch, foes of Spain. Safavid Iran, as much as Spain, was a consistent enemy of the sultan.[11] Though scholars have tried to play down the idea, Turks and Spaniards, like other early modern peoples, were conscious of living in civilizations defined by religion.[12] One implication of this broad civilizational awareness was that primacy in one's own "world" took precedence over battles with outsiders. Thus Spain was regularly preoccupied by the contest with France for hegemony in Christendom, just as the Ottoman state clashed with the Shi'ite Safavids of Iran. Still, a state that had the resources of the Ottoman Empire could accomplish a great deal. In Central Europe, the sultan's armies gained control of Hungary's heartland, the Danube basin. In the Mediterranean, his warships raided up and down the coastlines of Italy. For Europe, the Ottoman wars were not a secondary issue.

In both worlds, contemporaries looked for a climactic battle between Emperor Charles V (r. 1519–56) and Sultan Suleyman I the Lawgiver (r. 1520–66).[13] Charles saw himself as Christendom's chief defender against the infidels, and Suleyman sought to replicate in the infidel west the conquests made in the Muslim east by his father, Selim I (r. 1512–20). Belgrade and the island of Rhodes, long-time "bastions" of Christendom against the Turks, were captured in 1521 and 1522 respectively. In 1526, Suleyman led an army of about 70,000 to 80,000 men up the right bank of the Danube to Mohács, where his forces crushed the army of Hungary's King Louis II Jagellio, who died in the aftermath. By Ottoman law,

10 Holger T. Graef, "'Erbfeind der Christenheit' oder potentieller Bündnispartner? Das Osmanenreich im europäischen Mächtesystem des 16. und 17. Jahrhunderts – gegenwartspolitisch betrachtet," in Marlene Kurz, Karl Vocelka, and Thomas Winckelbauer (eds.), *Das Osmanenreich und die Habsburgermonarchie* (Vienna and Munich, 2005), 37–51.

11 James Tracy, "The Background War of the Early Modern Era: Christian and Muslim States in Contest for Dominion, Trade, and Cultural Preeminence," in Marguerite Ragnow and William D. Philips, Jr. (eds.), *Religious Conflict and Accommodation in the Early Modern World* (Minneapolis, 2011), pp. 9–54, 17.

12 One of the strongest arguments in Géraud Poumarède, *Pour en finir avec la Croisade. Mythes et réalités de la lutte contre les Turcs aux XVIe et XVIIe siècles* (Paris, 2004); James Tracy, "Asian Despotism? Mughal Government as Seen from the Dutch East India Company Factory in Surat," *Journal of Early Modern History* 3 (1999), 256–80.

13 Sanjay Subrahmanyam, "Sixteenth-Century Millenarianism from the Tagus to the Ganges," in *Explorations in Connected History: From the Tagus to the Ganges* (New Delhi, 2005), pp. 102–37.

Suleyman was now king of Hungary by right of conquest, and could dispose of the realm as he pleased. János Szapolyai, *Vojvode* of Transylvania, claimed the crown, and so did Charles V's brother, Archduke Ferdinand of Austria. After each got himself elected by a friendly diet, Ferdinand's forces won a battle, but Szapolyai played a more potent diplomatic card: Suleyman, loath to see a Habsburg on Hungary's throne, backed the *Vojvode*. On his next invasion of Hungary (1529), he paused, en route to Vienna, to install Szapolyai in Buda. Meanwhile, as Charles V remained in Italy, tending to other matters, Vienna held out against its besiegers, thanks to troops sent by the Imperial Diet. (See Figure 6.)

Ferdinand's efforts to gain ground in Hungary were undone by incompetent generals. He did not lose hope, because Szapolyai was childless and his supporters were nervous about Ottoman intentions. Throughout the late 1530s and 1540s, periods of direct conflict over control of Hungary and Transylvania alternated with dynastic negotiations and diplomacy, and in 1547 the Treaty of Istanbul was agreed between Suleyman and Charles.[14] By this and subsequent agreements, the sultan accepted Ferdinand's position in the parts of Hungary he controlled, in return for an annual "honorific gift" of 30,000 florins.

Charles V preferred to fight the Turks at sea, and a forward defense of Charles's Spanish and Italian shores became imperative after Suleyman extended his authority to the Barbary Coast: from 1534, Khair-ad-Din Barbarossa, bey of Algiers, was grand admiral (*kapudan paşa*) of the Ottoman fleet. In 1535 a Habsburg expeditionary force led by Charles stormed Goletta and entered Tunis without resistance. In 1537, an Ottoman attack on Venetian Corfu led the Republic to accept a papal proposal for a Holy League against the Turks. As commander of the combined Christian fleet, Andrea Doria, Genoese military entrepreneur and Charles V's galley-commander since 1528, managed to bottle up Barbarossa's galleys at Preveza, inside the Ambracian Gulf (September 1538). Barbarossa's escape, without damage, left Venice deeply suspicious of Spain's motives.[15] In 1541, Charles gave orders to mobilize an amphibious force for an assault on Algiers. But, still engaged in attempts to settle the religious crisis that had erupted in the Holy Roman Empire since the adoption of Protestantism by many German princes and cities, he put off departure until October, dangerously late in the sailing

14 Rhoads Murphey, "Suleyman I and the Conquest of Hungary: Ottoman Manifest Destiny, or a Delayed Reaction to Charles V's Universalist Vision," *Journal of Early Modern History* 5 (2001), 197–221.

15 John Guilmartin, *Gunpowder and Galleys: Changing Technology and Mediterranean Warfare at Sea in the Sixteenth Century* (Cambridge, 1974), pp. 15–56.

FIGURE 6. Siege of Vienna, 1529. Woodcut print by Hieronymus Ortelius. Getty Images / mikroman6.

season. Off Algiers, a ferocious storm sank half his supplies and doomed the expedition.

Charles V and Suleyman I now respectively turned their attention to France and the Safavid Empire. But corsair-expeditions against the infidel had become an honorable profession for both sides. The Knights of St. John, expelled from Rhodes in 1522, now embarked on their annual voyages from Malta, granted to them by Charles V in 1531. In North Africa, whole towns depended on the economy of privateering, supported by local investors. In 1551, Turgut Reis, Barbarossa's successor in Algiers, forced the surrender of the island of Gozo (a dependency of Malta) and took away some 5,000 captives. To the east, in ports like Durrës (Albania) and Modon (in the Peloponnese), captains of Ottoman galley squadrons made a lucrative side business of predatory landings in Spanish Naples. Near the head of the Adriatic, and especially in the small port of Senj, the *Uskoks* – mostly refugees from lands conquered by the Turks – made a living from booty. At sea, they preferred to plunder Ottoman vessels, but did not overlook those of Senj's old enemy, the Venetians. On land, their raids into Ottoman-held Dalmatia prevented the sultan's officials from repopulating a war-ravaged region.[16] Along the northern coast of the Black Sea, the sultan's Tatar allies regularly raided northward into Poland-Lithuania's southern provinces, and Zaporozhian Cossacks repaid the favor with attacks on Black Sea shipping, as far south as the Bosporus. Finally, along the 600-mile frontier between Habsburg and Ottoman land in Hungary and Croatia, treaties of peace did not prevent local commanders from doing what was needed to keep their men content.

Unremitting "small war" offered excuses for real war any time a major power made the choice. Neither Charles V nor his son Philip II (r. 1556–98) could respond to the assault on Gozo so long as Spain was heavily committed to war with France. But in 1560 Philip sent an expedition against Tripoli, now the base of Turgut Reis. Off the island of Djerba, where the army made camp, a surprise attack by Piyale Paşa, the Ottoman admiral, dealt Spain its worst Mediterranean defeat: nearly 10,000 men were killed or captured. In 1565, the Ottomans sought to build on this success by arranging for the fleets of Piyale Paşa and Turgut Reis to converge on Malta, with some 48,000 men. Ottoman guns battered down one fortress, but others held out, and the besiegers were about to break off when the viceroy of Sicily landed a relief force of 8,000; this

16 Suraiya Faroghi, "Trade: Regional, Inter-regional and International," in Suraiya Faroghi, Bruce McGowan, Donald Quataert, and Şevket Pamuk (eds.), *An Economic and Social History of the Ottoman Empire*, vol. 2: *1600–1914* (Cambridge, 1991–94), p. 482.

time it was the Turks who suffered heavy casualties. But if the corsairs of Malta escaped the sultan's wrath, the Hungarians would not. In 1566, Suleyman I marshaled his forces for a sixth and last campaign into Hungary, against Szigetvár, an anchor fortress in southwestern Hungary that had long frustrated Ottoman administration. With help from the Imperial Diet, Emperor Maximilian II (r. 1564–76) mobilized a large army near Győr (Raab), but for reasons best known to himself he chose not to use it for any action that might have relieved pressure on Szigetvár, which fell after a siege of five weeks.[17]

Sultan Selim II (r. 1566–74) was the son and grandson of great conquerors. His advisers debated whether the first campaign of his reign should be in Hungary or on the Mediterranean. Venice had forewarning, but no time to prepare: the attack came on Cyprus, the Republic's last major stronghold in the eastern Mediterranean, and a threat to vital traffic between Istanbul and Cairo. The 100,000 men who landed in June 1570 overwhelmed Nicosia's fortifications, leading to the slaughter of 20,000 Venetian subjects. But Famagosta held out. Venice joined Pope Pius V (r. 1566–72) in working to persuade a reluctant Philip II to enter a Holy League against the infidel. The agreement was not concluded until May 1571, and tensions between Spaniards and Venetians continued even after the combined fleet of more than 200 war galleys had sailed from Messina in September. Their encounter with a slightly larger Ottoman war fleet took place not at Lepanto, but off the northernmost of the Curzolaris Islands, on October 7. In the ensuing conflict, the Christians captured 117 Ottoman galleys and sank another 80 to 90. Had the bonds between Christian allies been stronger, they might have sought to exploit their victory by mounting another expedition to the eastern Mediterranean in 1572. Instead, Spain's commanders returned to their usual preoccupation with North Africa, and Venice signed a separate treaty of peace with the sultan in 1573. Meanwhile, the Ottoman state showed its mettle – eighty new war galleys were ready by January 1572, with more under construction. The Ottomans would have the last word: in 1574 they reconquered Goletta and Tunis, erasing Charles V's earlier victory. In 1578 the two sides concluded a one-year truce. Philip II had a war in the Netherlands, Murad III (r. 1574–95)

17 Pál Fodor (ed.), *The Battle for Central Europe: The Siege of Szigetvár and the Death of Süleyman the Magnificent and Nicholas Zrínyi (1566)* (Leiden and New York, 2019); Georg Wagner, "Maximilian II, der Wiener Hof und die Belagerung von Sziget," in Lajos Rúzsás (ed.), *Szigetvári emlékkönyo. Szigetvár 1566* (Budapest, 1966), pp. 237–68; Nicholas Vatin, "Introduction," in *Feridūn Bey. Les plaisants secrets de la campagne de Szigetvár* (Berlin, 2010).

was embarking on a campaign against the Safavids, and neither had the stomach for more great battles at sea.[18]

The privateering tradition continued. For example, the duke of Osuna, viceroy of Sicily (1610–1616) and then of Naples (1616–20), built with his own funds a galley fleet that had great success in the eastern Mediterranean.[19] From about 1580, the Mediterranean was more dangerous than ever for merchants. To establish a safer route to Constantinople, via Split and then overland, Venice overcame its historic anti-Semitism and struck a bargain with Spanish Jewish merchants based in the Ottoman Balkans. It was still profitable to import spices and silks from Cairo and Alexandretta (the outport for Aleppo), because ships on this route sailed in convoy.[20] But this trade too eventually faded, as the Dutch and English East India Companies gained a monopoly on traffic to Europe.

Meanwhile, along the Habsburg–Ottoman frontier, the reign of Rudolf II (r. 1576–1611) began with a new spirit of common effort. In the Empire, where Catholic and Lutheran princes now showed some willingness to collaborate, the Diet voted larger-than-ever subsidies for fortresses in Hungary. In the Austrian lands, governed by Rudolf's brothers, grants of toleration to a largely Lutheran aristocracy facilitated implementation of an ambitious plan whereby each province took responsibility for wages and provisions along a sector of the border.[21] The "military frontier," hardened by these improvements, was tested but not seriously breached during the Long Turkish War of 1593–1606. By the 1606 Peace of Zsitvatörök (at least as the Habsburgs understood it), Rudolf was no longer obliged to pay tribute for his lands in Hungary and Croatia. The emperor and the sultan would not fight another major war until the 1660s. This did not mean that the fighting stopped altogether. The *Uskoks* continued their raids, hoping to spark rebellion in the Ottoman Balkans. Venice, exasperated by Vienna's inaction, finally

18 Niccolò Capponi, *Victory of the West: The Story of the Battle of Lepanto* (London, 2006), ch. 10; María Antonia Garcés, *Cervantes in Algiers: A Captive's Tale* (Nashville, 2002).

19 Luis M. Linde, *Don Pedro Girón, duque de Osuna. La hegemonia Española en Europa a comienzos del siglo XVII* (Madrid, 2005), pp. 99–104; Domenico Ligresti, "L'organizzazione militare del Regno di Sicilia (1575–1630)," *Rivista Storica Italiana* 105 (1993), 647–78.

20 Renzo Paci, *La "Scala" di Spalato e il commercio Veneziano nei Balcani fra Cinque e Seicento* (Venice, 1971); James Tracy, "Il commercio italiano in territorio ottoman," in Franco Franceschini, Richard Goldthwaite, and Reinhold Mueller (eds.), *Il Rinascimento italiano e l'Europa*, vol. 4: *Commercio e cultura mercantile* (Vicenza, 2007), pp. 425–545.

21 Winfried Schulze, *Landesdefension und Staatsbildung. Studien zum Kriegswesen des innerösterreichischen Territorialstaates (1564–1619)* (Vienna, 1973).

hired Dutch mercenaries to force the Habsburgs to relocate their troublesome subjects (1615–17).

For European history, the chief result of these multiple wars was that the Ottoman advance to the west was halted. Behind a line made firm by about 1580, the Europe of privileged churches, privileged nobilities, self-governing towns, secular law courts, universities, and printing presses was secured for the future. Also, resources that might have gone to other purposes were committed to border garrisons, fortresses, and galleys. For the Austrian Habsburg monarchy and for Spanish Habsburg lands in Italy the costs of the Ottoman wars were vast; for the Holy Roman Empire, Venice, and Spain itself, they were substantial. Finally, for the Empire in particular, the Ottoman threat pushed this federation of Catholic and Protestant principalities to prove itself as a body politic.[22]

## Wars between France and the Habsburgs, 1499–1648: A contest for hegemony

Spain's engagements to the south and east reflected Aragon's long connection with Naples and Sicily. In 1503, Ferdinand the Catholic of Aragon sought to retake direct control of Naples from France. As ruler of Castile after Isabella's death (1504), Ferdinand sanctioned for-profit enterprises to establish military colonies in North Africa. In pursuit of the same priorities, Charles V (who ruled in Spain 1517–56) beat back another French attack on Naples (1528) and conquered Tunis (1535). But Spain's ambition was simultaneously drawn to the north, by unforeseen events that heaped crowns upon the head of Ferdinand and Isabella's grandson. Ruling in the Low Countries, Spain, Italy, and the Holy Roman Empire, Charles V exercised what was seen as a kind of hegemony in Europe. To be sure, contemporaries exaggerated his power, but their mere perceptions of what he could do allowed him to wield influence in far-flung parts of the Continent. Despite the flow of silver from the Spanish New World however, Charles's wars in France and Germany overburdened Castile's finances, and his successor, Philip II (r. 1556–98), had to decree a suspension of interest payments on the crown's debts in 1557. Philip's great struggle was in the Netherlands, where he was to achieve only partial success. After Spain finally reconquered Ostend (1603–6), at great cost, Philip III (r. 1598–1621) agreed to talks with the Dutch Republic. When the Twelve Years' Truce of 1609–21 expired,

22 Thomas Brady, *German Histories in the Age of Reformations, 1400–1650* (Cambridge, 2009), p. 355.

Philip IV (r. 1621–65), in alliance with his cousin, Emperor Ferdinand II, committed Spain to the path of war once again.

In France, as in all of Europe, lines of authority overlapped. The king's armed forces were led by great nobles, some of whom held hereditary offices giving them substantial authority over appointments, command, and control in the armies and navies. Many of the crown's revenues were alienated to bankers in Lyons, or to his own officials.[23] Still, no other European ruler had 20 million subjects, and no other king had such authority to raise taxes at will. In an age when Lilliputian princes could indulge Brobdingnagian fantasies, the ordinary ambitions of a French king would shake the earth. Indeed, it has been argued that a France that showed no ambition would have been in danger of breaking apart from within. In this light, the 1494 invasion of Italy was a "war of magnificence" meant to unite a potentially fractious aristocracy around the king. The second invasion (1499) led to a "war of prestige," because Spain now sought to check France's gains in the peninsula. Under Francis I (r. 1515–47) and Charles V, the struggle over Italy became a contest for hegemony, between Europe's largest kingdom and a dynasty whose lands bordered France on three sides.

Propagandists on both sides claimed that dynastic Europe, with its many independent states, was threatened by a looming *monarchia*.[24] But were the would-be tyrant's banners decked out in the double-headed eagle of the Empire, or the *fleur-de-lis* of France? Philip II assured all and sundry that his sole ambition was to preserve the lands bequeathed to him. Then in 1580 he made good his hereditary claim on the throne of Portugal, adding the profits of the Asian spice trade to the silver of the Americas. In hindsight, one can see that neither Charles nor Philip aimed for a universal *monarchia*.[25] But statesmen hostile to the Habsburgs found it harder to distinguish between hegemony and *monarchia*; the nearness of one idea to the other may also explain why French propaganda has had more success with historians than Habsburg self-justifications.

23 Robert Harding, *Anatomy of a Power Elite: The Provincial Governors of Early Modern France* (New Haven, 1978); Philippe Hamon, *L'Argent du roi. Les finances sous François Ier* (Paris, 1994). In the so-called *pays d'élection*, taxes were assessed by royal officials; in provinces that had estates – the *pays d'état* – this work was done by officials of the estates. Cf. James Collins, *Classes, Estates, and Order in Early Modern Brittany* (Cambridge, 1994).

24 Tracy, *Emperor Charles V*, pp. 27–28.

25 Ibid., pp. 305–7; Peter H. Wilson, *Europe's Tragedy: The Thirty Years' War* (London, 2009), p. 162: Spain's efforts to keep the peace after 1559 have wrongly been interpreted as "mere tactical withdrawals in a longer running struggle."

Italians were awed by the army that Charles VIII marched down the length of the peninsula (1494/95) in support of his dynastic claims on Naples, but they resisted him as he returned to France. With troops sent by his cousin, Ferdinand of Spain, the Aragonese king of Naples reclaimed the throne of Naples in 1496. But in 1499 Louis XII (r. 1498–1514) launched a new military campaign into conquered Milan. By reviving in addition his predecessor's claim to Naples, he provoked fears of French hegemony in the peninsula. Ferdinand struck a bargain with Louis XII for a Franco-Spanish partition of Naples, but then ordered his troops to expel the French. At Cerignola (1503), Spain's new *tercios* first showed what they could achieve against traditional French forces dominated by heavy cavalry.[26] Subsequent fighting in Italy once again culminated in a French military setback at Novarra (1513), leading to their expulsion from the duchy of Milan.

Francis I (r. 1515–47) began his reign by leading a French army across the Alps. The sudden appearance of his large force, supported by significant heavy guns, compelled France's rivals in Italy to come to terms. A large contingent of Swiss, newly arrived in Milan, refused to accept the agreement, but their furious assault at Marignano, outside Milan, was repelled by the French artillery. Meanwhile, Charles V was succeeding to his inheritances in the Burgundian Low Countries, Spain, and Austria and gaining election as Holy Roman Emperor (1519). The advent of these two rulers sowed the seeds for a conflict that would transcend the Italian theatre. Francis I struck first, through surrogate-led invasions of the Low Countries and of Spanish Navarre in 1521. Fuenterrabía was taken and held for a few years, but in the north Charles's commanders made a permanent change to the frontier by conquering Tournai. Despite the success at Marignano, which had given Francis control of Milan, the duchy fell to a Hispano-Imperial force, and French armies sent across the Alps in 1522 and 1523 failed to reverse the situation. Thus in October 1524 Francis I crossed the Alps a second time, at the head of an unprecedented French army of 40,000. He entered Milan without opposition, and began raising siege-works around Pavia, the headquarters for Spain's troops in Lombardy. At Charles V's behest, his brother Archduke Ferdinand engaged 15,000 *Landsknechte* for a march into Italy; in January they joined with *tercios* brought up by the viceroy of Naples. On February 24, 1525, Habsburg forces breached and charged through the walls of a large park

26 René Quatrefages, "L'Organisation militaire de l'Espagne, 1492–1592," Thèse de doctorate (Université de Paris-IV (Sorbonne-Panthéon), 1989); for the significance of the battle of Cerignola for military history, Jeremy Black, *War in the World: A Comparative History, 1450–1600* (Basingstoke, 2011), pp. 78–79.

outside Pavia, where Francis had placed most of his army. The king led his cavalry in a counterattack that showed some signs of turning the tide, until the French came under withering fire from Spanish *arcabuceros* posted in wooded areas of the park. More than 8,000 of the king's men were killed or wounded; Francis himself, now a prisoner, was taken off to Spain. (See Figure 7.)

The battle of Pavia had important consequences. While her son remained in captivity, Louise of Savoy, regent in France, made a first effort to reach out to the Sublime Porte. In Spain, Francis renounced his claims in Italy, and ceded the duchy of Burgundy to Charles V (Treaty of Madrid, 1526). Back in France, he repudiated the treaty, as did the estates of Burgundy. In Italy, Pope Clement VII (r. 1523–34), reflecting the Papacy's historic fear of a single dominant power in Italy, joined with France, Venice, and Medici Florence (ruled by the pope himself) in the League of Cognac (May 1526). The next spring Ferdinand's *Landsknechte* descended into Italy, but without the payments due from Charles V's bankers. Notwithstanding a truce with Clement VI concluded by the viceroy of Naples, imperial troops vented their wrath on the papal city; the eight-day *Sacco di Roma* (May 1527) was one of the worst

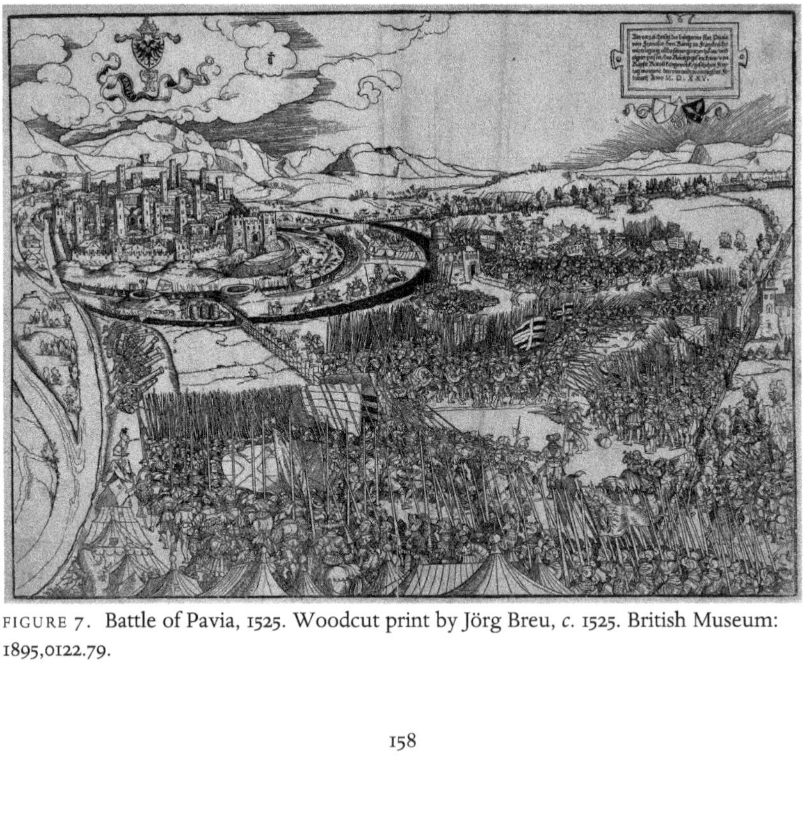

FIGURE 7. Battle of Pavia, 1525. Woodcut print by Jörg Breu, *c.* 1525. British Museum: 1895,0122.79.

military atrocities of the century. In Florence, foes of the Medici took the occasion to declare a new Republic. French forces now arrived in Lombardy, but they by-passed Milan and marched down into Apulia; as one town after another surrendered, they surrounded Naples from the landward side, Genoa's Andrea Doria, the naval *condottiero* who was at this time a French ally, blockaded Naples by sea. But Doria's sudden switch to the Habsburg side (July 1528) ended the blockade of Naples even as the French besieging army was decimated by plague. Meanwhile, in the Netherlands, the duke of Guelders, a French ally, had occupied the prince-bishopric of Utrecht; Habsburg troops expelled him from Utrecht and then from his own duchy. Both territories became Habsburg provinces. Hostilities ended with the Treaty of Cambrai (August 1529), recapitulating the Treaty of Madrid, save in regard to the surrender of Burgundy.

Milan became an issue once more in 1535, when Duke Francesco Sforza died without a male heir: might the duchy, an imperial fief, now be claimed by Charles himself? Francis I proposed a marriage alliance whereby Milan would go to his second son, the duke of Orléans, but Charles demurred. Then in February 1536, without apparent provocation, Francis invaded and occupied the lands of his maternal uncle, the duke of Savoy. Charles V responded by invading Provence (July 1536), but the defenses put in place by Anne de Montmorency, Constable of France, withstood attack.

Tensions continued to mount as it became clear that no compromise could be reached over the control of Milan. In July 1542, with a promise from the Ottoman sultan for operations in the Mediterranean to support the French, and with his armies in place for additional attacks in the Pyrenees and the Low Countries, Francis I declared war.

Despite France's preparations, Charles V had the best of things. In the Pyrenees, Perpignan held off a month-long siege. In the Low Countries, the Regent Mary of Hungary repulsed several incursions. In 1543, Barbarossa forced the surrender of Nice (a dependency of Genoa), but not its citadel. Since the *kapudan paşa* demanded a harbor for refitting over the winter, Toulon was evacuated for his benefit, letting Christendom see the Most Christian King's dependence on his infidel allies. Meanwhile, Charles joined German and Spanish units for a successful assault on the duke of Cleves, who had occupied Guelders. Charles's advance was hampered by dwindling funds, but Francis I had already signaled an interest in talks. The treaty of Crépy (September 1544) surprised contemporaries by restoring the territorial *status quo* of 1538.

Henri II (r. 1547–59) organized in 1553 a Franco-Ottoman assault on Corsica (a Genoese possession) that brought the island under temporary French control.[27] The army he sent in 1557 to aid the anti-Habsburg campaign of Pope Paul IV (r. 1555–59) occupied Siena, but only for a few months. Henri had better luck though renewing France's ties to Germany's Protestant princes. Elector Maurice of Saxony (r. 1541–53), an erstwhile ally of Charles V, was the leading figure in an alliance with France concluded in January 1552. Henri II marched into Lorraine and occupied three episcopal cities – Metz, Toul, and Verdun – that he claimed to rule as "imperial vicar," a title recognized by his German allies. Maurice advanced on Innsbruck, forcing Charles to flee across the Alps to Carinthia. But Archduke Ferdinand, hoping to limit the consequences of an Ottoman invasion of Transylvania, invited Maurice to peace talks: the two princes worked out a truce that Charles grudgingly accepted. Meanwhile, Habsburg military enterprise mobilized an army of more than 50,000. The target was Metz, where siege lines were opened in late October. But Duke Francis of Guise and his 6,000 men defended the city with confidence. Charles's artillery breached the walls twice, only to expose on both occasions new internal earthworks that were bristling with guns. As his army melted away in the bitter cold, Charles raised the siege on January 1, 1553. Broken in spirit, he lived *incommunicado* in Brussels for over a year.

Charles V abdicated in January 1556. Philip II, acutely conscious of Spain's fragile credit, had no wish to prolong the war which had begun in 1552. But the February 1556 Truce of Vaucelles was ruptured in January 1557, when Henri II sent one army to attack Douai and another to aid Pope Paul IV in Italy. From Brussels, Philip and his commander-in-chief, the exiled Duke Emmanuel-Philibert of Savoy, marched into Picardy, where they laid siege to St. Quentin. On August 10, 1557 a relief force led by Gaspard de Coligny, Admiral of France, was decisively defeated by Emmanuel-Philibert. This was to be the last important Franco-Habsburg battle of the sixteenth century. By the Peace of Cateau-Cambrésis (April 1559), France retained Calais (conquered from England in 1558) and the three bishoprics in Lorraine, but recognized Spain's rule in Milan and Naples, and returned almost all of Savoy-Piedmont to Emmanuel-Philibert. The treaty had critics in France, but neither kingdom had the will or the credit to keep fighting.

27 Antoine-Marie Graziani and Michael Vergé-Franceschi, *Sampiero Corso, 1498–1567* (Ajaccio, 1999).

For the next fifty years, the "Cold War" between France and Spain was more a matter of diplomatic *démarches* than of cannonades.[28] From 1562, France was torn by serial wars among three religious parties, each led by a magnate clan. To keep Spain occupied, Paris subsidized the rebel provinces of the Netherlands, albeit at a modest level.[29] Philip II intervened in France in 1584, when it became clear that the throne would pass to Henri of Bourbon, leader of the Huguenot (Protestant) party. Notwithstanding subsidies from Spain, the Catholic League, led by Duke Henry of Guise, failed to prevent the accession of a Protestant king. Henri IV (r. 1589–1610) converted to Catholicism in order to gain entry to Paris (1593), and then subdued the remnants of the League before turning his attention to foreign affairs. The 1598 Peace of Vervins recapitulated the terms of the treaty of 1559, save for unresolved claims to the marquisate of Saluzzo, which had been occupied by Duke Charles-Emmanuel of Savoy since 1588. Two years later French troops conquered Saluzzo, which Henri then exchanged for Savoyard lands west of the Alps, including a crucial bridge hitherto used by Spanish military units coming from Italy. Troops bound for the Netherlands henceforth had to cross the Alps farther east, through the Valtellina, north of Lake Como.

Louis XIII (r. 1610–43) became king at the age of eight, and in 1617 a faction around Louis seized control from the Regent, his mother. But both during and after the Regency, France was preoccupied with troubles within its borders caused by infighting among the aristocracy and Protestant rebellions. The only notable development in foreign affairs was that France made an ally of Duke Charles Emmanuel I of Savoy (r. 1580–1630), an erstwhile client of Spain. Then in 1624 Louis XIII named Cardinal Armand du Plessis de Richelieu as his *premier ministre*. In Italy, Richelieu picked up where Henri IV had left off. In 1624, and again in 1635, with more success, he sent an army to the Valtellina to block the passage of Spanish troops moving north from Lombardy. In company with Charles Emmanuel of Savoy, he launched an attack on Genoa that failed (1625). However, in 1628 a dispute arose over the succession to the two Gonzaga duchies of Mantua and Monferrato, strategically located on the western and eastern flanks of the Spanish duchy of Milan. Richelieu's decision to intervene in 1629 and then again in 1630 on behalf of one of the claimants to the two duchies, Charles de Gonzague-Nevers, was

28 For this term, see David Buisseret, *Henry IV* (London, 1984), p. 126; Michael Levin, *Agents of Empire: Spanish Ambassadors in Sixteenth-Century Italy* (Ithaca, 2005); for the French ambassador's role in promoting a separate treaty of peace between Venice and the Porte in 1573, Capponi, *Victory of the West*, pp. 310–13.

29 Geoffrey Parker, *The Dutch Revolt* (London, 1977), p. 298.

initially a high-risk strategy, threatening the prospect of all-out war with Spain. In the event, diplomacy, and the reluctance of the Habsburg powers to draw France into war, allowed Richelieu both to secure the succession for Charles de Nevers and to gain the strategic city and fortress of Casale-Monferrato for France.

The growth of Habsburg power in Germany thanks to the military successes of the Habsburg emperor and his Catholic allies was more concerning, but Richelieu could not address it so long as the *dévots* – those who favored a pro-Catholic foreign policy – sat with him on the king's council. Then in 1630, and following the success of Richelieu's Italian policy, Louis XIII dismissed Richelieu's rivals, leaving him a free hand to open discussions with the Protestant foes of Emperor Ferdinand II (r. 1619–37). Meanwhile Richelieu sent troops to occupy the lands of Duke Charles of Lorraine and invited the Empire's Catholic princes to sue for French protection – and accept French garrisons – against the threats of both the emperor and the now victorious Protestant Swedes. To his great satisfaction, Duke Maximilian I of Bavaria was among those who showed interest: one of Richelieu's main objectives was to detach Bavaria from its long-standing alliance with the Habsburgs. But the strategy of waging war by proxy had its limits. The entire French policy of exploiting a balance of military power between the Habsburgs and their Protestant enemies was overturned by the crushing defeat of the Swedish army at Nördlingen in south Germany (1634). Even before this, the activities of Spanish troops in Alsace had seemed increasingly threatening to France's tenuous occupation of Lorraine, and its various commitments to the Rhineland Catholic princes. The Spanish kidnapping of the archbishop-elector of Trier, and their occupation of his Electorate, provided a suitable *casus belli* for a conflict that had been looming for months. In 1635, France and the Dutch Republic signed an offensive alliance against Spain, and French forces mobilized for campaigns in Lorraine, in Alsace, and on the right bank of the Rhine. French armies would take the field every campaign season for the next twenty-five years.[30]

In the end, Richelieu's intervention in the Thirty Years' War brought a cluster of ten cities in the imperial province of Alsace, not including Strasbourg, under French rule. More importantly, the 1648 Peace of Westphalia reconfirmed the federal constitution of the Empire, thus preventing the Habsburg emperors from extending their authority, and opening the

30 David Parrott, *Richelieu's Army: War, Government and Society in France, 1624–1642* (Cambridge, 2001), pp. 84–109.

way for French influence in Germany, notably along the Rhine and in Bavaria. Against Spain, however, the fighting that occupied the balance of Richelieu's life (he died in 1642) achieved no enduring territorial gains, despite protracted sieges in the Spanish Netherlands, Catalonia's rebellion against Spain (May 1640), and the revolt of Portugal (December 1640). The outcome of battles gradually started to shift in France's favor, notably after its imposing victory at Rocroi (May 1643). Spain's Philip IV, concerned about separatist revolt and military setbacks, was prepared to open the negotiations that ultimately led to the 1648 Peace of Westphalia. War between France and Spain continued, in part because Richelieu's successor, Cardinal Mazarin, still hoped to acquire the Spanish Netherlands. The 1659 Peace of the Pyrenees formalized France's gains in the interim: Artois in the north, Cerdagne and Roussillon in the south.

France's success was not due to a modernization of French institutions, but it did have a good deal to do with geography. Both branches of the Habsburg dynasty had far-flung territories, and thus multiple foes, and multiple priorities. France had the luxury of giving single-minded attention to its borders to the north, east, and south, where French forces could both extend the territory subject to the king's writ and discomfit the Habsburgs. For European history in general, the implication of this outcome was that it made the nation-kingdom, not the dynastic state embracing multiple peoples, the template for the future.

## Inter-state religious wars: The Holy Roman Empire, 1530–1633

The emergence of permanent religious schism in Europe from the middle decades of the sixteenth century was a huge factor in stimulating European warfare for the next century. State power might be turned on the heterodox local inhabitants of a territory to enforce conformity, and this was the main reason why so many wars of religion were civil wars, with confessional differences mapping on to preexisting discords: France's Wars of Religion (1562–98), the Dutch Revolt (1566–1648), the English Civil War (1642–51), and less familiar conflicts in Switzerland, Sweden, and Italy.[31] The Holy Roman Empire was a special case, because the contending parties were

31 Christian Bäder and Olivier Bangerter, *Kappeler Kriege, 1529–1531* (Au, 2003); William Bouwsma, *Venice and the Defence of Republican Liberty: Renaissance Values in the Age of the Counter-Reformation* (Berkeley, 1968); Paul Lockhart *Sweden in the Seventeenth Century* (London, 2004), pp. 9–13.

semi-sovereign states, and the key struggles here were both between reformist and Catholic conservative rulers and territories, and between the emperor and those he saw as challenging his authority by their religious choices. Charles initially sought to hammer out some sort of religious compromise in Germany, seeking to persuade the Catholic authorities to make doctrinal and organizational changes as the basis for a unifying settlement. Only when all those options were exhausted did he move to war to compel the Protestants of the Schmalkaldic League to accept his terms for a religious settlement. Even then, those who best understood the balance of forces in the Empire warned against belligerence. But Charles hoped that, by defeating the key Lutheran princes, John Frederick of Saxony and Philip of Hesse, he would be able to humble the Schmalkaldic League, and deprive France of allies in Germany. His diplomats lined up two key partners: Duke William IV of Bavaria and Duke Maurice of Saxony. Charles was taken off guard in 1546, when League forces, divining his intentions, blocked the pass by which Spanish troops were expected. Falling back on Ingolstadt, he and his men had to withstand a fierce cannonade before reinforcements arrived from the Netherlands. In the customary waiting game that ensued, Charles "held camp" longer than his foes, forcing Philip to submit. In April 1547 he invaded Saxony, with Maurice and Ferdinand, and caught John Frederick ill-prepared at Mühlberg. For his services, Maurice gained the Saxon electoral title from John Frederick, and many of the lands of the branch now condemned for taking up arms against the emperor.

Maurice in fact was one of a group of German princes who grew concerned that Mühlberg had given Charles too much direct power in the Empire. In 1552 he entered into secret negotiations with Henri II of France to obtain French subsidies, and ultimately a guarantee of French military support, in preparing resistance against Charles. Maurice (d. 1553) did not live to see the fruition of his policy, by which French troops occupied Lorraine and boosted a German Protestant revolt that drove Charles back into his Austrian lands. As negotiated by Charles's brother, Ferdinand, the 1555 Peace of Augsburg gave Lutherans the same rights in the Empire as Catholics, not just until the next Diet, as Charles had demanded, but in perpetuity.

According to one school of thought, the Peace of Augsburg merely "put a temporary end to open confessional warfare in Germany."[32] The text did

32 Quote in Geoffrey Parker, *The Thirty Years' War* (London, 1988), p. 18; different perspective on Augsburg in Peter H. Wilson, *Europe's Tragedy: A History of the Thirty Years War* (London, 2009), p. 43.

contain the seeds of future conflicts, as in its reestablishment of the rights of Catholic minorities in the eight Protestant cities. More importantly, the "Ecclesiastical Reservation" decreed by Ferdinand was never accepted by Protestants as part of the agreement; it required prince-bishops and any other Catholic clerics who became Lutheran after 1555 to resign their ecclesiastical offices and any territorial rights that these conveyed. Also, since the Peace granted no rights to Calvinists, the conversion to Calvinism of the Palatinate Elector, Frederick III (d. 1573), created fresh contentions. Finally, the decrees of the Council of Trent (1547–63) combined with Jesuit preaching to stiffen the spines of Catholic rulers; many developed plans to re-Catholicize their lands. Yet recent scholars argue that the Empire was able to absorb these tensions. This was partly due to the Ottoman threat; despite unmet Protestant demands, the Diet voted its largest grants at times of confessional confrontation.[33] Also, Ferdinand I and Albert V of Bavaria formed a bi-confessional alliance to maintain the peace.[34] Finally, the Wettin–Habsburg entente prevented the strongest Protestant state, the Saxon Electorate, from joining any combination against the emperor.

The first real tremors of what was to become a major international conflict fought for both political and religious objectives came in 1609 with the death of Duke John William of Cleves-Jülich without an heir. French and Dutch troops marched in to block a possible land-grab by Rudolf II (d. 1612) and to prevent a Catholic, pro-Habsburg succession. Religious lines were drawn after the two claimants embraced, respectively, Calvinism and Catholicism. However, the assassination of Henri IV in Paris by a Catholic radical on the eve of his departure to join his army took the momentum out of the build-up to conflict. Faced with the reluctance of their backers to escalate the war, the two agreed to a partition of the disputed lands.

Vienna was again the capital of the Habsburg monarchy under Emperor Matthias (r. 1612–19) after the decades when Rudolf resided in Prague. His chief minister, Cardinal Melchior Klesl, kept up an amicable correspondence with Protestant princes, but he also favored Catholics for crown appointments, including the Council of Regents in Prague. Matthias was childless, and Spain's Philip III had a claim to succeed him. This matter was regulated

33 Winfried Schulze, *Reich und Türkengefahr im späten 16. Jahrhundert. Studien zu den politischen und gesellschaftlichen Auswirkungen einer äußeren Bedrohung* (Munich, 1978), pp. 80–87.

34 Wilson, *The Thirty Years War*, pp. 200, 209; Maximilian Lanzinner, "Der Landsberger Bund und seine Vorläufer, 1555–1599," in Volker Press (ed.), *Alternativen zur Reichsverfassung in der Frühen Neuzeit?* (Munich, 1995), pp. 65–74.

by treaty in 1617: Matthias would be succeeded by his nephew, Archduke Ferdinand, and Spain would receive Habsburg lands in Alsace. At the same time, Ferdinand visited Prague to claim the Bohemian portion of his inheritance: he confirmed Protestant privileges, and was elected king by the estates. But there was tension over Matthias's restrictive interpretation of the Letter of Majesty, the articles of toleration granted by his predecessor, Emperor Rudolf II, in 1609. On May 22, 1618, a radical party among the Protestant Bohemian nobility, led by Count Matthias Thurn, reenacted the deed that began the Hussite rebellion 200 years earlier: two Catholic Regents were "defenestrated" from the council chamber of Hradčany Castle. A Protestant assembly now declared itself to be the Diet, and raised an army for Thurn. But no major prince recognized the rebel government, save for the Calvinist Frederick V of the Palatinate. In the imperial election of April 1619, representatives of the Electors of Saxony and Brandenburg cast their votes for Archduke Ferdinand. In August, the five Bohemian provinces renounced the election of Ferdinand, and chose Frederick V of the Palatinate as their new sovereign. His reception in Prague (December 1619) was less than joyous: Calvinists in his entourage smashed statues that were respected by Lutherans as well as Catholics. But Ferdinand was paralyzed by simultaneous Protestant rebellions in Upper Austria and in Hungary, which was occupied by Gábor Bethlen of Transylvania (r. 1613–29). Maximilian of Bavaria came to his rescue, pledging to mobilize an army funded by Bavaria and other Catholic German states to reconquer Upper Austria and Bohemia; in return, he would receive the electoral title at present held by Frederick V and hold Upper Austria until his war expenses were paid. John George of Saxony undertook to occupy Lusatia (northern Bohemia) on similar terms. To forestall an attack on Bavaria from the west, and to strengthen Spain's communication routes through the Empire, Philip III was persuaded to send troops from the Spanish Netherlands into the Palatinate. At White Mountain, east of Prague, a Catholic army under Jean t'Serclaes, Count Tilly, charged uphill in the mist to rout the forces of Christian of Anhalt (November 8, 1620). In the aftermath, Prague opened its gates to the imperial troops, rebel leaders were put on trial, and Czech noble lands were redistributed to Catholic soldiers of many nations.

Protestant courts were unnerved by this Catholic–Habsburg triumph. But Christian IV of Denmark and England's James I (father-in-law of Frederick V) limited themselves to providing aid to Tilly's chief remaining foe, Count Ernest of Mansfeld, until he disbanded his men in 1624. In 1625 Christian IV entered the fray in his own right, and compelled the Lower Saxon Circle to raise additional troops. But Tilly was ill-prepared to fight, and John George of Saxony got both sides to begin negotiations. Then in 1626 Tilly was joined by Count Albrecht Wallenstein, a Bohemian military enterpriser who engaged to maintain an army at no direct cost to Ferdinand II, thus freeing the emperor from dependence on the Catholic League. Wallenstein and his Dutch financier, Hans de Witte, perfected the "contribution system": towns bought peace by promising payments in cash or in kind, and the inevitable time-lag was bridged by loans raised by de Witte's agents in cities across Europe. Mansfeld, back in harness, drew Wallenstein off by invading Silesia. But Tilly defeated the Danes near Goslar (August 1626), and when Wallenstein again joined him they had no difficulty occupying Holstein. Though Danish and Swedish action stopped Wallenstein from creating a Baltic naval base (1628), the Danes were eventually forced to terms (Treaty of Lübeck, January 1629).

Flush with new triumphs, Ferdinand took a step that gratified Catholic maximalists (like his Jesuit confessor), but alienated moderate Catholics and his key Lutheran ally, John George of Saxony. The Edict of Restitution (March 6, 1629) decreed that Church lands that had become Lutheran after 1555 (contrary to the Ecclesiastical Reservation) had to be returned to Catholic control. In Regensburg (1630), the Electors showed their displeasure by refusing to elect Ferdinand's son as king of the Romans, and by forcing him to dismiss Wallenstein.

Meanwhile, Gustavus Adolphus landed a new army in north Germany, having been extricated by French and Dutch diplomacy from an inconclusive struggle with Poland-Lithuania over control of Prussia. The manifesto he issued justified his invasion as a defense of the Empire's traditional constitution, not as a religious crusade; this was so as not to offend the sensibilities of a potential ally, France's Louis XIII. In fact, since Lutheran princes meeting in Leipzig declared their neutrality, Gustavus had to accept the offer of a lucrative subsidy from France (Treaty of Bärwalde, January 1631). The king and his capable chancellor, Axel Oxenstierna, used other means to rally support among co-religionists. The Protestant cause probably benefited even more from events at Magdeburg in May: Tilly's siege of the archiepiscopal city, which had declared for Sweden, ended with a brutal sack and a fire in which some 20,000 perished.[35] Needing the support of Brandenburg and Saxony, Gustavus threatened Berlin in June; Elector George William

35 Hans Medick and Pamela Selwyn, "Historical Event and Contemporary Experience: The Capture and Destruction of Magdeburg in 1631," *History Workshop Journal* 52 (2001), 23–48.

now took the side of Sweden, and John George of Saxony soon followed suit. At Breitenfeld, near Leipzig, a Swedish–Saxon army confronted Tilly. Gustavus's light artillery and the cohesion of his infantry led to a brilliant victory for "the Lion of the North."[36] Swedish forces now pushed south and west. Not even Maximilian I could defend his land, and Gustavus entered Munich in May 1632. Wallenstein, recalled by Ferdinand II, expelled the Saxons from Bohemia, and headed for Nuremberg, the main Swedish base. After the costly failure to dislodge Wallenstein from his armed camp outside Nuremberg in the summer of 1632, and in order to protect his communication with the Baltic, Gustavus struck out for Saxony. On November 16 his army engaged with Wallenstein at Lützen, near Leipzig. After an exceptionally hard-fought struggle, Wallenstein's men quit the field first, but Gustavus lay among the dead.

In the ensuing months, Ferdinand II signaled willingness to bend by revising the Edict of Restitution to make 1612 the starting date for Catholic reclamations. John George sought to form a party that would declare neutrality, but Oxenstierna preempted him by announcing (April 1633) a renewal of the Treaty of Bärwalde: with French subsidies, Sweden and its allies would fight on until assured of suitable compensation. From this point on, other powers too were mainly concerned to salvage something tangible for all the blood and treasure they had expended. In the end the Peace of Westphalia allotted to Sweden a large indemnity to pay off its troops, plus Western Pomerania and the prince-bishoprics of Bremen and Verden. Electoral Saxony gained the hitherto Bohemian province of Lusatia. Bavaria retained its electoral title, together with the Upper Palatinate. An eighth electoral title was created for Frederick V's son and heir, Charles Louis, who was restored to a devastated Rhineland Palatinate. Emperor Ferdinand III (r. 1637–57) had to be content with a loyal and now mostly Catholic Bohemia, and 1624 as the date beyond which any Church lands that had been secularized had to be returned to Catholic control.[37]

The settlement of 1648 has often been presented as "bringing peace by taking religion out of politics." But many recent studies have emphasized the presence of religion in European politics and diplomacy during the eighteenth and nineteenth centuries, not its absence. Still, Germany's religious

36 Wilson, *Europe's Tragedy*, pp. 470–77; Lockhart, *Sweden in the Seventeenth Century*, pp. 52–53; Parker, *The Military Revolution*, p. 24.

37 Konrad Repgen, "Negotiating the Peace of Westphalia: A Survey with an Analysis of the Major Problems," in Klaus Bussmann and Heinz Schilling (eds.), *1648: War and Peace in Europe*, 3 vols. (Münster and Osnabrück, 1998), vol. 1, pp. 355–72.

conflicts did have long-term consequences. In the 1540s, Mary of Hungary had warned Charles against deploying an army of mercenaries to achieve a religious goal: "I know not whether God wishes to reduce the heretics to obedience by men whose lives are so execrable." If proof were needed of how wars in the name of religion (like any other wars) unleashed the demons of unmitigated force, it was to be found in Wallenstein's protection-racket, fastened on great swaths of Germany with the backing of anonymous investors. It does seem that the Thirty Years' War "discredited the use of force to obtain confessional or political objectives within the Empire."[38]

***

From about 1500 to about 1650, most Continental states with a capacity to make war became involved in one or more of the patterns of conflict discussed here. There has not been space for more than a few references to connections between the various strands of war, but it should be clear enough that a state's decision to make an ally or to go to war often had implications in more than one sphere (for example, a government that aligned itself with France might also find it easier to get trade concessions at the Sublime Porte). If this chapter has given readers a more informed sense of the multiple priorities that early modern rulers and statesmen had to balance in their minds, it will have served its purpose.

38 For Mary of Hungary, Tracy, *Emperor Charles V*, p. 205. Wilson, *Europe's Tragedy*, pp. 758, 759 (the quotes). Cf. Dale van Kley, *The Religious Origins of the French Revolution: From Calvinism to the Civil Constitution of the Clergy, 1560–1791* (New Haven, 1996).

6

# The evolution of naval warfare, 1450–1650

NICCOLÒ CAPPONI

Alfred Thayer Mahan's *The Influence of Sea Power upon History* was undoubtedly a quintessential by-product of an age that believed in universally applicable rules, in this case that a navy's function was the same in the seventeenth as in the nineteenth century, the command of the sea its ultimate goal. Naturally enough, over the years Mahan's sweeping theoretical framework has received its share of criticism. In his 1911 book *Some Principles of Maritime Strategy*, Sir Julian Corbett argued that it was more important to deny to one's opponent the command of the sea, rather than seize it for oneself, so that 'the enemy can no longer attack our lines of passage and communication effectively, and that he cannot use or defend his own'.[1] Taking a leaf from Corbett, John F. Guilmartin underscored how Mahan's principles cannot be applied to the early modern Mediterranean, the physical conditions of the area defying the paradigms applied by Anglo-American naval historians to the oceanic world.[2] Besides, even if Mahan did acknowledge the importance of weaponry, Geoffrey Parker has pointed out that *The Influence of Sea Power upon History* 'contained no discussion of guns, sails or ship design, because the author did not believe that changes in these things could affect the application of strategic principles'.[3]

While such criticisms are justified, there is a risk of substituting Mahanian determinism with technological, economic, cultural, or philosophical determinism – to name just a few – in explaining the evolution of naval warfare between the mid-sixteenth and the mid-seventeenth centuries. Inevitable, perhaps, since, like Mahan, historians find comfort by working within

1 Julian Corbett, *Some Principles of Maritime Strategy* (London, 1911), p. 105.
2 John Guilmartin, *Gunpowder and Galleys: Changing Technology and Mediterranean Warfare at Sea in the 16th Century*, 2nd ed. (London, 2003), pp. 31–51.
3 Geoffrey Parker, *The Military Revolution: Military Innovation and the Rise of the West, 1500–1800*, 2nd ed. (Cambridge, 1996), p. 82.

a pattern. One should also add that all of the above-mentioned *-isms* were certainly factors, to a greater or lesser degree, in shaping the maritime strategies of different states during the early modern period.

Still, to understand what certain countries did, or wished to do, with their navies we need to consider what the latter could, or could not, accomplish. In 1540 the Ottoman grand vizier Lütfi Paşa would warn Sultan Suleyman I that 'however important may be the business of land affairs, those of the sea are far more important'; but, as much as they may have agreed with Lufti's argument, Suleyman and his successors had to deal with an essentially army-oriented administration.[4] In 1556 Piero Machiavelli could argue that 'the affairs of the sea are two thirds of the game'; yet in order to win, a nation needed to have the right naval cards and – more importantly still – know how to play them.[5]

By the end of the fifteenth century, guns had changed the face of European sea fights, with ships evolving into a weapons system and naval gunfire becoming increasingly effective. Shipbuilding had also considerably improved: sailing vessels were now bigger, sturdier and, thanks to full-rigging, more manoeuvrable than their predecessors. This meant smaller crews, while increased dimensions allowed for a greater number of firearms, thus augmenting the vessel's offensive and defensive abilities. However, the preferred warship remained the oar-driven galley, a lighter but more versatile craft not constrained by winds in its movements, although the low sides made it more susceptible to rough seas. A standard galley was conditioned by seasonal weather, its operative capacity limited to the months between March and November. On the other hand, a galley could mount big guns below its *arrembata* (*arrumbada* in Spanish, a low forecastle covering the main battery) and direct them against an enemy by use of oars. Therefore, in the sixteenth and seventeenth centuries it was not unusual for a galley to tackle a fully rigged sailing ship with a large artillery complement and emerge victorious from the match.

Whether driven by wind or by human muscle, at the end of the fifteenth century warships were seen as floating castles, as they appear somewhat fancifully in the Italian humanist Roberto Valturio's *De Re Militari*. Valturio backed his theories with the recognised authority of numerous ancient

4 Lütfi Paşa, *Das Asafnâme des Lutfi Pascha*, ed. and trans. Rudolf Tschudi (Berlin, 1910), pp. 31–2.
5 Biblioteca Nazionale Centrale, Florence, *Magliabechi*, E. B. 15.10 (Memorandum by Piero Machiavelli to Cosimo I de' Medici, *c.* 1556), f. 176r.

Greek and Roman authors.[6] As for any fortress, ships needed to be besieged and taken by storm or defended in an equally determined hand-to-hand struggle. Philippe de Cleves, writing around 1516, saw it this way, while emphasising the role played by artillery in 'softening up' the personnel of an enemy ship before boarding it.[7] Still, Cleves and others understood that, important as it may have been to take a ship, it was equally vital to avoid being boarded in turn. The solution to this problem was threefold: (a) try to eliminate the opponent's crew; (b) attempt to disable the adversary's ships; (c) keep the attackers at a distance by a superior display of weaponry.

The decision to adopt one, or all, of these options could be a matter of tactical doctrine or simply opportunity, triggered by factors sometimes as prosaic as a situation of numerical inferiority. At Jeddah, in 1517, the Ottomans used a combination of galley-mounted guns and artillery from coastal fortifications to repulse a numerically superior Portuguese fleet. In 1557 the interloper William Towerson and a French associate encountered a group of five Portuguese ships, including three caravels and one 'barke', off the coast of Guinea. In the ensuing battle the Portuguese managed to fend off the attackers by skilful use of artillery, targeting upper works, masts, and hulls. With smaller vessels gunfire could be devastating, as demonstrated by Vasco da Gama's 1502 encounter with an Indian fleet of substantial size off the coast of Malabar. On this occasion, da Gama's squadron of eighteen ships of various sizes managed to send a good portion of the enemy fleet to the bottom of the ocean, Portuguese artillery playing havoc with the lighter Indian craft. The recorded sinking of a vessel with gunshot had already happened at the battle of Zonchio in 1499, a Venetian ship of '700 barrels' (roughly 450 tons) being riddled with balls, the Venetian historian Marino Sanudo adding bitterly 'more from our bombards than the Turks', the shots intended for the Turks' – one of the first known examples of a loss at sea due to friendly fire. The clash, however, was a traditional hand-to-hand fight, all other losses were caused by the blaze that destroyed three carracks, two Venetian and one Ottoman, that had grappled together.[8]

In the early sixteenth century, the sailing ship appeared to be prevailing over the oar-powered galley as the primary war vessel. The former's alleged advantages were its holding capacity and the number of guns it could carry,

6 Roberto Valturio, *De Re Militari* (Paris, 1534), p. 297. Valturio's work, written in the 1450s, was first printed in Verona in 1472.

7 Philippe de Clèves, *L'Instruction de toutes manières de guerroyer sur mer*, ed. Jacques Paviot (Paris, 1997), pp. 49–55.

8 Marino Sanudo, *I diarii di Marino Sanuto*, ed. R. Fulin, 58 vols. (Venice, 1880), vol. 3, p. 15.

the invention, around the end of the fifteenth century, of the water-tight gunport allowing for artillery to be placed on the lower deck. The smaller crews of sailing ships resulted also in reduced operating costs: just the rowing element of a twenty-four-bench standard galley with the system *alla sensile* (requiring three rowers per bench, each pulling one oar) totalled 144 men, while a carrack (a three- or four-mast square-rigged ship) could operate with as few as one man for every 8 tons of burden. To make matters worse, a change in culture had turned the galley oarsman, once a respected professional, into a pariah. Even in Venice, where rowing on a bench did not carry the same social stigma as in other places (the Ottoman Empire was another exception to this trend), it was becoming difficult to find an adequate number of people willing to row in a galley, security concerns limiting the numbers of convicts and slaves that in increasing numbers were being sent to the oars. Sailing vessels, therefore, appeared a viable alternative.

However, for the best part of the sixteenth century, galleys held a number of crucial advantages. For one, the disposition of the main battery at the prow, with enough recoil room for the main central gun, meant that a galley could carry ordnance of considerable weight, heavier than what most other crafts could carry. Besides, a galley relied on human muscle as well as on the wind to move, whilst sailing ships necessarily needed at least a breeze. This would prove a serious handicap in working together with rowing vessels: in the campaign following Zonchio, the Venetian commander would complain about the inability of the sailing ships to keep up with the rest of the fleet. The unfeasibility of sailing ships operating in conjunction with galleys would also be underscored later by the Genoese Antonio Doria, who in a 1539 memorandum to the Emperor Charles V unequivocally stated that:

> Considering the expense of having a mixed fleet of galleys and carracks (*navi*), and given the difficulty, nay the impossibility, that they work in conjunction, and because the enemy [the Ottomans] is strong in galleys, it can retreat, spread or close its ranks employing greater opportunities than those given by the winds and the seas, I advise skilled captains to employ only galleys for an expedition.[9]

The Ottoman *kapudan-i deryâ* (roughly: Grand Admiral) Hayreddin Barbarossa shared the same opinion, there being little wind during the summer months in the Mediterranean. As late as the Cyprus War of 1570–2, Sebastiano Venier, leader of the Venetian contingent at the battle of

9 Bibliothèque National de France, *Ms Italien 210*, 'Discorso di Antonio Doria sopra le cose turchesche per via di mare' (1539), f. 189v.

Lepanto, would oppose the decision to add sailing ships to the Venetian fleet, believing them too vulnerable to enemy action if becalmed.

Venier's misgivings were grounded in bitter experience. At the battle of Preveza (1538), during the war of the first Holy League against the Ottomans, the allies sported several armed sailing ships, with two Venetian carracks placed in front of their array so as to disrupt the Ottoman formation with heavy artillery fire. Forced to retreat after an inconclusive blockade of the gulf of Preveza, the allies were caught by the enemy in a state of disorder and with their becalmed sailing ships well to the rear. In the ensuing fight their gunfire kept the assailants at bay, but two other Venetian ships went up in flames and sank, and two Spanish ones were taken after a fierce fight – the other sailing vessels saved only by a providential rising wind. Venice drew its own conclusions from the Preveza debacle. In 1545 the Venetian Senate made dispositions for the creation of a force in readiness of 100 normal galleys and 4 large ones (*galie grosse*); significantly, no sailing ships were included. A few years later Venice again considered adding them to its fleet; unfortunately, the heavily armed galleon completed in 1559 sank on its maiden voyage, when badly anchored artillery bunched together caused it to capsize.

Such events, and the unsatisfactory performance of the carracks at Preveza, pushed other Mediterranean states to concentrate on the galley as their preferred war vessel. In the mid-sixteenth century the Ottomans still employed sailing crafts, but, devoid of ordnance, their role was limited to trade, transporting siege guns and horses, or providing logistical support to their armies or oared fleets. Indeed, at this stage the ultimate triumph of the sailing ship was in no way a given, even in the Northern European countries traditionally associated with its development. The French made extensive use of galleys during the 1543–6 war against England, and in turn the English built several galleys of their own to counter the threat. Although more suited to the calm waters of the Mediterranean, in the right circumstances galleys could function equally well in the seas of Northern Europe, as witnessed during the War of the Spanish Succession by the French convict oarsman Jean Marteilhe:

> In the year 1704, we were stationed in the port of Ostend, to observe a Dutch squadron, who cruised at the mouth of the harbour. Whenever the weather was calm, we sallied out, and cannonaded them at a distance, as our guns carried much farther than theirs; and thus continued to harass them till the rising wind always compelled us to retire.[10]

10 Jean Marteilhe, *The Memoirs of a Protestant Condemned to the Galleys of France for the Sake of His Religion*, trans. James Willington, 2 vols. (Dublin, 1765), vol. 1, pp. 65–6.

But, manoeuvrable and heavily armed as a galley might be, it was handicapped by its reduced storage capacity, necessitating a string of logistic facilities – whether to replenish its water supply and food stocks or substitute empty gunpowder barrels – in order to function effectively. All this limited a galley's operational range, forcing it to stay in the vicinity of supply centres and fresh-water sources. Besides, despite its often formidable prow-mounted gun-battery, a galley could only carry a limited number of artillery pieces, technical constraints allowing, at best, for the mounting of small-calibre swivel guns to protect its flanks. These could be quite sufficient in fending off enemy combatants attempting to board, but not another vessel. Therefore, the galley was denied one of the chief advantages of the sailing ship: broadside artillery; although in the mid-sixteenth century the smaller guns mounted along sailing ships' sides, whether on the upper or lower deck, did not have the same destructive force as the larger pieces fitted at the stern or aft.[11] Naval engineers were therefore faced with a riddle apparently impossible to solve: building a vessel with an all-round firepower capacity, logistically self-sufficient and capable of moving in the absence of wind.

It would be up to the Venetians to square the circle. At the time of the battle of Zonchio, Venice had already converted large galleys – once freighters – into warships, their size allowing them to house bigger guns and a larger fighting contingent. Still, these were just overgrown galleys – later known as *bastarde* – with only slightly more punch than regular ones, and in the 1520s and 1530s the Venetians built a number of experimental vessels to replace these *galie grosse*. Despite high hopes, the results were meagre. Nevertheless, the *quinquireme* designed by the humanist Vettor Fausto, drawing on examples found in ancient texts, paved the way for future developments. The Venetians were not the only ones trying to blend the advantages of the galley with those of the sailing ship. Under Henry VIII the English built such hybrid vessels called galleasses, though, according to one author, they 'possibly suffered from the same disadvantages of the carrack, without much mitigation, and were rapidly becoming obsolete'.[12] Clearly, a different approach was needed to find a satisfactory solution and the Venetians had the obvious answer. But, as often happens in history, it took some time for them to realise it.

11 Nicholas Rodger, 'The Development of Broadside Gunnery, 1450–1650', *The Mariner's Mirror* 82 (1996), 304.

12 David Loades, *The Tudor Navy: An Administrative, Political and Military History* (Aldershot, 1992), p. 95.

Structural problems stood in the way of increasing a galley's ordnance. A bow chaser even of modest dimensions needed sufficient space to recoil. In a galley this problem was solved by mounting the piece on a sleigh, allowing it to travel backwards into the craft's hold and ending at the mainmast. The other ordnance below a galley's *arrembata* was fixed to a platform attached to the hull, with the vessel itself absorbing the recoil. This implied, though, that at most medium-sized pieces could be mounted on the ship's prow; moreover, these could only shoot balls, of stone or metal, requiring reduced charges, lest the guns burst, shatter the main structure, or slam the guns into the rowers' benches. The same principle applied to the artillery on the gunwales, swivel guns loaded with grapeshot for anti-personnel use. In round ships there was enough room to allow the stern-chasers to recoil along the deck; broadside guns, on the other hand, suffered from space constraints and it is not clear if they were fixed when firing and then pulled back, outboard loading being commonly practised at the time.

Which type of projectiles a gun would shoot was a matter of tactical philosophy. Shot could be made of stone, or cast metal (lead, bronze, or iron). Metal shot was easier to produce but, with a higher specific weight than a stone ball of the same calibre, necessitated heavier guns and more powder. While stone shot could be employed against vessels, it was preferable as an anti-personnel weapon. This was the tactic favoured by the Ottomans, 'considering that by shattering into pieces, smashing the timber and sending splinters flying through the air they [stone balls] do more damage than iron balls, that pass clean through the wood'.[13] Properly directed, a stone cannonball could have a devastating effect on a galley and on its crew. The historian Paolo Giovio's account of the battle of Capo d'Orso (1528) described in the following terms an encounter between Filippino Doria's Genoese squadron, supporting the French attempt to take Naples, and a Spanish force led by Don Ugo de Moncada:

> Count Filippo [Doria] who in this [artillery fire] placed great care, as well as being a crack shot, discharged against the enemy his large artillery piece called the *Basilisk*. The terrible ball, smashing through the ram and the forecastle [of Moncada's galley] with a horrific slaughter of men, travelled from bow to stern across the bridge with such violence that, having already killed more than thirty between soldiers and sailors, it slew many honourable

13 Luigi Firpo (ed.), *Relazioni di ambasciatori veneti al Senato. Tratte dalle migliori edizioni disponibili e ordinate cronologicamente*, vol. 13: *Costantinopoli, 1590–1793* (Turin, 1984), p. 622.

people on the poop deck . . . so that the Marques of Vasto and Don Ugo were spattered with their blood and guts.[14]

Because of the high quality of their soldiers, the Ottomans relied heavily on boarding and therefore deemed it important to thin the opponent's ranks with individual firearms, bows, and incendiaries. Emphasis on such weapons allowed them to equip their vessels with artillery of lesser quality – at least according to Venetian documents about captured Ottoman pieces – but perfectly adequate for the task of decimating enemy crews.[15] To counter the threat of being boarded, Venetian galleys carried numerous swivel guns, in part also to compensate for their usually small contingent of fighting men. Yet, anti-personnel artillery was only partially helpful against a numerically stronger opponent, committed to battle and far better trained in hand-to-hand combat; nor was the Venetian custom of making oarsmen fight alongside soldiers and marines much of an advantage, since Ottoman rowing crews were trained to do the same. In short, if the Venetians wished to redress the balance, they had to transfer the focus from the men to the vessels.

This represented a major cultural shift. For centuries captured enemy ships had been the accepted proof of success in naval combat. Besides, taking a craft and its crew meant a gain for the winner, whether financial or material: everything to the last nail could be sold for a profit, while prisoners were priced as slaves or ransomed. Minus enslavement – except in the Mediterranean, due to very particular circumstances – this type of warfare would be a constant feature of privateering up to the mid-nineteenth century, while prize money for crews who took an enemy ship was still being awarded by the Royal Navy in 1918. A Venetian galley owner similarly approached warfare with an entrepreneurial mentality. But in Venice the line between public and private interests was blurred, to say the least: the mantra that what was useful or detrimental in the private sector would likewise impact the public one was constantly echoed in Venetian council chambers. This attitude affected naval affairs, which is hardly surprising given that, although it possessed a large swath of the northern Italian mainland, sixteenth-century Venice still drew a substantial amount of its livelihood from the sea. The imminent change of perspective in naval warfare did not, therefore, affect war gains as much as it did how they were conceived: if private and public

14 Paolo Giovio, *Delle istorie del suo tempo di Mons. Giovio*, 2 vols. (Venice, 1564), vol. 2, p. 55.

15 Archivio di Stato di Venezia (ASV), *Consiglio dei Dieci* (*CX*), *Parti Comuni*, Reg. 30, f. 156v (28 Nov. 1572). ASV, *CX*, *Parti Comuni*, filza 122, n.n.f. (25 Feb. 1574).

interest coincided, then it was to everyone's profit if enemy ships went to the bottom rather than being taken.

The Venetians did not create something *ex nihilo*; rather, they improved on what they had at hand. It is unclear who, between 1567 and 1570, initially had the idea of rebuilding a number of large merchant galleys and turning them into gun-bristling platforms. But the real heroes of the whole enterprise are known: Domenego Grande, *sottoproto dei marangoni* (assistant foreman) of the Arsenal, who managed to 'make the forecastle part of the quickwork, while before it had belonged to the deadwork', thus allowing for the mounting of heavy guns; and Antonio Surian, who invented a way to place artillery pieces on the vessels' sides, without them recoiling into the rowing benches.[16] While uncertainty surrounds the exact workings of Surian's device, there is no doubt it gave the *galleasses*, as the refurbished crafts came to be known, a powerful broadside battery. The sort of punch that Venetian galleasses could pack would be revealed at the battle of Lepanto (7 October 1571), when their artillery fire disrupted the Ottoman formation in the opening phases of the engagement, sinking or disabling perhaps as much as a third of the enemy fleet through a combination of fire from the bow, stern, and broadside pieces. On at least one occasion during the fight the galleasses' presence helped to save the day for the Holy League; indeed, it was recognised at the time that, had it not been for the new Venetian craft, the outcome of this close-fought battle could have been very different.

Immediately after Lepanto there was a rush to copy the new Venetian wonder-weapon – with mixed results. In the months following the battle the Ottomans converted several *mavna* merchant vessels to carry an increased weight of ordnance and with all-round firing ability, yet they erroneously assumed that the Venetian galleasses were simply modified *galie grosse* and not the result of decades of experiments and trials. Consequently, the Ottoman efforts in copying the galleasses were dismissed, with reason, as ineffectual by informed Western commentators.[17] Likewise, Tuscany's attempts to imitate the Venetian craft ended in a miserable failure, despite all efforts, and crippling expenditure, including the building of a dedicated arsenal in Portoferraio on the Isle of Elba. Only the Spanish (in reality the Neapolitans, Naples being

16 ASV, *Collegio, Suppliche da dentro*, 5, f. 205r. ASV, *Senato Mar*, 226, n.n.f (Report of the *Provveditori alle Artiglierie*, 20 Aug. 1616).

17 Colin Imber, 'The Reconstruction of the Ottoman Fleet after the Battle of Lepanto', in *Studies in Ottoman History and Law* (Istanbul, 1996), pp. 87–8; Eugenio Albèri (ed.), *Relazioni degli Ambasciatori veneti al Senato raccolte, annotate ed edite da Eugenio Alberi*, 3rd series (Florence, 1855), vol. 3, p. 222.

a Spanish viceroyalty at the time) appear to have mastered the art of fabricating the new vessels: their seaworthiness, manoeuvrability, and firepower greatly impressed the English during the Armada campaign of 1588 to the extent that they considered adopting them, albeit in a modified form. The Spanish galleasses' impressive performance may have stemmed from the fact that they were square-, rather than lateen-, rigged and used the rowing system *a scaloccio* (four or more men in a bench pulling a single oar), which meant more muscle to drive the craft. On the other hand, available evidence suggests that Neapolitan galleasses had a less powerful battery on the broadside than the Venetian ones. It is possible that the Spanish were unable to copy Antonio Surian's invention, or that, like all those who imitated the original design, they saw the galleass simply as a glorified galley with more guns, overlooking the main reason behind the crafts' original conception.

Paradoxically, whilst everyone was busy copying the Venetians' new weapon, the Venetians themselves were the first to do away with it. For all its worth, the Venetian galleass had several drawbacks that limited its value as a fighting vessel. The galleass's original crew amounted to a mere 165 men rowing *alla sensile*, adequate for a twenty-six-bench *bastarda* galley but insufficient for a considerably bigger craft. This made the *Serenissima*'s galleasses slow, needing to be towed by other vessels when in a fleet (Neapolitan galleasses suffered from no such limitation, on one occasion during the Armada campaign actually towing a slower sailing ship). The Venetian galleass, despite its formidable hitting-power, was thus ill-suited to chase ships and depended upon enemy vessels coming within gun-range in order to be effective. Unfortunately for the Venetians, the Ottomans had learnt to stay away from the galleasses' artillery and, under the able leadership of the *kapudan-i deryâ* Uluç Ali Paşa, they successfully adopted a hit-and-run strategy in their naval operations in the year following Lepanto, turning tactical draws into strategic victories. It is therefore unsurprising that once the Venetians made peace with the Porte in 1573 they immediately mothballed the galleasses, leaving them there for the next thirty or so years, reverting to simple galleys for their war fleet.

Though introduced in the Mediterranean, the novel concept of destroying or disabling of enemy ships with long-range gunfire, rather than capturing them with boarding parties, would be honed in the seas of Northern Europe. Yet once again, the shift in mentality necessary to break the deadlock of a disadvantageous situation would take time to catch on, in the face of entrenched traditional practices. By the end of the sixteenth century long-range exchanges between ships were no novelty in naval warfare, as the above-mentioned 1557 encounter off the Guinea coast suggests. However, gunfire was still primarily intended as a deterrent, a prelude to boarding, or as an attempt to intimidate the opposing crew into surrender. Nor was the use of firepower to sink enemy vessels new. The Portuguese had first developed the technique of impressing a ricocheting effect on the ball by shooting into the sea at a certain angle and hit the enemy craft at the waterline – a technique the Italians called *il tiro di ficco* – although some contended that this caused the shot to lose too much speed and hitting power to cause damage of any significance to even medium-sized ships. By 1588 the English had perfected the skill of sinking vessels by hitting them at the waterline, so much so that king Philip II felt obliged to warn the duke of Medina Sidonia, in charge of the Armada, to take appropriate countermeasures.

As the prospect of a conflict between England and Spain loomed steadily larger in the two decades following Lepanto, the English were faced with an apparently insoluble problem. Given the fighting skills of Spanish soldiers they stood little chance in a traditional boarding-style maritime battle against the Spanish fleet, just as the Venetians had understood when facing the Ottomans. It is tempting to think that Elizabeth I's shipbuilding policy was influenced by the outcome of the sea fights during the Cyprus War – printed works describing them abounded and the Tudor regime kept a keen eye on what was being published in Europe. In fact, naval operations during the Armada campaign of 1588 contained elements that can be traced back to the previous conflict in the Mediterranean, the English employing a combination of hard-hitting artillery, long-range gunfire, and harassing tactics. This synergy was possible thanks to a well-planned shipbuilding programme, starting in 1573 with the launching of the *Dreadnought*: a 700-ton galleon (a four-mast sailing ship combining the low forepart of a galley with the high stern of a sailing ship), the first of several craft whose size, length, and sleekness allowed for faster sailing and greater manoeuvrability than any foreign vessel. Additionally, *Dreadnought*-type galleons' displacement and longer decks allowed for bigger guns and an increased weight of the broadside, even if the heavier guns were probably the ship-smashing bow and stern chasers.

Bigger and better-quality artillery pieces found their way onboard English galleons. A 1576 inventory of naval ordnance shows an abundance of large-calibre pieces, in particular bronze culverins and demi-culverins. Given that these types of guns had a superior range over cannons of the same calibre – as confirmed by research done by the Venetians in the seventeenth century – they were clearly intended to inflict as much damage as possible to an enemy ship from a distance. In 1591, the lower-deck broadside of the *Revenge* (Sir Francis Drake's flagship during the Armada campaign) included ten culverins and four demi-culverins; the total weight of shot delivered by these guns being as devastating as the Venetian galleasses' broadside at Lepanto.[18] The English had managed to combine strength with swiftness, and the heavy ordnance carried by their galleons during the Armada campaign would repeatedly allow them to pound the outgunned Spanish with all-round artillery fire. Though only one Spanish galleon was sunk during the battle of Gravelines (8 August 1588), the damage inflicted by English guns on Medina Sidonia's ships disrupted Philip II's plans to invade England for good.

Writing sometime in the 1630s, the Elizabethan sea-dog Sir William Monson would list the Armada campaign together with the battles of Ponta Delgado (1582) and Lepanto as the most important naval clashes in recent history. Significantly, the outcome of all three encounters depended on a judicious use of gunfire, or a lack thereof. It was probably at Ponta Delgada that the English learnt to keep the Spanish ships at a distance, after the flagship of Filippo Stozzi, commander of the French expeditionary force, succumbed to boarding, despite inflicting considerable artillery damage to enemy vessels. The English were quick to take notice and during the Armada campaign they used the superior seaworthiness of their ships to avoid being grappled. Some thirty years after Ponta Delgada, Sir Walter Raleigh commented that: 'To clap ships together, without consideration, belongs rather to a mad man, than to a man of warre: for by such an ignorant braverie was Peter Strossie lost at the Azores, when he fought against the Marquesse of Santa Cruz.' Raleigh praised Charles Howard of Effingham, Lord High Admiral of England, for refusing the advice of 'many malignant fools' to grapple with the Spanish during the Armada campaign.[19]

Boarding, however, remained the ultimate goal in naval engagements. Stressing the importance of having the advantage of the wind in a sea fight, the experienced Monson stated: 'For he that has the wind is out of danger of being boarded, and has the advantage where to board and how to attempt the enemy'; the Florentine seaman Alessandro Falconi expressed the same point.[20] Others shared this attitude; in the first half of the seventeenth century the Dutch combined a favourable wind gauge with gunfire as

18 Peter Earle, *The Last Fight of the Revenge* (London, 2004), pp. 144–145, 181.
19 Walter Raleigh, *The History of the World* (London, 1614), Bk 1, p. 297
20 William Monson, *The Naval Tracts of Sir William Monson*, ed. Michael Oppenheim (London, 1913), p. 98; Alessandro Falconi, *Breve instruzione appartenente al capitano de vasselli quadri* (Florence, 1612), pp. 20–1.

a prelude to taking vessels by storm. Capturing an enemy ship thus still remained the main objective of naval warfare, despite the destructive power of long-range gunnery devastatingly proved at Lepanto and during the Armada campaign. On these occasions, however, Venice and England were fighting for survival; once the peril had passed, and with it the need to destroy enemy vessels, it was business as usual. Can this be reduced to the simple fact that, as one author aptly puts it, 'a prize was worth a lot of money from which all the crew, from master to cabin boy, benefitted financially'?[21]

To answer the question, one needs to shift the focus from immediate gains to long-term losses. Today we perceive the sole purpose of warships as fighting enemy vessels; but in the early modern age substantial expenditure on warships was possible only with investments, which in turn required a profitable return. This was true of both state- and privately owned vessels; indeed, the distinction between the two was often blurry. Elizabeth I put a considerable amount of her own money into privateering, losing her investment, in the form of two royal ships, in John Hawkins's 1568 debacle at San Juan de Ulúa, but reaping an enormous profit from Drake's 1577 semi-piratical activities in the Pacific. Similarly, the Medici grand dukes of Tuscany profited from the endemic state of warfare in the early modern Mediterranean. In 1626, Gualterotto de' Bardi and Niccolò Gianfigliazzi, two Florentine knights of Malta, obtained from the Grand Duke Ferdinando II the authorisation to build a warship in the Pisa arsenal for a cost of 3,000 Tuscan *scudi* (a substantial sum), on the agreement that 10 per cent of all profits from their corsair enterprises should go to the sovereign, as well as a flat sum payment of 1,000 *scudi* even if they should fail to gain any booty. It was not unusual for corsairs to loan or buy ships from the grand duke, who would also provide artillery and victuals, in the form of hardtack biscuit, in return for a percentage of all profits.

Sailing ships had made a comeback into the Mediterranean thanks to the Dutch, followed by the English, both establishing merchant colonies in the various coastal cities in Italy, North Africa, and the Levant. The English, typically, combined trade with unofficially sanctioned piracy, operating primarily with *bertoni* – akin to galleons, but smaller according to contemporary descriptions. The seaworthiness of these ships left a long-lasting impression, as did their ordnance: cast-iron guns, though heavier, more prone to burst, and inferior in quality to brass ones, were far cheaper. As a result everyone tried, by fair means or foul, to obtain cast-iron guns, the

21 Robert Smith, 'Ships and Guns of the Tudor Navy 1495–1603', in Carlo Beltrame and Renato G. Ridella (eds.), *Ships and Guns: The Sea Ordnance in Venice and Europe between the 15th and 17th Centuries* (London, 2011), p. 115.

manufacturing technique for which the English managed to keep secret for several decades. The Dutch went a step further by creating cheaply built – some would say *too* cheaply built – ships to match this ordnance. In the 1630s, any Dutch vessel travelling to the Mediterranean had to follow a strict set of rules regarding the number of guns and crew members, depending on its freight and destination; in case of multiple stops, ships had to be of at least 360 tons with fifty crew members and carry twenty-four cast-iron guns firing at least a 5-pound shot. The Dutch were also aspiring to a certain degree of standardisation for the calibre of their naval ordnance – not always possible in a country heavily depending on imported hardware – useful when converging firepower in order to inflict maximum damage on a precise section of an enemy's hull. Such ships were as good for trade as for sea fights and were used extensively by the Dutch to augment their war fleets, a cheap way of having a ready-made navy without burdening the citizenry with the cost of building a large number of vessels whose sole purpose was combat and therefore considered a straight financial loss.

This policy paid dividends at the battle of the Downs on 21 October 1639, when a Dutch fleet of mixed war and merchant vessels under Admiral Maarten Tromp soundly defeated a Spanish one of similar composition under Antonio de Oquendo. In the preliminary encounter, on 16 September, an outnumbered Tromp employed the same tactic used by the English against the Armada, exploiting his ships' swiftness to avoid being grappled by the Spanish while raking them with shot. When the main clash occurred the next month, the substantially reinforced Dutch admiral went at the Spanish in the traditional way, the fireships sent to disrupt Oquendo's formation followed by surrounding, pounding, and boarding single enemy vessels. The Spanish lost more than half of their force: about forty Spanish warships and armed transports of the fifty-three engaged in the fight were, in Tromp's own words, 'stranded, sunk, burnt, or taken'.[22] To these casualties one should add about 5,000 infantry killed, captured, or seeking refuge on the nearby English coast. Numerically the battle of the Downs appeared to be a disaster of the highest degree for Spain, and anti-Spanish pamphleteers had a field day magnifying the Dutch success, with many later historians buying into this narrative.

However, the widespread rejoicing for the victory in the United Provinces overshadowed a number of naval and political shortcomings that would become apparent within a couple of decades. For one, each province had

22 Charles Boxer (ed. and trans.), *The Journal of Maarten Harpertszoon Tromp, 1639* (Cambridge, 1930), pp. 65–7.

its own admiralty, complicating the coordination of Dutch naval resources. Moreover, for war purposes the various admiralties hired merchant ships whose masters were loath to put the vessels' owners property at risk, often refusing to engage and ignoring orders from the fleet commander or his flag officers. Finally, the Dutch fleet at the Downs did not optimise the number and calibre of the guns carried by the constituent ships; Oquendo's flagship managed to escape despite taking 1,700 shots to the hull and repulsed several enemy boarding attempts. Still, the Dutch had beaten the Spanish in the boarding game, uniting shock with nimbleness.

'If you can't lick 'em join 'em', goes the saying. But the English decided that joining would be fruitless. The English response to Dutch naval developments was to construct bigger ships with greater hitting power. This would mean sacrificing agility to metal, a throwback to the galleass but with a notable difference. The Venetian vessel had been created with an emphasis on long-range fire to disable enemy galleys. To obtain the same result against new-generation galleons, however, would have implied gargantuan-sized guns and a severe strain on the structure of the ship carrying them. Galleys mounted short-range stone-throwing ordnance to kill oarsmen and crewmen, as well as to damage another galley's quickwork. Although stone-firing ordnance was on its way out in Northern Europe, a substitute was needed in order to smash a galleon's hull and destroy its propulsion system of masts and sails. Thanks to improved design and casting techniques, by the 1620s the English had developed a shorter, lighter weapon – called a *drake* – in cast iron or brass, that if fired with a reduced charge produced a lower muzzle velocity, with the same effect at close-range as the stone-shot artillery of old. Brass guns were more expensive but more prestigious than cast-iron ones, and so it is understandable why the former were chosen for Charles I's prize warship the *Sovereign of the Seas*, launched in 1637. With its 1,600 tons and 102 guns on three decks, the *Sovereign* was the most powerful ship of its time, and a sufficient deterrent against any maritime aggressor.

The launch of the *Sovereign* heralded the beginning of the final separation between ships designed for war and those conceived for commercial needs. It would still take a couple of decades for this division to be complete, but already at the beginning of the 1630s the parsimonious grand duke of Tuscany was investigating the possibility of buying two sailing ships 'entirely dedicated to war' from the United Provinces.[23] The drive behind the

23 Archivio di Stato di Firenze (ASF), *Depositeria Generale, Parte antica* (*DGA*), 659, n. 1318 (Mandate by the *Depositario Generale* Alessandro Caccini, 13 May 1633), n.n.f.

FIGURE 8. Idealised Mediterranean port with ships. Painting by Salvator Rosa, Palazzo Capponi. Permission granted by Capponi family for reproduction.

decision primarily lay in lightening the burden that six permanent galleys imposed on the Tuscan treasury. Between 1625 and 1630 their overall cost, between building, manning, arming, and maintenance, amounted to 15 per cent of Tuscany's annual revenue and one-third of the total military budget. In 1626 alone the expenses for the galley fleet amounted to nearly 155,000 *scudi*, whilst profits in return (booty and ransoms) stood at less than 14,000. At the same time the Tuscans were building and arming a twenty-eight-bench *bastarda* galley costing 10,715 *scudi*, to which one had to add thousands more for wages, victuals, and repairs. On the other hand, in 1630 the Tuscan government purchased a fully armed 520-ton galleon from the Van der Broecke company of Leghorn for less than 12,000 *scudi*, including twenty-two guns, two of them of brass and the rest cast-iron 'weight of Flanders' 8-pounders; in addition, the overall cost for running a thirty-or-more-gun galleon came to only about half that of a galley, with the added possibility of engaging in commerce to cover the expenses.[24] (See Figure 8.)

Thanks to its trading connections with England and the United Provinces a small maritime state with limited financial resources like Tuscany could keep abreast with naval developments. In all three navies, naval tactics stagnated as galleon squadrons were treated as if they were galleys: supposed

24 ASF, *DGA*, 655, n. 1165 (Purchase of the ship *S. Giovanni Battista* [Oct. 1630]), n.n.f.

to advance in line abreast, with each vessel poised to assist the others in attacking a single objective or repelling an assailant. Mastering this galley-style of warfare had allowed the Dutch to emerge victorious in their sea fights with the Spanish. But it also appealed to individual captains' craving for booty, leaving a fleet commander with little scope for a coordinated effort: the comte de Guiche, who witnessed the Dutch in action during the 'Four-Days Battle', would compare them to cavalry, 'the squadrons of which leave the formation and charge separately'.[25] The pendulum of naval warfare appeared to swing repeatedly between tactics to facilitate boarding and another set aiming to avoid being boarded, with no way to break this deadlock in sight.

It may seem paradoxical that one of the main changes in maritime warfare, destined to last a century and a half, principally stemmed from landsmen with no previous connection to naval affairs. In the aftermath of the English Civil Wars, the Commonwealth, seeking to combine military competence with political and religious reliability, turned to the New Model Army to put straight the disorganised navy. Several senior army officers, principally Robert Blake, Richard Deane, and George Monck, were appointed as Generals-at-Sea to reshape and lead the English fleet. The test for the new commanders arrived soon enough with the outbreak of the First Anglo-Dutch War in 1652. Predictably, in the first engagements confusion reigned, partly due to the inexperience of some captains and partly to the indiscipline of others, in particular the masters of impressed merchant vessels. In addition, a lack of commonly shared tactics meant that more than once the English were unable to vanquish their foes, even when victory lay within their grasp.

In reality, the English and Dutch navies suffered from the same problems; but while the Dutch saw the solution in pressing melee attacks with ever greater vigour, the English adopted the opposite approach. As they sought to bring discipline and order to the Commonwealth fleet, the Generals-at-Sea grasped the same fundamental truth that their forbears at the time of the Armada and the Venetians before Lepanto had: they could not beat the Dutch at the boarding game, and nor did they need to. Following the inconclusive and – for the English – disappointing three-day battle of Portland in February 1653, the following month Blake, Deane, and Monck issued their *Instructions for the better ordering of the Fleet in Fighting*, providing clear

25 Armand de Gramont, comte de Guiche, *Mémoires du comte de Guiche: Concernant les Provinces-Unies des Pais-Bas* (London, 1744), p. 267.

indications on how ships should proceed in formation and in battle ahead or astern of the flagship in one straight line. Moreover, fleets were to be divided into three squadrons, with the second-in-command in charge of the right 'wing' – a term that echoed the old formation of line abreast – while the third-in-command led the left 'wing', with these two squadrons preceded or followed by the central one.[26]

The line-ahead formation was no novelty in naval warfare. It had probably first been developed by the Portuguese as a defensive formation, as William Towerson witnessed in a battle near Guinea in 1557. The Dutch also used it in the running fight of 16 September 1639 to avoid coming to grips with the numerically superior Spanish, and it may have been employed by Monck's squadron at the battle of Portland. The English, however, understood that the line-ahead not only made boarding very difficult but also brought the full weight of a fleet's artillery to bear on an attacker advancing line-abreast: the same principle behind the conception of the Venetian galleass. In addition, the English had a superiority in ordnance, both in numbers and weight of shot: no Dutch ship had more than fifty-six guns, most had fewer than forty, while at the battle of Portland the English had twenty-six ships with forty or more guns, eight with between fifty and sixty-two pieces. Many of these guns were 24-pounders, far more powerful than anything the Dutch could muster on their broadsides.

The combination of big ships, powerful ordnance, and the line-ahead formation produced immediate results. In a long-range artillery duel at the battle of the Gabbard (2 June 1653), the English inflicted a terrible beating on the under-gunned and ill-supplied Dutch, who lost twenty ships with many more damaged. The same pattern was repeated the following year at the battle of Scheveningen (31 July 1654), when Monck managed to pass through Admiral Tromp's fleet four times in succession on opposite tacks, the artillery firing at almost point-blank range. The Dutch lost fifteen ships (the English claimed thirty), with many others suffering serious damage, forcing Admiral Witte de With (Tromp having been killed in the fight) to beat a hasty retreat.[27] However, it would take some time before the line-ahead became widespread: the Dutch fully adopted it during the Second Anglo-Dutch War (1665–7), and only after another major debacle at Lowestoft (13 June 1665). The French were slower to accept the new tactical

26 Julian Corbett (ed.), *Fighting Instructions, 1530–1816* (London, 1905), pp. 99–104.
27 David Davies, *Pepys' Navy: Ships, Men and Warfare, 1649–1689* (Barnsley, 2008), pp. 255–7.

formation, but by the end of the 1680s they had embraced it with a vengeance. (See Figure 9.)

Other countries lagged still further behind, often due to economic and geographical factors. To work properly, the line-ahead needed a certain number of ships, the more powerful the better; and this meant very deep pockets. Around the end of the 1660s, the Tuscan government calculated that 200,000 *scudi* would be necessary to build in England a first-rate man-of-war, and a quarter of that sum to build a 600–700 ton, forty-to-fifty-gun fourth-rater. Considering that 50,000 *scudi* was what the Medici grand duke disbursed yearly for the upkeep of three galleys (7 per cent of Tuscany's annual revenue), one can understand his reluctance to follow the new trend in naval warfare. Besides, even had Tuscany possessed the finances needed to build big ships, it lacked the necessary material resources to do so, a problem which beset other maritime countries. The Dutch and Spanish relied heavily on Baltic timber, hemp, flax, tar, and pitch for their shipbuilding industry, while the 3,000 aged oaks needed to build a first-rater meant that even a lumber-rich country like France had to import large quantities of wood for its fleet's construction and maintenance. Although they may not have shared the hyperbole, European leaders would surely have agreed with the Ottoman

FIGURE 9. Battle of the Sound fought between Swedish and Dutch fleets, 1658. The melee battle ended in victory for the Dutch. Painted by Willem van de Velde the Elder, 1665, Rijksmuseum, Amsterdam.

grand vizier Koja Sinan Paşa, who stated in 1589 that: 'One can launch a campaign on land by a mere command: everybody mounts his horse and sets off. A naval expedition is not like that ... however great the material investment and human efforts made, it can only be realized in seven to eight months.'[28]

As always, it was not just a question of material and human resources, as much as the ability to profit from them. The Ottomans certainly had the assets, such as hemp, wood, metal, tar, flax, and all the other resources needed for shipbuilding; they also possessed a pool of skilled mariners, and were also able – should the need arise, as it did after Lepanto – to tap into the human reserves of subject Muslim states of North Africa. However, during the first phases of the Cretan War against Venice (1645–69) the Ottomans were systematically outfought by the Venetians, who managed to combine at an operational level – albeit not always in a satisfactory manner – galleys, galleasses (much improved in their performance), and galleons rented from the Dutch and the English. Venice started building its own sailing warships after the end of the conflict, and also adopted the line-ahead formation. This meant not only abandoning galley-style tactics, but also relegating galleys to logistical and raiding roles. The Ottomans followed suit, although it took them some time to catch up with the new developments. Once this happened, they proved a match for their opponents; so much so that at the battle of Matapan (17 July 1717) a mixed Venetian–Portuguese–Maltese sailing fleet and an Ottoman one, both employing the line-ahead formation, fought each other to a standstill.

The Venetians and the Ottomans had come to realise what other naval-oriented states already understood: the line-ahead had been transformed from a defensive to an offensive formation. But once it was in widespread use the possibility of obtaining a decisive victory was largely negated; on most occasions naval battles were reduced to an artillery duel between two opposing fleets. Ships went on being taken, but boarding was largely relegated to piracy or privateering; rather, an enemy vessel would be forced to strike its colours after losing a gunnery fight with one or more opponents. Prize money was still sought by officers and sailors alike, but the line-ahead meant that now the interest of individual commanders had to coincide with those of a navy and, therefore, of a state. Personal maritime profit had given way to collective gain.

28 Quoted in Pál Fodor, 'Between Two Continental Wars: The Ottoman Naval Preparations in 1590–1592', in *In Quest of the Golden Apple: Imperial Ideology, Politics, and Military Administration in the Ottoman Empire* (Istanbul, 2000), p. 174.

Nevertheless, the line-ahead as a tactical method stifled the opportunities for fleets to deliver a knock-out blow. To break this impasse some proposed the building of even bigger ships with more ordnance, while one English maverick argued for the adoption of Venetian-style galleasses – despite the fact that the Venetians had decided they were obsolete – forcing the Navy Board to write an extensive reply explaining how such vessels were unsuitable for English waters.[29] In reality, the line-ahead suited everyone – except, of course, tactical geniuses – for the simple reason that it worked. Besides being a fairly easy concept to comprehend, it maximised firepower and allowed admirals more control over their fleet, discouraging individual commanders from having bright ideas involving even a moderate amount of risk.

Was the line-ahead revolutionary, as some have argued? Yes and no. *Per se* it certainly was not, although the stunning English victories against the Dutch after its initial adoption give the impression of a dramatic break with the past. Moreover, it tended to create a class of very conservative naval officers – of the type so abhorred by Mahan – perfectly content with following clearly outlined rules. Yet the line-ahead unintentionally tied the many loose threads – economic, technological, and theoretical – that had been dangling over European maritime affairs since the end of the fifteenth century. Furthermore, it gave navies an updated strategic role in controlling sea lanes by the constant presence of fleets-in-being. The possibility of inflicting a knock-out blow would only arise in Nelson's generation thanks to improved signalling, updated gunnery, and a change in mentality, allowing melee-centred tactics to return. Like Ebenezer Blackadder, during the Napoleonic Wars the Royal Navy understood, as had the Venetian, Dutch, and English navies in centuries past, what it had and also where to put it.

29 Brian Lavery, *The Ship of the Line*, vol. 1: *The Development of the Battlefleet, 1650–1850* (London, 1983), p. 59.

7

# Military finance, organisation, and state development, 1450–1650

PETER H. WILSON

For over half a century, discussion of the relationship between military finance, organisation, and state development has been dominated by the contested concept of a 'military revolution'; the belief that there were one or a few periods of fundamental change that transformed both war and wider European history. More recently, this has been supplemented by the idea of smaller, but more frequent 'revolutions in military affairs' (RMAs) as individual military organisations respond to, or anticipate, changes made by their likely opponents. Technology is generally considered to drive both forms of 'revolution', as innovative weaponry and institutional practice transform war, rendering older models ineffective and obsolete. Change flows through a series of chain reactions, as states adapt to new conditions, modifying their structures to sustain and direct altered armed forces, and revising their forms of interaction with society both to extract the necessary resources and to legitimate their use in war-making.[1]

This debate has operated on two levels. The more obvious one is the discussion of the character and significance of changes in the way wars have been fought. There is, however, a deeper level of argument about the nature of history itself, and particularly about the relative significance of developments in one area for changes in other aspects of human affairs. This broader issue cannot be resolved here, but it is possible to identify a technologically related transformation of European warfare between around 1450 and 1520 which fundamentally altered the relationship between military organisation, finance, and state structure.[2] The main

1 Clifford Rogers (ed.), *The Military Revolution Debate: Readings on the Military Transformation of Early Modern Europe* (Boulder, 1995); MacGregor Knox and Williamson Murray (eds.), *The Dynamics of Military Revolution 1300–2050* (Cambridge, 2001).

2 For a more detailed overview of this and its repercussions, see Jeremy Black, *European Warfare 1494–1660* (London, 2002). For changes in organisation and tactics, see

technological changes were improvements in metallurgy and gunpowder manufacture which significantly improved the effectiveness and reliability of firearms in both battle and siege warfare. However, these changes only had impact thanks to other developments, notably the demographic and economic recovery from Europe's mid-fourteenth-century crisis which supplied the additional human and material resources needed to sustain substantial growth in the scale, scope, and duration of wars. The varied pace of change itself proved a catalyst by upsetting the previous balance of forces across Europe. Uncertainty surrounding the new weaponry and methods encouraged great variety in organisation and tactics, creating opportunities for dramatic (if not necessarily decisive) victories as one force gained a temporary superiority over a differently armed and organised foe. These variations soon levelled out as proven methods were swiftly copied. This transfer of military practice and culture was facilitated by that other great early modern 'revolution': the rapid transformation and expansion in communications associated with the development of printing and postal services, such as that in the Holy Roman Empire after 1490. Subsequent changes in organisation and tactics across the next 150 years were largely refinements of the methods tested around 1500.

The improved firearms and other close-order weapons, like pikes, were only fully effective when used by large, disciplined bodies of men organised into standardised units subordinated to a clear hierarchy of command. Training needed to instil not only competence in weapons handling, but also precision and cohesion of movement. Driven by necessity and the Humanist rereading of ancient Roman military history, it became common to organise soldiers into regiments subdivided into companies, each commanded by a captain assisted by two to four junior officers. Discipline and control were gradually strengthened by reducing overall infantry company size from 300–400 men around 1500 to 100–200 by 1650, whilst retaining the same officer establishment.

These changes shifted the basis of military proficiency from individual prowess and experience to a new form of professionalism requiring technical expertise and theoretical knowledge. Such ideas were not necessarily incompatible with the warrior ethos of Europe's hereditary nobility. Matters like drill, tactical formations, and fortress design were seamlessly absorbed into

Christer Jörgensen, Michael F. Pavkovic, Rob S. Rice, Frederick S. Schneid, and Chris L. Scott, *Fighting Techniques of the Early Modern World* AD *1500*–AD *1763* (London, 2005).

princely and aristocratic education. The issues of subordination and responsibility proved more problematic. It was still an age of warrior kings, exemplified by Sweden's Gustavus Adolphus who died in the battle of Lützen in 1632. Other monarchs accompanied their armies and could take command, including French kings into the eighteenth century, and some Holy Roman emperors, like Charles V and (whilst king of Hungary) Ferdinand III. However, the new, more exalted ideal of monarchy required the ruler to distance himself from immediate responsibility in order to preserve the aura of power regardless of the actual outcome of operations. The same desire to display omnipotence complicated the choice of field and regimental commanders, however. Grandees had claimed high command by virtue of their status and ability to bring large bands of retainers into the field, and continued to do so after 1500, simply using their wealth and local influence to organise regiments of professionals in place of their earlier levies. While it often proved difficult to deny demands for preference, most monarchs avoided entrusting politically sensitive or militarily significant commands to independently minded grandees after about 1550, but status continued to influence appointment and promotion well into the nineteenth century in most countries.[3]

These difficulties are one aspect pointing to the changed relationship of war to politics after 1450, which is usually explored as an aspect of modernisation. Debates on whether there are singular or multiple paths to modernity generally conclude that European states either adopted an authoritarian 'coercion-intensive' solution to the pressures of war or followed a constitutional 'capital-intensive' route.[4] These two models have their theoretical antecedents in the historical sociology of the late nineteenth century, especially the works of Herbert Spencer and Otto Hintze. There is some discussion surrounding the speed of this process, with various countries (usually Britain and the Dutch Republic) considered more 'progressive' than others. More controversial still are the alleged merits of the different routes.

3 For detailed case studies, which also discuss differences in organisation and tactics, see James Wood, *The King's Army: Warfare, Soldiers, and Society during the Wars of Religion in France, 1562–1576* (Cambridge, 1996); David Parrott, *Richelieu's Army: War, Government and Society in France, 1624–1642* (Cambridge, 2001); Roger Manning, *An Apprenticeship in Arms: The Origins of the British Army 1585–1702* (Oxford, 2006); Fernando Gonzáles de León, *The Road to Rocroi: Class, Culture and Command in the Spanish Army of Flanders, 1567–1659* (Leiden, 2009).

4 Charles Tilly, *Coercion, Capital and European States,* AD *990–1992* (Oxford, 1992). Other important examples include Bruce Porter, *War and the Rise of the State* (New York, 1994); Brian Downing, *The Military Revolution and Political Change* (Princeton, 1992); Thomas Ertman, *Birth of the Leviathan* (Cambridge, 1997).

Previous assumptions of the superior war-making capacity of authoritarian monarchies have been overturned by research indicating that constitutional regimes taxed their subjects far more heavily and organised more efficient 'military-fiscal states'.[5]

The debate has concentrated on the period after 1650, and has been dominated by discussion of the Western European states, especially France, Britain, and the Dutch Republic, though research on war finance in the Scandinavian kingdoms is now quite extensive and the situation is improving for Spain, the Italian states, the Austrian Habsburg monarchy, and Central Europe.[6] Poland-Lithuania and the Balkan principalities remain under-studied.[7] The view remains broadly conflictual, seeing change emerging from antagonism between rulers and subjects rather than cooperation. Above all, research still concentrates on central state institutions as the primary agents of 'modernisation'.

These approaches explain the period 1450–1650 through the paradigm of the 'military enterpriser', or mercenary contractor, who is regarded as a temporary substitute until the development of state institutions caught up with the increased scale of warfare.[8] It is believed that hiring enterprisers allowed the state to circumvent the constraints and vested interests of the late medieval feudal order. However, mercenaries could only be an expedient, because they were allegedly motivated only by material interests and lacked 'higher ideals' like patriotism. Thus, while the scale of war increased, the quality of its organisation and conduct 'degenerated', reaching 'the nadir of brutality and pointlessness' in the Thirty Years' War (1618–48).[9] This process seems epitomised by Albrecht von Wallenstein, the most ambitious of all the enterprisers, who raised an entire army for the emperor before being

5 A concept developed by John Brewer, *The Sinews of Power: War, Money and the English State, 1688–1783* (New York, 1989) and subsequently backdated by James Wheeler, *The Making of a World Power: War and the Military Revolution in Seventeenth-Century England* (Stroud, 1999). See also Philip Hoffman and Kathryn Norberg (eds.), *Fiscal Crises, Liberty, and Representative Government 1450–1789* (Stanford, 1994); and Pepijn Brandon, *War, Capital and the Dutch State (1588–1795)* (Leiden, 2015).

6 Marjolein 't Hart, Pepijn Brandon, and Rafael Torres Sánchez, 'Maximising Revenues, Minimising Political Costs: Challenges in the History of Public Finance of the Early Modern Period', *Financial History Review* 25 (2018), 1–18.

7 For further references, see Robert Frost, *The Northern Wars 1558–1721* (Harlow, 2000).

8 For a critical engagement with this term and the broader debate on military contracting, see David Parrott, *The Business of War* (Cambridge, 2012).

9 Michael Howard, *War in European History* (Oxford, 1976), p. 37. This view is still deeply embedded in the political science literature that engages with history, e.g. Peter Singer, *Corporate Warriors: The Rise of the Privatized Military Industry*, 2nd ed. (Ithaca, NY, 2008), pp. 10–37.

murdered on his employer's orders in 1634 when it was feared he would change sides.[10]

This paradigm rests on the late nineteenth-century definition of the state as a sovereign 'monopoly of violence' over a given area, naturally drawing attention to the infrastructure of central institutions like war ministries and treasuries. Historical scholarship conventionally identifies the emergence of such institutions by the mid-sixteenth century, and traces their subsequent development as a process of extending authority into the localities through agents like the French *intendents*, appointed after about 1550 and used more systematically and extensively after 1630 to enforce compliance and promote the standardisation of administrative practice.[11] This perspective often rests, implicitly or explicitly, on top-down theories of change whereby the state first disciplines its own servants and then uses them to modify its subjects' behaviour.[12] The ability (and even the desire) of central authorities to affect change is frequently exaggerated. Central institutions played a limited role throughout this period. For instance, the Habsburg monarchy established a court war council (*Hofkriegsrat*) in 1556 to advise on strategy, process military appointments, maintain records, and deal with supply contractors. However, similar agencies already existed in several Austrian provinces, while much of the actual business of recruiting, accommodating, clothing, feeding, and arming soldiers remained in the hands of officers, provincial officials and the corporate assemblies known as the Estates well into the eighteenth century.[13]

While central institutions did become more powerful after about 1600, they remained only loosely connected to the largely self-governing world of villages and towns where most early modern Europeans lived. Politics remained fundamentally a matter of convincing local rural and urban elites to cooperate. For present purposes it is best to conceive of the state not as a set of central institutions, but as a framework to coordinate and activate legally enforceable claims on human cooperation and material resources. The social and political tensions generated by early modern warfare are

10 The extensive literature is summarised by Geoff Mortimer, *Wallenstein: The Enigma of the Thirty Years War* (Basingstoke, 2010).
11 Richard Bonney, *Political Change in France under Richelieu and Mazarin 1624–1661* (Oxford, 1978).
12 Gerhard Oestreich, *Neostoicism and the Early Modern State* (Cambridge, 1982).
13 Despite its focus on the period after 1650, much information can be found in Michael Hochedlinger, *Austria's Wars of Emergence* (Harlow, 2003); and W. D. Godsey, *The Sinews of Habsburg Power: Lower Austria as a Fiscal-Military State* (Oxford, 2018).

explicable as disputes over the legitimacy of these claims. Religious controversy following the Reformation complicated matters further, as did underlying economic change.

Demands for resources intersected with two related but competitive economic systems.[14] A substantial proportion of Europeans lived in households which were the primary units for production, consumption, and reproduction. Households were orientated towards risk minimisation, sustaining the survival of their members, preserving autonomy from external interference, and maintaining friendly relations with immediate neighbours. Production was directed towards self-sufficiency and immediate consumption, with only a small surplus set aside to meet obligations to secular and spiritual lords. The commercial, cash economy emerged in parallel with that of the household. It was geared towards producing a surplus for sale and thus had to integrate its members into wider market networks. Producers became dependent on the market as they no longer furnished all their own needs directly. However, commercialised markets also offered producers opportunities for greater individual autonomy from the household, community, and lords.

Early modern political development was never simply a matter of the state exploiting the commercialising economy to escape feudal constraints by extracting cash to hire mercenaries. Instead, cash was used when existing arrangements failed to deliver the human and material resources in the quantity and quality at the time and location required. Cash, not military enterprisers, provided a substitute for underdeveloped institutional infrastructure. It never entirely displaced contractual obligations, which were continually revised in line with experience and new ideas. For example, long-standing requirements for able-bodied males to assist repelling invasions and combating lawbreakers and natural disasters provided the basis for new forms of conscription and militia service from the 1570s in many countries. Conscription and other forms of extraction in kind often proved cheaper and more efficient than fiscal measures. The relatively inflexible character of the largely agrarian economy made fiscal extraction difficult, with people's ability to pay taxes dictated by the seasonal rhythm of production and markets.

This also affected recruitment. Compulsory systems targeted the landless and underemployed, who also supplied the bulk of 'volunteers'. It was easiest to recruit in winter when employment and food were scarce. War thus

14 Heide Wunder, 'Finance in the Economy of Old Europe', in Peter-Christian Witt (ed.), *Wealth and Taxation in Central Europe* (Leamington Spa, 1987), pp. 19–47.

followed a seasonal cycle, with the autumn and winter as periods of preparation and for recovery from previous operations. Campaigns began in the spring, as roads and rivers became passable and as fresh grass became available to feed the large numbers of cavalry horses and transport animals. While this imposed significant military constraints, it also provided political opportunities to seek alliances, or settle differences through negotiation during the autumnal and winter lull in operations. Variations in European society and ecology further affected warfare. For example, Polish, Hungarian, and Balkan forces operated in sparsely populated regions with low agricultural production, necessitating a high proportion of cavalry to move armies rapidly before they ate up all produce in their immediate vicinity. Such forces were good at covering ground, but poor at holding it, because they lacked sufficient infantry and artillery to capture and garrison strategic towns. In Sweden, the small scale of most agricultural production allowed for the substitution of female for male labour, permitting an effective conscription system to be developed from the 1540s. Indeed, women were an integral, if forgotten, element in all European armies, helping in the constant search for food. They also substituted for an effective logistical infrastructure, but they consumed additional resources and acted as a further constraint on military movement.[15]

Political considerations also hindered the full exploitation of the commercialising economy. All prevailing social philosophies and religions stressed the necessity for a firm guiding hand to promote thrift, piety, and obedience, and saw the household as the pillar of a stable hierarchical order. Far from promoting commercialisation, official measures sought to minimise its impact on householders whose interests were protected through legislation, such as that reinforcing parental control, stiffening masters' authority over servants, and securing household self-sufficiency by encouraging primogeniture.

Nonetheless, it is possible to identify a transition across Europe from a 'domain state' to a 'tax state'.[16] The domain state was rooted in the household subsistence economy and the ideal that each element in society, including the monarch, should 'live off their own' resources. Rulers were expected to fund their activities from crown assets and regalia / prerogatives, only drawing on their subjects' resources in exceptional circumstances. In

15 John A. Lynn, *Women, Armies and Warfare in Early Modern Europe* (Cambridge, 2008); Mary Ailes, *Courage and Grief: Women and Sweden's Thirty Years War* (Lincoln, NE, 2018).
16 Concepts first proposed by Joseph Schumpeter in 1918 and developed by Ladewig Petersen, 'From Domain State to Tax State', *Scandinavian Economic History Review* 23 (1975), 116–48.

contrast, subjects' resources provided the primary base for the tax state, which was clearly linked to the market economy, but also to a more active and inclusive ideal of the state with an expanded public role. This entailed a shift from patrimonial politics, based on multiple chains of personal ties of vassalage and patronage, to a broader impersonal commonwealth transcending the lives of any individual. The state was now defined by common subordination to an impersonal institution, the purpose of which was to promote the security and welfare of all. This transition required centralisation and consolidation to simplify the internal hierarchy by removing autonomous intermediary bodies between state and subjects and integrating the latter within a common legal framework.

It is important to remember that this transition represented state *formation*, not conscious 'state building'. Centralisation and consolidation were rarely desired and certainly not goals in themselves, but instead emerged as by-products of the never-ending scramble for resources as rulers and elites struggled for prestige, influence, and security, or responded to popular pressure. Other cultural factors contributed to the improvised and uncertain process of state development. Frugality and financial precision were contrary to the aristocratic ethos, to which rulers also subscribed, and which celebrated benevolent munificence and conspicuous consumption. The same applied to their servants, for whom money was simply a means for social advancement. Recent biographies of military enterprisers have downplayed the economic calculations emphasised by Fritz Redlich in his classic study.[17] Most enterprisers were minor nobles who saw military contracting and service as ways of enhancing their prestige and influence. Many proved poor businessmen, while the demands of service and the need to network at court left little time for efficient management or capital accumulation.[18] New forms of accounting were adopted from the early sixteenth century, but their purpose was primarily legal, not fiscal, and recorded liabilities and obligations to individual creditors or subjects. Planning and oversight were further hindered by administrative fragmentation, which directed each source of revenue through separate accounts attached to important areas of expenditure, as well as near perennial wishful thinking when estimating the actual likely costs of warfare.[19]

17 Fritz Redlich, *The German Military Enterpriser and His Workforce*, 2 vols. (Wiesbaden, 1964–5).

18 Stephanie Haberer, *Ott Heinrich Fugger (1592–1644)* (Augsburg, 2004); Frank Göse, *Otto Christoph Freiherr von Sparr 1605–1668* (Berlin, 2006); Robert Rebitsch, *Matthias Gallas (1588–1647)* (Münster, 2006).

19 Peter H. Wilson, 'War Finance, Policy and Strategy in the Thirty Years War', in Michael Rohrschneider and Anuschka Tischer (eds.), *Dynamik durch Gewalt? Der*

The growing scale of war prompted attempts to rationalise the domain state and make it more productive through new forms of management and the use of the royal prerogative to create additional sources of revenue.[20] Expediency usual dictated acceptance of long-term loss in return for short-term gain. Sweden's involvement in costly wars in Poland and Germany after 1621 obliged its monarchy to donate crown assets to nobles and contractors as rewards or in lieu of payment. The reduction of crown domains halved royal revenue between 1644 and 1653, while the nobility doubled their landholding to two-thirds of peasant farms. The remaining free peasants now bore the main burden of the country's other taxes, but their protests were blocked by the nobility until the crown initiated the resumption (*reduktion*) of alienated assets in 1680.[21] Similarly, the sale of newly created public posts in Spain and France saddled both countries' monarchies with large numbers of salaried, but largely honorific office-holders. Periodic reforms, such as those in Sweden, continued into the eighteenth century, but declined in significance in line with the overall place of crown assets in generating revenue.

Royal prerogatives also provided the main basis to expand indirect taxation, since they entitled the ruler to sales taxes and customs and excise duties. These were continually revised and adapted to expand the range of items liable, improve collection, and combat smuggling and other forms of evasion. For example, the Spanish *alcabala* was first used as a general sales tax in 1342, but was revised, notably in the 1490s, while further amendments doubled returns between 1570 and 1612. Such measures targeted the urban population who were the primary agents of the commercialising economy and who often pioneered indirect taxes to fund their own civic self-government. Not only were towns centres of wealth, production, and exchange, but it was also relatively easy to collect indirect levies by charging goods as they passed through the gates, thus directly linking urban fortifications and state finance. Excise taxes were scarcely revolutionary. Nonetheless, there were many false starts, and they were often 'farmed' to consortia who advanced the crown fixed sums in return for a free

*Dreißigjährige Krieg (1618–1648) als Faktor der Wandlungsprozesse des 17. Jahrhunderts* (Münster, 2018), pp. 229–50.

20 See generally Richard Bonney (ed.), *The Rise of the Fiscal State in Europe c. 1200–1815* (Oxford, 1999); Bartolomé Yun-Casalilla and Patrick O'Brien (eds.), *The Rise of Fiscal States: A Global History, 1500–1914* (Cambridge, 2012). Useful case studies include Marjolein 't Hart, *The Making of a Bourgeois State: War, Politics and Finance during the Dutch Revolt* (Manchester, 1993); Michael Braddick, *The Nerves of State: Taxation and the Financing of the English State, 1558–1714* (Manchester, 1996); Jan Glete, *War and State in Early Modern Europe: Spain, the Dutch Republic and Sweden as Fiscal-Military States, 1500–1660* (London, 2002).

21 Anthony Upton, *Charles XI and Swedish Absolutism* (Cambridge, 1998), pp. 51–70.

hand in collecting whatever they could from the population. In the longer term, though, excises proved important, especially where commercialisation increased the volume and value of goods in flow that could be taxed.

Direct taxes came in many forms, including levies on property, on land, and via poll taxes. Though late medieval rulers sometimes obtained new taxes ahead of operations, direct taxation mainly emerged retrospectively during the fourteenth century to amortise debts incurred during previous conflicts and relied on deals with local notables and corporate groups with vested interests in the state's survival. In return for political rights and other privileges, such groups agreed to raise money from their own dependants to enable the ruler to pay his creditors without having to alienate further domains. The intention was to stabilise the domain state, but the underlying tendency was to promote the emergence of the tax state. Meanwhile, such taxes hastened the development of Estates and other representative institutions as the only viable fora where rulers and politically important subjects could negotiate the distribution of the additional burdens. Rulers lacked both the legal right and the means to assess and collect direct taxes, and had to rely on nobles, clergy, magistrates, and others with local knowledge and jurisdiction to handle these tasks.

Three factors promoted the growth of additional direct taxation to fund current warfare during the sixteenth century. Political expediency encouraged rulers to establish new ways of taxing their subjects without having to bargain with notables in Estates or parliaments. The religious controversies of the Reformation encouraged this, since dissenters frequently used the power of the purse to widen their freedom of worship. For example, Protestant nobles bought religious concessions after the 1560s with taxes paid by their peasants to amortise the Austrian Habsburgs' war debts and fund defence against the Ottoman Turks. Social pressures provided a second factor via popular resentment at the unequal distribution of existing burdens and the exemption of nobles, clergy, and patricians from many direct taxes. Poll taxes were one early response to this pressure, since they were emergency measures which cut across corporate privileges and were linked to the language of necessity and imminent danger. Their potency was enhanced through the application of new techniques, again often pioneered in towns, which allowed taxes to be tailored more closely to people's ability to pay. The emergence of a new written and visual culture around 1500 allowed for more sophisticated tax registers (cadastres) and mapping of property ownership. The preparation and maintenance of these records reinforced the place of local officials like bailiffs and village and town mayors in the tax state.

Finally, the immediacy of military demands propelled change. Late medieval rulers were rarely able to pay and feed their armies directly, but the growing scale and duration of warfare made this much more problematic around 1500. A well-stocked farm might contain the equivalent of 3,000 daily rations; sufficient to feed its inhabitants for a year, or a quarter of an average medieval army for a day, but barely enough to provide for a single regiment of sixteenth-century infantry and its civilian camp followers. Armies needed a mechanism to tap resources from a wider area without having to divert scarce manpower to occupy it. 'Contributions', as this solution has been dubbed, was a form of obtaining money with menaces.[22] It is important to note that this was never intended as plundering, though it could degenerate into this. Instead, local elites were forced to redirect existing tax and tithe systems from their usual recipients and pay (usually at greatly inflated rates) money and supplies in kind direct to the army. The conversion of contributions into formalised, regular levies administered by civil officials answerable to the crown completed the introduction of the second form of direct war taxation and provided the principal means in most Continental countries to maintain armies by 1650.

The steady rise in taxes during the sixteenth century was partially absorbed by demographic growth and improved productivity, both of which levelled off around 1580 and even fell subsequently in some areas like Castile. Per capita tax burdens marched on, rising 75 per cent in Castile from 1621 to 1640 and more than doubling in France from 1610 to 1640, or three times faster than improvements in agricultural productivity. Consequently, no country managed to pay its troops on time or in full. Even the Dutch Republic, famed for the regularity of its payments to soldiers, had to rely on credit: company commanders borrowed each month from local businessmen (*solliciteurs-militair*) who received a fee and recouped their loan when the Dutch government eventually provided the wages.[23] This example illustrates how bankers provided one essential service in war finance by bridging time, advancing money when it was needed in return for repayment later. Less obvious, but just as important, was their ability to bridge distance by transferring currency

22 This has led some historians to characterise the early modern state as a protection racket: Charles Tilly, 'War Making and State Making as Organised Crime', in Peter Evans, Dietrich Rueschemeyer, and Theda Skocpol (eds.), *Bringing the State Back In* (Cambridge, 1985), pp. 169–91, developing ideas advanced earlier by F. C. Lane. Further discussion of contributions in Peter H. Wilson, *Europe's Tragedy: A History of the Thirty Years War* (London, 2009), pp. 399–407.

23 H. L. Zwitzer, *'De militie van den Staat'. Het Leger van de Republiek der Verenigde Nederlanden* (Amsterdam, 1991), pp. 91–9.

without the need to physically transport large quantities of cash to where the army was operating. The support of the Fuggers and other bankers with international connections enabled Charles V to be the first monarch to fight wars across the entire continent simultaneously since ancient Rome.[24]

International currency transfer was also essential to the development of subsidies around 1500. These were payments from one state to another in return for military assistance or to fight war by proxy. For example, France subsidised the Dutch for eleven years before openly joining them against Spain in 1635, and likewise backed Swedish intervention in Germany well ahead of an active alliance. Subsidies have attracted considerable attention, because they feature prominently in alliance negotiations, while payments can be traced relatively easily through diplomatic correspondence. In fact, they rarely covered more than 5 per cent of their recipient's war costs, and their real significance lies in the payment of relatively modest lump sums which could bridge delays in revenue collection and, above all, serve as collateral to raise much larger loans.

The expanded scale and duration of operations transformed borrowing into a structural feature of war finance. It grew in line with the development of the tax state, since loans were contracted to cover shortfalls in current revenue and dependent on subsequent receipts for repayment, locking governments into an endless cycle of borrowing and taxation. There was considerable reluctance to abandon the ideal of the domain state, where expenditure was supposed to be determined by income, with debts as temporary expedients. In practice, however, governments searched continually for cheaper and more reliable credit as the only way to cover their spiralling costs, creating national debts well ahead of the acceptance of the concept in the late seventeenth century. A national debt institutionalised deficit financing with borrowing covering a permanent shortfall between revenue and expenditure. What changed significantly in the two to three centuries after 1450 was how that borrowing was handled. Governments moved slowly from contracting individual loans tied to specific expenditure towards a consolidated debt serviced from general taxation. Interest-bearing government bonds were already used extensively by Spain in the sixteenth century, but were still linked to a single revenue source – in this case crown profits from New World bullion – to sustain government creditworthiness.[25] The Dutch Republic and later England were able to go further in the

24 James Tracy, *Emperor Charles V, Impresario of War* (Cambridge, 2002), p. 308.

25 Stanley and Barbara Stein, *Silver, Trade and War: Spain and America in the Making of Early Modern Europe* (Baltimore, 2000); Mauricio Drelichman and Hans-Joachim Voth,

seventeenth century, thanks to greater public confidence in their governments' general revenue raising capacity.

The development of consolidated national debts lessened governments' reliance on individual creditors, though the deaths of important bankers could still threaten state credit, as the Habsburg monarchy discovered as late as 1703 when it was plunged into crisis by the death of its main army contractor. Already in the sixteenth century, however, governments were obliged to change their treatment of creditors and no longer repudiated debts entirely as they had often done in the past. The bankruptcies which rocked Spain, France, and other countries in the sixteenth and seventeenth centuries in fact involved debt restructuring, including renegotiating repayment terms, off-setting obligations by transferring honours or assets, or by granting monopoly rights and other economic privileges.

Permanent indebtedness increased the ranks of those with a stake in the state beyond the nobles, clergy, and magistrates who handled taxation, though those groups were also prominent amongst government creditors. There was never a question of states following *either* a capital-intensive *or* a coercive route to modernity. Rather, the key variable was the institutional framework used to mediate the relationship between the government and those handling revenue and credit. Informal mechanisms, like court patronage, coexisted with formal representative institutions like Estates and parliaments, with the specific variant characterised by the balance between the two, rather than the absence of one or the other.[26] The courtly element generally predominated, growing with the strengthening of royal authority around 1620 which has entered history as 'absolutism'. Representative institutions were rarely eclipsed entirely and continued to provide arenas for fostering consensus with local elites essential in both revenue collection and loans. Single, national representative institutions were less common, declining in Denmark, surviving in Sweden, but dominating in England and the Dutch Republic. The significance of the English and Dutch experience, however, lies less in the vitality of their representative institutions, than in the motives of the dominant groups within them. The Dutch Revolt and Britain's mid-seventeenth-century civil wars and Glorious Revolution of 1688–9 represented victories for the groups controlling the Dutch States

*Lending to the Borrower from Hell: Debt, Taxes and Default in the Age of Philip II* (Princeton, 2014).

26 For a useful survey and excellent guide to the literature, see Wolfgang Neugebauer, 'Staat – Krieg – Korporation: Zur Genese politischer Strukturen im 17. und 18. Jahrhundert', *Historisches Jahrbuch* 123 (2003), 197–237.

General and the English Parliament. To achieve these victories and defend them subsequently, these groups needed to finance war efficiently and so had a vested interest in both agreeing taxation and ensuring its effective collection. Elsewhere, by contrast, elites were denied formal, collective influence over policy and often regarded their ruler's demands for taxes as unjustified claims on their assets and unwelcome intrusions into their local influence.

The rapid expansion in the scale and scope of war after 1450 accelerated the emergence of the tax state that had roots stretching back several centuries. The prolonged warfare engulfing much of Europe after 1560 ended hopes of retaining the late medieval ideal of the domain state. The mid-seventeenth century saw the tax state's decisive breakthrough, enabling it to support a cadre of regular, paid professional soldiers backed by a framework for mobilising supplementary militia or conscripts if needed. Central institutions were sufficiently developed to coordinate these activities, but actual delivery still largely depended on local elites who generally had little direct influence in formulating policy. The resumption of sustained warfare around 1670 shifted development to a second stage as central political authority overcame internal opposition from groups or regions reluctant to relinquish their traditional autonomy. The state concentrated on expanding central supervision and integrating local elites in more formal and uniform systems of administration. Developments entered a third phase around 1800 as internal upheavals, like the French Revolution, decided which groups controlled the more effective state apparatus, and as the last vestiges of older, autonomous military organisation were eliminated in the abolition of civic guards and widespread disarming of European populations by 1814.

# 8
# Fortification and siege warfare

SIMON PEPPER

The siege played such a dominant role in both medieval and early modern warfare that it is difficult to overstate the significance of the twin revolutions in gunpowder artillery and fortification which together transformed fortress warfare and the face of Europe. The continent was to be marked by chains of new frontier and coastal forts, expanded naval bases, and girdles of earth-filled ramparts and bastions around towns, often sweeping away picturesque medieval walls and towers and replacing them by much lower-profile works. The rash of urban citadels by which so many rulers sought to guarantee their security would be matched by what Parker called the 'demilitarised zone' of central France following the Wars of Religion and the Fronde. Ruined fortresses litter the continent. The demolitions following most civil wars, rebellions, or frontier adjustments speak to the importance of fortified places. Whether it was the local power projected by castle or border fort or the major concentration of resources protected by walled towns or major fortresses, fortified places were the main objective of offensive operations.

Many siege tactics, of course, predated by centuries the age of gunpowder. Strong places could be 'surprised', taken by 'deception', or stormed by 'escalade' (attackers swarming over the walls, regardless of casualties).[1] A 'blockade' could starve defenders into surrender – given time – or bring an enemy to battle by forcing an attempted relief. The threat of area 'bombardment' might terrify civilians into opening their gates. Bombardments became more destructive in the age of gunpowder, but unless incorporating incendiaries often failed to deliver decisive results.[2] Mines became more destructive too, allowing sieges to be sustained when solid shot ran short, and use of mines became an

1 The terms in inverted commas were used quite specifically in the early modern period for the kind of siege action indicated. I have tried to employ them consistently.

2 In the age of solid shot, of course, it was fire that destroyed towns; the damage from the missiles themselves was often limited. Although initially terrified, a population under bombardment quickly developed resilience, as did much more heavily bombed twentieth- and twenty-first-century civilians.

important siege tactic. However, it was in the most basic siege operations of all – the destruction of fortifications and the storming of breaches – that the most far-reaching changes would be seen following the introduction of effective gunpowder artillery. It will be argued that early modern fortress warfare hinged upon a shifting equilibrium between developments in fortification and the much less researched topic of artillery, its availability, performance, and tactics. The initial focus here will be on the long sixteenth century. However, recent research on the siege warfare of the early eighteenth century and the extraordinarily well-reported sieges of the Revolutionary and Napoleonic Wars provide a much broader context and an opportunity to ask how sieges figured in these eras of dynamic conflict. History with a capital 'H' and the history of technology often make uneasy bedfellows, but here the relationship is significant and it makes sense to begin with a review of the weapons and the walls.

Pedro Navarro, when surveying the walls of Florence with Machiavelli shortly before the siege of 1529–30, observed that 'more guns will always beat a lesser number'. A century and a half later Vauban said the same.[3] 'Guns beat guns' would become a maxim of twentieth-century warfare. How defensive guns could be mounted and artillery brought against them is fundamental to the modern siege. Logistics is also a key issue. Siege guns – whether the newest cast-bronze pieces firing iron shot, or the older wrought-iron guns firing stone balls – were heavy and difficult to move. A siege cannon of the mid-sixteenth century weighed with its carriage over 8,000 pounds and fired iron shot of from 40 to 60 pounds. Sieges consumed prodigious quantities of gunpowder and shot. As well as the gunners themselves, carpenters to repair gun carriages, blacksmiths, hauliers, forage providers, and the pioneers who dug trenches, levelled roads, and reinforced bridges were all essential to the movement and handling of a siege train, which in late sixteenth-century France counted 2,620 men on the royal payroll in support of twenty heavy guns.[4] Guns, ammunition, their costs, manufacture, manning, and movement together justify use of the term 'weapons system' for failure in any part was critical. When everything worked, however, guns delivered victory over heavily fortified states in the closing campaigns of the Hundred Years' War, the Ottoman conquest of the Balkans, and the *Reconquista* of Granada (to say nothing of the decisive edge

3 Navarro quoted in John Hale, 'The Early Development of the Bastion: An Italian Chronology *c.* 1450–*c.* 1534', in J. R. Hale, L. Highfield, and B. Smalley (eds.), *Europe in the Late Middle Ages* (London, 1965), pp. 466–94, 477; Vauban: 'superior artillery is always at an advantage'. Sébastian Le Prestre de Vauban, *A Manual on Siegecraft and Fortification*, trans. George A. Rothrock (Ann Arbor, 1968), p. 123.

4 James Wood, *The King's Army: Warfare, Soldiers, and Society during the Wars of Religion in France, 1562–1576* (Cambridge, 2002), p. 167.

firearms of all sorts gave to those spearheading overseas colonial enterprises). The best-known – and perhaps the most misunderstood – of these early artillery campaigns was the French invasion of Italy in 1494 by which Charles VIII sought to enforce his claim to the throne of Naples.

Two artillery trains accompanied the French army in 1494, one of them bombarding the Aragonese castles in Naples in April and March 1495. Contemporary Italians wrote breathlessly of the horse-drawn guns, moving fast enough to keep up with the mounted *gendarmerie*. 'Diabolical rather than human weapons' to Guicciardini; to Giovio their speed was 'incredible'. Talk of speed prompted Carlo Cipolla to reach for his 'pinch of salt'.[5] Siege trains (even French ones) did well to cover 10 miles a day. Giovio's description specified thirty-six guns, leading some to assume that all were siege cannon, which, if true, would have represented a prodigious weight of metal. Contemporary French *Bulletins*, however, made it clear that the heavy siege pieces (the eight or nine cannon serpentines, and the four gross culverins) formed only about one-third of the French train.[6] Guicciardini reported fortified places reduced by the French 'in a few hours . . . as formerly in the like number of days'.[7] From this he drew much wider conclusions. 'The French, in their invasion of Italy, infused so much liveliness into our wars that . . . whenever the open country was lost, the state was lost with it.'[8] These quotations support the narrative of the French shooting their way through Italy in 1494–5; as one distinguished historian put it, 'even the most majestic medieval walls melt[ed] before the blaze of such an assault'.[9] Yet the places which fell in 1494–5 were generally small and weak, and many were stormed without the use of guns. Strong fortresses were avoided until the French expedition reached Naples, where the Castelnuovo and two other castles had to be captured to secure the city. The Castelnuovo resisted bombardment for two weeks before its garrison surrendered following an accidental explosion in the castle magazine. The castle (medieval in origin and strengthened in the mid-fifteenth century with a low-level gun gallery) was defended by the French for many months later in 1495 before surrender

5 Carlo Cipolla, *Guns and Sails in the Early Phase of European Expansion 1400–1700* (London, 1965), p. 28.

6 Simon Pepper, 'Castles and Cannon in the Naples Campaign of 1494–95', in David Abulafia (ed.), *The French Descent into Renaissance Italy, 1494–95: Antecedents and Effects* (Variorum, 1995), pp. 263–93.

7 Francesco Guicciardini, *The History of Italy*, trans. Austin P. Goddard, 10 vols. (London, 1754), vol. 1, pp. 148–9.

8 Francesco Guicciardini, *Ricordi politici e civili*, No. 64 in *Opere inedite*, ed. Giuseppe Canestrini (Florence, 1857), p. 102.

9 Lynn White, 'Jacopo Aconcio as an Engineer', *American Historical Review* 72 (1967), 425–44.

following the detonation of a gunpowder mine beneath a key outwork, another event described in apocalyptic terms. Although probably not the first explosive mine, it reminds us that gunpowder gave the ancient art of undermining walls a new potency. The five-month siege of Rhodes in 1522 saw at least fifty Turkish mines. The final two years of the siege of Candia (1667–9) would see 1,364 Turkish mines exploded.[10]

The French certainly brought 'liveliness' to Italian warfare. Writing of the rapid collapse of any state defeated in the field, however, Guicciardini may have been thinking not of 1494–5 but of 1509 when, following the Venetian defeat at Agnadello by a largely French army, almost all of Venice's extensive *Terraferma* and its cities fell in two weeks to the forces of the League of Cambrai, most towns simply opening their gates. Similar wholesale collapses would be seen again, such as that of the heavily fortified Dutch Republic in 1672, invaded from all sides, its garrisons despairing of relief. The mythology of 1494–5 is deeply entrenched, however, and exaggerates both the power of the new artillery and the weakness of late medieval fortifications.

Guns in various forms had been around for at least 150 years by 1494 and all over Europe fortifications had evolved in response.[11] Medieval walls were packed with earth against incoming heavy shot, their towers reduced in height and filled. The fifteenth and early sixteenth centuries would see new artillery towers (often with massively thick walls as well as numerous tiers) stacking guns vertically; low external galleries mounting guns firing horizontally (their most effective trajectory); and immense earth banks (the *Wall* of greater Germany) built either in front of medieval walls or behind them, providing both a shield and a raised artillery platform. The most important of these early responses was the triangular bastion, which would become in Sir John Hale's words 'the International Style *par excellence*' of the Renaissance.[12] Originating in the pointed gun-tower of the late fifteenth century, the bastion evolved into a low earth-filled gun platform projecting from the walls. It allowed flanking fire along the curtain and the faces of adjacent works, while guns mounted on the faces fired outward to 'annoy the enemy'. Pointed bastions

10 Gregory Hanlon, *The Twilight of a Military Tradition: Italian Aristocrats and European Conflicts, 1560–1800* (London, 1998), p. 163.

11 Kelly DeVries, 'The Impact of Gunpowder Weaponry on Siege Warfare in the Hundred Years War', in Ivy D. Corfis and Michael Wolfe (eds.), *The Medieval City under Siege* (Woodbridge, 1995/99), pp. 227–44; and 'Facing the New Technology: Gunpowder Defenses in Military Architecture before the Trace Italienne, 1350–1500', in Brett Steele and Tamara Dorland (eds.), *The Heirs of Archimedes: Science and the Art of War through the Age of Enlightenment* (Cambridge, MA, 2005), pp. 37–71.

12 Hale, 'The Early Development of the Bastion', 466.

also eliminated the 'blind spots' which were nearly always found with semicircular or rectangular works, this becoming an obsession for designers despite impressive performances in action of old-fashioned circular works at Padua (1509), Rhodes (1522), and Famagusta (1570–1). Bastions were just as frequently irregular as symmetrical and adapted easily to different sites. This intrinsic flexibility and simplicity is often obscured by the seductive illustrations of fantastic exercises in star-shaped regular polygonal geometry, the drawings criss-crossed with literally hundreds of interlocking fields of fire. (See Figures 10 and 11.)

Since fortifications were not expected to hold out indefinitely without relief, the success of a defence was measured by delay. Even the earth-filled ramparts of the fully formed Italian bastioned system would fall to sustained battery at the close ranges (from 60 to 100 paces) urged in the treatises. In the early stages of a siege it was not extra thickness in the walls but effective counter-battery fire that extended a defence.[13] Once heavy siege guns were established behind gabions – wicker baskets filled with earth, the sandbags of early modern warfare – only other big (or at least medium-calibre) guns could match them. The suppression of defensive fire in the opening artillery duel allowed the heavy battering pieces to be advanced to the outer edge of the ditch, marking the beginning of the endgame.

The ditch itself was the other key defensive element: often a wet ditch in the north of Europe but nearly always dry in the south. To Francesco di Giorgio, 'a fortress without a ditch is like an animal without one of its limbs'.[14] Widening and deepening dry ditches made them more difficult to fill or to mine. By raising their outer bank (the counterscarp) older fortresses could be somewhat sheltered from gunfire. New fortresses would be built in deep ditches with only the upper works rising above ground level. Around the counterscarp ran the 'covered way', shielding infantry behind the sloping 'glacis'. Breaching artillery needed to be aimed at the lowest third of the rampart, requiring guns to be placed on the edge of the ditch, or 'crowning the covered way'. Mines were often used to blow-in the counterscarp, thus providing a slope into the ditch, where more trenches had to be excavated to provide shelter against fire from the *caponiers* (pill boxes) on its floor as well as from flanking artillery. Casualties in an assault, however, were just as likely to be caused by arquebus balls, and even by arrows or stones. The role of small arms in fortress defence is often overlooked. In 1502 the breach at Canosa was defended by 200 Spanish arquebusiers so effectively that after the surrender Jean D'Auton 'marvelled that so few should have been able to withstand so many'.[15]

13 Luigi Collado, *Pratica manuale di artiglieria* (Venice, 1586), f. 43.
14 Francesco di Giorgio Martini, *Trattati di architettura, ingegneria e arte militare*, ed. Corrado Maltese, 2 vols. (Milan, 1967), vol. 2, p. 474.
15 Cited in Frederick Taylor, *The Art of War in Italy, 1494–1529* (Cambridge, 1921), p. 43.

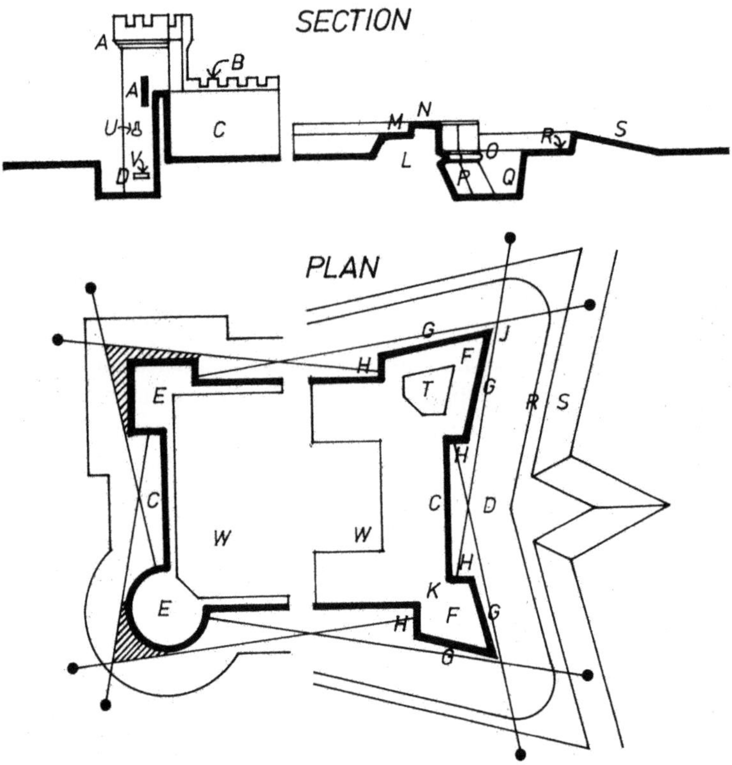


FIGURE 10. Medieval and early modern fortifications compared. The diagram contrasts the medieval wall and tower defences (left) with the bastion and rampart system (right) and provides us with a glossary of fortification terminology. A: Machicolations or machicoulis gallery, allowing defenders to drop heavy objects to the foot of the wall. B: Merlons, protective components between embrasures. C: Curtain, section of wall or rampart between towers or bastions. D: Ditch. E: Rectangular or circular medieval towers, with hatching indicating blind spots. F: Bastions. G: Face. H: Bastion flank. J: Salient or pointed tip of bastion (less acute angles, of course, gave more internal space for cannon). K: Gorge or throat. L: Rampart. M: Terreplein or gun platform. N: Parapet. O: Cordon, moulding dividing vertical and sloping sections of the scarp. P: Scarp, battered (sloping) lower part of scarp wall. Q: Counterscarp, outer wall of ditch. R: Covered way, protected infantry position outside the ditch. S: Glacis, gently sloping earth bank concealing the covered way and all but the uppermost defensive works. T: Cavalier, raised gun platform on bastion or rampart. U: Keyhole gunport. V: Letterbox gunport. W: Enceinte, circuit of fortifications surrounding a fortress (military usage).

Drawing by the author, reproduced from Simon Pepper and Nicholas Adams, *Firearms and Fortifications: Military Architecture and Siege Warfare in Sixteenth-Century Siena* (Chicago University Press, 1986), p. 4.

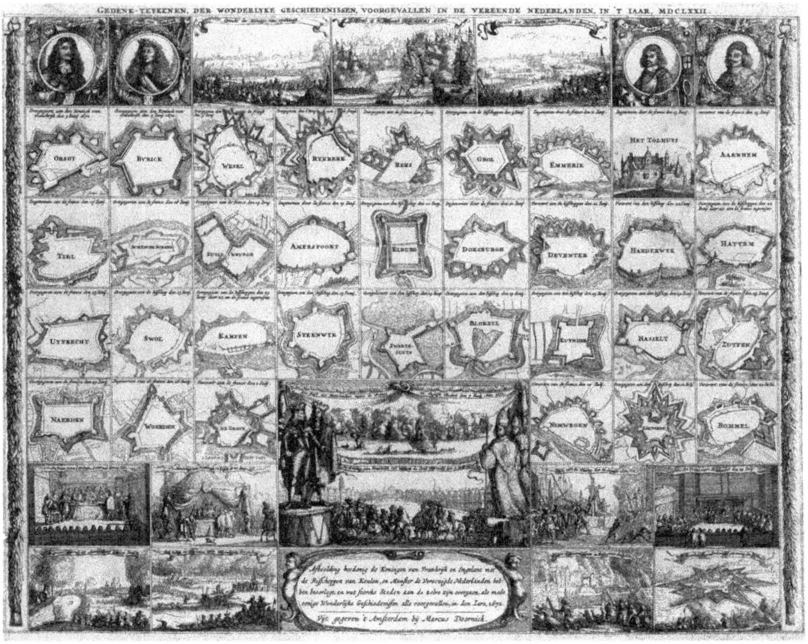


FIGURE 11. Fortifications in seventeenth-century Netherlands. Marcus Doornick's propaganda poster shows, in date order, the Republic's strongholds that were taken by siege, surrendered, or abandoned in May, June, and July 1672 when the Netherlands were attacked on land by France and her German allies (and at sea by England). Its political tone reflects the poor state of preparedness of the Republic and the many disasters that befell it until the recall of the prince of Orange as captain-general (February 1672) and his later installation as *stadtholder*. Vignettes illustrate the Republic's external enemies, events such as the lynching of the De Witt brothers and, of course, French atrocities. Under Orange leadership the tide of war finally turned at the end of the year with the recapture of Coevorden (shown bottom right-hand corner). Our interest in the poster is the snapshot it gives of the military architecture of an advanced European state by the second half of the seventeenth century. The bastioned system shows enormous flexibility.

*Source*: Rijksmuseum, the Netherlands.

The new fortifications dwarfed their medieval forebears, both physically and financially. The work beside Padua's Porta Codalunga in 1509 was defended by 1,000 men: 'questa fortezza' Bembo called it.[16] One new 'bulwark' at Arras (*c.* 1513) 'practically enclose[d] an entire football field'.[17]

16 Pietro Bembo, *Della historia vinitiana . . . volgarmente scritta, libri XII* (Venice, 1552), f. 128v.
17 Pieter Martens, 'An Early Sixteenth Century Drawing of Two Bulwarks at Arras', *Fort* 27 (1999), 67–92, 75.

Nettuno (1501–3), a four-bastioned coastal fortress, would have fitted inside a single bastion of the nine surrounding the Venetian border fortress of Palmanova (1593–8). For reasons of cost as well as complex property issues, big city defensive projects progressed slowly. Antwerp was the first major northern city to receive a completely bastioned urban *enceinte* (started 1543). The bastions of Lucca were planned in the 1520s, redesigned repeatedly, but completed only in 1650. The walls of many German cities besieged in the Thirty Years' War had by then seen only partial modernisation. Thus in many parts of Europe the first modern urban fortification to reach completion was a citadel. Utrecht's Vredenberg castle was started in 1528 on Charles V's orders but built at local expense. By mid-century new fortresses dominated Florence (Medici creations), Ancona, Perugia (both Papal), and Siena (Imperial). When Turin became the new capital of Savoy shortly after 1559, a five-bastioned pentagonal citadel was built and promptly copied for Antwerp. Six further identical citadels were then planned by Spain to secure the Low Countries. Not surprisingly, the Utrecht and Antwerp citadels were amongst the first to be destroyed in the early phases of the Dutch revolt, while the much resented Imperial citadel in Siena became the focus of revolt even before its completion. Control of a citadel and the strategic target it represented for jealous neighbours made it an asset of debatable value.[18]

For much of the sixteenth century the number of guns employed in sieges did not match the increasing size of fortifications. Tudor England was a pioneer manufacturer of cast-iron artillery, producing guns costing one-third as much as those cast in bronze and eventually contributing significantly to the growth of artillery strengths. But the English industry was still small and highly concentrated in Kent as the country drifted into Civil War in 1642, where it fell into Parliamentary hands together with the ordnance store at the Tower and the arsenal at Hull, leaving the Royalists critically short of guns.[19] France still struggled in Colbert's time to establish a modern heavy arms industry.[20] The Dutch Republic would become the most heavily armed and fortified state for its size and in 1589 started a central cannon foundry at The

18 For opposed views, David Parrott, 'The Utility of Fortifications in Early Modern Europe. Italian Princes and Their Citadels, 1540–1640', *War in History* 7:2 (2000), 127–53; Thomas Arnold, 'Fortifications and the Military Revolution: The Gonzaga Experience, 1530–1630', in Clifford Rogers (ed.), *The Military Revolution Debate: Readings on the Military Transformation of Early Modern Europe* (Boulder, 1995), pp. 201–26.

19 Stephen Bull, *The Furie of the Ordnance: Artillery in the English Civil Wars* (Woodbridge, 2008), pp. 54–80.

20 Cipolla, *Guns and Sails*, pp. 66–71.

Hague.[21] Lacking local raw materials, the Dutch developed iron and copper mines in Sweden, Germany, and Russia and by the early seventeenth century the de Geer's Swedish foundry had become 'the largest producer of cannon and cannon-balls not only in Scandinavia but possibly the whole of Europe'.[22] A watershed in the number of cannon employed in Dutch-led sieges around the year 1600 was noted by Geoffrey Parker. Grol (Groenlo) was besieged unsuccessfully with sixteen guns in 1595, and only fourteen in 1597; but in 1627 the Dutch took the place with eighty guns. The Dutch failed when only 22 guns were brought against 's-Hertogenbosch in 1601, but 116 guns achieved success in 1629.[23]

Sixteenth-century sieges, however, only rarely saw large concentrations of heavy cannon. Possibly as many as sixty heavy guns and many more small pieces were assembled for the siege of Padua (1509) by combining the resources of the League of Cambrai[24] and transporting guns over very long distances. Charles V's unsuccessful attack on Metz in 1552 saw a similar concentration, which was repeated for the emperor's capture and total destruction of Thérouanne in 1553.[25] Water was often the quickest way to move artillery, and the Dutch made notably good use of rivers and canals to achieve strategic concentrations.[26] But Padua, Metz, and Thérouanne all involved major power confrontations. Away from Europe's industrial and military heartland, the artillery shortages could not be disguised. Mid-sixteenth-century Italy saw siege trains generally no bigger than those of 1494–5 and a survey carried out in 1552 recorded

21 Olaf van Nimwegen, 'Maurits van Nassau and Siege Warfare (1590–1597)', in Marco van der Hoeven (ed.), *Exercise of Arms: Warfare in the Netherlands, 1568–1648* (Leiden, 1997), pp. 113–31, 122.

22 Jonathan Israel, *The Dutch Republic: Its Rise, Greatness and Fall 1477–1806* (Oxford, 1995), pp. 274–5; Jan Piet Puype and Marco van der Hoeven (eds.), *The Arsenal of the World: The Dutch Arms Trade in the Seventeenth Century* (Amsterdam, 1996); Cipolla, *Guns and Sails*, pp. 46–60.

23 Geoffrey Parker, *The Army of Flanders and the Spanish Road, 1567–1659*, 2nd ed. (Cambridge, 2004), p. 15.

24 The League, formed in 1508 at Cambrai, committed France, the Empire, Spain, the Papacy, and a number of other Italian powers to attack Venice. The Republic lost most of its mainland territorial holdings following defeat at Agnadello in 1509 but, after shifts within the League, Venice (by now an ally of France) regained all of this territory by 1517. The different contingents in the alliance make estimates of troop strengths and the number of guns difficult to calculate. For likely Venetian exaggeration, Michael Mallett and Christine Shaw, *The Italian Wars* (London and New York, 2012), pp. 94–5.

25 G. Zeller, 'Le Siège de Metz par Charles-Quint (octobre–décembre 1552)', *Annales de l'Est* 13 (1943), 1–270; B. de Salignac, *Le Siège de Metz par l'Empereur Charles V en l'an 1552* (Paris, 1850); Pieter Martens, 'La Puissance de l'artillerie de Charles Quint au milieu du XVIe siècle: Le siège de Thérouanne en 1553', in Nicolas Prouteau, Emmanuel de Crouy-Chanel, and Nicolas Faucherre (eds.), *Artillerie et fortification 1200–1600* (Rennes, 2011), pp. 119–42.

26 Van Nimwegen, 'Maurits van Nassau and Siege Warfare', p. 122.

that twenty-six Florentine fortresses held 625 guns, of which only 58 were heavy pieces, half of them in Florence. Most Florentine fortresses contained no heavy weapons.[27] By 1577 the Wars of Religion had so eroded France's early lead in artillery that the nineteen fortresses of the front-line province of Picardy in 1577 contained 360 pieces, only 89 of them heavy. Seven fortresses had no heavies.[28] Guilmartin identified a much wider Mediterranean shortage: 'there simply wasn't that much artillery around'.[29]

Artillery strength is at best a proxy for the resources deployed at a siege, albeit one widely employed by contemporaries. Other factors always need to be considered. Venice's successful defence of Padua in 1509 against the combined forces of the League of Cambrai is a case in point. Following Padua's recapture by Venice in a *coup de main*, the city's fortifications had been strengthened by a dry ditch excavated *behind* the extensively reinforced outer walls. The ditch was swept by fire from caponiers and backed by another massive rampart (probably inspired by the German *Wall*) mounting guns firing over the outer walls.[30] Although a large siege train was assembled by Maximilian I, Louis XII, and their Italian allies, their guns were roughly matched by Venetian defensive artillery (some of which had been removed from the Republic's galleys and moved by barge to Padua). Hard fighting, improved fortifications, and matched artillery therefore all played important parts in an operation which was certainly as significant as the events of 1494–5. But a rounded assessment would also include the unseasonable rain which flooded the besiegers' trenches, and the enormous size of the League army which magnified its supply problems in surroundings stripped bare by the defenders. Venice managed to get pay for its soldiers into the besieged city and – during the siege – the Venetian banker Priuli correctly predicted that the emperor would be unable for much longer to pay his troops.[31] Nearby Treviso had remained loyal in the aftermath of Venetian defeat at Agnadello. The triangle formed by Padua, Treviso, and the redoubts on the lagoon allowed

27 Arnaldo D'Addario, 'Burocrazia, economia e finanze dello stato Fiorentino alla metà del Cinquecento', *Archivio storico italiano* 121 (1963), 392–496 and table at 446.

28 Wood, *The King's Army*, pp. 168–70 and Table 6.4.

29 John Guilmartin, *Gunpowder and Galleys: Changing Technology and Mediterranean Warfare at Sea in the Sixteenth Century*, 1st ed. (Cambridge, 1974), p. 181. Also by Guilmartin, 'The Siege of Malta, 1565', in David Trim and Mark Fissel (eds.), *Amphibious Warfare 1000–1700: Commerce, State Formation and European Expansion* (Leiden, 2006), pp. 149–80.

30 Description in Francesco Guicciardini, *The History of Italy*, trans. Austin P. Goddard, 10 vols. (London, 1754), vol. 4, pp. 337–9.

31 Girolamo Priuli, 'I diarii', in *Rerum Italicarum Scriptores, XXIV*, 4 vols. (Bologna, 1912–38), vol. 3, pp. 286–7.

a defensive strategy that anticipated that of the later Austrian Quadrilateral. An attack on any one of them would bring support from the others. Guerrilla units and light cavalry threatened League communications. Outlying positions within the Venetian triangle, however, changed hands with bewildering frequency in a pattern which would be repeated in other Italian campaigns.[32]

This was also the pattern of the struggle which ended in the defeat of Siena and its French ally by Imperial and Florentine forces between 1553 and 1556. A single battle in the second year of the war, Marciano (2 August 1554), destroyed the French field force which until then had prevented the complete investment of Siena and had even briefly threatened Florence. The war of Siena was to be a war of many sieges.

The Imperialists planned in 1553 to isolate Siena from its hinterland by capturing Montalcino, 20 miles to the south. The plan stumbled on a classic frontier action at the small frontier town of Monticchiello which held out for eighteen days until its ammunition was exhausted. Although not a lengthy defence, it delayed the invasion for a critical month, including the capture of a dozen villages and castles to protect communications with Montalcino.[33] Montalcino was a tougher objective, with earthwork bastions covering the only narrow approach, backed by a second line of retrenchments. Montalcino held out for eighty days against an initial breaching battery (abandoned when the besiegers ran short of shot), mining (an impressive explosion but disappointing results), blockade (never completely water-tight), and finally an attack by sapping into an inadequately flanked bastion which had almost broken through when the Spanish troops were withdrawn to protect Naples from an approaching Turkish fleet (a product of the Franco-Ottoman naval alliance which scandalised Christian Europe by opening the French Mediterranean ports to the Turks).[34]

Phase II opened in January 1554 with a surprise night attack in severe weather which narrowly failed to take Siena but captured three outworks which marked the front-line until April 1555. Like other very lengthy operations, the siege was marked by a number of distinct phases. An inconclusive artillery duel was followed by four months (February to May 1554) of operations against outlying Sienese positions threatening Imperial communications; dozens of minor actions against medieval towers, fortified villas and monasteries, which

32 Other primary references in Simon Pepper, 'The Face of the Siege: Fortification, Tactics and Strategy in the Early Italian Wars', in Christine Shaw (ed.), *Italy and the European Powers: The Impact of War, 1500–1530* (Leiden and Boston, 2006), pp. 33–56, 41 n. 22.

33 Simon Pepper and Nicholas Adams, *Firearms and Fortifications: Military Architecture and Siege Warfare in Sixteenth-Century Siena* (Chicago, 1986), pp. 93–4.

34 Ibid., pp. 95–116.

were often abandoned when guns approached. Only after victory at Marciano were the Imperialists able to tighten their grip on the city and move against a number of the more important regional fortresses, which were often difficult to reach, but not to take once guns were positioned. Following another unsuccessful attempt in January 1555 to force a solution by concentrating artillery for a surprise battery, Siena – by then starving and despairing of French relief – surrendered in April.[35]

The practical limitations on the use of artillery were exposed in this campaign by the ammunition shortages which forced Monticchiello to surrender and cut short the Imperial bombardment of Montalcino, and by the difficulties experienced in bringing guns up to what were often primitively fortified positions. Much of the siege involved small forces, illustrating the trade off in any siege between numbers, costs, and the prospect of a timely outcome. Just before Marciano the Imperial strength outside Siena was given as 7,000. Three months later it was 3,500 'combat worthy infantry . . . the rest are all unfit'.[36] Only in the final weeks did Florentine-Imperial forces around Siena rise (perhaps to as much as 30,000), allowing a close blockade. These harsh realities of mid-century Italian siege warfare also prompted the development of fieldworks defended mainly by infantry with small arms. The zig-zag *tenaille trace* was reputedly first used for the forts around Mirandola (1552), allowing small units to blockade in security. Siege forts go back to Roman times, of course, often combined with fieldworks encircling a besieged town. Gunpowder artillery had now so greatly expanded the zone of siege operations that the much longer continuous rampart enclosing a besieged fortress required a much larger blockading garrison. Isolated siege forts, the experienced captain Castriotto explained, represented both a viable and a cheaper alternative.[37]

If artillery failed to deliver decisive results in mid-century Tuscany, the siege of Malta in 1565 has been called a 'gunners' siege'. The Ottomans brought sixty heavy battering pieces to Malta, where they were opposed by only thirty-six full cannon, twelve smaller modern guns, and three old-fashioned bombards firing stone balls, leading Guilmartin to conclude that the Knights had focused their procurement on heavy pieces. Just over half of the Knights' heavy guns – nineteen out of thirty-six – were placed in the

35 Ibid., pp. 117–39.

36 Marignano Correspondence with the duke of Florence in the Archivio di Stato, Florence, *Mediceo del Principato* 1854 (17 Jul. 1554); ibid., 1853 (5 Oct. 1554).

37 For Mirandola's forts, Girolamo Maggi and Jacopo Castriotto, *Della fortificatione delle città . . . Libri III* (Venice, 1564), ff. 90–94v. Castriotto served at both Mirandola and Montalcino before switching his allegiance from the Empire to France, becoming the king's director of fortifications.

outlying fort of St Elmo which covered the harbour entrances and was clearly going to be attacked first. This bold decision could have gone disastrously wrong. St Elmo's sharply pointed bastions restricted the internal space, making big-gun handling inconvenient, and the fort was overlooked by rising ground. In the event, its heavy guns were able to keep the stronger Ottoman batteries initially at a range of about 300 metres, much reducing the effectiveness of Ottoman fire.[38] St Elmo held out for twenty-eight days before being overwhelmed. Only then was the Ottoman artillery moved to the other side of Malta's Grand Harbour for the attack on the Knights' main positions. The principal siege started more than a month after fighting had begun at St Elmo, critically allowing time for relief to assemble in Sicily and to land when the Ottoman expeditionary force was already depleted, exhausted, and almost out of ammunition, with autumn gales threatening their fleet.[39]

Successful sieges and defences both hinged on logistics. The siege of Naples in 1527 ended in disaster for the League of Cognac when the maritime supply chain from France and Genoa to the French army outside Naples was broken (by the defection of Andrea Doria and his fleet to the Imperial camp), leaving the besiegers to waste away from starvation and disease before being cut to pieces in withdrawal.[40] La Rochelle was finally forced to surrender after a year-long siege in 1627–8 when Richelieu's army closed the channel into the port. For almost twenty years Venice defended Candia (Crete), resupplying the coastal fortress, defeating the Ottomans in naval engagements, and at times putting them under pressure by blockading Smyrna and the Dardanelles. Here and elsewhere the absolute essentials were ammunition and reinforcements. Food shortages caused misery, of course, but starvation (even for trapped civilians) was an elastic concept. Starving defenders would fight on given hope of relief or in fear of their fate should the place fall. Successful blockades were rarer than is often supposed.

Contested sieges, however, could be quite as bloody as battles. At La Rochelle's first siege of 1573 the Catholic army was effectively destroyed in four months before the Protestant stronghold. For this siege 'a supreme effort' had been made to assemble some forty-two cannon and great culverins and, in two weeks of battery, the Protestants' walls were demolished, or at least reduced to a high slope of loose rubble which had to be climbed against

38 Stephen Spiteri, *The Great Siege . . . MDLXV* (Privately published, Malta, 2005), p. 144, citing A. F. Cirni, *Commentari* (Rome 1567), p. 57: 'cento sessanta canne'.

39 Guilmartin, *Gunpowder and Galleys*, pp. 176–93.

40 Maurizio Arfaioli, *The Black Bands of Giovanni: Infantry and Diplomacy during the Italian Wars* (Pisa, 2005), pp. 98–161 for events in Naples in 1528.

fierce defensive fire.[41] Eight major assaults on the breach were then delivered throughout April and May, together costing more than 2,000 casualties. Brantôme (who was present) estimated 22,000 total casualties of which James Wood substantiates 6,000 combat losses, with about the same number sick and deserted. By the end of the siege – despite reinforcements – 'less than half of the original strength' remained fit for service.[42] Officers suffered particularly heavily, with 43 per cent killed and 30 per cent wounded. Of 1,500 pioneers, only 600 survived. The battering pieces had been brought up to the edge of the ditch, where the crews suffered crippling losses, mainly from small arms and light artillery. The Royal gunners were reduced to 27 per cent of their original strength. When hostilities resumed in 1574–5 the Royal army was unable to field more than one-third to one-half of the guns used at sieges during the first four civil wars. This would prove seriously limiting in the later civil wars, which would see many sieges but no more major battles until Coutras (1587). Overall, the besiegers of La Rochelle suffered between six and seven casualties for every one sustained by the defenders, with ratios of 3:1 for sorties and 10:1 for major assaults.[43]

Even higher total casualties would be suffered at Ostend (1601–4), which saw intense fighting in the opening and closing stages, and steady attrition for the three years and eighty days of the siege. The operations at Ostend illustrate the developing tactics of siege warfare; in particular, the use of artillery to enfilade defences. Here can be seen the beginnings of a more scientific approach to tactics, contrasting sharply with the crude frontal battery which killed so many gunners at La Rochelle.

Protected on its three landward sides by tidal flats and marshes, Ostend was fortified with bastions, an outer *enceinte*, and two wet ditches. A tidal creek on the west offered the only semi-dry approach. Guns mounted on dunes to the west of the creek reduced the Dutch, English, and Huguenot garrison to life in cellars and dugouts, but at 400–500 yards the range was too great to destroy fortifications. When guns on floating pontoons threatened to close the eastern harbour mouth, the defenders cut a new entrance channel to the North Sea. Their shipping supply line secured, the outcome at Ostend would turn on the defence of two key positions. (See Figure 12.)

41 Wood, *King's Army*, pp. 246–74 for La Rochelle (1573); and pp. 205–25 for the Royalist defence of Chartres (1568).

42 Ibid., pp. 269–71; Pierre de Bourdeille de Brantôme, *Œuvres complètes*, ed. L. Lalanne, 11 vols. (Paris, 1864–82), vol. 4, p. 90.

43 Wood, *King's Army*, pp. 258, 270–1, 273, also Table 10.2.

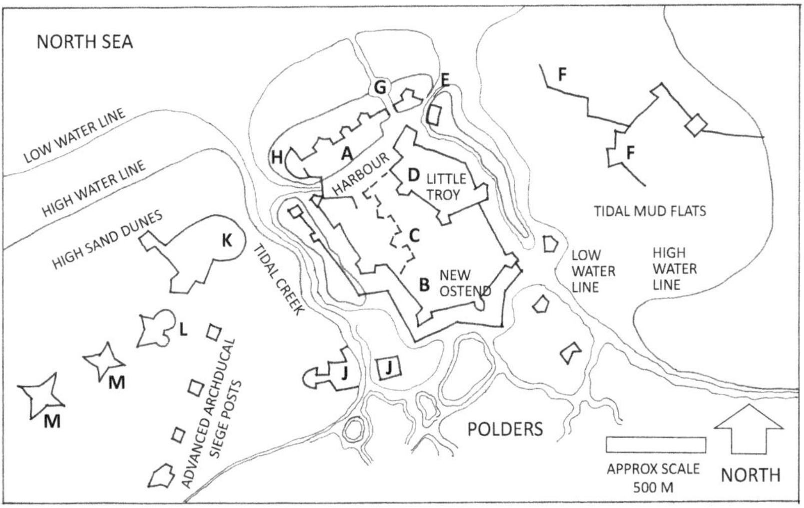


FIGURE 12. Siege of Ostend, 1601–4. Key: A: Ostend Old Town built on the inland slope of the Town Dune, defended by small bastions and beach obstacles on the sea-facing side. B: New Ostend (sixteenth century). C: Retrenchment built after loss of Polder positions. D: Little Troy, retrenchment built in final days of the siege. E: Original entrance to harbour closed to shipping by guns firing from F: the 'floating batteries'. G: New harbour entrance cut to secure access from the North Sea, later known as 'The Cut'. H: The Sandhill, battery commanding beach and tidal creek. J: The Polder (defensive) positions commanding the tidal creek and (after their capture) also sweeping the creek and rear of the western defences. K: Main Archducal artillery battery sited on top of the high coastal sand dune. L: Archducal battery on lower ground. M: Archducal forts (note use of tenaille trace).

*Source*: the plan drawn by author from the reconstruction in Edward Belleroche, 'The Siege of Ostend or the New Troy 1601–1604' reprinted in *Proceedings of the Huguenot Society*, Vol. III, no. IV (1892).

The Sandhill marked the western end of the old town dune and its guns in deep dug-outs swept the seaward end of the creek. If lost, the old town would be lost and with it the newly cut harbour entrance. The defenders also quickly grasped the importance of some polders beyond the south-west fortifications. On them new outworks were built, controlling the streams feeding the ditches and with a clear line of shot along the creek. The polder works were eventually lost in April 1603 (in a night attack in heavy rain) and efforts began to divert water away from the ditches. Under Ambrogio Spinola (who took command of the siege later in 1603) the approaches across the upper creek now became a marvel of sap and mine construction, some of the works being waterproofed and others pumped out between tides. The western *enceinte* was progressively breached by a series of mine explosions.

From the captured polders, Spinola's guns now enfiladed the western fortifications (shooting along and behind them), forcing withdrawal to a second line. When this was taken the Dutch fell back to their redoubt, Little Troy, enclosing perhaps 20 per cent of the original town. After the Sandhill (now well 'behind the front') was taken on 13 September 1604, Spinola's troops advanced quickly along the Old Town dune to the newly cut harbour mouth from where the ramparts of Little Troy could be enfiladed. A week later the defenders surrendered on surprisingly generous terms.[44]

If sixteenth-century Italy was the first school of modern siege warfare, its university would become the Low Countries and, later, France. The military engineers of the Italian Wars (whether soldiers or civilian technicians) learned largely by word of mouth and experience. The late sixteenth century saw the beginnings of a flood of printed treatises, fortification geometry taught at Padua by Galileo, and Maurice van Nassau's cadet engineers receiving initial classroom instruction at Leiden from Simon Stevin, before being attached to operational units or construction sites for practical experience.[45] Dutch experts were as likely to be found advising foreign powers as peripatetic Italians. It would be misleading to suggest that trained military engineers would now dominate the field because the three leading figures in late seventeenth-century siege warfare – Menno van Coëhoorn, Eric Dahlberg, and Sebastian Le Prestre de Vauban – all served as line officers before becoming siege specialists. This said, the century would see the establishment of cadres of trained Dutch and French engineers, increasingly systematic tactics, and considerable strategic vision in the planning of national defence systems.

The Dutch Republic's frontier defence line erected in the winter of 1605–6 comprised wooden blockhouses connected by ramparts running from the Zuider Zee along the IJssel to Arnhem, up the Neder-Rijn to the Waal, and thence to Tiel.[46] Later the Republic established forts controlling the 'waterline' – the floodable, zone protecting Amsterdam. Spain's military canal allowed artillery to be switched between the Rhineland and the southern Netherlands. The Habsburg 'military frontier' against the Ottomans now ran for hundreds of miles from Croatia, through Hungary and into Poland, with watchtowers, staked

44 Ostend: Christopher Duffy, *Siege Warfare: The Fortress in the Early Modern World 1494–1660* (London, 1979), pp. 85–9; for background Bentivoglio Giustiniani, *The History of the Wars of Flanders*, trans. Earl of Monmouth (London 1678); *A True Historie of the Memorable Siege of Ostend . . . Translated out of French into English by Edward Grimestone* (London, 1604).

45 John Hale, 'Military Academies in the Venetian Terrafirma in the Early Seventeenth Century', *Studi Veneziani* 15 (1973), 273–95; van Nimwegen, 'Maurits van Nassau and Siege Warfare', p. 126. On the treatises: Horst de la Croix, 'The Literature of Fortification in Renaissance Italy', *Technology and Culture* 6 (1966), 30–50.

46 Parker, *The Army of Flanders*, p. 11.

ditches, fortresses, and defended villages garrisoned mainly by armed civilians. Louis XIV's endless campaigns required Vauban to direct no fewer than forty-eight sieges, and to design or improve some 160 fortresses along the constantly changing frontiers of France. A central archive of fortress models was established in the Tuileries, where the king received weekly briefings and where foreign visitors could appreciate this aspect of 'la gloire'. Vauban himself inventoried the first 144 models.[47] What became known as the *frontière de fer* comprised two rows of fortresses and fortress towns behind the critical north-eastern frontiers, with others clustered on the Alpine and Pyrenean passes and around the key military ports on the Mediterranean, Biscay, and Channel coasts. Nine '*ex nihilo*' fortress towns housed civilians as well as garrison troops. The result when plotted on a map of France resembles a square with an empty centre (where many older fortifications had been sited): hence perhaps the name, the *pré carré* (literally 'square field', sometimes translated as 'duelling ground').[48]

Known today as a fortress builder, Vauban's contemporary fame was as a fortress taker. At his 'perfect siege' of Ath (1697) Vauban recaptured a fortress designed by himself after only fourteen days of open trenches, losing only 300 casualties, and consuming half the expected ammunition.[49] (See Figure 13.) Vauban's biographer, Sir Reginald Blomfield, called it *grande finesse*. Here Vauban had introduced 'parallels', trenches dug across the approaches – parallel to the besieged works – protecting saps and batteries from sorties. The parallels provided a defensible front at every stage. Saps zig-zagged forward to the next parallel, each aligned to deny the enemy a clear shot along their length. The final parallel formed a front line close to the enemy's covered way, from which the ditch would be gained, ravelins eliminated, breaches made, and the final assault launched. Zig-zag saps, of course, appear in the very earliest illustrated treatises, and something very like a parallel appears in a plan of La Rochelle in 1573. The progress of Vauban's approaches at Ath was so rapid that the third innermost parallel was omitted. Here Vauban had systematised 'ricochet fire' using small powder charges to skid or lob shot at low velocity into enemy works, destroying the palisades and staked entanglements on the glacis

47 Nicolas Faucherre, 'La Construction de la frontière. De l'usage stratégique des plans en relief', in Antoine de Roux, Nicolas Fauchere, and Guillaume Monsaingeon, *Les Plans en relief des places du roy* (Paris, 1989), p. 25.

48 For a modern overview see the numerous contributions to the exhibition catalogue, Isabelle Wormoes and Victoria Sanger (eds.), *Vauban, bâtisseur du Roi-Soleil* (Paris, 2007) marking the 300th anniversary of Vauban's death. Modern biography: Anne Blanchard, *Vauban* (Fayard and Paris, 1996).

49 Jamel Ostwald, *Vauban Under Siege: Engineering Efficiency and Martial Vigor in the War of the Spanish Succession* (Leiden and Boston, 2007), pp. 21–45.

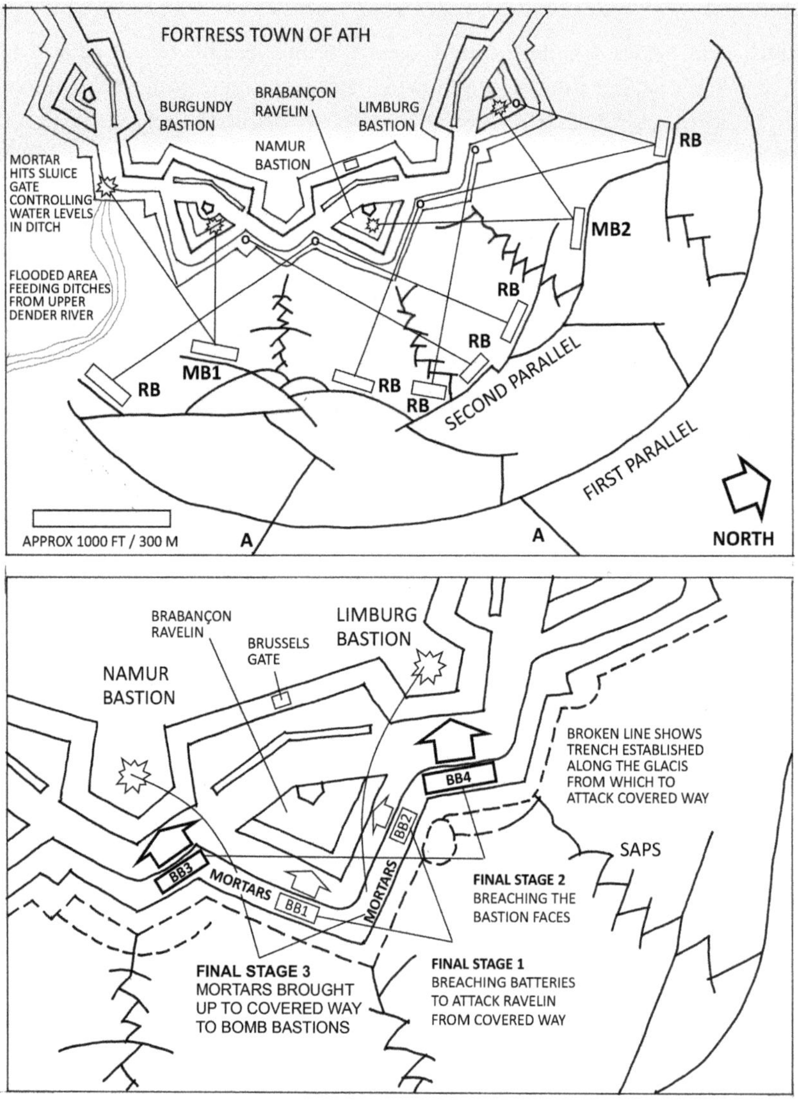


FIGURE 13. The perfect siege of Ath Maps drawn by author adapted from those illustrating the step-by-step description in Jamel Ostwald, *Vauban Under Siege: Engineering Efficiency and Martial Vigor in the War of the Spanish Succession* (Leiden and Boston, 2007), pp. 21–45.

Upper plan records Vauban's opening moves. Approaches opened on the night of 22–23 May with 2,500 ft of trenches (A) and the first parallel excavated was approximately 600 metres from the outer works. The second parallel was established on 24–25 May, batteries emplaced on 25–26 May, and fire opened on 27 May with thirty-six guns in six ricochet batteries (marked RB) scouring the covered way. Mortar batteries (MB1 and MB2) dropped bombs into the ravelins and destroyed the sluice gate controlling the flow of water into the ditches from the dammed river Dender (perhaps the only element of luck). The town was briefly flooded. Subsequently water levels in the ditches were lowered. The saps advanced to the glacis. The covered way (by then abandoned) occupied on the night of 29–30 May.

Lower plan records final attacks on the ravelin (final stage 1) and bastions (final stages 2 and 3). On 1 June, breaching batteries (BB1 and BB2) on the covered way brought down outer faces of the ravelin. Miners tunnelled from the glacis to the floor of the ditch and 'bridged' it with fascines. Storm troops captured the ravelin, assisted by ricochet fire from or just clearing the parapets to dismount guns and create a destructive passage through defenders (as well as saving powder). A combination of ricochet fire and mortars (and the increasing use of high-angle howitzers firing explosive shells) represented a serious threat to open bastions, which soon became cluttered with traverses thrown up to limit blast damage or the destructive passage of ricochet shot. For his last new fortress town at Neuf-Brisach (1698–1705) Vauban employed enclosed 'tower bastions' which can be seen as a revival of the artillery towers. The civilian houses and barracks were all built with bomb-proof basements. Vauban, it should be said, made no claims to originality, and wrote no promotional treatise on fortification; indeed, he demonstrated a pragmatic disregard for 'systems' of fortifications.[50] His tactical insights and clarity, however, left their mark. Siege parallels and ricochet fire would still be employed in formal sieges in the mid-nineteenth century.

Vauban's casualty-averse tactics proved less durable. With the support of Louvois and the ear of Louis XIV (present at many of Vauban's earlier sieges) what Ostwald has called 'engineering efficiency' made progress. When Vauban was promoted Marshal of France in 1703 and then promptly side-lined, French commanders as well as their Dutch and English opponents abandoned *finesse* for firepower and pressed their assaults remorselessly. Another step-change in resources seems to have occurred in the early eighteenth century. At twenty-five sieges from 1697 to 1712 the besiegers' artillery typically included some sixty cannon and thirty heavy mortars, excluding howitzers and the light Coëhoorn mortars. For Lille (1708) the besiegers brought 120 guns; for Tournai (1709) 130; for Douai (1710) 288 guns

CAPTION FOR FIGURE 13. (cont.)

BB1 and BB2. Mortars were brought forward to the covered way from MB1 and MB2 (upper drawing) to target the Limburg and Namur bastions. The final stage 2 started on 2 June, when breaching batteries (BB3 and BB4) opened fire on the bastion faces. Mortars, now sited on the covered way, bombed the bastion interiors. Breaches were judged 'practicable' by 5 June. Terms of surrender were negotiated on 6 June and Ath was evacuated on 7 June.

50 Internal technical manuals for his own engineering team are published in English translation: Sébastian Le Prestre de Vauban, *A Manual on Siegecraft and Fortification*, trans. George A. Rothrock (Ann Arbor, 1968). Extensive reports and notes are to be found in *Les Oisivetés de Monsieur de Vauban. Ou Ramas de plusiers mémoires de sa façon sur différents sujets*, ed. Michèle Virol for l'Association Vauban (Seyssel, 2007).

and 80 mortars.[51] Vauban was careful of soldiers' lives and was reputedly appalled by Coëhoorn's infantry assault across open ground to take the covered way at Namur, costing 2,000 Dutch and English casualties. Marlborough's storm of the Schellenberg, a Swedish-built fort commanding Donauworth, cost 5,500 Allied casualties – one-third of the attacking infantry – but in a day of fighting secured fort, town, and a key bridge over the Danube on the road to Blenheim.[52] Ostwald's analysis of the sieges of the War of the Spanish Succession reveals much 'martial vigour' as commanders increasingly short-cut regular sieges 'by the forms', using storm tactics, escalade, surprise, and even that tactic of last resort, the area bombardment. His dataset of eighty-five sieges in five theatres shows that most were short. Only 33 per cent lasted longer than a month, while 56 per cent lasted less than two weeks from investment to capitulation.[53] This is very far from the traditional picture of early modern sieges as ponderously slow, even 'leisurely' operations; although the time consumed by recovery and movement between sieges meant that campaigns may well often have become bogged down by the large number of fortified objectives that needed to be captured when operating in densely urbanised North-West Europe.

Were these patterns replicated in earlier wars? John Lynn assembled data on 135 French sieges from 1445 to 1714 to challenge aspects of Geoffrey Parker's 'military revolution' thesis; namely, that the growth of early modern armies was attributable in large measure to the lengthier sieges brought about by the proliferation of bastioned fortifications which demanded more manpower to besiege, defend, and garrison.[54] Lynn argued that siege lengths declined from 1500 to 1714, averaging sixty-five days (median sixty-one) in the sixteenth century, lengthening to seventy-six days (median thirty-four) in the years 1635–59, before settling to forty-three days (median thirty-seven) for fifty-eight sieges in 1702–14. The differences in the datasets for 1702–14 (eighty-five sieges in Ostwald, only fifty-eight for the much longer period covered by Lynn) illustrate obvious difficulties. Is it always the short sieges which fail to make the cut? What of the many places which 'changed hands under duress' but with no

51 Ostwald, *Vauban Under Siege*, Appendix F.

52 Richard Holmes, *Marlborough: England's Fragile Genius* (London and New York, 2008), pp. 269–76.

53 Ostwald, *Vauban Under Siege*, Appendix C.

54 John A. Lynn, 'The *Trace Italienne* and the Growth of Armies: The French Case', in Rogers (ed.), *Military Revolution Debate*, pp. 169–99 (first published in *Journal of Military History* 55 [1991], 297–330); Geoffrey Parker, 'The 'Military Revolution', 1560–1660 – A Myth?', *Journal of Modern History* 48 (1976), 195–214; Geoffrey Parker, *The Military Revolution: Military Innovation and the Rise of the West, 1500–1800*, 2nd ed. (Cambridge, 1996).

(or only minimal) siege operations?[55] Data from earlier periods are more difficult to measure accurately; but Lynn's dataset included only the sieges of Ravenna, Pavia, and Carignano from more than half a century of heavy major power involvement in the Italian Wars. These earlier wars now cry out for the analysis in depth given to those of Louis XIV. Is it possible, indeed, that the patterns observed for the War of Siena (above) or for the French Wars of Religion represent a separate category of lower-level conflict, with siege tactics developed to cope with the more limited resources available far away from the Low Countries, the 'heartland' of the military revolution?

Do later conflicts help us? Although not a central feature of the Revolutionary or Napoleonic Wars, sieges continued to figure in practically all theatres. Napoleon's eight-month blockade of Mantua in 1796–7 generated victories for France on the north Italian battlefields of Lonato, Castiglione, Caliano, Bassano, Arcola, and Rivoli as the Austrians attempted to relieve their prized Mantuan fortress – spectacularly reversing the balance of centuries between sieges and field engagements. Marlborough had also wanted to fight battles but fought only four in his entire military career against dozens of sieges. In the Peninsular War the French fought at least twenty-five exceptionally well-recorded sieges.[56] Most were offensive operations against Spanish-held positions, but there were eight French defences when the tide of war turned against the Empire in 1812–13. One-third lasted more than a month, including the six-month Spanish defence of Girona, the two sieges of Zaragoza, and Suchet's capture of Taragona, which won him a marshal's baton for what he called 'three sieges in one'. Another 30 per cent lasted less than two weeks. However, the main combat phases of the Peninsular sieges – the artillery duel, breaching battery, and assault – were pressed much faster (and in some respects, harder) than a century earlier. Once the siege artillery had arrived (often weeks after the initial investment), both French and British fire was notable for its accuracy and speed: typically, three days or less to create practicable breaches, and often only hours. Sir John May's slim volume focusing on the technical problems of rapid cannon fire at Cuidad Rodrigo and San Sebastian was translated as *Die Artillerie- oder Schnell-Belagerung* – and this said it all, for the gunners. For the generals it was the old problem of

55 Parker, *Army of Flanders*, pp. 7–9, responds to Lynn and provides further data on surrenders culled from contemporary analyses.

56 J. Belmas, *Journaux des sièges faits ou soutenus par les Français dans la Péninsule. De 1807 à 1814* (Paris, 1837) is a four-volume record, started in 1811 on Napoleon's orders and completed in the 1830s by an officer who had served in Spain and now had access to plans and documents from both French and British archives.

artillery and ammunition transport over long distances and bad roads that was the major source of delay.

Another aspect of the Napoleonic-era sieges is striking. Both civilians and soldiers *fought* in a way not seen since the confessional struggles of the sixteenth century. This is not to suggest that eighteenth-century sieges were not hard fought. But Spain saw a level of aggression that recalls the *Furia Francese* which so impressed the Italians in 1494–5. The nine British-led siege operations in the Peninsular War saw no fewer than twelve breaches assaulted and another twelve escalades (not all successfully in either case).[57] In Spain, moreover, resistance often continued long after the main *enceinte* had been breached and stormed, something which had practically disappeared a century earlier under conventions which saw surrender negotiations begin when the covered way was taken or, at the latest, when a breach was judged practicable. Breaches in Spain were now fiercely defended, generating fearful casualties. In 1808 the primitive walls of Zaragoza fell in hours to the French guns, but street-fighting inside the city continued for ten days before the invaders retired (following news of their defeat at Bailen). The city finally surrendered in 1809 after a second siege of fifty-two days, which included twenty-four days of fighting inside the walls as Spanish troops and civilians forced the French to advance, in General Rogniat's admiring words, by sap and mine 'from house to house, floor to floor, and room to room'.[58]

Spanish towns were widely spaced in poor country, placing almost as high a premium on logistics as the sea-sieges of the sixteenth century. Indeed, Wellington's sieges were often a race against relief and enormous risks were taken to achieve rapid results. Away from the coast Wellington was also often short of the appropriate ordnance (notably mortars and their bombs) as well as shot for the 24-pounder siege cannon which had now replaced the 40–60-pounders of the sixteenth century as the heavy weapon of choice.[59] Stronger powder and more accurate manufacture made

57 George Sydenham Clarke, *Fortification: Its Past Achievements, Recent Developments, and Future Progress* (London, 1907), pp. 285–7. For details of British operations, see John Thomas Jones, *Journals of Sieges Carried On by the Army under the Duke of Wellington in Spain during the Years 1811 to 1814*, ed. Lieut. Colonel Harry D. Jones, RE, 3 vols., 3rd ed. (London, 1846). Major-General Sir John Jones, KCB, RE, site engineer at Torres Vedras when a captain, wrote up his siege journals (first edition 1814) whilst recovering from a serious wound sustained at Burgos.

58 Joseph Rogniat, *Relation des sièges de Saragosse et de Tortose par les Français dans la dernière guerre d'Espagne* (Paris, 1814), pp. 28–9; Belmas, *Journaux des sièges* (1837), vol. 2, pp. 135–422.

59 Jones, *Journals of Sieges*, vol. 1, pp. 291–2: Wellington to Bathurst (27 Sept. 1812), pp. 315–16 for recovery and reuse of 24-pound shot, and use of French 16-pound shot in 18-pounder cannon.

smooth-bore artillery effective at hitherto unimaginable ranges.[60] Vauban had advised digging the first parallel at 600 yards, from which the relatively soft target of the defenders' artillery could be engaged.[61] Badajos's first abortive Allied siege saw breaches made at 800 yards in a single day.[62] Artillery positions could be engaged at twice this range (which meant that a besieger's batteries could be dispersed widely and many more guns brought into action). Sixteenth-century gunners struggled to sustain six rounds an hour per gun, although better times were recorded under test.[63] One battery served by Royal Navy crews at San Sebastian averaged 350 rounds a gun over 15.5 hours (22 rounds an hour).[64] The intensity of bombardments also escalated. In 1495 a French bombardment of 4,000 rounds in a month caused initial panic in Naples but surprisingly few casualties.[65] Gibraltar in 1782 received 6,500 shot and 2,000 shells in a day, showing the resilience of both civilians and soldiers under fire, a lesson which was to be repeated many times in subsequent wars and perhaps never really learnt.[66]

Improved range, accuracy, and intensity of artillery fire had so expanded the physical scope of siege operations that ditches, ramparts, and bastions no longer always seemed the best way to defend large areas. Artillery defences without obstacles had been the dream of the mid-sixteenth-century military engineer Francesco de' Marchi, whose treatise illustrated detached sea-forts defending an anchorage, the text suggesting that the same principle could apply to land-based fortification.[67] The guns that would make this possible were now available. Before the French Revolution the marquis de Montalembert had proposed multistorey artillery towers capable of

60 Brenda Buchanan, 'The Art and Mystery of Making Gunpowder: The English Experience in the Seventeenth and Eighteenth Centuries', in Steele and Dorland (eds.), *The Heirs of Archimedes*, pp. 233–74.

61 Vauban, *Manual on Siegecraft*, p. 28.

62 Jones, *Journals of Sieges*, vol. 1, pp. 43–4, firing at 120 rounds per day per gun (50 rounds per day from howitzers firing shells).

63 Girolamo Cataneo, *Dell'arte militare* (Brescia, 1571), f. 724v, reports 108 rounds in five hours from an Italian 50-pounder team in Brescia (1564) to settle a dispute over Italian or French supremacy in rapid fire. Francesco de' Marchi, *Della architettura militare, Libri tre* (Brescia, 1599), p. 6, estimates ninety rounds per twelve-hour shooting day for 20- pounders down to twenty-five rounds daily for 100-pounder cannon (the long culverin types being even slower because the length of the tubes made for lengthier reloading times).

64 Jones, *Journals of Sieges*, vol. 2, p. 34: 'never equalled at any siege', but see Belmas, *Journaux des sièges*, vol. 3, p. 444, reporting 300 rounds per gun in seven hours at Tortosa.

65 Pepper, 'Castles and Cannon', p. 284.

66 T. H. McGuffie, *The Siege of Gibraltar* (London, 1965), pp. 140ff.

67 De' Marchi, *Della architettura militare*, p. 105.

outshooting artillery brought against them. As Napoleon's empire crumbled, Lazare Carnot proposed another system of detached artillery forts, with the intervals swept by long-range artillery and the approaches saturated with mortar fire. To encourage aggressive sorties Carnot's ditches had gentle slopes, with a wall at the low point to block storm troops, the ditch itself swept with small arms fire from projecting caponiers. Both schemes abandoned the bastion: both would prove highly influential after 1815, when Europe's new frontiers were fortified.[68]

Another pointer to the future had been built outside Lisbon at the nadir of Allied fortunes in the Peninsular War. In 1810 the Lines of Torres Vedras protected Wellington's retreating army so effectively that the French declined to attack the 29-mile-long double chain of forts, redoubts, artillery batteries, and obstacles.[69] Behind it stood Wellington's army, which could be deployed rapidly along lateral military roads behind the two lines, with movements controlled by naval semaphore stations. Sir John Jones, who as a junior officer had supervised the construction of the lines, would later argue that properly supported field fortifications – a mobile support force was critical – had restored the defensive after a century dominated by increasingly mobile offensive warfare. He would be vindicated by a number of nineteenth-century operations (the defences of Sebastopol in the Crimea; Plevna in Bulgaria; and Vicksburg, Richmond, and Petersburg in the American Civil War) which suggested to some that the long-held distinctions between sieges and battles were becoming blurred.

There were still real differences, of course, not least in the (relatively) sophisticated command-and-control systems needed for sieges, and the divergent engineering and artillery skills of key personnel, in most armies now in different corps. Most basic of all, perhaps, battles were still fought in daylight while most sieges were nocturnal – darkness covering the opening of parallels and the establishment of heavy guns in battery. Soldiers paid a heavy price in daylight assaults. The famous line about war waged 'more like foxes than lions' is usually taken as a complaint about trench life but could equally have

68 E. M. Lloyd, *Vauban, Montalembert, Carnot: Engineering Studies* (London, 1887); also Steele and Dorland (eds.), *The Heirs of Archimedes*; Ken Alder, *Engineering the Revolution: Arms and Enlightenment in France, 1763–1815* (Princeton, 1997); Janis Langins, *Conserving the Enlightenment: French Military Engineering from Vauban to the Revolution* (Cambridge, MA and London, 2004).

69 Charles Oman, *A History of the Peninsular War*, 7 vols. (London, 1902–30), vol. 3, pp. 419–36 for detailed description (foldout map facing p. 432); Jones, *Journals of Sieges*, vol. 3, describes the lines in an Appendix. John Grehan, *The Lines of Torres Vedras: The Cornerstone of Wellington's Strategy in the Peninsular War 1809–12* (Staplehurst, 2000).

reflected a dislike of night operations. The crowds on the slopes above the Allied batteries at San Sebastian on 31 August 1813 thus witnessed a rare event, the daylight assault on a defended breach. Some 300 yards of the town's estuary wall had been brought down by Allied gunfire, the wall collapsing into a vast slope of broken masonry up which the storm troops had to climb. The British brigades approached along the estuary bank at low tide while a Portuguese unit drew gasps of admiration as it forded the stream waist-high to join the storm, which was supported by ricochet and howitzer fire to the rear of the French positions. There was carnage at the breach which saw 664 men killed and 1,047 wounded in about two hours as successive assaults were beaten off in an increasingly critical situation. Breakthrough was only achieved when guns in the siege batteries (the closest some 600 metres across the estuary) were ordered to target the French positions on the crest of the breach. By shooting closely over the heads of the storm troops as they scrambled up the debris, Colonel Dickson's guns enabled the infantry to gain the crest and to break into the town (further assisted by the explosion of a French magazine).[70] Wellington was appalled by the sack that followed, and for the loss of control which allowed the French to fight their way back to the citadel where they held out for another week. Although in many ways a shambles, the siege pointed to another future of close infantry support from artillery and an achievement in fire control which would have astonished the late-medieval artillerists.

Other aspects of the early nineteenth-century siege would not have astonished those who had taken part in the early developments discussed in this chapter. The gunners and fortification engineers of the fifteenth and sixteenth centuries would have had no difficulty in understanding the guns, the bastioned fortifications threatened by them, or the techniques employed by later siege tacticians. Performance had improved massively, to be sure, but until the mid-nineteenth century sieges (as well as battles) still employed mainly smooth-bore muzzle-loaded cannon firing solid round-shot propelled by gunpowder. Siege warfare in the period covered by this volume is practically defined by its reliance on these weapons. The heavy guns used at the end of the Hundred Years' War, the reconquest of Granada, or the French invasion of Italy in 1494–5 were little different in their fundamentals from those

70 Jones, *Journals of Sieges*, vol. 2, pp. 59–105 for the second siege of San Sebastian; Oman, *A History of the Peninsular War*, vol. 7, pp. 1–36, 59–62. For a modern professional soldier's evaluation: (Colonel) Nick Lipscombe, *Wellington's Guns* (Oxford, 2013), ch. 19, who sees in the close artillery support the origins of the 'rolling barrage' of World War I.

employed in the Crimea or the American Civil War. Mid-nineteenth-century advances in artillery represented real change, however, by making possible the ideas advanced by Montalembert and Carnot. Rifled artillery dramatically increased accuracy and range, while breech-loading increased rates of fire as well as the ability to conceal big guns in emplacements (even armoured turrets, like those of the latest warships). Rapid-firing small arms (soon to include machine guns) transformed local defence.[71] High explosives raised levels of destruction far beyond that achieved by gunpowder. Some of these ambitions had been first aired in the Renaissance, of course, well before the necessary technology existed. What would not have been expected by the early exponents of gunpowder siege warfare was the extent of what Yair Mintzker calls the 'defortification' of Europe's cities – much of it undertaken by threatened cities in time of war. Mintzker sees the beginnings of this process in the systematic slighting of urban defences in France's interior started under Richelieu and Louvois. France went on to export this policy. The defences of the Rhineland Palatinate were left devastated – and briefly demilitarised – by Louis XIV's army as it withdrew at the end of the Nine Years' War (1689–97). France's Revolutionary and Napoleonic armies repeated the exercise a century later. The years between 1800 and 1815 became a 'perfect storm' as many hundreds of ancient cities all over Germany were either coerced into dismantling their own fortifications by aggressive diplomacy or decided to neutralise themselves as wartime strategic targets after fierce internal disputes (which often echoed the civic debates centuries earlier when medieval walls and towers were replaced by early modern bastions and ramparts).[72] By the second half of the nineteenth century the continuous belts of ramparts, bastions, ditches, and outworks which had formed a seemingly impenetrable broad barrier between the European inner city and its spreading suburbs had already been largely replaced by ring roads, boulevards, and civic parks. New fortresses were still being built, but in smaller numbers in key strategic locations. Siege warfare had certainly not ended; one need only think of Leningrad and Stalingrad, or more recently Sarajevo and Mosul. But this form of warfare was taking on strikingly different forms and dimensions.

71 Quentin Hughes, *Military Architecture* (London, 1974), ch. 5 for post-Napoleonic Continental developments; Keith Mallory and Arvid Ottar, *Architecture of Aggression: A History of Military Architecture in North West Europe 1900–1945* (Architectural Press, 1973); Andrew Saunders, *Fortress Britain: Artillery Fortification in the British Isles and Ireland* (Liphook, 1989), chs. 9–11 for archivally based coverage.

72 Yair Mintzker, *The Defortification of the German City, 1689–1866* (Cambridge, 2012).

9

# European expansion in the Indian Ocean and Pacific, 1450–1850

LOUIS SICKING*

This chapter studies the history of European expansion in the oceans and the seas stretching east from the Cape of Good Hope. It aims to look at European violent activity here within the broader context of the history of the Indian Ocean, the Red Sea, the Persian Gulf, the Arabian Sea, the Bay of Bengal, the South Chinese Sea, and the Pacific Ocean. In this short chapter, only a few major developments can be traced. Roughly three phases can be distinguished: first, armed vessels – sea power – opened the door for later European success. Then overseas bases – factories – were consolidated by the construction of fortresses. Finally, the Europeans – the Portuguese, the Dutch, the English, and the French – became drawn into military enterprises inland. This chapter, though, focuses on the naval aspects of European expansion, more specifically on the use of warfare to support overseas trade or to prevent competitors from trading.

European attempts to gain access to the sea routes in the Red Sea and the Indian Ocean dated back to the late twelfth century when Reynald of Châtillon, lord of Transjordan in the crusader kingdom of Jerusalem, constructed five galleys on the Gulf of Aqaba, with which he captured the port of Eilat and also attacked Egyptian ports and ships. Saladin took his actions seriously enough to decisively suppress this activity. Other ambitious plans by the Genoese in the early fourteenth century to blockade Egyptian commerce with India by stationing galleys in the Gulf of Aden came to nothing.[1] These actions, though unsuccessful, were characteristic of much overseas European expansion for the centuries to come: maritime in focus and aspiring to the control of ports and sea lanes vital to

* I am indebted to Leonard Blussé, Jos Gommans, Gijs Kruijtzer, and the editors for their invaluable comments on an earlier version of this text. The usual disclaimer applies.

1 J. R. S. Phillips, *The Medieval Expansion of Europe* (Oxford, 1998), pp. 98–9.

long-distance trade.[2] Until the Portuguese entered the Indian Ocean at the very end of the fifteenth century, the European presence east of the Mediterranean remained limited to individuals travelling on local ships.

The most spectacular presence in the Indian and Chinese seas during the fifteenth century was Chinese: seven expeditions, consisting of heavily armed ships considerably larger than European seagoing vessels at the time, crossed the Indian Ocean under the command of the Muslim eunuch Zheng He. These voyages took in the Malay and Malabar coasts, Java, and Sumatra, reaching the Maldives, Hormuz, and Jiddah to the west, and the East African coast as far south as Kilwa. Undertaken between 1405 and 1433 along the monsoon route, the expeditions had a 'shock and awe' effect and gave a strong impression of what could have been possible. Nor did the Ming Empire's decision to refrain from maritime expansion after these expeditions bring an end to Chinese commerce and settlement. On the contrary, Chinese traders and settlers played a vital role in the economies of South-East Asia, as well as for the enrichment of China itself. China thus represented a major, if not *the* major, example of maritime commercial success in the wake of a powerful navy.[3]

## The Portuguese

It was not the Chinese, however, but the Portuguese who were the first to use seaborne violence and armed protection to build a maritime empire in the Indian Ocean for generating commercial profits. (See Map 6.) Portuguese kings tried to control the entire monsoon route and to impose taxes to provide armed protection for sea-traffic. Several factors worked in Portugal's favour. The kingdom took advantage of recent European innovations in sailing-ship technology and gunnery. It had also acquired extensive experience in long-distance sailing into unknown waters during the exploration of the Atlantic islands and the African west coast in the fifteenth century. Finally, the Portuguese elite inherited a tradition of crusades against

2 Louis Sicking, 'Funduq, Fondaco, Feitoria: The Portuguese Contribution to the Globalisation of an Institution of Overseas Trade', in Giampiero Nigro (ed.), *Maritime Networks as a Factor in European Integration* (Florence, 2019), p. 207.

3 Felipe Fernández-Armesto, *Pathfinders: A Global History of Exploration* (Oxford, 2006), pp. 109–16; Louise Levathes, *When China Ruled the Seas: The Treasure Fleet of the Dragon Throne, 1405–1433* (New York, 1994), pp. 20–1; Geoff Wade, 'The Zheng He Voyages: A Reassessment', *Journal of the Malysian Branch of the Royal Asiatic Society* 78 (2005), 37–58; Edward L. Dreyer, *Zheng He: China and the Oceans in the Early Ming Dynasty, 1405–1433* (New York, 2007); Tansen Sen, 'The Impact of Zheng He's Expeditions on Indian Ocean Interactions', *Bulletin of the School of Oriental and African Studies* 79:3 (2016), 609–36.

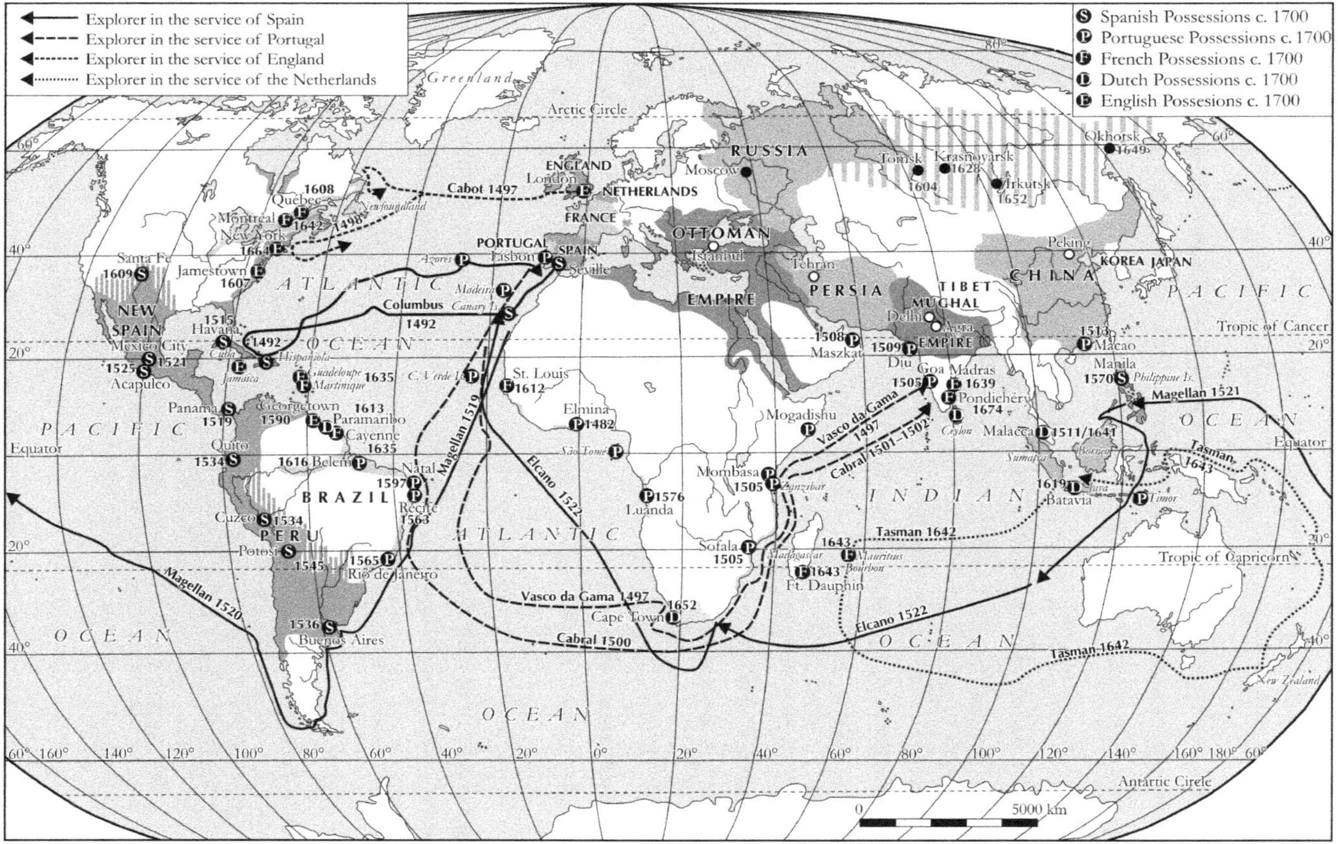


MAP 6. European exploration and empires, 1492–1700. Created by Béla Nagy.

Muslims, which was kept alive in Morocco, and played its part in the anti-Muslim dimension of Portuguese (ad)ventures in the Indian Ocean. In the Indian Ocean, Islam had acted as an integrating force among merchants from different areas for centuries, though it would be incorrect to assume that violence played no role here.[4] It had certainly prevented the first Europeans from developing peaceful commercial activities. The Portuguese did not believe themselves to be more violent than the people they met beyond the Cape of Good Hope.[5]

The Portuguese kings João II and Manuel I sent expeditions to the Indian Ocean to find Christian allies to fight against Muslims and to establish a direct sea route between Europe and Asia in order to create a profitable Portuguese-controlled trade in Asian goods, especially pepper, thus avoiding dependence on the Mamluk sultans who controlled the traditional trade routes through Egypt and Syria. Portuguese enmity towards Muslims, who dominated maritime trade in the Indian Ocean, greatly determined Portugal's Indian Ocean policy. Thus, during Vasco da Gama's first explorative voyage of 1497–9 and the second expedition sent out under Pedro Alvares Cabral in 1500, serious conflicts arose, mainly with the Hindu ruler of Calicut who was sympathetic to the Europeans, but rejected Portuguese demands that he ban Muslim traders. The Portuguese used violence against Indian ships and cities to enforce their will, learning that offensive actions paid off.[6]

Portugal's third expedition to Asia, again under da Gama, left in 1502 with orders to leave a squadron in the Indian Ocean. It was the start of a permanent Portuguese naval presence in the Indian Ocean, which was gradually built up by sending squadron after squadron to the region from 1503 onwards. The Portuguese confidently sought to control trading activity on the west coast of India through blockades, patrols, the bombardment of cities, and acts of terror. Asian ships now had to buy Portuguese licences (*cartazes*). A major attempt by the ruler of Calicut and Muslim traders to attack da Gama's fleet off Calicut in February 1503 failed, demonstrating the

4 Jan Glete, *Warfare at Sea, 1500–1650: Maritime Conflicts and the Transformation of Europe* (London, 2000), pp. 77–8; Michael Pearson, *The Indian Ocean* (London, 2003), pp. 124–5. For examples of non-European violence at sea before and after the arrival of Europeans in the Indian Ocean, see O. W. Wolters, *The Fall of Srivijaya in Malay History* (London 1970), pp. 15–16; Xavier de Planhol, *L'Islam et la mer. La mosquée et le matelot, VIIe–XXe siècle* (Paris, 2000), pp. 396–7.

5 Sanjay Subrahmanyam, *The Political Economy of Commerce: Southern India 1500–1650* (Cambridge, 1990), pp. 256–7.

6 Glete, *Warfare at Sea*, pp. 78–9; Pearson, *The Indian Ocean*, p. 124.

devastating effectiveness of the European combination of manoeuvrable sailing ships and heavy guns.[7]

Within fifteen years Portugal had established a power structure in the Indian Ocean, consisting of a permanent fleet and fortified entrepôts and naval bases, controlling the Ocean's major lines of trade. Goa, captured in 1510, became the capital of Portugal's *Estado da India oriental*. Melaka, taken in 1511, was the most important entrepôt of the South-East Asian spice trade. It enabled the Portuguese to control the most convenient route between the Indian Ocean and the China Seas. In 1515 Portugal took control of Hormuz at the entrance of the Persian Gulf. However, Alfonso de Albuquerque, governor-general from 1509 to 1515, who had led the campaign to create this maritime empire, did not achieve the success he anticipated. Crucially in 1513 his siege of Aden at the entrance to the Red Sea failed. The Portuguese were well aware that the traditional trade route through the Red Sea and Egypt was both faster and cheaper than the Cape route to India. Egypt was therefore at the centre of Portuguese strategic calculations during the first decades of the sixteenth century. In fact, an invasion of Egypt was on the Portuguese agenda, with its potential to make the expensive trade route around the Cape and the network of Portuguese bases in the Indian Ocean superfluous. Albuquerque had even dreamed of building a Suez canal once the conquest of Egypt had been effected.[8]

Portugal managed to establish a position in Asian waters largely thanks to the absence of strong opposition from the great continental states of the Indian subcontinent.[9] The most serious resistance the Portuguese encountered in the first instance came from the Ottomans.

## The Ottomans and the Portuguese

The *Estado da India* may have been the only power broker in the Indian Ocean in the sixteenth century, but the conquest of Mamluk Egypt in 1517 gave the Ottomans an enormous strategic advantage over the Portuguese, which would grow over the course of the century. The Ottomans posed the most serious threat to Portugal's position and together with the European powers they played an important role in maritime expansion through the Indian Ocean, the

7 Glete, *Warfare at Sea*, p. 79.

8 Ibid., p. 80; Giancarlo Casale, *The Ottoman Age of Exploration* (Oxford, 2010), pp. 33, 214 n. 88.

9 Geoffrey Scammel, *The First Imperial Age: European Overseas Expansion c. 1400–1715* (London, 1992), p. 78.

Red Sea, and the Persian Gulf. The Ottoman conquest of Mamluk Egypt and their subsequent expansion into Syria, Palestine, Mecca, and later Iraq (1534) gave the Ottomans a direct interest in both the Red Sea and the Persian Gulf, putting an end to Portuguese schemes to get Egypt into their grasp. Ibrahim Pasha, grand vizier from 1523 to 1536, instigated Ottoman policy in the Indian Ocean, resulting in a fully operational arsenal at Suez, the temporary occupation of Yemen, the opening of a new front against the Portuguese in the Persian Gulf, and Ottoman establishment of diplomatic ties with other Muslim powers across the Indian Ocean. No doubt the prestige the Ottoman sultan acquired as the protector of the holy cities of Mecca and Medina facilitated this process. The most striking expression of the Ottoman Empire's maritime aspirations was perhaps the three attempts to cut a channel – an Ottoman Suez Canal – from the Mediterranean to the Red Sea.[10]

Between 1538 and 1546 the Ottomans and the Portuguese fought each other all over the Indian Ocean. Both were preoccupied with control of the sea lanes. For the Ottomans, the sultan's sacred responsibility to guarantee the pilgrimage routes to Mecca and Medina was an important motivator, as was the recognition that protecting the interests of Indian Ocean merchants was in the interest of the Ottoman state. In 1538 the Ottomans sent a large galley fleet from Suez with a total crew of nearly 10,000 men and heavy artillery to lay siege to Portuguese-controlled Diu. It was the largest fleet they ever dispatched into the Indian ocean. However, the Ottoman fleet failed to win promised support from the sultan of Gujarat, and it was consequently repulsed by the Portuguese, demonstrating the importance of local relations to the outcome of external interventions. The Ottomans nevertheless managed to seize Aden and thus strengthen their ability to defend the Red Sea and to intervene in the Horn of Africa. In 1541 the Portuguese dispatched an expedition to the Red Sea aiming to destroy the Ottoman fleet at Suez, but the attempt was unsuccessful and the Red Sea remained an Ottoman waterway.[11]

In the early 1550s the Ottomans were active with fleets based at Suez, Aden, and Basra in the Persian Gulf. The Portuguese base of Muscat was sacked, Hormuz was besieged, and the island of Qeshm in the Strait of

10 John Darwin, *After Tamerlane: The Rise and Fall of Global Empires, 1400–2000* (London, 2008), p. 76; Jeremy Black, *War in the World: A Comparative History 1450–1600* (Basingstoke, 2011), p. 120; Casale, *The Ottoman Age*, pp. 33, 52, 82, 170.

11 Glete, *Warfare at Sea*, pp. 81–2; Black, *War in the World*, p. 121; Salih Özbaran, 'Ottoman Naval Policy in the South', in Metin Kunt and Christine Woodhead (eds.), *Süleyman the Magnificent and His Age: The Ottoman Empire in the Early Modern World* (London, 1995), p. 61.

Hormuz was plundered. An attempt to take Hormuz in 1554 failed. Ottoman raids on the East African coast during the second half of the 1580s resulted in most of the coastal cities between Mogadishu and Kilwa accepting Ottoman suzerainty, which forced the Portuguese to try to establish firmer control there. Consequently in 1589 a Portuguese fleet under Tomé de Sousa Coutinho defeated the Ottomans under Mir Ali Beg in the harbour of Mombasa. An imposing fortress, Fort Jesus, was built, confirming Mombasa's status as the most strategically important location on the coast. As on other occasions, local forces played a crucial role, as the surprising appearance of Zimba warriors from the interior of Africa forced Mir Ali to divide his forces and ships to defend the city against attack from both land and sea.[12]

The harsh physical environment of the Red Sea and Persian Gulf make Ottoman naval efforts in the Indian Ocean appear all the more impressive. The lack of locally available timber and the absence of a Suez canal made naval activity in the Indian Ocean highly complex and expensive. Moreover, Mediterranean-style galleys were unable to venture far from their bases in order to challenge the Portuguese sailing fleet. These logistical factors, as well as a lack of political will, go a long way towards explaining why Ottoman naval power was eventually restricted to the Red Sea and the Persian Gulf. The Portuguese had initially relied on the relatively small, but swift and seaworthy caravel, ideal for coastal exploration and navigation, as well as the *nau* or great ship, a very large carrack-type vessel. Around 1515 they introduced the galleon into the Indian Ocean. This vessel was longer and narrower than earlier carracks, and was faster, more manoeuvrable, and capable of carrying a heavier armament.[13]

It was not the Ottomans, but the island sultanate of Aceh, strategically located at the main trade route between South-East Asia and the Arabian Sea, which from the 1560s represented the most serious threat to the Portuguese in the Indian Ocean. Although the Ottomans and Gujarat both supported the Sultan of Aceh, who had converted to Islam in exchange for their aid against the Portuguese, neither the Ottomans nor Gujarat deployed the effective Portuguese combination of sturdy sailing ships armed with heavy guns. Aceh, though, used precisely this combination and managed to break Portuguese attempts to enforce naval blockades on Aceh and the Red Sea, attacking the Portuguese base in Melaka and establishing a trade network in

12 Black, *War in the World*, p. 122; Glete, *Warfare at Sea*, p. 82; Casale, *The Ottoman Age*, pp. 175, 177; Milo Kearney, *The Indian Ocean in World History* (London, 2004), p. 115.

13 Casale, *The Ottoman Age*, pp. 201–2; Black, *War in the World*, p. 123.

the Bay of Bengal, outside Portuguese control.[14] By the beginning of the seventeenth century, Aceh's new sovereign, Iskandar Muda (r. 1607–36) rapidly extended his rule over much of Sumatra and several neighbouring islands, and no longer depended on the strategic partnership with the Ottomans to keep the Portuguese threat under control. In 1629, however, the Portuguese permanently destroyed Iskandar's naval power when they repelled his attack on Melaka.[15]

Further east, Portugal was not faced with any major power, except Ming China, whose coast-guard defeated a Portuguese fleet in the early 1520s. Around 1557 the Portuguese were allowed to establish a trading post at Macao under strict Chinese control. Since the 1540s they had also been trading with Japan, and Portuguese merchants settled at Nagasaki in 1571. Acting as middlemen, the Portuguese developed a very profitable trade between Goa, Melaka, Macao, Nagasaki, and Manila. Each year a great *nau* made the voyage between Macao and Nagasaki taking Chinese silk to Japan, Japanese silver to China, and gold from these countries to India. This so-called 'silver ship' had its Spanish counterpart, a galleon annually crossing the Pacific from Acapulco in Mexico to bring American silver to Manila, after the Spanish colonisation of the Philippines in 1564. A flourishing exchange of silks for South American silver went on between Macao and Manila even after the prohibition of all trade between the colonial dominions of Portugal and Spain – between 1580 and 1640 joined in a personal union – by Philip II in 1581. In the China Seas the Portuguese acted as peaceful traders using arms only to defend their merchant vessels against pirates.[16]

However, after they had overcome the Ottoman and Aceh challenges and firmly established themselves in the Indian Ocean, the Portuguese would be challenged by newcomers from Northern Europe.

## The Dutch

Stimulated by Iberian embargoes denying trading through Iberian ports, from the 1590s the Dutch Republic, emerging from the Dutch Revolt, started trading directly with Asia, Africa, and America. The Dutch not only fought

14 Glete, *Warfare at Sea*, p. 82; Kearney, *The Indian Ocean*, pp. 115–16; Pearson, *The Indian Ocean*, p. 123.

15 Casale, *Ottoman Age*, pp. 180–1, 200.

16 Holden Furber, *Rival Empires of Trade in the Orient, 1600–1800* (Mineapolis, 1981), pp. 19, 22; Kearny, *The Indian Ocean*, p. 115; Pearson, *The Indian Ocean*, p. 132; Glete, *Warfare at Sea*, p. 83.

FIGURE 14. Sea battle between Dutch warships and Spanish galleys. Painting by Cornelis Bol, *c.* 1633–*c.* 1650, oil on panel, SK-A-4250, Rijksmuseum Amsterdam.

their Spanish enemy overseas, but also used their naval ability to establish an overseas trade network. (See Figure 14.) The Dutch penetration of the East Indies started in the Moluccas, the spice islands, rich in mace, nutmeg, and cloves. As the Malay Archipelago was windward of the Indian Ocean, it was easy for Dutch ships to project power into this area. Both trade and the use of violence to promote it became more focused with the unification of Dutch trading interests into the Dutch East India Company or *Verenigde Oost-Indische Companie* (VOC) in 1602.[17]

For the next two decades the Dutch fought the Portuguese and the Spanish. They conquered Amboina, Ternate, and Tidore in the Moluccas in 1605, and they blockaded Melaka in 1606–8. The moderately sized but well-armed Dutch East Indiamen managed to establish a fairly safe shipping service between the Indian Ocean and Europe and proved almost invulnerable to Portuguese counterattacks. The Dutch fought the Spanish off Canete in Peru in 1615, the first European fleet action in the Pacific, leading the Spanish to strengthen their American coastal defences. However, a Dutch blockade of Manila in 1616 was abandoned after the Spanish managed to defeat the Dutch fleet at Playa Honta in 1617.[18]

Like the Portuguese a century earlier, the VOC aimed at a monopoly in the spice trade and spared no means to achieve this. Through deportation or

17 Glete, *Warfare at Sea*, pp. 169–70; Kearney, *The Indian Ocean*, p. 120.
18 Glete, *Warfare at Sea*, pp. 170–1.

death, the Banda islands were systematically depopulated and then resettled with labourers from elsewhere, allowing the Dutch to achieve a monopoly in the cultivation and trade of nutmeg and mace by 1622. The monopoly on cloves was only acquired four decades later by concentrating its cultivation on Ambon whilst destroying it on other islands. Bitter guerrilla warfare was part of the process.[19] Once the local population had been subjugated, European competition had to be eliminated. The Portuguese had already left by 1606, the English withdrew from the Moluccas in 1623, and the Spaniards finally gave up their last fortress at Ternate in 1662.

The Dutch extended their trade networks to India and Savafid Persia, where they had to operate more cautiously than in the Malay Archipelago due to complex local circumstances: they were obliged to take account of the presence of the Moghul Empire and what remained of the Vijayanagar Empire, as well as the Portuguese with their power bases in Goa and Cochin. Gaining access to the Indian market was driven by the need for Indian textiles to exchange for spices. A Dutch intra-Asian trade network developed under military protection, with factories along the Coromandel coast with Pulicat as the VOC's headquarters, and further north in Masulipatnam, Hugli in Bengal, and Surat in the north-west. In Safavid Persia, the Dutch tried in vain to drive the export of silk from the Levant to the Cape route. Having more capital than its English counterpart, the VOC did manage to establish a profitable trade at Bandar Abbas and Mocha.[20]

With the foundation of Batavia (Jakarta) on Java in 1619, Governor-General Jan Pieterszoon Coen started to implement his grand strategy of making the VOC the dominant European force in Asian trade by establishing a network of trade factories, often fortified, stretching from Hormuz and Aden to Nagasaki. Batavia became the central entrepôt for trade between Europe and Asia. Dutch fleets blockaded Manila, Melaka, and Macao and cruised between Goa and Mozambique in 1621–2, interrupting Iberian trade. In 1624 the Dutch established a base at Formosa (Taiwan), Fort Zeelandia, for Chinese trade. In the same year, though, a similar effort to create a base on the west coast of America with a Dutch fleet crossing the Pacific failed thanks to improved Spanish defences.[21] Dutch military successes (against the Portuguese) were mainly determined by the VOC's maritime dominance: they not only had more vessels, but, ship for ship, also possessed superior fighting power and some of their vessels, such as the *jachten* (yachts), were

19 Femme Gaastra, *De geschiedenis van de VOC* (Zutphen, 1991), pp. 45–6.

20 Ibid., pp. 46–51; Furber, *Rival Empires*, pp. 58, 71.

21 Glete, *Warfare at Sea*, p. 172.

especially suited to war at sea. These vessels were lighter than the regular VOC ships, having a lower fore(castle) and stern, while their size could vary greatly, from 140 to 560 tons. The yacht was the most common type of ship used by the VOC in Asia between *c.* 1636 and *c.* 1680, a period which coincided with the intensification of the struggle with the Portuguese.[22]

At the initiative of Governor-General Antonie van Diemen, the Dutch began a new series of offensives, mainly against the Portuguese. Melaka and Goa were blockaded from 1633 and 1636 respectively. In two major battles at Goa the Portuguese attempted in vain to break the blockade. (See Figure 15.) The Dutch conquered Melaka in 1641. Thus, the strategic position around the Straits changed in their favour while they managed to conclude exclusive contracts for the delivery of pepper from Aceh, Palembang, and Jambi on Sumatra. Initially with the support of the king of Kandy, the Dutch conquered parts of Ceylon from the Portuguese between 1637 and 1656, acquiring a monopoly on cinnamon in the process. Possessions in India were extended across from Ceylon in the coastal area of Madurai and on the Malabar coast, where in 1663 the Dutch took Cochin from the Portuguese despite tenacious

FIGURE 15. A Dutch squadron commanded by Cornelis Simonsz van der Veer carries out a surprise attack on three Portuguese galleons in the Bay of Goa, 30 September 1639. Hendrick van Anthonissen, 1653, oil on canvas, SK-A-2126, Rijksmuseum Amsterdam.

22 Gerrit Knaap, Henk den Heijer, and Michiel de Jong, *Oorlogen overzee. Militair optreden door compagnie en staat buiten Europa, 1595–1814* (Amsterdam, 2015), pp. 121, 215, 221.

resistance.[23] Whilst the Portuguese were expelled from Japan in 1639, the Dutch were still allowed to trade with Japan from the tiny island of Deshima from 1641, which for some time became their richest factory in the east. A colony founded at the Cape of Good Hope in 1652 became a station to repair and replenish vessels on the route between Europe and Asia.[24]

Both ships and fortresses were key instruments of violence and force to promote and defend the VOC's trade and profit. Recent research has emphasised the great variety of ships used by the VOC. So-called 'homeward-bounders' (*retourschepen*) represented the largest-rated class of vessels, intended to carry Asian products homeward. These vessels dominated European–Asian shipping routes, but were also used in Asia as heavily armed traders to defeat Portuguese vessels in the Arabian Sea, or, in the case of older vessels, to transport soldiers from Batavia to the Spice Islands, where they were used as floating fortresses. The VOC also employed large men-of-war in the Arabian Sea and the Bay of Bengal. The aforementioned yachts represented the largest part of the VOC's fleet in Asia. They were adaptable vessels *par excellence*, combining both military and commercial functions. In 1653 the yacht *Concordia* transported a rich cargo from Surat to the Persian Gulf. When news arrived of the start of the first Anglo-Dutch War the *Concordia* was quickly unloaded and sent out to patrol the region, eventually capturing an English East-India vessel. Generally, the VOC adapted its fleet to circumstances, employing more defensible yachts in wartime, at the cost of efficient cargo carriers which were less able to protect themselves, like the bulk-cargo *fluits*.[25]

Fortresses and fortified settlements had formed a part of the European military tradition since the Crusades and offered a crucial strategic advantage for expansion. The medieval castle survived into the gunpowder age by adding bastions which transformed it into the artillery fortress. Fortifications allowed vastly outnumbered Europeans to project power overseas over local peoples and their armies. They were a condition *sine qua non* for a permanent presence overseas. With the arrival of the Dutch, the Portuguese, who had managed to withstand local powers by slowly introducing bastions and thick, low-lying walls, were forced to invest more heavily in

23 Glete, *Warfare at Sea*, pp. 175–6; Gaastra, *De geschiedenis van de VOC*, pp. 46, 52.

24 Kearney, *The Indian Ocean*, pp. 116–17; Furber, *Rival Empires*, pp. 62, 64.

25 Robert Parthesius, *Dutch Ships in Tropical Waters: The Development of the Dutch East India Company (VOC) Shipping Network in Asia 1595–1660* (Amsterdam, 2010), pp. 18, 68–9, 71, 78–9, 170; Herman Ketting, *Fluitschepen voor de VOC. Balanceren tussen oncostelijckheijt en duursaemheijt* (Zaltbommel, 2006).

expensive fortifications. The Dutch fortified the places they conquered – such as Galle, Colombo, and Negapatam – on a far more impressive scale than the Portuguese.[26] The VOC also built new fortresses, of which the 'castle' Oranje erected at Ternate in 1607 was the first. Dutch investment in fortress-building in Asia continued until the late seventeenth century when Dutch expansion lost momentum. In the eighteenth century most plans to modernise the fortresses of the VOC did not materialise. After the British conquered most of the VOC's empire, the Dutch concentrated on the defence of Java.[27] Heavy investment in fortification overseas took place mostly when and where other Europeans formed a threat. Fortresses allowed the Europeans to extend their power inland. Local rulers faced with a powerful local rival were often eager to allow Europeans to build fortresses and to conclude treaties with them, which improved their local importance and security. Fortresses also offered the advantage of security to local merchants and their trading stock, especially in times of war.[28]

The Dutch had imposed great costs and losses on the Iberian empires. They were also able to halt the Muslim spice trade to the Red Sea and the Levant, a goal which the Portuguese had never succeeded in achieving, and found a niche in the trade with Japan and China.[29] However, the Dutch were not universally successful. They lost a full-scale war on Formosa against the leader of the Ming in south-east China, Koxinga (Zheng Chenggong), whose troops successfully besieged the fortress of Zeelandia in 1662 (taking advantage of European military intelligence from deserters), which meant the end of the VOC as a political presence in East Asia.[30]

The Dutch conquest of Cochin dealt a deathblow to the Portuguese pepper trade. The *Estado da India* had been able to acquire a position amongst Asian powers, but it had never been compact and profitable enough, or sufficiently integrated, to withstand attacks from a powerful European opponent.[31] Naked force and diplomacy were the key factors behind Dutch success in Asia.[32] The VOC had been able to build up an empire with armed

26 Geoffrey Parker, 'The Artillery Fortress as an Engine of European Overseas Expansion, 1480–1750', in James Tracy (ed.), *City Walls: The Urban Enceinte in Global Perspective* (Cambridge 2000) pp. 386–416, 393–9.

27 Knaap, Den Heijer, and De Jong, *Oorlogen overzee*, pp. 225, 229.

28 Parker, 'The Artillery Fortress', pp. 415–16.

29 Glete, *Warfare at Sea*, p. 173.

30 Knaap, Den Heijer, and De Jong, *Oorlogen overzee*, pp. 111–12, 257.

31 George Winius, 'Luso-Nederlandse rivaliteit in Azië', in Gerrit Knaap and Gerke Teitler (eds.), *De Verenigde Oost-Indische Compagnie tussen oorlog en diplomatie* (Leiden, 2002), p. 129.

32 Sanjay Subrahmanyam, *The Portuguese Empire in Asia 1500–1700*, 2nd ed. (Oxford, 2012), pp. 223–4.

merchant vessels at a time when these were yet to be outclassed by specialised warships. In the eighteenth century, such vessels would give Britain a critical advantage vis-à-vis the Dutch.[33] Ironically, given that enormous expenditure on war and conquest would eventually lead to its stagnation and decline, the conquest of Cochin also provides a window onto the limitations of the VOC.

## The British

In reaction to the extremely high prices the Dutch demanded for pepper, a group of London merchants joined to create the English East India Company (EIC), which was granted a charter by Elisabeth I in 1600. When two English ships, lighter and better provided with heavy artillery, destroyed three Portuguese galleons off the coast of Surat in 1612/1613, the Moghul emperor Jahangir was so impressed that he granted the EIC the right to trade in Surat. Trading rights in Persia followed in 1622 as a result of English support for the Safavids in taking Hormuz from the Portuguese. Set up with less capital, and initially less warlike than the VOC, the EIC concentrated on peaceful trade with western India and the Safavid Empire. The English presence in the Indian Ocean was further strengthened with trading posts in Burma (Rangoon), Madras (1639), Bombay (1668), and the new city of Kolkata (1690).[34]

The Anglo-Portuguese alliance of 1662, sealed by the marriage of Charles II to the Portuguese princess Catherine of Braganza, gave the English some trading rights in Portugal's colonies. Bombay, a part of the dowry, was leased by Charles II to the EIC, and developed into a military and banking centre. Henceforth, the EIC's navy used Bombay as its base for protecting trading interests at Surat, although it faced serious problems with the European-inspired naval activities of the Sidis and the Marathas, rivals who each occupied a strategically located island at the entrance to Bombay harbour.[35]

In 1711, the Company established a trading post in Guangzhou in China, to trade tea for silver. By 1735 the textile trade from Bengal – which also provided opium, silk, and sugar – and the Coromandel coast was flourishing. Its grip on

33 Knaap, Den Heijer, and De Jong, *Oorlogen overzee*, p. 243.

34 Scammel, *The First Imperial Age*, pp. 18–19, 106; Glete, *Warfare at Sea*, p. 173; Pearson, *The Indian Ocean*, pp. 145, 147–8.

35 Kearney, *The Indian Ocean*, p. 123; Gijs Kruijtzer, *Xenophobia in Seventeenth-Century India* (Leiden, 2009), pp. 192–7, 210–11, 218–19; Ashin Das Gupta, *Indian Merchants and the Decline of Surat, c. 1700–1750* (Wiesbaden, 1979), pp. 260–72.

intra-Asian trade from the 1650s onwards ensured that the VOC remained the dominant European trading company until *c.* 1720. But the rising European demand for cotton and silk allowed the EIC successfully to challenge the VOC, better placed as they were to supply this demand in large quantities from their trading bases in Bengal. Rapid British expansion followed between 1740 and 1800, thanks to a vast rise in demand for tea in Britain and an increase in the opium trade to China. While the EIC concentrated on the trade link between Europe and Asia, a private English intra-Asian trade network developed from the late seventeenth century. This so-called 'country trade' expanded rapidly in the eighteenth century, making Indian merchants more and more dependent on the English for long-distance shipping.[36]

A large part of the European imports of Asian goods was paid for in silver bullion. On average, two-thirds of the exports of the VOC from Europe was bullion, which comprised almost 80 per cent of British exports to Asia between 1660 and 1720. This changed when the English interfered in inland Bengal production by acquiring revenue rights, coercing producers, and creating new trade networks from the 1760s onwards. As a result, British private traders came to dominate trade in the Indian Ocean, whilst the EIC rose to dominate trade to Europe.[37] Involvement in European wars with both continental and overseas dimensions led to the development, and military success, of British naval power. Consequently, the armed merchant vessels of the VOC proved no match for purpose-built British naval vessels.[38]

As the companies and other traders came to dominate the long-distance shipping and supersede overland trade routes, they provided the rulers of regional states with a valuable source of income. These states, which had emerged following the disintegration of the Mughal Empire, had an interest in the growing number of associations between English and Indian merchants. As a result, the EIC became more and more involved in Indian politics, both in Bengal and elsewhere. In south India, where the English and French trading companies fought each other between 1746 and 1754, the two companies allied themselves with competing contenders to the thrones of Hyderabad in the eastern Deccan and Arcot, capital of the Carnatic, in the Coromandel. The resulting Carnatic wars involved both the French and the

36 Pearson, *The Indian Ocean*, pp. 150–1; Femme S. Gaastra, 'Dutch, French and English Rivalry in Asia, 1688–1714', in *Guerres maritimes (1688-1713). IVes Journées franco-britanniques d'histoire de la marine* (Vincennes, 1996), pp. 227–43, 241–2; Catherine Asher and Cynthia Talbot, *India before Europe* (Cambridge, 2006), p. 282.

37 Pearson, *The Indian Ocean*, pp. 150–1, 153, 167.

38 Knaap, Den Heijer, and De Jong, *Oorlogen overzee*, p. 243.

English. In 1746 Joseph Dupleix, governor of the *Compagnie des Indes* on the Coromandel coast, besieged and conquered the British capital at Madras. In 1751 Robert Clive led British troops to take Arcot. When the new nawab of Bengal, Siraj ud-Daula, took Fort William at Calcutta in 1756, the British sent a fleet from Madras to Bengal. Calcutta was recaptured in January 1757 by Clive and Rear-Admiral Charles Watson. In 1757, the decisive battle of Plassey took place, during which the EIC was victorious over the nawab of Bengal.[39]

In return for their military assistance and political support the English and French companies each received rights to take control of revenues, while some of their employees profited through investments, loans, or purchase of revenue rights from rivals in Indian succession disputes. European military superiority, demonstrated during the south Indian wars, made Indian elites eager to enlist their help.[40] The Portuguese had already seen at the start of the sixteenth century 'that their naval superiority over land-based Asian empires gave them a means of sharing in the profits of the existing redistributive powers'. The *Estado da India* became a redistributive enterprise which served as a model for the Dutch and English East India Companies. The collective financial strength of the latter two companies allowed for a policy of armed trading which enabled them to play a (quasi-)political role.[41]

## The French, the Danes, the Swedes, and the imperials

Inspired by the example of the VOC, the Danes (1616) and the French (1664) had also founded chartered companies to penetrate the Asian markets. Unlike the VOC and the EIC, most of the capital of these companies came from the respective royal houses whose officials strictly controlled the companies' management. At Tranquebar in south-east India, the Danes settled at the invitation of the rajah of Thanjavur and built the fortress Dansborg. A year later, they opened a factory at Masulipatam. The Danish company, feeble because of lack of capital, went bankrupt in 1650 but continued to be involved in local trade at Tranquebar. In 1670 a new

39 Asher and Talbot, *India before Europe*, pp. 283–4; Nicholas Rodger, *The Command of the Ocean: A Naval History of Britain, 1649–1815* (London, 2004), pp. 253, 274–5; Philippe Haudrère, *Les Compagnies des Indes orientales. Trois siècles de rencontre entre Orientaux et Occidentaux* (Paris, 2006), pp. 205–11.

40 Asher and Talbot, *India before Europe*, p. 284.

41 Kirti Chaudhuri, 'Reflections on the Organizing Principle of Premodern Trade', in James Tracy (ed.), *The Political Economy of Merchant Empires: State Power and World Trade, 1350–1750* (Cambridge, 1991), pp. 438–9.

Danish East India Company was founded, again under royal control, which was quite successful around the turn of the eighteenth century. However, it too went bankrupt as a consequence of losses suffered during the Great Northern War (1700–21).[42] A third Danish company was set up in 1732, focused on China, sending at least one ship a year. Trade with Tranquebar and Hugli in India continued on an even smaller scale and was partly carried out by private Danish traders.[43]

The French established themselves on Madagascar, at Fort-Dauphin, hoping to develop it into a new Batavia, which could function as both an entrepôt for goods and an anchorage between Europe and Asia, and on Bourbon (Réunion), from where they moved to Mauritius in 1710, renamed Île de France. In India the French *Compagnie des Indes* set up trade in Surat, Pondichéry, and Chandernagore near Calcutta, encouraged by the Mughal emperor Aurangzeb (r. 1658–1707), who wished to counteract the British. After a period of Dutch occupation, the French built a fortress at Pondichéry in 1706, turning it into a strategic centre to secure French commercial activity in the region. From 1698 the French also started trading with China. Although clearly inferior to the English and Dutch companies in the seventeenth century, the *Compagnie des Indes* expanded rapidly in India during the first half of the eighteenth century and offered serious competition to its major rivals until the battle of Vandavasi in 1761 at which the British decisively defeated the French and took Pondichéry. After 1763 financial problems and high military costs eventually led to the liquidation of the *Compagnie des Indes*.[44]

In the southern Netherlands, some merchants attracted by the possibilities of a strong demand for Asian products just after the War of the Spanish Succession (1701–14) organised several expeditions from the Flemish port of Ostend. Stimulated by their success, in 1722 the Austrian emperor granted a monopoly for the trade between the ports under his dominion and those of Asia to the *Generale Indische Compagnie* (GIC), which operated from Ostend. Although the crews of all European ships in the Indian Ocean were mixed – that is, from various European and Asian countries – the character of the Ostend company was highly international at all levels, including its staff.

42 Haudrère, *Les Compagnies des Indes orientales*, pp. 69, 78–80; Sanjay Subrahmanyam, *The Political Economy of Commerce: Southern India 1500–1650* (Cambridge, 1990), pp. 182–3.

43 Furber, *Rival Empires*, pp. 211–16; Haudrère, *Les Compagnies des Indes orientales*, p. 88.

44 Haudrère, *Les Compagnies des Indes orientales*, pp. 69–78; Pearson, *The Indian Ocean*, p. 152; Kearney, *The Indian Ocean*, p. 125; Philippe Haudrère, *La Compagnie française des Indes au XVIIIe siècle* (Paris, 2005), pp. 820, 823.

Trade with Canton was quite successful; twenty-six vessels were sent there between 1718 and 1732. In India, after some pressure imposed on the nawab of Bengal, the imperials were allowed to establish themselves at Bankibazar, Balassor, Dhaka, and Bouroumpour for the silk trade. The Dutch and the British, wishing to eliminate yet another European competitor, joined to interdict the GIC's activities, a move which was formalised in an international agreement in 1731. However, during its existence the GIC was a lean organisation which managed to make profits in the tea trade, and to compete with the bigger, but less flexible, companies.[45]

The Swedish East India Company, founded in 1731, resembled its Danish counterpart in several ways, though a very limited amount of capital was the main parallel. Both companies took advantage of the vacuum created by the disappearance of the GIC. After some vain attempts to establish trade with India, the Swedes turned to China and became involved in the tea trade at Canton. This tea was mainly paid for with American silver acquired en route at Cadiz. It was meant not so much for consumption in Sweden but to be re-exported or – more in line with reality – smuggled into Britain.[46]

Neither the French East India Companies nor French naval efforts east of the Cape seem to have brought any financial benefit to the state. In the long run, strong state interference was detrimental to the companies' profits. The concerns of the French state were political. It had supported the companies to develop mercantile interests deemed essential to the state's political ends. Proportionally, the other, smaller companies seem to have performed best of all in terms of higher yields relative to the costs incurred.[47] However, they were frequently threatened with bankruptcy and, lacking the means to deploy enormous military expeditions, they remained far more limited in reach than their English and Dutch counterparts.

45 Haudrère, *Les Compagnies des Indes orientales*, pp. 84–5; Furber, *Rival Empires*, pp. 144–5, 221; Pearson, *The Indian Ocean*, pp. 185–6; Gijs Dreijer, 'Een strijd om erkenning. De onderhandelingen over de Oostendse Compagnie (1713–1723)', *Tijdschrift voor zeegeschiedenis* 17 (2018), 58.

46 Haudrère, *Les Compagnies des Indes orientales*, pp. 85–6; Christian Koninckx, 'The Swedish East India Company', in Jaap Bruijn and Femme Gaastra (eds.), *Ships, Sailors and Spices: East India Companies and Their Shipping in the 16th, 17th and 18th Centuries* (Amsterdam, 1993), p. 121.

47 Furber, *Rival Empires*, p. 334; Haudrère, *Les Compagnies des Indes orientales*, p. 822; Pierre Boulle, 'French Mercantilism, Commercial Companies and Colonial Profitability', in Léonard Blussé and Femme Gaastra (eds.), *Companies and Trade: Essays on Overseas Trading Companies during the Ancien Régime* (Leiden, 1981), p. 116.

## Britain's rise to domination, 1750–1850

The Seven Years' War (1756–63) heralded the onset of British domination in the Indian Ocean, though the French tried to counter the British on several occasions. While France successfully developed coffee production on Réunion and a sugar plantation economy on the Île de France, importing slaves from Mozambique, Britain focused on India. The defeat of Siraj ud-Daula of Bengal at the battle of Plassey in 1757 made the EIC the effective rulers of Bengal. Subsequently, all French possessions in India were taken, including Pondichéry. The Treaty of Paris in 1763 left Britain in control of India. Mauritius and Réunion remained the only French possessions in the Indian Ocean. Port Louis was, and remained, the *de facto* capital of the French trading empire and the key French naval base in the east, and an important centre for the European country trade, especially slave trading. The French now concentrated on East Africa, supporting the slave trade between Kilwa and Zanzibar.[48]

In western India, the EIC was forced to keep a small navy of its own, the Bombay marine, consisting mainly of small galleys to protect its shipping against the depredations of the Angrias, a ship-owning Maratha dynasty and their followers operating from at least ten fortresses. Their much-feared admiral, Kanhoji Angria, who had learned from the Europeans how to use violence to turn a profit, demanded 'protection-money' from both European and Asian ships in the 1720s and 1730s.[49]

In 1780 the British destroyed the fleet of Hyder Ali, *de facto* ruler of Mysore and a powerful French ally. French support under command of Le Bailli de Suffren, a former corsair, arrived from Mauritius in 1782, and he fought the British fleet under Edward Hughes on five occasions. These encounters have been considered 'unique in naval history for the intensity and balance of the fighting'. The naval actions of 1782 and 1783 may have created the heroic legend of Suffren in France, but the British position in India endured.[50]

48 Anthony Webster, *The Twilight of the East India Company: The Evolution of Anglo-Asian Commerce and Politics, 1790–1860* (Woodbridge 2009), p. 22; Haudrère, *La Compagnie française des Indes*, p. 752; Kearney, *The Indian Ocean*, pp. 126–7; Furber, *Rival Empires*, pp. 14–15, 138.

49 Rodger, *The Command of the Ocean*, pp. 274–5; Kirti Chaudhuri, *The English East India Company: The Study of an Early Joint-Stock Company 1600–1640* (London, 1965), p. 71; Furber, *Rival Empires*, p. 158; Kruijtzer, *Xenophobia*, pp. 219–20; H. C. M. van de Wetering, 'De VOC en de kapers van Angria. Een expeditie naar de westkust van India in 1739', *Tijdschrift voor zeegeschiedenis* 10 (1991), 117–30, 118–19; Sachin Pendse, 'Maratha Sea Power', in Christian Buchet and Gérard Le Bouëdec (eds.), *The Sea in History: The Early Modern World* (Woodbridge 2017), pp. 640–1, 644–5.

50 Rodger, *The Command of the Ocean*, pp. 356–7.

The Dutch Republic, which initially remained neutral during the Anglo-French wars after 1756, was thus able to conceal the VOC's military weakness for some time. The VOC continued to send the largest number of ships to Asia of all European companies until 1780, and on average these had a greater cargo-capacity than their EIC counterparts. But the company struggled against rising commercial competition, did not undergo institutional reform, and could not count on military support from the state to the same extent as the EIC. This weakness became painfully clear during the Fourth Anglo-Dutch War (1780–4) when the British took Negapatnam on the Coromandel coast from the Dutch and seized many homebound ships with valuable cargoes. Bankrupt, the VOC was nationalised in 1795.[51]

Napoleon's rise presented both new opportunities and challenges for Britain in the East Indies. When French troops occupied the Dutch Republic in 1795, the EIC took over most of the Dutch/VOC overseas possessions, including the Cape of Good Hope, which would not be returned. At the same time the EIC faced extensive French raiding from Mauritius, until the British conquered both Réunion and Mauritius in 1810, during which they suffered their only naval defeat at the hands of the French during the Napoleonic Wars among the reefs of Grand Port.[52] The American War of Independence had already compelled Britain to extend its overseas settlements elsewhere; control was tightened over possessions in India and new settlements were founded, like Penang on the Malay Coast, and settlement colonies in Australia and New Zealand, territories which had been discovered by the Dutch a century earlier, but were now claimed on behalf of Britain by James Cook.[53] Over the next fifty years Britain added a new series of strategically vital ports to its possessions: Colombo in 1796, Cape Town in 1806, Singapore in 1819, Melaka in 1824, Aden in 1839, and Hong Kong in 1842.[54] Meanwhile the Americans, to whom the West Indies had been closed off since the American Revolution, looked for other destinations for trade,

51 Gerke Teitler, 'De marine en de Verenigde Oost-Indische Compagnie', in Knaap and Teitler (eds.), *De Verenigde Oost-Indische Compagnie*, p. 79; Femme Gaastra, '"Sware continuerende lasten en groten ommeslagh". Kosten van de oorlogsvoering van de Verenigde Oost-Indische Compagnie', in Knaap and Teitler (eds.), *De Verenigde Oost-Indische Compagnie*, pp. 101–2; Femme Gaastra and Jaap Bruijn, 'The Dutch East India Company's Shipping, 1602–1795, in a Comparative Perspective', in Gaastra and Bruijn (eds.), *Ships, Sailors and Spices*, pp. 182–3.

52 Rodger, *Command*, pp. 435–6, 467, 546–7, 556–7, 560; John McAleer, *Britain's Maritime Empire: Southern Africa, the South Atlantic and the Indian Ocean, 1763–1820* (Cambridge, 2017), p. 238.

53 Kearney, *The Indian Ocean*, pp. 129–30.

54 Pearson, *The Indian Ocean*, p.191; McAleer, *Britain's Maritime Empire*, p. 241.

resulting in an enormous expansion of American commerce across the Pacific.[55]

Indian Ocean shipping had come to be dominated by one naval power. The EIC made enormous advances, extended its powers and jurisdiction, and was increasingly supported by the state. After the Great Indian Rebellion in 1857–8, however, the EIC was removed from any direct governmental control. All its assets passed directly to the crown, which took control of the former territories of the Company.[56] The Industrial Revolution, from which Britain was the initial beneficiary, was to have a profound influence on the shape of European presence in the Indian and Chinese seas. The transition from merchant capital to industrial capital stimulated the extension of British influence from port cities to territorial control, first and foremost in India. The introduction of the steamship reduced the time for a voyage from England to India from one year in the 1830s to between two and three months in the 1850s. The opening of the Suez Canal in 1869 revived the Red Sea route and annihilated distance yet further. Cape Town and several islands in the Indian Ocean, such as Madagascar and Mauritius, once significant way-stations for shipping, lost much of their importance.[57]

Further east, the Opium Wars (1839–41 and 1856–60) resulted in a series of agreements with China, under which over a dozen port cities, such as Shanghai and Canton (Guangzhou), were opened to Europeans, who were allowed to live under their own laws. Western gunboats began to police the Yangtse river. Coastal China was opened up to European industrial products.[58]

An evaluation of early modern European expansion in the Indian and China Seas rests on several different perspectives. First, the role of European shipping and military technology, including the building of bastioned artillery fortresses, has lately received renewed emphasis.[59] European overseas expansion and commercial success were dependent upon the constant use of force and the maintenance of military superiority.[60] Armed merchant vessels securing sea lanes between well-defended factories, fortresses, and fortified

55 Leonard Blussé, *Visible Cities: Canton, Nagasaki and Batavia and the Coming of the Americans* (Cambridge, MA, 2008), p. 61.
56 Webster, *The Twilight of the East India Company*, pp. 143, 145–9.
57 Pearson, *The Indian Ocean*, pp. 190–3, 196–7, 200, 210–11; McAleer, *Britain's Maritime Empire*, pp. 239–40.
58 Darwin, *After Tamerlane*, pp. 270–6.
59 Glete, *Warfare at Sea*, pp. 88–9; Darwin, *After Tamerlane*, p. 96; Knaap, Den Heijer, and De Jong, *Oorlogen overzee*, p. 243.
60 Geoffrey Parker, 'Europe and the Wider World, 1500–1700: The Military Balance', in Tracy (ed.), *The Political Economy of Merchant Empires*, pp. 161, 195.

cities – Melaka, Batavia, and Manila stand out – are the key to understanding Europe's long-distance trade in Asia. Some local rulers were well aware of these advantages and followed the European example by building fortresses in Bantam, Pati, Japura, and Surabaya, albeit in vain.[61] Instead of adjusting their armed forces to make the best use of European military technology, they attempted to adapt this technology to existing ideas of warfare. In this they mostly failed, though there were notable exceptions.[62]

Second, current global historiography has presented the impact of European expansion in the Indian and China Seas in relative terms. The European presence in Asia and East Africa involved small numbers of people, based in trading posts and forts whose purposes were overwhelmingly commercial, and were defended in order to maintain and safeguard overseas connections with Europe and other trading posts. Trade was dominated by low-volume, high-value goods, which have never been of decisive significance for the development of the world economy. Contact with local rulers was limited to the coast or along the fringes of major rivers, leaving great empires in the interior untouched. Consequently it is argued that early modern European overseas expansion can hardly be regarded as an integral part of the later, nineteenth- and twentieth-century European dominance in Asia and Africa. The Industrial Revolution thus represents a break in the history of the European presence in Asia, the chief beneficiary of which was the British Empire centred in South Asia. Yet the later colonies of British India and the Netherlands East Indies nevertheless evolved from bases established in the early modern era.[63] The principal continuity was the West's military superiority, which gave it a continuing advantage in the global 'military balance' in the nineteenth century.[64]

Finally, this early modern expansion cannot simply be regarded as a history of European exploitation of Asia. Although individual European merchants earned large fortunes in trade with Asia and the indigenous populations of the spice islands were largely wiped out or else brutally exploited, the effectiveness, efficiency, and profitability of European enterprise can be questioned. One need only look at the great rivalry between the East India companies. The fortunes of the EIC and VOC changed over time. While the VOC generated most of its profits in the seventeenth century, the most profitable

61 Parker, 'Europe and the Wider World', p. 171.

62 Tonio Andrade, *The Gunpowder Age: China, Military Innovation and the Rise of the West in World History* (Princeton, 2016).

63 Darwin, *After Tamerlane*, pp. 91, 93–5, 98, 110–11, 131, 153–5.

64 Parker, 'Europe and the Wider World', p. 195.

period for the EIC was the first half of the eighteenth century. Nevertheless, there were extreme fluctuations in profits throughout the early modern period, mainly due to the need to meet high protection costs. This was especially true for the French *Compagnie des Indes*, which may not have been a profitable venture at all. In the last decades of the eighteenth century the military and administrative expenses of all major trading companies increased greatly. It was ultimately the European taxpayer who had to pay for these protection costs because they could no longer be covered by the commercial profits from the trade with Asia; yet it was the European traders who continued to enjoy the companies' subsidised protection for their activities. It is hard, if not impossible, to draw general conclusions about whether the costs and benefits of military enterprise in overseas trade resulted in maintained net economic benefit. A recent study on the Dutch Republic has argued that the naval transformation around the middle of the seventeenth century – which meant that navies consisting of purpose-built warships became the decisive force in European naval warfare – led to a redefinition of the role of commercial companies. Until around 1650, fleets of ships that could interchangeably be used for trade or combat had been the backbone of Dutch commercial hegemony, and had enabled the Dutch Republic to expand both commercially and militarily overseas. Now, the Company bought off its naval obligations in European waters – leaving these naval tasks to the Republic's navy – whilst retaining full independence of action in Asia, where profit continued to be put first. The Company did not actively solicit for state support in Asia until the last third of the eighteenth century.[65] By then the British example had made it abundantly clear that regular naval strength was a key factor for overseas commercial expansion.

Some historians have suggested that the East India companies stimulated the economies of the lands surrounding the Indian and China Seas, rather than retarded them. But of course while local rulers, merchants, and bankers may have made fortunes from working with the European companies, many local workers involved in production for European markets were recklessly exploited under the severest of circumstances, and slave labour proliferated. To what extent the European presence in Asia made a difference in these respects may be debated since exploitation and slavery had always been present in the Indian Ocean and Chinese seas. Nevertheless, European expansion east of the Cape had wide-ranging social consequences, and whilst

65 Pepijn Brandon, *War, Capital and the Dutch State (1588–1795)* (Leiden and Boston, 2015), pp. 135–6.

the position of some groups of Asians improved, that of others deteriorated.[66] Perhaps, though, the clearest and most consistent point to be emphasised for this volume is that ships and military technology, including artillery and fortresses, were always the vital condition for European commercial presence and expansion in Asia.

66 Furber, *Rival Empires*, pp. 330–9; René Barendse, *The Arabian Seas: The Indian Ocean World of the Seventeenth Century* (London, 2002), p. 498.

10

# Conflict in Northeastern Europe and the emergence of Russia to 1721

CAROL B. STEVENS

Russia emerged as a European power in the early eighteenth century with a suddenness that alarmed its neighbors – and indeed some of its more distant potential supporters.[1] Russia's newfound prominence was in large part the outcome of a series of international conflicts often referred to as "the Northern Wars." Conflict over the fate of the eastern Baltic littoral had entered a new phase near the middle of the sixteenth century with the decline of the Livonian Order and the growing territorial ambitions of nearby states. Aside from the crusading Order itself, which had formally disbanded by 1561, the nearby states of Denmark, Sweden, Muscovy, Poland-Lithuania, and Brandenburg persistently battled one another over the fate of the littoral, in varying configurations but with surprisingly few intermissions until 1721. The more important of these multilateral conflicts are conventionally identified as the Livonian War (1558–83), the mid-seventeenth-century conflicts among Sweden, the Commonwealth, Muscovy, Brandenburg, and Denmark that included the Thirteen-Years' War (1654–67), and finally the "Great Northern War" (1720–21) which ended in Russian victory.[2] While the earlier conflicts remained relatively confined, in diplomatic and military terms, to Northern and Eastern Europe, the outcome of the last Northern War not only established the Russian Empire as the dominant Baltic state; it also led to Russia's broader recognition as a major force in the broader European diplomatic world.

These international conflicts, which had a dramatic bearing on the political and military development of the states involved, are examined here with a view to understanding Russia's eventual victory. Three themes emerge prominently, because of the role they played in framing the military and

1 British concerns in 1719 are voiced in Simon Dixon (ed.), *Britain and Russia in the Age of Peter the Great: Historical Documents* (London, 1998), pp. 204–6 (originally from *Sbornik Imperatorskogo Russkogo istoricheskogo obshchestva* 61 (1888), 561–67).

2 Robert Frost, *The Northern Wars, 1558–1721* (London, 2000).

institutional choices made by the combatants in the Northern Wars. First among these is the political relationship between monarchs and their nobilities: the intricate interactions of court elites to limit, negotiate with, or support their rulers are one facet of this relationship; the commitment of the nobility as a whole to a country's military adventures is another. Second is the devices that were available to combatant states to mobilize resources and manpower for war: those polities that relied exclusively on state administrative capabilities had quite different options from those which made use in varying degrees of private commercial, financial, and manpower networks. Access to, and willingness to experiment with, such networks opened access to greater resources, but could also result in unsustainable levels of debt.[3] Diversity in the size, prosperity, and ownership of lands were therefore as important in any given war effort as the nobility or the institutional manner of mobilizing resources. The third and final theme is the broader diplomatic and military context. Most immediately, Poland-Lithuania and Muscovy confronted the Ottoman Empire and an insurgent Ukraine, threats that imposed significant constraints on the Baltic theater.[4] However, the increasing importance of events in Eastern Europe to the Western powers also served to complicate the course of conflict in the Baltic, though it was also to establish Russia's growing influence within eighteenth-century Europe.

These three foci are intended to redirect attention from two more conventional approaches to military history. They de-emphasize the idea that advantages automatically accrued to a centralizing state and its bureaucracy *per se*, pointing instead to the diversity of arrangements that early modern rulers both used and had available to them. Nor, despite a large and related historiography concerning "early modern military revolution(s)," do the foci chosen here award explanatory primacy to changes in military technology, tactics, and strategy.[5] Rather, they draw attention to the reasons why governments and militaries adopted some such changes, and rejected or ignored others because they were inappropriate, too expensive, or politically inadvisable.

3 David Parrott, *The Business of War* (Cambridge, 2012).

4 Rhoads Murphey, *Ottoman Warfare, 1500–1700* (London, 1998), p. 108.

5 Geoffrey Parker, *The Military Revolution: Military Innovation and the Rise of the West, 1500–1800*, 2nd ed. (Cambridge, 1996); Clifford Rogers (ed.), *The Military Revolution Debate: Readings on the Military Transformation of Early Modern Europe* (Boulder, 1995); Jeremy Black, *War and the World: Military Power and the Fate of Continents, 1450–2000* (New Haven and London, 1998); Lars Bo Kaspersen, "How Denmark Became Democratic: The Impact of Warfare and Military Reforms," *Acta Sociologica* 47:1 (2004), 71–89, 75ff.

The Northern Wars emerged from a political vacuum which had developed at the eastern end of the Baltic during the early modern period. The Livonian Confederation, which had come into being in the early fifteenth century, at first ruled its territory on the basis of tense cooperation between the Church, the Livonian Order of crusading knights, secular German nobles, and the Hanseatic towns of Riga and Reval (Tallinn). By the sixteenth century the wealth and power of the Order, of Livonia's three landholding bishoprics, and of the Riga archbishopric had all been undermined by the decline of the Order and by the Protestant Reformation's impact on ecclesiastical rule. The established political relationships became increasingly fractious. Furthermore, the political troubles of the Hanseatic League, combined with burgeoning trade throughout the Baltic, made the Hansa city-states tempting targets for nearby powers.

## I

Various forms of "consultative monarchy" governed the principal combatants in the First Northern War. Denmark straddled the straits controlling the entrance to the Baltic Sea, a position which necessarily involved its monarch in international politics and conflict. By the sixteenth century, Denmark had established a tradition of elective monarchy and noble rule. Fredrick II's oath of accession (1559) confirmed that the Estates (the *Rigsdag*), and more importantly the small, elite Council (the *Rigsråd*), could check the authority of the king. In this context, it was the king's foreign ambitions that proved a primary source of tension between the monarch and the noble council throughout much of the period. The Sound Dues, levied on the passage of ships through the straits, were paid directly to the Danish crown, granting the monarch considerable financial independence from the nobility. Furthermore, as Danish kings were also the dukes of Schleswig and Holstein, they were often able to avoid purely Danish political constraints, allowing them to develop a strong navy, hire mercenary troops, and even undertake ambitious projects to protect and expand Denmark's political sway along the Baltic (or into the North Sea) with considerable independence from the Estates and Council.[6]

6 John Maarbjerg, "*Regimen Politicum* and *Regimen Regale*: Political Change and Continuity in Denmark and Sweden," *Scandinavian Studies* 72:2 (2000), 141–62, esp. 155; Paul Lockhart, *Denmark, 1513–1660* (Oxford, 2004), pp. 37–41, 47.

The political arrangements instituted in newly independent Sweden, under its first king, Gustav Vasa (r. 1523–60), differed only somewhat from those of Denmark. Whilst the government was also broadly consensual, Sweden's Estates (the *Riksdag*) played a much more prominent role, offsetting the power of the nobility in Council (the *Riksråd*). While the *Riksdag*, in particular, supported the Vasa kings' growing venturesomeness, including the building up of their navy, the Swedish monarch lacked the Danish king's financial and political independence. Furthermore, Sweden was relatively poor and sparsely populated. But its isolation provided a measure of military security, and its control over important military resources such as timber, copper, and iron, a degree of financial stability.[7] This enabled the Swedish crown, though less wealthy than its Danish counterpart, to raise an army largely of native recruits and supported the beginnings of a navy.

Both territorially and economically larger than Denmark and Sweden, Poland-Lithuania and the Grand Duchy of Muscovy were two kingdoms sitting at opposite ends of the spectrum of Eastern European consultative monarchies. Poland and Lithuania had been joined in dynastic union since the late fourteenth century. In political terms, Poland proper was dominated by its nobility or *szlachta*, who played leading roles in the central *Sejm*, a bicameral Diet composed of an elite Senate and a Chamber of Envoys made up of elected representatives from local councils of the nobility (*sejmiki*). The Diet ruled a prosperous Poland headed by its elected king. Frequently suspicious of monarchical power, the central *Sejm* was careful to limit the monarchy's independence in key financial and military matters. As a result, numerous military decisions were made on the local level by the *sejmiki*. Poland's partner in dynastic rule, Lithuania, had political and social systems that were initially closer to Russia's: it was ruled by a grand duke and a small advisory council of high nobility into the sixteenth century. However, given Lithuania's need for Polish military support and the attractions of Polish noble liberties, it drifted closer to the Polish political model, even before the 1569 unification of the two countries into the Commonwealth. Thereafter, the Commonwealth's military could draw remarkably successfully on the combination of a kind of standing army, locally recruited and supported troops, and mercenaries.[8]

On the other end of the spectrum, by the early sixteenth century, Muscovy was characterized by a centralized court and a largely noble army. The

7 Paul Lockhart, *Sweden in the Seventeenth Century* (London, 2007), pp. 3, 7–8.

8 Andreas Plakans, *A Concise History of the Baltic States* (Cambridge, 2011), p. 93; Daniel Stone, *The Polish-Lithuanian State, 1386–1795* (Washington, 2001), pp. 38, 44–45.

leaders of the greatest noble families were members of an advisory council, the *Boyar Duma*, whose power rested in networks of kinship and informal political influence rather than in institutional structures. The monarch, whose power was taken by much of Europe to be despotic and uncontrolled, was nonetheless limited by his need, amongst other considerations, to maintain a delicate balance amongst these court networks. At another level, the recent practice of awarding estates (*pomest'e*) in return for military service to the crown maintained the loyalty of a large nobility and supported a bow-and-arrow cavalry army at little cost. This noble cavalry was largely seasonal and militia-like; its support and wellbeing depended upon the management of individual landed estates. To these forces, Tsar Ivan IV (r. 1533–84) added a standing force of infantry and expanded and refurbished the army's artillery early in his reign.[9]

The First Northern War began in 1558 with a two-pronged incursion by Muscovite armies into Livonia, following a dispute over tribute owed from Dorpat (Tartu). One Russian army moved from Ivangorod against Narva, which fell in May, before moving along the Baltic toward Reval. Meanwhile, a second army crossed the border southwest of Lake Pihkva (Peipus) and marched northward to Dorpat, which fell in July. Muscovite cavalry, with significant Tatar contingents, as well as Muscovy's new infantry and updated artillery, featured prominently in these forces.[10] Muscovy's successful campaigns continued into the early 1560s. As well as besieging larger fortified cities, Muscovite forces also inflicted great damage on their Livonian foes by capturing smaller fortifications and pillaging the surrounding countryside to sustain their cavalry. At the battle of Ermes in 1560, they inflicted a devastating defeat on the Livonian Order, which was dissolved the following year. Only the distant fortresses of Riga, Reval, and Pernau (Pärnu), all of which could be reinforced from the sea, remained unconquered.

Livonia's slow response to the Muscovite invasion reflected several problems. First, Livonia's elite faced difficult political choices in the wake of the initial Muscovite invasion, choices which were complicated in some cases by local rebellions against German overlords. Their military and political problems were compounded by calamities vividly described in contemporary chronicles and hostile broadsides: Muscovite atrocities, bouts of epidemic

9 Carol Stevens, *Russia's Wars of Emergence, 1460–1730* (London, 2007), pp. 60, 69–70.
10 Janet Martin, "Tatars in the Muscovite Army during the Livonian War," in Eric Lohr and Marshall Poe (eds.), *Military and Society in Russia, 1450–1917* (Leiden, 2002), pp. 365–87. For the Livonian War, see Stevens, *Russia's Wars*, pp. 85–92.

disease, local marauders, and looting by the mercenary companies the Livonians had hired but failed to pay.[11] A strictly Livonian response became increasingly implausible, as population, leadership, and economy failed to recover.

Rather, as Muscovy advanced, members of the Livonian Confederation sought support, and were offered help by nearby states. By the early 1560s, Livonia was divided and occupied by four powers. Sweden became protector of Reval, Pernau, and much of modern Estonia; Poland-Lithuania of Riga, Courland, and some central Livonian fortresses; and the bishops of Courland and Saaremaa sold territory to the king of Denmark's younger brother, Magnus.

This turn of events in practice led to stalemate after the early 1560s. Periodic truces interrupted the fighting, which allowed the combatants to turn their military attention to other matters – in Muscovy's case, Tatar attack from the south. But the conflict dragged on. Events in Livonia exacerbated the Danish–Swedish hostility that had existed since the dissolution of the Kalmar Union in 1523. Yet, although the resulting Nordic Seven Years' War (1563–70) stemmed from the conflict in Livonia, and involved all of the principal combatants there, it was largely fought in Scandinavia or at sea. The conflict depleted the resources of both principals, proving both "costly and futile."[12] Unlike in Livonia, however, wartime destruction was reasonably limited, and recovery was quite rapid. Meanwhile, neither the enlistment of Poland-Lithuania on the side of Denmark nor Sweden's hopes of support from Muscovy proved fruitful.

But the Livonian conflict was by no means over. As one truce between Muscovy and Poland-Lithuania drew to a close, Muscovite forces marched into Lithuania itself and captured Polotsk in 1563, which helped to push Lithuania into formal Union with Poland at Lublin in 1569. The Union of Lublin had repercussions for the other powers in the region. Military reversals and the defection of Muscovite soldiers to Poland, amongst other factors, exacerbated the already harsh nature of Tsar Ivan's rule, undermining Muscovite consultative traditions. In 1565, Ivan abruptly divided the country into two unequal parts, each with its own military and administrative apparatus; within his directly governed portion, the *oprichnina*, Ivan

11 Johannes Renner, *Livonian History, 1556–1561*, trans. J. C. Smith, W. Urban, and J. W. Jones (Lewiston, 1997), pp. 37–39, 113–15; Plakans, *Baltic States*, p. 89; Alexander Filjuskin, *Ivan the Terrible: A Military History* (Barnsley, 2008).

12 Frost, *The Northern Wars*, p. 36.

unleashed a reign of terror.[13] Meanwhile, Magnus, the maverick brother of Frederick II of Denmark (r. 1559–88), directed Muscovite affairs in Livonia, albeit not very effectively, as Ivan's vassal.

Muscovy's direct intervention in Livonian affairs resumed in the mid-1570s, following the dissolution of the *oprichnina*. By late 1577, Muscovy again controlled nearly all of Livonia apart from the strongholds of Reval and Riga. Once again, though, Russian armies were unable to consolidate their gains, and a joint effort by the Commonwealth and Sweden retook fortress after fortress, even pushing into Muscovy itself. In 1583, an exhausted Muscovy signed a truce at Iam Zapolskii with Poland-Lithuania, and the Truce of Pliussa with Sweden; it lost no territory but retained none of its wartime gains, except a narrow strip of the Baltic shoreline.

The First Northern War starkly illustrates some of the fiscal and organizational difficulties that sixteenth-century states encountered when they embarked on prolonged, if often seasonal, wars. Given its relative wealth and independence, the Danish crown predictably hired a substantial number of mercenary troops from northern Germany for the Nordic War. Despite their undeniable success on the battlefield, however, these troops were unable to deliver a knockout blow. Even after severely cutting back the size of its mercenary forces, the war plunged the Danish crown deep into debt. But when peace negotiations failed, the Danish Estates and Council were nonetheless persuaded to continue supporting the war, by introducing reforms and new taxes designed to pay off the debts accrued in the war's course.[14] Nevertheless, after peace finally came in 1570, the king, and certainly his Council, evinced much more caution over subsequent military involvements, essentially pulling back from conflict in the eastern Baltic. The war had provoked a financial crisis, but eventually political consensus.

By contrast, Sweden's gains in Estonia as a result of the war represented a major expansion of its commitments following independence from Denmark. Given its relative poverty, Sweden raised a national army – comprised of noble cavalry assisted by peasant infantry – instead of hiring mercenaries. Recruitment was initially based on traditional military obligations, but during the course of the war developed into something close to conscription. The army was large, inexpensive, and reasonably effective in a defensive war. King Erik XIV (r. 1560–68) is much praised for his efforts to build up the Swedish navy, as well as reforming the training and organization

13 Sergei Bogatyrev, "Groznyi tsar' ili groznoe vremia?," *Russian History* 22 (1995), 285–308.
14 Frede Jensen, "Fredrik II.s militære politik, 1569–1570," *Historisk Tidsskrift* 12 (1973), 59–92, 90–91; Frost, *The Northern Wars*, p. 38; Lockhart, *Denmark*, pp. 106, 113–17.

of his infantry forces. Nevertheless, Swedish troops essentially remained a semiprofessional militia. Their battlefield operations underscored the greater efficacy of the Danish professional mercenary forces, even as the Danes suffered from their unacceptably high cost. Furthermore, the war aggravated King Erik's unstable mental condition. In 1568 the nobility rebelled, and Erik was deposed and imprisoned. However, the transfer of power to Erik's half-brother John was accompanied by no fundamental political realignment. Meanwhile, the Nordic war dragged on until the Peace of Stettin in 1570, with Sweden conceding little. The Swedes also managed to retake Narva before the Livonian War finally ended in 1583.[15]

Muscovy had emerged least well from the first round of the Northern Wars. Some elements of its military had clearly performed effectively. Ivan's new standing infantry regiments, the *strel'tsy* (musketeers) – who were entirely supported by the state treasury – first doubled, and then tripled, their numbers during the war as their year-round usefulness became clear. Russian artillery had also been revamped to good effect, whilst the Tatar allies had also been successful.[16] However, noble cavalry forces, who expected to be deployed for a season or for a campaign only, still comprised the bulk of the Russian army. Initially the *pomest'e* system had displaced nearly the entire cost of arms, supplies, and training onto the servicemen themselves. Ivan IV's military reforms included an arrangement of cash subsidies for the cavalrymen.[17] Yet even with this additional support, given the Livonian War's duration, its increasingly year-round character and escalating demands on manpower, by the war's end many noble cavalrymen found that their estates were unable to sustain their military service. The crown's apparent inability to consider other forms of compensation only confirmed the shortcomings of the *pomest'e* system in these new military conditions. By the 1580s, peasants and provincial nobles alike were deserting Muscovy's heartland for the kingdom's southern frontier.[18]

15 Jan Glete, *War and the State in Early Modern Europe, 1500–1660* (London, 2002), pp. 201–6; Frost, *The Northern Wars*, pp. 33–34; Clifford Rogers, "Tactics and the Face of Battle," in Frank Tallett and David Trim (eds.), *European Warfare, 1350–1750* (Cambridge, 2010), pp. 220–21.

16 Stevens, *Russia's Wars*, pp. 74, 97.

17 Decrees on Service, 1556. [Translation in George Vernadsky (ed.), *A Source Book for Russian History from Early Times to 1917*, 3 vols. (New Haven, 1972), vol. 1, pp. 141–42].

18 Janet Martin, "The Economic Effectiveness of the Muscovite Pomest'e System," in Brian Davies (ed.), *Warfare in Eastern Europe, 1500–1800* (Leiden, 2012), pp. 19–34; Marshall Poe, "Elite Service Registry in Muscovy, 1500–1700," *Russian History* 21:3 (1994), 251–88.

The political fallout from the hardship and disarray of war, the terror and land exchanges of the *oprichnina*, and the failures of the (*pomest'e*-based) military were intense and lengthy. Disputed succession followed the death of Ivan's heir; foreign intervention, famine, and rebellion in the south tore at the state's fabric. Only after 1613 was a new royal family, backed by a powerful court nobility, installed to lead a restoration. The status of impoverished noblemen, ruined by war and civil strife and no longer militarily effective, emerged as a central political issue for the new Romanov dynasty. In 1649, the legal institution of serfdom and the opening up of new lands and opportunities seemed to reinforce the economic underpinnings of provincial noble status. Military reforms, replacing both the *pomest'e* cavalry and the *strel'tsy*, whose qualities had also been undermined by the years of civil strife, necessarily came more gradually as successive governments experimented with military forms that could coexist with noble expectations. Experiments with hired mercenaries after 1613 were not only expensive but ineffective, and many of those who had raised and led these units were dismissed following the unsuccessful Smolensk War, 1632–34.[19] After 1649, Muscovy resumed the hiring of individual foreign officers (or groups of them), but to train and command new "foreign-style" regiments made up of domestic conscripts and recruits as infantry troops. These officers effectively lost their mercenary status as they were transformed into direct and (often unwilling) permanent Muscovite employees. The (largely serf) conscripts also received wages, food subsidies, and weapons while on duty. These arrangements moved Muscovy toward a heavily "state-administered" military; in this respect Muscovy resembled the Ottomans. But these new "foreign-style" troops remained for the time being subordinate to Muscovy's still reluctant *pomest'e* light cavalry forces.[20]

Despite prolonged disputes over royal elections in the middle of the conflict, the decentralized Polish-Lithuanian Commonwealth arguably emerged in the strongest position from the First Northern War. Throughout the sixteenth century the Commonwealth army was a diverse combination of the relatively unimportant noble levy (*pospolite ruszenie, służba ziemska*), a Quarter Army paid by the Polish crown to defend the southern frontier, and mercenary troops supported by taxes voted by the *Sejm*. The latter included Cossacks from the Registry and mercenaries, often Germans or Scots, with leaders appointed by the Polish king. The famous

19 Richard Hellie, *Enserfment and Military Change in Muscovy* (Chicago, 1971), pp. 129ff., 170–71.

20 Stevens, *Russia's Wars*, pp. 132, 135–38. Parrott, *The Business of War*, p. 200, defines "state administered" militaries.

Polish hussars, unusually well attuned to the character of fighting in Eastern Europe, also counted as hired troops; they were raised by noble captains who bore much of their costs, which might be subvented by booty. Military service in these and other regiments served as a pathway to advancement, particularly for the petty nobility. In the late sixteenth century, King Stefan Báthory raised more artillery and more infantry, both foreign and domestic, for his forces and limited his use of the Polish noble levy. More importantly, he convinced the *Sejm* to raise unprecedented taxes to hire many more troops. This army, with its larger infantry contingent and excellent hussar cavalry, enjoyed a clear advantage over both the Swedes and the Muscovites. By distributing both the cost of, and responsibility for, the army to magnates, mercenaries, and local and Commonwealth institutions, Poland could endure considerable military pressure, both politically and financially, though the payment of higher taxes proved difficult for Lithuania to sustain.[21] The alienation of royal lands proved a notable means for funding Báthory's military. The Commonwealth grew, and its kings briefly claimed the Russian throne as well as trying to regain the Swedish throne from the other branch of the Vasa family. However, the Commonwealth's victory over Muscovy during the Smolensk War proved to be the high water mark of its military power.

## II

It is not easy to extract the second round of the Northern Wars from other conflicts in and around the Commonwealth in the mid-seventeenth century. By the late 1630s, the political and economic climate of the Commonwealth had changed, even if its ruling institutions had not. Poland and Lithuania had been at war for some eight decades almost without interruption – against Sweden, Muscovy, Denmark, the Tatars, and the Ottoman Empire. Any renewed conflict would require higher taxes to fund the military. The consequent Polish reluctance to engage in foreign wars effectively suspended military and other appointments that supported the minor nobility, and by the 1640s many of them felt ill-used and ill-disposed to cooperate with the crown, whose own financial status had been undermined by the alienation of its lands. Despite the efforts of Commonwealth kings from Báthory on, Polish institutions were still predicated upon close cooperation and agreement with the *szlachta*. These and other tensions were brought to a head by

21 Dariusz Kupisz, "The Polish-Lithuanian Army in the Reign of King Stefan Báthory," in Davies (ed.), *Warfare in Eastern Europe*, esp. pp. 66–70, 76; J. Teodorczyk, "Wojsko," in *Polska w epoce odrodzeniu* (Warsaw, 1986), pp. 294–97.

the Orthodox Cossacks of Ukraine, who also had religious grievances against the crown.

The following Thirteen Years' War (1654–67) began in southern Poland and Ukraine. The Ukrainian Cossacks' initial assault on Royal Poland in 1648, led by Bohdan Khmelnytsky, was astonishingly successful, in part because the Commonwealth was unprepared and the *szlachta* were reluctant to resume fighting. When Khmelnytsky's early successes proved unsustainable, the Cossack leader enlisted Muscovite protection for Ukraine in 1654. This arrangement sparked a long and intense international conflict over the fate of Ukraine, which drew in the Ottoman Empire and its Crimean vassal in a confrontation involving the Commonwealth, Ukraine, and Muscovy that lasted at least until 1681.[22]

The fighting did not remain in the south, however, but rapidly spread into the Baltic, developing into the Second Northern War. When Muscovy joined battle against the Commonwealth in 1654, ostensibly in defense of Khmelnytsky, it deployed many of its troops in the north near Smolensk and Polotsk. This effort proved remarkably successful. By August 1655, the Muscovites occupied much of the Commonwealth as far west as Minsk and Wilno (Vilnius).

Unsurprisingly, a Muscovite military presence so close to Sweden's Livonian holdings quickly attracted its attention. Expanding the size of its army using military enterprisers, Sweden hastily assembled an army and entered the Commonwealth in July 1655, from both Swedish Pomerania and Livonia. Thus the conflict usually referred to as "the Second Northern War" began. Sweden intervened both to protect its gains from the Thirty Years' War and to cut short unpleasant domestic disputes. Its forces in western Poland were quickly successful, occupying Warsaw in September; King Jan II Casimir retreated southward, where he was followed, defeated, and then exiled by October. In the east, Janusz Radziwiłł, hetman of Lithuania, demoralized by his military losses to Muscovy and Sweden's victorious activities, and encouraged by his disagreements with Jan Casimir, negotiated Swedish protection for Lithuania, as Muscovite and Cossack forces continued their advance. Shockingly, nearly the entire Commonwealth suddenly lay in the hands of occupying armies. In 1655, the elector of Brandenburg therefore hastened to organize the defense of his Polish fiefdom, the duchy of Prussia. When Sweden nonetheless quickly occupied the duchy, the elector briefly

22 For further elaboration on the following narrative, see Brian Davies, *Warfare, State and Society on the Black Sea Steppe* (London, 2007), chs. 4 and 5; Frost, *The Northern Wars*, ch. 7; Stevens, *Russia's Wars*, ch. 5.

became a Swedish vassal and then ally in Prussia. Such dramatic changes raised some international concern: from the Habsburgs (fearful of attack from a Swedish Poland), from the Danes and the Dutch (fearful of Swedish domination over the Baltic), and from the Crimeans and Transylvanians, conscious of the resulting power shift from the south. Active responses to the war remained largely regional, however, except for Dutch support for the Danes.[23]

The Commonwealth remained the battleground for the armies, though Sweden's advantageous position there did not last. Muscovy signed a truce with the Commonwealth, declared war on Sweden, and invaded Livonia. A resurgent Poland joined with Lithuania to attack Ducal (eastern) Prussia. Seeing Sweden beset, Denmark too attacked. Carl Gustav of Sweden bolstered his forces with a degree of success, reinforcing his connection to Brandenburg and enlisting Transylvania as another ally. This support faltered, however, as Carl Gustav abandoned the morass in the Commonwealth to launch a daring attack on Denmark's home territories. A growing anti-Swedish alliance included the Habsburgs, Brandenburg (abandoning its previous, uncomfortable alliance with Sweden), Denmark, the Dutch, and the Commonwealth. Carl Gustav's death in 1660 and growing disagreement among the allies led to a welcome peace.[24]

While the Second Northern War largely ended there, continuing disputes in Ukraine kept both Muscovy and the Commonwealth fighting for two more exhausting decades in Ukraine and sometimes in the north. After Hetman Khmelnytsky's death, the Cossack leaders' dissatisfaction with their Muscovite allies and their attempts to negotiate a new relationship with Poland reignited fighting between the two powers. Poland's complex internal politics and growing exhaustion on both sides finally led to the signing of a thirteen-year truce at Andrusovo in 1667. A triumphant Muscovy gained Smolensk, Chernihiv, part of Vitebsk in the north, the Left-Bank Ukraine, and – ostensibly for two years only – the city of Kiev. The Cossacks were not consulted, though, and their fate remained unresolved. Petro Doroshenko, the new hetman of the Right-Bank Ukraine, then allied himself with Crimea and the Ottomans in an attempt to reunite Ukraine, but Muscovite forces frustrated his efforts. Ottoman troops then marched north to try to restore their influence, fighting Poland (1672–76) and then Muscovy (1676–81). The Turkish intervention shifted the focus of East European

23 Parrott, *The Business of War*, pp. 280, 284; Glete, *War and the State*, p. 188; Frost, *The Northern Wars*, pp. 178–82.

24 Lockhart, *Sweden*, pp. 94–100.

competition away from the Baltic and had a profound impact on several of the belligerents. The strain of fighting the Ottomans exacerbated Polish exhaustion and accelerated Russian military reform. For Ukraine, the period 1657–87 would be remembered as "the Ruin": an era of strife, destruction, and depopulation.

The lengthy and continuous, as opposed to seasonal, campaigning that had become an ever greater part of seventeenth-century conflicts reinforced the pressures on the combatant states. For one thing, in the aftermath of the Thirty Years' War, there were considerable changes in systems of military enterprise that had important political consequences. Sweden's was a particularly dramatic example. Often brilliant military leadership, unpredictable but effective tactics, and seasoned troops had produced a succession of astounding military victories in the Thirty Years' War and a powerful militarized empire in north Germany. Similar great victories followed during the Second Northern War as the Swedes attempted to defend their newly won empire. The demands of these wars on manpower led Sweden to supplement its renowned native troops with a heavy reliance on military contractors-creditors – many of them Germans with connections to the contractors of the Thirty Years' War, others Swedish.[25] The fiscal and personnel cost of these activities left Sweden with a crushing domestic burden after 1648, further exacerbated by the wars in the Commonwealth and against Denmark, where mercenaries were again used. This became a focus of political conflict, leading to considerable fiscal and political pressure from the *Riksdag* against the Swedish nobility in 1650 and an aristocratic reaction in the 1660s.

By contrast, it was only in the second third of the seventeenth century that Denmark overcame its reluctance to use military contractor-creditors or enterprisers from amongst its subjects, instead of foreign mercenary regiments. As it prepared to join the Second Northern War against Sweden in the late 1650s, however, it did supplement its native troops (which were recruited by similar means to Sweden's but without their effectiveness) with German enterprisers.[26] As in Sweden, this incurred significant debt, the repayment of which was made all the more difficult by the occupation of Danish soil by Swedish troops. Nonetheless, when the *Riksdag* met in 1660, the higher nobility resolutely defended its immunities. Outraged, the other estates offered the new king "absolute and hereditary" power, and the nobility was

25 Parrott, *The Business of War*, pp. 126–34, 293–300 and private communication; Frost, *The Northern Wars*, pp. 204–8; Lockhart, *Sweden*, pp. 80–85, 107–14.

26 Parrott, *The Business of War*, p. 111.

unable to stop them. Partly as a result of this offer, the Danish nobles' position was gradually redefined after 1660. The upper reaches of the military and the bureaucracy began to be filled through an increasingly meritocratic system, while a table of ranks ennobled those reaching the upper levels of the hierarchy, revitalizing the links between nobility and the military. In short, a "bloodless but dramatic royalist revolution" – leading to political and military restructuring and resumed commitment by the Danish nobility to its army – took place, stemming at least partly from the levels of debt incurred during the Second Northern War.[27] Denmark resumed its predominantly westward orientation, as confirmed by its reluctance to join battle over the eastern Baltic.

For Brandenburg-Prussia, a much less powerful state just beginning to gain prominence in the aftermath of the Thirty Years' War, this was not a viable option. Like several other German rulers, Elector Fredrick William was developing a standing army of trained native troops. The threat of Swedish invasion and guarantees of noble privileges overcame the Brandenburg Estates' initial reluctance in 1652–53 to collect taxes to support such troops. The Polish–Swedish War, Fredrick William's success in pitting his subjects' interests against one another, and his use of force against opposition in Ducal (eastern) Prussia allowed him to continue raising both taxes and the size of his army. The costs of a growing army – still in large part made up of hired mercenaries – were further subvented by Fredrick William's loaning his troops for the defense of the Holy Roman Empire. Brandenburg and Denmark were among those with lasting grievances against Sweden; the Scanian War of 1674–79 eventually pitted Sweden (and France, which subsidized the Swedish army) against Denmark and its north German allies. Brandenburg-Prussia's new military forces proved successful in campaigns against the Swedish, albeit at great cost and without a permanent and dependable native noble officer corps such as Sweden and Denmark enjoyed.[28]

The Swedish army did not perform to its usual standards in the 1670s. The loss of a small Swedish army to Fredrick William's forces at Fehrbellin was followed by a series of failures in northern Germany. Yet, as the battle of Lund (1676) indicated, the Swedes had not permanently lost their edge. In the

27 Leon Jespersen, "Court and Nobility in Early Modern Denmark," *Scandinavian Journal of History* 27 (2002), 129–42, 130–31; Kaspersen, "How Denmark Became Democratic," pp. 78–79.

28 Karin Friedrich, *Brandenburg-Prussia, 1466–1806: The Rise of a Composite State* (London, 2012), pp. 27, 31–32, 88; Peter H. Wilson, *German Armies: War and German Society, 1648–1806* (London, 1998), pp. 36–37.

end French pressure, and disputes among Denmark's allies, helped Sweden to retain nearly all of its north German lands at the Peace of Lund (1679). Sweden's existing military system, however, was bankrupt; the ensuing negotiations restored alienated royal lands (*reduktion*) and standardized the *indelningsverk*, a new allotment system that spread the cost of maintaining the army and ensured that a service-based nobility officered it. Though not without serious strains and some adjustment, military reorganization also helped to gradually restructure the agricultural economy, and the army offered a new path to social advancement to young men who joined the military. The Swedish commoner and nobleman alike thus remained linked to, and supportive of, the Swedish military system in the late seventeenth century. As a result, there was considerable social support for greater royal power – which helped to launch Swedish absolutism.[29]

The Commonwealth's implosion in the Second Northern War, "the Deluge," has often been attributed to the failures of its state structures compared with those of other military-fiscal states, like Muscovy and Sweden. The military-fiscal state (and the associated fiscal-military state) are usually understood as variants of European centralized bureaucracies focused on funding the expansion of land and naval forces, as became common in the seventeenth century. But recent research suggests that European states funded their military forces with a bewildering variety of privatized and informal arrangements as well as the more obvious use of developing public institutions; indeed, individual states' abilities to exert central administrative control over these arrangements could be quite limited.[30] Given this, creating a sharp explanatory distinction between centralized and decentralized governments seems unhelpful.

While it is true that the Commonwealth's military effectiveness rested on a negotiated consensus both between king and nobility and within the nobility, such a consensus was quite remarkably productive, drawing on local, public, national, and private resources in unique combinations. It has been suggested that, by the mid-seventeenth century, this system was losing

29 Jan Glete, "The Swedish Fiscal-Military State in Transition and Decline, 1650–1815," www2.historia.su.se/personal/jan_glete/Glete-Swedish_Fiscal_Mil_State_Trans.pdf (2006); Anthony Upton, *Karl XI and Swedish Absolutism, 1660–1697* (Cambridge, 1998), pp. 71–98; Lockhart, *Sweden*, pp. 139–42; Alf Åberg, "The Swedish Army from Lützen to Narva," in Michael Roberts (ed.), *Sweden's Age of Greatness* (New York, 1973), pp. 268–69.

30 Glete, *War and the State*, pp. 39–40, 60–61; Jan Glete, "War, Entrepreneurship, and the Fiscal-Military State," in Tallett and Trim (eds.), *European Warfare*, p. 307; Parrott, *The Business of War*, p. 12.

the flexibility that had given it such resilience, especially as the *liberum veto* developed.[31] Still, King Jan Sobieski managed to introduce some institutional change after the Deluge – and raised an army that he famously led to victory against the Turks at the siege of Vienna in 1683. On the other hand, these arrangements undeniably failed to provide the Commonwealth with a permanent military structure.[32]

No less significant was the broad decline in military resources available to the Commonwealth during the seventeenth century, just as many European armies were growing. Efforts to restore alienated crown lands, which had supported earlier royal military initiatives, were unsuccessful. The alternative, imposing high levels of taxation in the Commonwealth, had been difficult for Lithuania to sustain even before 1600. The impoverishment and simmering discontent of the petty nobility, which followed from the shortcomings of service landholding, was a further complication. Finally, the Second Northern War, the Thirteen Years' War, subsequent confrontations with the Ottoman Empire, and fighting over Ukraine all took place in large part on Commonwealth territory. It is not difficult to attribute at least some of the Commonwealth's problems to *economic* decline following the exactions of foreign troops and to the loss of population, agricultural land, and productive resources after the peace. Towns in Lithuania lost 40–90 percent of their population, often as a direct result of Muscovite action. In Ukraine, a side theater of the Baltic wars, the population was decimated and fled, often into neighboring Muscovy. In addition, Muscovy, Sweden, and the Ottomans annexed some 15–20 percent of Commonwealth territory. Under these circumstances, the impact of tax and fiscal reforms was necessarily limited. The yield of the 1649 hearth tax dropped by 40 percent after 1667. While the Commonwealth's revenues eventually approached pre-1655 levels, they did not increase any further.[33] For these reasons, the Commonwealth found it difficult to field large armies after the mid-seventeenth century. A sharp decline in resources, less flexibility in the *Sejm*, and the politicization of the

31 Stone, *The Polish-Lithuanian State*, pp. 181–83.

32 Robert Frost, "The Death of Military Culture? The Citizen Army and the Military Failure of the Polish-Lithuanian Commonwealth, 1648–1717," in Winfried Speitkamp (ed.), *Gewaltgemeinschaften. Von der Spätantike bis ins 20. Jahrhundert* (Göttingen, 2013), pp. 209–30.

33 Anna Filipczak-Kocur, "Poland Lithuania before Partition," in Richard Bonney (ed.), *The Rise of the Fiscal State in Europe, c 1200–1815* (Oxford, 1999), pp. 443–79; Juliusz Bardach, Bogusław Leśnodorski, and Michał Pietrzak, *Historia ustroju i praw polskiego* (Warsaw, 1994), pp. 180–81; for Polish reluctance to serve outside home territories, see Jan Ch. Pasek, *Memoirs of the Polish Baroque: The Writings of Jan Pasek* (Berkeley, 1980), p. 8.

military seem as important in this deterioration as the presence or absence of any particular state structures. Interaction among the various decentralized levels of the Commonwealth did little to offset these problems. The Commonwealth was therefore unprepared for what proved to be another devastating onslaught.

By contrast, the Second Northern War and the subsequent conflict in Ukraine dramatically catalyzed Muscovy's military transformation. By the late 1650s, much of its existing military establishment was stretched to the limit. For political reasons, in 1654 the army still relied on the large, inexpensive contingents of noble *pomest'e*-based cavalry. However, while the *pomest'e* cavalry were adept at defending Muscovy's southern frontier, they were poorly suited to waging a siege-based and nearly continuous war in Lithuania. The seasonal cavalry's reluctance to fight was made quite clear by their repeated failure to answer calls to service. Russian manpower losses during the fighting were also heavy. As conscription and recruitment produced new troops for the front, military restructuring took place under cover of war. The numbers of new, foreign-style regiments, which were easier to assemble and were led by foreign officers and conscripted troops, increased. Under the guise of military necessity, the poorest of *pomest'e* holders, as well as the *strel'tsy,* were incorporated into these formations.

The creation and rapid build-up of new military-administrative machinery at the regional level – to recruit, train, pay, feed, arm, and clothe troops – is a feature of Muscovy's military transformation commented upon less frequently by historians. By 1667, southern army groups provided much of the institutional framework for a more permanent army on a regional level; they continued to grow during the fighting over Ukraine.[34]

Such arrangements were perhaps more immediately costly than some alternatives, such as the combinations of foreign loans, mercenary officers, military enterprisers, and the drafts on private resources that other states resorted to. Moscow does not appear to have seriously considered these options, although Russia did try to borrow on the international financial market. Instead, it stretched its resources by in-kind collection and delivery of items such as food supplies. When higher taxation did not relieve a shortage of specie exacerbated by the payment of military salaries, Moscow debased its silver coinage in 1662. This led to riots in Moscow, as well as demands from foreign officers at the front to be

34 Stevens, *Russia's Wars*, pp. 160–72.

released from their obligations, and the silver coinage was restored in 1663.[35] Nonetheless, Muscovy's administrative systems from the 1670s onwards largely avoided (or at least avoided acknowledging) the crushing levels of direct debt that plagued states with different arrangements. The adaptability of the Russian system in the midst of war also emphasizes the commitment of successive monarchs, as well as the Russian elite, to restructuring the country's armed forces.

As with the Commonwealth, Muscovy's confrontation with the Ottomans (1676–81) profoundly shaped its later behavior in the Baltic theater. Further military reforms undertaken during, and just after, the Ottoman war abolished the precedence-ranking system and plans were made for foreign-style troop structures to be introduced throughout the field army. These changes were particularly striking in that they were introduced with the broad support, and indeed often under the leadership, of powerful noble families. No less important was the replication of military-administrative regions throughout the country, which laid a national institutional framework for the military. However, the decision to retain a very large fighting force, clearly prompted by the Ottoman threat, ensured that the army did not entirely move to standing troop structures, but overwhelmingly retained semipermanent forces with limited training, regardless of their new formation structure.[36] This new state-institutional capacity to muster, train, supply, and field a massive army with a growing number of Russian officers did not yield battlefield victories in the last decades of the seventeenth century. Consequently, the large territorial transfers of 1667 and Muscovy's increased organizational capacities are rarely given their due importance in presaging what was to become a sustainable military infrastructure.

## III

The next and final round of the Northern Wars provided a resolution in the contest for power in Northeastern Europe. It began simply enough. Peter I, newly ascended to the Russian throne, tried to extend Russia's activities as

35 William Reger, "In the Service of the Tsar," PhD diss. (University of Illinois, 1997), p. 128; Carol Stevens, *Soldiers on the Steppe* (DeKalb, 1996), ch. 3; Hellie, *Enserfment*, p. 228.

36 There is no direct evidence of Ottoman influence, but the presence of large and well-supplied Ottoman armies in the area strongly suggests a connection: Rhoads Murphey, "Ottoman Military Organization in South-Eastern Europe, *c.* 1420–1720," in Tallett and Trim (eds.), *European Warfare*, pp. 135–58; Virginia Aksan, "Locating the Ottomans among Early Modern Empires," *Journal of Early Modern History* 3:2 (1999), 103–34, 116–17.

a member of the anti-Ottoman alliance, the Holy League. But despite Peter's own victories against Ottoman Black Sea forts, and Habsburg gains in the Balkans, most Europeans had lost interest in the face of the upcoming War of the Spanish Succession. Peter found a like-minded ally in Augustus, the newly elected king of Poland-Lithuania. Augustus was also the elector of Saxony, and whatever the vagaries of Polish politics, the very respectable Saxon army was his to command.[37] Together with the Danes, who were unable to resist the opportunity for revenge against Sweden, Peter and Augustus renewed the Northern Wars, launching an attack on the new young monarch of Sweden, Karl XII, and in particular on Sweden's undergarrisoned eastern Baltic provinces. Their initial efforts were disastrous. Karl unexpectedly blockaded Copenhagen with naval support from the Dutch and the British, who were as usual anxious to protect their interests in the Baltic; this led to Denmark's immediate withdrawal from the war. Meanwhile Augustus, whose siege of Riga had been compromised by logistical failure and a lack of naval support, headed for winter quarters early. Karl then landed a relief force near Russian-besieged Narva and attacked the Russians in their trenches. The result was a rout. Sweden was not as vulnerable as it had seemed; the third, the Great, Northern War would last for twenty-one years.[38] (See Map 7.)

Karl's victories after 1700 provoked considerable concern in Western Europe. At issue were not only the Maritime Powers' naval interests in the Baltic. Karl's military style, which included successful campaigns waged over huge expanses of territory, threatened to involve the Ottoman Empire.

Russian military activity was focused in a different direction. The comprehensive round of military reforms that Peter launched after the defeat at Narva is often credited with "making the Russian military modern." But it would be more accurate to say that early eighteenth-century reforms built upon the framework of the previous century to create an army that was specifically crafted to defeat Sweden: Peter copied Swedish military training and tactics quite directly; indeed, he is reputed to have toasted captive Swedish officers as "my teachers" after the Russian victory at Poltava in 1709. Swedish institutions, too, were studied, copied, or adapted as effective means of supporting a "state-administered" army.

37 Augustus's actions also owed something to his concern for Saxony's standing in the Empire. Wilson, *German Armies*, p. 130.

38 For more detailed narratives of the Great Northern War, see Frost, *The Northern Wars*, chs. 9 and 10; Stevens, *Russia's Wars*, Part III; Wilson, *German Armies*, pp. 129–45.

MAP 7. The Great Northern War, 1700–21. Created by Béla Nagy.

Most significantly, much of the Russian army quickly, and quite visibly, became a permanent force. This change was reinforced by war itself. In the early years, Russian armies were persistently on the move, fighting against small Swedish fortresses and garrisons in the eastern Baltic. Not only did the ongoing conflict serve as a training ground for officers and men alike, but also Russian victories at the eastern end of the Baltic led to the foundation of Peter's emblematic new capital, St. Petersburg, in 1703. Meanwhile, in the army, peasant conscripts became soldiers, and soldiering became a new, separate, social identity; noble servitors too had little opportunity to return home. Peter worked to create national-level military institutions to serve them, by 1705 replacing or superseding the regional army-group administrations of the preceding century.

Peter's reforms traveled a bumpy road. At first, the tsar's restless energy unleashed a series of remarkable efforts by his inner coterie, who intervened ferociously and in defiance of protocol to ensure some degree of organizational effectiveness. Peter was not easily satisfied; ineffective or incomplete mechanisms were hastily replaced, as their royal progenitor impatiently sought remedies. These abrupt, and ferocious, displays of royal power more than once produced confusion, rather than improvement. Nonetheless, the military reforms of the early 1700s moved food supply and other military systems onto a cash basis, relying on a growing commercial economy, direct contracts with merchants, and stronger commercial networks. But rapid and substantive reform carried significant political risk, as Russian court elites resented their displacement by the tsar's favorites, who included influential Europeans and commoners, as well as fellow Russian aristocrats.

Meanwhile, the Commonwealth under Augustus, judged by Sweden to be the more fearsome enemy, constituted the main theater of war until 1706–7. Things did not go well for Poland-Lithuania. In addition to older, ongoing problems, there was an immediate political crisis, given that the Commonwealth's nobility did not universally recognize Augustus's election in 1698. Although the *Sejm* cooperated with its king to muster a reasonable army in 1703, there was a notable lack of enthusiasm for this path to war.[39] Furthermore, in Augustus's early campaigns, his well-trained and generally effective Saxon infantry failed to distinguish itself against the rather unpredictable fighting style of the Swedes. Polish-Lithuanian civil conflict rapidly complicated the international confrontation, with Commonwealth

39 Wilson, *German Armies*, p. 131; Stone, *The Polish-Lithuanian State*, pp. 249–50.

confederations each offering their own solutions to problems military and political, and proffering support variously to Augustus, Sweden, Russia, and Polish noble democracy. There seemed no solution in sight to the Commonwealth's declining population and continuing problems. The result was a political quagmire; the Great Northern War became at once a civil war and a multisided international struggle, with the Commonwealth serving as the battleground for both.

Karl XII's victories in the Commonwealth rested on his own remarkable generalship and justly famed army, whose support remained integral to the Swedish state and economy. He benefited from refinements in the organization and training of the army undertaken by his father, Karl XI. The standardized and regularized allotment system (*indelningsverket*) could rapidly mobilize a native cavalry and infantry of over 40,000 men. This system rested on in-kind support for the army during peacetime. It permitted intensive training while leaving Sweden's cash income for other purposes. For officers as well as for the men, military service remained a reliable, and often rewarding, lifestyle. In Sweden's Baltic and Pomeranian outposts, as in the field during the Great Northern War, native troops were augmented with mercenaries. Karl XII's army was in many respects an unusual one: it was, by comparison with other armies of the era, particularly mobile and aggressive in attack. It boasted effective heavy cavalry, which built upon the model of the Polish hussars, who attacked at a full gallop. Karl XII famously distrusted firepower; he may also have minimized his use of artillery during the Great Northern War because of lengthy supply lines. The battle of Kliszów (1702), at which the Swedes surprised and defeated a well-trained, larger, and well-entrenched Saxon-dominated force, was a testament to Sweden's enduring military prowess.[40]

Battlefield victories notwithstanding, Karl was unable to gain political control of the Commonwealth and his efforts to do so alienated some of his initial Polish support. Further difficulties beset the Swedish force. First, the Commonwealth's economic and demographic base continued to suffer from the adverse effects of occupation and epidemic. Despite the army's excellent organization, Sweden's small size and population required its armies to be quartered overseas and supported by external subsidies to support any prolonged fighting. Second, Karl compelled Augustus to abdicate the Polish throne and installed Stanisław Leszczyński in his stead, but he did

40 Comments by Austrian diplomat in Otto-Anton Pleier, "O nyneshnem sostoianii gos. upravlenii v Moskovii v 1710," in V. Naumov (ed.), *Iust-Iul, O.-A. Pleier. Lavry Poltavy* (Moscow, 2001), p. 398.

so hastily and without appropriate attention to Polish law and custom. Nonetheless, by 1706, Sweden's continuing victories in Poland and Saxony imperiled the stability of north Germany and reinforced the allies' concern that the Northern War might invite French intervention or have an impact on the role of Austria in the War of the Spanish Succession, threatening to intertwine the two wars. Having earlier promised Karl subsidies and loans, the allies now requested that Karl remain neutral in the western war, in return for guarantees of Swedish holdings in Germany; they were further relieved by his decision to turn eastward.[41]

Frustration factored into Karl's decision to attack Russia in 1708. The move has been criticized as rash and arrogant, but there were good logistical and strategic reasons for moving out of Poland, while the Swedish army was in excellent condition. The Russians' military situation also appeared less than optimal. The army's fighting quality was certainly much improved since 1700, but Russia still lacked some key features of a modern army, in particular specialized military education. Furthermore, Peter faced serious resistance, and rebellion, at home. He therefore ordered his army to withdraw in the face of the oncoming Swedes, destroying supplies as it retreated. Upon passing Mogilev, the Swedish army turned southward toward Ukraine, which promised better supplies as well as support from disgruntled Cossacks under their hetman, Mazeppa.

Sweden's military success did not persist. General Lewenhaupt, moving southward to meet the main Swedish forces with more men and supplies, was intercepted and defeated by Russian forces at Lesnaia in the fall of 1708. Karl's forces were thereafter reinforced by more men, fleeing the field of Lesnaia, but not with the supplies to sustain them. As Sweden tried to overcome the Russian advantages of proximity and a larger population, it lengthened its supply lines and imposed new strains on its home territory. Sweden's wartime casualties had imposed a progressively heavier demographic burden since the 1650s which peaked in the Great Northern War.[42] On the ground, Russia's scorched-earth policy, an unusually cold winter, and Peter's skillful use of irregular cavalry to harass the Swedes compounded their difficulties. The Russian destruction of the Cossack capital at Baturin

41 Stone, *The Polish-Lithuanian State*, pp. 250–53; Hamish Scott, "The Great Northern War and the Integration of the European States System," in John Hattendorf, Åsa Karlsson, Margriet Lacy-Bruijn, Augustus Veenendaal, Jr., and Rolof van Hovel tot Westerflier (eds.), *Charles XII: Warrior King* (Oxford, 2019), pp. 16–18.

42 John Landers, "The Destructiveness of Pre-Industrial Warfare: Political and Technological Determinants," *Journal of Peace Research* 42:4 (2005), 455–70, 458; Glete, *War and the State*, p. 206.

prevented the Cossacks under Mazeppa from providing substantial support to the Swedish army, nor was Polish aid forthcoming.

As it moved south, the Swedish army laid siege to the fortress-town of Poltava and turned to attack the Russians early in the morning of June 27, 1709. The Russians were relatively well supplied and were certainly the larger force; in particular, their artillery was in full working order and their munitions more abundant than Sweden's. Russian commanders used their troops cautiously; the soldiers stood firmly in redoubts or in their fortified camp against the attaching Swedes, but did not initially undertake more complex offensive maneuvers. Most importantly, Russian forces took advantage of Sweden's multiplying difficulties. Their enemy, while confident, was far from home and suffered from depleted supplies, food, and numbers; Karl, who was wounded in the foot, also lost contact with nearly one-third of the Swedish infantry at the redoubts before the main battle was joined. The Russian victory annihilated much of the Swedish army, left many Swedish officers in captivity, led to the deaths of Karl's Cossack supporters, and sent Karl himself across the Dnieper river to seek sanctuary in the Ottoman Empire. (See Figure 16.)

FIGURE 16. Battle of Poltava, 1709. Pierre-Denis Martin, 1725, painted to a commission by Peter the Great, but delivered after his death. Currently housed in the Catherine Palace, St. Petersburg. Courtesy of Museums of Russia.

The powers of Europe were both astounded and worried by the resounding defeat. The immediate result of Poltava was the reemergence of an anti-Swedish alliance. Denmark and Augustus, who was returned to the Polish throne with Russian support in 1709, were joined by others interested in the fate of Sweden's trans-Baltic lands. Sweden's neutrality vis-à-vis the Spanish Succession had been carefully arranged by the allies. But now the presence in northern Germany of Swedish troops defending a defeated empire could easily provoke, even if they did not initiate, fighting in the region. Such a conflict could easily lead to a corresponding withdrawal of troops and disengagement from the War of the Spanish Succession. In 1710, Britain, the Dutch, and the Empire joined the Danes, Saxony, Hanover, Brandenburg-Prussia, Russia, and others in the Neutrality Convention. A remarkable product of widespread concern, it offered to protect the neutrality of the southern Baltic shore, including Sweden's overseas provinces. Karl XII, however, refused to sign it.[43]

Peter's signature was a recognition that, despite Poltava, the Great Northern War was far from over. As Peter strove to consolidate his gains along the Baltic, his efforts were undermined first by a severe outbreak of the plague and subsequently by an Ottoman declaration of war. The major Russo-Ottoman battle on the river Prut (1711) was a bitter defeat for Peter, whose army was weakened by low numbers and the very long march into Moldavia; Russia relinquished its Black Sea gains of the 1690s.[44] Thanks to his recent institutional reforms, however, Peter's army was able to regain military momentum relatively quickly. Even as the war dragged on, Peter continued to engage in substantive political reforms, permanent institutions replacing the personalized, more *ad hoc*, rule of his early reign.

Yet domestic conflicts persisted for Peter. Rebellion had already erupted against the stringencies of his reign along the edges of the empire, though it was subsequently contained. Of greater significance, during the late 1710s, court elites lost patience with Peter's failure to acknowledge or consult them, and a number entered into negotiations with Peter's heir, Alexei. The entire episode – Alexei's flight to the Habsburg court, his return, and the investigations that followed – did not end with Alexei's death in prison in 1718.

43 Serhii Plokhy, "The Battle and Its Aftermath," in Serhii Plokhy (ed.), *Poltava 1709: The Battle and the Myth* (Cambridge, MA, 2012), pp. 81–189; Scott, "The Great Northern War," pp. 20–21.

44 Stevens, *Russia's Wars*, pp. 265–66; Ia. E. Vodarskii, *Zagadki Prutskogo pokhoda* (Moscow: 2004), pp. 180f.; Aksan, "Locating the Ottomans," p. 106, on the army's technical shortcomings at Prut.

Some structural concessions to the high nobility helped to stabilize Peter's government, while noblemen who had won preferment or regained status during the Northern Wars gradually provided a political counterweight to maintain the reforms.[45]

Meanwhile, Sweden remarkably fought on, despite the devastating loss of experienced troops and equipment at Poltava. By the time Karl had escaped his Ottoman refuge in 1714 to command his armies, however, the War of the Spanish Succession was over and diplomatic interests in Eastern Europe had shifted. By 1716, when Karl died before the Norwegian fortress of Fredrikshald, Sweden's Baltic, German, and Finnish possessions had been lost, and Karl's heirs hoped in vain for (previously reliable) help from the Dutch or British. Instead, new issues took primacy. Russian aspirations for Mecklenburg and the presence there of Russian troops aroused concern among contestants for Sweden's former holdings, such as Bremen and Verden, recently acquired by Hanover, whose elector was now the British king George I. To Peter I's great pleasure the Russian navy, the recipient of significant state investment, had won the battle of Gangut (Hanko) against Sweden in the Baltic in 1714. Western Europe was becoming vividly aware of an emergent naval and proven military power whose reach stretched beyond the Baltic. At long last, however, a Russo-Swedish peace was concluded at Nystadt in 1721. Peter became emperor of all the Russias and his new capital at St. Petersburg was assured.

Russian victory, in one respect, seemed almost inevitable given its size, prosperity, and population. Despite Sweden and Denmark's efficient mobilization of public and private resources, and the military talents of several Swedish kings, these advantages could not hope to match Russia's vast human and economic power and its new, if imperfect, administrative machinery. The precipitous decline of the Commonwealth, now dominated by Russia and devastated by prolonged Ottoman and Northern Wars on its own territory, eliminated Russia's only obvious rival in terms of size and wealth. The prospect of a solution to the internal political problems of the Commonwealth was undermined by a collaboration between King Frederick William I (r. 1713–40) of Brandenburg-Prussia and Peter I, cemented in the Treaty of Potsdam (1719).[46]

45 Paul Bushkovitch, *Peter the Great: The Struggle for Power, 1671–1725* (Cambridge, 2001), pp. 428–34, 442–43.

46 Robert Frost, "Everyone Understood What It Meant," in Plokhi (ed.), *Poltava 1709*, pp. 159–76; Friedrich, *Brandenburg-Prussia*, p. 88.

The strategies and devices Russia used to mobilize its resources were less self-evident factors in Moscow's triumph in the Baltic. Its increasingly state-administered military allowed the state to minimize long-term debt and to make use of in-kind resources at crucial moments. Nevertheless, the system itself was inherently expensive – requiring considerable stockpiling and duplication and plagued by corruption – which was somewhat offset by the increased use of paper and copper money, this time with relative impunity.[47] Despite its Scandinavian origins and its undeniable flexibility under pressure, the Petrine military model was autocratic and reliant upon continual draconian policing organized by a powerful monarchy to ensure its efficacy. It was not a model that would be widely imitated.

Perhaps most importantly, however, Russia made the transition to a new form of military organization as it garnered the political approval of its nobility. That is, it had moved in the late seventeenth century, with the participation of its high nobility, toward a permanent, trained military force with numerous infantry and a large native and foreign officer corps. The Scandinavian-oriented reforms of the Petrine period did not entirely deflect it from its earlier adaptation to Ottoman fighting styles, and the new structure proved relatively capable of adapting to further military innovation. And, despite the crises of the late 1710s, by the end of the reign the nobility's participation in the new military system had been thoroughly consolidated.

Russia's victory also came at great cost. The Russian Empire's military consolidation depended upon a regime of serfdom, put in place to satisfy and maintain noble status, and expanded by Peter to supply the military, among other things.[48] With its reinvigorated insistence on a lifetime of noble service to the crown, often as military officers, Petrine bureaucratization explicitly transferred renewed power to court elites. Finally, the fiscal underpinnings of the army were based upon militarization of that bureaucracy. This was a new military age; communities were losing their long-cherished autonomy; the higher human and financial demands of permanent and regular armies and navies were felt by all countries that sent them into the field. Life under military control where these forces fought and won (or lost) was another thing altogether, and it often took generations to recover. It was a costly change for all.

47 Boris Mironov, "Consequences of the Price Revolution in Eighteenth-Century Russia," *Economic History Review*, New Series, 45:3 (1992), 457–78, 472–73.

48 Evgenii Anisimov, *Vremia Petrovskikh reform* (Lenizdat, 1989), pp. 296–98.

But the end of the Northern Wars also created a new situation across Europe. That the Russian Empire would dominate the Baltic for years to come was self-evident. But beyond the boundaries of that old conflict, Europe had been forced to pay increasing attention to events in the east, and particularly to Russia's emergent power. Russia had not been an outsider in European diplomacy for the previous half-century.[49] Nonetheless, the events of the Great Northern War forced Europe's powers to adopt a more far-reaching perspective that marked Russian inclusion in a consolidating European diplomatic community.

49 Jan Hennings, *Russia and Courtly Europe: Ritual and the Culture of Diplomacy, 1648–1725* (Cambridge, 2016).

11

# The military in Safavid Iran, 1501–1736

RUDOLPH MATTHEE

## Introduction

For all of the obvious importance of warfare in the period when the Safavid dynasty (1501–1736) was in power in Iran, relatively little has been written on the topic of the military in this formative phase of Iranian history. Jamil Quzanlu, Vladimir Minorsky, and Laurence Lockhart were the first to address the organisation and development of the army in the Safavid period.[1] Their studies have since been supplemented by the work of Yahya Dhuka, Khanbaba Bayani, Masashi Haneda, Richard Tapper, Willem Floor, Walter Posch, Giorgio Rota, and the present author. Much work remains to be done, though, and what follows is therefore a preliminary survey.[2]

Warfare in pre-modern and early modern Iran remained confined to activity on land. Land-based in orientation and dependent on tribal military power, Iran's ruling dynasties were little familiar with the sea and made no effort to build a navy in either the Caspian Sea or the Persian Gulf until the eighteenth century. In this context it is useful to frame Safavid military history as essentially the story of the clash between the

1 Jamil Quzanlu, *Tarikh-i nizami-yi Iran*, 2 vols. (Tehran, 1936–7); Vladimir Minorsky, *Tadhkirat al-Muluk: A Manual of Safavid Administration* (Cambridge, 1943, repr. 1980), pp. 30–6; Laurence Lockhart, 'The Persian Army in the Safavid Period', *Der Islam* 34 (1959), 89–98.

2 Y. Dhuka, 'Tofang wa pishina-yi an dar Iran', *Barrasiha-yi Tarikhi* 2:3 (1967), 281–94; Y. Dhuka, *Artish-i shahanshahi-i Iran az Kurush ta Pahlavi* (Tehran, 1971); Khanbaba Bayani, *Tarikh-i nizami-yi Iran -Jangha-yi dawra-yi Safavi* (Tehran, 1974); Masashi Haneda, *Le Châh et les Qizilbâš. Le système militaire safavide* (Berlin, 1987); Rudi Matthee, 'Unwalled Cities and Restless Nomads: Gunpowder and Artillery in Safavid Iran', in Charles Melville (ed.), *Safavid Persia: The History and Politics of an Islamic Society* (London, 1996), pp. 389–416; Willem Floor, *Safavid Government Institutions* (Costa Mesa, 2001), pp. 124–280; Willem Floor, 'The Safavid Army. Continuity and Change', in Rudi Matthee (ed.), *The Safavid World* (London and New York, 2022), pp. 224–43; Walter Posch, *Osmanisch–safavidische Beziehungen 1545–1550. Der Fall Alḳâs Mîrzâ*, 2 vols. (Vienna, 2013); Giorgio Rota, 'Safavids against Turkmen in the Early Seventeenth Century: Warfare against the Nomads on the Caspian Steppe', *Iran: Journal of the British Institute of Persian Studies* 55 (2017), 35–43.

centralising ambitions of the ruling dynasty and the constraints of Iran's geography as well as the desire of a predominantly tribal periphery for autonomy. This periphery, it is important to note, constituted all of the territory outside of the direct orbit of the shah and his retinue, though – given the absence of a fixed capital – this was not geographically specific. (See Map 8.) Put differently, the entire military history of the eastern part of the Middle East after the Seljuq and Mongol conquests may be seen as a quest for ways to harmonise sedentary–nomadic relations.[3]

In more ways than one, the Safavids were an anomaly in the history of Iran. Whereas all other Iranian dynasties until the Qajars were of Turkic origin, their make-up reflecting their tribal origins, the originally Kurdish Safavids came to power with the assistance of tribal, (semi-)nomadic forces but were themselves non-tribal. Rather, they were the hereditary rulers of a Sufi order, whose initial religious persuasion is unclear, but who in due course gravitated towards (an ill-defined form of) Twelver Shiʿism. The Safavids, moreover, were not a conquering dynasty sweeping in from Central Asia, like the Seljuqs, the Mongols, and the Timurids, but rulers over an assemblage of several ethnically diverse tribes, Turkmen, and other remnants of Turkic peoples inhabiting the eastern Anatolian and Caucasian interstices of a succession of states which had emerged in the aftermath of the Mongol period. Like the military of their Mughal neighbours, the Safavid army long remained 'essentially, an aggregation of individual heroic warriors'.[4] They were overwhelmingly composed of ethnic 'Turks', as opposed to the 'Tajiks', ethnic Iranians, who mostly staffed the bureaucracy. This division, though real, was not fixed and is ill-defined in the sources, which do not spell out a 'functional specificity' for either group, though the best soldiers in the Safavid army were accordingly said to hail from (predominantly Turkic) Azerbaijan.[5]

3 Shahin Mustafayev, 'Between Nomadism and Centralization: The Ottoman Alternative in the History of the Aqqoyunlu State', in Jürgen Paul (ed.), *Nomad Aristocrats in a World of Empires* (Wiesbaden, 2013), p. 144.

4 Geoffrey Parker, *The Military Revolution: Military Innovation and the Rise of the West, 1500–1800* (Cambridge, 1988), p. 130.

5 M. B. Dickson, 'Shah Tahmasb and the Uzbeks (The Duel for Khurasan with ʿUbayd Khan: 930–946/1524–1540)', PhD diss. (Princeton University, 1958), p. 9. Fars was a great source of soldiers as well, at least in the early seventeenth century, when the troops led by Imam Quli Khan were said to be the best in the country. See Jean-Baptiste Tavernier, *Les Six Voyages de Jean Baptiste Tavernier . . . en Turquie, en Perse et aux Indes*, 2 vols. (Paris, 1676), vol. 1, p. 594.

MAP 8. The Safavid Empire

## Organisation and composition of the army and manner of fighting

Unable to count on their own tribal military force, the Safavids depended for military power on Turkoman troops called *Qizilbash*, 'Red Heads', so named after their red caps with twelve gores symbolising the twelve Imams of Shiʿism. The *Qizilbash* were recruited from the Turkoman tribes, frontier freebooters, a 'sort of residual category of a type of Turco-Mongolian frontier tribes that lived independently just beyond the control of state power'.[6] It was the *Qizilbash* who brought the Safavids to power and who, as cavalry troops, long remained the mainstay of their armed forces. They were recruited and fought according to tribe and clan, but frequent references to territorial origins suggest that the town or the region from which they hailed had some importance as well. Until the seventeenth century, almost all Safavid forces were recruited from the tribes, each of which was supposed to supply a certain number of troops. The tribal chief kept a record of the number to be mobilised from each tribe according to their size and means. These were to report for duty in the spring of each year, the beginning of the campaigning season.[7] Recruits had to bring their own weapons and equipment. The *Qizilbash* fought with 'bow and arrow, lance and buckler' and wore 'coats of mail and light breastplates with helmets of thin steel plate'.[8] (See Figures 17 and 18.) Great swordsmen and moving fast on horseback, they were redoubtable warriors, experts in a style of fighting marked by speed and manoeuvrability: the ambush at a mountain pass, the quick dash followed by retreat. The *Qizilbash* did not enjoy a salary from the shah; predatory looting was their reward, and additional costs were paid from the fiefs, *tuyul*, allocated to their leaders.[9]

The *Qizilbash* had a reputation for lacking any devotion to the Safavid clan, still less to an abstract concept of Iran. They were, however, known to be fiercely loyal to their spiritual leader, the shah, whom they revered as an incarnation of God, at least until his defeat by the Ottomans in 1514. Their blind obedience to him was shown in the

6 Thomas Barfield, 'Turk, Persian and Arab: Changing Relationships between Tribes and State in Iran and along the Frontier', in Nikki Keddie and Rudi Matthee (eds.), *Iran and the Surrounding World: Interactions in Culture and Cultural Politics* (Seattle, 2000), p. 70.

7 Eskandar Beg Monshi, *History of Shah ʿAbbas the Great*, trans. and ed. Roger Savory, 3 vols. (Boulder, 1978), vol. 2, p. 861.

8 Don Juan of Persia, *Don Juan of Persia: A Shi'ah Catholic 1560–1604* (London, 1923; repr. 2005), p. 51.

9 Jean Aubin, 'L'Avènement des Safavides reconsidéré', *Moyen Orient & Océan Indien* 5 (1988), 1–130, 39.

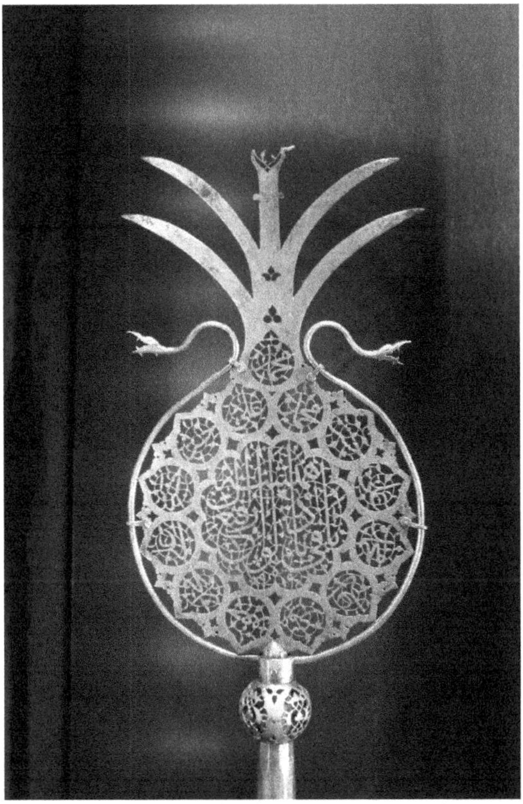

FIGURE 17. Army battle standard – Safavid era. TKS.1664, Topkapi Museum, Istanbul.

recklessness with which they threw themselves into battle without armour, ready to attain martyrdom.[10] Combined with a lack of discipline, this fanaticism cost them dearly in several decisive confrontations. Moreover, they suffered from the same drawbacks as previous tribal forces, which would lead later rulers to marginalise them. Their segmented tribal structures created perennial factionalism and seditiousness; they were difficult to mobilise and proved to be unreliable, either on the battlefield – where they often engaged in looting rather than fighting and were wont to flee – or as questionable allies. Nor could they

10 'The Travels of a Merchant in Persia', in Charles Grey (ed.), *A Narrative of Italian Travels in Persia in the Fifteenth and Sixteenth Century* (London, 1873), p. 206.

FIGURE 18. Helmet – Safavid era. TPM 1–348, Topkapi Museum, Istanbul.

be relied upon to stay the course of a campaign, since they could not leave their flocks and families for too long.[11]

Concerns about loyalty and discipline early on prompted the Safavids to search for alternatives to the unreliable *Qizilbash* forces. Initially, these were found in the so-called *qurchis* (literally, 'quiver bearers'): the shah's praetorian

11 For the Seljuq period, see David Durand-Guédy, 'Goodbye to the Türkmen? The Military Role of Nomads in Iran after the Saljuq Conquest', in Kurt Franz and Wolfgang Holzwarth (eds.), *Nomad Military Power in Iran and Adjacent Areas in the Islamic Period* (Wiesbaden, 2015), pp. 113–21.

guard, who from an early point in Safavid history formed the heart of the standing army and were always in the shah's immediate vicinity, or could be mobilised at short notice. Their numbers fluctuated, from 3,000 at the battle of Chaldiran (1514), dropping to 1,000 in the aftermath, and increasing to 5,000 in later times. They were paid and equipped by the shah. Originally recruited from the ranks of the *Qizilbash*, most *qurchis* were tribal, but, especially in the seventeenth century, one finds *qurchis* recruited from the ranks of the slave soldiers, *ghulams*, as well as the eunuchs. The head of the *qurchis*, the *qurchibashi*, would attain high rank in the same period.[12] Safavid efforts to create a loyal following are also reflected in their creation of 'artificial tribes', the best known of which are the *shahsivan*: the so-called 'lovers of the shah'.[13]

The most important manifestation of the Safavid search for military reliability, though, was a variant on the Middle Eastern tradition of slave soldiers, imported from abroad and moulded into a fighting force loyal solely to the ruler. In the Safavid state, this took the form of Georgian, Armenian, and Circassian war captives who were (nominally) converted to Islam before being introduced into the ranks of the army as *ghulaman-i khassa-yi sharifa* (slaves of the royal household). Beginning with the reign of Shah ʿAbbas I (r. 1587–1629), these troops came to form the mainstay of the Safavid army, though they never entirely supplanted the *Qizilbash*.

In addition to the *Qizilbash*, irregular forces, recruited from the local population, were frequently called upon to defend their own town or district. These were organised as city militias who were routinely equipped with firearms.[14]

The predominantly tribal and semi-nomadic character of the Safavid army had several consequences. One was that military expeditions, involving thousands of horses, were shaped by the need for grazing grounds, and preparations for combat were frequently interspersed with (and interrupted by) royal hunting and feasting. Campaigns involved the entire ruling elite and their retinue. The character of the Safavid army camp, *urdu*, reflects this. The

12 Jean-Louis Bacqué-Grammont, *Les Ottomans, les Safavides et leurs voisins. Contribution à l'histoire des relations internationales dans l'Orient islamique de 1514 à 1524* (Istanbul, 1987), p. 180; Qazi Ahmad b. al-Qomi, *Khulasat al-tavarikh*, 2 vols., 2nd ed. (Tehran, 2004), vol. 1, pp. 203–4; Mahmud b. Hidayat Allah Afushta'i Natanzi, *Naqavat al-athar fir zhikr al-akhyar*, 2nd ed. (Tehran, 1984), p. 185; Muhammad Yusuf Vala Isfahani, *Khuld-i barin* (Tehran, 1993), p. 768; Pietro Della Valle, *Viaggi di Pietro Della Valle*, 2 vols. (Brighton, 1843), vol. 1, p. 355; Haneda, *Le Châh et les Qizilbâš*, pp. 154–62, 191; Masashi Haneda, 'The Evolution of the Safavid Royal Guard', *Iranian Studies* 22:1 (1989), 57–86.

13 Richard Tapper, *Frontier Nomads of Iran: A Political and Social History of the Sahahsevan* (Cambridge, 1997), pp. 39–57, 72–91, 130–2, 139.

14 Posch, *Osmanisch–safavidische Beziehungen*, pp. 200, 223, 539, 733.

ambulant *urdu*, which effectively functioned as a capital during campaigns, was an elaborate affair. One observer claimed that at least 400,000 non-combatants – servants, foragers, water carriers, camel jockeys, and runners – were needed for the 50,000 troops available to the shah to function effectively, and since no one marched on foot, a great number of horses, camels, and luggage-carrying mules accompanied this vast host. The main commanders brought their harems with them as well.[15] The army camp, including the mobile *urdu-bazaar*, thus resembled a city; in effect, during campaigns it was the main city of the Safavid realm, according to Pietro Della Valle, who adds that the Iranians even brought portable bathtubs and kitchens on camelback on their campaigns.[16] Different observers give different perspectives on its layout and organisation. According to Della Valle, the Safavids had a rather disorderly way of setting up camp, which resembled more a sprawling bazaar than an army camp, making them vulnerable to night attacks.[17] Chardin, by contrast, presents an image of a camp based on a carefully chosen site and organised into a well-ordered ensemble of tents resembling a city divided into rectangular streets lined with spacious mobile mansions.[18]

Another consequence of the tribal nature of the Safavid army was that since its strength lay in guerilla warfare, marked by mobility and speed, it always sought to avoid open battles, which were consequently rare. The battle of Chaldiran of 1514 and the battle of Jam, fought against the Uzbeks in 1528, where the two parties confronted each other directly, are two prominent exceptions. Shah ʿAbbas I, for instance, habitually retreated before armies he knew were superior to his, and only fought four open battles during his forty-year reign.[19] Following a pattern that goes back at least to the Mongols, the preferred fighting method of the *Qizilbash* was to scout out the terrain for the position of the enemy and their strength, to shadow the opponent rather than confront him in open battle. The Iranians also made great efforts to prevent their enemies from living off the land, especially in the Ottoman borderlands. This often meant resettling the population, especially nomads, who were seen as unreliable. This might even take the form of evacuating and dismantling all

15 Francis Richard (ed.), *Raphaël du Mans missionnaire en Perse*, 2 vols. (Paris, 1995), vol. 2, pp. 118–20, 286–9.
16 Della Valle, *Viaggi*, vol. 1, p. 866.
17 Richard (ed.), *Raphaël du Mans*, vol. 2, pp. 118–20, 286–9.
18 Jean Chardin, *Voyages du chevalier Chardin en Perse, aux Indes*, ed. L. Langlès, 10 vols. and atlas (Paris 1810–11), vol. 5, pp. 483ff.
19 Giorgio Rota, 'The Mask of Shah ʿAbbas I', in Maria Szuppe, Anna Krasnowolska, and Claus V. Pedersen (eds.), *Medieval and Modern Iranian Studies: Proceedings of the 6th European Conference of Iranian Studies* (Paris, 2011), p. 172.

fortresses within reach of the enemy to prevent their capture.[20] No less destructive was a scorched-earth policy, depriving the enemy of food by destroying fields, bridges, and irrigation infrastructure, and poisoning wells, which was routinely deployed by the Safavids as a form of 'asymmetric' warfare against the numerically and logistically superior Ottomans.[21]

Such mobile warfare was hard to reconcile with the logistics and equipment needed for sieges, at which the Iranian army consequently never excelled – even though they are said to have had expert miners and sappers.[22] Cities anyhow were as often lost due to betrayal as through sieges, as exemplified by the Ottoman capture of Yerevan in 1635 and the loss of Qandahar to the Mughals in 1638.[23] But warfare was not necessarily a matter of conquering cities. Nomadic war-making was often more about finding suitable grazing grounds and opportunities for plunder than the long-term occupation of urban centres. Few cities were walled, and those that were, such as Tabriz, Yerevan, and Qandahar, were mostly located on the fringes, at the points where the vast empire was 'defended'.[24] Only the inner fortress, *qalʿa*, the origins of which go back to Seljuq times, might be fortified and served as a refuge for the elite in times of attack, leaving the urban population largely at the mercy of the enemy. In conditions where the boundaries between a city and its verdant surroundings were vague, fortified cities were not important as urban centres to be defended.

The relative absence of fortified cities also meant that the trajectory of siege warfare in Europe, whereby ever more formidable strongholds led to increasingly elaborate sieges resulting in still stronger fortifications, was not repeated in Iran. Conversely, just as it made sense for the Poles to strengthen their cavalry against the Cossack 'horsemen of the south', so it made sense for

20 [H. Chick ] (ed.), *A Chronicle of the Carmelites and the Papal Mission to Persia*, 2 vols. (London, 1939; repr. 2012), vol. 1, p. 53; Posch, *Osmanisch–safavidische Beziehungen*, p. 459.

21 Hans Dernschwam, *Dernschwam's Tagebuch einer Reise nach Konstantinopel und Kleinasien (1553/1555)*, ed. Franz Babinger (Munich and Leipzig, 1923; new ed. Berlin, 1986), p. 211; Richard, *Raphaël du Mans*, vol. 2, pp. 118–20, 289; Chardin, *Voyages*, vol. 5, p. 319.

22 Ruy Freyre de Andrada, *Commentaries of Ruy Freyre de Andrada*, ed. C. R. Boxer (London, 1930; repr. Abingdon, 2005), p. 46. For the Safavid lack of knowledge about siege warfare during their siege of Tabriz in 1585–6, see S. Sarıçiçek, 'Askerî ve Siyasi Yönden 1585–1586 Tebriz Kuşatması', *Savunma ve Savaş Araştırmaları Dergisi* 34:2 (2024), 91–109.

23 H. Dunlop (ed.), *Bronnen tot de geschiedenis der Oostindische Compagnie in Perzië* (The Hague, 1930), p. 616.

24 Rudi Matthee, 'Unwalled Cities and Restless Nomads: Gunpowder and Artillery in Safavid Iran', in Charles Melville (ed.), *Safavid Persia: The History and Politics of an Islamic Society* (London, 1996), pp. 389–416. Even Ardabil, sacred for the Safavids for being their ancestral town, was without walls, prompting Shah ʿAbbas I to contemplate removing the bones of his ancestors before the arrival of the Turks. See Della Valle, *Viaggi*, vol. 1, p. 779.

the Safavids to rely on the mobility of their cavalry-based armies as their principal means of defence.[25]

The natural environment set limits to both the scope of expansion and the extent of enemy penetration into the Iranian interior. Campaigns followed the rhythm of the seasons. Long-distance Safavid expeditions never started until well into the spring and generally fizzled out in the autumn, insufficient time to penetrate much further into Ottoman territory than Iraq or eastern Anatolia, or to venture much beyond Qandahar in the east. To the east and north-east of the Iranian heartland stretched vast barren, largely waterless deserts, separating Iran from the Mughal Empire and the great oases of Central Asia, and making conquest and control impossible, not to mention pointless. Descending from the Iranian plateau into the Mesopotamian lowlands was an option, and even a religiously sanctioned objective given the location of Shiʿism's holy shrines in Iraq, but defending the plains located beyond the Zagros Mountains was a difficult, if not impossible, task in light of the climate, the heat and the humidity, which was as alien to the *Qizilbash*, used to the bracing climate of the plateau, as was the local, Arab culture.[26] Moving due west also posed problems, for the first region to traverse on the way to the Ottoman heartland was mountainous eastern Anatolia with its bitterly cold winters. The Safavids were not in a position to ship ordnance via the Black Sea, as the Ottomans sometimes did. Conversely, Iran's heartland was shielded from direct Ottoman attack by distance and physical conditions. The Ottoman army generally had to winter in Aleppo or Mesopotamia before reaching western Iran, and although they managed to attack, seize, and (temporarily) hold on to Tabriz several times, they never penetrated beyond Zanjan in the north and Hamadan in the west.

The only external region that was easily accessible, as well as of sufficient economic interest to the Iranians, was the southern Caucasus: the fertile lowlands of Armenia and Georgia. The Safavids accordingly directed their armed forces towards this region early on and managed to subjugate and maintain part of it as a client state until the end of the dynasty. As had been the case with Turkic tribal bands in the Seljuq period, the raids conducted by the *Qizilbash* into the predominantly Christian territories of Armenia and Georgia in the sixteenth century counted as *ghazi* warfare: a form of jihad against non-Muslims.

The Safavid army followed the Mongol decimal system in its subdivisions. It comprised so-called *min-bashis*, commanding 1,000 troops, *yuz-bashis*

25 Parker, *The Military Revolution*, pp. 37–8. 26 Della Valle, *Viaggi*, vol. 2, pp. 195–6.

(centurions) in charge of 100 men, and *un-bashis*, who led units of ten.[27] Inheriting another Turco-Mongol tradition, the Safavid army from the outset consisted of two wings enclosing a centre: the right wing, *maymana*, and the left wing, *maysara*, with specific tribes occupying either, whilst the shah was usually to be found in the centre. The Shamlu tribe were always in the front rank of the right wing, whereas the Ustajlu invariably commanded the left wing. Other major *Qizilbash* tribes generally occupying the right wing were the Turkoman, the Rumlu, the Takallu, and the Kurds. The left wing, by contrast, was held by the Dhu'l-Qadr, the Afshar, and the Qajar. This division of labour was never institutionalised, yet the custom persisted until the end of the dynasty, even though, following the reforms of Shah ʿAbbas I at the turn of the seventeenth century, non-tribal forces – *ghulams* and *tufangchis* (musketeers) – inserted themselves into this arrangement as well.[28]

The number given for the Safavid army varies widely and fluctuated over time. According to a late sixteenth-century European estimate, the number of soldiers that the tribal leaders could provide was capped at about 50,000 to 60,000.[29] A clearly exaggerated 250,000 is offered as the highest number. One observer claimed that the Safavid army never had more than 30,000 men in the field.[30] That may be a low estimate, but the regime never seems to have been able to field more than 100,000 troops.

The limited number of Safavid soldiers naturally stemmed from Iran's small population of perhaps 8 million which, coupled with scarce natural resources, made for an equally small resource base. As a result, the Safavid army was perennially underfunded. This was manifest from the very beginning of the dynasty and even the reforms of Shah ʿAbbas I, which were partly designed to raise new crown revenue for military reforms, were unable to fundamentally alter this condition.

Both numerically and regarding equipment, the Safavids were inferior to their arch-enemies, the Ottomans, a problem that persisted over time. In 1618, for instance, Shah ʿAbbas I reportedly sought and failed to get hold of

27 Adam Olearius, *Vermehrte newe Beschreibung der muscowitischen und persischen Reyse* (Schleswig, 1656; facs. repr. Tübingen, 1971), p. 667; Chardin, *Voyages*, vol. 5, p. 314; Muhammad Husayn Mustawfi, 'Amar-i mali va nizami-yi Iran dar 1128', *Farhang-i Iran-zamin* 20 (1975), 399.

28 Monshi, *History of Shah ʿAbbas*, vol. 2, pp. 755–7, 885; Vala Isfahani, *Khuld-i barin*, p. 255; Hasan Bayk Rumlu, *Ahsan al-tavarikh* (Tehran, 1978), p. 281; Haneda, *Le Châh et les Qizilbâš*, pp. 48–61.

29 Ajuda Library, Lisbon, 'Della guerra tra Persiani e Turchi', 46-X-10, n.d., f. 374r–v.

30 Ugo Tucci, 'Una relazione di Giovan Battista Vechietti sulla Persia e sul regno di Hormuz', *Oriente Moderno* 35:4 (1955), 149–60, 156.

a type of long-range musket that the Ottomans had access to.[31] The Iranians consequently feared the military might of their main adversaries, whose skills were constantly honed on the battlefields of South-Eastern Europe. This sense of inferiority is expressed as grudging admiration of the Ottoman military wherewithal in the Persian-language sources.[32] Its most powerful manifestation was an instinctive Safavid fear of becoming entangled in a two-front war.[33] As a consequence, almost all the armed conflicts fought between the two parties were launched by the Ottomans – usually at a time when they had their hands free in the Balkans, or saw the Safavids as weak, divided, or preoccupied with their eastern foes. The best examples of this are the so-called War of the Two Iraqs of 1534–5, when the Ottomans launched an offensive after making peace with the Habsburgs, and the Ottoman expedition against Georgia and Shirvan in 1578, following another truce on the European front and coinciding with weak rule and bitter divisions among the Safavids.[34]

Vis-à-vis the Mughals, on the other hand, the Safavids felt, and seem to have been, superior. In keeping with what was perhaps a general prejudice towards Indians as unwarlike, Shah ʿAbbas is said always to have mobilised fewer troops against the Mughals than they sent to fight the Iranians.[35] Still, Qandahar, the stronghold at the intersection of the two realms, went back and forth between them several times, until it definitively fell to the Safavids in the war of 1645–8. This was also the period when the Iranians, finally free from their fear of encirclement by their Sunni enemies and of facing armed conflict on multiple fronts, gave up on war, refusing to accede to multiple requests by Europe's Christian powers to join them in an anti-Ottoman alliance.

The Ottomans may have outdone the Safavids militarily, but they much preferred to fight the Europeans rather than campaigning in the east, primarily for logistical reasons.[36] In the Balkans they would find abundant food and

31 Della Valle, *Viaggi*, vol. 1, p. 666.

32 Tilmann Trausch, *Anpassung und Abbildung. Das Türkenbild in safawidischen Chroniken des 16. Jahrhunderts* (Berlin, 2011), pp. 71ff., 90ff.

33 See Rudi Matthee, 'Safavid Iran and the "Turkish Question" or How to Avoid a War on Multiple Fronts', *Iranian Studies* 52:3–4 (2019), 513–42.

34 For the War of the Two Iraqs, see Walter Posch, 'Der Osmanische Feldzug in die beiden Iraq (940–942/1534–1535)', Master's diss. (University of Vienna, 1991). For the war of 1578, see Bekir Kütükoğlu, *Osmanlı–İran siyâsi münasebetleri (1578–1612)* (Istanbul, 1993); and Rudi Matthee, 'The Ottoman–Safavid War of 986–998/1578–90: Motives and Causes', *International Journal of Turkish Studies* 20:1–2 (2014), 1–20.

35 Rudi Matthee, *Persia in Crisis: The Decline of the Safavids and the Fall of Isfahan* (London, 2012), p. 125.

36 Dernschwam, *Tagebuch*, p. 31.

water, whereas during campaigns against the Safavids they might not be able to obtain water for several days.[37] The Ottoman specialty in siege warfare was of little use in territory with few fortresses and where traditional steppe tactics, emphasising lightning cavalry attacks and feint retreats, prevailed.[38] Furthermore, the Safavids repeatedly ravaged the area between Tabriz and Van several times under Shah Isma'il I (r. 1501–24) and again during the wars fought during the reigns of Sultan Süleyman (r. 1520–66) and Shah Tahmasb (r. 1524–76) with scorched-earth tactics, employed in part to make up for their inferiority in firepower. Ultimately, the perception in Istanbul was that 'one square foot of land in Hungary would be of more worth to the Sublime Porte than all of the vast, empty deserts of Persia'.[39]

## Sixteenth-century developments

The troops that initially accompanied Shah Isma'il in the early years of the sixteenth century numbered no more than 7,000.[40] In the course of his campaigns, this number grew, perhaps to as many as 20,000.[41] With these, he managed to conquer large swaths of western Iran between 1501 and 1514, including Azerbaijan, western Iran (Iraq-i 'ajam), Fars, Diarbakr, central Iraq (1508), Shirvan, and Khurasan (1513). An eastern campaign resulted in the capture of Herat from the Uzbeks in 1510.

Shah Isma'il's triumphs came to an abrupt halt at the battle of Chaldiran on 23 August 1514, which followed a series of Safavid incursions into Ottoman territory and the humiliating treatment of Ottoman ambassadors at the Safavid court. The result was a crushing defeat for the Safavids, who were faced with an Ottoman force equipped with the kind of rifles and gun carriages that they were able to effectively deploy. Following their victory at Chaldiran, the Ottomans momentarily halted the advance of the Safavids into Anatolia. Yet it was only a partial triumph, for, faced with the same environmental and seasonal limitations as the Iranians, they quickly gave up Tabriz and retreated westward following their victory. For the Safavids,

37 Stephan Gerlach, *Stephan Gerlachs deß Aeltern Tage-Buch. Der von zween glorwürdigsten römischen Kaysern, Maximiliano und Rudolpho, ...* (Frankfurt am Main, 1674), p. 219.

38 Virginia Aksan, 'War and Peace', in Suraiya N. Faroqhi (ed.), *The Cambridge History of Turkey*, vol. 3: *The Later Ottoman Empire, 1603–1839* (Cambridge, 2006), p. 100.

39 Phillip Williams, 'The Lords of All Muslims? The *Avisos de Constantinopla* and Christian Perspectives on the Ottoman–Safavid Rivalry', in Enrique García Hernán, José Cutillas Ferrer, and Rudi Matthee (eds.), *The Spanish Monarchy and Safavid Persia in the Early Modern Period: Politics, War and Religion* (Madrid, 2016), p. 160.

40 Haneda, *Le Châh et les Qizilbâš*, p. 63.

41 Aubin, 'L'Avènement des Safavides', pp. 32–3.

though, the defeat at Chaldiran was decisive: Isma'il went into mourning and subsequently withdrew from state affairs. He did procure new weapons in the aftermath of his defeat, but he never waged war again. Instead, he descended into a welter of drinking that would end his life ten years later.

Shah Tahmasb, Shah Isma'il's eldest son and successor, is known as a war-wary, rather reclusive ruler. But, faced with a massive rebellion in the ranks of the *Qizilbash*, he could ill afford to stay aloof from armed conflict. He thus spent the first ten years of his reign subduing the fractious *Qizilbash*. Before the end of this operation, Iran was targeted by near-simultaneous invasions by the Uzbeks in the east and the Ottomans in the west, necessitating a two-front war. The former in 1528 took Mashhad and Astarabad in the north-east. The Safavid–Uzbek confrontation that followed later that year, culminating in the battle of Jam, demonstrated that the Iranians had learned from their past mistakes. Just as the Ottomans had prevailed over the Iranians in 1514 by using the *Wagenburg* tactic of chaining baggage carts together to form a defensive wall, so the Safavids now employed that technique to best the Uzbeks.[42]

In 1534, having made peace in Europe, the Ottomans moved east and took Baghdad and Mesopotamia from the Safavids, briefly occupied Tabriz, and advanced as far as Zanjan. This conflict, known as the War of the Two Iraqs, was followed by two more Ottoman invasions, the first in 1547, when they sought (and failed) to turn Iran into a vassal state by supporting Tahmasb's renegade half-brother, Alqas Mirza. This time, too, the Ottomans (briefly) took Tabriz, prompting Shah Tahmasb to move his capital to Qazvin, located 450 kilometres inland and thus far less exposed to external attack. Having learned from the experience of his father's crushing defeat at Chaldiran, Shah Tahmasb chose to retreat from Tabriz, sacrificing his capital to bottle up his enemy within its confines. He also resorted to the scorched-earth tactic that the Safavids were known for, burning crops and poisoning wells between the Ottoman border and Tabriz, thus compounding the problems facing the Ottoman invaders. When the Turks entered Tabriz in the summer of 1548, they were exhausted and quickly succumbed to hunger and thirst. The frustrated Ottomans, worn out by the Safavids' tactics and faced with *Qizilbash* warriors on either side of the border, soon had to give up on Tabriz, and retreated in the direction of Van, which they managed to take from the Safavids. The Iranian troops responded by penetrating deep into Ottoman territory, marauding, killing, and gathering booty as far as Erzincan.[43]

42 Dickson, 'Shah Tahmasb and the Uzbeks', pp. 129, 133, 136.
43 Posch, *Osmanisch–safavidische Beziehungen*, pp. 481–95.

The campaign did not result in significant territorial loss or gain, and in 1555 the two parties, exhausted and demoralised, signed the Peace of Amasya, dividing Armenia and Georgia evenly between the two parties, making the Iranians cede control over Azerbaijan to the Ottomans, and allowing the latter to turn their attention to the European front.[44] This accord would hold until 1578. Shah Tahmasb, a recluse in his later years, would wage no more external wars after 1555. Intent on sidelining the *Qizilbash*, he did stage various incursions into the Caucasus, though, which became a reservoir of war captives.

Then, as later, Iran's perennial penury affected its military capacity. Shah Tahmasb had remunerated his military in cloth, and, upon the accession of his successor, Isma'il II (r. 1576–7), Iran's soldiers had not been paid for fourteen years.[45] Isma'il's short-lived and bloody reign, followed by that of the purblind and weak Shah Muhammad Khudabanda (r. 1578–87), only aggravated this turmoil. The Ottomans took advantage of Iran's manifest weakness in this period by launching a new war against their eastern neighbour, capturing Shirvan and Georgia. The hostilities lasted until 1590 and caused a great deal of damage to the lands straddling the two empires. The Mughals seized Qandahar in 1590, which had been in Safavid hands since 1537.

## The question of firearms

Contrary to a hoary myth that was laid to rest long ago by Roger Savory, firearms were not introduced to Iran under Shah 'Abbas I at the behest of the envoy-adventurer Anthony Sherley, but as early as the fifteenth century, in other words before the rise of the Safavids as a state and shortly after the first recorded use of gunpowder weapons in India and the Ottoman Empire.[46] The first incontrovertible use of firearms and gun-casting in Iran dates from the reign of the Aqqoyunlu ruler Uzun Hasan (r. 1453–78). The *Qizilbash* used cannon from the early days of their formation to besiege eastern Anatolian cities, and Shah Isma'il introduced a (small) regiment of musketeers directly after his disastrous defeat at of Chaldiran.[47] Terms such as *khumpara* and *mushak*, as well as *qal'a-kub*

44 Zahit Atçıl, 'Warfare as a Tool of Diplomacy: Background of the First Ottoman–Safavid Treaty in 1555', *Turkish Historical Review* 10:1 (2019), 3–24.

45 [Chick], *Chronicle of the Carmelites*, pp. 47, 53; Vincentio d'Alessandri, 'Narrative of the Most Noble Vincentio d'Alessandri', in Grey (ed.), *Narrative of Italian Travels*, p. 218; Rumlu, *Ahsan al-tavarikh*, p. 623.

46 Roger Savory, 'The Sherley Myth', *Iran: Journal of the British Institute of Persian Studies* 5 (1967), 73–81.

47 Bacqué-Grammont, *Les Ottomans, les Safavides*, pp. 146, 157, 165.

(fortress-smasher) and *badlij* or *balmiz* (siege-cannons) occur with great frequency in sixteenth-century Safavid texts.[48] The use of handguns (*tufang*) is also mentioned repeatedly in early Safavid sources.[49]

From their early days, though, the Safavids often sought to acquire guns and gunners from abroad, suggesting the difficulty they had keeping up with technological advances and the expertise needed for handling armaments. Shah Isma'il thus sought to acquire gunners and cannon founders from Venice as early as 1502 and again in 1509, as well as from the Portuguese in 1515.[50] Shah Tahmasb received a large quantity of arquebuses from Emperor Charles V, and Shah 'Abbas I in 1595 purchased a new type of 'quick-firing' arquebus in Russia and also received weaponry during subsequent missions.[51] Safavid shahs continued to request cannon and expertise until the dynasty's end. Finally, as is true of the Ottomans and the Mughals in this period, the Safavids employed foreign, mostly Venetian and Portuguese, mercenaries, gunners, and cannon-masters.[52]

Yet firearms spread far and wide during the sixteenth and seventeenth centuries. The people of Kirman are said to have used them against marauding Uzbeks in 1509–10.[53] In the 1660s the inhabitants of the Iraqi borderlands would not sell anything but for firearms.[54] One observer in 1694 called the Baluchis, not previously known for their familiarity with firearms, 'good shots'.[55] Another, reporting from Shamakhi in north-western Shirvan in 1711, claimed that, whereas formerly the locals had only had arrows and lances, they were now armed with pistols and sabres.[56]

48 Examples in Rudi Matthee, 'Firearms, History', *Encyclopaedia Iranica* 9 (1999), 620–1. For casting, see Parviz Mohebbi, 'Firearms, Production of Cannon and Muskets', *Encyclopaedia Iranica* 9 (1999), 628–31.

49 For examples, see Rumlu, *Ahsan al-tavarikh*, pp. 205, 207, 208.

50 Bianca-Maria Scarcia-Amoretti (ed.), *Šāh Ismā'īl nei 'diarii' di Marin Sanudo* (Rome, 1979), pp. 9–10, 162–3; and Gugliemo Berchet (ed.), *La repubblica di Venezia et la Persia* (Turin, 1865), p. 25.

51 Srećko M. Džaja and Günter Weiss (eds.), *Austro-Turcica 1541–1552. Diplomatische Akten des Habsburgischen Gesandtschaftsverkehrs mit der Hohen Pforte im Zeitalter Süleymans des Prächtigen* (Berlin, 1995), p. 224; P. P. Bushev, *Istorii posol'stv i diplomaticheskikh otnoshenii russkogo i iranskogo gosudarstv v 1586–1612* (Moscow, 1976), pp. 226–7, 347, 360.

52 Jürgen Andersen and Volquard Iversen, *Orientalische Reise-Beschreibungen* (Schleswig, 1669; repr. Tübingen 1980), pp. 149–50; Lajos Fekete, *Einführung in die persische Paläographie. 101 persische Dokumente* (Budapest, 1977), pp. 529–33; Nationaal Archief (Dutch National Archives, The Hague), VOC 1501, Diary Van Leenen, f. 706.

53 Ahmad Vaziri Kirmani, *Tarikh-i Kirman*, ed. M. I. Bastani-Parizi, 2 vols., 3rd ed. (Tehran, 1985), vol. 2, p. 598.

54 Tavernier, *Les Six Voyages*, vol. 1, p. 180.

55 Anne Kroell (ed.), *Nouvelles d'Ispahan 1665–1695* (Paris, 1979), p. 65.

56 S. J. Monier, 'Mémoire de la province de Sirwan', in [Père Fleuriau (ed.)], *Lettres édifiantes et curieuses* (Toulouse, 1810), vol. 4, p. 35.

Yet, following Geoffrey Parker's dictum that the 'war machine had to adapt to it as well as to adapt it to local circumstances', the integration of methods and weaponry into existing structures was no easy process.[57] It proved difficult to 'merge the mobility of the horseman with the firepower violence of the arquebusier'.[58] In some cases such fusion was seamless. The *zamburak*, a swivel gun mounted on the back of a camel, is an example of imported technology easily adapted to local circumstances of mobile warfare. Probably of eastern origin, the *zamburak* is first recorded at the turn of the seventeenth century and remained popular, in Iran as well as Afghanistan, until the nineteenth century.[59]

Yet overall, integration proved difficult. Several factors impeded a quick and comprehensive assimilation of firearms into the Safavid army. This was in part a matter of logistical problems. Iran was much further removed from the primary source of technological innovation at the time, Europe, than the Ottoman Empire, or even Mughal India with its comparatively easy oceanic access. The manufacture and use of cannon also posed problems. Unlike the Ottoman Empire with its Mediterranean ports, or Russia with its many navigable rivers, Iran was effectively landlocked, making the transportation of heavy ordnance difficult. Transport also hampered production, as did the lack of available materials. Iran had few areas where necessary ingredients such as saltpetre and sulphur were available; though the availability of the latter facilitated a firearms industry in Azerbaijan and Shirvan.[60] Nor did the Safavids lack cannon foundries.[61] The best cannon was reportedly manufactured to the south in Lar, where sulphur and saltpetre were found.[62] Cannon were also cast in Mashhad.[63] Safavid artillery is said to have been better than that of the Mughals.[64]

The difficulty of integrating firearms was also a matter of culture. Although firearms became widespread, their introduction into the Safavid army did not follow a smooth path. The 'anonymous death' of firearms

57 Parker, *The Military Revolution*, pp. 142–3.

58 Weston F. Cook, *The Hundred Years War for Morocco: Gunpowder and the Military Revolution in the Early Modern Muslim World* (Boulder, 1994), p. 71.

59 Mulla Jalal al-Din Munajjim, *Tarikh-i ʿAbbasi (ruznama-yi Mulla Jalal)* (Tehran, 1987), p. 407; Audrey Burton, *The Bukharans: A Dynastic, Diplomatic and Commercial History 1550–1702* (New York, 1997), p. 297; Matthee, 'Firearms', pp. 620–1, with image.

60 M. Kh. Geidarov, *Goroda i gorodskoe remeslo Azerbaidzhana v XIII–XVII vekov* (Baku, 1967), pp. 80–2.

61 Chardin, *Voyages*, vol. 4, p. 91.

62 Matthee, 'Unwalled Cities', p. 396.

63 Andersen and Iversen, *Orientalische Reise-Beschreibungen*, pp. 153, 155.

64 Iqtidar Alam Khan, 'Indian Response to Firearms [1300–1750]', *Proceedings Indian History Congress* 58 (1998), fn. 98.

radically altered the individual and 'heroic' form of traditional combat.[65] Mounted soldiers thus deemed the use of firearms, which necessitated dismounting, beneath their dignity.[66] The musketeers, the *tufangchis*, were inferior to the *Qizilbash* mostly because they fought on foot and were not tied to the tribal aristocracy.[67] Sixteenth-century Persian chronicles mention *tufangchis* with some frequency as of *c.* 1530 but do not accord them any special place.[68] By all accounts, their status was low. Called 'men taken from the plough', their reputation was that of cannon fodder, *'alaf-i shamshir* (literally, 'sword fodder').[69] Recruitment also faced obstacles, for people were loath to see their sons disappear into a force of such low standing, and whoever could offered money to be exempted from the obligation.[70]

Artillery exemplifies the dictum that the adoption of new technology is of limited value unless it is accompanied by cultural changes.[71] Cannon, hard to cast and hard to move over land, remained even more marginal than handguns to an army that remained cavalry-centred and was less intent on conquering land and urban centres than on subduing enemies for tribute. Two years after Chaldiran in 1516, Shah Isma'il I reportedly had 50 carriage-drawn cannons, a number that had increased to 100 a year later.[72] But artillery never played a central role in war. The *tupchibashi* (cannon master), who also served as the admiral of the (non-existent) navy, thus had a light workload since the Iranians had only a few substandard pieces in the borderlands, and the guns in Isfahan were covered with dirt.[73] The French Capuchin Father Raphaël du Mans, a long-time resident of Iran, insisted that the Iranians were expert at tunnelling under walls, not to place mines but to weaken the foundations, hastening their collapse.[74] In 1684 the German Engelbert

65 Sanjay Subrahmanyam, 'Warfare and State Finance in Wodeyar Mysore, 1724–25: A Missionary Perspective', *The Indian Economic and Social History Review* 26:2 (1989), 203–33, 228.

66 John Cartwright, 'Observations of Master John Cartwright', in S. Purchas (ed.), *Hakluytus Posthumnus or Purchas His Pilgrimes*, 20 vols. (Glasgow, 1905), vol. 8, p. 503.

67 Jean de Thevenot, *Voyages de M. de Thevenot au Levant*, 3 vols., 3rd ed. (Amsterdam, 1727), vol. 3, p. 343.

68 Posch, *Osmanisch–safavidische Beziehungen*, pp. 222–3.

69 Della Valle, *Viaggi*, vol. 1, p. 759; Tavernier, *Les Six Voyages*, vol. 1, p. 593. The translation is from the 1678 English translation of Tavernier's work, p. 224; Thevenot, *Voyages de M. de Thevenot*, vol. 3, p. 343; Engelbert Kaempfer, *Exotic Attractions in Persia, 1684–1688*, trans. and ed. Willem Floor and Colettte Ouahes (Washington, DC, 2018), p. 63; Cornelis de Bruyn, *Reizen over Moskovie en Perzie* (Amsterdam, 1714), p. 157.

70 Tavernier, *Les Six Voyages*, vol. 1, p. 565.

71 Jeremy Black, *War in the World: A Comparative History, 1450–1600* (New York, 2011), p. 674.

72 Scarcia-Amoretti, *Šāh Ismāʿīl*, p. 398; Bacqué-Grammont, *Les Ottomans, les Safavides*, p. 168.

73 Tavernier, *Les Six Voyages*, vol. 1, p. 594.

74 Richard, *Raphaël du Mans*, vol. 2, p. 119.

Kaempfer claimed that movable artillery was hardly available in the Safavid army and that no artillery and grenades were used in official battles. Kaempfer also observed that the cannon left by the Portuguese upon abandoning Hormuz sixty years later now served as decoration for the front hall of the royal palace in Isfahan, and that the ordnance sent by Duke Frederick III of Holstein-Gottorp in 1637 now decorated the entrance to the royal garden in Qazvin.[75]

## Shah ʿAbbas I's reforms

As in so many other ways, the reign of Shah ʿAbbas I represented a military high point for the Safavids. Shah ʿAbbas was not just a good strategist and an excellent tactician but was also distinguished by his concern to minimise the harm done by his troops to the civilian population. Unlike most of his contemporaries, he did not billet troops in people's houses, and he made his soldiers pay for goods furnished and services rendered by the populace.[76]

Shah ʿAbbas's strategic acumen first manifested itself in 1590. In that year, he concluded a peace with the Ottomans, leaving them in control of Azerbaijan, Georgia, Shirvan, and parts of Luristan. The concession was humiliating, but it was also deliberate, for ʿAbbas needed the lull in warfare against his most formidable foes so that he could undertake a series of internal reforms. A few years earlier, the shah had come to power amidst a welter of *Qizilbash* seditiousness; and the war of 1578–90 had highlighted their shortcomings. Shah ʿAbbas's reforms aimed to strengthen royal power by way of redirecting loyalty away from the tribe to the person of the ruler. His policy of marginalising the *Qizilbash*, whilst maintaining the army's strength, has been described as a transformation from a 'companionship-in-arms' into a 'hierarchically structured' force.[77]

The most important component of this transformation was the intensification of a policy dating back to Shah Tahmasb, that of elevating the *ghulams* to high administrative and military rank by appointing them as provincial governors overseeing newly converted crown (*khassa)* lands and as military commanders loyal to the shah. The conversion of land to crown domain was

75 Kaempfer, *Exotic Attractions*, pp. 63–4.

76 Pietro Della Valle, *Delle conditioni di Abbàs Rè di Persia* (Venice, 1628; repr. Tehran, 1976), pp. 35–6; P. Pacifique de Provins, *Le Voyage de Perse* (Assisi, 1939), p. 258.

77 The terminology here is borrowed from Isenbike Togan, 'Ottoman History by Inner Asian Norms', in Halil Berktay and Suraiya Faroqhi (eds.), *New Approaches to State and Peasant in Ottoman History* (London, 1992), p. 197.

precisely meant to raise the revenue needed to pay the new soldiers – whose wages were met from the royal coffers. The principal military recruits were Georgians, who were seen as faithful and trustworthy and instinctively anti-Turkish, who thus served as a counterweight to the *Qizilbash*.[78] By the late seventeenth century, Georgia is said to have been obliged to furnish 40,000 soldiers to the shah in times of war.[79]

More specifically, the shah created a *qullar* corps composed of *ghulams*, elevating their leader, the *qullar-aqasi*, to high rank.[80] The Safavid *qullar* corps had many of the features of the Ottoman Janissaries, although its members seem to have lacked the same organised training and the attendant *esprit de corps*. This also prevented the Safavid *ghulams* from becoming a praetorian guard, capable of overwhelming the state, as slave soldiers would do in Ottoman Iraq and Egypt.[81] About half of the total number of *ghulams*, anywhere between 8,000 and 18,000 men, served in a military capacity.[82]

Shah ʿAbbas also created or, rather reorganised, a corps of musketeers, *tufangchis* (mounted infantry) making this corps a part of his reconstituted military apparatus, one in which joining the army was no longer a privilege, but an obligation. Even though the *tufangchis* were entered in the ranks of the *ghulams*, their creation may have been intended to provide a counterweight to the *qurchis* and the *ghulams*.[83] The size of the musketeer corps was variously estimated at between 8,000 and 20,000 men.[84] It seems that the corps was similar to the *sekban* in the neighbouring Ottoman Empire, who may have served as a model for Shah ʿAbbas I.[85]

How much power the *Qizilbash* had lost is visible by the fact that, by the end of ʿAbbas's reign, they had ceased to be a separate force, serving with all

78 Sir Thomas Herbert, *Travels in Africa, Persia, and Asia the Great*, ed. John Anthony Butler (Tempe, 2012), pp. 358–9.

79 G. A. Ezov (ed.), *Snosheniia Petra Velikogo s armianskim narodom* (St Petersburg, 1898), letter Israel d'Ori, p. 39.

80 Vala Isfahani, *Khuld-i barin*, p. 768; Chardin, *Voyages*, vol. 5, pp. 304–7; Floor, *Safavid Government Institutions*, pp. 172–4.

81 G. Rota, 'Fighting with the Qizilbash: Preliminary Remarks on Safavid Warfare', in K. Franz and W. Holzwarth (eds.), *Nomad Military Power in Iran and Adjacent Areas in the Islamic Period* (Reichert, 2015), pp. 238–9.

82 Floor, *Safavid Government Institutions*, p. 170.

83 Iskandar Big Turkaman, *Tarikh-i ʿAlam-ara-yi ʿAbbasi*, ed. Iraj Afshar, 2 vols. (Tehran, 1971), vol. 2, p. 1106; Rota, 'Fighting with the Qizilbash', p. 241.

84 Della Valle, *Viaggi*, vol. 1, pp. 354–5; Chardin, *Voyages*, vol. 5, pp. 297, 304–6; Tavernier, *Les Six Voyages*, vol. 1, p. 593; Thevenot, *Voyages de M. de Thevenot*, vol. 3, p. 343; Mirza Naqi Nasiri, *Titles and Emoluments in Safavid Iran: A Third Manual of Safavid Administration*, trans. Willem Floor (Washington, DC, 2008), p. 34.

85 Matthee, 'Unwalled Cities', pp. 391–3; Rota, 'Fighting with the Qizilbash', pp. 239–41.

the other army forces under the *qurchibashi*.[86] Neither the *ghulam* corps nor the *tufangchi* regiment fundamentally altered the structure of the army, though, and the Safavids held on to their traditional *Qizilbash*-like fighting tactics, even if they adopted firearms and modified the structure of their army. ʿAbbas's reforms buttressed his 'absolutism' yet failed to effect a truly profound social and political transformation, such as a greater degree of state autonomy.

Once he had laid the basis for his reforms, Shah ʿAbbas first turned his attention eastward, to Khurasan, where he retook Herat in 1598. In anticipation of the resumption of hostilities with the Ottomans, he next transferred his capital from Qazvin to Isfahan. In 1603, he took advantage of the Ottoman conflict in Hungary, as well as internal revolts, to attack and retake Tabriz and Yerevan. Two years later, the Safavid army defeated the Ottomans in the battle of Sufiyan, a rare direct confrontation that the Safavids won, albeit less through superiority than through Ottoman complacency and Kurdish betrayal.[87] The wars that followed were concluded first with the Treaty of Istanbul of 1612 and next, in 1618, with the Peace of Sarab. These accords, while including the payment of tribute in silk by Iran, marked the failure of the Ottomans to recover the parts of western Iran that they had lost in 1603. In 1622 Shah ʿAbbas, having extended his control towards the Persian Gulf, managed, with English naval support, to expel the Portuguese from the isle of Hormuz. Three year later he recovered Mesopotamia by seizing Baghdad from the Ottomans.

Shah ʿAbbas's reforms paid great dividends for maintaining Iran's territorial integrity, but they did not put his military on a solid financial footing. Its financial woes were at times so severe that troops remained without pay for years on end, sapping the strength of the military and causing widespread desertion. At times his soldiers received their pay in the form of leather strips or broadcloth. If anything, Shah ʿAbbas exacerbated these problems by transferring the army's centre of gravity away from the self-financed *Qizilbash* towards the state-paid *ghulams*. These circumstances naturally increased incidences of desertion.[88]

86 Floor, *Safavid Government Institutions*, pp. 160–1.

87 Colin Imber, 'The Battle of Sufiyan, 1605: A Symptom of Ottoman Military Decline?', in W. Floor and E. Herzig (eds.), *Iran and the World in the Safavid Age* (London and New York, 2012), pp. 91–102.

88 Della Valle, *Delle conditioni*, p. 40; Della Valle, *Viaggi*, vol. 2, p. 301; Dunlop, *Bronnen*, pp. 391–2.

## Later developments

Shortly after the accession of Shah Safi I, Shah ʿAbbas's grandson and successor, in 1629, the Ottomans, taking advantage of their neighbour's presumed weakness, attacked again. Though reluctant to react with a full campaign and dissuaded from undertaking one by his great-grandmother, Shah Safi did mount a counter expedition.[89] Yet Safavid defeat at Marivan in 1630 allowed the Ottomans to penetrate western Iran as far as Hamadan. Yerevan changed hands several times in 1635–6. Once again the Ottomans took and destroyed Tabriz; in 1638 they managed to recapture Baghdad and Mesopotamia and, in the same year, Qandahar was lost to the Mughals.

The Safavids at this point continued to fight the Mughals over Qandahar, but in 1639 resigned themselves to the Ottoman control over Baghdad by concluding the Peace of Zuhab. The decision to cut back on the military was deliberate, the outcome of fiscal exigencies coupled with pressure from the forces that came to prominence following ʿAbbas I's reign, principally women and eunuchs.[90] The accession of Shah Safi in 1629 had been accompanied by an infusion of (a clearly exaggerated) 200,000 *tumans* into the armed forces.[91] But this did little to improve the state of the army and in 1635 Grand Vizier Mirza Taqi asked the Dutch for an 8,000-*tuman* loan to pay the army.[92] Meanwhile debates at the court pitted different parties against one another, the *ghulams*, represented by Grand Vizier Mirza Saru Taqi, who short-changed the army in an attempt to undermine the *Qizilbash*, against Jani Khan, himself a *qurchibashi* who represented their interests.[93] Ultimately the pacific forces of the inner palace prevailed, leading to Jani Khan's assassination at the shah's orders in 1645. While it was not the sole factor at play, the Treaty of Zuhab had significant repercussions for the status and the quality of the Safavid military. It foreclosed armed conflict with the Ottomans and drew as fixed a boundary between the two powers as the logistics of the period permitted, reducing the reliance of the Iranians on the Arab and Kurdish tribes in their western borderlands and, conversely, the ability of these to play off the two powers.[94]

89 Dunlop, *Bronnen*, pp. 747–9. 90 Chardin, *Voyages*, vol. 5, p. 324.

91 Muhammad Maʿsum b. Khajigi Isfahani, *Khulasat al-siyar. Tarih-i ruzgar-i Shah Safi Safavi* (Tehran, 1989), p. 41. A *tuman*, a money of account, equalled 40 Dutch guilders and 15–16 Ottoman piastres in the 1660s. This sum would have amounted to one-third of the state's annual income.

92 Dunlop, *Bronnen*, p. 547.

93 Dutch national Archives (NA), Coll. Geleynssen de Jongh 283, Daghregister Winnincx, f. 204.

94 Rudi Matthee, 'Relations between the Center and the Periphery in Safavid Iran: The Western Borderlands v. the Eastern Frontier Zone', *The Historian* 77 (2015), 431–63.

Following Zuhab, the defensive strategy of the Safavids lay in seeking protection behind the natural barriers of the vast deserts and mountain ranges that surrounded the Iranian plateau. They were also lucky that their old rivals were, temporarily, occupied on other fronts. The Mughals were too weak and divided or, in the case of Sultan Aurangzeb (r. 1658–1707), too preoccupied with war in the Deccan, to try to recapture Qandahar – after the Safavids seized the city in 1648. Meanwhile the Ottomans became engaged in new wars in Europe. The mid-century Candia War over control of Crete sapped their already depleted treasury, while further conflict in South-East Europe led to territorial concessions at the Peace of Karlowitz (1699) following defeat at Vienna in 1683.

In these circumstances, it is not surprising that, without battlefield training and with repeated budget cuts, the Safavid army deteriorated. The security of the court was strengthened with the creation of a 600-strong corps of personal guards, *jaza'iri*, in 1654.[95] Yet in the same period the posts of *sipahsalar* (army commander) and *tupchibashi* were (temporarily) abolished.[96] Soldiers' pay was also reduced by a quarter: in 1652 each soldier reportedly received a mere one *tuman* for himself, his servant, and his two horses for the entire year. Soldiers also had to fend for themselves for victuals; hence the enormous size of the *urdu* bazaar in the later seventeenth century.[97] The result was a greater burden on the local population, which now had to pay the soldiers' salaries in a process accompanied by much theft and violence.[98] Merchants, too, in this period were forced to contribute to the upkeep of soldiers.[99]

The effects of the crisis became dramatically visible during the war over Qandahar in 1645–8. Various observers estimated the effective numerical strength of the Safavid army at 50,000 soldiers at that point, virtually unchanged from a century earlier.[100] Yet, underfed and underequipped, many soldiers deserted from the battlefield. That the Iranians nevertheless

95 Chardin, *Voyages*, vol. 5, p. 310.

96 Ibid., pp. 312–13; Savory, 'The Office of Sipahsālālar'; and Floor, *Safavid Government Institutions*, p. 19–20.

97 Chardin, *Voyages*, vol. 5, pp. 324–6; Cornelis Speelman, *Journaal der reis van den gezant der O.I. Compagnie Joan Cunaeus naar Perzië in 1651–1652*, ed. A. Hotz (Amsterdam, 1908), p. 212 ; Poullet d'Armainville, *Nouvelles Relations du Levant. Avec une exacte description du royaume de Perse*, 2 vols. (Paris, 1668), vol. 2, p. 148.

98 Zakʿaria of Agulis, *The Journal of Zakʿaria of Agulis*, trans. George A. Bournoutian (Costa Mesa, 2003), pp. 84–5.

99 NA, VOC 1215, Gamron to Isfahan, 11 Dec. 1655, f. 810.

100 NA, VOC 1178, Gamron to Batavia, 16 Oct. 1649, f. 628v; Richard, *Raphaël du Mans*, vol. 2, p. 119.

managed to take the stronghold arguably owed less to the strength of their army than to Mughal weakness.

Matters further deteriorated under Shah Sulayman (r. 1666–94), whose reign, after an initial period of ruinous panoply, saw state expenditure, including that towards the military, contract. No longer warriors on horseback, both he and his successor, Shah Sultan Husayn (r. 1694–1722), became sedentary rulers, engaged in private pleasure within the confines of their palace and poorly informed about the true state of the country, including the deteriorating state of the army. Chardin in the 1660s still speaks of the orderly way of paying soldiers, who were not forced to live off the land at the expense of the peasantry.[101] But he also observed that during military parades the same (limited number of) soldiers kept marching in front of the ruler.[102] Frequent musters were ordered to combat this corruption but, as Kaempfer maintained, the officials sent by the court were easily fooled or bribed.[103] And from at least the reign of Shah Sulayman soldiers were paid not in cash but in *barat*, promissory notes that had often lost part of their value by the time soldiers wanted to use them. Once again arrears in pay, of up to seven or eight years, became commonplace. Dutch sources indirectly confirm these dire conditions, reporting in 1681 that the soldiers around Bandar ʿAbbas had not been paid for five years.[104]

The Iranians at that point are said to have retreated into the interior of their country, feeling safe behind the rugged mountains and the desolation of the deserts flanking the plateau, counting on strategic depth. Of course, the same depth worked against the army. For the vast lonely emptiness also meant a lack of victuals, leaving soldiers with little to plunder and thus without means of subsistence.[105]

On paper, the numerical strength of the army was still quite impressive at the turn of the eighteenth century. A statistical overview speaks of a total of 180,000 soldiers – 120,000 cavalry and 50,000 infantry, plus 10,000 engaged in the operations of the *tupkhana*.[106] The Georgians alone were held to furnish 40,000 troops to the Safavid army in times of war. Foreign estimates reduce

101 Chardin, *Voyages*, vol. 9, pp. 222–3. 102 Ibid., vol. 5, p. 322.

103 Kaempfer, *Exotic Attractions*, pp. 64–5.

104 Richard, *Raphaël du Mans*, vol. 2, p. 393; Kaempfer, *Exotic Attractions*, pp. 64–5; G. A. Ezov (ed.), *Snosheniia Petra Velikogo s armianskim narodom* (St Petersburg, 1898), p. 39; Chardin, *Voyages*, vol. 5, p. 427. The Dutch confirm that soldiers were not paid. See NA, VOC 1355, Gamron to Batavia, 5 Mar. 1681, f. 399v.

105 Gabriel de Chinon, *Relations Nouvelles du Levant* (Lyon, 1671), pp. 69–70, 72.

106 Mustawfi, 'Amar-i mali', p. 397.

this to 100,000 men, all cavalry and 10,000 royal guards, with no infantry to speak of.[107]

The forces that had come to prevail in government under Sultan Husayn, mostly eunuchs and clerics, did not deem it necessary for salaries to come out of the royal treasury and ordered they be paid from *mulk* (private landed property) income. Half of the soldiers thus would not get paid for months.[108]

The result of this neglect became clear as external threats and internal revolts threatened the territory once again at the turn of the eighteenth century. A lack of troops made it difficult for the regime to defend Darband against Cossack raids, resist ongoing Uzbek incursions into Khurasan and Baluchi depredations in the Qandahar region, or quell internal Luri and Bakhiyari tribal revolts.[109] In 1696 it took six months for the Safavid army to be mobilised.[110]

The real danger, though, came from the east, and the Safavid army proved unequal to the challenge. In 1716 massive desertion was reported in the vicinity of Herat.[111] In the same year, the shah undertook a half-hearted effort to strengthen the country's financial base by removing all the silver and gold from the coffins of his forebears to pay the troops.[112] But it did not make much difference, for money was not the only problem. In 1717 one commentator, pointing to an age-old underlying problem, called the army in want of 'neither men nor money' but said that Iran's soldiers were 'undisciplined' and, above all, had no 'officers of sufficient abilities to command them'.[113] In late 1718 the Dutch reported that, faced with the conquest of Bahrain by Masqat, the loss of almost all of Khurasan, and bad news arriving from Astarabad and Yerevan, the Iranians were finally mounting a campaign. Yet a lack of funds severely hampered the recruitment of soldiers and even forced the ones who had been mobilised to return to the cities.[114] In the same year, a 'lack of discipline' led to the cancellation of a campaign against the Afghans. In 1719 there were reports of Iranian soldiers plundering villages.[115]

The intelligence about the weakness of Iran's defence forces provided to Moscow by the Russian envoy Artemii Volynskii in 1720 no doubt

107 Ezov, letter Israel d'Ori, p. 39. 108 Mustawfi, 'Amar-i mali', p. 399.
109 N. Sanson, letter 13 Sept. 1691, in Kroell (ed.), *Nouvelles d'Ispahan*, pp. 48–9.
110 Muhammad Ibrahim b. Zayn al-ʿAbidin Nasiri, *Dastur-i-shahriyaran* (Tehran, 1992), p. 104.
111 NA, VOC 1862, Gamron to Heren XVII, 15 Feb. 1716, fol. 2553.
112 P. P. Bushev, *Posol'stvo Artemiia Volynskogo v Iran v 1715–1718 gg.* (Moscow, 1978), p. 205.
113 John Bell of Antermony, *Travels from St Petersburg to Various Parts of Asia* (Edinburgh, new ed., 1805), p. 76.
114 NA, VOC 1913, 2nd ser., Report Ketelaar to Swols, fols. 489–90.
115 Bushev, *Posol'stvo Artemiia Volynskogo*, pp. 201, 265–6; Matthee, *Persia in Crisis*, p. 238.

contributed to Tsar Peter's decision to attack Iran and occupy the Caspian provinces as part of his project to control the region's silk production and gain a trade connection with India.[116] A similar report about Iran's lamentable defences presented in Istanbul by ambassador Dourry Efendy likewise probably informed the Ottoman decision to invade Safavid territory from the north-west.[117]

The Safavids' defeat by the forces that would bring them down in 1722, the Afghans, was less due to the strength of the latter's army – the Afghans only numbered 10,000 warriors who, in keeping with their tradition of mobile warfare, were ill-equipped for siege warfare – than to the lack of organisation and discipline in the Safavid army.[118]

116 Bushev, *Posol'stvo Artemiia Volynskogo*, passim.
117 Ahmad Dourry Efendy, *Relation de Dourry Efendy, ambassadeur de la Porthe Otomane auprès de roy de Perse*, ed. L. Langlès, trans. M. de Fiennes (Paris, 1810); and Salahsur Khassa Kamani Mustafa Agha, *Fathnama-yi Iravan. Ravabit-i Iran va ʿUthmani dar astana-yi bar uftadan-i Safaviyan 1132–1137 h.q.*, ed. Mehmed Munir Aktepe, trans. Nasr-Allah Salihi and Safiya Khadiv (Tehran, 2015).
118 Judas Thaddeus Krusiński, *The History of the Late Revolutions in Persia*, 3 vols. (London, repr. 2018), vol. 1, p. 151.

12

# Wars of dynasties, wars of empires: The nature of European conflicts, 1648–1792

HAMISH SCOTT

## I

War was a regular feature and, at times, a dominant characteristic of international relations between the Peace of Westphalia in 1648 and the beginning of monarchical Europe's struggle with Revolutionary France in 1792. At least until the Enlightenment, contemporaries viewed it not merely as an acceptable way of pursuing international rivalries, but as a more normal and natural state of affairs than peace. Periods of open conflict, during which diplomatic representatives would usually be withdrawn, were assumed to be inevitable and, indeed, were frequently the anticipated outcome of the policies adopted by rulers and their advisers. The eighteenth century was significantly more pacific than its seventeenth-century predecessor had been, though in turn much more bellicose than its nineteenth-century successor. According to one calculation, the European 'great powers' were engaged in warfare for eighty-eight years of the century 1600–1700, sixty-four years from 1700–1800, and twenty-four years from 1800–1900.[1] During the shorter period between 1700 and 1790, Russia was at war at some point during all nine decades; Austria, France, and Britain during eight; Spain and Sweden during seven; Prussia during six; and the Ottoman Empire during five: figures which underline the ubiquity of armed struggle even during the less bellicose eighteenth century.[2]

Whilst warfare remained a near-permanent feature of international relations, the motives which underlay such conflicts were changing, albeit slowly. Three principal sources of hostilities can be identified: religious divergence, dynastic rivalry, and competition for resources within Europe

1 Jack Levy, *War in the Modern Great Power System, 1495–1975* (Lexington, 1983), esp. pp. 139–43 and Table 6.5, p. 141; ibid., ch. 2 passim, provides a classification of which states can and cannot be considered 'great powers', by which Levy means essentially the leading states at any period.

2 Tim Blanning, *The Origins of the French Revolutionary Wars* (London, 1986), p. 37.

and, increasingly, overseas. Though most wars involved two, and sometimes all three, of these factors, which tended to reinforce each other, in the account which follows they will be considered separately to facilitate analysis. Their relative importance would change during this period. The middle decades of the eighteenth century witnessed the appearance of the term 'great power', together with political developments which produced the European international system which would prevail at least until 1914, and this has encouraged an emphasis upon state power and competition for economic resources as dominant motives for international conflict.[3] Yet deeply traditional factors such as dynastic rivalry, monarchical prestige, and religious faith long continued to spark conflict alongside more obviously modern causes of this rivalry and the wars which resulted.

## II

Religion was one traditional source of conflict which continued to be important, though its significance did decline. Its very survival as a factor in relations between states has often been doubted, particularly by international relations theorists of a realist persuasion who portray the Westphalian settlement as a watershed. They believe it ushered in an international system comprising sovereign states of approximately equal strength dedicated to the pursuit of enhanced status, and so inaugurated a century and a half when new, more secular forms of politics took hold.[4] *Realpolitik* principles came to guide foreign policy, the resulting struggle for territory and resources, and the pursuit of a balance of power. This influential viewpoint left no space for such a deeply irrational factor as religion, which was believed to have ceased influencing national foreign policies after 1648. It fitted in with an established and wider interpretation of the seventeenth and eighteenth centuries as a period of significant progress towards a more secular world-view, which had its roots in the German sociologist Max Weber's influential notion of the 'disenchantment of the world'.

From this perspective, the Thirty Years' War brought down the curtain on the age of religious conflicts, which dated back to the Crusades and had been intensified first by the fifteenth-century Ottoman advance and then by the sixteenth-century Reformation, which fractured the unity of medieval Christendom. The Westphalian settlement appeared to domesticate religious

3 Hamish Scott, *The Emergence of the Eastern Powers, 1756–1775* (Cambridge, 2001).

4 Though see the important qualifications in Derek Croxton, 'The Peace of Westphalia and the Origins of Sovereignty', *International History Review* 21 (1999), 569–91.

conflict, removing it from the sphere of international politics. It contained important religious provisions which applied within the *Reich*: the principle of *cuius regio, eius religio*, initially applied to the Catholic and Lutheran faiths by the religious Peace of Augsburg (1555), was now extended to Calvinism. It also created a more robust legal framework to deal with German religious conflicts.

The Papacy's own diminished political influence appeared to confirm this verdict. Its exclusion from the negotiations at Münster and Osnabrück symbolised its decline, as did the agreement between the signatories to the peace settlement to disregard Rome's anticipated protests. That this appears to have been proposed by the Catholic monarchies seemed further confirmation of the Papacy's declining importance. Though this reduced significance could also be accommodated within a wider secularising perspective, once again decline rather than extinction appears the most satisfactory description of Rome's trajectory after 1648: a generation later, a papal diplomat would mediate the Peace of Nijmegen (1678–9). The Papacy retained a sizable diplomatic corps until the French Revolution and far beyond, though representatives were sent only to Catholic courts, limiting its influence. European international relations continued to be dominated by the Catholic monarchies of France, Spain, and Austria, until the rise of Orthodox Russia and Protestant Britain and Prussia during the eighteenth century.

By the mid-seventeenth century, Europe's Catholic monarchies were becoming dismissive of Rome's political role and quite prepared to bully the pope of the day. They increasingly intervened both in papal elections, to secure a candidate favourable to their views, and in the selection of ecclesiastical figures, particularly in the *Reich*, where the Rhineland territories of Mainz, Trier, and Cologne were each ruled by an archbishop-elector. This new attitude was exemplified by the notably high-handed, and certainly disproportionate, behaviour of Louis XIV after a violent clash in a Rome street in 1662 between the pope's unruly Corsican guard and members of the French ambassador's household, some of whom were killed. The French king immediately broke off diplomatic relations and occupied Avignon and the Comtat Venaissin (two papal fiefs within France), menaced the Papal States and, after a confrontation which lasted for eighteen months, forced Pope Alexander VII (1655–67) to accept a humiliating settlement.[5] The efficacy of such tactics obviously depended upon the pope's personality: the formidable Innocent XI (1676–89) resisted much more vigorously and effectively

5 Paul Sonnino, *Louis XIV's View of the Papacy (1661–1667)* (Berkeley and Los Angeles, 1966).

than most. But the trend was clear and also irreversible. One striking illustration was the way in which, during the third quarter of the eighteenth century, the leading Catholic powers – France, Spain, and Portugal, with Austria's tacit support – first expelled all Jesuits from their own territories and then forced a reluctant Papacy to suppress the Society of Jesus, thereby legitimising their seizure of its property.[6]

The disappearance of religion from international rivalries after 1648 therefore appears both a simplistic and a premature verdict. Factors other than faith had always been important, and at times decisive, during the previous century, while there was never a clear-cut antithesis between 'religion' and 'politics' as a motivation for the actions of rulers and statesmen. The Thirty Years' War itself is now recognised to have been less of a purely religious conflict than was once believed. Sweden's long-serving Chancellor, Axel Oxenstierna, reflecting on his country's motives for intervening in Germany in 1630, famously declared that 'it was thus a question not so much of religion, as of public tranquillity [*status publicus*] in which religion was comprehended'.[7] This was a much more accurate guide to the significance of religious factors, both during the fighting and for many decades thereafter. Though there might no longer be 'wars of religion', that did not mean that the international system had become secularised overnight.

Religious issues continued to be a major source of conflict in the later seventeenth century. Europe's monarchies saw themselves as divinely ordained communities and, although considerations of political interest undoubtedly became more influential during this period, this in itself helps to explain the continuing significance of religion for international relations. Foreign policies were strongly influenced by the Catholic, Protestant, or Orthodox world-view of the ruler in question and were articulated employing religious language at least until the early eighteenth century. Confessional uniformity was still viewed as a foundation of internal stability and coherence: among the European political elite, only the Dutch Republic, Poland-Lithuania, and, after 1689, the British state were prepared to allow some religious minorities a degree of toleration, and even that fell significantly short of full civic equality.

6 Hamish Scott, 'Religion and *Realpolitik*: The duc de Choiseul, the Bourbon Family Compact, and the Society of Jesus, 1758–1773', *International History Review* 25 (2003), 37–62.

7 Quoted by Michael Roberts, 'The Political Objectives of Gustav Adolf in Germany, 1630–2', in *Essays in Swedish History* (London, 1968), 84.

The tensions created by these domestic priorities periodically washed over into international rivalries. Friction was overwhelmingly between Europe's major religious blocks: Catholic, Orthodox, Protestant, and Islamic. Yet within Protestantism itself there was significant tension between different confessions and primarily between Lutheranism, long unreconciled to the admission of the Reformed or Calvinist Church to the Religious Peace of Augsburg in 1648, and Calvinism. One illustration of the tension which could result was the dispute between Lutheran Sweden and Calvinist Brandenburg-Prussia in the later 1680s, leading to a formal, though brief, rupture in relations in 1691.[8] Most religious tension, however, arose between rival faiths: Catholic against Protestant, Catholicism and Orthodoxy against Islam.

Throughout this period religion remained particularly important in relations between Christian monarchies and their Islamic neighbours. Following the relief of the siege of Vienna in 1683 Austria spearheaded a successful counter-attack against Ottoman power, which had hemmed in and periodically menaced Europe during the two centuries since Constantinople's fall in 1453. Two periods of warfare (1683–99, 1716–18) restored to Habsburg control large areas of South-Eastern Europe and especially the greater part of Hungary (occupied by the Ottoman Empire since the first half of the sixteenth century), and so transformed relations. Though subsequent fighting during the 1730s led to some gains being handed back, Europe no longer stood on the defensive against Islam.

The fighting after 1683 had been waged by the 'Holy League': an alliance of Austria, Poland-Lithuania, Venice, and, latterly, Russia, largely constructed by papal diplomacy. Innocent XI had, as a young man, unsuccessfully aspired to go on a crusade, and as head of the Church always took very seriously his responsibility for Christianity in the East. He himself gave significant sums of money to support the military effort, instructed the clergy in Southern Europe to contribute as well, and repeatedly permitted Church wealth to be taxed.[9] In the Habsburg monarchy, such taxation raised significant sums to help fund the war of 1716–18 and, during the later 1720s and 1730s, to repair fortresses in Hungary and then to defray some of the costs of the campaigns of 1737–9.[10] The anti-Ottoman offensive amounted to an undeclared final

8 Daniel Riches, 'The Rise of Confessional Tension in Brandenburg's Relations with Sweden in the Late-Seventeenth Century', *Central European History* 37 (2004), 568–92.

9 Ludwig von Pastor, *The History of the Popes from the Close of the Middle Ages*, English translation, 40 vols. (London, 1891–1953), vol. 32, pp. 160–245, passim.

10 Peter G. M. Dickson, *Finance and Government under Maria Theresia 1740–1780*, 2 vols. (Oxford, 1987), vol. 2, pp. 265–6.

Crusade, and underlined how relations with Islam continued to be viewed through a religious prism. This was equally true where Spain's relations with Islamic communities in North Africa, and defence of its enclaves there, were concerned. It also influenced Madrid's desire to limit the depredations of North African corsairs, inspiring a series of punitive expeditions during the eighteenth century against Algiers, long seen as the principal pirate lair.

Both on the Mediterranean frontier and in South-Eastern Europe, religion remained central to the continuing rivalry between Christianity and Islam, as it did in a different way to the growing confrontation between Orthodox Russia and the Ottoman Empire. Yet Spanish, like Austrian, policy could follow religious dictates much more easily when it faced outwards. Within Europe by contrast, both states needed Protestant allies to achieve political goals and particularly to sustain rivalry with France. In practice this made it impossible for their policies to run precisely along confessional fault-lines, qualifying but not overturning generalisations about religion's enduring importance.

The repercussions of internal conflicts for the rivalries of leading states were a further way faith could still be significant. This was especially true within the *Reich*, where the confessional geography was peculiarly complex and even intractable. Co-religionists rallied to the defence of beleaguered minorities, while the fact that several major rulers within the *Reich* also held crowns elsewhere – primarily the Hanoverian dynasty, which ruled in Britain after 1714, the Saxon Wettins who were kings of Poland-Lithuania (1697–1763), and the Hohenzollerns of Brandenburg, 'kings in Prussia' after 1701 – also helped to bring these episodes to wider international attention.

In larger polities too the presence of religious minorities could be a source of strife, and these domestic conflicts periodically spiralled out into the international sphere. The generation after Westphalia was relatively untroubled, though the rising in Ukraine (1648–54) had an important religious dimension, as Orthodoxy clashed violently with Catholicism. During the 1680s, however, a renewed Catholic advance began against Protestant minorities who had hitherto enjoyed a degree of toleration, but now came under renewed pressure. This had significant international repercussions. In Hungary, the Habsburg reconquest was accompanied by a Catholic offensive (which had begun during the 1670s) against the Calvinist and, to a lesser extent, Lutheran communities which had thrived during the long Ottoman occupation. In 1685 the Catholic Pfalz-Neuburg line of the ruling family succeeded to the Electorate of the Palatinate, traditionally the leader of militant Protestantism within the *Reich*, reviving tensions there. In the following year the duke of Savoy with his army invaded the Alpine valleys of the Vaudois (west of the capital, Turin) in an attempt to extirpate the Protestant communities – the descendants of the medieval Waldensians – who lived there, killing or imprisoning many of them. Louis XIV encouraged and assisted this last initiative, and his own actions did most to revive religion as an international issue. Royal pressure on France's sizable Huguenot minority had increased sharply after the Peace of Nijmegen (1679) and culminated in the celebrated Revocation of the Edict of Nantes (October 1685), when French Protestants were deprived of their right to worship according to their own faith, enjoyed since the very end of the sixteenth century. Around 200,000 Huguenots were driven into exile, principally in the Dutch Republic, Brandenburg, and England.

Anti-French propaganda highlighted their fate, and portrayed the Revocation as exemplifying violent Catholic persecution. It was the final way in which religion still influenced international rivalries. Monarchies were coming to make more and more use of pamphlets and newspapers to criticise rivals' conduct, justify their own actions, and seek support for their cause. William III made particularly extensive use of this medium in his struggle with France. The *stadtholder*-king was himself a religious moderate who both favoured and practiced toleration. Yet for a generation beginning in the 1670s, Williamite propaganda was based upon an identification of Dutch and, subsequently, British interests with international Protestantism and European liberty, with William cast in the role of providential divine deliverer from the universal monarchy to which Louis XIV was said to aspire.[11] The Revocation came to play a central role in this campaign, which was spearheaded by *émigré* Huguenot writers. It was given a degree of plausibility by the fact that – in France, Savoy-Piedmont, Hungary, and Britain under James VII and II (r. 1685–8) – absolutism and Counter-Reformation Catholicism appeared to march hand-in-hand. But this approach also had real practical limitations: Emperor Leopold I (r. 1657/8–1705) might be a persecutor of Hungarian Protestants, but for William III he was also a highly desirable ally against Louis XIV.

Religion's propaganda value, though, exactly like its wider international role, tended to diminish over time. Occasionally it could still play an important role. In 1719–22 and again in the later 1730s there were rumours of

11 Tony Claydon, 'Protestantism, Universal Monarchy and Christendom in William's War Propaganda, 1689-1697', in Esther Mijers and David Onnekink (eds.), *Redefining William III: The Impact of the King-Stadholder in International Context* (Aldershot, 2007), pp. 125–42.

a Protestant league, though this did not materialise; while as late as the Seven Years' War, Frederick the Great (r. 1740–86) sought to play the confessional card, pronouncing the fighting to be a religious struggle in which Protestant Prussia and Britain faced Catholic France and Austria who were allied with Orthodox Russia. The Prussian king's notorious contempt for revealed religion, coupled with the more evidently secular spirit of the mid-eighteenth century, ensured that his claims were met with a sceptical response. Yet more traditionally inclined rulers were still influenced by religious considerations. The celebrated Franco-Austrian *rapprochement*, the Diplomatic Revolution of 1756, was facilitated by Louis XV's belief that the end of the traditional rivalry between Europe's two leading Catholic monarchies could improve his own spiritual standing. Fearful that his Jesuit confessors would force him to end his liaison with Madame de Pompadour, he hoped that such a religious coup might be sufficient to save his soul while retaining his mistress. Even at the very end of the period, religion was far from moribund as a factor behind international rivalry: the excesses of the French Revolutionaries against the Catholic Church and their promotion of the Cult of Reason added something of the character of a crusade in support of Christianity and against disbelief to the warfare of the 1790s. But by the mid-eighteenth century, if not actually earlier, it was becoming less influential than it had been hitherto as a driving force in conflict.

## III

Dynasticism, an equally traditional factor, proved even more resilient and remained an influential, and at times the dominant, animus at least until the end of the *ancien régime*. Though the importance of dynastic considerations in international politics slowly declined, particularly from the mid-eighteenth century onwards, as the search for new lands and resources within and beyond Europe assumed ever greater importance, they always remained influential. In addition to influencing actual policy, dynasticism provided the framework within which relations took place, together with a language in which international rivalry could be conducted. (See Map 9.) Europe's ruling houses believed that they constituted an international monarchical society, while their informal interaction provided a second, separate level for political aims to be pursued at.[12]

12 Emphasised in Lucien Bély, *La Société des princes, XVIIIe siècle* (Paris, 1999).

MAP 9 Europe after the treaties of Utrecht and Rastatt (1713, 1714). Created by Béla Nagy.

Individual state policy decisions primarily took account of the family interests of the ruling house, rather than more modern concepts such as *Realpolitik*, 'national security' or even the 'balance of power'. In this respect dynastic concerns corresponded both to the nature of political authority and to the structure of the individual polities which comprised the European states-system in this period. Rulers regarded their own territories as a reservoir of economic and human resources with which to pursue their ambitions. They had an essentially proprietary conception of monarchy, and viewed themselves as the temporary owners of their possessions, entrusted with the responsibility of defending and, if possible, extending the patrimony which they had inherited and passing it on to their successor in good order.[13] The family's claims to particular territories or other successions formed part of this inheritance and had to be pursued whenever possible. The intensity with which dynastic claims were remembered and pursued was remarkable. In the aftermath of Emperor Charles VI's death in autumn 1740, Spain's Bourbon King Philip V (r. 1700–46) seriously considered a bid for the Austrian Habsburg throne, on the grounds that the Spanish monarchy had long been ruled by the House of Habsburg.[14]

This was particularly important in Western Europe, where territorial units claiming sovereignty were packed densely together and dynastic claims were therefore more difficult to resolve within a strongly legalistic international culture. One reason why the Italian peninsula was so central to international rivalries during the first half of the eighteenth century was that it was a region of fluid political and dynastic geography, where territories might be acquired or exchanged. A different pattern prevailed in the continent's eastern half, largely due to the disruptive impact of successive waves of conquest, the last of which had been the Ottoman advance during the fifteenth and sixteenth centuries. That, together with the elective nature of the Polish-Lithuanian throne and the absence of accepted rules of succession in Russia, ensured that dynastic disputes assumed different forms and, overall, were less dominant there.

Most countries of any political standing were ruled by monarchs claiming an authority that was absolute, at least in theory. In these polities the interests of the sovereign and the immediate ruling family took priority in

13 Herbert Rowen, *The King's State: Proprietary Dynasticism in Early Modern France* (New Brunswick, 1980).

14 See Edward Jones Corredera, 'The Memory of the Habsburg Monarchy in Early Eighteenth-Century Spain', *Global Intellectual History* 6 (2021), 443–62.

decision-making. These men, and less frequently women, were usually involved directly, and often actively, in government, and foreign policy and warfare were everywhere the dominant issues in informal discussions at court and in council meetings. Concern for their own and their dynasty's reputation, before both contemporaries and posterity, dominated these rulers' approach to international issues. A decade into his personal rule, Louis XIV frankly admitted to his 'extreme desire to augment the glory of my crown', declaring that the objective all monarchs pursued in their foreign policies was 'our glory and the greatness of our state'.[15] This view was echoed two generations later by Frederick the Great, who declared whilst still crown prince his hope that the Prussian monarch would one day 'cut a fine figure among the great powers of the world and play a significant role'. What Prussia's king termed 'the honour of the dynasty and the country' remained central to international rivalries until the end of the *ancien régime*, and far beyond.[16] During the 1780s the veteran Habsburg foreign minister Wenzel Anton von Kaunitz-Rietberg concurred, confirming that glory and reputation remained central objectives of monarchical policy. A concern with honour was central to this outlook: Hugues de Lionne, Louis XIV's first important foreign minister, had declared in 1662 that exactly like most rulers and noblemen, the king was 'sensitive to the last degree in anything which affects his honour', and this remained true at least until the French Revolution.[17]

Describing the polities which comprised Europe's *ancien régime* as 'states' derives from convenience – it is difficult to provide a persuasive alternative – and relies on the familiarity which arises from frequent usage, but does not accurately capture the realities of eighteenth-century administrative structures. Monarchs did not think of their possessions in such terms and whilst there was undoubted progress during the later seventeenth and eighteenth centuries towards the kind of bureaucratic, and therefore impersonal, state idealised by Max Weber, it fell far short of the theoretical model he constructed 100 years later. The ruler of a major power who came closest to viewing his territories as a 'state' in the modern sense was probably Joseph II, Holy Roman Emperor (1765–90), co-regent in the Habsburg

15 Paul Sonnino (ed.), *Mémoires for the Instruction of the Dauphin* (New York, 1970), pp. 170, 221–2.

16 Both quoted by Gerhard Ritter, *Frederick the Great: An Historical Profile*, English translation (London, 1968), p. 33, from the 'Natzmer letter' of 1731.

17 Quoted by Derek McKay and Hamish M. Scott, *The Rise of the Great Powers 1648–1815* (London, 1983), p. 16.

monarchy (1765–80) and later its sole ruler (1780–90), and his efforts to create a centralised, uniform polity out of the widely dispersed and disparate territories and societies which he had inherited ended in failure. Early modern Europe consisted of a series of composite monarchies, made up of different territories, political societies, and administrative and legal systems, which had been imperfectly integrated and sometimes even spoke different languages. The conflicts which arose between these polities not only reflected the interests of the ruling families but centred upon issues arising from their marriages and successions.

This was reinforced by the considerable role played by rulers in decision-making. The Dutch Republic, especially during the periods of Regent ascendancy (1650–72, 1702–47), was an important exception to these generalisations, and commercial and economic considerations always exerted far greater influence upon its external policy. Yet there were equally extended periods (1672–1702, 1747–95) when the House of Orange ruled as quasi-monarchs and the *stadtholder* was able to exert real initiative and even direction, whether overtly or covertly. William III, who became *stadtholder* in the emergency of the French invasion of 1672, was granted exceptional authority, which he retained and carefully defended. He operated through a small number of close advisors and established his own secret diplomacy which operated side-by-side with the Republic's official representation. Around twenty favourites of the *stadtholder* carried out important private missions on his behalf – some seventy such embassies took place during his reign – and only once these secret negotiations had been concluded were the States-General informed of what had been agreed. William III's personal control over the Republic's policy increased after the creation of the Anglo-Dutch political union in 1689.

The nascent British state which came into existence after the Revolution of 1688–9, when a constitutional monarchy replaced the would-be absolutism of the Stuarts, followed a different pattern again, since ministers and the two Houses of Parliament (through their control of the purse strings) came to play a significant part in policy-making. Yet, even in Britain, the king's role remained very considerable and decisions about war and peace were clearly part of the royal prerogative. William III was used to exercising personal direction over Dutch policy, and, as British monarch after 1689, continued to act as his own foreign minister, negotiating personally with ambassadors, conducting the important correspondence himself, and taking decisions without much reference to British politicians. Royal influence survived the accession of the Hanoverians in 1714. Attention has focused upon the extent and the ways in which George I (r. 1714–27) and George II (r. 1727–60) were able to influence British policy in the interests of their German homeland, the Electorate of Hanover. Yet both monarchs also played a full part in the making of British policy, reading diplomatic dispatches and discussing policy with their ministers – exactly as George III (r. 1760–1820) was subsequently to do.

This monarchical control was not seriously weakened and may actually have been strengthened by the emergence of more specialised departments for foreign affairs after 1700. France led the way, with the creation of a much larger and more elaborate foreign office under the marquis de Torcy, who was Louis XIV's final foreign minister (1698–1715), and this provided a model for other rulers. In Russia the old *Posol'skii Prikaz* ('Ambassadorial' or 'Foreign Affairs Chancellery'), which had become increasingly important and independent, came to be replaced by a College of Foreign Affairs in 1719. In Brandenburg-Prussia – where Frederick William I (r. 1713–40) carried through an extensive administrative reorganisation – a Department of External Affairs (*Auswärtiges Amt*) was set up in 1728; five years later it began to be known as the *Kabinettsministerium*. During the first half of the eighteenth century, Spain, Sardinia, and the Habsburg monarchy all acquired more modern and professional foreign offices. These initiatives, however, may actually have strengthened royal control over policy-making, by providing a more extensive and efficient infrastructure able to implement a monarch's wishes.

Until the end of the *ancien régime* and far beyond, rulers were accustomed to playing a prominent, and frequently dominant, part in discussions about policy and especially diplomatic and military strategy. The abilities or otherwise of the individual monarch, and sometimes the gender too, were important in determining the extent of this role. A weak ruler might exert less influence, though – France's Louis XV (r. 1715–74) would be a case in point – he or she could disrupt and weaken policy by meddling behind the scenes. A female sovereign would often be less influential within what long remained a strongly masculine political society. When Queen Anne (r. 1702–14) succeeded to the British thrones, Louis XIV immediately began to conduct negotiations with her ministers rather than the ruler herself; he had hitherto negotiated directly with her predecessor, William III. There were, however, important contrary examples: Philip V's second wife, Elisabeth Farnese, and Maria Theresa, as ruler of the Habsburg monarchy (r. 1740–80), both exerted real influence upon policy, as did the Russian empresses Elizabeth (r. 1741–62) and Catherine II (r. 1762–96).

The impact of rulers, and other members of the royal family, upon the making of political decisions exemplified the dynastic nature of international relations. They viewed the defence and (where possible) enhancement of their country's reputation, status, and resources as their primary duty. Questions of external security and decisions about war, peace, and alliances enjoyed a pre-eminence in council discussions. Not until the generation after the destructive Seven Years' War (1756–63) would domestic policy and the tasks of reconstruction and reform begin to eclipse international rivalry as rulers' principal concerns, and then temporarily and due entirely to the legacies of the fighting.[18]

The leading ministers of *ancien régime* Europe, in the major states at least, owed their position to their expertise in foreign policy: figures such as Kaunitz in the Habsburg monarchy, who loomed large in Vienna for almost four decades after he became *Staatskanzler* ('State Chancellor') in 1753, or Heinrich Rüdiger von Ilgen, who was influential in Brandenburg-Prussia under Frederick III/I (r. 1688–1713) and dominant under his successor Frederick William I, dying in office in 1728. There was a series of such powerful figures in France: the minister for war Louvois, the son and successor of Le Tellier (the secretary of war), and the long-serving foreign minister Colbert de Torcy under Louis XIV; Cardinal Fleury (1726–43), the duc de Choiseul (1758–70), and the comte de Vergennes (1774–87) under his two Bourbon successors. Eighteenth-century Russia could boast almost as many influential advisors: Alexander Danilovich Menshikov under Peter I, Andrey Ivanovich Ostermann under Anna (r. 1730–40), Alexey Petrovich von Bestuzhev in the reign of Elizabeth, and first Nikita Panin and then Grigory Aleksandrovich Potemkin under Catherine II. Even states of the second rank followed this trend: men such as the two Bernstorffs, Johan Hartwigg Ernst and his nephew Andreas Peter, in Denmark, Graf Heinrich von Brühl in Saxony-Poland, and the marquis Carlo Francesco D'Ormea in Savoy-Piedmont.

The continuing importance of dynasticism was most evident in the nature, and even the names, of many major conflicts, as well as in the provisions of the treaties that concluded them. The Peace of Westphalia itself had an important dynastic dimension, though this is often neglected. It was most evident in the

18 Hamish Scott, 'The Seven Years War and Europe's Ancien Régime', *War and History* 18:4 (2011), 1–37.

widespread backing given to the Hohenzollern claim to Pomerania. The Brandenburg elector, Frederick William (r. 1640–88), had the strongest title based on dynastic affiliations to the previous rulers, and was supported by other states in the face of Sweden's powerful political and military position and determination to retain the entire duchy, ensuring that in the final treaty Brandenburg secured Pomerania's larger, if less populated and economically less advanced, eastern territories. It was also evident in the creation of an eighth Electorate for Bavaria, which resolved a long-standing Wittelsbach family rivalry. Even Vienna's strengthened control over Bohemia was ultimately dynastic: in 1627, after the defeat of the Bohemian rebellion, the territory was declared definitively a hereditary Habsburg possession and this settlement secured international recognition at Westphalia.

To a greater or lesser extent, dynastic ambitions underpinned most *ancien régime* conflicts. This was exemplified by the War of Devolution (1667–8), with which Louis XIV's personal rule opened. Louis attempted to enforce his Spanish wife's claims to part of the southern Netherlands, to which he believed she had a right under the will drawn up by her father, Philip IV (r. 1621–65). The conflict between 1689 and 1697, usually known as the Nine Years' War or the War of the League of Augsburg, was also, and to an important extent, a 'War of the British Succession': a struggle to determine whether the House of Stuart should be restored to its British thrones, effectively abandoned by James VII and II in 1688, or whether the Dutch *stadtholder* William III, of the House of Orange, and his English wife Mary, herself a Stuart and daughter of the fallen ruler, should be set up in their place. In both cases Louis XIV demonstrated his support for the principle of hereditary right.

Dynastic issues continued to be prominent. Within the *Reich*, fighting after 1689 was a struggle for the succession both to the Palatinate and to the archbishopric of Cologne. The War of the Spanish Succession (1701–14) was, on one level, exactly what its name proclaimed it to be: a struggle to control the Spanish monarchy, with its extensive territories and rich resources within and beyond Europe. Louis XIV and his great rival Emperor Leopold I (r. 1657/58–1705) were first cousins. Their mothers were both children of Philip III (r. 1598–1621). In addition, each was married to a daughter of Philip IV, and advanced claims on that basis, despite the renunciation agreements which each princess had accepted on her marriage. The eventual succession of the Bourbon Philip V came about only after a decade of intensive warfare and was widely seen as threatening the balance of power. The fighting, moreover, was also an implicit struggle over the British succession, since a French victory may well have led to the return of the House of Stuart, a point that ministers in London keenly appreciated.

Dynastic issues dominated three of the final four continental wars of the *ancien régime*. This was particularly true of the Polish Succession conflict (1733–5/38), which in fact incorporated three dynastic questions. The most obvious, if not the most significant, dynastic consequence was the election of a second Saxon elector, Augustus III (r. 1733–63) to the throne of Poland-Lithuania, now a Russian client state, after the death of his father Augustus II (r. 1697–1733). The struggle also revolved around the ambitions of Elisabeth Farnese, the Italian-born second wife of Philip V, to secure territories in her native Italy for her two sons, together with the titles and incomes which would go with them. This coincided with Philip's ambition to restore Spain's dominance over the western Mediterranean, severely weakened by the loss of Gibraltar and Minorca to Britain in the Utrecht settlement.[19] The expected succession of Philip's son by his first wife, Ferdinand, would leave Farnese's own sons excluded from the Spanish throne. The strong-willed queen herself had a claim both to Parma (as niece of its duke) and more distantly to Tuscany. She achieved one part of this aim during the 1730s, establishing her elder son Don Carlos as duke of Parma (1731/2–1735) by diplomatic pressure and then – as a result of the fighting in the peninsula in which Spanish troops played a part – translating him south to Naples and Sicily (the 'Kingdom of the Two Sicilies'), where he was to rule as King Carlos VII (r. 1735–59).

The Polish War's final dynastic dimension would anticipate the fighting over the Austrian Succession which would break out after the death of Holy Roman Emperor Charles VI (r. 1711–40). Charles – who had previously been the Habsburg candidate for the Spanish throne as 'Carlos III' – was survived by only two daughters. In accordance with the famous Pragmatic Sanction (1713), the eldest, Maria Theresa, would succeed to the Habsburg family lands, but as a woman could not be elected emperor. Her husband's identity would be all-important, especially when the choice was revealed to be Francis Stephen, duke of Lorraine and Bar since 1729. France was unwilling to see its eastern frontier, successfully strengthened during Louis XIV's reign, undermined by the passing of Lorraine to the new rulers of Austria, long-standing rivals since the late fifteenth century. Louis XV went to war to avert this and succeeded in evicting Maria Theresa's intended husband from his

19 Christopher Storrs, *The Spanish Resurgence 1713–1748* (New Haven, 2016).

homeland in Lorraine. Though Francis Stephen harboured deep and permanent resentment towards France, he was compensated with Tuscany in the Italian peninsula, where the Medici dynasty had conveniently died out and which he was to rule as grand duke (r. 1737–65). On his death it came to be ruled by his younger son Leopold (r. 1765–90), thus passing to a cadet branch of the Austrian Habsburgs.

The repercussions of Charles VI's own death in October 1740 were even more momentous. He had secured Maria Theresa's succession to the Habsburg hereditary possessions through extended negotiations with the various territories ruled from Vienna, but international reactions were much more difficult to predict, let alone control. Political and territorial concessions made by Charles to rival rulers, in return for undertakings to respect the Pragmatic Sanction, now proved worthless. A series of would-be predators joined Prussia's Frederick the Great, who invaded Silesia a mere six weeks after the old emperor's death, in seeking to exploit the new ruler's gender and, to a lesser extent, youth and inexperience. Simultaneously an attempt was made to break the Habsburg grip on the imperial crown, which had endured for over three centuries. Though the elector of Bavaria, Charles Albert, was elected emperor as 'Charles VII', his reign was short (1742–5). Upon his death, Francis Stephen was elected emperor (1745–65) and Habsburg continuity restored. Maria Theresa demonstrated considerable resolution in resisting the coalition seeking to despoil her of her hereditary territories and, with the exception of the important loss of the rich province of Silesia (to Prussia), avoided major territorial cessions in the peace of Aix-la-Chapelle (Aachen) (1748).

Once again minor alterations were made to the Italian dynastic map, where Farnese's ambitions were fully realised. Her younger son, Don Philip – by now also the son-in-law of France's Louis XV, whose favourite daughter he had married – was installed as duke of Parma and Piacenza (1748–65), though with only limited rights of succession. The French king's support was important; one reason for his signature of an Austrian alliance in the mid-1750s was a wish to secure a more appropriate sovereignty for his daughter and son-in-law than Parma and Piacenza. The second Treaty of Versailles (May 1757) provided for most of the Austrian Netherlands to pass to the young couple (who would also receive a royal title) if and when Austria recovered Silesia, but Prussia's survival during the Seven Years' War ensured that this did not come to pass.

The Bavarian electorate itself provided the occasion for the *ancien régime's* final dynastic dispute: the brief and indecisive War of the Bavarian Succession

(1778–9). The death of the last male Bavarian Wittelsbach, Max Joseph, at the close of 1777, left the Palatine line of the family in the person of Charles Theodore as heir to the Electorate. This threatened to establish a strong Wittelsbach position in southern Germany, and to weaken Habsburg interests in a region where the two families had long been rivals. It thus encouraged Austria to seek the Electorate for itself, acquisition of which would round out Habsburg possessions, in exchange for its own distant territories in the southern (Austrian) Netherlands. Opposition to such a scheme both within the Empire and more widely, coupled with Vienna's mishandling of the issue, prevented both this, and a subsequent attempt in the mid-1780s, to acquire Bavaria for the Habsburgs. Yet the failure of these, and other projected exchanges, highlighted the attachment of rulers to territory which had long formed part of their ancestors' patrimony, demonstrating the continuing strength of dynasticism.

All four eighteenth-century succession struggles were precipitated by the death of a reigning sovereign. This underlined the dangers, as well as the difficulties and complexities, of royal inheritance. Early modern political society was wary of an heir who was female or a minor, believing this to be a fertile source of weakness and instability, and such assumptions could all too quickly translate into reality. Yet only the French royal crown formally barred a woman from succeeding, or the crown passing through a female, and even this principle rested on the questionable authority of the early medieval Salic Law, which had been consistently applied since the 1380s. The imperial dignity was also reserved for men, as Charles VI's difficulties over Maria Theresa's succession highlighted. Most rules of royal inheritance simply gave sons precedence over daughters and assumed the primacy of intergenerational succession, and such provisions were not proof against the vagaries of either royal demography or frequent warfare. One consequence had been that, shortly after 1600, all Europe's royal thrones had been in the hands of as few as five dynasties: the Habsburgs, Bourbons, Stuarts, Oldenburgs, and Vasas.[20] This situation was quite exceptional and proved transient, as new dynasties – notably the Romanovs in Russia and the Braganzas in Portugal – emerged, while one or two of the composite

20 Robert Oresko, 'The House of Savoy in Search for a Royal Crown', in Robert Oresko, Graham Gibbs, and Hamish Scott (eds.), *Royal and Republican Sovereignty in Early Modern Europe: Essays in Memory of Ragnhild Hatton* (Cambridge, 1997), p. 302 n. 52. By contrast there had been eight major dynasties in 1500 and thirteen in 1300: Daniel Nexon, *The Struggle for Power in Early Modern Europe: Religious Conflict, Dynastic Empires and International Change* (Princeton, 2009), p. 85.

monarchies created earlier did not survive. But the number of ruling monarchical families remained relatively small until the end of the *ancien régime*: during the eighteenth century there were no more than eighteen royal crowns in Europe, held by fifteen separate dynasties.[21]

The resulting tensions were exacerbated by widespread dynastic intermarriage. Finding suitable royal brides and grooms had always been a potential source of difficulties, and these had been increased by the sixteenth-century Protestant Reformation, since only a minority of princes and princesses were willing to change their religion for the uncertainty of a foreign throne. The division of Europe into confessional camps had compounded the problems which resulted from biological accident. The result was that more and more marriages were between a relatively small number of families. To give only one example, the Spanish infanta Margareta Theresa was both Leopold I's wife and his niece.

These intermarriages frequently created politically awkward claims, and were thus a further source of potential instability. To deal with this danger, the practice had developed of inserting renunciation clauses restricting future gains by inheritance when dynastic marriages were being arranged. A daughter or, much less frequently, a younger son being married formally gave up their claim to a particular piece of territory or even to a throne, in an attempt to avert future problems. Yet such renunciations could always be ignored or overturned if it became expedient to do so. In 1646, when a Franco-Spanish marriage was being considered, Cardinal Mazarin – the French foreign minister – voiced a prevalent attitude when he declared that 'once the Infanta marries His Majesty, we can hope for the succession to the Spanish thrones, whatever renunciations she has to make'.[22]

The succession of Carlos II in Spain was a case in point. Both the Habsburg emperor Leopold and the French king Louis XIV had married daughters of Philip IV, and in addition were themselves sons of Spanish infantas. Their wives, before setting out for Vienna and Paris, had formally renounced all claims to the Spanish monarchy. During the later seventeenth century, as the issue of the Spanish Succession became increasingly urgent, these renunciations were cast aside and claims to the whole inheritance vigorously advanced on the basis of their legal rights. The failure to pay marriage portions and dowries, whether in whole or part, could furnish a pretext.

21 Calculation from Daniel Schönpflug, 'One European Family?: A Quantitative Approach to Royal Marriage Circles 1700–1918', in Karina Urbach (ed.), *Royal Kinship: Anglo-German Family Networks 1815–1918* (Munich, 2008), p. 27.

22 Quoted by McKay and Scott, *Rise of the Great Powers*, p. 8.

Moreover, it was always possible to argue that the rights of royal succession were fundamental laws, and as such could never be renounced.

A further problem was that dynastic intermarriage created a significant number of overlapping and competing claims to territories and thrones, and these could be revived at moments of political crisis. The authentic voice of the Enlightenment, Voltaire, commented acidly on the central role of dynastic rights as a primary cause of eighteenth-century wars:[23]

> A genealogist proves to a prince that he is a direct descendant of a count whose relatives had made a family pact, three or four hundred years ago, with a house [*maison*] of which not a trace remains in living memory. This house had remote claims to a province whose last owner died of a stroke. The prince and his council conclude without difficulty that this province belongs to him by divine right. This province, which is a few hundred miles away, protests in vain that it does not know him, and that it has no desire to be ruled by him, that to impose laws on people, their consent at least is necessary. These speeches do not even reach the ears of the prince with the incontestable rights. At once he finds a large number of men with nothing to lose; he dresses them up in blue serge at a hundred and ten *sous* a yard, edges their hats with thick white piping, drills them and marches them off to glory.

Jean Rousset de Missy's *Les Intêrets presens des puissances de l'Europe* (1733), an influential contemporary guide, distinguished between enduring strategic interests and the historically validated claims (*prétensions*) which every ruler maintained for use whenever an appropriate opportunity arose, and which were supported by treaty provisions and laws of dynastic succession.[24]

A further source of difficulties was the existence of multiple sources of sovereignty and overlapping authorities within a particular area. This was especially acute within the *Reich* and throughout the northern half of the Italian peninsula, where the Empire also exercised overlordship across an area known as *Reichsitalien*, which was a legacy of medieval political developments. Seventeenth- and eighteenth-century international relations were marked by a series of disputed claims, great and small, dormant and active, as rulers waited for the opportune moment – usually when a rival was weak and

23 In the article on 'Guerre [War]' in the *Dictionnaire philosophique* (1764): quoted and translated by David Williams (ed.), *Voltaire: Political Writings* (Cambridge, 1994), p. 8.

24 Jean Rousset [de Missy], *Les Intêrets presens des puissances de l'Europe, fondez sur les Traitez conclus depuis la Paix d'Utrecht, et sur les Preuves de leurs Prétensions particulières*, 2 vols. (The Hague, 1733), 'Preface'. A good example would be the claim to Jülich and Berg by the Hohenzollern rulers of Brandenburg-Prussia, for which see ibid., vol. I, pp. 201–73.

vulnerable, or a throne vacant or contested – to assert these claims and to pursue them with every prospect of success.

This was exactly what took place in the aftermath of Charles VI's death in late autumn 1740, as a series of rulers came forward with claims to parts of the Habsburg inheritance to which Maria Theresa had just succeeded. The gender of the new ruler was a further incentive to such political opportunism. These claims were formally drawn up in manifestoes couched in the legal and genealogical language of dynasticism. Even Prussia's Frederick the Great was careful to produce a worked-out claim to Silesia, though he himself did not take it very seriously. When one of his advisers, Heinrich von Podewils, submitted a draft of a Hohenzollern manifesto, the king cynically commented in the margin: 'Bravo: the work of an excellent charlatan.'[25] His predecessors had taken such legal rights more seriously. Until the 1740s the Hohenzollerns were rulers of a second-rank state, but this did not weaken the tenacity with which they had upheld their claims for more than a century both to the Silesian duchy of Jägerndorf and to the Lower Rhenish territories of Jülich and Berg.

The principal conflicts which lacked a major dynastic dimension were the three Anglo-Dutch Wars between the 1650s and 1670s, the wider Dutch War (1672–9), the Great Northern War (1700–21), and the Seven Years' War (1756–63). The fighting around the Baltic during the first two decades of the eighteenth century had an implicit dynastic element, since the Saxon elector Augustus would have lost his new Polish-Lithuanian throne if Sweden had been victorious. The Seven Years' War also had a minor dynastic dimension in the proposals to create a kingdom in the southern Netherlands for the French king's son-in-law.

War did not invariably arise from dynastic concerns, which could prevent hostilities as well as bring about fighting. The best illustration was the generation of relative peace in Western and Central Europe following the settlements of 1713/14 and 1721. There would be no serious fighting between more than two leading states before the 1730s, and a majority of these states remained at peace throughout most of the period. One significant reason was the financial burden and material cost of the wars of 1689–1714: all governments faced significant deficits and growing debts, as well as essential reconstruction after the preceding generation's intensive and destructive warfare. But widespread dynastic uncertainty was also important. Both the Hanoverians in Britain and the Bourbons in Spain were busy establishing

25 Quoted by Ritter, *Frederick the Great*, p. 82.

themselves. In France Louis XIV's remarkable longevity, together with the unexpected death of several potential heirs, meant that after the old king died in September 1715 the French succession rested on the shoulders of a sickly child of five: his great-grandson, the future Louis XV. In the composite Habsburg monarchy, Charles VI's lack of a male heir and its repercussions shifted attention to domestic affairs, particularly during the 1720s. In Sweden too, dynastic upheaval was central to the political changes which followed the death of the childless Charles XII (r. 1697–1718). The succession of his younger sister, Ulrika Eleonora, and, after her abdication (1720), of her husband, Frederick of Hesse, was one factor facilitating peace around the Baltic throughout the 1720s and 1730s. Another was the dynastic uncertainty in Russia following Peter I's death in 1725, and that of his only surviving grandson five years later.

The continuing importance of royal marriages provided a further, traditional means by which dynastic concerns continued to influence international relations. Finding suitable wives and husbands for royal children remained significant at least until the very end of this period. During the 1780s, Joseph II – who was simultaneously pushing through a series of modernising reforms intended to transform his far-flung territories into a unitary state – devoted considerable time and energy to securing a suitable bride for Francis, the son of his younger brother Leopold, and his own eventual successor both in the Hereditary Lands and in the *Reich*.[26] The emperor began the search early in 1781, shortly after the beginning of his personal rule, when Francis was only thirteen, but the marriage did not take place for another seven years and then largely as a result of Joseph's tenacity. The union, moreover, served a wider political purpose: Francis's bride was Elisabeth of Württemberg, the sister-in-law of the Russian grand duke Paul (later Paul I [1796–1801]), and so it would strengthen Austria's alliance with St Petersburg, which was central to Habsburg foreign policy throughout the 1780s.

Royal marriages were employed to reinforce established alliances and to foster and strengthen new ones throughout the *ancien régime*, as well as to maintain and consolidate dynastic prestige. After the Seven Years' War, France's leading minister, Choiseul – probably at Louis XV's instigation – sought to strengthen the Austrian alliance concluded in 1756, through a dynastic marriage of the traditional type. This eventually took place in

26 See Derek Beales, *Joseph II*, 2 vols. (Cambridge, 1987–2009), vol. 2, pp. 124–32, 167, 359, 565.

1770 when the Habsburg archduchess Marie Antoinette married the dauphin, the future Louis XVI (r. 1774–93). Choiseul had earlier drawn a revealing distinction between 'political' and 'family' alliances.[27] The former, exemplified by the treaty with Vienna, were less stable and secure than the latter, embodied in the contemporaneous alliance with Madrid, signed in 1761. This was the third Family Compact to be concluded during the eighteenth century and proved much the most enduring, surviving until the early years of the French Revolution.

The Bourbon axis, established by means of the succession of a junior line of the French royal family south of the Pyrenees, was unique among leading monarchies. The potential importance of such family-based political links became clear at the beginning of the 1750s when Louis XV concluded a secret agreement with the future king of Spain to renew these ties after the latter succeeded to that throne. The Bourbon alliance had been severed when Ferdinand VI (r. 1746–59) succeeded his father Philip V. The Anglophone political sympathies of the new Spanish king's influential Portuguese wife were the main reason for the disruption of the Bourbon alliance. Ferdinand, however, was childless and his heir was his step-brother Carlos, then ruling Naples and Sicily. At the beginning of the 1750s Louis XV and Carlos exchanged private letters agreeing to renew the Family Compact when the latter succeeded in Madrid, and this was duly done.[28] The French king may even have invented the nomenclature for the Franco-Spanish alliance, the 'family compact' (*pacte de famille*), during the detailed negotiations in 1761, highlighting the extent to which international relations continued to be dominated by the dynastic interests of the ruling families. Indeed, as late as 1787 the new Prussian king Frederick William II (r. 1786–97) intervened militarily in the Dutch Republic in order to rescue his sister, Wilhelmina. The princess was married to the Dutch *stadtholder*, William IV (r. 1751–95), and had been detained by the Patriots in the course of a continuing struggle for political control. An invasion by Prussian forces quickly rescued her and restored her husband's authority.

## IV

The more sophisticated mechanisms for administrating foreign policy which had emerged in the earlier eighteenth century were but one dimension of

27 See the 1765 'Mémoire au Roi', in Fernand Calmettes (ed.), *Mémoires du duc de Choiseul, 1719–1785* (Paris, 1904), pp. 389–90.

28 Rohan Butler, 'The Secret Compact of 1753 between the Kings of France and Naples', in Oresko, Gibbs, and Scott (eds.), *Royal and Republican Sovereignty*, pp. 551–79.

a wider expansion of government at this time. More elaborate and efficient departments of state emerged, better able to formulate and implement policy and to collect taxes, conscript soldiers, and extract matériel to support warfare. These agencies were primarily concerned within central administration. Until the end of the *ancien régime* and far beyond, local government remained almost universally deeply traditional and was overwhelmingly a matter of informal initiatives instigated by local noblemen, often unsalaried or poorly and irregularly paid, limiting the effectiveness with which central government aims could be achieved at the provincial level. Yet whatever view is taken of the reality or otherwise of 'absolutism', there is little doubt that the ability of the monarchical state to extract resources from its subjects did increase, and may have significantly increased, during this period.

This contributed to the third defining characteristic of international conflict in this period: that these were wars consciously, and increasingly, fought for control of human and material resources, and by the eighteenth century they were increasingly being fought beyond Europe. Two factors reinforced this trend. Governments came to think of resources in new and rather different ways, eventually believing that these were not fixed but might be increased through trade, which could be reciprocally advantageous. Second, the colonial possessions and imperial aspirations of Europe's leading states came to play a greater role in international rivalries. By the mid-eighteenth century, the wars of dynasties were also becoming – and perhaps more significantly – wars of empires.

The search for additional demographic and economic resources had always been an implicit, and sometimes explicit, aim of monarchical foreign policy. Intangible factors such as prestige, reputation, and status were important, but they were measured in terms of the gain – or loss – of territory in peace settlements, which provided a convenient, if approximate, index of international success or failure. During the seventeenth century, however, this traditional approach began to be overlaid, though never replaced, by the idea that rulers could and should take positive steps, in peacetime and during hostilities, to secure the largest share of the available resources and so enhance their own and their subjects' prosperity. These ideas – which it is convenient, if not altogether accurate, to describe as 'mercantilist' – were predicated upon the conviction that all Europe's monarchies were engaged in a permanent struggle for available resources, which were more-or-less fixed, and therefore amounted to an early form of economic nationalism.

One of the leading advocates of mercantilist policies was Jean-Baptiste Colbert, economic, finance, and naval minister during the first two decades of

Louis XIV's personal rule. Colbert famously declared that 'one nation could improve its commerce, its merchant marine, or its manufactures, only by taking away something from the trade, the shipping or the industry of another country'.[29] In peacetime this was pursued through protectionist policies, which imposed high tariffs in an attempt to exclude a rival's natural produce and especially its manufactured goods. The Bourbon monarchy, exactly like its competitors, sought to foster trade by establishing chartered companies and also to acquire overseas possessions. Simultaneously the French government attempted to encourage domestic production and to develop luxury manufacturing which reduced the need for high-cost imports and generated goods which could in turn be exported, securing further resources and, above all, precious metals for France in payment. One important dimension of mercantilism was the acquisition of bullion, not because it was equated with wealth and so was something to be hoarded, but for the important reason that precious metals were needed to put people, and resources, to work and to lubricate international trade, as contemporaries recognised.

This approach assumed that a country's wealth – exactly like its political power – was relative to that of its rivals. One late seventeenth-century theorist, the Austrian cameralist Philipp Wilhelm von Hörnigk, precisely expressed the basis of this philosophy:

> whether a nation be today mighty and rich or not depends not on the abundance or scarcity of its powers and riches, but principally on whether its neighbours possess more or less than it. For power and riches have become a *relative* matter, dependent on being weaker and poorer than others.[30]

Such thinking blurred the distinction between peace and war. Hörnigk's contemporary, the famous English theorist Sir Josiah Child, described commerce as 'a kind of warfare'. Colbert went even further, declaring that trade was a source of 'perpetual strife, both in time of war and peace, to decide which should have the greater share'.[31]

These views were most influential in the three leading countries of North-Western Europe: the Dutch Republic, France, and England, and

29 Evan Luard, *The Balance of Power: The System of International Relations, 1648–1815* (Basingstoke, 1992), p. 204, citing Charles Woolsey, *Cole, Colbert and a Century of French Mercantilism*, 2 vols. (New York, 1939), vol. 1, p. 342.

30 Quoted by Eli Heckscher, *Mercantilism*, 2 vols., 2nd ed. (London, 1955), vol. 2, p. 22, from Hörnigk's celebrated *Österreich über alles, wann es nur will* (1684).

31 Luard, *The Balance of Power*, p. 215.

they influenced both internal policies and international rivalries. The three Anglo-Dutch Wars fought during the third quarter of the seventeenth century (1652–4, 1665–7, and 1672–4) exemplified this kind of economic conflict. The Dutch were Europe's leading commercial and maritime power, with a political system uniquely receptive to the interests of the Dutch merchant elite, the Regents, and their dominant commercial city, Amsterdam. This was especially important during the first two wars, when the stadtholderate was in abeyance after William II's death in 1650, and the regime of Johan de Witt was in power (1653–72), which favoured the interests of the urban Regent oligarchy in the maritime provinces. Even so, the three Anglo-Dutch wars were not purely economic conflicts but had significant political and naval dimensions.

France was at least as great an economic rival of the Dutch Republic as England, particularly during Colbert's ministry. The Republic's wealth and consequent international power were envied, being viewed as the main obstacles to the further enhancement of France's own prosperity and political dominance. In 1672 Louis XIV launched a carefully prepared invasion of Dutch territory, instigating four decades of near-continuous conflict, which contributed significantly to the Republic's political and, less certainly, economic decline. Paradoxically, however, the attack on the Dutch Republic in 1672 had much less to do with economic rivalry than it did with political resentments and even religious divergence. Louis XIV detested the Dutch as republicans and Calvinists, and resented their intervention in the War of Devolution, which he believed had limited his gains from Spain. Urged on by Louvois, the rising star of the French court, and by his leading commander, Turenne, the king had begun to prepare militarily and diplomatically for an attack almost as soon as the Peace of Aix-la-Chapelle (1668) had been signed with Madrid. Colbert viewed the Republic as a deadly economic rival whose wealth, and therefore power, must be reduced, and he had introduced anti-Dutch tariffs in 1664, 1667, and 1670–1. Yet the minister argued against the projected invasion because he feared – correctly, as it turned out – that war would undermine the notable, but precarious, improvement he had achieved in French finances during the first decade of Louis XIV's personal rule.

One significant source of Dutch economic and commercial power was the Republic's global colonial empire and the trade and resources it generated. France, like England, envied these colonies and aspired to acquire overseas possessions of its own. The mercantilist orthodoxy was that a country's prosperity depended upon securing such territories. Lands beyond Europe were an important source of raw materials and, in particular, strategic items such as timber and other naval stores, essential if a powerful fighting navy were to be built. Colonial trade, in theory at least, could be reserved to the mother country. This was the purpose of the English Navigation Acts of 1651 and 1660–3, which also sought to prevent costly imports into England itself. Spain's attempts to uphold a monopoly of imperial trade had identical aims. During the eighteenth century, as the populations of Europe's colonies increased in number and gained in economic maturity, such territories also became captive markets for manufactured goods produced in the mother country.

Europe's three leading colonial powers during the second half of the seventeenth century were Portugal, Spain, and the Dutch Republic. The English and the French certainly had the appetite to acquire overseas possessions, but territorial gains were difficult to achieve within the parameters of existing geographical knowledge and the prevailing political and economic balance of power. That was one reason why the impending extinction of the Habsburg dynasty in Madrid had been so important from the 1660s onwards, and why it proved impossible to settle the Spanish Succession peacefully. Spain was Europe's foremost colonial power, with a vast land empire in Central and South America and on islands scattered across the Caribbean, together with a foothold in the Far East, centred on Manila in the Philippines. The American territories generated significant wealth for the Spanish crown, in the form both of the *quinto* (the celebrated 'royal fifth') on the silver mined in the mountains of Bolivia and Mexico, and of a variety of taxes and duties levied throughout the overseas possessions. Although the sums generated for the crown had declined from their earlier peak, they remained an important source of royal revenue.

Madrid's extensive possessions in Europe were actually more central to the frenzied and, eventually, unsuccessful diplomacy which tried to arrange a negotiated settlement of the Spanish Succession at the end of the 1690s. This was an important reminder that contemporaries still underestimated the significance of overseas territories for international rivalries. Yet an appreciation of the value of such possessions was growing: as early as 1709, in the midst of the Spanish Succession conflict, Louis XIV had declared unequivocally that 'the principal object of the present war is the trade of the Indies and the wealth it produces'.[32] By the beginning of the eighteenth century, the potential value of colonies was coming to be recognised in many of Europe's capitals. Simultaneously, both the French monarchy and the new British state

32 Quoted by Ragnhild Hatton, *Europe in the Age of Louis XIV* (London, 1969), p. 215 n. 8.

were able to increase their possessions overseas: the former in Canada and Louisiana, the latter along North America's eastern seaboard in the Caribbean.

Britain's dynamic empire progressively encroached on Spain's vulnerable colonial structures, the latter being both a geographical neighbour and a deadly rival. During the eighteenth century metropolitan control over events at the periphery was at best limited, and periodically this proved to be an important source of tension. Friction frequently arose due to the unauthorised actions of subordinates, both officials and military commanders, in the colonial sphere, and governments in Europe were forced to react to situations which they had not created and from which they often wished to remain detached. One notable example was the complex origins of the Anglo-French Seven Years' War in North America, which arose principally from rivalries and clashes between settlers in the Ohio Valley during the 1750s. Another was the way in which, twice during a single generation, Spain and Britain were taken to the brink of open hostilities by the provocative actions of subordinates: in 1770–1 over the Falkland Islands, which lay off the extreme south-east of South America, and in 1790 over an incident at Nootka Sound, on the distant north-west coast of North America.

Behind such accidental clashes, however, stood Madrid's efforts to uphold its colonial trade monopoly and to defend the Pacific Ocean as a Spanish *mare clausum*. One important dimension of three successive phases of Anglo-Bourbon conflict during the eighteenth century – 1739–48; 1756–63; 1778/9–83 – was mounting British pressure upon the Spanish imperial system, leading to territorial losses, particularly in the Peace of Paris signed in 1763. Britain's expanding empire pressed equally hard upon its French rival during what has been termed the 'Second Hundred Years' War' (1689–1815), but was more accurately a 'Seventy Years' War' extending from the 1740s to the end of the Napoleonic Wars in 1815. The decisive phase was the Seven Years' War, when France was overwhelmingly defeated and suffered huge territorial losses. It was excluded from the North American mainland, losing Canada and handing over French Louisiana to Spain as compensation for Madrid's losses, in an attempt to strengthen the Third Family Compact of 1761 after the severe Spanish defeat in the war with Britain. The French position in the Indian subcontinent was all but destroyed, establishing British pre-eminence there, while Louis XV's monarchy also lost territory in West Africa.

Britain's extensive gains overseas during the third quarter of the eighteenth century, and the considerable land empire now ruled from London, promoted a significant adjustment of official attitudes, which became more consciously imperialist. During the 1760s ministers began to talk, for the first time, of 'the British Empire', an attitude which contributed to the clashes with the thirteen North American mainland colonies and provoked the War of American Independence. An identical, if slightly later, change was evident in Madrid. By the 1780s the territories which comprised Spain's American empire were coming to be known for the first time by officials as 'colonies' (*colonias*) and not as hitherto, 'kingdoms' (*reinos*). These semantic shifts reflected the changing way in which metropolitan governments viewed their overseas territories, as well as their increasing importance for European rivalries.

Within Europe an analogous process was under way, with the acquisition of new territories and subjects acquiring an enhanced, and more precisely articulated, importance. Cameralism, a distinctive variant of mercantilism, developed in Central Europe and particularly its German-speaking regions. It conferred an even greater role upon territorial governments and imposed upon them more extensive social responsibilities. Cameralism as an economic theory had first been elaborated during the second half of the seventeenth century by writers such as Hörnigk and Johann Joachim Becher. It incorporated two dominant assumptions: that a wealthy, and therefore powerful, state required prosperous subjects, and that governments could and should seek to promote prosperity, both by the kind of protectionist policies characteristic of mercantilism and by the internal control and regulation of trade and manufacturing.

These were the origins of an important change in the way international relations were conceptualised during the middle decades of the eighteenth century as a new notion of power took shape, particularly in Central and Eastern Europe. Policy decisions had always taken account of available resources, but hitherto this had been done in fairly general and approximate terms: geographical extent, population, military and naval strength, potential and actual wealth. Such calculations now became more precise and came to depend upon detailed estimates of economic, demographic, and even geographical resources. The collection of reliable quantitative information was characteristic of the long eighteenth century and an aim of government; the word 'statistics' – originally meaning 'the study of the state' – was first used at this time. The information assembled by administrative bureaus across Europe facilitated calculations of relative power and these slowly replaced the established juridical framework of international relations.[33] Two

33 Harm Klueting, *Die Lehre von der Macht der Staaten. Das außenpolitische Machtproblem in der 'politischen Wissenschaft' und in der praktischen Politik im 18. Jahrhundert* (Berlin, 1986), which should be read in the light of the notable review by Grete Klingenstein (*English*

influential cameralist writers elaborated this approach during the 1750s and early 1760s: Johann Heinrich Gottlob von Justi, especially in his *Staatswirtschaft* (1755) and *Die Chimäre des Gleichgewichts von Europa* (1758), and Jakob Friedrich von Bielfeld in his *Institutions politiques* (1760). Both emphasised the centrality of economics and highlighted the significance of internal organisation to calculations of power, and their ideas had a considerable and surprisingly rapid impact upon practical statecraft. This was first evident during the detailed negotiations which determined the initial round of annexations in Poland between Austria, Prussia, and Russia from 1772 to 1773, when an attempt was made to equalise the gains each secured.[34]

In the eastern half of the continent, the acquisition of new lands and peoples had always been particularly valued by rulers as the most promising way to enhance their own power. Lacking the dense populations, developed trade networks, and mature fiscal systems to be found further west, the rulers of Sweden, Brandenburg-Prussia, and Russia – which all emerged as leading states during the seventeenth and eighteenth centuries – depended less upon tax revenues and more upon levies of men, animals, grain, and matériel to support armies and facilitate expansion. This was particularly evident in the distinctive systems of conscription which each developed during this period: the *indelningsverk* (literally 'allotment system') in Sweden after 1680, the Russian system of territorial recruitment which was finalised under Peter I, and the cantonal system in Brandenburg-Prussia which had matured by 1733. All three successfully reconciled relative economic backwardness – at least by comparison with the leading countries of North-Western Europe – with the provision of a powerful army. The success of these initiatives, and especially Prussia's cantonal system, was evident in Austria's belated and incomplete adoption of territorial recruitment during the 1770s and 1780s.

The search for new provinces and new subjects was central to the eighteenth-century international rivalries of Northern, Central, and Eastern Europe. The acquisition of more precisely conceptualised human and economic resources overlaid, but did not replace, more traditional motives for conflict and expansion. This was facilitated by the relative decline of the two previously dominant polities, Poland-Lithuania and the Ottoman Empire. Together with the collapse of Sweden's Baltic Empire in the earlier eighteenth century, this created potential scope for considerable annexations. This

*Historical Review* 103 [1988], 134–8), which rightly insists upon the crucial importance of books IX and X of Montesquieu's *Spirit of the Laws* (1748).

34 Scott, *Emergence of the Eastern Powers*, ch. 7.

was in sharp contrast to the situation further west and south where political geography – with the exception of the Italian peninsula – was much more stable and new provinces correspondingly more difficult to acquire. In Southern and Western Europe, after the peace settlements of 1713/14, major territorial exchanges were few: France purchased the island of Corsica from the declining Italian republic of Genoa (1768); two years earlier it had secured Lorraine upon the death of Stanisław Leszczyński, by now Louis XV's father-in-law. Leszczyński had earlier been given the duchy when Francis Stephen was driven out and forced to accept Tuscany.

By contrast all the rising states of Central and Eastern Europe made substantial gains. Austria secured extensive lands from the Ottoman Empire in the wars of 1683–99, 1716–18, and 1788–91, and also annexed the Bukovina (1775). Prussia seized the wealthy province of Silesia from the Habsburgs in the 1740s and defended it successfully in the Seven Years' War. Russia, however, gained the most, annexing considerable territory around the Gulf of Finland when the Swedish Baltic Empire was broken up in 1721, and further Finnish lands after a bilateral war with Sweden in 1741–3. To the south, after a false start under Peter I, Russia secured huge gains at the expense of the Ottomans during the wars of 1768–74 and 1787–92, together with the important peacetime annexation of the Crimea (1783–4). The three eastern powers together partitioned Poland-Lithuania out of existence (1772–95), thereby securing for each state the largest territorial gains of all without major hostilities, besides suppressing Polish insurgency.[35] (See Map 10.)

The brutal cynicism of the three Polish partitions attracted the censure of the *philosophes*, who viewed it as typical of *ancien régime* rivalries. The Enlightenment's economic ideas made a more positive and enduring contribution, however, to the way international rivalry was conceptualised. The assumption that the volume of economic resources was fixed and unchanging, and that one monarchical state could grow in wealth, and therefore power, only at the expense of a rival, had underlain warfare and diplomacy for more than a century after the Peace of Westphalia. International rivalry, in other words, was a zero-sum game. This attitude, which was fundamental both to a mercantilist and to a cameralist outlook, had made international rivalry appear near inevitable. The economic ideas of the physiocrats offered the first important and systematic challenge to the established outlook. By asserting that increased wealth would result from an

35 Austria of course did not participate in the second partition of 1793.

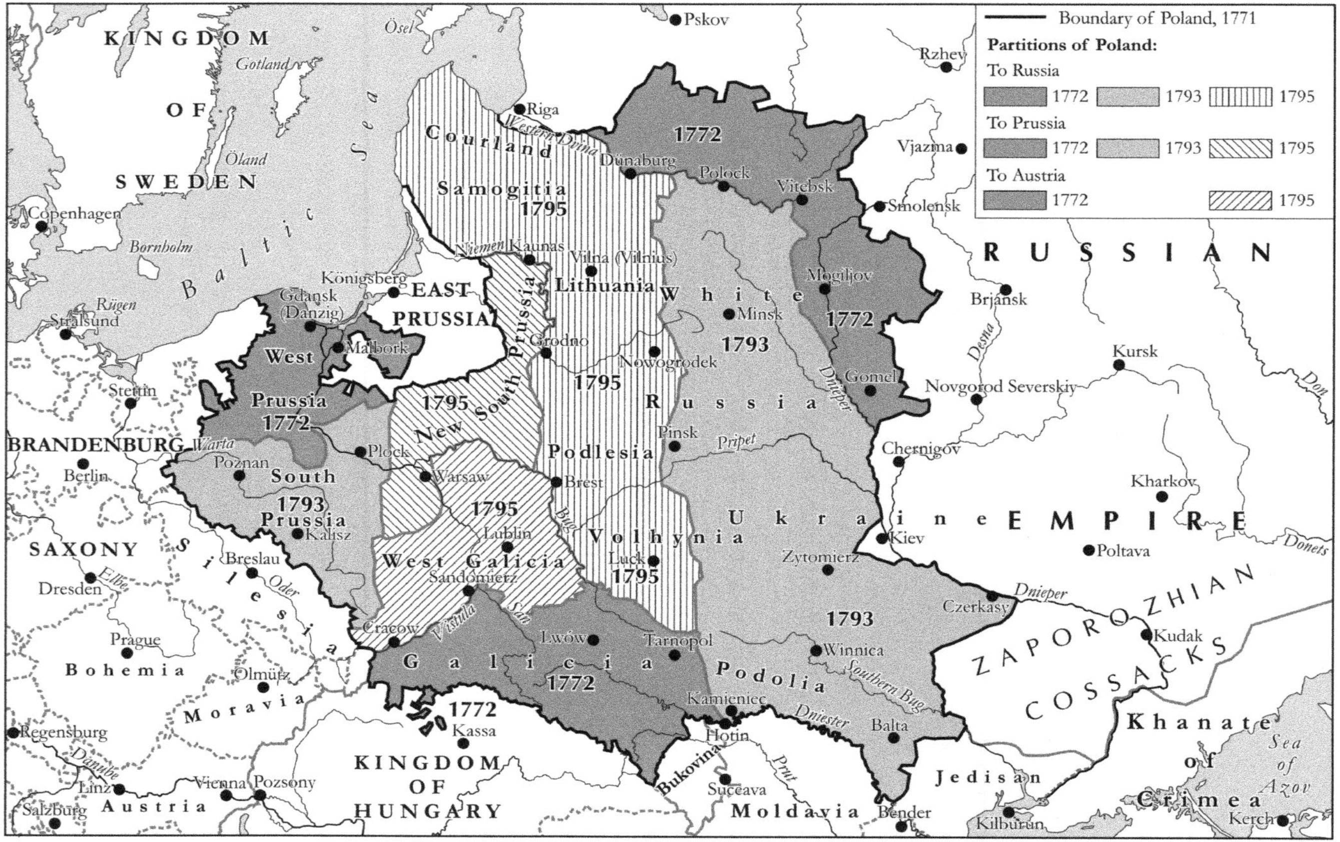


MAP 10. The partitions of Poland, 1772–5. Created by Béla Nagy.

expansion of trade, and that the volume of economic resources could be increased in this way, it powerfully challenged established assumptions.

This was one foundation of an important – if not at first widely influential – critique of the perceived inevitability of international rivalry. The so-called 'new diplomacy' of the eighteenth century, rooted in the economic theories of the *philosophes*, was the first systematic challenge to established ways of conceptualising such conflict.[36] Yet such challenges should not be exaggerated. Religious and dynastic factors and, above all, a drive for honour and reputation now often operating at a national level, ensured that many sources of international conflict remained deeply traditional long into the nineteenth century and even until 1914.

36 See Marc Bélissa, *Fraternité universelle et intérêt national (1713–95). Les cosmopolitiques du droit de gens* (Paris, 1998), esp. pp. 179–97; and Felix Gilbert, 'The "New Diplomacy" of the Eighteenth Century', reprinted in Gilbert's collected essays: *History: Choice and Commitment* (Cambridge, MA, 1977), pp. 323–50, esp. 345.

13

# The standing army: France and Brandenburg-Prussia – a comparison

BERNHARD R. KROENER

## The state of research

Generations of historians have seen the interplay between the early modern state and its armed forces, and between warfare and state formation, as key factors in the process of modernisation.[1] The creation of the modern state was most powerfully expressed through the supposed symbiosis between absolute regimes and standing armies. The image of geometric order and discipline generated by formations of infantry drawn up in kilometre-long battle lines; the authorities' direct involvement in provisioning, equipping, and uniforming its soldiers; central government's reach into every aspect of warfare and military planning. All of these have been regarded as defining traits of the interconnection between the standing army and the state.[2] Research on the inner structures of early modern military society has, until recently, been coloured by preconceptions about functioning hierarchies and chains of command, an increasingly effective military administration, rigid discipline, and corresponding efficiency in the waging of warfare.[3]

1 Otto Hintze, 'Staatsverfassung und Heeresverfassung' (1906), in *Staat und Verfassung. Gesammelte Abhandlungen zur allgemeinen Verfassungsgeschichte*, ed. Gerhard Oestreich (Göttingen, 1970), pp. 52–83; Samuel Finer, 'State- and Nation-Building in Europe: The Role of the Military', in Charles Tilly (ed.), *The Formation of National States in Western Europe* (Princeton, 1975), pp. 84–163; J. Kunisch (ed.), *Staatsverfassung und Heeresverfassung in der europäischen Geschichte der frühen Neuzeit* (Berlin, 1986).

2 Harald Kleinschmidt, *Tyrocinium Militare. Militärische Körperhaltungen und-bewegungen im Wandel zwischen dem 14. und dem 18. Jahrhundert* (Stuttgart, 1989); Henning Eichberg, 'Geometrie als barocke Verhaltensnorm. Fortifikation und Exerzitien', *Zeitschrift für Historische Forschung* 4 (1977), 17–50; William McNeill, *The Pursuit of Power: Technology, Armed Forces and Society since* A.D. *1000* (Chicago, 1982).

3 Louis André, *Michel Le Tellier et l'organisation de l'armée monarchique* (Paris, 1906); Curt Jany, *Geschichte der Preußischen Armee*, 4 vols. (Berlin, 1928–33); Gerhard Papke, *Handbuch zur deutschen Militärgeschichte*, vol. I: *Von der Miliz zum stehenden Heer. Wehrwesen im Absolutismus* (Munich, 1979).

Such a top-down view remained unchallenged as long as researchers relied almost exclusively on sources derived from governmental and/or legal provenance, leaving an impression of overwhelming state authority reaching right down to the level of the common soldier.

This perspective has recently started to evolve as military historians – strongly influenced by trends in the social history of everyday life, and increasingly by cultural history – have shifted their focus towards symbols, rituals, instruments, and agents of warfare in history. Following from this, new studies have turned their attention to the larger economic, social, and infrastructural preconditions underlying warfare.[4] Yet despite this shift of perspective, many of the assumptions still reflect a late nineteenth-century tendency to view modernisation primarily through the framework of the intersection between state power and military growth.

Examining the military structures of France and Brandenburg-Prussia, both viewed as 'prototypical absolutisms' of the eighteenth century, this chapter will attempt to trace the different stages of reform and stagnation in their military systems, and their development in relation to the respective legal, governmental, and social contexts of the two states. The authorities' monopoly on violence may have put an end to the political independence of armed forces, but it did not substantially impact the many 'privatised' elements of early modern warfare.[5] The traditional tendency to focus on the military aims and purposes of the central state must be set against consideration of the individual and collective aims, interests, and intentions of the army itself, as well as those of local and regional authorities. This perspective is necessary in order to understand the institutions of regimental proprietorship and company management, the recruitment of soldiers and officers, quartering, and the decentralised acquisition and maintenance of military equipment. Private entrepreneurs found employment in all aspects of military logistics required for campaigning, including transport, supply,

4 See the series of articles edited by Daniel Hohrath and Klaus Gerteis in *Aufklärung* 11:2 (1999), 3–97 and 12:1 (2000), 3–79, concerned with eighteenth-century warfare and the Enlightenment.

5 Stig Förster, Christian Jansen, and Günther Kronenbitter (eds.), *Rückkehr der Condottieri? Krieg und Militär zwischen staatlichem Monopol und Privatisierung. Von der Antike bis zur Gegenwart* (Paderborn, 2010); David Parrott, *The Business of War: Military Enterprise and Military Revolution in Early Modern Europe* (Cambridge, 2012); Matthias Meinhardt and Markus Meumann (eds.), *Die Kapitalisierung des Krieges. Kriegsunternehmer im Spätmittelalter und in der Frühen Neuzeit / The Capitalisation of War: Military Entrepreneurs in the Late Middle Ages and the Early Modern Period* (Münster and Berlin, 2021).

and hospital services. On a smaller scale, individual sutlers sold provisions directly to the troops, and provided care for the wounded. In twenty-first-century terminology, eighteenth-century networks of military logistics resembled a 'private–public partnership', the key 'private' element to this partnership being the regimental and company proprietors. In the face of the constraints imposed by such enterprise, the degree to which eighteenth-century militaries were centralised under governmental authority was limited.

## The French army

In the French armed forces, the shift from military units as independently managed economic entities to a 'semi-entrepreneurial system' occurred during the Thirty Years' War.[6] The situation changed gradually as the generation of officers whose experiences were formed by that conflict died out or left the service. For another thirty years, the French marshals Condé and Turenne, like their counterparts – the imperial commander-in-chief and military writer Montecuccoli; Brandenburg's Derfflinger and Sparr; and the legendary Swedish commander Carl Gustaf Wrangel – shaped the image of military forces.[7] Yet practices of recruitment and management changed only gradually. Monarchs functioned as military employers, though the sovereign as the *roi de guerre* (warlord) also embodied the nobility's *vertus guerrières* and shaped the identity and behaviour of his officers, both through his presence on the battlefield and by means of his public representation.[8] Rather than being seen merely as a contracting party in mutually beneficial financial negotiations, the warlord monarch was looked up to by the nobility as the lasting embodiment of their social aspirations. Yet the sovereign also stood at the centre of a patronage network, granting material favours, privileges, and heightened social status.[9] During the second half of the seventeenth century, the interests

6 John A. Lynn, *Giant of the Grand Siècle: The French Army 1610–1715* (Cambridge, 1997), p. 222, who for this concept draws upon David Parrott, *Richelieu's Army: War, Government and Society in France, 1624–1642* (Cambridge, 2001).

7 John A. Lynn, *The Wars of Louis XIV 1667–1714* (London, 1997), p. 33; relevant material on Condé in Henri d'Orléans, duc d'Aumale, *Histoire des princes de Condé pendant les XVIe et XVIIe siècles*, 8 vols. (Paris, 1885–96); Jean Bérenger, *Turenne* (Paris 1987); Gerd-Ulrich Herrmann, *Georg Freiherr von Derfflinger* (Berlin 1997). Frank Göse, *Otto Christoph Freiherr von Sparr 1605–1668* (Berlin 2006).

8 Joel Cornette, *Le Roi de guerre. Essai sur la souveraineté dans la France du Grand Siècle* (Paris, 1993); Guy Rowlands, *The Dynastic State and the Army under Louis XIV* (Cambridge, 2002).

9 William Beik, 'The Absolutism of Louis XIV as Social Collaboration', *Past and Present* 188 (2005), 195–224; Roger Mettam, 'The French Nobility, 1610–1715', in Hamish Scott

of the martial nobility and the monarch had started to align, consolidating the former's social and political position within the monarchical state, and providing scope to defend the status of the nobility against competing elites up to the end of the *ancien régime.*[10]

## *The generals*

Challenging the French commanders' traditional rights to appoint officers from out of their circle of confidants and clients, Cardinal Richelieu abolished the high military offices of *connétable de France* (1626) and *amiral de France* (1627).[11] The reason given by Louis XIV in 1661 for abolishing the *colonel général de l'infanterie* – to prevent him 'establishing his clients in every infantry unit' – indicated a consistent policy of suppressing, or reducing the status of, the highest military offices.[12] Another aspect of this policy was that generals were no longer permitted to hold their rank in addition to the proprietorship of a regiment. The generals' status in office primarily reflected their court connections, and the monarch's status as the dominant source of military patronage. So long as they were members of the court aristocracy, the *noblesse presentée*, generals had considerable social capital – far more than the regimental proprietors – capital which they used to support and sponsor their *clientèles*. Their primary goal was not exerting influence over the army's regimental structures, but rather securing their existing patronage networks outside of the military.[13]

(ed.), *The European Nobilities in the Seventeenth and Eighteenth Centuries*, 2 vols. (London, 1995).

10 Ronald Asch, 'Ständische Stellung und Selbstverständnis des Adels im 17. und 18. Jahrhundert', in *Der europäische Adel im Ancien Régime. Von der Krise der ständischen Monarchie bis zur Revolution (1600–1789)* (Cologne, 2001); Steven Gunn, David Grummitt, and Hans Cools, 'War and the State in Early Modern Europe: Widening the Debate', *War in History* 15 (2008), 371–88; Guy Rowlands, 'The Ethos of Blood and Changing Values? Robe, Epée and the French Armies 1661–1715', *Seventeenth-Century French Studies* 19 (1997), 95–108.

11 Parrott, *Richelieu's Army*, pp. 469–70; Rowlands, *The Dynastic State*, pp. 188–9; Bernhard Kroener, 'Législateur de ses armées. Verstaatlichungs- und Feudalisierungstendenzen in der militärischen Gesellschaft der Frühen Neuzeit am Beispiel der französischen Armee im Zeitalter Ludwigs XIV.', in Ronald Asch and Hans Duchhardt (eds.), *Der Absolutismus – ein Mythos? Strukturwandel monarchischer Herrschaft in West- und Mitteleuropa (ca. 1550–1700)* (Cologne, 1996), pp. 311–28.

12 Jean Chagniot, 'La Rationalisation de l'armée française après 1660', in Association des Historiens Modernistes des universités (ed.), *Armées et diplomatie dans l'Europe du XVIIe siècle. Actes du colloque de 1991*, Bulletin No. 16 (Paris, 1992), pp. 97–108, 100.

13 André Corvisier, 'Les Généraux de Louis XIV et leur origine sociale', in *Les Hommes, la guerre et la mort* (Paris 1985), pp. 145–75, Lynn, *Giant of the Grand Siècle*, pp. 305–10; Rowlands, *The Dynastic State*, pp. 282–5.

### *The colonel-proprietors*

The goals and motivations of a colonel were rather different. Traditional interpretations of the reorganisation of the French armed forces after 1660, which opened up colonels' commissions for purchase, argue that this was an attempt to gradually, but deliberately, undermine the operations of informal chains of command which drew on established social networks. More recently, though, it has been shown that such networks and connections among groups sharing provincial and local identities, linked by distinct traditions and utilising established patron–client networks, played a vital role in creating regimental *esprit de corps*. Drawing on the patronage networks of the high nobility to recruit units established links that connected the captains and subaltern officers and even extended down to the ordinary soldiers, providing units with coherence and acting as a stabilising factor, even under extreme battlefield conditions. Ties of regional, local, or even family affiliation, and common ground in dialect or religious denomination, were not simply factors conducive to integration amongst foreign regiments in royal service. They were no less powerful as forces for army cohesion amongst French units, factors that were more immediate and stronger than loyalty to the crown or royal authority.[14]

The purchase of a commission and the rights of ownership associated with it provided the proprietor with a limited degree of financial security. The returns gained from regular pay and the prospective resale value of the unit had to be set against the considerable expenses required for the colonel-proprietor to cut an appropriately aristocratic figure in public life, and the obligations of distributing patronage to his client-officers.[15] Yet, as with the generals, the social status associated with a colonel's commission was highly esteemed, and in the eyes of the holder could amortise the financial risks of the position. John Lynn's assessment that 'from the point of view of a captain or colonel, the French semi-entrepreneurial system combined the worst of two worlds' appears exaggerated; it overlooks contemporary ideals of the aristocracy as well as the colonels' aim of using the military in order to consolidate

14 Kroener, 'Législateur de ses armées', pp. 318–23; Rowlands, *The Dynastic State*, pp. 166–71; Lynn, *Giant of the Grand Siècle*, p. 223; Jean Chagniot in Anne Blanchard, Gilbert Bodinier, Jean Chagniot, Philippe Masson, and Jean Meyer, *Histoire militaire de la France*, vol. 2: *De 1715 à 1871* (Paris, 1992), p. 41; Bernhard Kroener, 'Deutsche Offiziere im Dienste des "allerchristlichen Königs!" (1715–1792). Aspekte einer Sozialgeschichte der Elite deutscher Fremdenregimenter in Frankreich im 18. Jahrhundert', in Jean Mondot, Jean-Marie Valentin, and Jürgen Voss (eds.), *Deutsche in Frankreich; Franzosen in Deutschland 1715–1789* (Sigmaringen, 1992), pp. 53–72.

15 Rowlands, *The Dynastic State*, p. 201.

their existing patronage networks.[16] The colonel-proprietor acted as an intermediary between the sovereign's interests and those of his officers, whose aim was essentially to use their military service to establish or consolidate their own rank within the French nobility. To the crown, selling a commission was primarily an opportunity to skim off capital from the resources of wealthy nobles. It is doubtful, given the relatively small number of commissions actually offered for sale, that this practice constituted a systematic, concealed attempt to tax the nobility.[17] Still, the dramatic increase in the number of regiments, from 30 in 1660 to 250 by 1730, presented the crown with a very real opportunity to sponsor and manage individual members of the aristocracy in the military. The danger of disbandment was considerable: approximately half of the established troops were typically disbanded during the short periods of peace under the reign of Louis XIV, and these tended to be the most recently commissioned units. Nonetheless, the number of regiments permanently maintained during peacetime rose from 30 in 1660 to 100 in 1700.[18] The king's decision not to disband a regiment after a peace settlement raised the financial value of the commission. This could motivate relatively less wealthy regimental commanders to keep their units' equipment, as well as the number and quality of serving troops, in good condition. The *ancienneté des régiments*, a European-wide means of establishing the hierarchy of regiments according to their date of establishment, thus became indicative of their respective military quality as well as their ranking.

The position of colonel, or *mestre de camp*, was first and foremost an honorific rather than a functional rank, and was closely linked to ideals of social status. In the regiments of the *maison du roi* (the king's hugely extended bodyguard) a first-generation noble would have been considered an upstart. Yet, in the more recently established regiments, it was not unusual for a colonel's commission to be sold to wealthy members of the bourgeoisie. The diversity of commission prices, reflecting the regiment's rank and whether it was a cavalry or infantry unit, created internal differentiations among the military elites corresponding to birth and wealth. These distinctions tended to favour 'nobles of the robe', those whose noble status had come via the purchase of senior legal, administrative, or financial office, usually reflecting a successful family background in mercantile or financial

16 Lynn, *Giant of the Grand Siècle*, pp. 222–3. 17 Ibid., p. 223.

18 André Corvisier, 'Louis XIV, la guerre et la naissance de l'armée moderne', in *Histoire Militaire de la France*, vol. 1, pp. 383–414, 388; Kroener, 'Législateur de ses armées', p. 317.

dealings. The access to money and influence enjoyed by the 'robe' put members of the less prosperous provincial nobility at a serious disadvantage in the quest for military office. Fuelled by the enlightened military literature of the eighteenth century, this situation generated two diametrically opposed views.[19] Traditionalists considered military office to be first and foremost a position of honour, one closely linked to traditional noble qualities. Opposing this, reformers argued for the unconditional recognition of professional military achievement, regardless of the social rank of the holder.

Yet this debate came too late: regarding military service as an expression of elite culture had facilitated, even under Louis XIV, a gradual decline in the military effectiveness of the French army that became progressively more marked after 1715. Long periods of peace, and the accession of rulers less interested than the Sun King in public displays of the *vertus guerrières*, exacerbated the social and financial exclusivity apparent in acquiring commissions. It effectively prevented any professionalisation of the French army, while the tensions it generated provided the basis for several attempts at reform after 1763.[20] This was compounded by an inflationary increase in senior military offices, particularly at the level of colonels and generals, fuelled rather by external criteria than as a reward for military merit. The increasing social exclusivity of senior military office drove the price of military commissions, as well as the expenses of service, to unprecedented heights, adding to the regime's financial woes.[21]

### *The captains as proprietors*

A captain's commission, like that of a colonel, included managing an economic as well as a tactical entity and was purchased, even if the colonel-proprietor retained the right to nominate candidates to the captaincies of his regiment. Both the company and the regiment were business enterprises, and proprietors hoped that they would be able to recover the investment made at

19 Michael Sikora, '"Ueber die Veredelung des Soldaten". Positionsbestimmung zwischen Militär und Aufklärung', in Hohrath and Gerteis (eds.), *Die Kriegskunst*, vol. 1, pp. 25–50; Claudius Fischbach, *Krieg und Frieden in der französischen Aufklärung* (Münster, 1990).

20 Jean Chagniot, 'Mobilité sociale et l'armée (vers 1660–vers 1760)', *XVIIe siècle* 122 (1979), 37–49, 39; Rafe Blaufarb, *The French Army 1750–1820: Careers, Talent, Merit* (Manchester, 2002); Claudia Opitz-Belakhal, *Militärreformen zwischen Bürokratisierung und Adelsreaktion. Das französische Kriegsministerium und seine Reformen im Offizierkorps von 1760–1790* (Sigmaringen, 1994).

21 Bernhard Kroener, 'Herrschaftsverdichtung als Kriegsursache. Wirtschaft und Rüstung der europäischen Großmächte im Siebenjährigen Krieg', in Bernd Wegner (ed.), *Wie Kriege entstehen. Zum historischen Hintergrund von Staatenkonflikten* (Paderborn, 2000), pp. 145–73, 168; Lee Kennett, *The French Armies and the Seven Years War: A Study in Military Organisation and Administration* (Durham, NC, 1968), p. 97.

purchase, or at least keep the subsequent financial burden to a minimum. Commanders of regiments and companies were each responsible for the recruitment and outfitting of their soldiers; soldiers' weapons were provided by the monarch in times of war, but their maintenance was the responsibility of the company captains.[22] Operational costs incurred by the officers on campaign were settled subsequently by the crown, usually according to rates that failed to cover the actual expenses. As a general rule, the proprietor would expect to bear around 90 per cent of the initial expenses in establishing his company, while their share of subsequent maintenance costs was approximately 10 per cent.[23] Yet the consequent elevation, or preservation, of wider social status acquired through the holding of military command provided ample compensation for this private investment in the eyes of the proprietors.

A proprietor's authority over his unit partly expressed itself in appointment rights over all those subordinate positions that were not for sale.[24] In contemporary thought, both regiments and companies formed a *république* distinguished by regulations that were to be enforced both by the acting senior officer and by the colonel-proprietor. Ownership could express itself in other, less legal ways. To the modern observer French military uniforms, an innovation that Louis XIV systematically introduced, give an impression of regularity and subordination to the authority of the crown. Yet a closer examination reveals any number of distinctly individual elements in these uniforms, and similarly distinctive features in regimental equipment, flags, and trophies. The proprietors spent money and effort imposing distinctive family heraldry, particular colours, and monogrammed details and insignia on the uniforms and equipment of their regiments.

Relations of patronage – including mutual support among the officers and financial favours, passed down from the highest positions – evolved into an integral part of social career management. Thus, it appeared reasonable for subordinate officers to leave military service after their leader had been killed or resigned.[25] After the Seven Years' War, the reforms of the war minister, the duc de Choiseul, attempted to curtail such patron–client structures within regiments by appointing lieutenant-colonels and majors without consulting

22 Lynn, *Giant of the Grand Siècle*, pp. 180–3.

23 Hervé Drévillon, *L'Impôt du sang. Le métier des armes sous Louis XIV* (Paris 2005), p. 133.

24 Kroener, 'Législateur de ses armées', p. 321; Jean Chagniot, 'The Ethics and Practice of War amongst French Officers during the Seventeenth Century', *War and Society* 10 (1992), 19–36, 28.

25 Sharon Kettering, *Patrons, Brokers and Clients in Seventeenth Century France* (Oxford, 1986), p. 139; Chagniot, 'The Ethics and Practice of War', p. 23.

colonels or captains. This led to protests in several regiments, some of which escalated into outright mutiny.[26]

## *Military administration and recruitment*

In return for their financial commitment, the regimental proprietors controlled the recruitment, outfitting, and supply of their units, and were able to delegate aspects of these functions to their captains in order to strengthen existing patronage ties. In peacetime, recruitment, acquisition of clothing and equipment, supply and forage, and the disbursement of funds for the troops' pay offered the greatest avenues for colonels and captains to make a profit. This permitted a limited degree of commercial agency and allowed the regimental proprietors to profit from interest on loans made to their subordinates to meet some of these costs. Yet the relative freedom granted to officer-proprietors to manage their units spawned an ever more elaborate network of administrative institutions designed to exert more effective control over proprietors' commercial activities. It was with some justice that Louis XIV bore the epithet *Louis administrateur*, just as Frederick William I, dubbed the 'Reformer King of Prussia', would do a generation later.[27]

Louis's reign linked membership of the *conseil d'état du roi*, the central administrative organ, not simply to birth, but also to administrative experience and qualification. This initially confronted the *noblesse d'épée* – the traditional military nobility of old-established families without such training or experience – with insurmountable obstacles. Families of the professional *noblesse de robe*, like the Colberts, the Le Telliers, and the Phélypeaux, dominated the *conseil du roi* and occupied key positions in the administration. But to secure these positions, and to affirm their own social status at the top of society, they developed relationships of patronage and family connecting them to members of the great *noblesse d'épée*. A clear example of this was the close association of François-Michel Le Tellier, marquis de Louvois and Louis XIV's war minister, with the marshals Lorge and Duras. This form of dual loyalty was designed to mutually benefit both *robe* and *épée*; it was not hostile to the crown, but rather combined aspirations for power and mutual support

26 Jean Chagniot, 'Les Faux-semblants du désarmement collectif dans la France des Lumières', in *Histoire Militaire de la France*, vol, 2, pp. 41–5.

27 Albert Cremer, 'Der Strukturwandel des Hofes in der frühen Neuzeit', in Rudolf Vierhaus (ed.), *Frühe Neuzeit – Frühe Moderne? Forschungen zur Vielschichtigkeit von Übergangsprozessen* (Göttingen, 1992), pp. 75–89, 88; Peter Baumgart, 'Friedrich Wilhelm I. (1713–1740)', in Frank-Lothar Kroll (ed.), *Preußens Herrscher. Von den ersten Hohenzollern bis Wilhelm II* (Munich, 2000), pp. 134–59, 134.

with the social ambitions of the *robe*. During the second half of Louis's reign, it also helped to alleviate those tensions between military leaders and administrative institutions which had been a notable feature of the 1660s and 1670s, seen for example in the conflicts between Louvois and confident military commanders like Turenne and Condé. The integration between robe and sword nobility continued into the eighteenth century, and the previously distinct spheres of military and administrative leadership were tied together through close-knit networks of corruption and favouritism.[28] By the end of the Seven Years' War, the regulatory system had been thoroughly undermined by all-pervasive social relationships between the officer corps and those military administrators supposedly holding them to account.[29]

Le Tellier's administrative reforms of the 1660s and the rigid control of his son and successor, Louvois, mostly restrained the military elites from fraudulent practices and abuses on French soil. However, cases of embezzlement, blackmail, pilfering, and bribery accompanying French campaigns abroad were frequent and readily apparent throughout the eighteenth century. In part, these reflected supply bottlenecks and financial delays, and not all colonels and captains were guilty of active involvement or complicity. But, for many commanding officers and for the military elites in general, operations on enemy territory offered the opportunity to make profits – even personal fortunes – well above any normal opportunities provided within the military system.[30]

The crown's inability to provide adequate or timely funds to support the immense costs of the army inevitably led to situations in which military operations could only be sustained if food- and munitions-contractors, military administrators such as the *commissaires des guerres*, and indeed the generals made considerable financial advances on their own part. But this state of affairs was widely seen as unsatisfactory. The ineffectiveness of military suppliers and wholesalers, and the immense profits made by the financial consortia behind them, led the French crown to repeatedly shift responsibility for army subsistence between contractors and direct

28 Kroener, 'Législateur de ses armées', p. 324.

29 André, *Michel Le Tellier et l'organisation*; Louis André, *Michel Le Tellier et Louvois* (Paris, 1942); Lynn, *Giant of the Grand Siècle*.

30 Myron Gutman, *War and Rural Life in the Early Modern Low Countries* (Assen, 1980); Jean-Paul Rorive, *Les Misères de la guerre sous le Roi Soleil* (Lüttich, 2000); Rolf Kießling, 'Armeen auf dem Durchmarsch – Kriegserfahrungen der Bevölkerung', in Johannes Erichsen and Katharina Heinemann (eds.), *Die Schlacht von Höchstädt / The Battle of Blenheim. Brennpunkt Europas 1704* (Ostfildern, 2004), pp. 68–79; Horst Carl, *Okkupation und Regionalismus. Die Preußischen Westprovinzen im siebenjährigen Krieg* (Mainz, 1993), pp. 201–21.

administrative control throughout the eighteenth century. In the latter case, it was no less regularly discovered that the crown's officials simply lacked access to the financial resources needed to secure and arrange regular supply contracts, and that personnel were not sufficiently qualified or well enough connected into supply networks to build effective support systems. The response was once again to fall back on the practice of outsourcing supply and support facilities to contracting companies. Those responsible were aware that while this might be the least bad option, it did nothing to lower the costs of army supply, nor did it ensure a constant level of performance. Suppliers sought to reduce the financial risks involved in contracting through underhand practices, reneging on supply obligations, and making inadequate provision for anticipated needs. These practices were justified as compensation for payments made in advance or for inevitable additional expenses incurred in the course of the campaigns.

The problems were well entrenched by the outbreak of the Austrian Wars of Succession, although the military success of the *maréchal* de Saxe – an exceptional war-profiteer in his own right – masked some of the concerns about their damaging consequences. By the beginning of the Seven Years' War, however, corruption in the logistical and support services had started to impair the basic functioning of the military.[31] Pamphlets produced by Enlightened critics – the so-called *soubisiades* – targeted members of the high nobility and military elites, like the marshals Richelieu and Rohan-Soubise, for their involvement in these corrupt practices. Such criticism reached further into the management of the war and even the monarchy did not entirely escape it. Yet the majority of officers escaped criticism and continued to profit – albeit on a smaller scale – from the same machinations.[32]

Money spent on the purchase of a commission and the upkeep of a unit could be recovered, provided of course that the officer survived the war. In case of death, the crown automatically reclaimed the commission without offering compensation. At best, the officer's family could look to discretionary grants offered from the king's bounty. More than 300 French captains

31 Carl, *Okkupation und Regionalismus*, pp. 166–71; Kennett, *The French Armies*, p. 99; Charles-Nicolas Dublanchy, *Une intendance d'armée au XVIIIe siècle. Étude sur les services administratifs à l'armée de Soubise pendant la Guerre de Sept Ans* (Paris, 1908), pp. 35–6.

32 Kennett, *The French Armies*, pp. 99–119; Carl, *Okkupation und Regionalismus*, pp. 213–21; André Dussauge, *Études sur la Guerre de Sept Ans. Le Ministère de Belle Isle. Krefeld et Lütterberg 1758* (Paris 1914), p. 47; Bernhard Kroener, 'Militärischer Professionalismus und soziale Karriere. Der französische Adel in den europäischen Kriegen 1740–1763', in Bernhard Kroener (ed.), *Europa im Zeitalter Friedrichs des Großen. Wirtschaft, Gesellschaft, Kriege* (Munich, 1989), pp. 99–132.

were killed during the Seven Years' War, 1,000 more were wounded and dubbed *héros subalternes* by contemporary writers.[33]

A fundamental concern in seeking to balance the proprietor-officers' economic independence and initiative with the crown's administrative authority was establishing real or effective troop numbers. In this, the interests of senior military commanders and of the crown and its administration were aligned in seeking to gauge as exactly as possible the actual size of the corps at their disposal. At the same time, the administration wished to know the precise number of soldiers for whom the colonel-proprietors had been guaranteed pay. These joint concerns generated similar responses across early modern Europe: the formation of systems of commissioners whose positions, though initially temporary, became the core of governmental military administration. The task of a military commissioner consisted primarily of verifying the soldiers' physical presence and compiling lists and registers – *montres* – of troop strengths.[34] Regular inspections and reviews certainly contributed to containing excessive fraud among the colonels and captains. Together with the monarch's increasing control over the military this led to a marked improvement in accountability in the second half of the seventeenth century. Yet, as with other aspects of the administration, embedded interests, complicity between *robe* and *épée*, and tolerance of corruption all ensured that these successes did not endure.[35]

France's comparatively high population density made it possible to adhere to the principle of voluntary recruitment, at least during times of peace. As recruitment remained in the hands of the proprietors until the end of the eighteenth century, the usual turn of phrase in any written exchange before 1740 was *'soldat du sieur de Bellefonds au régiment de Navarre'*. The designation *soldat au régiment de Navarre* from the second half of the century onwards finally emphasised the position of the crown as lord and employer.[36] A new and centralised system of recruitment, introduced in France at the middle of the century, had the same

33 André Corvisier, 'Les "héros subalterns" dans la littérature du milieu du XVIIIe siècle et la réhabilitation du militaire', *Revue du Nord* 66 (1984), 827–38.

34 André Corvisier, *Les Contrôles de troupes de l'ancien regime*, vol. 1: *Une source d'histoire sociale* (Paris, 1968–70); André Corvisier, *L'Armée française de la fin du XVIIe siècle au ministère de Choiseul. Le soldat*, 2 vols. (Paris, 1964).

35 Bernhard Kroener, 'Wirtschaft und Rüstung der europäischen Großmächte im Siebenjährigen Krieg. Überlegungen zu einem vergleichenden Ansatz', in Militärgeschichtliches Forschungsamt (eds.), *Friedrich der Große und das Militärwesen seiner Zeit* (Herford, 1987), pp. 143–75.

36 André Corvisier, 'Clientèles et fidélités dans l'armée française aux XVIIe et XVIIIe siècles', in *Les Hommes*, pp. 191–214, 198–201.

rationalising goal.[37] King and crown, however, remained remote to large parts of the population. The voluntary decision to enrol was usually dependent on the rumours of better living conditions, or on personal contacts and established local identity within the unit. Central assembly points for the recruits of an entire region and anonymous deployment to regiments would prove a strong disincentive to voluntary enlistment.

In the case of the existing system for regional defence, the provincial militia, changes in recruitment procedures caused even greater problems. Recruits for the *régiments provinciaux* of the militia were randomly selected by drawing lots, and those chosen forcibly conscripted or deployed as substitutes for those who had bought themselves out of militia service. Moreover, these conscripts served as a potential reservoir for the regular armies during the wars of the eighteenth century. Even if this practice became less and less common, those concerned still saw it as anonymous and arbitrary, while enlightened critics of the contemporary military presented it as a symbol of despotism.[38]

However, those who subscribed to rationalist arguments for greater government and administrative efficiency, especially in the period of reform after 1763, saw conscription by lot and anonymous allocation to regiments both as a means to reduce inefficiency and corruption in recruitment and as a way to tackle wider administrative weaknesses extending down to company level. The failure of those efforts shows how vital personal networks remained as the means to reform traditional structures. These reforms could only be permanently implemented, though, if they offered a minimum of benefit and participation to those concerned.[39] That said, French military reforms after 1763 did have the effect of modernising and professionalising the technological, tactical, and operational aspects of the military, particularly the subaltern officers and the *armes savantes* (artillery and engineers). It was the latter who would benefit most from the more meritocratic promotional opportunities offered in the armies of the Revolution and Empire after 1790. Among generals and majors, the conflicts between the traditional court aristocracy and wealthy members of the *noblesse de robe* to establish their respective social position continued and

37 Chagniot, 'Les Faux-semblants', p. 36.

38 Georges Girard, *Racolage et milice* (Paris, 1922); Arlette Farge, *Les Fatigues de la guerre XVIIIe siècle – Watteau* (Paris 1996), pp. 19–37; Stéphane Perréon, *L'Armée en Bretagne au XVIIIe siècle. Institution militaire et société civile au temps de l'intendance et des États* (Rennes, 2005), pp. 270–1.

39 Matthew Anderson, *War and Society in Europe of the Old Regime* (Leicester, 1988), p. 48.

intensified.[40] Admission into the highest military ranks remained first and foremost a process based on social exclusion and social connections. This defeated all those initiatives and efforts since the reign of Louis XIV that had aimed to establish the counter-principle of military professionalism based on seniority and performance.

## The Brandenburg-Prussian army

From 1652, and in the aftermath of the Nuremberg 'Execution-Congress', the majority of troops that had been kept under arms since the end of the Thirty Years' War were demobilised, with the notable exception of French and Spanish troops still at war with each other. Of the 120,000–150,000 soldiers previously in the Empire perhaps as few as 20,000, and no more than 50,000 men, remained in the service of Sweden, the emperor, the States-General, or the armed Estates of the Holy Roman Empire. Whether or not this reduction led to a seamless transition from mercenary forces to state-run standing armies is open to question.[41] That said, the fate of the politically ambitious military entrepreneurs of the Thirty Years' War after 1648 clearly reveals the limits of their power in relation to an early modern state that was prepared to assert control over the armed forces and over the previous sources of income of their military proprietors. The interests of this latter group after 1648 were primarily their own financial needs, seeking to recover their military investment in more challenging political circumstances. They no longer sought to transform their proprietary contingents of troops into independent political and military power centres.[42]

Yet not all states moved to exclude their officer-proprietors from a distinct role in military management: Brandenburg's Great Elector retained recruitment under contract, practised during the previous thirty years, until the mid-1680s, thus reinforcing the position of the proprietors in directly recruiting their military personnel.

40 Opitz-Belakhal, *Militärreformen*, pp. 261ff.; Bernhard Kroener, '"Des Königs Rock". Das Offizierkorps in Frankreich, Österreich und Preußen im 18. Jahrhundert – Werkzeug sozialer Militarisierung oder Symbol gesellschaftlicher Integration?', in Peter Baumgart, Bernhard R. Kroener, and Heinz Stübig (eds.), *Die preußische Armee zwischen Ancien Régime und Reichsgründung* (Paderborn, 2008), pp. 72–95, 80.

41 Johannes Burkhardt, *Der Dreißigjährige Krieg* (Frankfurt am Main, 1992), pp. 213ff.

42 Bernhard Sicken, 'Der dreißigjährige Krieg als Wendepunkt: Kriegführung und Heeresstruktur im Übergang zum miles perpetuus', in Heinz Duchhardt (ed.), *Der Westfälische Friede. Diplomatie, politische Zäsur, kulturelles Umfeld, Rezeptionsgeschichte* (Munich, 1998), pp. 581–90.

### *The captains and colonel-proprietors*

Unlike in France, the generals in Brandenburg-Prussia were always proprietors of at least one regiment, and Field Marshal von Derfflinger even accumulated ownership of three units.[43] In his capacity as commander-in-chief, the sovereign entered into a contract – known as a capitulation – with each officer-proprietor, determining the regiment's deployment and the material support it would receive. Ideally, the sovereign also provided the necessary funds for recruiting and equipping the troops. The regiment as a whole was not the property of its commanding officer, but he was assigned the first company in the regiment as his personal unit, and he took responsibility for its recruitment, outfitting, and management. In the face of the territory's precarious financial situation, the Great Elector was repeatedly forced to ask his officers to advance funds for pay, recruitment, weaponry, and ammunition.[44] These ongoing financial commitments were substantial, and a number of officers hoped to gain favour through their commitment to the elector's service so as to retain their commissions when the state reduced troop numbers. Others who had invested not only their own resources in recruitment, but also those of relatives, remained hopeful that frequent campaigning would allow them the opportunity to collect an adequate return on their initial and ongoing financial commitments.[45]

Regular mustering and inspections of the troops conducted by the commanding officer and the war commissioners were, unlike the initial situation in France, characteristic of the Brandenburg army from the outset. Yet such corporate responsibility had its limits. Contemporaries complained about the exaggerated claims, random exactions, and manifold attempts at profiteering made by the proprietors.[46] Yet in fact the state support received by proprietors was substantial. If a unit had entered a campaign at full strength, refunds for any material losses could be claimed from the elector. The transfer of a company into the hands of another captain was a financial transaction, and the elector compensated those proprietors whose companies were decommissioned after a peace agreement. During the reign of Frederick I (r. 1688–1713), still more that of Frederick William I (r. 1713–40), cutbacks in troop numbers were less common and the expansion of the Brandenburg-Prussian army from 18,000 men in 1688 to 80,000 in 1740 left none of the proprietors in fear of

43 Curt Jany, *Die alte Armee von 1655 bis 1740 (Formation und Stärke)*, (Berlin, 1905), pp. 82–3.
44 Jany, *Geschichte der Preußischen Armee*, vol. 1, p. 157. 45 Ibid., p. 158.
46 Ibid., pp. 168–9.

losing their commissions without reason.[47] In this process of expansion and maintenance, capitulations lost more and more of their explicitly contractual nature and were reduced to a one-sided agreement that initially gave the colonel the right to assemble a military unit through his own initiative, and finally became an official act by which the colonel had an existing regiment 'consigned and conferred' upon him by the elector's authority.

Unlike the situation in France, the actual commissions of Brandenburg-Prussian colonels and captains were not put up for sale. Officers were appointed to serve in a regiment, even the individual company captains, by direct commission from the sovereign, and on the recommendation of the colonel-proprietor. The officers had to commend themselves in the first instance as worthy candidates for selection. It was this process that distinguished client–patron relations in the Brandenburg-Prussian army. A regiment's proprietor was responsible for the selection of all his officers, including the captains, but not their direct appointment. Parallel to developments in France, Brandenburg-Prussian authorities gradually implemented a standardised uniform across the whole army. But here again, variations in uniform, colours, and the addition of heraldic symbols and badges on sleeves and lapels were tolerated, signifying the distinctiveness of the regimental proprietor, and his claims to ownership of the unit.[48]

By the mid-1680s the last veterans of the Thirty Years' War had disappeared from the Brandenburg-Prussian army.[49] At the same time, the influx of French Protestant officers increased after the Potsdam Edict of Toleration in 1685, bringing with them the influence of the French military reforms. By the end of the decade approximately 300 of the 1,000 Brandenburg-Prussian officers were of French origin, but, with the increasing integration of Huguenot families and a trend towards self-recruitment among the military elites, their relative numbers dropped from the beginning of the eighteenth century and levelled out at approximately 8 per cent of officers.[50] Unlike their fellow officers from Brandenburg-Prussia, they could not rely on networks of patronage rooted in the aristocratic society of the region. Thus, their loyalty was first and foremost to the sovereign, on whose benevolence and support their further social advancement depended. This group established a specific

47 Christopher Duffy, *The Army of Frederick the Great*, 2nd ed. (London, 1996).

48 Jany, *Geschichte der Preußischen Armee*, vol. 1, pp. 341–6.

49 Ibid., p. 313; Peter-Michael Hahn, 'Aristokratisierung und Professionalisierung. Der Aufstieg der Obristen zu einer militärischen und höfischen Elite in Brandenburg-Preußen von 1650–1720', *Forschungen zur Brandenburgischen und Preußischen Geschichte N.F.* 1 (1991), 161–208, 182.

50 Göse, *Sparr*, pp. 313–14.

concept of service and duty in the Prussian army, which – under wartime conditions and Frederick II's character and personality – gradually developed into an *esprit de corps* relating to professional performance.[51]

At first glance, proprietors of regiments and companies held a position similar to the officers of equivalent rank in the French army at the time. In contrast to noble society in France with its fissured social topography, the legal and social status of the ranks of the Brandenburg-Prussian nobility may appear remarkably homogeneous. Yet when it came to their traditions, cultural influences, and religious denominations, the Brandenburg-Prussian nobility were highly heterogeneous. The landed gentry's economic position – especially in the monarchy's central provinces – was strained due to agricultural crises and the devastation of the Thirty Years' War, helping to ensure that a political alternative to the initiatives towards state centralisation in military and administration could not develop. In a process of gradual amalgamation that occurred with different intensity in the elector's individual provinces, military service established itself as an opportunity for the nobility to access social networks and aim for an enhancement of status and their economic position through patronage and strategic marriage alliances.[52] Yet, although a core of career-officers came into being, considerable numbers of subaltern officers served only for a limited period of time and – following the French example – retired quickly to take up other pursuits or to manage their estates.[53]

Control over the company and its resources enabled the captains to maintain a lifestyle appropriate to their social standing. With appointment to the company command, the captain became a part of the military network of patronage centred upon the colonel-proprietor. In turn, the crown sought ways of generating loyalty and obligations and strengthening its influence, particularly in the remote parts of the dynastic territories.[54] During the eighteenth century several wealthy and influential members of the old aristocracy

51 Hahn, 'Aristokratisierung und Professionalisierung', p. 182; Ernst Opgenoorth, *'Ausländer' in Brandenburg-Preußen als leitende Beamte und Offiziere, 1604–1871* (Wiesbaden, 1967); T. Schmidt and H. Schnitter, 'Die Hugenotten in der brandenburgisch-preußischen Armee bis zum Jahre 1807', *Militärgeschichte* 24 (1985), 233–9.

52 Hahn, 'Aristokratisierung und Professionalisierung', pp. 161–208; Edgar Melton, 'The Junkers of Brandenburg-Prussia, 1600–1806', in Scott (ed.), *The European Nobilities*, vol. 2, pp. 118–69; William Hagen, *Ordinary Prussians: Brandenburg Junkers and Villagers 1500–1840* (Cambridge, 2002); Frank Göse, *Rittergut – Garnison – Residenz. Studien zur Sozialstruktur und politischen Wirksamkeit des brandenburgischen Adels 1648–1763* (Berlin, 2005).

53 Göse, *Rittergut*, pp. 104–10.

54 Hahn, 'Aristokratisierung und Professionalisierung', p. 194.

joined the ranks of colonels and generals, utilising their uncontested position as intermediaries between the central state and agencies in the provinces. Yet these high-ranking elements in the officer corps could generate disputes over precedence and rank, and even incite military insubordination.[55]

After Frederick II's accession, due to his role as king-in-arms, and the structural changes that occurred among the officers during long years of war, the Prussian army gradually changed. While 95 per cent of officers were still members of the nobility and thus aristocratic values remained central, their hierarchical mindset gradually merged with a more specifically military *esprit de corps* that also recognised professional abilities as a crucial prerequisite for advancement.[56] After 1763, service in the Prussian army became increasingly appealing to officers from all over the Holy Roman Empire, and between 1762 and 1800 only a minority of officers came from Brandenburg-Prussia, Pomerania, and East Prussia. To these foreign officers, the king became the focus for their concept of duty, a mindset that was later diffused amongst all officers.[57] This placed the king at the centre of a military network of patronage, while a caste of military noblemen, whose status and wealth was largely independent of inherited landed property, established itself.

It is here that a number of contrasts with the French situation in the second half of the eighteenth century become apparent. The reforms to the French army after 1763 improved military equipment and established training schools to create a new generation of professional subaltern officers. But the promotion of such 'new men' continued to be hampered by conflicts for influence between the court nobility and the *noblesse de robe* and real progress towards promotion based on professional merit became possible only after the Revolution. Meanwhile, Brandenburg-Prussia believed that it had achieved a recipe for military success through the right blend of traditional and meritocratic input into the army. Yet in reality Frederick II's long reign resulted in stagnation, only allowing for minute modifications to over-rigid structures. The practice of promoting according to seniority alone significantly raised the average age of captains in the following years, and strengthened their reluctance to subject their material security to the changing fortunes of war in the hope of getting an early promotion.[58] Moreover, neither Frederick William II (r. 1786–97) nor Frederick William III (r. 1797–

55 Ibid., pp. 178–81. 56 Hagen, *Ordinary Prussians*, pp. 300–7.
57 Melton, 'The Junkers', p. 147.
58 Bernhard Kroener, 'Die Armeen Frankreichs und Preußens am Vorabend der Schlacht von Jena und Auerstedt', in Eckhardt Opitz (ed.), *Gerhard von Scharnhorst. Vom Wesen und Wirken der preußischen Heeresreform – Ein Tagungsband* (Bremen, 1998), pp. 12–30.

1840) succeeded in securing a central position in the networks of patronage the way Frederick II had achieved.

### *Recruitment*

Brandenburg-Prussia's system for territorial defence had failed during the Thirty Years' War, and, faced with economic crisis in the years following the war, the territorial Estates were unable to restore the traditional form of militia defence against the sovereign's aim to establish a *miles perpetuus* in Brandenburg.[59] Elector Frederick William favoured employing professional soldiers rather than recruiting a local militia; the European mercenary market offered extensive sources for supply of experienced troops after the end of the war. All capitulations made by the elector allowed for recruitment at the levels of both company and regiment, and thus provided proprietors with an opportunity to make considerable profit.[60] In addition to recruiting experienced troops from foreign territories, the officers' local bases provided excellent opportunities to draw on their personal connections to facilitate recruitment.

Frederick I was to fall back on the traditional deployment of the militia, partly out of financial necessity and partly out of confidence that the system could generate militarily effective units. While the standing troops, financed through subsidies, were deployed abroad, the militia should secure and defend the territory. During the War of the Spanish Succession, members of the militia were repeatedly and forcibly drafted to serve as a reservoir of personnel to compensate for the heavy losses sustained in the war. Thus, forms of recruitment that had become customary in other territories, especially in France after 1690, were also employed in Brandenburg-Prussia.

### *The cantonal system*

Immediately after his accession to the throne in 1713, Frederick William I disbanded the by-then discredited militia. But he maintained the established militia practice of recruiting natives from his territories for the army, continuing the procedure of drawing up full lists of candidates fit for military service. Each regiment was assigned a so-called canton, an area close to its garrison-base, from whose population it could replenish its troops.[61] Because

59 Papke, *Handbuch*, vol. 1, pp. 100ff.; Peter H. Wilson, 'Social Militarization in Eighteenth-Century Germany', *German History* 18 (2000), 1–39, 32–3.

60 Hahn, 'Aristokratisierung und Professionalisierung', p. 191.

61 Martin Winter, *Untertanengeist durch Militärpflicht? Das preußische Kantonssystem in brandenburgischen Städten im 18. Jahrhundert* (Bielefeld, 2005), pp. 62ff.; Curt Jany, 'Die Kantonsverfassung des altpreußischen Heeres', in Otto Büsch and Wolfgang Neugebauer (eds.), *Moderne preußische Geschichte*, 3 vols. (Berlin, 1981), vol.

of the soaring numbers of soldiers recruited in the early 1720s, there were frequent irregularities in regimental recruitment, which in turn led to a corresponding increase in desertions.[62] The escape of farmers' sons and artisans to neighbouring territories increasingly impaired the Prussian economy, and even led to local riots among the populations.[63] It is therefore understandable that even the colonels urged the king to find a more permanent solution.[64]

The cantonal system was progressively modified up to 1733. The initial procedure was based on the way the militia had previously been recruited, but it did not involve the cooperation of the territorial Estates to gather recruits into local, independent defence units. Rather, those drafted were auxiliary personnel for the standing armies, assigned to a specific regiment and forced to serve for two, later three, months per year. Captains and colonels no longer acted as independently operating entrepreneurs, but were now agents of the crown, negotiating the muster of a certain number of recruits with local authorities on behalf of their sovereign. Approximately half the head count of a company was comprised of so-called natives or cantonists, men drafted from the respective canton; the other half was recruited by the captain in the traditional way from among those not liable to military service, usually subjects of other princes inside and beyond the Holy Roman Empire.[65] The funds necessary to attract recruits were gained from retained pay, provision funds, and tax-money originally intended for cantonists currently on leave. Thus, the captain had an immediate financial incentive to grant as much leave as possible to those drafted within the cantonal system.

Otto Büsch's influential study proposed that the cantonal system created a broad-based militarism within Brandenburg-Prussian society.[66] Yet, on the basis of a maximum military service of eight to twelve weeks per year, this

2, pp. 767–809; Hartmut Harnisch, 'Preußisches Kantonssystem und ländliche Gesellschaft', in Bernhard Kroener and Ralf Pröve (eds.), *Krieg und Frieden. Militär und Gesellschaft in der Frühen Neuzeit* (Paderborn, 1996), pp. 137–65.

62 Michael Sikora, *Disziplin und Desertion. Strukturprobleme militärischer Organisation im 18. Jahrhundert* (Berlin, 1995), 69–75.

63 Ralf Pröve, 'Zum Verhältnis von Militär und Gesellschaft im Spiegel gewaltsamer Rekrutierungen (1648–1789)', *Zeitschrift für Historische Forschung* 22 (1995), 191–223, 208ff.; Jörg Muth, *Flucht aus dem militärischen Alltag. Ursachen und individuelle Ausprägung der Desertion in der Armee Friedrichs des Großen* (Freiburg, 2003), pp. 41–4.

64 Max Lehmann, 'Werbung, Wehrpflicht und Beurlaubung im Heere Friedrich Wilhelms I', *Historische Zeitschrift* 31 (1891), 254–89.

65 Winter, *Untertanengeist durch Militärpflicht?*, pp. 167–83.

66 Otto Büsch, *Military System and Social Life in Old Regime Prussia 1713–1807: The Beginnings of the Social Militarization of Prusso-German Society* (Atlantic Highlands, 1997). See the critical appraisal of Büsch's thesis in Wilson, 'Social Militarization', pp. 1–39.

was unlikely to occur. It was true that a canonist on leave remained a soldier, and as such subject to the jurisdiction of his regimental officers. But again, Büsch's work rests on the mistaken assumption that senior officers held lands and tenants within the particular canton assigned to their regiment. In fact, Frederick William and his successors were concerned to ensure that officers did not hold land in the particular canton assigned to their respective units in order to avoid conflicts of interest.[67] Although the lords of these estates and the captains moved in the same social circles, their immediate economic aims were opposed, even diametrically so. A landowner sought to increase the yields of his property, utilising the workforce of peasant labourers to cultivate his land. An officer, however, was charged with a military assignment from the king based on the canton: ensuring that his unit remained at full strength and overseeing the training, equipment, and operational capacity of his troops.[68] In fact, regimental officers frequently exploited these divergent interests to negotiate advantageous deals for themselves by giving the landowners more access to their labour force. Similar opportunities for profit were available when cantonal officers recruited 'foreign' troops. The latter were not allowed to leave the garrison town but were granted enough free time to employ themselves as handymen, craftsmen outside of the regulatory guilds or '*Freiwächter*' – soldiers with no military responsibilities and available for casual labour. In these ways the captain could make considerable deductions from their pay to his own profit.

### *The 'business of the company'*

The company as an economic unit was at the centre of the eighteenth-century Prussian military, but, since the Prussian military archives were destroyed, source-based insight into its inner workings and structure is not easily attainable. Contrary to the situation in France, where a captain's commission was purchased outright, in Prussia the commissions of both colonels and captains were bestowed according to conditional seniority: a combination of years of service and military performance. This practice underpinned the principle of a community of socially equal officers and as such corresponded to the structures of a relatively 'flat' noble society as well as the requirements of experienced

67 Wilson, 'Social Militarization', p. 25.

68 Harnisch, 'Preußisches Kantonssystem', pp. 162–3; Jürgen Kloosterhuis, 'Zwischen Aufruhr und Akzeptanz: Zur Ausformung und Einbettung des Kantonssystems in die Wirtschafts- und Sozialstruktur des preußischen Westfalens', in Kroener and Pröve (eds.), *Krieg und Frieden*, pp. 167–90, 183.

military professionalism.[69] But a commission brought additional benefits so that skilful management of the company could yield profit and thereby improve the officer's personal financial circumstances. In the eighteenth century, company management also concerned the colonel-proprietor, as he owned the regiment's first company.

The starting point for the business of managing a company came with the financial transaction by which the new captain paid a stipulated sum of money in compensation for purchases of uniforms and regimental equipment bought by his predecessor. Parts of the subsequent business were the profits to be made through recruitment and in exploiting recruits' leave from service.[70] The cantonal system set strict limits to illegal or excessive recruitment as the captain had to consult with the local notables and representatives before recruiting from the assigned canton.

Nonetheless, these restrictions were abused, and the reign of Frederick II saw the captains' rights to recruit limited. In the face of the territorial Estates' continued complaints about random and illegal recruitment by the captains, enrolment and recruitment were given over to the colonels. For this purpose, an additional position among the regimental staff officers was created, similar to the majors of French regiments and with the same purpose of bypassing the captains. Another decision of the king was even more decisive in the decline of the captain's powers of recruitment. During the Seven Years' War the crown's central administration had taken charge of all troop recruitment across 'foreign' territories, and this state of affairs continued after the war as a response to tight budgetary constraints. The loss of control over recruitment cost the captains up to 50 per cent of their previous revenues, a loss that could not be compensated for by an increase in income elsewhere.

Yet not all opportunities for captains to exploit their companies for profit had disappeared. Wages could still be deducted from those soldiers on leave, and those temporarily laid-off from military duties. Captains could still do business in the acquisition and distribution of items of clothing like shoes, stockings, and waistcoats. A further source of income was derived from legal fees – for instance, for marriage licences, which required the candidate to pay a certain sum to the captain in order to demonstrate his ability to support a household. Perhaps most significant of all was the captain's discretion in the allocation of quarters for his troops.

69 Papke, *Handbuch*, vol. 1, pp. 181–2.
70 Pröve, 'Zum Verhältnis von Militär und Gesellschaft', p. 199 n. 34.

In the course of the reforms that followed Frederick II's death, captains ultimately lost control of the disposal of money accrued from the cantonists on leave. In return, they regained control over recruitment carried out in foreign territories from the central administration. Experience had shown that in a centralised system of recruitment, neither the recruiting agent nor the captains took any particular interest in the recruits' qualities. Personal connections, and loyalties based on them, still formed the basis of any given individual's performance in the eighteenth century. The 1792 reform of the cantonal system and its accompanying measures did not entirely obliterate the practice of company management, which continued to offer opportunities for legal and illegal profit until its final abolition in 1808.[71]

## Conclusion

This chapter has outlined striking similarities in the structural conditions of seventeenth- and eighteenth-century European warfare using the examples of France and Brandenburg-Prussia, but has also shown a number of differences between the individual systems.

While France's comparatively high population density allowed its army to replenish its troops from volunteers and only to fall back on drafting members of the militia during times of war, the Prussian authorities were compelled both to recruit abroad and to rely on organised conscription of subjects through the cantonal system to make good numbers. To the nobility, the essential element of the military elite, the concept of duty was a central part of their self- and public-images in France as well as Brandenburg-Prussia. The many and complex internal nuances within the French nobility ensured that military rank was chiefly determined by position within that hierarchy. Tension between members and groupings within the nobility and deliberate dissociation from an increasingly confident bourgeoisie ensured the failure of any efforts to professionalise the military elites. The practice of selling captains' and colonels' commissions, reinforcing the interdependence of military functionality and financial resources, blocked promotion based solely on merit in France.

In comparison, the Prussian nobility appears more homogeneous both in terms of its economic situation and with regard to legal status. This facilitated promotion based on conditional seniority, a combination of years of service and merit. It also encouraged military professionalism in times of war, but the

71 Papke, *Handbuch*, vol. 1, pp. 274–81.

concept had reached its limits by the end of the eighteenth century, by which time it had generated a corps of overly aged officers. Although captains' and colonels' commissions could not be bought in Prussia, the system of company management offered sufficient financial incentives to attract competition. The system worked so long as Prussia's territorial economy, catering to the needs of the military, provided the necessary means. The Prussian system plunged into crisis at the end of the eighteenth century in the face of economic decline and worsening state indebtedness. If military service could not be combined with a simultaneous improvement of status or financial gain, it was merely a transitory rite of passage for members of the French and Prussian nobilities. In France, the positions of captain or colonel were marks of existing status. In Prussia, on the other hand, officerships had been the primary mechanism for social advancement.

As long as the sovereign was surrounded by the nimbus of the *vertus guerrières* and thereby acted as king-in-arms, it was possible to merge aristocratic ideology and military professionalism within the limits of the social constitution of the *ancien régime*. If the ruler failed to fulfil that martial role, strictly military values were replaced by the competing values of the aristocracy outside of the military.

The importance of technical military skills and military professionalism for ordinary subaltern officers in both territories developed only towards the end of the eighteenth century. Officers and subalterns in the artillery and engineers may have yearned for the day when wealth and birth would no longer be the key criteria for advancement in the military, but such aspirations could only be fulfilled after fundamental reforms in the system. Meanwhile, Prussian reformers wrestled with the impossibility of altering the military system installed by Frederick II without simultaneously toppling the constitution of the state. In France and Prussia alike, reform-orientated senior officers and administrators sought to redistribute the intermediate powers and authority held by captains and colonels in recruitment and company management, seeking to place these responsibilities instead in the hands of what they hoped would be an efficient and centralised administration. Yet, in both cases, the intermediary agencies constitutive of the social system of the *ancien régime*, and the consent of individuals and social groups required for the exercise of any form of authority, proved to be insurmountable obstacles.

14

# Navies, technology, and organisation, 1648–1815

ROBIN BRIGGS

Naval warfare changed out of all recognition from the late sixteenth century onwards through the rapid development of large square-rigged warships carrying heavy broadside gun batteries. A whole series of developments followed, with a long (if far from smooth) evolution in ships, equipment, strategy, and tactics continuing down to the last sailing navies of the early nineteenth century. It was clearly no accident that this naval revolution coincided with a great age of global European empires, which would have been impossible to create or maintain without effective naval power. Galleys and other oared craft became largely obsolete, except for some amphibious operations in the Mediterranean and for use in shallow waters around the innumerable Baltic islands. The crushing Dutch victory over a Spanish fleet at the battle of the Downs (1639) marked the first occasion when the full power of broadside gunnery became evident. Then the three Anglo-Dutch wars between the 1650s and 1670s saw a series of savage and bloody engagements between the fleets of two nations that were coming to be known as the Maritime Powers. The combination of imperial and trading ambitions, new financial arrangements, and relatively open societies enabled first the Dutch, and then the British, to develop naval power to new heights, in turn allowing them to punch well above their weight on the international stage. Under Louis XIV, France did mount a serious challenge to the Dutch and English, and for a time possessed the largest navy in the Western world. However, by the 1690s the French, and more gradually the Dutch, were finding the costs of maintaining this level of power at sea, as well as on land, to be too great.

Navies based on specialised warships and extensive port facilities were immensely costly long-term investments, which also demanded new levels of bureaucratic organisation. These financial demands were very considerable even in peacetime, because wooden ships typically lasted only ten or twelve years between rebuilds that might equal their original building cost (with less satisfactory vessels simply being scrapped), while ports silted up and their facilities became inadequate for new generations of larger ships. A navy that did not keep substantial reserves including stocks of cannon, hull and mast timber, sails, and cordage would hardly be able to sustain the briefest campaign. The dockyards, with thousands of skilled workers, rapidly became the largest industrial enterprises in the Western world. In wartime, navies made voracious demands for men, armaments, and foodstuffs, on a scale that had significant effects on national economies. The international trade in naval supplies was valuable in itself, and strategically vital to the contending powers. In many ways, then, this was a precociously advanced sector of the early modern world, at the cutting edge of technology and administration. On the other hand, this precocity created major problems for the statesmen and admirals who tried to work out effective versions of both strategy and tactics. The distinctly unimpressive writings of the time on naval affairs suggest that key issues had barely been identified, while the absence of anything resembling naval staffs reflected this conceptual poverty. Arguably it was only at the very end of this period, from the 1790s, that British ministers and the Royal Navy finally discovered ways to unlock the full potential of naval power, in the context of what became virtually a worldwide conflict. Others might consider that the Seven Years' War marked the watershed, but no earlier war looks a serious contender.

In some respects the Anglo-Dutch wars were exceptional, in respect of both the strategic situation and the attitudes of the combatants. Dutch dependence on maritime trade left the Republic dangerously vulnerable to a rival power that was ideally placed for operations in both the North Sea and the Channel; only the crucial Baltic trades were not under immediate threat. Early successes had also encouraged the Dutch to believe that their fleet was capable of defeating all comers, so they felt they had no choice but to confront the English challenge directly. Since the English could only achieve their war aims by gaining supremacy at sea the result was a series of set-piece battles in relatively confined waters, perhaps the only time over the whole period when two large navies opted to fight it out in this fashion. In these formative decades the line of battle emerged as the tactic that allowed ships to use their main broadsides effectively, whilst also protecting them from being picked off and attacked in isolation. Sets of fighting instructions were formulated as the only practical way to preserve some kind of order, as the relative impotence of admirals to direct fleets in battle became apparent. Meanwhile a distinction was being drawn between heavier ships capable of standing in the line of battle and smaller cruisers, the lesser ships being used for scouting and – in a defensive and offensive capacity – trade warfare. The Dutch pioneered new convoy systems, though the Spanish had been deploying convoys since the sixteenth century. In practice there were great difficulties in protecting merchant shipping, even when owners and captains could be persuaded to comply with orders, so losses on both sides were very high. On the whole, the English seem to have been the net gainers, because their extensive captures in the 1650s greatly expanded their stock of efficient ships and kick-started the merchant fleet's continuous growth. This was crucial for military capacity at sea, because seamen were an essential and finite resource; unless they were available in sufficient numbers no amount of shipbuilding could make a country into a serious maritime power.

The English Republic, having created a highly effective fleet, proceeded to use it for operations in the Mediterranean and the New World. The Western Design of 1655, which saw an amphibious operation in the Caribbean capture Jamaica, was in many respects a portent of things to come. Its implications, though, like the results of this expedition, were far from wholly positive. Significant groups among the British ruling elites were developing an attachment to a 'blue-water' strategy that was premised on illusions about the fragility of the Spanish American empire, and grossly underestimated the difficulties of long-range naval operations. Perhaps in the end this particular (and remarkably enduring) set of rose-tinted spectacles worked to Britain's advantage, but unfounded optimism along these lines would lead to many expensive disasters, not least through the agency of that notably lethal enemy, the mosquito. In the shorter run, Oliver Cromwell and his advisers had miscalculated on another front: naval operations that were naïvely expected to be self-financing proved cripplingly expensive, greatly contributing to the debts that brought down the Republic and restored the Stuarts.

Over the next half-century a new configuration of forces restructured European affairs on both land and sea, as Louis XIV's France emerged as the predominant power against which other states increasingly ranged themselves. The king's minister Jean-Baptiste Colbert and his son the marquis de Seignelay set out to create the greatest fleet in Europe, with an administrative structure and a set of naval bases that would largely endure until the Revolution and even beyond. This was an astonishing achievement, with lasting results, for, despite a temporary blip in the early eighteenth century, France had now moved permanently into the ranks of the major naval powers. At great expense Colbert had established an infrastructure that, up to a point, functioned; inevitably, though, he had also left unsolved some major problems that the increasingly beleaguered French royal state of the next century never wholly overcame.

None of the late seventeenth-century navies seems to have made significant improvements in ship design, although ships tended to get bigger and there was a modest amount of incremental progress. The French, with the highest budget for a time, were naturally dependent on foreign experts and craftsmen in the early stages, and many of their first-generation ships had short and rather unsatisfactory careers. Even their successors tended to be fragile and drew too much water to operate safely in several key campaigning areas. Dutch shipbuilders were always affected by different problems, usually building down to a restrictive budget, while shallow waters in and around their harbours limited ship size and draught, so that Dutch warships tended to sail poorly in open waters. Largely thanks to Charles II's intelligent personal interventions, the English probably built the best warships of the 1670s, but then blundered by replacing them with some of the worst performers of the 1690s. All the navies failed to generate satisfactory designs for the smaller cruisers, whose stubby hulls, rendering them slow and leeward, proved a persistent defect. But, for several reasons, these inadequacies attracted less notice than they might have done. Naval policy was in most cases conservative, even when not thoroughly confused, so fleets operated for relatively short periods close to home, when more ambitious campaigns might have exposed their limitations. When battles did occur, the outcome could usually be attributed to numerical superiority, or in single-ship actions to the bravery and professionalism of the officers and crews, rather than to the technical qualities of the ships. Naval architecture was far more of a craft skill than a scientific activity; indeed, for a very long time to come most of what passed for theory in this field was largely erroneous and better avoided. Nor had anyone really worked out just what roles specific ship types should be expected to perform, so design briefs were rare and generally unhelpful. Nor did naval weaponry evolve significantly: limited metallurgical understanding and poor quality control both contributed to a situation where the mass-produced iron cannon of 1700 may well have been significantly inferior to their predecessors of 1650.

At the outset of Louis XIV's Dutch War (1672–9) an ephemeral Franco-British alliance had some grandiose and unrealistic plans, rapidly frustrated by Dutch defensive tactics, but this was essentially a land war in which naval operations played only a secondary role. In 1688, on the other hand, the opening of a much greater set of wars was marked by the greatest naval feat of the age, William of Orange's invasion of England. Complete success cannot hide the fact that this was an astonishing gamble, made possible only by the 'Protestant wind' that penned the English fleet in the Thames estuary and carried the invaders down to Torbay. The French, who had committed themselves to a land campaign and made minimal preparations at sea, looked on impotently as William's triumph dramatically altered the European balance of power. For the next quarter-century the two great Maritime Powers were to be linked in an uneasy and quarrelsome alliance, mainly held together by fear of France through two long, complex, and immensely expensive conflicts (the Nine Years' War of 1688–97 and the War of the Spanish Succession, 1702–13). In some respects the most important developments of the period were only indirectly military, because the adoption of many Dutch financial practices in Britain had enormous implications for both present and future war. European wars were becoming wars of attrition – not even the great victories of the duke of Marlborough or Prince Eugene of Savoy proved to be decisive – ending in peace settlements determined as much by the financial endurance of the combatants as by their military achievements. In this context, the great cost of navies relative to their effectiveness was bound to raise questions. This would first lead to some interesting strategic choices, followed by an era when only the British would remain committed to a predominantly maritime strategy – a choice made by a political elite with little real understanding of naval affairs and largely motivated by rooted aversions to standing armies and Continental commitments.

The opening campaigns of the Nine Years' War saw some staggeringly inept performances by both sides, although these can be attributed to kings and ministers even more than to the admirals. A dispersal of allied forces in 1690 allowed the French to achieve temporary superiority in the Channel, then an absurd decision by the earl of Nottingham and Queen Mary compelled Admiral Torrington to fight a battle he did not want off Beachy Head, leading to heavy losses in the Dutch squadron. With no invasion force ready to follow it up, however, this French victory proved completely sterile. When an army was prepared in 1692 the Toulon squadron failed to arrive in time, but Louis XIV forced Admiral Tourville to fight all the same, resulting in heavy defeat at Barfleur. Although the French soon replaced the lost ships, their inadequate dockyard facilities and their manning problems, exacerbated by intolerable financial pressures, steadily reduced their capability. The most effective naval intervention in this war came when William III insisted that the allied fleet winter in Cádiz in 1694, thereby enabling it to remain in the Mediterranean until very late in the year and totally disrupt French strategy. Otherwise, though, the British navy failed to exploit its opportunities across the Atlantic, with disease, as usual, proving to be the greatest enemy, while in the Channel it tried (and failed) to meet public expectations by a series of futile raids on French ports. Meanwhile the French increasingly turned to commerce raiding, using small detachments of warships. Despite some striking successes, however, these raids rarely paid their way and their overall effect on allied commerce was surprisingly small. It was the main French fleet that captured the Smyrna convoy in 1693, while, once shortage of money had restricted French fleet operations, more British and Dutch warships were available for trade protection. With their large crews, privateers were costly to fit out and too many voyages failed to encounter any ships worth capturing in the vastness of the ocean. Moreover, the practice of spreading ship-ownership across many hands and the development of marine insurance distributed the risks to shipowners quite effectively, so that losses of several per cent per annum did not have much of a deterrent effect on trading ventures.

When hostilities resumed in 1702 the French naturally concentrated their efforts on protecting the Spanish treasure fleets in the Atlantic and in the Mediterranean. This policy succeeded in bringing most of the bullion safely across the Atlantic, but also led to a shattering defeat at Vigo Bay in 1702, leading in the long term to an enduring Anglo-Portuguese alliance. Two years later the allies took Gibraltar, then a battle off Malaga, which ended in a draw, ensured that the British would keep this strategic fortress. In 1707 a combined operation against Toulon, marked by the use of recently developed bomb vessels, alarmed the French into sinking their ships and their intention to refloat them later was largely unrealised. In any case, financial exhaustion soon compelled the French to revert to trade warfare, with the same mixed results as before; whatever the potential of this strategy, it was never going to work without a main fleet. When peace finally came in 1713 the French navy was almost ruined, while the Dutch had greatly reduced their naval commitment, by default leaving the Royal Navy as a virtual winner. This dominant status was reinforced when a Spanish fleet was heavily defeated off Cape Passaro in 1718, frustrating Spain's ambitions in Italy. The two following decades were unsurprisingly a period of technical stagnation since, in the absence of serious challengers, the British continued to build over-gunned and rather unseaworthy warships, whose limitations were brutally revealed in a war with Spain in 1739. This conflict soon merged into the general War of the Austrian Succession (1740–8), during which the British discovered that the Spanish American empire was anything but the easy target they had anticipated. From 1744 Britain was at war with France as well; after some dismal performances the navy redeemed itself with two important victories against French squadrons in 1747, whilst in 1745 a colonial force with naval support took the crucial Acadian fortress of Louisbourg. A calamitous French expedition to retake it in 1746 demonstrated that the French navy, slowly reconstituted since the 1720s, was still quite incapable of operating effectively at long distance. This disaster stemmed from a lethal combination of ineffective command, poor hygiene leading to crippling epidemics among the crews, faulty navigation, and inadequate provisioning.

Yet this rather unexciting war marked a turning point in several respects. The French government learnt that the country's commercial and imperial ambitions could only be realised with the aid of naval strength; down to the Revolution the Bourbon monarchy would try, at ruinous expense, to be a great power both on land and at sea. In Britain worries about French intentions, and the defects shown up after 1739, inspired a major effort to improve the navy, by developing better command and administrative structures, investing in dockyards, and building improved modern ships. Soon Spain, the Baltic powers, and others would join in what amounted to a naval race, as national warship tonnages soared to unprecedented levels. In this new world of competition three interlinked factors gradually transformed naval warfare, by rendering sailing navies far more efficient, so that they could operate on a global scale to unprecedented effect. The first was a revolution in ship design instigated by the French in the 1740s, apparently through the efforts of a couple of talented naval designers rather than deliberate planning by the authorities. The two basic ship-types that would dominate fleets until 1815 emerged: the frigate and the larger two-decker. (See Figure 19.) The dimensions of ships were naturally dictated by the size of human beings, the guns to be carried, and the availability of large timbers for construction. Within these limits there were huge gains from building ships that were long relative to their height out of the water and could carry the heaviest effective guns on their main gun-decks. This greatly improved a vessel's seagoing performance and enabled it to fight even the biggest three-deckers on nearly equal terms. For most situations, the ideal compromise was the seventy-four-gun ship of the line, which carried twenty-eight heavy guns, yet was relatively economical in terms of construction and manpower costs. For scouting, trade protection, and other general duties the frigate, with its single gundeck and a notable combination of speed and range, was the ideal craft. Fortunately for the British, the French navy does not seem to have realised quickly enough what its constructors had achieved, so made no effort to build the new ships in great numbers. Following a curious delay after

FIGURE 19. Model of the French eighty-gun two-deck warship *Foudroyont*. Launched at Toulon in 1750 and captured by the British off Cartagena in 1758, serving subsequently in the Royal Navy. Musée d'histoire de Nantes, 294.

several examples were captured, British dockyards started to produce their own versions from the mid-1750s.

The second revolution was one in the health and wellbeing of sailors at sea, the more remarkable given the limitations of contemporary medical science. Here the British, from the time of the unprecedentedly long cruises made by the Western Squadron off Brest in 1758 and 1759, comfortably outstripped all rivals. The key factors were high standards of hygiene and cleanliness, adequate supplies of good food (including fresh food whenever possible), and the gradual adoption of lemon juice as an anti-scorbutic. This last development only became standard practice in the Royal Navy around 1800, but scurvy had already ceased to be a major problem before this date. Since seamen were always a scarce resource, keeping them healthy was evidently a major gain, while the ability to make very long voyages and remain efficient opened new strategic possibilities.

The final crucial trend involved a whole range of technological changes, which rapidly gained pace over the final decades of the eighteenth century. The coppering of hulls was the most dramatic single improvement, but there were also developments in guns and gunpowder, blocks and rigging, ventilation, pumps, anchors, and much else. Here too British industrial and artisanal strength, notably in the metallurgical trades, was a huge advantage, reflected in a widening gap between the Royal Navy and all its rivals. French espionage was rather good at following British advances, but in the absence of an appropriate industrial base it often proved impossible to copy them effectively. Even without these developments, Britain had a crucial, permanent advantage in securing naval supplies, because British firms and traders had come to dominate the Baltic markets for iron, timber (notably fir for masts and spars), tar, and other vital commodities.

These were not the only respects, though, in which historians can now see, as contemporaries often could not, how the dice were loaded in favour of the British. The British seafaring population, always larger than the French, seems to have grown much faster across this period. Indeed, it is somewhat mystifying that an impressive growth in the value of French maritime trade down to 1790 generated only a trivial increase in the numbers of registered seamen. Although the Royal Navy envied France's formalised conscription method, and would have imitated it but for parliamentary opposition, in practice the relatively chaotic British recruitment methods were more productive. For all the notoriety of the much-misunderstood press gang, there were high proportions of volunteers in British crews, and the navy turned landsmen into useful seamen whereas the French system tended to consume the existing stock of trained men. Over successive wars the numbers of men serving at sea in the Royal Navy rose to staggering levels (142,000 at the peak in 1810), while rates of death, disablement, and sickness fell considerably. Amongst the other naval powers, Spain had substantial numbers of experienced seamen, yet for some reason a registration system based on French principles proved unable to draft enough of them to man Spanish fleets properly. The Dutch always relied on a straightforward practice of hiring seamen, which seems to have been quite effective although costly and perhaps ultimately capped the size of their fleets. The Baltic navies employed various forms of conscription, with Denmark and Sweden able to call on significant seafaring populations, in contrast to Russia. Smaller Mediterranean powers that built ships for prestige never seem to have had the capacity to man them efficiently, apart from the privateering vessels of the Barbary States on the North African coast. In wartime, both France and

Spain tended to draft large numbers of soldiers onto their ships to make up for the shortfall in seamen, with predictably unfortunate results. From 1755 the British followed a more practical route by creating a Marine Corps, whose members might lack specific nautical skills, but were accustomed to sea service and provided a striking force for amphibious operations.

There was a large private shipbuilding industry in Britain, capable of building all but the largest warships on contract, so that in wartime the royal dockyards could give priority to maintenance and repair without too much effect on new construction. In France, by contrast, virtually everything had to be done in the three great bases of Brest, Rochefort, and Toulon, where bottlenecks in supply and problems with the labour force (which included many convicts) were endemic. Like everything else concerning the navy, the French dockyards were at the mercy of the ramshackle financial practices of the state, which in a highly technical service could have particularly destructive consequences. Matters were made even worse because late and unreliable payment of bills and wages combined with a disastrous failure to check overspending by departmental ministers. Contractors who supplied the French navy had to reckon on bankruptcy in the medium term, a powerful inducement to charge heavy premiums and cheat on quality wherever they could. Spanish shipbuilding followed a mixture of British and French practice to produce ships that the British rated highly, and the Havana dockyard made effective use of tropical hardwoods. By 1796, however, financial stringency put an end to new building and seriously reduced even regular maintenance operations. The defeat at Trafalgar, combined with the chaos generated by the 1808 revolt against the French, brought about the virtual demise of the Spanish navy, the prelude to the collapse of the Spanish Empire in the early nineteenth century.

Britain's far more robust financial arrangements allowed governments to borrow at levels that sometimes terrified the ministers themselves, while until 1796 a curious anomaly enabled the navy to convert its short-term debt into interest-bearing negotiable bonds. However, the anxiety of the Admiralty and the Navy Board not to draw attention to this legally suspect arrangement limited their recourse to it. Parliament would periodically absorb the shortfall into the National Debt, thereby extending the life of this flexible device. On this basis efficient bodies like the Victualling Board and (in later years) the Ordnance Board were able to establish long-term relationships with contractors, enforce quality controls, and secure excellent value for money. It appears that the Royal Navy was never seriously affected by financial constraints, while from the 1750s it had liberated itself (by somewhat devious means) from the intrusive parliamentary surveillance that had previously hampered effective ship-design. By contrast, all of Britain's major naval rivals – the French, Spanish, and Dutch – experienced periods of penury which seriously reduced their operational efficiency. The French navy was also increasingly plagued by Colbert's administrative system and its legacy, which gave rise to endemic conflicts between officers of the pen and the sword, without achieving the financial control that was supposedly its *raison d'être*. These quarrels reflected the enduring tensions between robe and sword that dogged the *ancien régime* on many fronts. Although few naval administrators in the dockyards could claim the status of *noblesse de robe*, sea officers drawn from the traditional sword nobility seem to have missed few chances to criticise or humiliate colleagues whose help they really needed, but whom they saw as upstart pen-pushers. The faults for which the officers of the pen took the blame usually resulted from financial problems quite outside their control, while it is far from clear that any of the later eighteenth-century attempts at reform by giving sea officers more power in the dockyards did much to improve this situation. On paper British naval administration might look distinctly incoherent, but in practice its reliable financial base usually encouraged pragmatic solutions and cooperation. It also permitted the development of naval bases both at home and abroad, a crucial trend.

The principal French fleet bases, Brest and Toulon, were excellent natural ports possessing some real advantages. Toulon was particularly well placed to obtain timber and food supplies from a wide region, was virtually impossible to blockade effectively, and gave a head start for Mediterranean operations. Brest, at the extremity of the poor province of Brittany, was much harder to supply by land. On the other hand, though, its strategic location made it very hard for the British to stop the French from operating in the Atlantic. There were, unfortunately, three countervailing drawbacks. The bases were separated by a long and hazardous voyage, so the French fleet rarely operated as a single body. This was further exacerbated by rivalries between the officers and men of the Atlantic and Mediterranean squadrons, representing as they did very different cultures and even languages. Finally, since the French had no Channel port capable of housing large ships, French officers were understandably very reluctant to enter these dangerous waters, so the British never had to face any serious interference with their own bases. In the eighteenth century the whole focus of British activity moved from the Thames estuary to the 'western' bases of Portsmouth and Plymouth, in which very large sums were invested, although it was only in 1817 that a new breakwater rendered the anchorage off Plymouth truly satisfactory.

In the process the British gained a crucial lead in the number of dry docks, transforming the maintenance capacities of their dockyards. By 1793 there were nineteen of these, against France's eight, some of which did not function properly. The Spanish navy enjoyed good facilities at Cadiz, Cartagena, and Ferrol, as well as the notable overseas dockyard at Havana. However, it was only the British who managed to develop a global network of naval bases including at various times Gibraltar, Minorca (later replaced by Malta), Antigua, Jamaica, Halifax, the Cape, and Bombay. After the loss of the American colonies there was indeed a major shift in the whole British attitude to empire, with the focus now moved away from territorial expansion to the development and protection of trade, of which these nodal points were one of the clearest expressions.

It was inevitable that all navies relied on the larger pool of seamen to provide the essential skilled element amongst their crews. It was a different story where officers were concerned, partly because there was a natural expectation that military command could only be exercised by men of some social standing. The French navy reckoned that its officers should be drawn exclusively from the nobility, although in practice some non-noble merchant officers were recruited in wartime. A similar situation developed in Spain, and both countries had large numbers of poor nobles who might be attracted by such employment, especially when systems of half-pay gradually evolved to guarantee officers an income even when they were not serving. France developed an elaborate shore-based training system for naval officers, with particular emphasis on mathematical and navigational skills. The British had followed a different route, stemming from the remarkable 1677 decision by King Charles II that aspiring lieutenants must pass an examination in seamanship, after they had served a minimum of three years at sea, at least one of these as a midshipman. In practice this led to teenage boys going to sea under the patronage of captains, to learn practical skills (largely from warrant officers and experienced seamen) before submitting themselves for examination. As a result, British officers often lacked the polish normally associated with a gentleman's education, but there is little doubt that they were better seamen and managers of men than those of rival navies. But whatever the system, promotion inevitably relied on some combination of seniority and influence, though by the later eighteenth century the British navy had evolved various methods for rewarding gallantry in battle by promotion. A notable innovation came in 1747 whereby senior captains could be promoted as admirals 'without distinction of squadron', and thus effectively retired on half-pay. This made it possible both to select the most talented and energetic officers as admirals who actually commanded fleets, and to make room for younger captains; this was arguably a crucial factor in the vastly improved performance of the Royal Navy after that date. Although one should not underrate the *ancien régime* French officers, among whom a fair proportion were highly skilled, a more rigid promotion system led to elderly captains and even older admirals, until the Revolution changed everything. There were jealousies and even hatreds amongst senior British officers, notably during the American War, but it is hard to imagine any First Lord of the Admiralty concurring with the French naval minister of the 1750s who lamented that 'in the navy they all hate one another': a bitter (albeit exaggerated) reflection which is easily evidenced.

Officers and seamen had to be paid, and this could lead to serious practical difficulties, especially when peace was made and large numbers of men were eligible for accumulated pay just when treasuries were empty. One disastrous occasion in France saw the main fleets completely disabled in 1760 – at the climax of the Seven Years' War – because there was no money available to pay that year's recruits or feed them, so the port officials decided they were legally obliged to send them home. In Britain, the Admiralty's failure to increase seamen's pay (static since 1652) in a period of inflation was a major cause of the great naval mutinies of 1797. Yet such extreme cases were very rare, and even in 1797 it appears that perceived inequities in pay and treatment were as important as the absolute level of pay. A remarkable, and largely forgotten, feature of this episode was the extent to which officers (including Nelson) admired and respected the responsible conduct of the mutineers in the Channel Fleet, as opposed to the more radical group who briefly took control at the Nore. This is a salutary reminder that warships were not the floating prisons of imagination; survival at sea, and even more success in battle, depended on cooperation and high morale. Naval discipline could be harsh, and there were bad, even sadistic officers in every navy, but the reaction to such men (who often ended up being dismissed from the service) suggests that the general position was very different. Of course, warships were enormously overmanned in all situations other than battle, because so many men were needed to operate the guns; work-rates were therefore rather low, although men paid for this by living in very cramped conditions, an area where the British insistence on shipboard cleanliness and hygiene came into its own. It is also clear that the British style of naval warfare after the 1750s, which saw an exceptionally high ratio of ships at sea year on year, was only possible because crews were healthy and well-fed (and allowed virtually unlimited quantities of small beer or other alcoholic drink). Though the French navy managed something similar during the War of American Independence, partly because the colonists provided supplies, this was something of an exception.

Within shipboard society an ordinary seaman could rise to a position of considerable authority, as a warrant officer or a technical specialist, and would be respectably paid. However, there is little indication that men normally saw naval service as a route to social advancement, and sailors were notorious for the feckless ways in which they disposed of the quite large sums they could receive on discharge. This also applied to prize money, taken from preying on enemy trade ships. In the seventeenth and early eighteenth centuries the Dutch, French, and British navies all offered considerable opportunities for individual enrichment by such means, heavily skewed towards officers, especially senior officers, by the regulations for sharing out the proceeds. From 1740 onwards the British navy became the great predator on a global scale, far outstripping all rivals. Hopes of making a fortune stimulated recruitment at all levels, but were particularly important for the officers, who in the British case were often younger sons or other men of very modest means. There was an obvious risk that the system would distort the priorities of admirals and captains, and on some occasions it patently did. The Admiralty's solution was to reward captors of enemy warships by routinely valuing and purchasing these ships at a fair price which was paid as a bonus ('Head Money') to the captor. Such activity had profound implications, raising very difficult legal issues about the rights of neutrals and what constituted a blockade, while also creating major diplomatic problems for belligerents. The most celebrated instance of the latter was the 1780 Armed Neutrality of the North, when an alliance of Baltic naval powers posed a deadly, if ephemeral, threat to British strategy and supplies. The Law of the Sea was virtually a creation of the post-1600 era, and, on the whole, a surprisingly effective one; the Admiralty Courts seem to have set rather high standards in distinguishing between legitimate captures and others. From the perspective of the British this helped defuse potential diplomatic tensions, though it frustrated those whose claims to prize money were disallowed, as so many major powers inevitably felt that their trade and prosperity were threatened by British naval dominance.

Until the 1790s, though, that British dominance was far from assured, primarily because the 'family compact' between the Bourbon kings of France and Spain meant that these two large navies repeatedly combined forces against Britain. This had little effect on the Seven Years' War (1756–63), because Spain only entered the conflict in 1762, by which point French naval power had virtually collapsed. From the start of this war the French government had appreciated that, given the disparity of forces, there was no chance of winning a purely maritime conflict against the Royal Navy, and so had settled on a strategy of waging war on land against Britain's isolated European ally, Prussia, to counterbalance inevitable colonial losses, presumably envisaging a repeat of the peace of 1748. When this strategy started to go disastrously wrong, French ministers resorted to a second familiar plan, mounting a cross-Channel invasion. In response the British developed a powerful Western Squadron to blockade Brest, a strategy that was to have enormous future consequences. Yet it was only possible because of a new generation of ships with superior sea-keeping qualities and the rapid improvement in health on board, backed up by a very effective (if partly improvised) system of supply at sea. In 1759, the 'year of victories', France was unable to send any significant help to her embattled colonies, while the invasion scheme was frustrated over the whole summer and autumn. When bad weather finally allowed the Brest fleet to get to sea in November, it led only to a decisive British victory at Quiberon Bay. The loss of Canada and most of France's positions in India, confirmed by the peace settlement of 1763, would turn out to mean an irreversible shift in the global balance of power. At the time this was less evident, because it was the Caribbean sugar islands, most notably Saint-Domingue, which formed the vital element in the dynamism of the French Atlantic seaboard, through the 'triangular trade' that revolved around the slave economy of the plantations. Meanwhile the rapid capture of Havana and Manila demonstrated both the new-found capabilities of the Royal Navy and the problems in making any kind of long-term gain at the expense of the Spanish Empire. The Spanish colonists in the New World had long realised that they could not expect much military help from Spain, and so had developed their own systems of defence, drawing on what were now large local populations. At such distances the British could never land and supply large enough armies to take proper control, when the colonists displayed little if any enthusiasm for revolt against the Spanish crown, as the invaders naïvely persisted in expecting. The fashion in which the British took conscious advantage of their maritime domination to deplete France's limited stock of seamen was another aspect of the Seven Years' War; this was largely possible because desperate French shipowners and seafarers turned from trade to privateering, with the result that many thousands of French sailors become prisoners for the duration of the war.

The decade following the Seven Years' War saw a vast collective expansion of navies across maritime Europe, since many powers with economic or strategic positions to defend and, if possible, advance drew their own conclusions from the war. Beyond the three major powers already mentioned, there was much activity in the Netherlands, Denmark, Sweden, and Russia, whilst the Ottoman Empire and some Italian states played more marginal roles in this maritime expansion. Spain was becoming acutely aware of the vulnerability of its empire, while French ministers combined a rational wish to protect the lucrative Atlantic trades with vaguer aspirations to reverse the humiliating defeats of the recent past. In purely numerical terms, Britain's relative position inevitably declined over the decades down to the 1790s, but there were significant compensations. Between them, the earl of Sandwich (First Lord of the Admiralty 1771–82) and Sir Charles Middleton (Comptroller of the Navy 1778–90) carried through a range of reforms designed to modernise the Royal Navy, placing it and its dockyards on a much sounder footing. Adequate stocks of seasoned timber were built up and kept under cover, so that the ships, built to the excellent designs made, or inspired, by Sir Thomas Slade in this period, proved exceptionally long-lasting. There is a persistent myth that scientific French ship design of this epoch was superior to artisanal British efforts, already propagated by some contemporary observers, which sits uncomfortably with the outcomes of naval wars. French ships did have one potential, if ambiguous, advantage, in that they were usually larger, and this ought to have conferred some performance advantages. The basic reason for this discrepancy was that, in a striking early case of 'path dependence', the French navy was locked into the use of a heavier lower-deck gun, a 36-pounder as against the British 32-pounder; because the French pound was heavier, the weight of the ball was actually 39 British pounds. Since there was a persistent shortage of cannon it was impossible ever to reverse this fortuitous choice, even though the big guns were too heavy for rapid fire, and needed larger crews as well as larger ships. French designers, unfortunately, concentrated on speed, producing light and highly stressed hulls that deteriorated quickly and offered poor protection in battle, so larger ships were in practice weaker ships unsuited for extended sea service; the British routinely rebuilt those they captured to make them usable. This even applied to frigates, where the British also copied French hull designs but made them far stronger and gave them greatly extended storage capacity for long voyages. The greatest single technical advance was the use of copper plating on the underwater section of hulls after 1780, which protected them against shipworm and also greatly reduced encrustation with barnacles and weed, necessitating less frequent cleaning, and thus making it much less likely that fleets would be held up by slow ships with foul bottoms. This innovation nearly proved disastrous, because electrolytic action underwater destroyed the iron bolts holding the hull timbers together, but the British copper manufacturers quickly developed a cold-rolled copper bolt to fix the problem. They caused some dismay by selling the bolts to the French as well, although it looks as if the standard of French coppering was always much inferior, a problem exacerbated after 1793 when the French were thrown back on their own resources.

In the War of American Independence, the French navy was for once able to fight the type of war it was best suited for, one in which simply holding one's ground at key points and avoiding major battles was enough to undermine Britain's already shaky efforts to fight a major war 3,000 miles from home. Indeed, large-scale French operations in the Caribbean and along the American seaboard were probably only possible at all because the colonists were able to provide most of the necessary port facilities and supplies. In addition, the Spanish alliance meant that the allied navies enjoyed numerical superiority, although the attempt to mount an invasion of England in 1779 proved a fiasco, with the enormous Franco-Spanish fleet ravaged by an epidemic of typhus. Admiral Rodney's victory at the Saintes in 1782 was taken by all parties as the signal that the British navy was regaining the ascendancy. The French government also turned out to have miscalculated hopelessly at a geostrategic level, as the loss of the thirteen colonies did no serious harm to Britain's great power potential. After their crippling expenditure on the war, the French had very little success in penetrating the American market, while Anglo-American trade boomed at unprecedented levels. Worse still, after 1783 the naval minister, the marquis de Castries, embarked on a lavish campaign of expansion that paid no heed at all to the growing financial plight of the French monarchy, and was a major contributor to the fiscal and political collapse of 1787 that opened the way to the Revolution. Castries would hardly have taken much pleasure in learning that his investment enabled the French navy to perform better than expected for several years after the Revolution. This was despite the incredible disarray in the dockyards (culminating in the occupation of Toulon by the British in 1793), and the emigration of most of the noble officer corps. British hopes for the rapid disintegration of their old enemy quickly proved to be over-optimistic, but there is no real doubt that French operational capacity was seriously diminished by the Revolutionary upheavals, or that the ensuing disparity proved impossible to reverse.

Naval war developed on an unprecedented scale between 1793 and 1814, with just a brief interval in 1801–3. Over this period the British navy not only inflicted repeated defeats on the French; it also virtually eliminated the navies of several other powers – principally the Spanish, Dutch, and Danes – that were absorbed by the French or were forced into alliances with them. By 1810 the total tonnage of the Royal Navy was greater than that of all other European navies combined, a remarkable situation that was achieved by capturing and destroying enemy ships, not by outbuilding its rivals. The gulf in fighting capacity was greater yet, as British ships routinely spent most of their lives at sea, whilst other fleets spent most of their time anchored in ports and roadsteads. The ratio in terms of sinkings and captures in actual conflict over the whole period has been estimated at an astonishing 60:1 in favour of the British, all the more remarkable as in most cases opponents fought with great courage, testified by appalling casualty rates. The Nile (1798) and Trafalgar (1805), in particular, were battles of annihilation (see Figure 20); in both cases the few French ships of the line to survive the battles later fell into British hands, so that not a single large vessel ever made it back to

FIGURE 20. Opening stage of the battle of the Nile, 1798. Oil on canvas, Thomas Whitcombe. National Maritime Museum, Greenwich.

France. Tens of thousands of seamen were imprisoned, and even had Napoleon's efforts to rebuild his fleet after Trafalgar produced seaworthy ships (as they too rarely did), there would not have been enough skilled men to put any significant number of them to sea. Blockade and trade warfare destroyed the long-range commerce of France and her allies, with previously booming Atlantic ports like Nantes and Bordeaux falling into precipitous decline. In earlier conflicts there had been forewarnings of such possibilities, but not even the Seven Years' War had inflicted permanent damage on French colonial and international trade, which bounced back immediately after the peace settlement. This time an interruption for twenty years made such a recovery impossible, while simultaneously the great slave revolt on Saint-Domingue had destroyed the linchpin of the old French Atlantic system. This generation of maritime ascendancy saw Britain achieve a dominant position in world trade that would last for well over a century; when the navy reached its peak of 142,000 men at sea in 1810, there were another 164,000 registered as merchant seamen, as new markets opened up around the globe.

British dominance was not easily won, and it was only with the victories of 1797–8 that a decisive gap clearly opened up. The Glorious First of June (1794) was certainly a major victory, but also highlighted significant defects; some ships and captains performed poorly, and cautious leadership allowed a beaten enemy to escape, while the French fleet put up impressively stiff resistance. In a sense, it was French success on land, leading to the serious possibility of an invasion of England or Ireland, which led the Royal Navy to a policy of aggressive close blockade that became routine from 1800 onwards. The merits of this strategy were less obvious than they appear with hindsight; holding the sea through all seasons, close to what was normally a lee shore, made huge demands on both the ships and on the seamanship of their officers and men. Those admirals who argued against this policy had a plausible case, which might have been supported by citing the considerable number of ships lost or damaged over the long years of blockade. In the end, though, this cost was clearly outweighed both by the strategic success of blockade and by the way hard sea service worked to improve the quality of the navy. In the last decade of the war, smaller ships added a further dimension to the blockade, using their boats for cutting-out operations that devastated coastal trade all the way from the Adriatic to the North Sea. A further argument in support of a close blockade was the huge scare in 1799, when the Brest fleet took advantage of the loose watch on the port to escape and enter the Mediterranean, only to miss several opportunities to overwhelm scattered and outnumbered British forces. This was by far the best chance the French navy ever had to alter the course of the war; the failure to exploit it had much to do with the ingrained French practice of assigning commanders specific missions, rather than seeking to exploit situations as they arose. The British faced another, though lesser, crisis in 1803–5, when the earl of St Vincent (unwisely installed as First Lord of the Admiralty 1801–4) indulged all of his numerous prejudices in a headlong 'reform' campaign that badly disrupted the dockyards, leaving the navy very short of seaworthy ships. This was something of a blip, quickly put right by Trafalgar; in general, this period saw the British take advantage of their technical and industrial capabilities to advance ship design and construction, improve weaponry with better guns and the formidable short-range carronades, and further increase their lead in efficient systems of rigging. By the end of the wars, they were poised to build a new generation of big sailing warships that decisively outperformed all their predecessors.

From the start of the wars in 1793 Britain had naturally reverted to a familiar blue-water strategy, combining European alliances and subsidies with colonial conquests. As usual this proved difficult and slow to execute, but by the Peace of Amiens (1801) extensive gains had been made, only for nearly all of them to be given back. After 1803 the job had to be done again, this time even more thoroughly, including the takeover of the Dutch possessions in the Far East. Although this policy certainly helped provide the resources to keep the war going and to subsidise Britain's allies, while naval power thwarted Napoleon's invasion plans, on its own a blue-water policy was never sufficient to win the war. Admittedly from 1809 onwards the navy was able to sustain the British and allied forces involved in the Peninsular War, which the duke of Wellington himself acknowledged could only have been won on the basis of maritime supremacy. In diplomatic terms, however, that supremacy actually worked against Britain, because most of her potential allies in Europe were just as hostile to British domination of the wider world as they were to French aggression on the Continent. This was one reason why it took so many years for the rulers of Europe to join in the general alliance against France they eventually formed in 1813. That alliance would never have come about but for Napoleon's ultimate effort to defeat Britain by economic warfare, through the unpopular and largely unenforceable Continental System, which led to the breach with Russia and the disastrous campaign of 1812. In 1815 the Prussian Field-Marshal Gneisenau declared that

> There is no mortal to whom Great Britain has greater obligations than to this blackguard, for it is the events which he has brought about which have raised England's greatness, security and wealth so high. They are lords of the sea, and neither in this dominion nor in world trade have they any rivals left to fear.

When Napoleon resorted to economic warfare, he was acting on the ancient French belief that the British economy was inherently fragile because it was based on trade rather than land. It was in this respect that the great wars of 1793–1815, in which the Franco-British conflict represented the ultimate confrontation between land and sea power, saw the eventual triumph of the latter. The Continental System did admittedly cause some major trading and industrial disruption in Britain, despite the innumerable leaks that allowed large, if reduced, quantities of British goods to continue reaching European markets. The compensation was a major reorientation of British maritime trade, most notably after the Spanish revolt of 1808 opened extensive new opportunities in Latin America. The 164,000 seamen in the merchant fleet at the end of the wars were indeed as significant as the slightly smaller number at sea on warships, embodying as they did a huge expansion in long-distance trade, and the rich financial rewards it brought. This interlocking maritime domination demonstrates just how far both the human and the material aspects of war at sea had developed since 1650: the great wind-powered warships of the 1800s were truly formidable weapons, capable of turning a European conflict into a global one. This conferred a peculiar advantage on Britain, whose unusually loose and flexible economic, social, and political structures may have provided a crucial advantage. In the maritime sphere a complex and integrated state of this type ultimately outperformed its traditional absolutist rivals, with their comparatively rigid structures, to a decisive extent. British success rested on the financial, administrative, and logistical advances of many decades, even centuries, without which Nelson and his colleagues would never have had the means to achieve their stunning victories.

15

# Warfare in early modern Africa, c. 1450–c. 1850

RICHARD REID

## Introduction

Sub-Saharan Africa was on the threshold of a new and violent era in the second half of the fifteenth century. The ensuing four centuries would see innovative forms of military organisation, novel cultures of militarism underpinning such systems, and new wars, as well as new ways of fighting them. There were often different factors at work in different regions; the presence of external drivers was a key distinction between Atlantic Africa and the rest of the continent, for instance. However, warfare across early modern Africa had much in common, in terms of the aim to control factor endowments, to maximise population, and to construct enduring ideological systems, whether territorially or culturally defined. In some ways – certainly in terms of the underlying trends and broad contours of Africa's military history – the existence or absence of external intrusion is a distraction, however significant it was in particular places at particular times. The outcome of the processes in motion between *c.* 1450 and *c.* 1850 was an expansion in military scale, the professionalisation of soldiery, the adoption of new weaponry, and the militarisation of the polity – whether 'state-based' or otherwise. The militarisation of African polities and societies was an ongoing process between the fifteenth and the nineteenth century, a period which in many ways witnessed the laying of the foundations of modern African political systems; this would culminate in a veritable military revolution in the nineteenth century, a transformation in the organisation and culture of violence, without which Europe's later partition of the continent cannot properly be understood.

We need to briefly consider Africa's enormous diversity – geographical, demographic, cultural, economic, political – in order to appreciate the contexts within which war evolved. (See Map 11.) To begin with, an understanding of different physical terrains is clearly critical. In the open grasslands of the great northern savannah, for example, armies tended to be combinations of cavalry and infantry, ranging over wide areas; the key challenge facing war-makers and would-be state-builders was to keep control of far-flung frontiers, and of their own horsed regiments which often resisted centralisation projects. South of the savannah, the presence of the tsetse fly – which spread trypanosomiasis, or sleeping sickness – meant that it was generally impossible to keep horses and other pack animals, with the exception of the far south (modern South Africa) and a handful of other pockets. Thus most sub-Saharan armies were dominated by infantry, with all the impediments associated with war waged exclusively on foot, not least in terms of supply and communication. In thick forest and woodland, armies generally had a narrower operational range. Nonetheless, in markedly fertile agricultural zones – the Ethiopian Highlands, the West African coastal forest, or the Great Lakes area of Uganda – large, centralised state-building projects were possible, and such states were able to field comparatively large armies over longer distances. In sum, agricultural surplus and political centralisation supported armed expansionism. In drier savannahs and grasslands – such as are found in parts of east and north-east Africa – pastoral societies dedicated to the keeping of livestock developed distinctive military systems based not on hierarchy and territorial conquest but on age-sets, in which young men were both warriors and herdsmen on behalf of the wider community. Clearly, distinct environments produced particular political forms, and thus in turn military systems which made best use of terrain and manpower. Across much of the continent, however, it was the concern to maximise population which drove a great deal of conflict: underpopulation in many areas led to continual processes of fission and fusion, as communities split and reformed, and as local migration led to the creation of new and frequently militarised frontiers between states and societies. It was the drive to master those frontiers – and conversely, the periodic march of the frontier on the centre – which was at the heart of much African warfare, in the early modern period as in more recent times.[1]

## The changing patterns of African warfare, from the mid-fifteenth to the late eighteenth century

There are only a handful of regions for which we have anything approaching detailed source material on the nature of warfare for the fifteenth century, including the Ethiopian Highlands, Egypt, and the western savannah. In Christian Ethiopia, large (and often cumbersome) armies – comprising

1 Richard Reid, *Warfare in African History* (New York, 2012), esp. ch. 1.

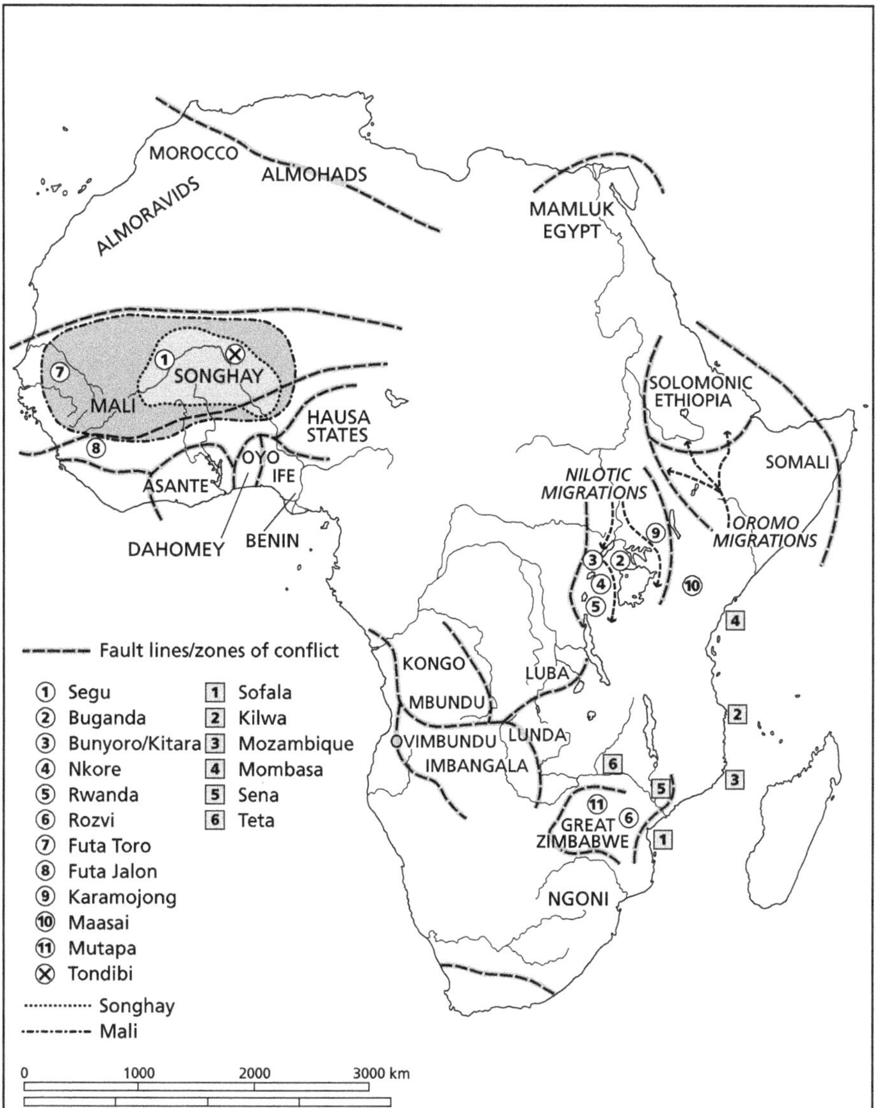


MAP 11. Warfare in early modern Africa, *c.* 1450–*c.* 1850.
Source: Richard Reid, *Warfare in African History* (Cambridge, 2012), map on p. xviii (military states in Africa, *c.* 1000–*c.* 1800). Reproduced with permission.

a royal core bolstered by infantry and some cavalry forces raised in the provinces – waged a series of wars with the forces of the Muslim sultanate of Adal from the late fifteenth century onward. In the most devastating of these conflicts, between 1529 and 1543, an Adal army equipped with Turkish guns, composed of Somali convert soldiers, and inspired by the religious leadership of Ahmad ibn Ibrahim brought the Christian state to the brink of destruction. A rump Ethiopian force adapted quickly, emulating the mobility and the compact size of Adal units, and in the meantime benefited from the timely intervention of some Portuguese musketeers, part of a force recently arrived in the Red Sea area.[2] Ethiopia survived, and in the decades which followed underwent something of a military reorganisation – driven in part by the retrenchment necessary in the wake of the Somali assault, and in part by the large-scale immigration of groups of Cushitic-speaking Oromo pastoralists. The Oromo were noted horsemen, while their system of age-grade regimentation made them a readily mobilised armed force – as highland Ethiopians discovered to their detriment. Yet, while some groups clashed with the latter, other Oromo became absorbed into highland society and, in time, became key political players and reinvigorated the imperial military.[3]

By the end of the sixteenth century, the Ethiopian army had adopted firearms to some extent, but guns remained very much a secondary weapon for some time to come. Instead, infantry were predominantly equipped with spears and swords, and were increasingly deployed in conjunction with Oromo cavalry groups. Elsewhere in the same period, however, guns were deployed rather more effectively in actual combat. Egypt had for the previous two centuries been governed by one of the world's most impressive military aristocracies, the Mamluks; yet their horsed regiments, sufficient to see off Mongols and Christian crusaders alike in the thirteenth century, were overwhelmed by Ottoman firepower in 1517 in what was perhaps the earliest engagement using firearms on African soil. It is tempting to see this as representative of a fundamental shift in available technology. The same could be claimed for the battle of Tondibi, where in 1591 a Moroccan army used muskets and light cannon to send the Songhay army scurrying from the field. Songhay – its power resting on archery and the rapid deployment of cavalry – had for some time been the pre-eminent power of the West African savannah. It now entered a spiral of – prolonged – decline, which would eventually see the great empire consumed by the desert, although the

2 Mordechai Abir, *Ethiopia and the Red Sea: The Rise and Decline of the Solomonic Dynasty and Muslim–European Rivalry in the Region* (London, 1980).
3 Donald Crummey, *Land and Society in the Christian Kingdom of Ethiopia* (Oxford, 2000).

invading Moroccan force did not, in the end, fare much better. In the years that followed, remnants of the army severed links with Morocco itself, and operated as mobile bands of raiders through the early decades of the seventeenth century before they were gradually absorbed by adjacent societies.[4] Meanwhile Songhay's own demise may have marked the temporary end to the great cavalry empires of the western savannah, but it did not denote the arrival of the firearm as the region's weapon of choice. Horsemen equipped with lances and, less commonly, bows remained the predominant military formation over the next 200 years or more, and it was only in the nineteenth century that guns had a significant impact on savannah warfare.

Meanwhile Atlantic Africa – defined here as reaching between the Senegambia region and Angola, and encompassing a sizable hinterland – arguably became the most overtly militarised zone anywhere on the continent during this period, owing in large part to the expansion of the slave trade during the seventeenth and eighteenth centuries. Yet a process of militarisation across Western and Equatorial Africa, owing to gradual population increase and an attendant expansion in political and economic scale, is discernible over several centuries prior to this. In West Central Africa, notably, the available evidence points towards greater military cooperation across wider areas, the fielding of ever larger armies, and in general ever more destructive patterns of warfare. By the fifteenth century, the intensification of organised violence involved the organisation of young men into age-sets (discernible in central and coastal Angola, for example) and the use of an increasingly diverse array of weaponry: throwing and stabbing spears, body armour, throwing knives, and battle axes are all in evidence in the period prior to the escalation of the Atlantic slave trade.[5]

Yet the external slave trade unquestionably led to heightened levels of violence and militarisation across a vast area. In Kongo, an early participant in the slave trade following the arrival of the Portuguese in the 1480s, slave-raiding armies increasingly equipped with muskets – and sometimes accompanied by contingents of Portuguese musketeers – scoured ever wider areas in search of captives for export. Kongolese political culture incorporated a strand of militarism which had its echo in rebellious provinces, where

4 Lansine Kaba, 'Archers, Musketeers and Mosquitoes: The Moroccan Invasion of the Sudan and Songhay Resistance (1591–1612)', *Journal of African History* 22 (1981), 457–75.

5 Jan Vansina, *Paths in the Rainforests: Toward a History of Political Tradition in Equatorial Africa* (London, 1990); Jan Vansina, *How Societies Are Born: Governance in West Central Africa before 1600* (Charlottesville, 2004).

violence escalated as the outcome of both demand for slaves from the coast and the predations of royal armies. In the course of the sixteenth century, Kongo descended into civil war. In the deeper interior, the trend was towards 'private' armies under entrepreneur warlords. In the area of modern Angola, for example, the Imbangala emerged out of slave-raiding violence in the late sixteenth century. Mobile bands of professional warriors, frequently serving as mercenaries in others' armies, they may have originated from a subversive faction of a local Ovimbundu army; in any case, they had soon overrun much of the Ovimbundu area, and in the 1570s and 1580s invaded the Atlantic coast. By the middle of the seventeenth century, there were several areas under the control of Imbangala bands stretching across central Angola. Imbangala armies had a ferocious reputation – theirs was a form of total war, involving widespread pillaging as well as the capture of people for sale – and they were organised along complex lines of command; boys were taken when young and trained up to be fed into the system. In the early seventeenth century, both the Portuguese and the kingdom of Ndongo used Imbangala mercenaries; famously, the defeated Ndongo queen Njinga Mbande underwent initiation herself as an Imbangala, and modelled a new community which replicated Imbangala militarism.[6]

While evidence relating to fighting formations is scant for the sixteenth century, it is clear that in the West African forest infantry forces fought in relatively tight formation – especially those which regularly encountered cavalry armies. Further south, in West Central Africa, somewhat looser formations are suggested by the evidence. But a general trend across the Atlantic zone was the deployment of archers as skirmishers in support of soldiers equipped with short spears, clubs, and knives whose main aim was to close on the enemy and engage the latter in hand-to-hand combat. There is no evidence of significant European influence on local tactics, at least in the sixteenth century, although some Portuguese were hired as mercenaries in Central Africa. Once again the impact of firearms was limited at this stage, as when they were deployed it was as missile weapons alongside bows, which were not displaced for some decades yet.[7] In any case, sixteenth-century guns – especially early 'trade guns', assembled specifically for African

6 John Thornton, 'The Art of War in Angola, 1575–1680', *Comparative Studies in Society and History* 30 (1988), 360–78; Linda Heywood, *Njinga of Angola: Africa's Warrior Queen* (Cambridge, MA, 2017).

7 R. Kea, 'Firearms and Warfare on the Gold and Slave Coasts from the Sixteenth to the Nineteenth Centuries', *Journal of African History* 12 (1971), 185–213; John Thornton, *Warfare in Atlantic Africa 1500–1800* (London, 1999).

commerce – were less than reliable in tropical conditions. The damp heat caused gun barrels to rust, and powder and matches were of little use in the rainy season.

The Imbangala exemplified a new kind of restless, predatory military culture, one generated by the slaving violence of the age. But other, no less predatory, political forms emerged with direct links to the external slave trade, in particular new territorial states with increasingly complex military systems. The process is exemplified by the polities of the West African coast and immediate hinterland, encompassing modern-day Nigeria, Benin, and Ghana, during the seventeenth and eighteenth centuries. Most of the major centralised and territorial states of the region deployed armies in a similar way and towards similar ends.[8] In Asante, which emerged in the course of the seventeenth century, the 'national' army was composed of contingents supplied by tributary chiefdoms. A large, permanent force, it underpinned further Asante expansion in the eighteenth century, during which time it was brought under the control of the king, the *Asantehene*, to an ever greater degree. The kingdom of Oyo, located in savannah land just north of the coastal forest, was capable of fielding armies of some 100,000 men – and its tsetse-free location meant that Oyo armies made extensive use of cavalry. While some cavalry units may have represented a 'standing army' – the professionalism invariably involved in the use of horses implies as much – most of Oyo's soldiers were part-time, drawn from the various provinces of the kingdom, and deployed as a 'national' force. The command structure was split between the *basorun*, commander of the metropolitan army, and the *kakamfo*, in charge of the provincial forces. Both office-holders represented considerable political authority, too, and each vied with the other as well as with the king himself, the *alafin*, for influence. By the late eighteenth century, it was the *kakamfo* who had emerged as the leading political as well as military figure.[9] More generally the army wielded enormous power in Oyo politics, no doubt owing to the fact that the slave-trading empire was dependent on military success for its commercial strength.

The army played a central role in the politics of Dahomey, too. Although Oyo drew tribute from Dahomey from the 1730s onward, the kingdom was a dynamic and expansionist state in its own right, its roots in the slave trade and organised around a predatory militarism. Dahomey fought wars for slaves, large numbers of which were earmarked for export, but slaves

8 Robert Smith, *Warfare and Diplomacy in Pre-Colonial West Africa* (London, 1989).
9 Robin Law, *The Oyo Empire: A West African Imperialism in the Era of the Atlantic Slave Trade* (Oxford, 1977).

intended for internal use were no less important. Dahomey's army comprised a series of well-drilled regiments increasingly equipped with firearms, in particular flintlock muskets. Each regiment made use of emblems and was sustained by a system in which young boys were assigned to soldiers for rigorous training. The head of the army, the *agau*, was assisted by two deputy commanders, the *zohenu* and the *fosupo*, and the army as a whole was divided into 'left' and 'right'-hand wings. As in Oyo, Dahomey's military leaders wielded considerable political influence, again underlining the central significance of the army in matters of state.[10] Further north, the Wolof states in the Senegal area likewise grew up in response to external demand for slaves, and organised themselves around the deployment of armies of royal slaves, or *ceddo*. Slave soldiers were prevalent elsewhere, too. In the early eighteenth century, on the upper Niger river, a young warrior named Biton Kulibali transformed his age-regiment into a highly effective force of armed slaves known as the *tonjon*. The kingdom of Segu resulted, embodying the increasingly professional organisation, the armed entrepreneurialism, and the ever more sophisticated cultures of militarism of the age.[11]

Across the region, the size of armies themselves doubtless varied from time to time even within particular polities. But along the Atlantic seaboard, armies were probably fairly compact: some of the smaller states fielded no more than a few hundred men at any given time, while Dahomey's forces probably ranged between 5,000 and 20,000. Oyo and Asante were capable of larger-scale mobilisation, and might field armies of between 50,000 and 100,000 – the latter figure representing, in all probability, an absolute upper limit, for few polities possessed the logistical or technological capacity to support such huge forces on the move. Logistical problems became more acute the further armies were from home. Most campaigns involved support systems of women and slaves carrying food and other supplies, but armies ranging further afield depended on foraging and requisitioning in enemy territory. These limitations meant that warfare was often outsourced: slaving states were frequently served by client polities which raided on their own accounts and passed slaves into the systems of the former. Beyond the forest, cavalry were crucial: in Oyo and Segu, for example, expeditions comprised horsed raiders supplemented by infantry, organised as the need or opportunity

10 Robin Law, *The Slave Coast of West Africa, 1550–1750* (Oxford, 1991).
11 Richard Roberts, 'Production and Reproduction of Warrior States: Segu, Bambara and Segu Tokolor, *c.* 1712–1890', *International Journal of African Historical Studies* 13 (1980), 389–419.

arose. In the savannah, these would have been more sweeping manoeuvres, characterised by somewhat looser formations. Ambush and more compact formations were favoured in forest and woodland. In Dahomey, carefully planned annual campaigns were favoured, involving the dispatch of armies towards well-established hunting grounds, especially to the north. The use of mercenaries, meanwhile, was increasingly common across the Atlantic zone through the seventeenth and eighteenth centuries, another indication of growing professionalism. Smaller coastal polities such as Allada and Whydah, for example, made use of Akwamu soldiers from the neighbouring Gold Coast region in the late seventeenth century.

Of course, the most important tactical innovation of this era was the increasingly widespread adoption of firearms. The fusil, or flintlock musket, was standard in European armies by the late seventeenth century, and by the early eighteenth century it was likewise displacing older matchlocks among the armies of Atlantic Africa. However, an immediate qualification is needed here: the flintlock's impact was very much limited to littoral societies and those just a little way inland. Savannah armies north of the West African forest generally eschewed guns for much of this period, cavalry forces preferring javelins, swords, and bows. And, although guns did reach the markets of the Central African interior, they were not generally deployed in battle. Extant armouries and tactics remained intact, with clubs, throwing knives, and short stabbing spears being used in hand-to-hand combat. Elsewhere, however, fusils had a major impact. In some areas, bows and arrows were almost completely displaced over the course of the eighteenth century. Along the Slave Coast – dominated by the Dahomeans, who adopted flintlocks with especial skill and vigour – guns were widely used, as they were too along the adjacent Gold Coast, evidenced by the hiring of mercenaries from that area on account of their shooting abilities. Firearms influenced battlefield formation, as the close formations of the seventeenth century sometimes gave way to more open, fluid deployments in the eighteenth, often involving smaller and more mobile units. Several African armies now sought to avoid close combat altogether, and amassed musketeers at vantage points from which fields of fire were designed to overwhelm enemies at distance. Exceptions to this loosening up of battlefield formation were to be found in the frontier zones between forest and savannah, where infantry armies were accustomed to facing enemies on horseback. Dahomean and Asante armies, for example, maintained close order and concentrated rates of fire, having observed that horses (such as those of Oyo) tended not to charge down such hostile obstacles.[12]

Similar, if not identical, processes of militarisation can be discerned in other parts of the continent, notably in East and South Africa, over much the same period, even if the effects appear somewhat less dramatic than in the Atlantic zone. In the south-east, Mutapa – the successor state to Great Zimbabwe – dominated the central Zambezi valley and adjacent coastal lowlands, under the rule of the *Mwene Mutapa*, a chiefly title indicating 'conqueror'. Its armies drew in tribute from a wide area, and protected the export of gold to the coast. In East Africa, several small chiefdoms had merged by the middle of the fourteenth century to form the kingdom of Kitara, soon to be succeeded by another military power, Bunyoro, in the foundation of which an immigrant clan from the north played a key role. These were the compact, militarised monarchies characteristic of the Great Lakes region of East Africa, capable of fielding sizable armies of part-time soldiers.[13] In the course of the sixteenth century, Bunyoro was joined by other states – Buganda, Toro, Nkore, Rwanda, and Burundi – which took advantage of the fertile arc between lakes Victoria and Tanganyika. While there was variation in the detail, each boasted a capacity for rapid military mobilisation and built traditions of martial prowess into their social and moral edifices. Kings needed to be war leaders, capable of protecting their peoples from outside attack but equally able to command regular campaigns aimed at territorial expansion and resource extraction.[14]

Across East Africa between 1500 and 1700, then, entrepreneurial leaders built up networks of clients – often through the distribution of cattle – which could be mobilised for military service, service which underpinned expanding political authority.[15] Military expeditions were probably longer-range than previously, involving men equipped with spears, bows, and the long shields characteristic of the region, and were aimed at both the extension of political territory and the extraction of neighbours' resources. Such campaigns were essentially predatory raids, but raids which sustained ambitious and increasingly complex polities. These were centralised political systems, but there were other models of militarism in the area, including those of decentralised

12 Kea, 'Firearms and Warfare'; Thornton, *Warfare in Atlantic Africa*.

13 David Schoenbrun, *A Green Place, a Good Place: Agrarian Change, Gender and Social Identity in the Great Lakes Region to the 15th Century* (Oxford, 1998).

14 Roland Oliver, 'Discernible Developments in the Interior, *c.* 1500–1840', in Roland Oliver and Gervase Mathew (eds.), *History of East Africa*, 2 vols. (Oxford, 1963), vol. I.

15 Schoenbrun, *A Green Place*.

pastoral peoples – the Karamojong and Maasai, notably – whose age-regiments facilitated a distinctive form of military service. Young men equipped with spears as tall as themselves could be mobilised over wide areas in defence of cattle and community.[16]

Ever more enhanced military states emerged during the seventeenth and eighteenth centuries, from the Ethiopian Highlands to the South African Highveld. Ethiopia had survived the trials of the sixteenth century and now consolidated its position around the new capital of Gondar from the 1630s: Gondar's lack of fortification suggested, perhaps, a greater sense of security than had been the case for many decades. Oromo units (both cavalry and infantry) were increasingly co-opted into the imperial army, and Oromo chiefly lineages wielded ever greater political power, rooted in their command of large numbers of soldiers. In particular, emperors created new Oromo units under their direct command in the attempt to counteract the military power of regional magnates. Ethiopian armies comprising contingents from several provinces might number between 50,000 and 100,000 infantry and cavalry, further swollen by enormous logistical corps. But the perennial problem of centrifugalism remained, as provincial overlords sought to utilise their own military strength to challenge central authority.[17] In the course of the eighteenth century, armed districts and hostile frontiers successfully eroded imperial authority to the point – by the 1770s and 1780s – that Ethiopia as a unified polity had effectively ceased to exist. A cycle of wars had begun which would last to the middle of the nineteenth century, fought by sizable provincial kingdoms competing over both resources and imperial ideology. Guns, meanwhile, were certainly available, especially in the northern districts which were more closely connected to the Red Sea trade, but they were only gradually having a significant impact on battle formations.[18]

Further south, Buganda's economic growth enabled it to make a bid for hegemony over the northern Great Lakes region, by fielding relatively large, well-fed armies of part-time soldiers recruited by provincial chiefs with an interest in a successful expansionist state. That expansionist state was directed by an increasingly powerful kingship which, in the course of the eighteenth century, was able to reward military loyalty and prowess with land and

16 See various contributions in Thomas Spear and Richard Waller (eds.), *Being Maasai: Ethnicity and Identity in East Africa* (Oxford, 1993).

17 Crummey, *Land and Society*.

18 Mordechai Abir, *Ethiopia: The Era of the Princes* (London, 1968); Richard Reid, *Frontiers of Violence in Northeast Africa: Genealogies of Conflict since 1800* (Oxford, 2011).

political appointments. Ganda armies equipped with spears and body-length shields were dispatched to raid for livestock and slaves, or were deployed to annex new districts blessed with especially desirable factor endowments. In this way Buganda incorporated tracts of good farmland and pasture, occupied areas rich in iron ore and wood for weapons, and began to reach southward to meet the global trade networks emerging towards the end of the eighteenth century. War, meanwhile, underpinned a thriving military culture at home, and a set of martial values according to which 'good citizenship' was measured. Yet there were constraints to the reach of Ganda armies, which became increasingly cumbersome owing to the large logistical corps which attended them – corps which ironically were designed to enlarge Ganda military capacity. And as its soldiers were mostly part-timers, whose labour was critical to the domestic economy, the kingdom waged war according to agricultural timetables.[19]

In Southern-Central Africa, similarly militarised state-formation processes are discernible, although the Maravi state at the south end of Lake Malawi proved rather less durable than those in the northern lake region. Between *c.* 1600 and *c.* 1650, under Masula of the Kalonga dynasty, this dynamic polity drew revenue from the ivory trade to sustain large and readily raised armies; but, in a pattern often replicated, the state barely outlived its charismatic leader. After Masula's death, politics became fragmented, leaving the region open to the predations of small-scale private armies. Something similar transpired south of the Zambezi, where in the seventeenth century the Mutapa state deteriorated to be replaced by a new, highly militarised polity under the leadership of the *Changamire*, whose army, the *Rozvi* ('destroyers'), dominated the northern part of the Zimbabwe plateau. Like the Mutapa forces before them, the *Rozvi* were not a standing army, but there is evidence of military reform in terms of the organisation of distinct regiments and ever greater use of firearms, imported from the Portuguese on the Mozambican coast. Yet it was their disciplined battle order rather than new technology which gave the *Rozvi* their decisive edge. They were able to rapidly mobilise relatively small forces of around 2,000 men, who carried out punitive and extractive expeditions against outlying districts, and protected the increasingly important coastal commerce. This was a highly professional outfit; and an ethos of professionalism was in evidence in the nearby Zambezi valley, too.

19 Richard Reid, *Political Power in Pre-Colonial Buganda: Economy, Society and Warfare in the Nineteenth Century* (Oxford, 2002).

Here, Portuguese *prazos* – land grants, originally created for white settlers – were increasingly dominated by Afro-Portuguese warlords, or *prazeiros*, the product of local intermarriage and a hardy, militarised frontier culture. In the course of the eighteenth century, these warlords led armies of slave-soldiers (known as *Chikunda*) not dissimilar in form and ethos to those of the West African hinterland, who spent much of their time feeding off burgeoning trade, raiding neighbours for tribute, and hunting. Like the *Rozvi*, the *Chikunda* made increasing use of muskets.[20]

East African pastoralists such as the Maasai, meanwhile, developed a somewhat different system whereby youthful aggression might be harnessed in the interests of the larger community: age-grades enabled the speedy mobilisation of small units of spearmen (perhaps numbering no more than a few dozen at any given time) who were deployed to raid neighbours' cattle and secure good pasture. A distinctive culture of militarism resulted, whereby young men were simultaneously herdsmen and warriors: the roles were indelibly intertwined. There were clear similarities with the Oromo in the Ethiopian region, and indeed with the Ngoni in the South-East African lowveld. Yet there was an important difference, for while the Maasai eschewed 'state'-formation in the conventional sense, Ngoni age-regiments underpinned a process of political centralisation. Among the Ngoni, age-regiments equipped with assegais and trained in rapid manoeuvre and close combat were increasingly harnessed by political elites who sought armies to consolidate political authority and protect economic resources. By the late eighteenth century, several centralised polities had emerged in the Ngoni region – highly militarised, and founded on a distinctive, professional military ethos – paving the way for a remarkable political-military transformation in the early nineteenth century involving the Zulu, as we will see below.[21]

Meanwhile, across the three centuries under review here, there was a gradual expansion of European military activity in Africa, mostly involving the Portuguese. From the early 1500s, Portuguese galleons armed with cannon moved along the Indian Ocean coast and subdued one Swahili city-state after another, culminating in the submission of Mombasa in 1589. The

20 Edward Alpers, *Ivory and Slaves in East Central Africa: Changing Patterns of International Trade to the Later Nineteenth Century* (London, 1975); Richard Gray, 'Portuguese Musketeers on the Zambezi', *Journal of African History* 12 (1971), 531–3.

21 John Wright, 'Turbulent Times: Political Transformations in the North and East, 1760s–1830s', in Carolyn Hamilton, Bernard Mbenga, and Robert Ross (eds.), *Cambridge History of South Africa*, vol. 1: *From Early Times to 1885* (Cambridge, 2010).

defence policy of the Swahili states, insofar as one can be discerned, had long focused on the land; attack from the sea had never been seriously contemplated. Having established a garrison at the mouth of the Zambezi, the Portuguese looked inland, capturing the Swahili trading posts of Sena and Tete in the 1530s. But beyond this their ambitions were thwarted by African military capability and malaria. Attempted Portuguese invasions of Mutapa in 1571 and 1574 were calamitous failures, demonstrating in the starkest terms that Europe's military technology proffered no clear advantage. They had a rather greater impact in Kongo – where Portuguese musketeers sometimes attached themselves to local slaving expeditions, and were sufficiently admired for their combat skills to be hired as mercenaries – and in Ethiopia, where their intervention against Ahmed ibn Ibrahim's Somali forces in the 1540s was ultimately successful. Elsewhere, Cape Colony at the Cape of Good Hope – founded in 1652 – was rapidly becoming one of the most violent and militarised of Afro-European frontiers. Here, Dutch settlers' seizure of Khoisan cattle prompted Khoisan counter-raids, and led to significant conflicts in 1659–60 and in the mid-1670s. Armed initially with matchlocks and then with flintlocks, the Dutch, or Boers ('farmers'), expanded onto Khoisan land from the late seventeenth century onward, compelling dispossessed Khoisan into a life of hunting and raiding. The Boers themselves would counter the growing Khoisan threat by forming militia units known as commandos, in what represented an escalation of frontier violence. Across the eastern borderlands of Cape Colony, too, the Boers were engaged in cyclical conflict with the Xhosa, who represented a rather more robust obstacle to their seemingly inexorable advance than the Khoisan. Prolonged warfare, waged between Boer and Xhosa raiding parties, was fought over the *Zuurveld*, Xhosa grazing land. This began in the 1770s and 1780s, and periodically erupted throughout the nineteenth century. The Boers would increasingly look to the British, in charge of Cape Colony from the 1790s, for assistance in these frontier wars; but British reluctance to expend money and manpower in pursuit of the Xhosa would give rise to new tensions within the territory.[22] Overall, the European military presence was clearly increasing – in West Central Africa, in Cape Colony, and along the eastern seaboard; and although its significance during this era should not be exaggerated, the nature of the encounter offered enough harbingers to suggest a rather more violent relationship in the decades to come.

22 Noel Mostert, *Frontiers: The Epic of South Africa's Creation and the Tragedy of the Xhosa People* (London, 1992).

## The transformation of warfare in the nineteenth century

The trend across the continent was clearly towards more overt and more sophisticated militarisation, and the increasingly efficient organisation and deployment of armed force on the part of polities of various hues. There had been, in sum, a marked expansion in the scale of warfare. But these dynamics became all the more intense in the first half of the nineteenth century, making it possible for us to talk of an African military revolution in this era. In West Africa, for example, the collapse of the Oyo Empire led to several decades of military reform among the Yoruba. Fortified city-states, the most powerful of which was Ibadan, were dominated by armed entrepreneurs who commanded private and highly professional armies of young men (known as 'war boys') armed with matchlocks and, increasingly, flintlock muskets. They demonstrated the primacy of military might in Yoruba politics. Drilled, disciplined, and trained in battlefield manoeuvre and in the use of firearms, these forces hired themselves out to the highest bidder and attracted the displaced and uprooted as they did so, symbolising a socioeconomic as much as a military revolution.[23] In this era of the 'illegal' slave trade (which had been abolished, at least at the European end, in the early years of the nineteenth century), there were plenty of commercial opportunities for well-armed, well-led, and cohesive bands of men, seeking material, social, and sometimes political advancement. The privatisation of violence and the social, political, and economic transformation it involved can be discerned across the continent. The Ovimbundu and Chokwe in west-central Atlantic Africa developed similarly entrepreneurial armies, again dedicated to the capture of slaves, whether for export or for domestic usage. Trading-and-raiding polities led by powerful, charismatic figures grew up in the Zambezi valley, too, a region in which the slave trade was actually expanding. Among the Yao of Malawi, Mozambique, and southern Tanzania, opportunistic but highly effective professional militaries emerged on the back of the Indian Ocean slave trade, with use often being made of slave soldiers. A distinctive, but not dissimilar, harnessing of youthful aggression by warrior leaders in search of economic and political opportunity can be seen among the pastoralist Maasai and Turkana in the East African grasslands.[24]

23 Ade Ajayi, 'Professional Warriors in Nineteenth-Century Yoruba Politics', *Tarikh* 1 (1965), 72–81; Robert Smith and Ade Ajayi, *Yoruba Warfare in the Nineteenth Century* (Cambridge, 1971).

24 John Lamphear, 'Brothers in Arms: Military Aspects of East African Age-Class Systems in Historical Perspective', in Eisei Kurimoto and Simon Simonse (eds.), *Conflict, Age and Power in North East Africa* (Oxford, 1998).

In some areas, men-at-arms were coupled to new state-building projects. Later in the nineteenth century, in the 1860s and 1870s, East Africa witnessed a transformation in the organisation and uses of violence. Among the Nyamwezi and the Kimbu, notably, dramatic economic change brought about by the growth of long-distance commerce, the predations of the slave trade, and the resultant erosion of older social ties and forms produced new forms of military leadership and private armies of displaced young men who formed the backbone of new polities. These new communities were sometimes transient, but they demonstrated remarkable military dynamism. Known among the Nyamwezi of northern Tanzania as *ruga ruga*, young fighters might be bandits and criminals, preying off commercial highways, or they might be harnessed by charismatic leaders such as Mirambo, who was known for his military skill. Mirambo's armies fought with flintlock muskets, and were noted for their speed of deployment and manoeuvre; as communities became increasingly 'urban' and fortified their settlements for protection, Mirambo developed siege tactics accordingly, and sought to absorb the defeated into his 'empire'. The state he created was unstable – it disintegrated shortly after his death in 1884 – but it was a remarkable political and social experiment while it lasted, representative of a broader trend across the region; namely, a marked expansion in political scale underpinned by new forms of warfare and military organisation. Among the neighbouring Kimbu, the war leader Nyungu-ya-Mawe sought to do much the same thing.[25]

Arguably, however, the most noted instance of this military revolution was among the Ngoni of South-East Africa, where heightened economic competition, a tendency towards political centralisation, and the increasing militarisation of the age-regiment system produced the imposing Zulu state. In the course of the 1810s, a remarkable commander named Shaka – who was in some ways exceptional, but in other ways represented a broader trend across much of sub-Saharan Africa – developed a highly militarised polity with a view to using war for both economic ends and social reconstruction. The Zulu regiments (*amabutho*) honed skills and tactics which had been developed in the second half of the eighteenth century, and fought in close formation with *assegais*, using rapid flanking manoeuvres to crush ponderous and unprepared enemies. Young men served in these regiments for many years before being released and permitted to marry, giving rise to a ferociously martial identity which succeeded in absorbing many non-Zulu. Yet the resultant turmoil – known as the *mfecane*,

25 Norman Bennett, *Mirambo of Tanzania, 1840?–1884* (New York, 1971); Richard Reid, *War in Pre-Colonial Eastern Africa: The Patterns and Meanings of State-Level Conflict in the Nineteenth Century* (Oxford, 2007).

alluding to violent upheaval – forced many communities to migrate away from the epicentre of the Zulu revolution. Some of these were scattered through modern-day Zimbabwe and Malawi by the 1830s and 1840s; other offshoots had reached northern Tanzania by the 1850s, where they undoubtedly influenced the Nyamwezi and Kimbu, noted above. Wherever they went, they brought with them Zulu military tactics and ethos; and many of those who came into contact with them sought to emulate the Zulu in turn. Thus for example the Swazi, Ndebele, Sotho, Bemba, Hehe, Bena, and Sangu – stretching from modern KwaZulu Natal to central Tanzania – became highly militarised polities in the original Ngoni mould, albeit with some local flavour. From mid-century, especially, these armies adopted firearms, which had been unavailable to (or, at any rate, had been eschewed by) Shaka himself.[26]

Across the West African savannah, religious ideologues harnessed local military cultures in pursuit of theocratic states, and reorganised their armies accordingly in the name of Islam. This was a pattern which in fact dated back to the early and mid-eighteenth century, with jihadist movements in Futa Jalon and later in Futa Toro making particularly effective use of archers. In both cases, early victories bred confidence and provided momentum. An even more spectacular jihad erupted in 1804 among the Fulani and Hausa of present-day northern Nigeria, led by the reformist cleric Uthman dan Fodio. It initially involved amassed infantry, which achieved what may have been a 'chance' victory at the battle of Tabkin Kwotto, where the jihadist forces' archers formed what seems to have been an impromptu square and drove the enemy cavalry from the field. Thereafter the insurrection grew in confidence, and the square became its defensive cornerstone; but cavalry forces were quickly developed as the offensive wings of the jihad, with horsemen wearing padded quilts and chainmail and equipped with lances and swords. Infantry, meanwhile, continued to use bows, and only in the mid-nineteenth century were firearms adopted with any vigour. By 1810, what became known as the Sokoto Caliphate had been established – and it would prove a durable creation.[27] Others across the region would draw inspiration from this militarily and ideologically dynamic entity. In the 1850s, the Tukolor jihad of al-Hajj Umar Tal made use of a highly regimented army equipped with firearms and

26 Elizabeth Eldredge, 'Sources of Conflict in Southern Africa, *c.* 1800–30: The "Mfecane" Reconsidered', *Journal of African History* 33 (1992), 1–35; Carolyn Hamilton, '"The Character and Objects of Chaka": A Reconsideration of the Making of Shaka as "Mfecane" Motor', *Journal of African History* 33 (1992), 37–63.

27 Joseph Smaldone, *Warfare in the Sokoto Caliphate: Historical and Sociological Perspectives* (Cambridge, 1977).

some small artillery pieces seized from the French in the Senegal area; Tukolor infantry and cavalry operated in tandem, and the army was noted for its capacity to undertake large-scale manoeuvres in formation. A little later, Samori Ture in the 1870s and 1880s would play down the jihadist element in his own state-building project, but would make effective use of professional soldiers on horseback as well as on foot, carefully organised into regiments. The Sokoto Caliphate had been launched in the 1800s by archers, and consolidated by horsemen with no interest in (or knowledge of) firearms; but, by the 1880s, Samori's men were increasingly equipped with modern guns, and they could repair guns, too, and make ammunition.[28]

Longer-established states, too, embarked on programmes of military reform, or expanded their military operations significantly. Following the collapse of Oyo, Dahomey expanded its slaving activities eastward into Yoruba territory, and thrived on the 'illegal' slave trade down to the 1850s, famously making use of an apparently new regiment of female fighters. Even after it embraced so-called 'legitimate commerce' (i.e. exporting agricultural produce rather than human beings), Dahomey's was one of the most deeply rooted military cultures anywhere in the continent; and war, of course, was necessary too to supply the slaves needed for the domestic economy.[29] In East Africa, Buganda had militarised its political establishment by the early nineteenth century, creating a series of new military chieftaincies which further underpinned a sophisticated and pervasive militarism which had much in common with that of Dahomey. Ganda ambition expanded significantly as long-distance commercial traffic built up around Lake Victoria, and the kingdom aspired to control of both land and lake trade routes. With those ends in view, Buganda became one of the largest importers of firearms in the region from the 1860s, although the 'gun culture' at the royal capital was arguably detrimental to actual battlefield performance: untrained but privileged young men armed with muskets increasingly enjoyed a prominence in Ganda armies which was disproportionate to their actual skill. Meanwhile, in order to offset the limitations of war on land, the Ganda developed a fleet of war canoes on Lake Victoria, designed to extend the kingdom's commercial as well as political reach. It was only partially successful, but the navy represented one of the most dramatic exercises in military reform anywhere in the region.[30] It was neighbouring Bunyoro, however, which arguably undertook even more effective military reforms by creating *barusura*,

28 Martin Legassick, 'Firearms, Horses and Samorian Army Organisation, 1870–1898', *Journal of African History* 7 (1966), 95–115.

29 Smith, *Warfare and Diplomacy*. 30 Reid, *Political Power*.

regiments of young men equipped with rifles. During the reign of Kabalega in the 1870s and 1880s, Bunyoro enjoyed something of a resurgence as a result.[31]

Firearms and military professionalism characterised Ethiopia's resurgence, too, although here violence was honed as a result of several decades of almost continual war between the 1770s and the 1850s following the collapse of the medieval Christian state. Entrenched cultures of militarism, new forms of military leadership, a greatly swollen military 'class', and ever more efficient deployment of armed force characterised the fierce commercial and ideological competition between the ethno-regional polities of the highlands, until a particularly gifted tactician, Kassa Hailu, won a decisive series of pitched battles between the mid-1840s and 1850s. Kassa had encountered the deadly efficiency of modern firepower while fighting the Egyptians on the north-west Ethiopian frontier in the 1840s, and dedicated his career to professionalising his army and acquiring that weaponry. In 1855, he was crowned Emperor Tewodros of a putatively 'reunified' Ethiopia, but spent most of his reign on campaign against his own subjects. He had sought to regularise his army, to pay his soldiers regular salaries, and to curb the natural inclination of highland troops to raid and ravage; in this he was unsuccessful, for the repeated challenges to his legitimacy led to Tewodros resorting to widespread, often shockingly punitive, violence against rebel soldiers and suspect communities alike. He killed himself in 1868, faced with a British expedition which had been dispatched to rescue some Europeans whom Tewodros had taken hostage; but his short reign, and those of his successors – Yohannes and Menelik – laid the foundations for a highly militarised polity, modern Ethiopia, and demonstrated the transformative power of violence in pursuit of political and economic goals. Later in the nineteenth century, Menelik in particular used Ethiopian military might – and an enormous arsenal of modern firearms acquired over many years of trade – to famously see off an Italian invasion, and to expand the boundaries of his empire to the south and east of the highland plateau.[32]

Tewodros, Yohannes, and Menelik each sought to acquire foreign military technology, and to some extent emulate the military modernity exhibited by European and Egyptian armies. It was a pattern discernible elsewhere, too, notably across North Africa, where exposure to European armaments was all

31 Shane Doyle, *Crisis and Decline in Bunyoro: Population and Environment in Western Uganda, 1860–1955* (Oxford, 2006).

32 John Dunn, '"For God, Emperor, and Country!" The Evolution of Ethiopia's Nineteenth-Century Army', *War in History* 1 (1994), 278–99; Sven Rubenson, *The Survival of Ethiopian Independence* (London, 1976); Reid, *Frontiers of Violence*.

the more intense. Egypt is both exceptional, in terms of the scale of its attempted military revolution, and also the exemplar of a larger process. Under Muhammad Ali, who from 1805 governed the ostensibly Ottoman province as an autonomous state, Egypt embarked on a remarkable series of reforms – economic, political, and military – which were at least partly inspired by Europe, while also aimed at preventing further European incursion. Building up an army of some 200,000 well-armed and disciplined troops, Muhammad Ali expanded southward into Sudan and the Red Sea, and seized control of the Hijaz; it was an exercise in military expansionism which was partially financed by a greatly expanded cotton industry. Yet some of the money also came in the form of European loans, ironically, while Muhammad Ali also sought to make use of European expertise as he saw fit. But when he invaded the Levant and Syria in the 1830s, an increasingly nervous European coalition intervened to cut him down to size. Defeated, Muhammad Ali's forces were reduced to some 18,000 men, while the great economic projects designed to sustain the Egyptian renaissance were placed under close foreign supervision.

Events in Egypt demonstrated that Europe would respond decisively and aggressively to perceived threats emanating from the continent, or to protect local interests, but more broadly European armed incursions in the first few decades of the nineteenth century were piecemeal and not supported by any great political enthusiasm. By mid-century, the French were waging aggressive, if 'low-level', war in Algeria, and the British were sporadically engaged in frontier wars in South Africa; but elsewhere, armed encounters between European and African forces were relatively rare. The British fought the Asante in the 1820s and again in the 1860s, in the attempt to protect and stabilise their 'Gold Coast' possessions; the 1867–8 campaign against Ethiopia was something of an anomaly. From their base in Senegal, the French were engaged with Umar Tal in the 1850s and 1860s. Usually these 'local' wars, dealing with recalcitrant 'tribes', as Europeans saw it, were a nuisance, and were entered into with great reluctance. This fact notwithstanding, however, by the 1850s and 1860s such armed encounters did indeed demonstrate growing European military power, and the widening gulf between African and European capability. By the middle of the nineteenth century, the rough parity between European and African militaries (and even, in some places at certain times, African superiority) was being superseded by clear European technological advantage. Firepower was obviously important: by the 1860s, Europeans had access to breech-loading rifles, which few Africans were able to get hold of, although there were exceptions. But just as significant was

Europe's organisational power: in the realms of communications and logistics, Africans could not compete. And although in the middle decades of the nineteenth century the idea of a 'scramble for Africa' was unthinkable – both in military and, more importantly, in political terms – the technological, organisational, and tactical advantages necessary to make it happen were already in place.[33]

## Conclusions: Towards the age of high imperialism

It is not within the scope of this chapter to discuss the 'scramble' for Africa in any detail.[34] Nonetheless, three points can be made in this respect by way of conclusion. The first is that the African military revolution described above involved, self-evidently, a dramatic escalation of armed competition between polities across the continent, providing small European-led forces with critical leverage over a number of states and societies. In many respects, accordingly, their influence was disproportionately heightened, vis-à-vis their actual size and military efficacy. In other words, the transformation of violence in Africa in the nineteenth century, and the expansion in the scale of warfare across the continent, facilitated Europe's own armed incursions in the later part of the century – which can in turn be considered something of an extension of Africa's military revolution.

Second, and related to this, the upheavals of the nineteenth century produced the swell of manpower, and the skills and knowledge of warcraft, on which European officers absolutely depended. War had generated, across the continent, pools of uprooted and dislocated young men (and women), armed entrepreneurs with a range of skills and local knowledge, which could be tapped into by Europeans, using a combination of luck and dexterity – just as they had been periodically tapped into by local warlords. The nineteenth century had engendered cultures of mercenary professionalism in violence, and it was these cultures which ultimately facilitated the European partition of the continent. Again, it is important to stress that the 'scramble' for Africa would never have been possible without African agency: for all its military advantages, Europe had neither the will nor the ability to impose itself on Africa without utilising local troops and expertise. And thus again, the era of

33 Douglas Porch, *Wars of Empire* (London, 2001); Bruce Vandervort, *Wars of Imperial Conquest in Africa, 1830–1914* (Bloomington and Indianapolis, 1998).

34 John Lonsdale, 'The European Scramble and Conquest in African History', in Roland Oliver and George Sanderson (eds.), *Cambridge History of Africa*, vol. 6: *From 1870 to 1905* (Cambridge, 1985).

the 'scramble' needs to be seen as at least partly the outcome of Africa's own prolonged period of revolutionary upheaval.

Finally, the violence in Africa described in this chapter – during the nineteenth century, in particular, but also throughout the era under examination – was consistently misunderstood by European observers and was used, ultimately, to justify European imperialism. Perceptions of African warfare became increasingly racialised; outside observers did not see the socio-political creativity, or the developmental potential, of African war, but only violent disorder which was symptomatic of African backwardness and needed to be brought to an end. As European discourse around the continent became increasingly couched in the language of race, and as the European 'civilising mission' took shape, centred on ideas about development and primitiveness, the 'savage barbarity' which was espied everywhere in Africa – not least the persistence of the 'illegal' slave trade – was used to validate armed intervention. It may not have been a direct *cause* of the partition, but it provided the essential cultural and moral framework within which imperialism was deemed not merely possible but necessary. This was a chronic misapprehension of what had been momentous events across Africa; and it is a view which has persisted to the present, in various ways. The late nineteenth-century interpretation of much African violence as fundamentally barbaric and illegitimate has proven robust, and has frequently prevented more nuanced understanding of the role and practice of warfare in the continent's early modern history.

16

# Warfare in India, 1450–1850

KAUSHIK ROY

## Introduction

The dominant interpretation of warfare in the Indian subcontinent before the establishment of British rule is that it was comprised of unorganised melees by forces of undisciplined militia. This stemmed from the fact that pre-British Indian states were weak polities with divisible sovereignty; they were – to use the terminology of Burton Stein – segmentary states, lacking any concept of frontiers and standing armies. The divisive caste system of India further debilitated the pre-British indigenous states and armies. The argument goes that the rise of British power in the second half of the eighteenth century resulted in a sea change in warfare. The British introduced a bureaucratic state with standing armies capable of waging decisive battles and conclusive sieges in India. This interpretation dates back to two nineteenth-century British scholars of colonial India. They argued that Indians were incapable of constructing stable states and structured armies due to their racial failings.[1] And at the beginning of the twenty-first century, historians may have substituted a racial analysis for a cultural one, but otherwise they argue along more or less the same lines, that the limited scale of organised inter-state violence reflected the constraints upon the states of pre-British India.[2]

More recently, two historians – Jos Gommans and Dirk Kolff –have attempted to explain the unique political and military culture of India in terms of South Asia's environment. They claim that the teeming millions of India were dependent on the unpredictable monsoon-fed agriculture. When rainfall failed, marginal peasants were eager to join forces raised by power brokers, who could readily arm these recruits and

1 George-Bruce Malleson, *The Decisive Battles of India* (1885; repr. Jaipur, 1986); William Irvine, *The Army of the Indian Moghuls* (1903; repr. Delhi, 1994).

2 Stephen Rosen, *Societies and Military Power: India and Its Armies* (New Delhi, 1996); Pradeep Barua, *The State at War in South Asia* (Lincoln and London, 2005).

provide them with horses. This generated a martial culture among certain peasant communities, the net result of which was the creation of a dynamic military labour market. The presence of so many warriors did not encourage rulers to maintain permanent armies. Rather, in times of emergency, *ad hoc* forces could be raised from the market of free-floating, part-time peasant mercenaries. Moreover, this vigorous military labour market prevented any Indian state from establishing a monopoly over organised violence and enabled power brokers easily to challenge the authority of a centralising state apparatus. Only once they had demilitarised this labour market could the British establish a centralised state structure. This is an interesting thesis, but one that begs the question of why Indian powers were unable to demilitarise this military labour market as the British did. Gommans and Kolff imply that this was because the Weberian notion of state building was exclusive to Western Europeans and was absent among the Indian rulers whose political culture accepted the concept of fragmented and multiple layers of sovereignty.[3]

In contrast, this chapter argues that the military situation in South Asia was much more complex than the picture painted above suggests. Warfare in pre-British India was not in a state of *stasis* but was continuously evolving. Decisive battles and sieges waged by standing disciplined armies occurred in the subcontinent long before the arrival of the British. This is not to minimise the importance of the transformation in state building and warfare initiated by the British during the eighteenth and nineteenth centuries. Here, 'India' refers to South Asia as a whole, including Afghanistan. (See Map 12.) This chapter is divided into three sections. The first deals with the nature of warfare before the rise of the Mughal Empire; the second section focuses on the Mughal era; whilst the last section discusses the rise of British military power in India.

## Warfare in pre-Mughal India: *c.* 1000–1520s

### *Turkish and Rajput systems of warfare*

From the late tenth century CE onwards, Turks from Afghanistan started attacking India. By 1206, they had established the Delhi Sultanate, which

3 Dirk Kolff, *Naukar, Rajput and Sepoy: The Ethnohistory of the Military Labour Market in Hindustan, 1450–1850* (Cambridge, 1990); Jos Gommans, 'Warhorse and Gunpowder in India, *c.* 1000–1850', in Jeremy Black (ed.), *War in the Early Modern World: 1450–1815* (1999; repr. London and New York, 2004), pp. 105–27.

MAP 12. Map of India

endured until the establishment of Mughal rule in South Asia. The Turks were able to repeatedly defeat the *Rajputs* – who had established regional principalities in north and west India – due to their superior technology and military doctrine. While the *Rajputs* used simple bamboo bows, the Turks used longer-range composite bows which were also much easier to handle from horseback. Moreover, the *Rajput* straight swords proved less versatile than the lighter *tulwars* (curved swords) of the Turks, which had a longer reach and could easily, and repeatedly, be used from horseback to slash at enemy horsemen. Turkish iron stirrups also provided a superior platform than the wooden stirrups of the *Rajputs* for shooting arrows from horseback. Lastly, in terms of speed and stamina, the *Rajput* horses were inferior to the steppe-bred horses of the Turks. Throughout the pre-modern period, the sedentary monarchies of India and China were unable to breed good-quality horses in the fertile river valleys dedicated to crop cultivation. Both regions lacked adequate amounts of grassland required for breeding good horses whilst the hot and humid climate of India is especially ill-suited to horse breeding.[4]

The *Rajput* way of warfare was dominated by the heavy cavalry charge, led by the king seated on an elephant. Turkish horse archers would repeatedly dislocate the command of *Rajput* forces at the beginning of a battle by killing such exposed leaders. Whereas the *Rajput* culture of warfare prohibited night attacks, the Turks excelled in using the cover of darkness to surprise the enemy. The *Rajput* code of conduct also prohibited pursuing a defeated enemy. At the first battle of Tarain (1191 CE), the numerically superior *Rajput* cavalry force was able to blunt the Ghurid cavalry, but the *Rajput* king Prithviraj Chauhan (r. 1177–92) allowed the Turks to withdraw. In contrast, the Turks pursued the defeated enemy to their destruction. They utilised light-horse archers to soften their enemy from a distance, before delivering the *coup de grâce* with a heavy cavalry charge. This classic Turkish tactic succeeded at the second battle of Tarain (1192 CE), which resulted in the defeat of the *Rajput* Confederacy and the establishment of the Turkish Ghurid dynasty in Delhi, which ultimately flowered into the Delhi Sultanate.[5]

The *Rajputs* were unable to stand up in set-piece battles against the Turks and resorted to static defensive warfare, erecting massive stone forts. However, from the thirteenth century onwards, the expansionist Delhi Sultanate was able to demolish these forts using traction trebuchets, which they learnt to use from West Asia. Between the second half of the thirteenth century and the first half of the fourteenth century, the Mongol Jagatai khanate launched repeated attacks against the Delhi Sultanate. The

4 Simon Digby, *War Horse and Elephant in the Delhi Sultanate* (Karachi, 1971).

5 Rashmi Upadhyaya, 'Military Strategy of the Rajputs during the Early Medieval Period', in Shahabuddin Iraqi (ed.), *Medieval India 2: Essays in Medieval Indian History and Culture* (New Delhi, 2008), pp. 173–90; Jagadish Narayan Sarkar, *Military History of India* (1960; repr. Bombay, 1970), pp. 32–7.

introduction of market reforms allowed Sultan Alauddin Khalji (r. 1296–1315) to extract revenues adequate for maintaining a large, well-trained cavalry force and disciplined *paiks* (infantry). The core of the army was comprised of *ghulams* (slave soldiers), a technique which had spread from the Middle East into India. The Sultanate cavalry was trained in *furusiya* exercises, like the Mamluks of Egypt. Following the Mongols, the sultans conducted large hunts to train the cavalry in conducting intricate manoeuvres. Under Sultan Muhammad bin Tughluq (r. 1324–51), the Delhi Sultanate's cavalry numbered 600,000 men. Medieval Western European armies were tiny by comparison, and only in China was a pre-modern polity able to maintain larger armies. All these measures enabled the Sultanate to contain the Mongols. The success of Timur's invasion of India (1398) did not reflect any Tartar technological or doctrinal superiority, but rather stemmed from the fact that the Delhi Sultanate had by then broken up into several competing parts.[6] Not for the first time, politics proved to be the crucial driver in shaping military affairs.

### *Gunpowder warfare in the Deccan*

Richard Eaton and Phillip Wagoner argue that the introduction of gunpowder weapons in the Deccan in the second half of the fifteenth century instigated a military revolution.[7] In fact, the ancient Hindus had used sulphur for making fire arrows whilst the Arabs in Sind, and then the Turks, had used naptha fire. Whilst campaigning in India, Mahmud of Ghazni (r. 971–1030) had used explosive mines in siege warfare. The knowledge of gunpowder entered India in the late thirteenth century with the Mongols.[8] The Delhi Sultanate thus possessed pyrotechnic devices. Indeed, the term 'military revolution' is coming under challenge from historians of Western warfare. Whether or not a military revolution occurred in the Deccan between 1450 and 1550, gunpowder weapons certainly, and decisively, changed the face of battle in the Deccan during this period.

During the second half of the fifteenth century, the Bahmani Sultanate – which had become independent of the Delhi Sultanate – used cannons (wrought-iron breech-loaders) in siege warfare. They had acquired gunpowder technology from the Mamluk Sultans of Egypt. In response, Krishna Raya (r. 1509–29) of the Vijayanagara Empire, the principal opponent of the Bahmani Sultanate,

6 Peter Jackson, *The Delhi Sultanate: A Political and Military History* (Cambridge, 1999).

7 Richard M. Eaton and Phillip B. Wagoner, *Power, Memory, Architecture: Contested Sites on India's Deccan Plateau, 1300–1600* (2014; repr. New Delhi, 2015), pp. 241–2.

8 Iqtidar Alam Khan, *Gunpowder and Firearms: Warfare in Medieval India* (New Delhi, 2004), p. 1.

accessed gunpowder technology from the Portuguese. To control the lucrative spice trade of the Indian Ocean, the Portuguese supplied gunpowder technology to Vijayanagara for use against the Bahmanis, who were in turn supported by Portugal's rivals, the Mamluks of Egypt and later the Ottomans. In 1510, Bijapur – a province which had become independent of the Delhi Sultanate – used field artillery and bombards against the Portuguese. Yet the adoption of gunpowder weapons in India was not merely a case of copying external powers. The gunsmiths of the western Deccan amalgamated the Ottoman and Portuguese technologies and built upon them further. So much so, that in 1513 the Portuguese reckoned that the Deccani matchlocks were superior to their own, and on an equal plane with Bohemian firearms. On 19 May 1520 at the battle of Raichur Doab, Vijayanagara's combined infantry–cavalry troops armed with matchlocks were able to overwhelm the cavalry and field artillery force of Bijapur. The two armies were technologically matched, and the Bijapuri defeat was due to the inept tactical handling of their field artillery.[9] The point to be noted is that, despite the use of gunpowder weapons, the mainstay of Indian warfare, cavalry, did not disappear from the battlefield but continued to play an important role.

As a result of their defeat at the hands of Hindu Vijayanagara, the Muslim sultanates of the Deccan scrambled to modernise their arsenals. The Bidar Sultanate produced forge-welded iron cannons. Trunnions were introduced by the Ottoman engineers. From the Portuguese, the swivel fork was adopted, allowing the lateral movement of the barrels. The upgraded gunpowder technology of the Muslim sultanates of the Deccan was an important factor behind the defeat at the battle of Talikota (5 January 1565) of the Vijayanagara Empire, which had grown complacent after its victory at Raichur Doab.[10] But the victory of the Muslim sultanates proved to be short-lived in the face of the expansion of the Mughals in north India.

## The Mughal military revolution: 1526–*c.* 1730s

Following Irvine, Jos Gommans has claimed that Mughal warfare was spectacularly theatrical. The Mughal force was all pomp and show, while its combat effectiveness was minimal: it was a machine for the political co-option of potential enemies. It was a sort of disorganised, non-cohesive

9 Richard Eaton, 'India's Military Revolution: The View from Early Sixteenth-Century Deccan', in Raziuddin Aquil and Kaushik Roy (eds.), *Warfare, Religion, and Society in Indian History* (New Delhi, 2012), pp. 85–108.

10 Sarkar, *Military History of India*, pp. 91–9; Eaton and Wagoner, *Power, Memory, Architecture*, pp. 243–87.

political umbrella, ill-suited for fighting battles of annihilation. Like the Mughal army, the Mughal Empire itself was, as Douglas Streusand puts it, a theatre state.[11] In fact, the Mughal army was a bureaucratic, combat-effective, and technologically proficient force. In addition to gunpowder weapons, the armed forces of the Mughal Empire continued to depend heavily on cavalry.

### *Babur's revolution in military affairs*

Having established himself at Kabul, Zahiruddin Muhammad Babur (1483–1530), a military adventurer from Ferghana, conquered north India and founded the Mughal Empire. After defeating Sultan Ibrahim Lodhi (r. 1517–26) at the first battle of Panipat (21 April 1526), he was able to establish Mughal supremacy in Delhi. By combining Uzbek *taulqama* tactics with gunpowder weapons acquired from the Ottomans, Babur initiated a revolution in military affairs in India. *Taulqama* tactics involved attacks on the flanks and rear of an enemy force with horse archers. Whilst the horse archers attacked the flanks and rear of the enemy, Babur's field artillery (light leather cannons, similar to those used by Gustavus Adolphus, and light brass mortars) and infantry (equipped with matchlocks supported on tripods) directed their firepower against the enemy's front. Babur constructed field fortifications – digging ditches and deploying wagons, like the Hussite leader Jan Žižka – as an added protection for his gunpowder infantry. Babur's deployment of gunpowder infantry and field artillery was influenced by the Ottoman order of battle at Chaldiran (23 August 1514).

Babur's victory was also facilitated by the political disintegration of the Lodhi Sultanate. When Babur attacked India, Punjab declared independence against Ibrahim, and the *Rajputs* under Rana Sanga (1482/4–1528) were already autonomous in west India. Deccan and Bengal had been lost long ago. The Indian rulers were dependent on the supply of Central Asian horses through Afghanistan, but Babur severed this supply. In desperation, Ibrahim raised a levy of Afghans and fell back on the use of elephants. With his rag-tag, undisciplined army, Ibrahim launched a frontal charge on Babur's force at Panipat. Ibrahim depended on sheer numerical superiority (80,000 men against Babur's 12,000 trained troops) to overwhelm the Mughals. While Mughal field artillery and matchlock men tore huge gaps in Ibrahim's

11 Jos Gommans, *Mughal Warfare: Indian Frontiers and High Roads to Empire, 1500–1700* (London and New York, 2002); Douglas Streusand, *The Formation of the Mughal Empire* (Delhi, 1989).

assaulting troops, the horse archers attacked the Lodhi force from the flanks and rear. Babur noted in his autobiography: 'His right and left ... were massed in such a crowd that they could neither move forward against us nor force a way for flight.'[12] This battle was a classic example of the *Kesselschlacht* (cauldron battle) at the tactical level. Some 16,000 Lodhi soldiers died in the confrontation. The same tactical formula enabled Babur to defeat Rana Sanga at the battle of Khanwa (17 March 1527). The frontal charge of *Rajput* armoured cavalry equipped with swords and lances – similar to the Frankish cavalry of the Crusades – was wiped out by an army of matchlock infantry and field artillery (cannons and stone-throwing mortars used as anti-personnel weapons) at the front and horse archers at the rear and both flanks.[13] At the battle of Ghagra (1529), Babur defeated the Afghans of Bihar. The Afghans used *bans* (rockets which came from south China across the Himalayas to Bengal and Bihar), but they were no match against the field artillery, matchlock men, and horse archers of the Mughals.

### *Akbar's* mansabdari *system*

Babur's grandson Jalaluddin Muhammad Akbar (r. 1556–1605) initiated a true military revolution by integrating the Mughal revolution in military affairs with broader social constituencies. Akbar introduced the *mansabdari* system to systematise the Mughal force. The *mansabdars* were organised in thirty-three ranks. Officers were promoted and demoted according to their performance and were rotated throughout the empire. In accordance with their ranks, they had to maintain a fixed number of cavalry and baggage animals. For raising and maintaining the cavalry contingents, the officers were granted *jagirs* (land grants). The *jagirs* were not hereditary but were dependent on military service. To broaden the military, the *Rajput* chieftains were also granted *jagirs* and were integrated within the *mansabdari* system. Akbar also upgraded his military's weaponry. Matchlocks were lightened to enable the infantry to better manoeuvre on the battlefield, and brass and bronze field guns were also manufactured. The Mughal gunpowder infantry fought in open order, emphasising skirmishing and marksmanship. The presence of horse archers, elephants, and camel-mounted swivel guns – equivalent to the horse artillery which emerged in Western Europe much later – made the adoption of a linear battle order impracticable. The Mughal revenue

12 *Babur-Nama (Memoirs of Babur), Translated from the Original Turki Text of Zahiruddin Muhammad Babur by Annette S. Beveridge*, 2 vols. (New Delhi, 1989), vol. 2, p. 474.

13 Ibid., p. 547.

system was able to extract 50 per cent of gross agricultural produce, and about 80 per cent of the state budget went towards maintaining the *mansabdars*.[14] Such a polity could hardly be characterised as a minimalist state, as some Western historians have done.[15] The population of Mughal India was about 150 million and Akbar's army was able to absorb 10 per cent of the male population of his empire.[16] Under Akbar's grandson Shah Jahan (r. 1628–1658), when the empire was at its zenith, the Mughal army had on its payroll some 600,000 cavalry and 500,000 infantry. The gargantuan size of the Mughal army could be matched only by the Ming and Manchu empires in China.

### *Decline and fall of the Mughals*

Andrew de la Garza claims that the Mughals initiated a military revolution between 1500 and 1600, but it was an unfinished revolution. Somewhat like Paul Kennedy and Tonio Andrade, Garza argues that the absence of a serious military threat decelerated the rate of Mughal military-technical advances.[17] Conversely, I argue that the Mughal failure to come up with cast-iron field artillery and flintlock muskets reflected the nature of the hostile forces facing the Mughals. Emperor Aurangzeb (r. 1658–1707), Shah Jahan's son, directed his energies against the two surviving polities within the subcontinent: the Bijapur and Golkunda sultanates. Both polities relied on massive fortifications to contain the advancing Mughal force.[18] Consequently, the focus of the Mughal Empire was on constructing bigger siege cannons, rather than quick-firing light field artillery suited for use on the battlefield. The Mughals continued to adapt and improve their siege artillery, shifting from stone to iron cannonballs.

After the Bijapur and Golkunda sultanates were conquered in 1686 and 1687 respectively, the principal threat which emerged in the Deccan was the Marathas. Under their charismatic leader Shivaji (1627/30–1680), the Maratha peasants started an insurgency against the Mughals. After Shivaji's death, the Maratha insurgency gained further momentum. The Maratha way of war

14 Jos Gommans and Dirk Kolff, 'Introduction: Warfare and Weaponry in South Asia: 1000–1800 AD', in Gommans and Kolff (eds.), *Warfare and Weaponry in South Asia: 1000–1800* (New Delhi, 2001), p. 23.

15 Ibid. 16 Streusand, *The Formation of the Mughal Empire*, p. 41.

17 Andrew de la Garza, *The Mughal Empire at War: Babur, Akbar and the Indian Military Revolution, 1500–1605* (London and New York, 2016).

18 Jadunath Sarkar, *History of Aurangzeb*, 5 vols. (1919; repr. Bombay, 1972), vol. 4, pp. 83–139, 259–326.

involved *ganimi kava*: the word *ganim* means a horseman, and *kava* refers to the forward and backward movement of light cavalry. The Maratha cavalry conducted hit-and-run attacks against the Mughal line of communications, implementing ambushes, establishing roadblocks, and conducting raids far behind the Mughal frontlines. Against this guerrilla warfare, it made no sense to invest in capital-intensive weapon systems like cast-iron field artillery and flintlock muskets. Moreover, in the mountainous terrain of western Maharashtra the Mughals found it difficult, if not impossible, to drag their guns with them. This region also lacked navigable rivers for transporting the heavy guns. Further, the Marathas did not wait to be attacked, and retreated when the Mughals advanced.[19] Therefore, the Mughals needed a manpower-intensive, rather than a capital-intensive, conventional, force to conduct counter-insurgency operations against the Marathas. When the Mughals had their hands full against the swarming tactics of the Maratha light cavalry, the Persian storm broke over their heads.

Nadir Shah, emperor of Persia (r. 1736–47) probably had the most powerful conventional army in Asia at that time. The 'Napoleon of Asia' had already fought successfully against the Russians and Ottomans. In 1603–7, 1620–3, and 1649–53, the Mughals and the Safavids had fought over Kandahar. The Mughal failure before Kandahar was due to their lack of adequate siege artillery and weak logistics. The Mughal road from Lahore to Kabul and thence to Kandahar failed to adequately sustain the Mughal forces in southern Afghanistan just as the Spanish Road in Flanders could not adequately maintain the Spanish army in the Low Countries. However, the Safavids were unable to advance east of Kandahar.[20] In 1738, though, Nadir Shah moved into Mughal-occupied Afghanistan. Kabul and Ghazni fell easily to the Persian conqueror. This was partly because the pay of the Mughal field force in Kabul was in arrears. The ongoing counter-insurgency in the Deccan and the governor of Bengal's declaration of independence (revenue from this province was used to supply the Kabul garrison) prevented the Mughals from sending any funds and soldiers to strengthen their frontier defences in Afghanistan.[21]

19 Jadunath Sarkar, *A Short History of Aurangzeb* (1930; repr. New Delhi, 1979), pp. 154–201, 239–95.

20 Kaushik Roy, *War and Society in Afghanistan: From the Mughals to the Americans, 1500–2013* (New Delhi, 2015), pp. 35–48.

21 Kaushik Roy, *Warfare in Pre-British India: 1500* BCE–*1740* CE (London and New York, 2015), pp. 142–4.

The Mughal imperial force advanced to Karnal, where Nadir's Persians confronted them. On 24 February 1739, the battle of Karnal was fought. Nadir enjoyed superiority both in the fields of software (tactics, doctrine, training, leadership) and hardware (weaponry). The quick-firing Persian field guns destroyed the slower-firing heavy Mughal guns, which lacked mobility and flexibility. The handguns of the Persian infantry were also superior in terms of range and rate of fire to the muskets of the Mughal infantry. Lastly, the Persian horses were much better than the Mughals' Indian-bred horses, and – due to financial difficulties – the Mughals were unable to import Central Asian horses in large quantities from Kabul. As crucial, the leadership of the Mughal emperor Muhammad Shah (nick-named *Rangila* for his excessive addiction to wine, women, music, and dancing; r. 1719–48) proved to be highly deficient, and the Mughal high command during the battle also fragmented. The Mughal army was divided into three battle groups. While Nizam, the semi-independent governor of the four Mughal provinces in the Deccan, commanded the left battle group, Muhammad Shah himself commanded the Mughal centre and the semi-autonomous Mughal governor of Awadh Saadat Khan (1680–1739) commanded the right battle group. Muhammad Shah failed to coordinate the actions of the three battle groups, and the troops under Nizam and the Mughal emperor did not even participate in the battle. Consequently, the Persian forces were able to annihilate Saadat Khan's battle group.[22] Karnal dealt a death blow to the Mughal Empire. Nadir Shah plundered the Delhi treasury after the battle, and the Mughal Empire passed into the dustbins of history. By this time, the British East India Company (EIC) was making inroads in South Asia.

## British conquest of India: *c.* 1740s–1849

The scholarly consensus is that the EIC, by introducing the Western way of warfare (which emerged in Western Europe out of the military revolution that occurred between 1500 and 1750), was able to overwhelm the feudal cavalry force of the segmentary Indian states. The argument (influenced by John Brewer) runs that the British created a military-fiscal apparatus which enabled the EIC to raise and sustain permanent bureaucratically sustained armies. At the same time, fragile indigenous polities lacked the sustained

22 Michael Axworthy, *The Sword of Persia: Nader Shah, from Tribal Warrior to Conquering Tyrant* (London and New York, 2006), pp. 175–208.

cashflow necessary to keep their forces in the field throughout the year.[23] Conversely, here it is argued that the British adapted to the Indian terrain by adopting certain indigenous elements and fusing them with imported Western military techniques. The net result was a hybrid military organisation. In this situation Indian powers were not static but attempted to modernise their state structures and armed forces. However, the indigenous powers failed to maintain a stable balance between the imported Western elements and indigenous military techniques.

### *The East India Company's hybrid military system*

Rather than trade, land revenue was the chief source of income for the EIC. The demand for more land revenue resulted in continuous wars and expansion, which in turn required ever larger armies. From the very beginning, the EIC realised that it lacked adequate white manpower to build a field army of the magnitude required to conquer India. Bringing Europeans from home and transporting them back to Britain was not only time-consuming, but also costly. The only option for the EIC was to absorb Indian manpower, and train and equip these troops in the Western style of warfare. This meant recruiting them into the infantry and drilling and disciplining them to Western standards. These troops were referred to as sepoys. A sepoy was four times cheaper than a British soldier in India. Furthermore, European soldiers frequently got sick in the dusty, hot, and humid climate of India, where the life of a sepoy was substantially longer than that of a European private. In 1801, the EIC had 120,000 sepoys.[24] The British realised that to campaign over long distances in India – the distance between Calcutta and Delhi, for instance, is equal to the distance between London and Saint Petersburg – and to counter the cavalry-heavy forces of their indigenous opponents, the EIC needed a large cavalry force. The British were unwilling to depend on the supply of horses from Afghanistan, which was under the Durrani monarchy, and so they imported horses from Australia. However, this proved to be

23 Dirk Kolff, 'The End of an *Ancien Regime*: Colonial War in India, 1798-1818', in Johannes A. de Moor and H. L. Wesseling (eds.), *Imperialism and War: Essays on Colonial Wars in Asia and Africa* (Leiden, 1989), pp. 22–49; Douglas Peers, *Between Mars and Mammon: Colonial Armies and the Garrison State in Early Nineteenth-Century India* (London, 1995); Channa Wickremesekera, *'Best Black Troops in the World': British Perceptions and the Making of the Sepoy, 1746–1850* (New Delhi, 2002); Christopher Bayly, 'The British Military-Fiscal State and Indigenous Resistance: India 1750–1820', in *The C. A. Bayly Omnibus* (2009; repr. New Delhi, 2010), pp. 238–75.

24 Kaushik Roy, 'Military Synthesis in South Asia: Armies, Warfare, and Indian Society, *c.* 1740–1849', *Journal of Military History* 69 (2005), 685.

immensely expensive, and the required number of horses and European cavalrymen were just not available. Therefore, the British started recruiting Indian cavalry troopers (*sowars*), equipped with lances and swords in the Indian style. In terms of horsemanship, the British officers accepted that the Indians were better than the Europeans. To manage the Indians, the British absorbed certain indigenous organisational formats such as the *panchayats* – which comprised a board of five *panches* (non-commissioned officers) who were responsible for the internal administration of the regiment and meting out disciplinary punishments – instead of introducing their own court-martial mechanism. However, the British were smart enough not to induct the Indians into the commissioned officer ranks and did not train the Indians in commanding large formations. To bridge the linguistic and cultural gulf between the British officers and Indian privates, some Indians were trained as non-commissioned officers. Ultimately, the EIC's armies were a brown force officered by the British. In 1830, the number of Indians in the service of the EIC's land force was five times higher than the number of Europeans.[25]

For logistics, the EIC utilised the *banjaras* (mobile grain traders), who had supplied the armies of South Asia from the early medieval era onwards. For a host of non-combatant duties, the EIC recruited Indians as coolies, tailors, armourers, blacksmiths, *beldars* (for making roads), and so on. The British copied the indigenous practice of using elephants and camels as baggage animals for operations in eastern and western India, respectively. Mughal news writers were pressed into service for espionage and intelligence gathering. The ordnance factories set up by the British in India employed Indian labourers and used saltpetre from Bihar. The British also adopted the *bans*, which they later developed into Congreve Rockets.[26] While the Royal Navy deployed frigates and ketches for additional firepower, the Bombay Marine of the EIC utilised the indigenous *grabs* – a shallow-draft boat with three masts and a prow like a Mediterranean galley – and *gallivats* – which utilised both oars and sails – for operating in the coastal water off the Malabar Coast. The British adapted to Indian waters by adopting indigenous crafts to supplement their Western war vessels. The Western regimental system, howitzers, and battleships were thus fused with indigenous combat and non-combat techniques. It was this hybrid military machine

25 Peers, *Between Mars and Mammon*, p. 63.
26 Kaushik Roy, 'Rockets under Haider Ali and Tipu Sultan', *Indian Journal of History of Science* 40 (2005), 646–50.

that enabled the British to overwhelm not only the French in India but also the indigenous powers.

### *Incomplete modernisation and the collapse of the Indian regimes*

The Mysore kingdom, the Maratha Confederacy, and the *Khalsa*/Sikh kingdom were the three principal Indian powers that challenged the British military in the subcontinent. The accession of Haider Ali (1720–82) in 1761 heralded the rise of Mysore as a threat to the British. He and his son, Tipu Sultan, fought four wars with the EIC. During the First Anglo-Mysore War (1767–9), Haider had light cavalry, infantry equipped with muskets and bayonets, and artillery which was serviced by the French. Haider's light cavalry was able to ravage the EIC's Mysore presidency. Since the slow-moving EIC infantry could not come to grips with Haider's light cavalry, who in turn were unable to defeat the former in a set-piece battle, the war ended in a draw. During the Second Anglo-Mysore War (1780–4), Haider requested that the Marathas join the war against the British, but the Marathas made peace with the EIC. What would have happened had Mysore and the Maratha Confederacy joined forces remains one of the great counterfactuals of Indian history. After Haider's death in 1782, his son and successor Tipu carried on the struggle for another two years. This war also ended in a draw, with the parties signing the Treaty of Mangalore on 11 March 1784.[27]

Tipu Sultan (r. 1782–99) realised that Mysore's light-cavalry-centric force was able to pillage and plunder the British-controlled Indian countryside from which the EIC derived revenues, but he could not destroy the EIC's armies on the battlefield. He therefore started raising Western-style infantry and field artillery on a large scale. He realised that the maintenance of a capital-intensive army would require greater financial resources, which in turn demanded the wholesale restructuring of the political fabric of Mysore: transforming Mysore into a true fiscal-military state. The *amils* – officials responsible for assessing and collecting revenue from cultivated land – conducted a survey of the entire kingdom. Agricultural land was brought under the direct management of the state. A department of finance and revenue was established, maintaining records in Persian, Marathi, and Kanarese. The ordnance department not only supervised the stores and dealt with the manufacture of arms and ammunition, but was also

27 Kaushik Roy, *War, Culture and Society in Early Modern South Asia: 1740–1849* (London and New York, 2011), pp. 80–7.

responsible for supplying the army with provisions and war stores. The muskets manufactured in its factories were equivalent in quality to the smoothbore muskets used by the EIC's infantry. Iron foundries were set up for manufacturing cannons. A bank and mints were established to pay the troops in cash. Tipu launched campaigns of pacification to centralise the state apparatus. For instance, large parts of Mysore were under the control of *poligars*, semi-autonomous chieftains who ruled their domains from fortified strongholds with their own feudal armed retainers and quasi-bureaucratic civilian administrations. The *poligars* never paid revenues to the state and, when the central government was weak, conducted private wars amongst themselves. In response, Tipu launched repeated punitive campaigns against the *poligars*, demolished their fortified bases with cannons and forced them to disband their private forces. Lands previously under the control of the *poligars* were brought under the direct administrative control of the Mysore monarchy.[28]

Anxious to nip in its bud the threat that Tipu posed, the EIC attacked Mysore in 1790, resulting in the Third Anglo-Mysore War. Lacking an adequate number of light cavalry, the EIC was able to convince the Maratha chieftains to supply them with their own. Lord Cornwallis (1738–1805) laid siege to Seringapatnam, the capital of Mysore. Although logistical failure prevented Cornwallis from capturing the fort, Tipu was nevertheless humbled. He lost one-half of his kingdom when the war ended in 1792. Tipu realised that the British could be defeated if their sea lines of communications were cut. He also decided to raise revenue from maritime trade. Therefore, he started building an ocean-going fleet, sending emissaries to Constantinople and Paris for shipwrights able to manufacture warships.[29] With the aid of French engineers, Tipu started converting the Seringapatnam fort in accordance with *trace italienne* design. The British could not ignore this threat. In 1799, they instigated the Fourth Anglo-Mysore War. Once again, British diplomacy triumphed; the EIC was able to forge an anti-Mysore alliance with the ruler of Hyderabad and the Marathas. In a short sharp campaign, the numerically superior British-led alliance categorically defeated Mysore. Tipu died fighting when Seringapatnam was stormed.

Before Tipu, Shivaji had realised that a navy comprised the first line of defence. In 1659, he had established a Maratha navy. The fleet of the most famous Maratha sea lord, Kanhoji Angre (1669–1729), who became the

28 Roy, *War, Culture and Society in Early Modern South Asia*, pp. 72–9.

29 Philip Macdougall, *Naval Resistance to Britain's Growing Power in India, 1660–1800: The Saffron Banner and the Tiger of Mysore* (Suffolk, 2014), pp. 149, 152.

admiral of the Maratha navy in 1697, comprised *pals*, *gurabs*, and *gallivats*. These vessels were built to operate in shallow waters and lacked the heavy guns to take on the British capital ships. In many ways, the Maratha ships fought like the classical Greek and Roman navies. Instead of depending on firepower generated by the broadsides, as did British ships, Maratha vessels utilised the tactics of grappling and boarding. Kanhoji's ships were fast and manoeuvrable in the coastal waters where the British man of war's deep draught placed it at a disadvantage. Kanhoji's naval guerrilla tactics (somewhat equivalent to the French *guerre de course*) in the coastal waters off Maharashtra inflicted huge losses on the British merchantmen. However, British diplomacy and divisive Maratha politics destroyed the Maratha navy, which was not a public enterprise of the state but rather a private venture of powerful chieftains. Manaji Angre accepted the authority of *Peshwa* (prime minister of the Maratha Confederacy) Balaji Vishwanath, but Manaji's half-brother, Tulaji Angre, challenged his authority. In 1755, the *Peshwa* ordered Ramaji Pant to attack Tulaji with the aid of the British in Bombay. Manaji's fleet transported Ramaji Pant's army to Sevendurg, the principal base of the Maratha navy, and the EIC's fleet provided support to the Maratha ground force. The surrender of Sevendurg marked the end of the Maratha navy.[30] When the decisive Second Anglo-Maratha War occurred (1803–5), the Maratha navy was a force of the past.

The Maratha Confederacy fought three wars with the EIC. The First Anglo-Maratha War lasted from 1775 to 1782. The Marathas, influenced by the French in the Deccan, had raised gunpowder infantry (*gardi* corps) for fighting with Ahmad Shah Durrani (1723–72) – a former Afghan general of Nadir Shah's who, after Nadir's death in 1747, had proclaimed himself the ruler of an independent Afghanistan and founded the Durrani dynasty – during the third battle of Panipat (14 January 1761). While the *gardi* corps was equipped with French-pattern muskets, Ahmad's *Rohilla* infantry muskets had greater range and penetrative power.[31] After their defeat at the hands of Ahmad, the Marathas lost faith in gunpowder infantry and reverted to their traditional *ganimi kava*. Further, the political economy of the *Peshwa* was in shambles, and he could not maintain such a capital-intensive force. In 1780, the EIC's Bombay army under General Thomas Goddard advanced along the Western Ghats. However, the Maratha light cavalry, equipped with lances, swords, and matchlocks, and the camel-mounted infantry, equipped with

30 Kaushik Roy, *The Oxford Companion to Modern Warfare in India: From the Eighteenth Century to Present Times* (New Delhi, 2009), pp. 258–60.

31 Hari Ram Gupta (ed.), *Marathas and Panipat* (Chandigarh, 1961).

wheel-lock muskets, utilised their established swarm tactics and forced Goddard's withdrawal.[32] Since the EIC was also fighting Mysore, it decided to make peace with the Marathas.

After this war, Mahadji Sindia (1730–94), the leading member of the Maratha Confederacy, realised that Maratha light cavalry could not destroy the EIC's gunpowder army with their time-honoured tactics of harassment. Therefore, he decided, at roughly the same time that Tipu Sultan was assembling such a force in Mysore, to raise a Western-style infantry army. Mahadji hired 260 European officers in order to westernise his army. The most famous among them was the Savoyard Benoit de Boigne, who in 1793 started raising *campoos* (brigades) of ten infantry battalions each. The recruits were drilled for six months, including twenty-eight days of musketry training, forty days of gunnery practice, and a three-month period over which they were taught military manoeuvres. Regular rates of pay and wound pension in cash – much higher than those available in the EIC's armies – were introduced. Both the EIC and the *campoos* recruited from the peasants of Rohilkhand and Awadh. Boigne set up numerous foundries and arsenals for casting brass guns and grapeshot. The *campoos* were provided with 4-, 8-, and 12-pounder field guns. The Maratha artillery was based on the Gribeauval system and had elevating screws, technology derived from both Ahmad's army and the French.[33]

After destroying Mysore, the EIC turned its attention towards the Maratha Confederacy. Randolf G. S. Cooper asserts that British victory in the Second Anglo-Maratha War (1803–5) reflected the EIC's domination of South Asia's political economy and the successful use of espionage.[34] But the breakdown of the Maratha high command also played a crucial role in the British victory. The most crucial battle of this war was fought at Assaye (24 September 1803). Just before the engagement, the Maratha command structure disintegrated because all the European officers deserted the *campoos* and joined the EIC.[35] Still, the *campoos* put up a spirited resistance. One British officer warned Major General Arthur Wellesley (later the duke of Wellington, b. 1769–d. 1852) before the battle: 'I tell you general . . . their infantry and guns will

32 Madhav Ramchandra Kantak, *The First Anglo-Maratha War, 1774–1783: A Military Study of Major Battles* (Bombay, 1993).

33 Kaushik Roy, 'Firepower-Centric Warfare in India and the Military Modernization of the Marathas: 1740–1818', *Indian Journal of History of Science* 40 (2005), 611–13.

34 Randolf Cooper, *The Anglo-Maratha Campaigns and the Contest for India: The Struggle for Control of the South Asian Military Economy* (Cambridge, 2005), p. 284.

35 John Pemble, 'Resources and Techniques in the Second Maratha War', *Historical Journal* 19 (1976), 375–404.

astonish you.' Wellesley went on record saying that he was more uncertain of victory at Assaye than at Waterloo. Jaswant Rao Holkar (1776–1811) entered the fray after Daulat Rao Sindia (Mahadji's successor) was humbled due to the internecine rivalry still wracking the Maratha state. Jaswant Rao by following *ganimi kava* was able to defeat Colonel George Monson's force near Chambal. If Jaswant Rao and Daulat Rao had cooperated, then things would have been far more difficult for Wellesley. Finally, in 1805, the EIC's commander-in-chief, Lord Lake, pursued Jaswant Rao to the borders of Punjab and forced him to surrender.[36]

After the Second Anglo-Maratha War, the Marathas were down but not out. The *Peshwa*, irritated by continuous British meddling in his internal affairs, declared war against the EIC in 1817, though this final war, the Third Anglo-Maratha War, would last for less than two years. The *Peshwa* was supported solely by the Maratha chieftain of Berar. Holkar and Sindia, remembering their defeat at the hands of the British in the preceding war, remained neutral. After 1803, the EIC had prevented the further westernisation of the Maratha forces by barring Maratha chiefs from employing European mercenaries and by occupying the seaboard of India, thus denying indigenous powers the means to forge maritime links with any outside power hostile to the British. Hence, the *Peshwa* and his ally Berar Raja fought with only light cavalry and were ultimately defeated in a grand counter-insurgency campaign conducted by the EIC in central India.[37] The Maratha Confederacy was abolished, and the EIC turned its attention towards the last surviving indigenous power in the subcontinent: the *Khalsa* kingdom.

After the British victory in the Second Anglo-Maratha War, Ranjit Singh (r. 1801–39), the founder of the *Khalsa* kingdom of Punjab, realised the importance of Western-style combat. He started recruiting into *Dal Khalsa* (*Khalsa* army) the personnel of the *campoos* which had been disbanded by the British after 1805. Ranjit also employed European officers – especially those who had served in Napoleon's *Grande Armée* – to train recruits in Western warfare. To strengthen the loyalty of these foreign officers, they were forced to marry Indian women and were barred from doing anything which went against the tenets of the Sikh religion. Many of them wore *pugris* and grew beards and all were required to take an oath that they would fight, even against their mother country if so required. The Sikhs sneered at marching in step and called it *rusq-i-lulooan* (bloody fool's

36 Roy, 'Firepower-Centric Warfare in India', pp. 619–20.

37 Reginald Burton, *The Maratha and Pindari War* (1910; repr. Delhi, 1975).

ballet). They only wanted to fight as light cavalry. Since the Sikhs were unwilling to serve as infantry soldiers, Ranjit recruited the north Indians, namely Gurkhas and Kumaunis. From the British, the *Dal Khalsa* adopted horse-drawn artillery guns which replaced the Mughal camel-borne swivel guns. To maintain the westernised military establishment Ranjit, like Tipu, transformed the apparatus of his state. The troops were paid regularly in cash and a finance ministry, as well as a pay office for the troops, were established. Mints and banking facilities were introduced. The craftsmen at the gun foundries built upon the cannon technology provided by French artificers. Qualitatively, the muskets manufactured in Punjab were on a par with, if not better than, the Brown Bess used by the EIC personnel.[38]

After Ranjit Singh's death in 1839, though, civil–military relations disintegrated. The European officers dabbled in the succession politics of the *Khalsa* kingdom. Resentful of the privileges and power that they had amassed, the Sikh chieftains banded together to force the European officers to leave Punjab. However, the personnel of the westernised battalions of *Dal Khalsa* refused to accept the authority of the Sikh chieftains who were not trained in Western-style combat. The *Dal Khalsa* thus became a headless giant like the *campoos* in 1803. The EIC decided it was time to destroy such a force. On 21 December 1845, the *Dal Khalsa* clashed with the EIC at Ferozeshah. The Sikh field guns were quantitatively and qualitatively superior to the EIC's artillery park. General Hugh Gough's frontal assaults resulted in heavy casualties for the infantry brigades of the EIC. Gough's troops had spent all their ammunition by the day's end. The next day, Tej Singh arrived from Ferozepur with fresh troops and 80,000 pounds of gunpowder, but instead of attacking the British, he retreated. This reflected the fierce tensions within the Sikh high command, and the deep mistrust between the *Khalsa* court and the commanders of the *Dal Khalsa*.[39] Palace intrigue, rather than military superiority, thus ensured that the EIC survived to fight another day. By 1846, the EIC's force was able to defeat the *Dal Khalsa*. The British then started demobilising the westernised battalions of the *Dal Khalsa*. The disgruntled Sikh chieftains made a last attempt to fight the British, which resulted in the Second Anglo-Sikh War (1848–9). Like the Maratha chiefs in the Third Anglo-Maratha War, though, the Sikh chiefs were forced to fight the EIC with only light cavalry. Drilled and

38 Roy, *The Oxford Companion to Modern Warfare in India*, pp. 86–90; Kaushik Roy, 'The Hybrid Military Establishment of the East India Company in South Asia: 1750–1849', *Journal of Global History* 6:2 (2011), 195–218, 201.

39 Roy, 'Military Synthesis in South Asia', pp. 683–4.

disciplined EIC's infantry equipped with Brown Bess muskets and supported by field artillery and *sowars* were able to destroy the Sikh force.[40] By 1850, the whole subcontinent had come under British control.

## Conclusion

The political landscape of India prior to the rise of the EIC was not covered with fragile political entities, fighting each other with undisciplined, unorganised militias; pre-British warfare was not 'a flower war'. The battles which were fought before the rise of British power were certainly bloody. However, the rise of the British radically transformed the politico-military landscape of South Asia. The mid-eighteenth century could be taken as 'zero hour', if such a concept can be applied to history. Despite the introduction of gunpowder weapons in the thirteenth century, horse archers remained a crucial element in warfare. But their dominance ended with the rise of hybrid British military organisations in the mid-eighteenth century. The British combined Western gunpowder weapons with Indian manpower and animal resources, resulting in the emergence of hybrid military organisations under the auspices of the EIC. By this time, quick-firing field artillery and infantry equipped with flintlock muskets and organised in regiments were able to outrange the traditional horse archers. However, the superiority of gunpowder was episodic rather than intrinsic, and the hybrid military organisation allowed the EIC to project power from its coastal enclaves to India's interior.

Both Mysore and the Marathas attempted to check British naval supremacy in the Arabian Sea. However, both powers had only riverine and coastal navies, and these regional navies failed to pose a credible threat to the British. However, the greatest failure of Mysore, the Maratha Confederacy, and the *Khalsa* kingdom was in the sphere of military software, not hardware. While technologically their forces reached a level equivalent to that of the EIC, they were unable to create an indigenous officer corps. None of the Indian powers attempted to set up officer academies to train their aristocrats and their sons as Western professional officers. Dependence on the European mercenaries to lead their westernised units proved to be a deadly error. Drilled and

40 Kaushik Roy, 'Technology and Transformation of Sikh Warfare: *Dal Khalsa* against the *Lal Paltans*, 1800–49', *Indian Journal of History of Science* 41 (2006), 387–410.

disciplined infantry, quick-firing field artillery, and cavalry – both heavy cavalry, who acted as a battering ram, and light cavalry, acting as scouts and skirmishers – and a fiscal-military organisation that supported these armies represented a paradigm shift in warfare. It was one that remained dominant until the dawn of the twentieth century.

17

# Japan and Korea, 1450–1850

EUGENE Y. PARK

The Japanese archipelago and the Korean peninsula are neighbouring regions, with histories of similarities and contrasts. Currently inhabited by a population of about 123 million in an area slightly bigger than Germany, Japan has been relatively isolated throughout much of the last two millennia. In the late nineteenth century, Japan reinvented itself from a land on the margins of the Sinitic cultural zone to a world power. In contrast, Korea – a landmass a little larger than Great Britain and inhabited by about 78 million people – was an active participant in the China-centred world order during much of its 2,300 year-history before losing sovereignty to Imperial Japan in 1910. (See Map 13.)

To an extent, the history of war in early modern Japan and Korea from about 1450 to 1850 explains the two regions' contrasting fortunes by the turn of the twentieth century. In 1860 when Japan dispatched its first diplomatic mission to the United States, which had been the first Western nation to establish relations with both Japan and Korea, its members were quintessential samurai: each man with his front pate shaved, wearing the distinctive *kamishimo* dress, and, of course, bearing a set of long and short swords (*katana, tantō*). When the first Korean mission visited America in 1883, its personnel were aristocratic scholar-officials donning a horse-hair top hat and a voluminous high waist-line robe (*top'o*).

As of the mid-nineteenth century, the most powerful among Japanese and Korean elites alike were learned bureaucrats versed in a cannon of classics in literary Sinitic ('classical Chinese'), but the military's place in these two societies was markedly different. In early seventeenth-century Japan, almost incessant warfare among various political actors came to an end with the accession of the stable, centralised Tokugawa Shogunate (1603–1868) dominated by a bureaucratic warrior elite who upheld a code of martial virtue, though Buddhism, Confucianism, and Shintoism were also influential ideologies. In contrast, in Korea under the Chosŏn dynasty (1392–1910), a centralised

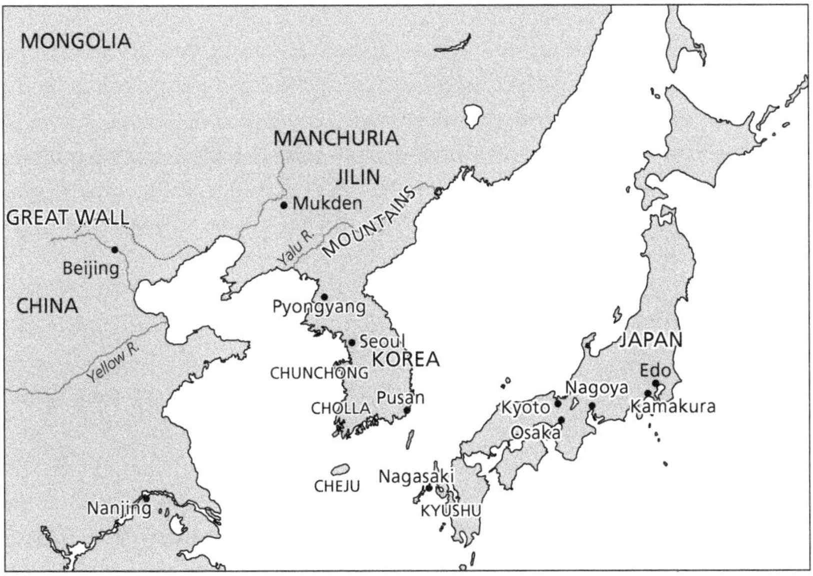


MAP 13. Map of Japan and Korea

monarchy governed through an officialdom dominated by hereditary *yangban* ('two orders', civil and military officials), in which a civil aristocracy held precedence over a military aristocracy, and Confucianism remained the official ideology.

The following discussion of the history of war in Japan and Korea divides the subject period into two parts: an era of sweeping transition (*c.* 1450–*c.* 1650) followed by one of relative stability. Constant fighting in Japan among various armed groups during this earlier phase – referred to as the Sengoku ('Warring States') Period (1467–1573) – ended with the political reunification effected by the Tokugawa Shogunate. In Korea an earlier military organisation based on military service by males aged from sixteen to *sixty se* (lunar years), except slaves (*nobi*), disintegrated, and was replaced by a semi-professional army. Between 1650 and 1850 in Japan, the samurai – who evolved into a warrior elite with bureaucratic functions whilst upholding the *bushidō* ('the way of the warrior') code – dominated a centralised *bakuhan* system maintained by the shogunal government (*bakufu*, military government), comprising domains (*han*), each governed by a daimyo (hereditary military governor). In Korea a civil aristocracy backed by the muscle of a military aristocracy continued to control the state. A relatively stable social hierarchy in both Japan and Korea did not entail much upward social mobility, at least not in the sense of allowing any non-elites to join the existing hereditary elite. Nonetheless, non-elites with wealth, influence, and status aspiration increasingly performed local security duties or pursued swordsmanship in Japan, whereas in Korea the government-administered military examination (*mukwa*) and martial popular tales of vernacular narrative fiction (*kundam sosŏl*) appealed to *chungin* ('middle people'), who were government-employed experts in various areas, and commoners.

## From war to peace in Japan, 1450–1650

From the mid-fifteenth to the early seventeenth century, a decentralised Japan experienced cataclysmic upheaval. The troubles began with the Ōnin War (1467–77), which massively reduced the power of both the imperial court and the Ashikaga Shogunate (1338–1573) in Kyoto. The war saw the rise of the daimyo as regional strongmen, each supported by forces of professionalised samurai, and other armed groups, which fuelled the subsequent incessant warfare.[1] This lengthy Sengoku Period finally gave way to a period of reunification initiated by the capture of Kyoto (1568) by Oda Nobunaga (1534–82), the first of the three Great Unifiers. Progress towards reunification was seemingly undone with a renewed round of civil war following the death of the second Unifier, Toyotomi Hideyoshi (1536?–98). But when Tokugawa Ieyasu (1543–1616) emerged victorious from this conflict after the battle of Sekigahara (1600), he managed to establish a stable, centralised polity that would rule Japan from Edo (present-day Tokyo) for the next two and a half centuries.

During the Sengoku Period – when both the imperial court and the Ashikaga Shogunate exercised only nominal authority – disputes were overwhelmingly militarised.[2] Efforts to achieve and improve military capability were widespread not just among the major daimyo, smaller-domain lords (*kokujin*, *dogō*), and lower-status warriors, but also among various other autonomous groups and individuals.[3] Achieving *de facto* political

1 Mary Elizabeth Berry, *The Culture of Civil War in Kyoto* (Berkeley, 1994), p. 29; David Spafford, *A Sense of Place: The Political Landscape in Late Medieval Japan* (Cambridge, MA, 2013), pp. 123–257.

2 Tatsusaburō Hayashi, 'Kyoto in the Muromachi Age', in John W. Hall and Toyoda Takeshi (eds.), *Japan in the Muromachi Age* (Berkeley, 1977), p. 22; Kenneth Grossberg, *Japan's Renaissance: The Politics of the Muromachi Bakufu*, 2nd ed. (Ithaca, 2000), pp. 52–63.

3 David L. Davis, '*Ikki* in Late Medieval Japan', in John W. Hall and Jeffrey P. Mass (eds.), *Medieval Japan: Essays in Institutional History* (Stanford, 1974), pp. 222–46.

independence, religious organisations such as the *Shin* Buddhist (Jōdo Shinshū) *Hongan-ji* defended monasteries large enough to function as bases from which to mobilise local peasants as fighting forces; others maintained defensive installations, such as fortress ramparts and walls. *Rōnin* – the majority of whom were warriors of peasant status unattached to a lord or master – served as mercenaries, and some became *ninja*, known for their specialised combat skills. Chronic violence varied in scale from armed demonstrations to the massive battle of Sekigahara with over 80,000 troops fighting on each side, a battle which practically mobilised the entire archipelago and divided it into two camps. In practice, though, most armed conflicts were smaller disputes confined to a daimyo domain.[4]

Oda Nobunaga initiated the process of reunification, bringing roughly a third of Japan under his control. Beginning his career as a minor warlord, in 1573 Nobunaga formally ended the Ashikaga Shogunate. Apart from skilled leadership and the ability to mobilise larger forces than his enemies, Nobunaga's success also rested on his use of infantry (*ashigaru*) firearms. First introduced in 1543 by Portuguese merchants, by the 1560s the Japanese were mass-producing muskets. In 1575 at the battle of Nagashino, Nobunaga revolutionised musket tactics in Japan with a division of labour between loaders and shooters, assigning three guns to each shooter. The crushing victory obtained by Nobunaga over Takeda Katsuyori (1546–82), his most formidable enemy, encouraged other daimyo to rush to acquire muskets, but ensured that Nobuaga retained his military advantage. By the time of his death in 1582, at the hands of a treacherous general, Nobunaga had forced the submission of the daimyo of most of central Japan, even if he had not defeated them outright.[5] He also subjugated other autonomous groups, including religious organisations. In an effort to break the power of the Buddhist establishment, he supported the Christians, whose number had been growing since the arrival of European missionaries and merchants in 1542.[6]

Nobunaga's *de facto* successor, Toyotomi Hideyoshi, completed the reunification of Japan, although the political breakdown after his death meant that the archipelago redivided. Hailing from a peasant background and beginning his career as a servant of the Oda clan, Hideyoshi rose to

4 Anthony J. Bryant, *Sekigahara 1600: The Final Struggle for Power* (Oxford, 1995), p. 25.

5 George Elison, 'The Cross and the Sword: Patterns of Momoyama History', in George Elison and Bardwell L. Smith (eds.), *Warlords, Artists, and Commoners: Japan in the Sixteenth Century* (Honolulu, 1981), p. 66.

6 Marius B. Jansen, *The Making of Modern Japan* (Cambridge, MA, 2000), p. 11.

become one of Nobunaga's trusted generals. Bent on further securing his hegemonic position, Hideyoshi not only tore down the defensive installations of subjugated foes, but also, in 1588, forbade ordinary peasants from possessing weapons. Although this measure was enforced through a 'sword hunt', the measure did not stop peasant revolts.[7] With the successful Siege of Odawara in 1590, Hideyoshi eliminated the last meaningful resistance to his authority. Hideyoshi's peasant origins precluded him from proclaiming himself Shogun, which required descent from the Minamoto clan, a scion of the imperial family that had established the Kamakura Shogunate (1185–1333). Nonetheless, in 1585 Hideyoshi secured the prestigious position of first secretary and regent combined, *kampaku*, from the court.[8]

Having secured a degree of domestic stability, Hideyoshi's ambition extended outward, and in 1592 he launched a massive, ultimately futile invasion of Korea. The resulting Imjin War (also known as the East Asian War, 1592–8) pitted roughly 160,000 Japanese invaders against an alliance of up to 84,500 Korean troops and 60,000 Chinese on the peninsula.[9] Hideyoshi's ultimate ambition beyond the conquest of Korea was to relocate the imperial court to Beijing – which would serve as the capital of a Japanese empire stretching to India. But a more realistic aim was to overturn the East Asian trading network through which the governments of Chosŏn Korea and Ming China continued to stifle Japan's growing demand for trade.[10] Another pragmatic advantage was that the enterprise channelled the warlike energy of a highly militarised society outward. Furthermore, since the bulk of the invasion force was drawn from the heavily Christianised population of western Japan, casualties facilitated Hideyoshi's anti-Christian persecution, which reflected his suspicion that Christians were agents of European powers with designs on Japan. Besides the numerical superiority that the Japanese could bring to bear, their style of fighting – which combined the close-quarter combat of deadly

7 Bernard Susser, 'The Toyotomi Regime and the Daimyo', in Jeffrey P. Mass and William B. Hauser (eds.), *The Bakufu in Japanese History* (Stanford, 1985), pp. 130–47.

8 Mary Elizabeth Berry, *Hideyoshi* (Cambridge, MA, 1982), pp. 8, 38–96, 102–6, 168–86, 223–5.

9 Stephen Turnbull, *Samurai Invasion: Japan's Korean War, 1592–98* (London, 2002), p. 42, 109; Kenneth Swope, 'Beyond Turtleboats: Siege Accounts from Hideyoshi's Second Invasion of Korea, 1597–1598', *Sungkyun Journal of East Asian Studies* 6 (2006), 177–206, 186.

10 Berry, *Hideyoshi*, 208; Jurgis Elisonas, 'The Inseparable Trinity: Japan's Relations with China and Korea', in John W. Hall (ed.), *Cambridge History of Japan*, vol. 4: *Early Modern Japan* (Cambridge, 1991), pp. 235–300, 265–71; Yi Tae-jin, *Dynamics of Confucianism and Modernization in Korean History* (Ithaca, 2007), pp. 90–1.

swordsmen with large contingents of conscripted peasants equipped with muskets – initially overwhelmed the poorly prepared Koreans, who fielded only a few thousand experienced and battle-hardened troops. Before long, though, Chinese intervention and increasingly better organised Korean resistance ultimately kept the Japanese from making any lasting gain from the war – aside from 50,000 to 60,000 Korean captives, including potters, other artisans, and Neo-Confucian scholars, who made meaningful contributions to early modern Japanese culture. On the orders of the dying Hideyoshi, the Japanese withdrew from Korea.[11]

In the course of the subsequent power struggles among major political players, Tokugawa Ieyasu triumphed as the new master of Japan. Following his victory at Sekigahara, Ieyasu secured the position of shogun and inaugurated the Tokugawa Shogunate. He further solidified the new order by eliminating Toyotomi Hideyori (1593–1616), the son and successor of Hideyoshi, and other potential adversaries at the battle of Osaka (1614–15). Ieyasu's son and successor, Tokugawa Hidetada (r. 1605–23), further stabilised the *bakufu*. The third Tokugawa shogun, Tokugawa Iemitsu (r. 1623–51), established a precedent of requiring his chief shogunal retainers in Edo to immerse themselves in a study of Confucianism as a suitable ideology for the now peaceful country. In 1635 a new law backed up this Confucian rhetoric with coercion by imposing the *sankin-kōtai* ('alternate attendance') system on the daimyo, obliging them to divide their time between their domain and the shogunal capital.[12]

Tokugawa Ieyasu's success against his rivals had been accomplished against the background of new military technology. At Sekigahara the combatants deployed nearly 20,000 muskets: a number comparable to, if not greater than, the numbers employed on contemporary European battlefields.[13] While the Tokugawa Shogunate continued to oversee the manufacture and inventory of its own muskets, the *bakufu* pursued its general policy of maintaining peace. Not only prohibiting daimyo from manufacturing or stockpiling firearms, the Tokugawa uniformly banned anyone other

11 The estimate of Korean captives is according to Naitō Shunpō, 'Jinshin-teiyū eki ni okeru hiryo Chōsenjin no sakkan mondai ni tsuite' [The repatriation of Korean captives of Hideyoshi's invasion], parts 1–3, *Chōsen gakuhō* 29 (1963), 33 (1964), and 34 (1965) as cited by Yasunori Arano, 'The Formation of a Japanocentric World Order', *International Journal of Asian Studies* 2:2 (2005), 185–216, 197.

12 Conrad D. Totman, *Early Modern Japan* (Berkeley, 1993), pp. 50–6, 105–13.

13 On the widespread use of muskets in Japan at the time, see Samuel Hawley, *The Imjin War* (Seoul, 2005), pp. 3–7.

than the samurai from bearing any weapons.[14] The decisiveness with which Edo closed off the country to the West – except for one or two Dutch trading ships permitted to enter Nagasaki each year, and the establishment of amicable relations with Korea – also helped to ensure that the non-elite Japanese population remained unarmed.

## Warfare and Korea's government army in transition, 1450–1650

In contrast to Sengoku–Tokugawa Japan, in Chosŏn Korea the court exercised command over military forces that were evolving from a universal conscript army into a semi-professional organisation. Fully established by the mid-fifteenth century, the early Chosŏn army was based on the service of all adult males except slaves, who constituted perhaps 30 per cent of the population. For the most part, Korea's military organisation was effective in campaigns against the raiding Jurchens (ancestors of the Manchus), in the north and multiethnic pirates, the *Wakō*, in the south. However, during the decades of relative peace from the late fifteenth century, the government began to allow tax payers to substitute their military obligations with a military cloth tax (*kunp'o*) that covered the expenses of the increasingly downsized standing army. During the Imjin War at the turn of the seventeenth century the record of Korean land forces against the Japanese and the Manchus was at best mixed. In response, the government sought to improve military capability with a semi-professional army comprising new military divisions that featured troops trained in swordsmanship and musketry, among other weapons.

Extolling the imagined ideal society of early China where peasants served rotation duties in defending the community, early Chosŏn Confucian scholar-officials instituted an army guided by the principle of universal military service. But while all were theoretically liable to provide military service, in practice each active-duty serviceman was assigned two or three support taxpayers (*pongjok, poin*) by the government to provision him. Nor was military service restricted to the commoners and *chungin*: members of the aristocracy, the *yangban*, too, performed military service, albeit in less

14 For the most part, the Tokugawa Shogunate sought to exercise social control of violence, even when samurai honour was involved. Eiko Ikegami, *Taming of the Samurai: Honorific Individualism and the Making of Modern Japan* (Cambridge, MA, 1995), pp. 241–60.

onerous ways in special elite or ceremonial guard units.[15] Maintained along these lines, this developed early Chosŏn military organisation, which relied more on archers, cavalry, and heavy artillery than on swordsmanship, and was relatively effective in dealing with its foes, the Jurchens and the *Wakō*.

During a stretch of relative peace from the late fifteenth to the mid-sixteenth century, several developments contributed to the collapse of the early Chosŏn military organisation. *Yangban* males began to shun military duties as a part of their class privilege of exemption from taxation: a prerogative which they justified with their ostensible service to the state as full-time Confucian students (*yusaeng*) or scholar-officials. The phenomenon was part of an overall socioeconomic trend, according to which most taxpayers stopped performing actual military duties. The government increasingly allowed commoners drafted as infantrymen (*chŏngbyŏng*) or marines (*sugun*) to remit bolts of cloth – collected from their support taxpayers – for release from their active duties by hiring substitutes. Usually a private slave or a vagabond, such a substitute was likely to perform the military obligation instead. Further institutionalising the practice, the government imposed a military cloth tax on every non-slave adult male in 1537, but both the state's finances and the army's effectiveness deteriorated, with corruption spreading among agencies and tax-contractor merchants involved in the levy of the various troops and their substitutes.[16]

The breakdown of the early Chosŏn military organisation accounts for Korea's generally poor battlefield performance at the turn of the seventeenth century. In response to mounting Jurchen military pressure in the late sixteenth century, the government resorted to frequent large-scale military examinations and conditionally allowed previously excluded illegitimate sons (*sŏŏl*) and slaves to compete. Instituted in 1402 to recruit military officials, the military examination – which, as a way of recruiting military officials, originally stressed a candidate's knowledge of the classics and martial skills centred around polo, lance-wielding, and various forms of archery – became more solely focused on archery and also a stopgap measure for recruiting rank-and-file soldiers from more modest backgrounds to man border garrisons, since the overall troop strength of the standing army dwindled.[17] Under the command of Yi Sunsin (1545–98) the Korean navy performed brilliantly against the Japanese, defeating

15 Eugene Y. Park, *Between Dreams and Reality: The Military Examination in Late Chosŏn Korea, 1600–1894* (Cambridge, MA, 2007), p. 24.
16 James B. Palais, *Confucian Statecraft and Korean Institutions: Yu Hyŏngwŏn and the Late Chosŏn Dynasty* (Seattle, 1996), pp. 70–7.
17 Park, *Between Dreams and Reality*, pp. 26–9.

the numerically superior adversary in all twenty-six engagements save one draw, and Kwŏn Yul (1537–99) ultimately provided diligent, effective leadership for land troops. All the same, throughout the conflict both commanders constantly struggled with manpower shortages as well as partisan bickering at the court.[18]

During the Imjin War, the Chosŏn state implemented measures to strengthen the military organisation, which was increasingly centred upon a core of better trained, more professionalised troops, especially those recruited from the capital, Hansŏng (present-day Seoul). Its army acquired these soldiers in addition to conscripts, although the government still supported all of them with revenues from the military cloth tax and other sources. The war persuaded otherwise conservative policy makers of the power of new technology, and in 1593 the government set up the Directorate-General for Military Training (*Hullyŏn Togam*) to train the Three Skills Army (*Samsubyŏng*), which comprised musketeers, archers, and close-combat 'killers' (*salsu*, comprising swordsmen, pikemen, and spearmen) – in accordance with the military strategies of Ming China's Qi Jiguang (1528–88), who had earlier designed this military system to counter *Wakō* raids in eastern China.[19] Also, among various changes, the military examination began to feature a musket test, mainly intended by the Directorate-General for Military Training for those among the troops already trained in the new weapon.[20] Rather ironically, though, Korea found little use for its new army after 1637. Not only did the Manchus subjugate Korea in that year before conquering China, but also the Tokugawa Shogunate, which prioritised internal stability, normalised its relations with Korea – giving each Korean embassy a grand reception that was both lavish and costly.[21]

## Bureaucratic warrior elite and martial virtue in Japan, 1650–1850

Tokugawa Japan maintained peace and stability through a rigid social hierarchy and relative isolation. The prohibition of Christianity as a dangerous religion accompanied the so-called *sakoku* ('locked country') policy, which

18 On Yi's success during the war, see Turnbull, *Samurai Invasion*, pp. 82–91.

19 Palais, *Confucian Statecraft and Korean Institutions*, pp. 85–9.

20 Ibid., pp. 84–6; Park, *Between Dreams and Reality*, p. 51.

21 For a classic discussion of Japanese reception of Korean embassies, see Ronald P. Toby, 'Carnival of the Aliens in Edo-Period Art and Popular Culture', *Monumenta Nipponica* 41 (1986), 415–56.

kept out all foreigners except authorised visitors from China, Korea, the Ryukyu Islands, and Holland. Internally, the *bakufu* promoted Confucianism as an ideology suitable for education and social harmony. A clearly defined status hierarchy recognised 'four occupations', namely: (1) gentry scholars, (2) peasant farmers, (3) artisans and craftsmen, and (4) merchants and traders. Peace facilitated transportation, booming commerce, the use of money, and the rise of urban centres. No longer engaged in warfare and also suffering from a shortage of bureaucratic positions, an increasing segment of the samurai population reflected on their callings as practitioners of *bushidō*, which they articulated more strongly and perfected as a moral code.

While the warrior ethos persisted among the Japanese elite, the shogunate and daimyo domains maintained civil governance through samurai retainers who were essentially bureaucrats. The overall cultural spirit of the era inspired the samurai to pursue scholarship and refinement. For example, Tokugawa Mitsukuni (1628–1701), who was a prominent daimyo from a collateral Tokugawa branch, pursued such undertakings as compiling the *Great History of Japan (Dai Nihon shi)*.[22] Both Buddhism and Shinto undeniably retained influence in a by-then largely secular society, but Confucianism found strong favour among the elite. The influence of Chinese language and classics was at its strongest since the formative centuries of early Japan, though institutional manifestations were not as ubiquitous as was the case in Chosŏn Korea.[23] In fact, *kokugaku* – which articulated more nativist cultural sensibilities – gradually gained strength, acting to de-centre China in the world-view of both Japanese elite and lower-status scholars.[24]

By the eighteenth century, the Tokugawa system could no longer ignore the growing economic weight of townsmen and merchants, who were even making loans to those among the warrior elite who faced financial hardship. Until 1862, the *sankin-kōtai* system of enforced residence in Edo remained a tremendous financial burden on the daimyo, as each trip demanded a large train of goods and an imposing entourage.[25] With the revenue surplus from the prosperity of the late seventeenth and early eighteenth centuries depleted by the mid-eighteenth century, the shogunate devalued gold and silver, pocketing the difference in actual value between newly minted coins and the old. The

22 On the *Great History of Japan*, see John S. Brownlee, *Japanese Historians and the National Myths, 1600–1945: The Age of the Gods and Emperor Jimmu* (Vancouver, 1997), pp. 29–42.

23 Marius B. Jansen, *China in the Tokugawa World* (Cambridge, MA, 1992), p. 4; Ikegami, *Taming of the Samurai*, pp. 302–8.

24 David Earl, *Emperor and Nation in Japan* (Seattle, 1964), pp. 66–7.

25 Jansen, *The Making of Modern Japan*, pp. 127–41.

devalued currency and resulting inflation only further increased the financial burden on both the *bakufu* and ordinary peasants. At times, Edo increased the tax levy on the peasantry of the area it directly administered. Meanwhile, the conservative economic policy of increasing rice production led to oversupply, and the decreased grain price created financial hardship not only for the peasant cultivators but also for samurai retainers who received their salaries in a predetermined amount of rice according to each retainer's rank.[26]

In the nineteenth century, the Tokugawa Shogunate finally lost its footing. Expanding cultivation in Japan, as well as Korea, fuelled by advances in agricultural technology since the thirteenth and fourteenth centuries, reached its limit with denuded mountains, flooding, harvest failures, famine, and epidemics.[27] In the course of the Tenpō Famine (1832–7), which the *bakufu* was largely powerless to ease, lower-level samurai, especially in domains such as Chōshū and Satsuma which had traditionally been associated with discontent towards the centre, began to pursue their own reform programmes and agendas that increasingly challenged Edo's prerogatives.[28] At the same time, Western powers such as Britain and Russia began to pressure the shogunate to open its doors for diplomatic relations and trade, though these initial efforts were unsuccessful.[29]

In spite of these socioeconomic changes, the samurai continued to play a central role in governing the country throughout the Tokugawa period, and *bushidō* continued to guide their endeavours. Somewhat ironically it was during the prolonged period of peace that *bushidō* won widespread recognition as the code of moral principles for all samurai to observe. *Bushidō* literature of the era shows that the samurai were seeking more general applications of martial principles and experience in peacetime, as well as reflecting on a long history of war. In an early eighteenth-century work, *Hagakure (Hidden by the Leaves)*, Yamamoto Tsunetomo (1659–1719) relates many of the sayings of Nabeshima Naoshige (1537–1619), who was a major daimyo and an ally of Ieyasu, to Naoshige's grandson, whom Tsunetomo served as a retainer. Naoshige boldly declares: 'The Way of the Warrior . . . is to find a way to die. If a choice is given between life and death, the samurai must choose death. There is no more meaning beyond this.'[30] The *rōnin*

26 Totman, *Early Modern Japan*, pp. 236–40, 269–70, 296–315, 328–32, 341–7.

27 Conrad D. Totman, *Pre-Industrial Korea and Japan in Environmental Perspective* (Leiden, 2004), pp. 154–60.

28 Totman, *Early Modern Japan*, pp. 241–2.

29 Ibid., pp. 275–8, 493–501.

30 *Hagakure* in Chūo Kōronsha, *Nihon no Meicho* [*Great Books of Japan*], 50 vols. (Tokyo, 1969), vol. 17, pp. 58, 76–8, 98–9, as cited in David J. Lu, *Japan: A Documentary History*, vol. 1: *The Dawn of History to the Late Tokugawa Period* (New York, 1997).

scholar and strategist Yamaga Sokō (1622–85) – who wrote extensively on *bushidō* and *bukyō* ('teachings on military affairs') – presented a more practical interpretation of martial virtue and conceptualised a more general *shidō*, the 'way of the gentlemen', which was applicable to all stations of society. Sokō sought to codify a kind of 'universal *bushidō*' with an emphasis on pure Confucian values while stressing the singular and divine nature of Japan and Japanese culture. These radical concepts – including ultimate devotion to the emperor, who essentially had become a mere political figurehead in his capacity as supreme priest since the fifteenth century – put him at odds with the shogunate, which exiled him to the Akō domain, incidentally the later setting of the Akō Incident (1701–3).[31]

Also known as the Revenge of the Forty-Seven *Rōnin* or the Akō Vendetta, the Akō Incident revealed the tensions between a stable, centralised socio-political order and the idealised code of *bushidō*. It began when the daimyo of Akō assaulted a shogunal official for an insult and subsequently was forced by the *bakufu* to commit *seppuku*, an honourable, ritualised suicide. Forty-seven of his retainers, now masterless *rōnin*, avenged his honour after patiently waiting and planning for two years, ultimately killing the official. While recognising that the samurai had followed the precepts of *bushidō* by avenging the death of their lord, the befuddled shogunate could not overlook the fact that the *rōnin* had defied the authorities by taking justice into their own hands and committing a murder. In the end, after receiving a volume of petitions pleading on behalf of the perpetrators, the *bakufu* sentenced the *rōnin* to death for the murder but ordered them to commit *seppuku* instead of submitting to execution as common criminals.[32] The increasingly generalised appeal of *bushidō* throughout all society was such that, by the nineteenth century, the shogunate was allowing some non-samurai personnel serving the *bakufu* to wear a crested kimono and even carry swords.[33] But the socio-cultural efficacy of martial virtue across status boundaries in peacetime was not unique to Tokugawa Japan.

31 It was in fact not until after the 1899 publication of Nitobe Inazō's *Bushido: The Soul of Japan* that the term '*bushidō*' came into common usage in Japan and the West. From then on, the Japanese government and military increasingly manipulated *bushidō* to suit their needs. Karl F. Friday, 'Bushidō or Bull? A Medieval Historian's Perspective on the Imperial Army and the Japanese Warrior Tradition', *History Teacher* 27 (1994), 339–49.

32 G. Cameron Hurst III, 'Death, Honor, and Loyalty: The Bushidō Ideal', *Philosophy East and West* 40 (1990), 511–27, 523–5; Ikegami, *The Taming of the Samurai*, pp. 223–41.

33 Anne Walthall, 'Peripheries: Rural Culture in Tokugawa Japan', *Monumenta Nipponica* 39 (1984), 371–92.

# Confucian scholar-officials and martial virtue in late Chosŏn Korea, 1650–1850

The late Chosŏn state maintained an army no bigger than that dictated by the needs of ensuring internal security, and the rank-and-file were commoners. Besides maintaining an army of paid soldiers, allowing formerly excluded slaves to serve was another salient characteristic of the later Chosŏn military organisation. The army, for the most part, reflected the rigid social hierarchy in the sense that ascriptive status largely determined how far a soldier or an officer could advance in rank. Yet, in spite of this, the military enabled non-elites to enhance their social status to an extent, as well as inspiring the popular imagination with the aspiration that martial success could bring social advancement.

The military organisation of late Chosŏn Korea was a by-product of internal changes in the levy system, as well as the earlier wars with the Japanese and the Manchus. Besides the Directorate-General for Military Training, the subsequent establishment of four new special military units completed the so-called Five Military Divisions (*O kunyŏng*) – which collectively functioned as the capital army – by 1682. Unlike the troops reporting to the Directorate-General for Military Training, most of whom were professionals on the government payroll, the other four divisions maintained rotating-duty conscripts provisioned through the support of taxpayers: a continuation of earlier Chosŏn practice. In the provinces, the government set up the *sogo* army based on Qi Jiguang's previous military proposals, which incorporated small units, the *shuwu* – a hierarchical organisation incorporating men trained in firearms – as well as archers and swordsmen. Self-supported, commoner recruits tended to be those too poor to buy their way out of military obligation, whereas illegitimate sons and slave recruits were attracted by certain incentives. After his military service, an illegitimate son became eligible to sit a government service examination, whereas a government slave received manumission; in contrast, the *sogo* army recruited private slaves by compensating their owners with court ranks or substitute slaves, and the recruits themselves received no incentive.[34] Among the new units, musketeers, in particular, impressed the new masters of China: the Manchu Qing dynasty (1616–1912). Upon requests for military aid from the Qing in 1654 and 1658, the Chosŏn government dispatched hundreds of musketeers to fight the Russians near Khabarovsk, and the combined Qing–Chosŏn troops drove them back from the borderlands.

34 James B. Palais, *Politics and Policy in Traditional Korea* (Cambridge, MA, 1975), p. 322 n. 42; Palais, *Confucian Statecraft and Korean Institutions*, pp. 88–91, 417–20.

More than two centuries of peace, however, took a toll on Korea's military capacity. Seeing little need to maintain a large standing army, the government allowed much of the population to substitute active-duty service with the military cloth tax, as had been the case for decades preceding the Imjin War. In a letter addressed to the Catholic bishop in Beijing to seek foreign military assistance against the government's anti-Catholic persecution, Hwang Sayŏng (1755–1801) succinctly, yet accurately, summarised Korea's military capacity: 'Korea does not have a strong military. It never was a major military power, but over the last couple of hundred years, when Korea has been at peace and has had no need for a strong army, it has grown even weaker. The people do not know anything about weapons or fighting.'[35]

While its military capability was limited, as we have seen the late Chosŏn army allowed a degree of upward social mobility and influenced expressions of status aspirations in popular culture. At the top, increasing domination of the civilian branch of the central government by a small number of *yangban* families based in Hansŏng and its vicinity led to the political marginalisation of other aristocratic families. Through the military examination, some *yangban* family lines evolved into a military aristocracy who became junior political partners of the civil aristocracy, while most *yangban* lineages became detached from central officialdom, settling for a status as local aristocrats. Despite the differentiation, the central civil official, central military official, and local *yangban* families continued to recognise one another as *bona fide* members of the aristocracy. Meanwhile, non-elites began to participate *en masse* in the military examination, as passing it satisfied their status aspirations and helped them win recognition in their social circles, albeit without securing membership in the aristocracy.[36] The late Chosŏn military examination assumed the dual role of a political institution that guaranteed a place in the power structure for a segment of the aristocracy, but also a socio-cultural mechanism that released tension engendered by the rigid status hierarchy. At the same time, such cultural genres as vernacular narrative fiction, *p'ansori*, which is vocal-percussional music featuring extended narratives, masked dance (*t'alch'um*), and shaman rituals celebrated various historical military men as gods.[37] (See Figure 21.)

35 'An Appeal for Aid', from *Hwang Sayŏng paeksŏ*, pp. 16–17, 107, cited in Yŏngho Ch'oe, Peter Lee, and Wm. Theodore de Bary (eds). *Sources of Korean Tradition*, vol. 2: *From the Sixteenth to the Twentieth Centuries* (New York, 2000), p. 137.

36 Park, *Between Dreams and Reality*, pp. 86–177. 37 Ibid., pp. 168–77.

FIGURE 21. Contrasting depictions of Korean military men. (Left) A seventeenth-century portrait of Kwŏn Ŭngsu (1546–1608), a high-ranking military official donning court dress with an embroidered rank plate showing a tiger symbolising his status as a military man. Jinju National Museum, South Korea. (Right) Im Kyŏngŏp (1594–1646), a quixotic military commander who died as a scapegoat amidst an international political intrigue, inspired popular imaginations for his dream of avenging Korea's humiliations by the Manchus and is depicted as a warrior god in a nineteenth-century shaman painting. Chungminsa, Yeonpyeong-myeon, Ongjin-gun, Incheon. Sources: (Left) 'Kwŏn Ŭngsu yŏngjŏng (Pomul che 668–1 ho), kaktae wa pyŏru (Pomul che 668 ho) – Kungnip Chinju pangmulgwan' [Portrait of Kwŏn Ŭngsu (Treasure number 668–1), ceremonial dress belt and inkstone (Treasure number 668) – Jinju National Museum], navykanggu (blog), last modified 23 August 2017, https://blog.daum.net/navykanggu/2516. (Right) 'Han'guk ŭi shamŏnijŭm: musindo' [Korea's shamanism: shaman god paintings], Han'guk yŏksa munhwa sinmun [Korean History and Culture News], 22 February 2019, www.ns-times.com/news/view.php?bIdx=1898.

## Closing thoughts

Comparing the history of war in Japan and Korea between 1450 and 1850 highlights both similarities and differences. As of the mid-nineteenth century, a relatively stable, centralised political system, guided by Confucianism and monopolised by a hereditary-elite status group, exercised power in both regions, though each faced growing internal and external pressures. In Tokugawa Japan, the samurai, who became a bureaucratic warrior elite honouring *bushidō*, dominated the centralised systems of the shogunate and daimyo domains whilst also dealing with an increasingly influential class of merchants and townsmen. In Chosŏn Korea, the Confucian scholar-official elite, the *yangban*, maintained a rigidly hierarchical system while minimally satisfying the demands of social newcomers with status aspirations. The position occupied by the military differed in Japan and Korea, as did the martial ethos of each society, but in both societies the realm of military activities – real or imagined – offered some advancement opportunities to status-seeking non-elites.

Post-Imjin War Japanese and Korean societies shared a common attitude towards war. In Korea, as well as in Japan, the expanding influence of Confucianism not only stressed morality over legalism or militarism as the ruler's preferred means of governance, but also acculturated most aristocrats to distance themselves from martial virtue.[38] Physical combat skills did not constitute a defining element in the self-identity of the *yangban*, except for members of the military aristocracy based in the capital and some local *yangban* who took military examinations. In this light, the centuries of pre-Tokugawa internal warfare and its warrior elite make Japan seem like an anomaly in East Asia, but the Japanese code of warriors, *bushidō*, came of age only during the peaceful Tokugawa period when the samurai class reflected on its calling. In fact, official portraits of Tokugawa shoguns uniformly depicted them wearing court dress rather than armour (albeit with the hilt of a sword protruding from the robe), as if these supreme generals had to legitimise their positions as the emperor's chiefs-of-staff rather than as warriors *per se*.[39] (See Figure 22.) These practices put Japan more in line with Korea or China than Europe, where, even in the late modern era, rulers, aristocrats, and statesmen have donned military dress and carried swords, especially for ceremonial reasons.

In the latter half of the nineteenth century, a beleaguered Tokugawa Shogunate under internal and external pressures gave way to a restored imperial Japan bent on achieving equality with Western powers. Eager to

38 John K. Fairbank, 'Introduction: Varieties of the Chinese Military Experience', in Frank A. Kierman, Jr and John K. Fairbank (eds.), *Chinese Ways in Warfare* (Cambridge, MA, 1974), pp. 6–9.

39 Andrew Goble, 'Visions of an Emperor', in Jeffrey P. Mass (ed.), *The Origins of Japan's Medieval World: Courtiers, Clerics, Warriors, and Peasants in the Fourteenth Century* (Stanford, 1997), p. 119.

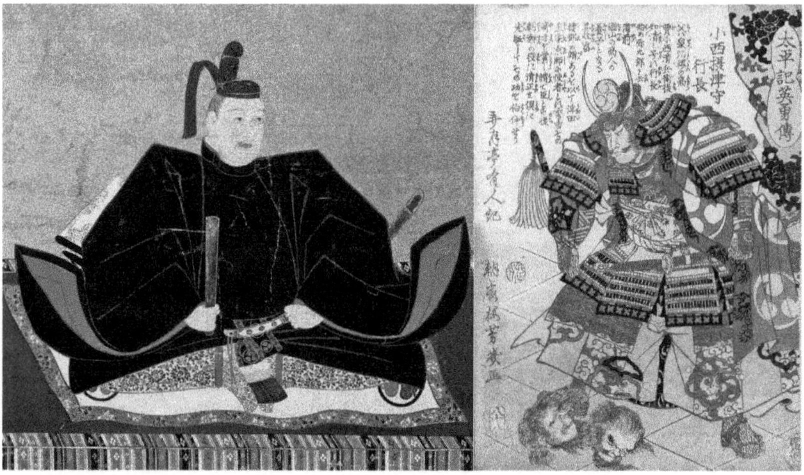


FIGURE 22. Contrasting depictions of Japanese military men. (Left) A portrait of Tokugawa Hidetada (1581–1632), the second Tokugawa shogun. Matsudaira Saifuku-ji, Tokyo, Japan. (Right) Depicted as a fearsome warrior in a woodblock illustration (*ukiyo-e*) by Utagawa Yoshiiku (1833–1904) is Konishi Yukinaga (1558–1600), a Christian daimyo who was on the losing side that fought against Tokugawa Ieyasu in the battle of Sekigahara. He chose death by execution rather than sinning with suicide. *Konishi Settsu-no-kami Yukinaga Ukiyo-e [Ukiyo-e of Konishi Yukinaga the God of Setsu], Taiheiki eiyūden [Biographies of the Heroes of the Grand Pacification]* (Edo [Tokyo], 1867), series number 88. Sources: (Left) 'Tokugawa Hidetada no shōjōga' [Portrait of Tokugawa Hidetada], Edo gaido [Edo Guide], https://edo-g.com/images/view/2. (Right) 'Konishi Yukinaga no ukiyo-e (Taiheiki eiyūden, Konishi Setsu-no-kami Yukinaga)', History Guide, https://history-g.com/eng/images/view/59.

secure a reliable supply station for its ships on their way to the China market as well as assistance for shipwrecked sailors, in 1853, the US government dispatched a fleet under the command of Commodore Matthew C. Perry (1794–1858) to negotiate a treaty with Japan. Helplessly watching the Americans put on a peaceful yet deliberate display of military force and technology, the *bakufu* did not respond and Perry departed, promising to return. When he did the following year, the shogunate had no choice but to sign the Treaty of Kanagawa (1854), the first among a series of unequal treaties signed with Western nations. As their agents and goods entered Japan, growing popular resentment against the foreigners also undermined the regime, which was now perceived as powerless. Increasingly subversive, the lower-level samurai of the traditionally restless Chōshū and Satsuma domains began to encroach on the court in Kyoto. Rallying in the name of the new, young monarch, the Meiji Emperor (r. 1867–1912), they defeated the shogunal armies. The Meiji Restoration of 1868 inaugurated ostensibly direct imperial rule and ushered in a revolution that, as it unfolded, would transform Japan from an isolated island country on the margins of Sinitic world order into an imperial power pursuing its strategic interests in neighbouring regions such as Korea.[40]

At the time, Korea was pursuing a conservative reform programme geared towards controlling rampant corruption and social unrest through a stronger monarchy while improving military defences against Western incursions. Along with famines and epidemics rooted in environmental disasters due to centuries of over-cultivation, an oligarchy of consort families had been monopolising political power from 1800 to 1864 as a series of kings ascended the throne either unprepared or as minors. The corruption spreading among local officials fanned discontent, sparking uprisings. Upon gaining power as the regent through backdoor negotiations, the Hŭngsŏn Taewŏn'gun (1820–98), the father of a young, new monarch, King Kojong (r. 1864–1907; emperor from 1897), ended the oligarchy of consort families, returned military command to the throne, offered more opportunities for advancement to men from diverse backgrounds, abolished most of the thousands of tax-exempt Confucian private academies (*sŏwŏn*) that had been exacting resources from local population for centuries, and rebuilt a royal palace that the Japanese had burnt down during the Imjin War of 1592–8.[41] While pursuing domestic reforms, the Taewŏn'gun rejected Western demands to establish diplomatic relations, fighting off French and American incursions (1866, 1871). Upon commencing his personal rule (1873), Kojong pursued treaty relations with all powers with strategic interests in Korea. But Kojong's efforts to establish peaceful relations with Japan were unsuccessful. After deliberately ignoring his gestures to establish relations as equals, an increasingly aggressive Japan resorted to gunboat diplomacy and became the first nation to impose an unequal treaty on Korea, the Treaty of Kanghwado (1876).[42] Although Kojong pursued a sweeping modernisation programme, Korea's geopolitical position deteriorated as ensuing internal turmoil and disturbances invited interventions by China and Japan.

40 Andrew A. Gordon, *A Modern History of Japan: From Tokugawa Times to the Present.* (Oxford, 2003), pp. 50–69.

41 In Chosŏn Korea, *taewŏn'gun* was an honorific title for a king's father not in the line of succession. Unlike three previous honorees, the Hŭngsŏn Taewŏn'gun was not only alive when his son ascended the throne but also subsequently wielded power as the regent until pressured to step down in 1873.

42 Palais, *Politics and Policy in Traditional Korea*, pp. 24–57, 110–59, 176–201, 238–51.

In hindsight, any attempt to utilise the military history of Japan and Korea from 1450 to 1850 to explain how the two regions fared differently in the mid-nineteenth century vis-à-vis the challenges of Western imperialism is ill-advised. At first glance, it may be tempting to view Japan's samurai whose culture honoured *bushidō* – rather than Korea's military *yangban* whose martial expertise the Chosŏn state subordinated to literary pursuits – as better adapted to the social Darwinism of a nineteenth-century world dominated by Western empires. However, the fact that the Tokugawa Shogunate collapsed decades before the Chosŏn state should at the very least caution students of modern East Asian history against reading too much into the history of war in early modern Japan and Korea.

18

# Warfare in South-East Asia, 1448–1851

MICHAEL CHARNEY

The mid-fifteenth century saw the slow emergence of new states across mainland and island South-East Asia after a period of substantial political decentralisation and fragmentation, on the one hand, and the withdrawal or ejection of Chinese armies, on the other. The new or reinvigorated states of the region – including Ava (Upper Burma), Pegu (Lower Burma), Ayudhya (today Thailand), and Dai Viet (today Vietnam) – stimulated a new period of martial vigour from the 1450s as they expanded at the expense of their neighbours. In successful campaigns in the 1450s and in 1471, Dai Viet conquered Champa on its southern frontier twice, leaving the Vietnamese state as the permanent hegemon over mainland South-East Asia's eastern littoral, relegating Champa to a mere tributary shadow of its former self. (See Map 14.) Ava and Pegu waged a bitter war for decades for dominance over the Irrawaddy valley, a contest that spilled over into Arakan, on the Bay of Bengal, mainland South-East Asia's thin, western littoral. Ayudhya, in the central mainland, made itself the dominant political and military power in the Chao Phraya river valley. In maritime South-East Asia, Melaka, founded in the early fifteenth century, revived an archipelagic, commerce-oriented empire on the scale of its predecessor, the maritime trading empire of Sri Vijaya. Melaka's power partially depended on arrangements with local seafaring people, the Orang Laut, who aided the assurgent state in controlling trade in the Straits. The rise of Melaka and the extension of its influence into Sumatra and into the Java Sea confirmed the ongoing decline of the Javanese kingdom of Majapahit, which saw its access to the sea gradually whittled away.

The rise of new states coincided with new weapons and the arrival of foreign mercenaries. The fifteenth century provided the first specific evidence of rudimentary firearms of Chinese or Indian manufacture used in South-East Asia. Amongst the first references for soldiers using firearms are those relating to Indian mercenaries at Bassein in 1406. Nevertheless, the

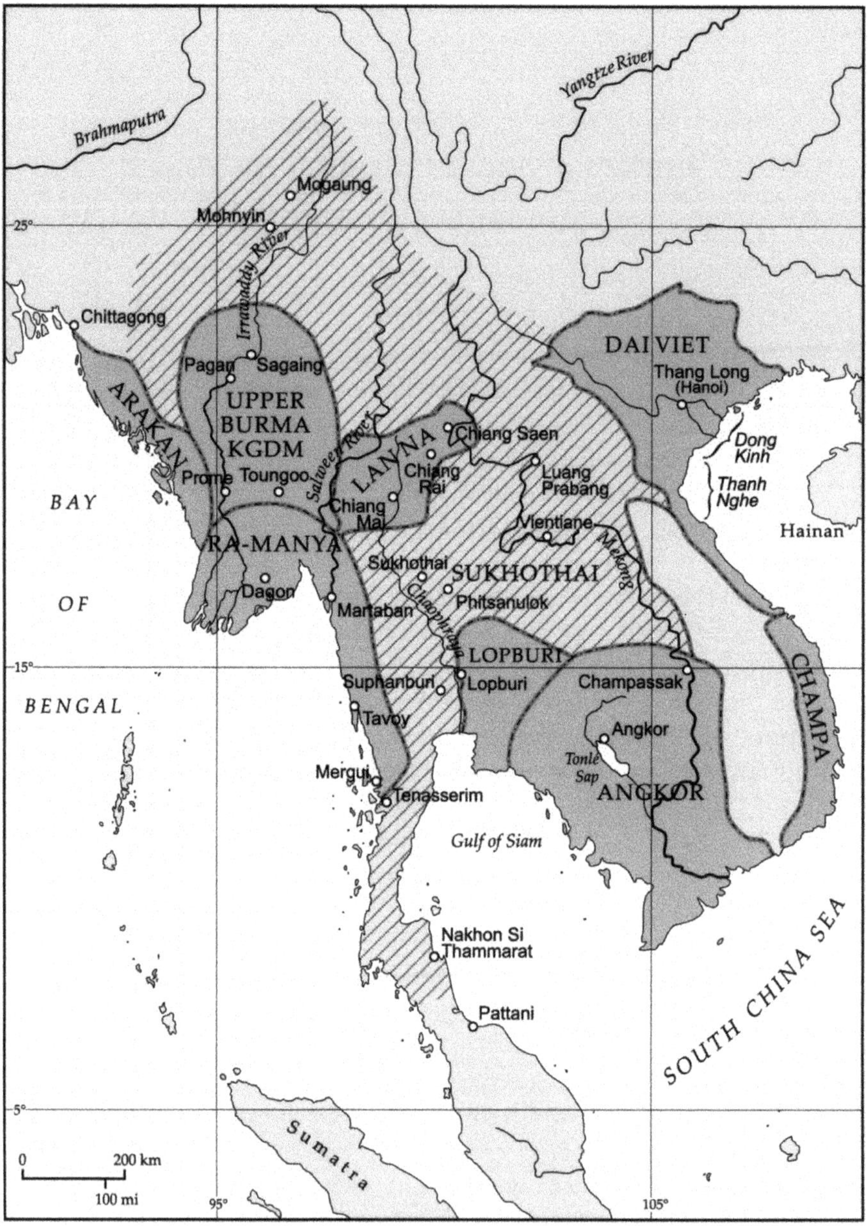


MAP 14. Mainland South-East Asia, *c.* 1340. Source: Victor Lieberman, *Strange Parallels: Southeast Asia in Global Context, c. 800–1830* (Cambridge, 2003), p. 13. Reproduced with permission.

quality of these weapons meant that while they might add to the repertoire of military commanders in the field, their tactical contribution was extremely limited. Greater reliance continued to be placed on old 'tools of war' such as elephant mounts in combat, which was possible because Asian elephants were trainable, unlike their African counterparts. As in the classical era, elephants were important because of their efficacy in crossing jungle and hilly terrain, but also because they provided a better line-of-sight for archers mounted on saddleback platforms. Elite riders also relied on elephants partly for this reason, allowing them to spot opposing elite riders with whom they aimed to engage in personal combat in the midst of battle. Engaging with an elephant-mounted opponent in personal combat on anything but another elephant in this period would have been foolhardy. As a result, an important part of preparations for war was securing elephants and, for large campaigns, capturing more and training them.

On the periphery of the mainland, in the highlands, and in many of the outer islands of the archipelago, smaller-scale political formations and a more detached relationship with maritime trade often meant that ritualised warfare, especially headhunting, would continue unabated into the twentieth century. As this kind of warfare was of a smaller scale and did not characterise the major polities with which Europeans in the archipelago and lowland indigenous states of the mainland were concerned, there was little external intervention in these conflicts, technological or otherwise.

## The mid-sixteenth to mid-seventeenth century

The century or so that straddled the middle of the sixteenth century to the mid-seventeenth century saw one of the most intense periods of warfare, in terms of range, human cost, and intensity, in South-East Asian history prior to World War II. Larger states consolidated more human and material resources that could be directed with greater intensity against larger-scale enemies. Armies and navies of a magnitude not seen previously in the region fought against, and sometimes with, Europeans who entered the region for the first time when the Portuguese visited Melaka in 1509. By the measures of indigenous chronicles and European accounts, armies were estimated to reach, with considerable exaggeration, into the millions. However, this hyperbole, in the absence of reliable statistics, reflected just how dramatic this escalating scale of warfare was for contemporaries.

Several factors loomed large in determining the outcome of warfare in the region during this period. First, access to the sea was important both as an objective of war and as a means to attaining victory. As maritime commerce increased from the mid-fourteenth century, controlling coastal ports provided lucrative revenues that could supplement those already available from agriculture and overland trade. Control of ports thus became the main goal of many campaigns. Second, maritime trade connections also aided access to mercenaries, who needed to be paid with cash, but were more likely to be loyal to the central ruler because they depended on that income from the royal coffers. Indigenous warriors, who were mobilised through patron–client hierarchies built on other loyalties, tended to shift loyalty according to the changing status quo. Third, from the sixteenth century, along with numerous Portuguese mercenaries came Portuguese firearms and cannon, more powerful and more accurate than those provided earlier by the Chinese, and thus advantaging whoever controlled the ports where these first arrived.

Fourth, one of the transformations brought by the Portuguese in the sixteenth century and then the Dutch in the seventeenth was the projection of substantial external military challenges across vast areas of sea and coasts. In the early fifteenth century, the Chinese admiral Zheng He (1371–1434) had made seven voyages through the region, some even reaching Africa, but it was the Portuguese in the sixteenth century who tried to establish permanent control over the trade in the region by seizing control of ports. The first was Melaka in 1511 and, from there, they controlled the flow of trade through the strategic Straits of Melaka. The Spanish in turn had taken Manila but were satisfied by a focus on their Chinese trade, exchanging silk for silver across the Pacific from Spanish America. In South-East Asia, the ability of the Europeans to strike quickly by sea with a powerful arsenal of cannon fired from gundecks within ships, permitted by Western ship structure, wreaked havoc on indigenous shipping and in the seventeenth century ruined indigenous coastal ports across archipelagic South-East Asia. On the mainland, control of the river entrances, often aided by natural bars that blocked the entrance of deep-draft ships without the aid of local pilots, would effectively keep the Europeans out until the mid-nineteenth century and the deep interior safe from the devastation experienced by the coastal areas of the archipelago.

On the mainland, the two main states to take advantage of the new resources available were Burma in the west and Dai Viet in the east, but with different results. In Burma, the first Toungoo dynasty rulers moved south from the upriver town of Toungoo to bring the coastal port of Pegu under their control in the 1540s and made it the royal capital. There they began to draw upon all the benefits that direct access to maritime trade offered, hiring Portuguese mercenaries, buying Portuguese guns and cannon, and paying for it all with revenues derived from maritime commerce. The first Toungoo dynasty (1531–99) undertook campaign after campaign, erecting the largest territorial empire in the history of the mainland in less than half a century, reunifying all of the Irrawaddy valley under one regime for the first time since the Pagan dynasty (1044–1287) and also conquering Ayudhya and the Lao kingdoms. The emphasis on maritime trade had been so strong, however, that the first Toungoo dynasty rulers undermined their own demographic and agricultural base. Poor administrative arrangements meant they depleted the manpower under their control relative to outlying lords, and soon the latter turned on the central court and besieged Pegu, reportedly driving the defenders to cannibalism, leading to its fall in 1599.

The ability to control the port, and thus access up the river system to the interior, was more challenging for Dai Viet, for as it expanded southwards it came to include two major river delta systems at the two extreme ends of the country. Dai Viet's powerbase had centred on control of the Red River delta for centuries, but under the Mac dynasty (1528–92), the progenitor of the Nguyen clan, Nguyen Hoang (1525–1613), exploited his role as governor of a district on the southern frontier to begin developing a family base there, drawing upon the resources of the Mekong river delta, as had the Chams and the Khmer previously. Although Dai Viet remained, theoretically, a single empire under the restored Lê dynasty (1592–1788), it effectively divided into two kingdoms, with the Trinh family dominant in the north, where the Lê Emperor was hosted, and the Nguyen family dominant in the south. Each relied not only on a different delta, but also on different European suppliers of firearms, the Trinh turning to the Dutch and the Nguyen to the Portuguese. They fought each other to a stalemate, unable to breach the walls erected at the narrowest point in the mid-section of the littoral, in a civil war lasting nearly fifty years (1627–73).

In the maritime world, which included both the archipelago and the coastal regions of the mainland, indigenous states were fast disappearing as a result of European efforts to dominate the region. The Portuguese and then the Dutch East India Company (VOC) in South-East Asia were more present in the region than the Spanish, who were principally a Pacific, rather than a South-East Asian, power until the 1820s, although the Spanish governor of the Philippines did sponsor a failed expedition to Cambodia in the 1590s. The Portuguese empire in Asia, the *Estado da India*, governed by the viceroy at Goa, sought to dominate the trade routes across the Indian Ocean and through South-East Asia into the South China Sea, and taking Melaka in 1511 gave them an important strategic position on this route. This threatened the trade revenues of Muslim states in the western archipelago and they struck back by repeatedly sending fleets that blockaded and besieged the fortress there. The most dogged of these states was Aceh on the northern coast of Sumatra, which benefited from mercenaries and weaponry lent by the Ottoman Empire, whose rulers saw the Portuguese as no less of a threat to the Ottoman presence in the Red Sea and the Persian Gulf.

Around the *Estado da India* hovered men, outcasts and desperadoes, who formed throughout the region what George Winius has labelled the *Estado da India*'s 'Shadow Empire' in the area around the Bay of Bengal.[1] These men occupied an intermediary zone between European and indigenous political space and served and cooperated with both sides at different times depending on the profits to be made. Their military impact on the region was important, because when they were hired out as mercenaries they brought current firearms and, crucially, gunpowder-making technology to their South-East Asian employers and provided their skills in using them in indigenous warfare. Although the first Toungoo dynasty in Burma deployed large numbers of indigenous musketeers, Burmese chronicles and even the account of Mendes Pinto make it clear that the Portuguese formed the elite of the musketeers in Burmese employ. The danger with using such men was that they were notorious opportunists who were quick to seize a better opportunity if it presented itself.

The best-known example is the case of the Portuguese mercenary Philip de Brito, who rebelled against his Arakanese employers after the fall of the first Toungoo dynasty in 1603 and built a kingdom centred upon his fortress at Syriam in the Burmese delta that endured for a decade. After a three-month siege the Burmese king of the restored Toungoo dynasty based in the north at Ava captured the fortress, impaled de Brito, and sent the Portuguese captives north to serve as hereditary artillerymen for the Burmese army. A key feature of this contest was the supply of gunpowder. De Brito depended on supplies from the *Estado da India*, having married the viceroy's niece, revealing the collusion of members of Portugal's formal and informal empires in the region. When gunpowder failed to arrive, the Burmese were able to get close enough to the fortress to mine the walls of the fort, a new tactic they developed using their own gunpowder. When the British fought the

1 George Winius, 'The "Shadow Empire" of Goa in the Bay of Bengal', *Itinerario* 7 (1983), 83–101.

Burmese 200 years later in the First Anglo-Burmese War, they would compliment the Burmese on their skill in applying gunpowder in this way.

The Dutch joined the Portuguese as a ubiquitous European presence in the region from 1600. The VOC sought to monopolise the spice trade. The brunt of the fighting to realise this ambition in the first half of the seventeenth century took place on Java, which saw the reversal of Mataramese power following Sultan Agung's (r. 1613–45) failure in both 1628 and 1629 to take Batavia (modern Jakarta) from the Dutch. In its role as a 'terrestrial power' in the Indonesian Archipelago, the VOC developed a substantial cavalry, increasing its size from 24 horsemen against Mataram's 200–300 in 1628 to 400–500 (including its allies) against Trunajaya's 600 in 1679.[2]

But the VOC developed broader interests in the archipelago, in part to supply the workers and the strongholds involved in the spice trade and partly to supplement cargoes. As a trading company that was also at war with Spain and Portugal for parts of this period, the VOC became heavily involved in local politics, finding local rulers in the archipelago eager to make alliances against the Iberians who had drawn Maluku into a virtually constant state of war from 1570. Through the treaties the Dutch drew up, sultanates such as that of Ternate committed themselves to a Dutch monopoly in the purchase of spices, relegating autonomous Javanese traders to the status of smugglers in Dutch eyes and subject to a violent Dutch interception policy. Local allies, virtually now vassals of the VOC, found themselves attacked when they sought to go their own way.[3] Another important area of conflict was Dutch involvement in the struggles between the Bugis and Makassarese in South Sulawesi, largely through the VOC's alliance with a Bugis mercenary captain, Arung Palakka (1634–96), who eventually became a local king with their help. The wars resulted in the Dutch cementing a foothold in yet another part of the archipelago.

The Spanish colonial project in the Philippines was again more circumspect than the Portuguese and the Dutch in their activities and aspirations in the Malay Archipelago. Nevertheless, the Dutch blockade of the Philippines for much of the seventeenth century forced the Spanish to expand beyond Luzon and further south in order to survive. Although the Dutch eventually

2 Peter Boomgaard, 'Horse Breeding, Long-Distance Horse Trading and Royal Courts in Indonesian History, 1500–1900', in Greg Bankoff and Sandra Swart (eds.), *Breeds of Empire: The 'Invention' of the Horse in Southeast Asia and Southern Africa 1500–1950* (Copenhagen, 2007), p. 40.

3 Gerrit Knaap, 'Crisis and Failure: War and Revolt in the Ambon Islands, 1636–1637', *Cakalele* 3 (1992), 1–26.

abandoned their blockade, deeper involvement in the archipelago brought the Spanish into conflict with the Muslim sultanates in the south. Although the Spanish waged numerous campaigns to secure the region, the Muslim south remained effectively out of the control of the Catholic north until well into the nineteenth century.

European and indigenous warfare in the archipelago contributed to severe population dislocation. The warfare in South Sulawesi, for example, pushed both Bugis and Makassarese to move elsewhere in the archipelago. Continual warfare offered opportunities for migrants from outside the region as well, especially in the case of Japanese Christian migrants pushed out of Tokugawa Japan, who now joined the courts of Ayudhya, Arakan, and even the Spanish Philippines as mercenary bands. Europeans continued to offer their services as mercenaries amongst other roles and Makassarese and Bugis groups in their role as mercenaries destabilised some local courts and propped up some others.

## 1630s to mid-eighteenth-century transitions

By the 1630s, however, the dramatic military campaigning of the previous century or so had begun to die down. Warfare in South-East Asia changed in three fundamental ways over the course of the seventeenth and early eighteenth centuries. First, while archipelagic states trudged on, in most areas of the archipelago (and the coastal mainland) naval warfare diminished in scale and range. While the mainland remained beyond the reach of Europeans, at least after the de Brito episode, European military superiority on the seas made the archipelago extremely vulnerable to European expansion. The huge battle fleets that had besieged Melaka in the sixteenth and early seventeenth centuries largely disappeared after the sinking of Iskander Mudah of Aceh's (r. 1607–36) huge fleet of several hundred ships in 1629. However, local naval expeditions and even regional raiding, particularly in the Sulu Zone and elsewhere in the eastern archipelago, would continue until the nineteenth century. This was partly due to the diminishing resources of these states due to mid-seventeenth-century commercial decline, especially since, unlike the mainland, the island world could not compensate for lost maritime revenues via overland trade with China. Increasingly, maritime states disappeared altogether, broken by the aggressive VOC, which reacted to declining revenues militarily by eliminating local competition from coastal indigenous states in the archipelago. Anthony Reid has argued that this period ushered in the beginning of South-East Asian poverty; the turning

FIGURE 23. Naval cannon. Source: author's photograph.

point came with the 'struggle between the rising absolutist states of South-East Asia and European maritime power', conflict heralded by the defeats of the rulers of Aceh and Mataram in 1629.[4] It also saw the beginning of the end of indigenous rule, which was first demonstrated by declining indigenous vitality in naval warfare. (See Figure 23.)

Second, although firearms had been a novel element of warfare in the sixteenth century, and perhaps earlier in a rudimentary form, they became sufficiently accurate and powerful by the early eighteenth century – as South-East Asian armies shifted from the matchlock to the flintlock – to end the role of the elephant as a tactical combat animal. The elephant would not disappear from the battlefield and indeed would be used highly effectively by both indigenous and even invading European armies in the nineteenth century as a baggage animal, able to negotiate difficult terrain with heavy loads. However, as elephant-mounted princes began to be felled on the battlefield (at an admittedly close distance) even by the late sixteenth century, their utility in man-to-man combat was clearly waning. South-East Asian

4 Anthony Reid, *Southeast Asia in the Age of Commerce, 1450–1680*, vol. 2: *Expansion and Crisis* (New Haven, 1993), p. 274.

commanders and princes now mounted local ponies, which provided a much lower line-of-sight, but also greatly diminished their vulnerability to snipers.

Third, less visibly, but in the long run more importantly, the increasingly important place of firearms as the determining factor in real and expected military contests required careful management of the weapons, the accurate production of gunpowder for them, and effective tactics and training to make full use of them. As a result, this period saw the development of standing armies for the first time in South-East Asia. The demonstrable unreliability of European rulers ended battlefield dependency on their troops and military resources. In their place, indigenous soldiers, often made up of resettled groups of war captives, were formed into units that specialised in gunpowder and other weaponry.[5] In the case of Burma, for instance, Indo-European captives taken at the aforementioned siege of Syriam in 1613, and other European captives taken later, formed a particular group of artillerymen, married to Burmese wives, and their descendants would continue to serve in that capacity until the end of the dynasty, alongside indigenous musketeer regiments. As royal servicemen whose careers and fortunes stemmed from supporting the throne rather than the intermediary agency of a tributary prince who might have his own pretensions to the throne, they possessed the reliability that had originally been hoped for in the European mercenaries. They were also available in all seasons, unlike levies who would have to leave for home to tend their fields for long periods of time.

Siam was amongst the last states to move away from reliance on foreign military expertise, as the state continued to host perhaps the most cosmopolitan array of foreign traders, mercenaries, and other guests of any mainland polity. The outward-looking King Narai (r. 1656–88) made a Greek, Constantine Phaulkon, a trusted minister, who conspired with King Louis XIV to bring the country under French control. An expedition of French soldiers was sent, who managed to take control of the forts controlling the Chao Phraya river, but in 1688 the court of the dying King Narai was overthrown by a powerful and politically alienated faction, leading to the expulsion of the French and Phaulkon's execution.

The peace, however, also brought further problems that seem to have stifled other martial developments that should reasonably have occurred in response to such dramatic shifts from the core elements of earlier warfare in the region. During the 'long peace' that preceded these upheavals, 'kith and

5 Victor B. Lieberman, *Strange Parallels: Southeast Asia in Global Context, c. 800–1830*, vol. I: *Integration on the Mainland* (Cambridge, 2003), p. 166.

kin' personal connections meant that the most capable individuals did not usually reach the highest levels of military command. Without major wars of suitable duration, military tactics could only be tested in venues such as the palace grounds where combat and warfare could be simulated. Here dominant status groups and other cliques determined who won mock battles and held military commands rather than offering recognition to those with real-world experience and demonstrable martial ability. Consequently, lethargy, indifference, and even resistance to change, as well as incompetence, slowed and confused the trajectory of martial developments in the region. From the beginning of this post-1630 period, the scene was set for revolutionary change, but it would take the convergence of economic, political, social, and other problems to provoke the crisis that would instigate this revolution.

## The last great indigenous wars of the mainland and Java

From the mid-eighteenth century to the early nineteenth century a series of major wars, that were almost exclusively fought between indigenous polities, swept through the region. These wars brought down almost all of the old dynasties of the region or saw their permanent fracture. More importantly, they were not confined to either the mainland or the archipelago, but mainly experienced in those parts of the region with the largest populations and ruled over by the most powerful of the remaining indigenous states. By the end of the period, South-East Asia would be dominated by only a handful of states that had finally swept up the myriad of petty kingdoms which had populated the region for much of the early modern period. Although the timing of these upheavals suggests that at least some of these conflicts were interrelated, their fundamental causes were much more complicated. The coincidence of the upheavals was partly due, however, to administrative arrangements that had stagnated over the course of a long-term peace that followed the resolution of wars across the region in the sixteenth and early seventeenth centuries. As dynasties grew older, the ministerial networks that managed the kingdoms and their riches became embedded and the royal courts proved unable to reform in response to local problems that crept in over time. These local problems built up until they reached a breaking point: Lower Burma in 1740 and central Vietnam three decades later (1771) saw the outbreaks of rebellions that initially met with enormous success, although they proved relatively short-lived. Mataram on Java collapsed into a civil war that ushered in Dutch domination of the Javanese heartland. Administrative problems in Ayudhya facilitated the polity's fall to Burmese armies in 1767. As a result of decades of military contests across the region, by the beginning of the nineteenth century, South-East Asia's political map had been transformed. (See Map 15.)

These wars had three major impacts on warfare and military evolution in the region. First, the wars were so sweeping in scale and their political changes so dramatic and sudden that they unleashed a huge amount of raw manpower and material resources. This windfall often gave the victors the freedom to organise new kingdoms and to draw on the lessons they had learnt in this most trying of times. Second, the wars decimated a generation of court and military elites, many of whom had risen on family ties alone and subsequently failed to excel in an uncontrolled military environment. This paved the way for a new generation of men who had risen on the basis of their own military abilities to assert themselves. These new elites owed little to the past and were open to new ideas adopted on the basis of efficacy rather than tradition. Third, the wars provided an intense arena in which military technologies, tactics, and organisational schemes could be tested. These developments coincided to generate the most widespread, violent, and innovative campaigning the region had seen since the sixteenth century.

Political expansionism now fuelled decades of military conflict in the western and central mainland. Burma's aspirations regarding the Chao Phraya river valley were revived under King Alaunghpaya (r. 1752–60) and his sons, in particular Hsinbyushin (r. 1763–76) and Bodawhpaya (r. 1782–1819). Having defeated the forces of the southern court at Pegu by 1757, Alaunghpaya launched a massive invasion, nearly taking the city of Ayudhya, but he was mortally wounded in the process. Hsinbyushin succeeded where his father had failed and took Ayudhya in 1767, devastating the country. Later, his younger brother, Bodawhpaya, sent successful military expeditions westwards and northwards against Arakan and Assam respectively as he sought to expand the kingdom, perhaps even into India. The campaigns of all three of these kings led to tens of thousands of deportees being sent to the Irrawaddy valley as captive labour, while those sent to the court would leave an important cultural imprint.[6] Siam's rulers, rebuilding the country after the fall of Ayudhya, also acquired manpower through similar expansionary wars in the central mainland that brought the Chao Phraya river valley and much of the Mekong river valley under their control.

6 Bryce Beemer, 'The Creole City in Mainland Southeast Asia: Slave Gathering Warfare and Cultural Exchange in Burma, Thailand and Manipur, 18th–19th Centuries', PhD diss. (University of Hawaii, 2013).

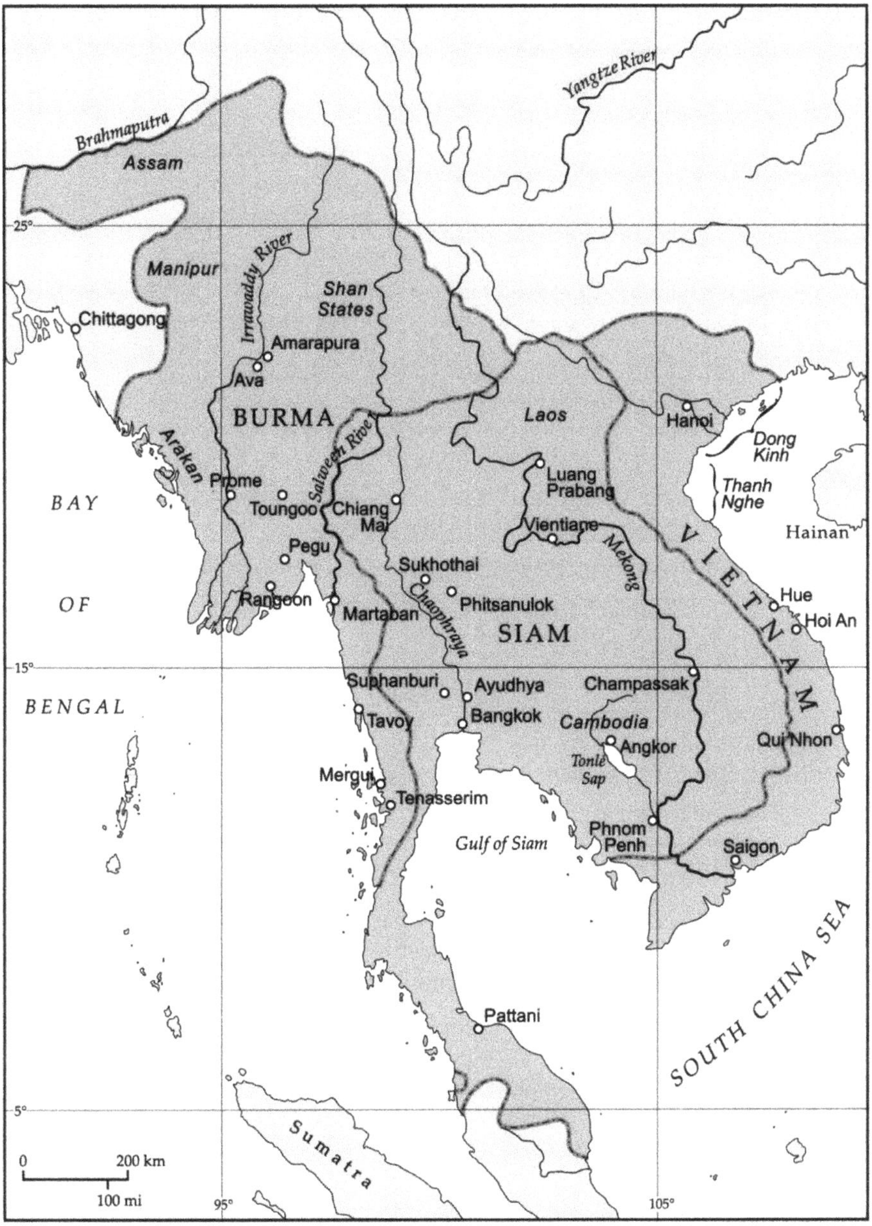


MAP 15. Mainland South-East Asia in 1824. Retrieved from Victor Lieberman, *Strange Parallels: Southeast Asia in Global Context, c. 800–1830* (Cambridge, 2003), p. 32. Reproduced with permission.

In the eastern mainland, an extremely complicated civil war repeatedly reconfigured the political landscape. A regional uprising in the south broke out in 1771, led by the three Tay Son brothers and boosted by heterogeneous forces that included large numbers of Chinese soldiery financed and raised though support from Chinese merchants settled in Vietnam. The uprising coincided with another breakdown in relations between the Trinh and the Nguyen rulers.[7] In the fighting that followed, the Trinh defeated the Nguyen in 1774, then the Tay Son defeated the Siamese who were attempting to restore the Nguyen in 1785. This was followed in 1786 by the Tay Son conquest of the Trinh, while in 1788–9, the Tay Son established their own dynasty (1788–1802) and defeated a Qing Chinese army that had invaded to restore the Lê dynasty. Yet this apparent military success by the Tay Son was reversed in 1802, when Nguyen Anh, a refugee prince of the defeated Nguyen family with backing from Siam, defeated the Tay Son brothers and established the Nguyen dynasty that would rule Vietnam until 1945.

After the mainland was completely divided amongst these three states they then turned on each other. However, as each state was better organised than those they had succeeded, and all sides having had recent military experience and considerable military success, stalemate quickly ensued. Bodawhpaya dispatched yearly campaigns against Siam, draining his manpower resources without much success. For the first time since the early seventeenth century, a major indigenous battle was fought in the seas in the northern part of the Straits of Melaka between the Burmese and a Malay fleet under Thai command for control of Phuket Island. Such relentless campaigning with little military success placed considerable strains upon upstart rulers whose legitimacy partially rested on battlefield success, contributing to a major religious crisis in the Burmese kingdom from 1812. Vietnam and Siam fought each other for control of Cambodia, passing the country back and forth until the signing of a joint protectorate over the country in 1846.

Unable to make headway militarily and permanently hemmed in by natural barriers, each of the three mainland dynasties settled into the same kind of military inertia that had damaged their war-making capabilities from the mid-seventeenth to the mid-eighteenth century. Warfare devolved back into mock battles on the royal parade grounds and the odd punitive expedition against rebellious villagers or uncooperative local governors. In Vietnam, for example, indigenous generals focused on static defence,

7 George Dutton, *The Tay-son Uprising: Society and Rebellion in Eighteenth-Century Vietnam* (Honolulu, 2006), pp. 200–1.

ignoring ongoing changes in European military technology that increased range and mobility; tactically, the Vietnamese army focused more on the need to suppress rebels than on potential European threats.[8]

## The renewed European expansion

While the remaining indigenous armies slipped into military conservatism, the Europeans increasingly became the dominant military force in the region, first in the archipelago and then on the mainland. At first, this probably seemed irrelevant to local security on the mainland. Until the 1820s, Europeans in the region predominantly fought each other as external wars spilled over into South-East Asia. The British, for example, occupied Manila for eighteen months (1762–4) during the Seven Years' War and occupied Java during the Napoleonic Wars. Such contests were fleeting episodes which usually had no more than a small impact on the overall nature of war in the region. However, from the 1820s and up to 1899 with the outbreak of the Spanish–American War, Western military activity in South-East Asia was almost exclusively directed at indigenous states. This was partly due to the establishment of a new status quo amongst the major European commercial powers in the Straits of Melaka. The Dutch, who had replaced the Portuguese as the dominant power in the Straits of Melaka in the mid-seventeenth century, were displaced from there in turn by the British East India Company. In 1786 the British had secured Penang, now renamed Prince of Wales Island, the port of Singapore in 1819, and, by the Anglo-Dutch Treaty of 1824, Melaka, surrendering Bencoolen on western Sumatra in exchange. This treaty effectively divided the archipelago into two zones of influence and expansion and also marked a point at which Europeans in the region stopped fighting each other and turned instead to bringing what remained of indigenous-ruled South-East Asia under their control.

Further to the north, the pursuit of rebels on its periphery led the Burmese court indirectly into a war with the British East India Company. When Burma took Arakan and Assam, both these kingdoms had long traditions of independence and perceived rivalry with Burma, on the one hand, and on the other, long open borders with the Company, which maintained it had an inviolable sovereign right to keep the Burmese from transgressing over those borders. Elites and others who did not want to submit to the Burmese thus merely migrated to the British side of the border and then commenced raiding the

8 Thomas Hodgkin, *Vietnam: The Revolutionary Path* (New York, 1981), p. 111.

villages of their former homeland. When Burmese punitive missions proceeded to violate the border, in pursuit of these bandits or rebels, the Company tried to stop them, leading to violence and then a declaration of war.

The First Anglo-Burmese War (1824–6) was amongst the Company's most serious engagements in Asia. The war required invading Burma overland from Bengal into Arakan and then across the mountains into the Irrawaddy valley, with a maritime landing force aiming to take Rangoon and the Lower Burma Delta. The Company's forces suffered heavy losses not only due to Burmese weapons, but also to diseases for which they were unprepared. The Burmese defended themselves by stockading and resisting the British advance. In the delta, the British were able to bring up heavy fire by using their gunships, including the *Diana*, the first time a steamship was used in warfare. Ultimately, the Burmese court sued for peace and signed the Treaty of Yandabo in 1826, removing control of Assam, Manipur, Arakan, and Tenasserim from the Burmese and ceding these territories to the British.

While the British were heavily engaged in Burma, the Dutch found themselves involved in a major rebellion – the Java War (1825–30) – when a Javanese prince emerged and led peasants in an uprising against Dutch rule. Fought against a people rather than against a state, the war had some of the same features of the kind of guerrilla war that the British would face in Burma after 1885 and the Americans in the Philippines after 1899: combat dominated by small arms; a blurring of the lines between fighters and non-combatants; and the absence of a front line. In response, the Dutch army applied the fortress-system known as *benteng stelsel*, which involved light columns operating independently and moving quickly out of an expanding network of fortified positions aided by local inhabitants. This anticipated the kind of decentralised response to an equally decentralised opponent that other colonial forces would apply in pacification campaigns in Burma, French Indochina, and the Philippines from the late nineteenth century onwards.[9]

The fact that existing kingdoms were unable to expand against each other, and the new European military presence in the region, should have made it clear that there was a new martial context in South-East Asia. Why there was a prolonged delay among the remaining indigenous states in adopting Western military doctrines remains unclear. Part of the answer lies in the conservatism of the royal courts. But it could also be argued that as the Europeans had been a threat in the sixteenth and early seventeenth centuries which had passed after a few decades, so in the nineteenth century South-East

9 Merle Ricklefs, *A History of Modern Indonesia since c. 1200* (Stanford, 2001), p. 152.

Asian societies, accustomed to a cyclical view of history, believed that once again the European assurgence would prove temporary. Certainly, in some cases, European attacks were mere punitive strikes, such as the French attack on the Vietnamese navy at Da Nang in 1847 in reaction to anti-Catholic policies, which would not be followed up by another French attack on Vietnam for ten more years. In the case of Burma, the Company would not return militarily for another twenty-six years.

It is true that Burmese court chroniclers minimised the magnitude of the defeat in the first war to allow the court, and with it the country, to forget what had happened. As reflected in the chronicles, the royal courts appear to form exclusive intellectual communities, each of which looked to their own pasts and their own experience in warfare for lessons on how to deal with new challenges, rather than benefit from sharing their experiences with others. However, through royal orders and other contemporary documentation, it is clear that indigenous rulers were aware of and wanted to know more about some of the problems their counterparts were having with particular military challenges, such as those of the Portuguese and the Dutch, or from the firearms they brought. Indigenous courts thus had a clear desire to adapt and learn from the experiences of others facing comparable challenges, but slow communications, and the lack sometimes of a shared prior text, made it difficult to put the information into any sort of wider historical or geographical context. Instead of relying on military adaptation, Siam turned to diplomacy. After seeing the outcome of the First and Second Anglo-Burmese Wars and realising it was less likely to win a military contest against the British than their stronger neighbour who had just been defeated, the Siamese adopted a concessionary approach to the British and then to the French in the last half of the nineteenth century that saw them sacrifice territory in the southern Malay peninsula and in the Mekong river valley, contributing to Siam's success as the only South-East Asian state that would permanently avoid European rule.

19

# War in the era of Qing imperial consolidation and expansion, 1587–1804

PAMELA CROSSLEY

The Qing Empire's military drew from the traditions of bodyguards and booty warfare in North-East Asia (primarily what is now southern Jilin province) in the late sixteenth century. The foundations of imperial expansion were built during the long war with Ming China, from 1618 to 1644, which allowed the Qing to absorb the central features of Ming military technology. Patterns of human management and technology application established in this period persisted over the next forty years as the Qing completed their conquest of China and Taiwan. After 1685, Qing expansion spread out to Mongolia, Qinghai, Tibet, and what is now the province of Xinjiang. These wars, against less densely populated, sometimes nomadic zones, changed Qing campaigns significantly. By the nineteenth century, the century and a half of focus on the continental frontiers left the Qing poorly prepared for seaborne challenges, and from some technologies that the Qing had previously regarded as less relevant to their military needs.

## Qing consolidation and expansion

The wealth of the Qing emperors' ancestors was founded on commerce, primarily the trade of gathered goods, textiles, and livestock in the markets of the Ming province of Liaodong and the Korean trading posts on the Yalu river; trading monopolies were frequently maintained through warfare. In 1616 Nurhaci (1559–1626), then the *beile* or chieftain of the Jianzhou Jurchens, declared himself khan of the Aisin khanate (Chinese Jin 金, gold). In 1618 the khanate declared war on Ming China, and by 1621 had seized Shenyang, the Liaodong provincial capital. After Nurhaci died in 1626, the war continued under his successor khan, Hung Taiji (1592–1643). Nor did he solely fight the Ming to prise away the western territories of Liaodong as far as the Great Wall. The Aisin also fought against the formidable Chakhar state of eastern Mongolia, which collapsed amid internal troubles after the death of its ruler,

Lighdan, in 1634, and forced Korea to acknowledge Aisin suzerainty.[1] The structure of the Aisin state was transformed. Lighdan's symbols of state, capital, populations, and family were all taken by Hung Taiji, who used them to craft a new form of rule in Liaodong. In 1635 Hung Taiji banned the name 'Jurchen', ordering that the term 'Manchu' be used instead.[2] In 1636 the state was reorganised when Hung Taiji declared himself the Qing emperor. The Qing Empire continued the fight against the Ming to control western Liaodong. In 1643 Hung Taiji died and was succeeded by his minor son, with a regency of princes ruling on his behalf. Less than a year later, the leader of a huge rural uprising in China captured Beijing and drove the last Ming emperor to commit suicide. In the resulting chaos, the Ming general Wu Sangui requested that the Qing armies enter China to defeat the rebels. This was accomplished, but rather than restoring the Ming, the Qing established themselves at Beijing and proceeded to consolidate their control over northern China, entrusting the pacification and administration of south China to a number of Chinese power brokers of north-eastern origin, including Wu Sangui.

Resistance to the Qing in the Yangtze river basin and southern China was led by the local gentry, by Ming pretenders, and by the Zheng family pirate confederacy based on Taiwan. Most Ming pretenders were quashed early. However, as campaigns against Zheng Chenggong (Koxinga) proceeded in the 1670s, the governors who had been given power over the south rose in rebellion, presenting the first existential challenge to the teenage Kangxi Emperor (r. 1661–1722), at the outset of his personal rule. In 1683 war against these rebellious governors – the so-called Three Feudatories – and against the remaining members of the Zheng family in Taiwan concluded.[3] Thereafter, challenges to the Qing would come from the south-west (Sichuan and Yunnan provinces), the west (Qinghai and Tibet), the north-west (Xinjiang), and the north (from Russia and the Mongols).

1 Gertraude Roth Li, 'The Manchu–Chinese Relationship, 1618–1636', in Jonathan Spence and John Wills (eds.), *From Ming to Ch'ing: Conquest, Region and Continuity in Seventeenth-Century China* (New Haven and London, 1979), pp. 1–38; Gertraude Roth Li, 'State Building before 1644', in Willard Peterson (ed.), *The Cambridge History of China*, vol. 9 pt 1: *The Ch'ing Empire to 1800* (Cambridge, 2002), pp. 9–72; Pamela Crossley, *A Translucent Mirror: History and Identity in Qing Imperial Ideology* (Berkeley, 1999), pp. 206–15.

2 Evelyn Rawski, *The Last Emperors: A Social History of Qing Imperial Institutions* (Berkeley, 1998), pp. 35–7.

3 Crossley, *A Translucent Mirror*, pp. 110–18.

The Kangxi Emperor was aware that Mongols had been a perennial thorn in the side of the Ming. As we will see below, some Mongol groups had been incorporated into the Qing forces, aiding Qing attempts to incorporate Mongolian leadership and neutralise challenges from ambitious Mongols. Romanov Russia had periodically attempted, without success, to penetrate Qing territories in Manchuria (particularly the present-day province of Heilongjiang on the Amur river), but Mongolia remained a possible theatre of subversion. As soon as the problems in south China had been resolved in 1685, the emperor pressed the war (1685–9) against the few Russian troops stationed on the northernmost bend of the Amur river. The conflict concluded with the Treaty of Nerchinsk, which demarcated the border between the two empires, with penalties for its violation, and established terms of diplomatic equality between them. The Kangxi Emperor proceeded to incorporate the populations of Mongolia into the empire. He had early success with the largest group, the Khalkha, but was opposed by Galdan, the Oyirad leader of the Zunghar khanate in the west: a large federation controlling most of modern-day Xinjiang, as well as parts of Tibet and Qinghai. In 1690 the Qing forced Galdan's retreat from Khalkha, and in 1696 they destroyed the greater part of Galdan's army. Galdan died the next year while in retreat from the Qing armies.[4]

Survivors from Galdan's armies retreated westward and leaders of the Zunghar khanate occupied Lhasa in 1718. The Qing sent forces to Tibet and Qinghai to attempt to drive the Zunghars from the region, taking Lhasa in 1720. Succession disputes in the Zunghar khanate during the 1750s created an opportunity for the Qianlong Emperor (r. 1736–95) to press Qing conquests towards western Mongolia.[5] By 1757 the Xinjiang territory had been occupied and sporadic resistance suppressed. Qing commanders proceeded to deliberately exterminate the Zunghars, killing males, and distributing women to Qing soldiers. Combined with an ongoing smallpox epidemic, the effect was the virtual annihilation of the Zunghar population. Their destruction finally stabilised Qing control over western Mongolia, Xinjiang, Qinghai, and Tibet. (See Map 16.)

In 1765 the Qing took advantage of the Burmese invasion of Siam to invade Burma itself, though this was successfully countered by the Burmese

4 Peter Perdue, *China Marches West: The Qing Conquest of Central Eurasia* (Cambridge, MA, 2005).

5 For background, see James Millward, *Beyond the Pass: Economy, Ethnicity, and Empire in Qing Central Asia* (Stanford, 1999), esp. pp. 29–45.

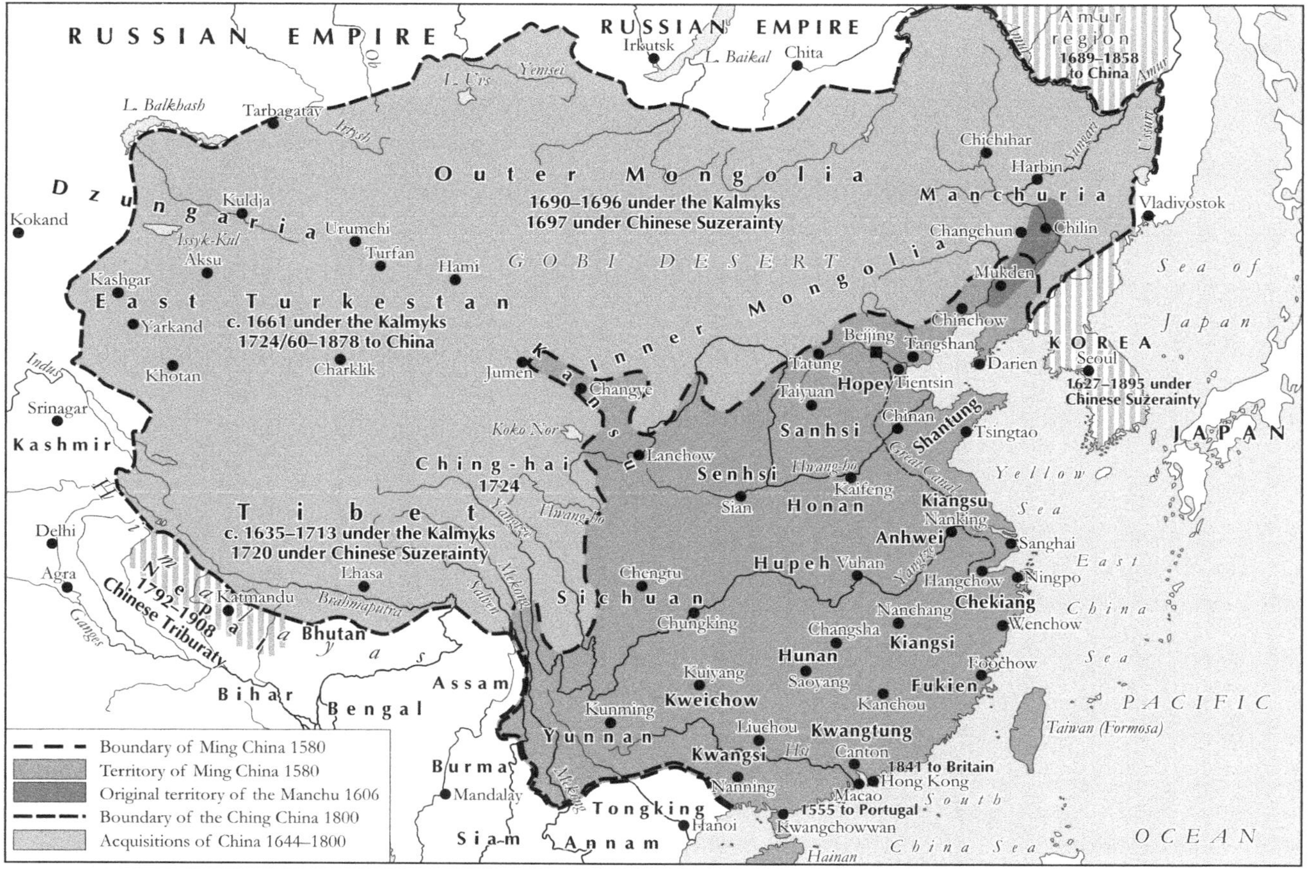


MAP 16. China under the Manchu (Qing) Empire, 1644–1912. Created by Béla Nagy.

in 1769.[6] In 1788, an opportunity to occupy Annam (Vietnam) appeared when the last ruler of the Lê dynasty was overthrown and requested Qing military assistance to restore him. They agreed and retook enough Annamese territory to claim that the emperor had been restored, but Annam's military leaders quickly drove them out of the country. As Qing forces withdrew from Annam, Nepal attempted to seize part of Tibet. The Nepalese expeditionary force was repelled, but in 1790 they mounted a much larger campaign in Tibet. However, the Qing, relying upon traditional troops and local militia to combat this incursion, were victorious in 1792.

By this time, the Qing were embroiled in repeated uprisings from the Tibetan population settled in Sichuan province. Defeating such internal uprisings would dominate Qing military thinking for more than fifty years to come.[7] Neither the heavy weapons deployed in the original wars against the Ming nor the cavalry warfare used against the nomadic populations of Mongolia and Xinjiang proved effective in these new theatres of war, which were often dominated by humid, mountainous terrain. Meeting the challenges of these uprisings preoccupied the empire at the end of the eighteenth century.

## Human management: The Eight Banners and the Green Standard armies

In the mid-fifteenth century, the Jurchen lineage leaders of southern Manchuria lived alongside their bands of bodyguards in fortified villages. This was clearly the origin of the military forces surrounding the proximate Qing founder, Nurhaci, whose bodyguards and their settlements are attested by various late sixteenth-century documents. It appears that by 1601 Nurhaci was ruling the Manchu primarily through socio-military organisations known as *gūsa*, a possible reference to the 'banners' (*qi* 旗) of Ming military organisation. In theory, the organisation of the banners allowed the Qing state to administer the entire population, since each soldier was supported by his family, the whole family was enrolled in the banner, and membership of a banner family was hereditary. The continuity with long-standing North

6 For more, see Alexander Woodside, 'The Ch'ien-Lung Reign', in Peterson (ed.) *The Cambridge History of China*, vol. 9 pt 1, pp. 260–8; Peter Lorge, *War, Politics and Society in Early Modern China, 900–1795* (London, 2005), pp. 166–8.

7 Yingcong Dai, *The Sichuan Frontier and Tibet: Imperial Strategy in the Early Qing* (Seattle, 2010), pp. 117–47.

Asian traditions of governing the population by way of estates, or patrimonies (*ordo*) – which included land, livestock, and people (in some degree of dependent status) under the command of an aristocratic lord – was apparent.[8]

Prior to the creation of the khanate in 1616, each of the four previous banners had been split into two, each designated by a colour (yellow, white, blue, and red) together with a tag for 'bordered' (*kubuhe*) or 'plain' (*gulu*). Each banner was owned by one of the eight members of the princely council. The consolidation of rulership after accession of the second khan, Hung Taiji, in 1626 was largely coterminous with the consolidation of control of the banners; by 1631 Hung Taiji was the outright owner of three of the banners, which were afterwards called the 'Three Superior Banners' (*san shang qi*, *ilan dergi gūsa*) and privileged in all matters of residence, appointment, and budget. By the time he had transformed the state into an empire in 1636, Hung Taiji controlled all eight banners.

The onset of war with Ming China in 1618 marked the beginning of profound changes in the relationship of the Eight Banners to actual warfare.[9] Early descriptions of the mobilisation of the Jurchen forces make clear that the bannermen were not originally full-time soldiers. They were working villagers – generally engaged in hunting, fishing, gathering, animal husbandry, farming, goods processing, or peddling – who were eager to join campaigns against neighbouring villages or more distant enemies in the hope of returning with booty to be shared among the household. In return, household members supplied the soldiers with uniforms, armour, cash to purchase weapons and food, and grain and dried meats for the campaign. As more villages were conquered, more adult men were coerced into joining new campaigns, or were motivated by the prospect of booty. Mobilising for war cost Nurhaci comparatively little, and the returns to him in territorial aggrandisement and the lord's share of the booty were enormous. The relationship between the bannermen and the khan was in-kind, based upon direct enhancement of the khan's ability to distribute goods and status.

The creation of the banners meant the gradual absorption of able-bodied men over the age of fourteen and under the age of sixty into the reserves of the standing forces, with 25–30 per cent of them appointed to active service at

8 Karl Wittfogel and Chia-Sheng Feng, *History of Chinese Society: Liao, 907–1125* (Philadelphia, 1949), p. 510.

9 Pamela Crossley, *Orphan Warriors: Three Manchu Generations and the End of the Qing World* (Princeton, 1991); Edward Rhoads, *Manchus and Han: Ethnic Relations and Political Power in Late Qing and Early Republican China, 1861–1928* (Seattle, 2000); Mark Elliott, *The Manchu Way: The Eight Banners and Ethnic Identity in Late Imperial China* (Stanford, 2001).

any given time. In essence, it transformed the adult male population from farmers who sometimes went to war into warriors who sometimes performed other chores. This was not usually an individual process; records of this early period suggest that whole villages were enrolled in companies called 'arrows' (*niru*). Captaincies were largely hereditary, and affiliation with specific units remained to a marked degree genealogically stable into the nineteenth century. Amalgamation of the *niru* into the new banner organisations embodied the administrative absorption of regional populations under Nurhaci's government. The declaration of war against the Ming in 1618, and subsequent occupation of territory within the Ming province of Liaodong, was the beginning of the process by which the Eight Banners became a military occupation force living among the immensely larger civilian population of Liaodong, rather than the medium for administering a principally nomadic population. It was the beginning of the monetisation of the support and rewards that were supposed to flow from the khan to his followers.

In 1634 the process began of constructing parallel banner organisations for the Mongols, incorporating both the easternmost Mongols who had been with Nurhaci since the early part of the century and the newer populations from Chahar. Each of the Eight Banners subsequently was composed of both a Manchu and a Mongol variant (e.g. Manchu Bordered Yellow, Mongol Bordered Yellow). Groups and individuals who had been the regime's earliest adherents from eastern Mongolia were for general administrative purposes thereafter nominalised as Mongols of the Eight Banners. A few of the early Mongol adherent lineages were enrolled in the Manchu banners and afterward were considered Manchu.[10]

This was parallel to the creation of the Chinese-martial banners in 1642, when some politically favoured lineages of the Chinese-martial were also made Manchu.[11] This group were known as the *Hanjun baqi*, or the 'Chinese-martial Eight Banners' (*nikan cooha*) in the eighteenth century. The region of Jilin, where the Qing ancestral state had formed, was settled in pockets by Chinese and Korean farmers, whose surnames often became part of the toponymy of the countryside and, eventually, part of the names of Jurchen lineages that were formed in the area in the fifteenth and sixteenth centuries. Many local Jurchen magnates had made use of Chinese speakers – whether they were by ancestry Chinese, Korean, or Jurchen – for certain tasks, including taking their livestock

10 See also Pamela Crossley, 'Making Mongols', in Pamela Kyle Crossley, Helen F. Siu, and Donald S. Sutton (eds.), *Empire at the Margins: Culture, Ethnicity, and Frontier in Early Modern China* (Berkeley, 2006), pp. 58–83.

11 Crossley, A *Translucent Mirror*, pp. 89–128.

and finished goods to the markets inside Liaodong province and handling written communications with Chinese and Korean officials. In 1596 Nurhaci was accompanied by a group of Chinese-speaking bodyguards following a black flag. It is probable that a descendant of this organisation formed the portions of Nurhaci's armies that by the later 1620s were called *ujen cooha*, literally 'heavy troops' or 'cherished troops'.[12] Among their functions was the use of firearms and cannon. Even after these units were absorbed into the new Chinese-martial banners in 1642, *ujen cooha* remained as a sobriquet for the Chinese-martial troops as a whole. By the time that Qing troops marched into Beijing in 1644, the Chinese-martial constituted well over half of the entire banner force. Thereafter, the state acted intermittently to lower the numbers and the influence of the Chinese-martial bannermen, primarily by racialising their identities. Examination quotas and officer appointments for Chinese-martial bannermen were steadily decreased over the course of the eighteenth century and they were eventually thrown into competition with civilians; their histories were revised to suggest that they were actually a segment of the subjugated Chinese population. By the later nineteenth century, the Chinese-martial bannermen had thus been reduced to a minor component of the banner population.[13]

After the Qing established their capital at Beijing in 1644 the Eight Banners were settled in segregated communities. By law, they were required to live in these compounds and adult males were expected to train as soldiers and await military appointments. About half of the entire banner population that entered China was settled in, or in close proximity to, Beijing, with the remainder dispersed to the Chinese provinces and provincial capitals to enforce the law and suppress disorder. Soldiers were supposed to be paid in monthly stipends of grain and silver that were designed to support a household of roughly five members. But the finances of the empire, always under strain because of the demands of repeated military campaigns between 1621 and 1804, were never sufficient either to provide the garrison populations with adequate food and housing or to allow the empire to field very large armies of Eight Banner troops.[14]

12 Pamela Crossley, *The Manchus* (Hoboken, 1997), pp. 203–5; Pamela Crossley, *A Translucent Mirror*, p. 96.

13 Pamela Kyle Crossley, 'The Conquest Elite of the Ch'ing Empire', in Peterson (ed.), *The Cambridge History of China*, vol. 9 pt 1, 310–59.

14 For an overview of the economy of the garrisons, see Elliott, *The Manchu Way*, esp. pp. 116–62; and Crossley, *Orphan Warriors*, esp. pp. 47–75.

Good information on the populations of the companies is only available for the period after the creation of the Grand Council in 1729 to centralise military management, and particularly after 1735, when the government began to improve its systems of rewards and status in the banners, and after 1785 no new companies were created. It is probable that in the late seventeenth century the male adult population of the garrisons in China was something on the order of 150,000–200,000 men, and by the mid-eighteenth century may have risen to about 250,000–300,000. But active soldiers with military stipends totalled less than 100,000 in the mid-seventeenth century, increasing in the early eighteenth century to 120,000 regular appointments (and a small number of training or cadet appointments). The inconstancy with which the government actually paid the stipends had marked effects upon the morale of the Eight Banners. The government attempted to ameliorate discontent within the banner garrisons by occasional disbursement of compensatory sums, some of them specially designated for relief of bannermen debt.[15] Nevertheless, the court was determined for ideological reasons to retain the banner populations as an occupation caste, and for practical reasons was unwilling to release a million unemployed, disgruntled, undisciplined former warriors into society. Many remedies were attempted, none of which worked, apart from ignoring bannermen escapes from the garrison compounds to live as civilians. Before the nineteenth century, the obligatory annual military budget never fluctuated much below 13,000,000 or much above 17,000,000 ounces of silver, regardless of whether or not the empire was at war.[16] This is evidently because well over 90 per cent of the annual expenditures went not to salaries of officers or common soldiers, or costs of mobilisation, but to provide support to the non-combatant and unsalaried population that made up 80–90 per cent of each garrison community.

As the Eight Banner armies entered China, the Qing leaders realised that means other than enlarging the Chinese-martial banners had to be found in order to absorb the huge numbers of deserters from the Ming forces, as well as willing local recruits. It was for this purpose that the armies of the Green Standard were created; appointment was not hereditary but rather came from recruitment.[17] Green Standard soldiers were salaried, and after the later seventeenth century were rarely deployed away from their homes; when extended war took them further afield, there was a limit of three to five years

15 Crossley, *Orphan Warriors*, pp. 54–5; Elliott, *The Manchu Way*, 307.
16 Elliott, *The Manchu Way*, 195–205. 17 Ergang Luo, *Luying bingzhi* (Beijing, 1984).

before the soldiers were reassigned to their areas of residence. They were rarely gathered in garrisons and were normally dispersed throughout a district in small stations. The primary role of Green Standard troops was as enforcers of local order, but occasionally they were the closest available troops to respond to a conflict; for instance, they were the first to be mobilised to strike at Burma in 1765. Like the Eight Banners, the number of Green Standard armies rose steadily until the mid-eighteenth century, followed by a protracted decline until the end of the dynastic period. The numbers of the Green Standard were three times that of the Eight Banners forces in the mid-eighteenth century, which would probably put their numbers near a million, and even after the Taiping War in the mid-nineteenth century they probably still totalled well over 400,000 troops.[18] In the mid-eighteenth century, expenditure on Green Standard troops ran to perhaps 30 per cent of the imperial budget in any given year, compared with the 25 per cent spent on the Eight Banner populations.[19]

## Arming soldiers: Manufacture and use of personal weapons

The population of China – perhaps 350 million in the middle of the eighteenth century – did not produce an enormous infantry force. The Three Feudatories Rebellion – the most formidable military threat to the Qing state before the nineteenth century – was based on a force of somewhere between 10,000 and 20,000 soldiers, most of them defectors from the Eight Banners and the Green Standard armies.[20] Instead of overwhelming infantry or cavalry forces, the empire relied upon technology to meet military challenges.

Manchu bannermen on foot normally wore cotton padding, perhaps reinforced by leather, silk, or woven paper on the inside and rivets on the outside, along with a lacquered helmet. They defended themselves with shields of woven bamboo, rattan or cane, and on occasions when they left off their padded clothing because of the heat, the shield could be their only protection. Musketeers typically wore loose jackets and trousers, and caps

18 Kwang-ching Liu, 'The Military Challenge: The North-West and the Coast', in Denis Twitchett and John K. Fairbank (eds.), *The Cambridge History of China*, vol. 11: *Late Ch'ing 1800–1911* (Cambridge, 1980), pp. 202–6.
19 Elliot, *The Manchu Way*, p. 310.
20 Nicola di Cosmo estimates between about 7,000 and just under 18,000. See *The Diary of a Manchu Soldier: 'My Service in the Army' by Dzengseo* (London, 2009), p. 113 n. 27.

with upturned brims and red fringes falling from the top button. Mounted soldiers in the Manchu banners sometimes wore metal helmets, though most were lacquered, and some in the Mongol banners preferred a leather cap. Their armour might be lacquered wood covered by silk or dyed cotton padding, and some are clearly depicted in finely woven mail. Officers usually wore a surplice in the banner colour. Green Standard officers often wore padded armour, but foot soldiers were usually without armour and defended themselves with shields.

The distinctive composite bow, used whilst standing on the ground, was the primary feature of traditional Jurchen hunting, diverging from its Mongolian and Korean cousins, which had evolved at an early date to be used from horseback.[21] Consequently, they were lighter than the Manchu bow and had a draw strength that did not go much above 150 pounds. The Manchu bow did not have tremendous range (about 500 feet on average, compared with about 900 feet for the Mongol bow) or speed, but its force was remarkable. Equipped with the long hunting arrow, in the hands of a skilled hunter it was capable of impaling two deer at once and stopping elk and bear in their tracks. Before the decimation of Siberian tigers in Manchuria, this was the favoured bow for the tiger hunt. Early Qing troops were equipped with such a bow, and even after it was marginalised in the banner forces by lighter and more accurate bows and specialised arrows (*sirdan*), it remained the standard for the banner troops' training and testing; the greatest champions were credited with drawing the equivalent of 240 pounds.[22] It is possible that such bows were the standard weapons among the 'New Manchu' (*ice manju*) populations drawn from the tribal peoples north and east of the original Manchu area. Their primary advantage in close contact was their armour-piercing capabilities. Illustrations of Nurhaci's campaigns show that they were used in the front lines against other Jurchen armies; against Ming forces their use was less common and always occurred at a late stage of the battle, during close combat. Though its practicality on the battlefield was limited, the bow remained the symbol of bannerman identity and an indispensable prop in elite portraiture.

In its last decade, the Ming Empire became interested in the importation, manufacture, and development of hand cannon and muskets. Imperial

21 Peter Dekker, 'Rediscovering Manchu Archery', paper presented to the World Traditional Archery Festival, Busan, Korea (2008).

22 Dekker quotes Stephen Selby, *Chinese Archery* (Hong Kong, 2000) on the declining distances required for qualification in the military examinations: about 308 feet *c.* 1640, about 193 feet in 1693, about 115 feet in 1760.

records state that as early as 1558 the court had commanded the manufacture of 10,000 guns – adapted from the Portuguese arquebus – for distribution to its armies and for sale to neighbouring states. Muskets, usually called 'bird guns' or 'fowling pieces' (*niaoqiang*) may have first been obtained from Japanese pirates – who had themselves acquired them from the Portuguese – in the mid-sixteenth century; it is also possible that the first muskets were acquired from the Ottomans via the overland trade of the earlier sixteenth century.[23] Though the Chinese were well aware of 'gunpowder', and may have used a more combustible form than the Europeans of the period, they were unused to the manufacture of the powder in the form of the standard small cakes suitable for loading muskets, though the Chinese learnt this technique after some unimpressive trials of the new weapons.

The Ming were challenged on several fronts, all of them continental, nomadic zones. Consequently, Ming military innovation focused on war against mounted archers and swordsmen. Volleys were used in China, as in Europe, to interrupt the onslaught of horsemen, but for unknown reasons neither the Ming, nor the Qing after them, employed precision volley squads against mounted enemies. Instead, they focused on technological responses to the need for rapid individual reloading. Precision was not a great object; the greatest value of Ming handguns was probably their ability to startle enemy horses with loud noises and noxious smells. The Ming were particularly fond of the 'three-eyed gun', originally a hand cannon, and of later muskets that could fire multiple shots (as many as ten) in sequence through separate barrels (with separate triggers) before having to be reloaded. They also developed a matchlock device with a detachable breech; soldiers carried multiple breeches with them, all pre-loaded with gunpowder, and switched between them as needed. Ming gunmakers also designed a firearm loaded with cartridges on a delivery belt that dropped each into the breech behind the barrel. The gunman's job was to light the fuses on each cartridge on the belt so that they would fire sequentially as they dropped into position behind the barrel.[24]

When the Aisin went to war against the Ming in 1618, they knew well the spectrum of handguns actually *deployed* (rather than *designed*) by the Ming. They had witnessed the full complement of Ming weaponry during the Imjin Wars (1592–8) in which Jurchens – and Nurhaci personally – had been recruited by the Ming court to help repel the Japanese invasion of Korea,

23 Kenneth Chase, *Firearms: A Global History to 1700* (Cambridge, 2010), pp. 144–5.
24 Ibid., pp. 146–50.

and had later encountered some of these weapons in their own war in Liaodong. Once the Qing had entered Liaodong province and initiated decades of war against the comparatively heavily armed Ming troops, they began to develop musket units of their own. Since the cavalry were largely on their own side, the Qing could stick to the development of mechanically reliable, relatively accurate single-shot muskets. Their dedication to these weapons seems to have been reinforced in later wars against the Zunghar khanate, whose horsemen were supplied with muskets (*niaoqiang* again); in at least one confrontation with Qing troops, the Zunghars dismounted, had their camels lie in a row, covered them with felt, and used the spaces between them as gun-ports.[25] The Qing court aggressively sought the cooperation of Jesuits in the transfer of musket technology, and so the adoption of wheel-lock, and then flintlock, muskets followed soon after their dissemination to ground troops in Europe and the Ottoman Empire. As the Qing pursued their wars against the Mongols beginning in the later seventeenth century, they experimented with advanced artillery, but preferred single-shot guns that were easy to use and repair, and relatively light to carry. The first official history of the Qing (1927) stated that a Qing soldier-turned-scholar of the seventeenth century had invented a repeating gun of the sort the Europeans introduced in the nineteenth century, but no corroborating evidence for this claim has been found.[26] Certainly, the Qing during the eighteenth century had access to any existing gun technology. But the circumstances of their engagements meant that they continued to prefer lighter muskets, and the designs approached the minimum weight that could withstand the force of combustion and recoil.

## Horses

The Jurchens were horse-breeders, and for decades before declaring war on the Ming had been supplying horses to the markets of Liaodong, where many were purchased by the Ming military. These horses were certainly derived to some major degree from the Mongolian steppe ponies, but they were slenderer and perhaps taller, suggesting some derivation from the larger horses of Central Asia. The Ming typically fed their horses 'beans' (*dou*, most likely alfalfa) with a minor addition of grass or hay. For the entire dynastic period the Qing would be troubled by the necessity to provide grain

25 Perdue, *China Marches West*, p. 150.

26 Zhao Erxun (ed.), *Qingshi gao* (Beijing, 1927), p. 292 provides a detailed description, but what it is based on is uncertain.

for their horses, a problem that the Mongolian pony rarely encountered. Like the Ming, the Qing prized quartermasters – who often scoured the countryside with cash seeking grain for horses and men – as highly as their finest troops in prolonged engagements, and rewarded them with equal honours.[27]

In the sixteenth century, sale of horses through Ming Liaodong greatly enriched the Jurchens of Nurhaci's home area, and his family federation in particular. They were also increasingly familiar with a style of military horsemanship that between 500 CE and 1500 became generalised across Mongolia, China, Korea, and Japan. Its distinct material feature was the use of thick, long (often ovoid) saddle fenders made of several layers of hide glued or stitched together. Such fenders prevent the rider from making close contact with the flank of the horse when riding. Training requirements for horses and riders were minimal, allowing maximum mobilisation of the riding warriors. The horse's movements were especially suited to the needs of a mounted archer, since a horse will tend to ride in long sweeping movements unless steered from the bit, allowing the archer time and stability to take aim. However, the rider must sacrifice the conventional movements of cavalry fighting, in which guns, swords, or spears are used against infantry. Mongol attacks, and those of the Manchu horse units adapting their practices, tended to rely upon swift mounted archery attacks against standing lines of enemy troops, veering off, re-nocking, and attacking again. Steadiness and endurance were consequently highly prized virtues in a horse; lateral nimbleness and a high degree of training were not.[28]

Horses were essential for moving troops; like their European counterparts, early Qing soldiers relied upon horses to carry them swiftly to the battlefield, after which they would dismount to shoot arrows or fight with swords and lances. The common depiction of Mongol and Manchu horsemen equipped not only with bows and arrows but also with muskets slung over their shoulders suggests that at certain times they dismounted from their horses to shoot from cover. At the outset of the war against the Ming, the Jurchens were sure to be seriously outgunned, and so an emphasis upon horses in war was essential. In the battle of Sarhū in 1619, Nurhaci prevented a combined Ming and Korean force of perhaps 160,000 troops from laying siege to his home fortress.[29] Utilising a variety of guerrilla tactics, sabotage, and lightning

27 Perdue, *China Marches West*, p. 183; Crossley, *Orphan Warriors*, pp. 135–6.

28 Pamela Crossley, 'Flank Contact, Social Contexts, and Riding Patterns in Eurasia, 500–1500', in Morris Rossabi (ed.) *How Mongolia Matters: War, Law and Society* (Leiden, 2017), pp. 129–46.

29 Frederic Wakeman, *The Great Enterprise: The Manchu Reconstruction of Imperial Order in Seventeenth-Century China* (Berkeley, 1985), pp. 62–6.

attacks by mounted archers, Nurhaci's 1,000 horsemen are reported to have killed nearly half the attacking force and sent the rest home in defeat.

The importance of riding well enough to hurry to a battle, as well as to fight by one means or another from horseback, demanded that every bannerman – not only Mongols or Manchus – be able to ride, and be equipped with a horse. Indeed, the regulations for the garrisons of the later seventeenth century designated over half of all soldiers in the Manchu and Mongol banners, and over three-quarters of the Chinese-martial, as horsemen.[30] These regulations, it should be noted, do not distinguish between mounted archers, cavalry, or mounted infantry but rather refer generally to 'horse soldiers' (*mabing* 马兵). There is every reason to assume that to the Qing government, designating a certain number of men in any unit as horse soldiers meant (a) that when needed the unit should be able to field that many men able to ride at least to the point of engagement and (b) that the unit should be able to provide sufficient horses for them. It is possible that the very high proportion of 'horse soldiers' expected from Chinese-martial units was an acknowledgement of the reality that soldiers able to perform as mounted infantry would be plentiful even if soldiers able to function as mounted archers were not. The proportion of horsemen demanded by the regulations was an expression of the value the Qing assigned to horse soldiers in the history of their wars against the Ming and subsequent conquest of southern China, and of their expectations for the roles of horse soldiers in the wars of expansion.

Yet, by the end of the eighteenth century, adequate horsemanship skills were not common in the banners, and a decreasing number of bannermen had access to even a single horse, let alone the five remounts stipulated in the regulations. Despite their history as horse-breeders and traders, once the Qing transferred to China they encountered the same difficulties as the Ming had before them in acquiring sufficient horses for the army. The earlier dynasty had found that as the agricultural economy developed, open land suitable for grazing diminished. Grain needed by the horses was hard to find in some years, as the human demand for oats, barley, and sorghum made the feed prohibitively expensive. Earlier dynasties had monopolised the special tea and horse trade in Sichuan, using it as a conduit for horses from Tibet and Central Asia into the military stables. By the time of the Ming, though, this monopoly had

30 Guanjie Niu, 'Qingdai mazheng chucao', *Yanshan Daxue xuebao* 2 (2006), 43–7, 45. For background, see also S. A. M. Adshead, 'Horse Administration under the Ch'ing', *Papers on Far Eastern History: Ming and Early Ch'ing China* 17 (1978), 71–9.

broken down, and the number of horses that could be acquired was negligible. The Ming had found that their principal source of horses was the Mongols, with whom they were often at war. Even in times of peace, the cost of Mongol horses was high and the quality could be low. In 1600 the Ming were buying 3,000 horses a year, a significant expense, but hardly a large enough number to make a difference to the military.[31] From 1736 the Qing established an independent supply of horses through a network of stud farms in Gansu province, close enough to supply campaigns in Mongolia but firmly under their control. Occupation of Xinjiang after 1755 opened up new venues for horse breeding, as well as a possible line of supply from the Kazakhs. By the end of the eighteenth century the supply of horses from these sources was approaching the 100,000 minimum needed if the Qing were to provide their cavalry with good horses and remounts. But by then the venues of war had shifted southward and the need for horses had diminished. The stud network fell into disarray, and horses were sold to ameliorate budget shortfalls.[32]

Imperial records indicate that the number of horses attached to the Eight Banners may have peaked in the earlier eighteenth century (roughly corresponding to the reign of the Yongzheng Emperor, 1723–35), at a little over 95,000 horses for the Eight Banners and about 115,000 horses for the Green Standard armies.[33] These difficulties in raising and keeping horses were replicated in almost every banner garrison that the Qing established. Many of the garrisons faced serious problems of overcrowding and a shortage of space within a decade or two of the conquest; some were forced to build on their horses' pastures or else to use them for military practice. Often the land was sold in order to make up for deficits in the garrison budget. When the land was disposed of, the horses themselves were sold to the meat markets in nearby cities. In 1723 cash allotments to buy local grain for horses were less than 3 per cent of the Eight Banners budget for that year, and it was probably understood that most of that would quickly be diverted to other expenses.[34] By the nineteenth century, riding a horse on the battlefield was, in reality, the preserve of an elite in the Mongol and Manchu banners.

31 Wakeman, *The Great Enterprise*, pp. 202–3. 32 Perdue, *China Marches West*, pp. 353–6.
33 Niu, 'Qingdai mazheng chucao', p. 45; Luo, *Luying bingzhi*, p. 396.
34 Elliott, *The Manchu Way*, pp. 308–10.

## Institutional weapons: Fortification, cannons, and cartography

Like other Jurchen leaders of his time, Nurhaci passed more than once through the gates of the Liaodong provincial defences of the Ming, and even into the capital at Beijing. He would have noted that the Chinese methods of fortification were somewhat different from the pattern that loosely characterised Korea and Manchuria. Whereas the Chinese preferred a single face to their fortifications from ground to battlements, interrupted by heavy gates and 'arrow towers' where heavy guns could be placed, the pattern to the east was a base of fortified material, topped with wooden structures. The materials used could vary widely, from dressed stone to compressed earthen blocks to 'rammed earth' (*hangtu*) faced with brick or field stone. At Nurhaci's first capital, a rammed earth base about a yard high was topped by a wooden palisade, which was strengthened later with a second palisade inside it. Later, larger fortifications were built at a new capital nearer to Liaodong, this time of rammed earth. Since much warfare in East Asia in all periods was siege warfare, whenever possible fortifications were built around at least one well, and usually encompassed enough exposed ground to allow for the repair of the rammed earth walls as necessary. Fortifications were fundamentally built to provide high observation points and to defend the population within from fire and arrow shot, as well as the ingress of mounted warriors. In China, where artillery – with or without gunpowder – had been used for centuries before 1600, the ability to withstand prolonged bombardment was also prized.

It can be assumed that Nurhaci's participation in the Imjin War was highly formative in the early Qing approach to cannon.[35] The Manchu would already have been familiar with the type of smaller bronze cannon, as well as cannon firing clutches of – sometimes burning – arrows often placed on the walls of Ming fortifications in Liaodong, but it is unlikely that they had previously experienced the gigantic cannons brought to the war front by the Ming armies. Chinese metallurgists produced cannon at the beginning of the fifteenth century, but they were not effective in warfare. As with their handguns, the Ming first acquired tactical cannon from the Portuguese. Afterward, the Jesuits aided in the publication of manuals and design plans.[36] They impressed upon Ming soldiers and officials that mathematics

35 Kenneth Swope, 'Crouching Tigers, Secret Weapons: Military Technology Employed during the Sino-Japanese–Korean War, 1592–1598', *Journal of Military History* 69 (2005), 11–41, 13.

36 Joanna Waley-Cohen, 'China and Western Technology in the Late Eighteenth Century', *The American Historical Review* 5 (1993), 1525–44, 1531.

and cartography were essential aids to correctly aim cannon. At the Qing court, Jesuits combined their roles as masters of mathematics and astronomy with cartography and cannon design in their service to the emperors.

The Ming had a preference for extremely large cannon, and by the late sixteenth century their ordinance is believed to have equalled the largest artillery used in Europe.[37] They adhered closely to the original *hongyi pao*, or Red Barbarian Gun, that they had been aware of since the 1570s. In the war in Korea, they deployed gigantic cannon of this type, at least one of which was 20 feet in length, over 3,000 pounds in weight, and reportedly made the ground quake for 'ten *li*' (about 3.5 miles).[38] It may have had a range of 2 miles. These were complemented in the defence of fortifications and trenches by the small *folangji* (the Frank), a cannon under a yard long and a little over 30 pounds, which fired hefty caches of scattershot. These guns were all transported to Korea by ship (sometimes equipped with *folangji* on their flanks), the last major naval deployment by an imperial power based in China before the late nineteenth century.[39]

However, little of the cannon technology Nurhaci witnessed in Korea was translatable to the war in Liaodong. They were able to adapt the Korean 'firecart' (*hwacha* 火車), a sort of pushcart with small guns mounted on the front to propel more than 100 steel-tipped or burning arrows.[40] Illustrations of Nurhaci's battles against neighbouring Jurchens, inked three decades after the engagements, show his soldiers pushing the weapons towards the gates of enemy fortifications. After 1618, the Jurchens rapidly captured not only Ming cannon but also the troops to operate them, many of whom went into the Chinese-martial banners. From then until the onset of the Three Feudatories Rebellion in 1673, the Qing used Ming cannon – most of it directly inherited, rather than newly cast, pieces – against Chinese cities and defenders using bows and arrows, muskets, and a few poorly maintained cannons of their own. Wu Sangui and his fellow Three Feudatory Commanders were energetic in scrounging iron and bronze for the casting of new cannon, and stimulated Qing cannon production.[41] The Kangxi Emperor demanded lighter and more accurate weapons, threatening his Jesuit astronomy master with execution if he did not design and cast better big guns.[42] The Jesuits produced nearly 500 weapons, providing the material infrastructure for the artillery component of almost all subsequent Qing campaigns.

37 Ibid. 38 Swope, 'Crouching Tigers, Secret Weapons', p. 21.
39 Ibid., p. 34.
40 Ibid. 41 Wakeman, *The Great Enterprise*, pp. 1110–11.
42 Waley-Cohen, 'China and Western Technology', p. 1531.

The campaigns in Mongolia – first against the Russian Empire in 1685, and soon after against the various Mongol federations – presented serious problems of transport and deployment for the larger cannon. Consequently, Jesuit designers lengthened the barrels for accuracy, used slightly lighter materials (including wood) in the manufacture, and emphasised the production of smaller cannon. Nevertheless, cannons were a constant source of frustration in the campaigns against the Zunghars in Mongolia and later in Xinjiang.[43] Their tactical value was also open to question. At the battle of Ulan Butung in 1690, where the Qing prevailed and forced Galdan's withdrawal from eastern Mongolia, the great cannon of the Qing Chinese-martial general Tong Guogang became hopelessly stuck in a marsh and, according to legend, it actually sank beneath the surface. Qing commanders were frequently punished – often by whipping – for abandoning their cannon in mud or mire as enemies dismounted from their horses and swarmed towards them with arrows flying and swords drawn.[44]

The Qing had even more reason to be sceptical about heavy cannon as they turned in the late eighteenth century to the suppression of internal rebellions in the mountainous and humid areas of the upper Yangtze delta, the south-eastern shoulder of the Tibetan plateau, and the highlands of Yunnan. When the Tibetans of Sichuan (Jinchuan) built piled stone forts that the Qing could not destroy from afar with cannon, the Qing attempted to work around the impassable terrain by carrying ingots of copper and tin up the mountainsides and trying to cast cannon while camped outside the rebel stronghold. Few of these cannons bore up for the time needed to reduce the stone walls; several exploded, inflicting more casualties than the rebels themselves. The emperor ordered a Jesuit to Jinchuan to cast mortars for the army. He eventually not only made the weapons, but improved their calibration, making them devastatingly effective.[45] Mortars – in no way a new technology for the Qing – hereafter found increasing employment in such terrain.[46]

Declining interest in heavy artillery increased the Qing court's attraction to muskets, mortars, cavalry, and the always-reliable bow and arrow. Even without the extensive use of cannon, the Qing were technologically superior to their enemies – internal and external – in the seventeenth and eighteenth centuries. By the end of the eighteenth century, declining capacity in mining and in supplies of wood would have interfered with continued cannon manufacture, even if the Qing had been fighting in terrain where the cannon could have been advantageous.

43 Perdue, *China Marches West*, pp. 188, 194, 184. 44 Ibid., pp. 157, 159.
45 Waley-Cohen, 'China and Western Technology', p. 1538. 46 Ibid., p. 1539.

## Conclusion: The Qing military at the dawn of the nineteenth century

The White Lotus rebellion of 1795–1804 was a turning point in Qing military financing, deployment, and infrastructure. The war ranged across some of the most populous centres of central China, and for the first time the empire leveraged its huge population to raise militia forces supplementary to its small professional force. Qing commanders also became more adept at defeating the rebels, who resorted to the same tactics that their ancestors had used in the Ming wars in southern Jilin and Liaodong in the early seventeenth century. Nevertheless, suppressing the rebellion took eight years, and at its conclusion knowledgeable Qing officials had no illusions that the networks of rebel organisation had actually been wiped out. Although tactical innovations for retaining control of areas reclaimed from the rebels had been effective, the basic Qing practice – used against smaller insurrections through the eighteenth century – of achieving a local victory, disbanding the militia, and then moving on to the next trouble spot endured. Rebel communications, supplies, and covert participation were rarely dismantled.

After the mid-eighteenth century, the Qing emphasis was upon defence, primarily of their northern borders from Xinjiang to the Amur river in Manchuria that were under constant threat from Russia or from Mongol apostates. Responses even to internal disorder were slowed by the declining number of horses available to the military. By contrast, the coasts were imperilled only by the long-standing problem of piracy; the Qing had a few very large cannon mounted along the coast on rigid platforms, mostly to discourage pirate vessels that might pass by. Though Qing commanders in the White Lotus war were well aware of the deficiencies in their strategies, they also knew that they had no option but to continue their smash-and-move-on practices, because the money to sustain local occupation was not available. Perhaps as much as 120 million ounces of silver had been spent on the war: a sum that exceeded the empire's regular annual budget by 30–50 per cent.[47] The poverty of the Eight Banner garrisons seriously impeded the maintenance of any serious military function among the population, and the poverty of the state meant that the Green Standard armies could neither be enlarged nor improved. A great deal of the resources of the imperial treasury had been siphoned off by an insidiously destructive corruption network in the high ranks of the military, a product of decades of Qing campaigns to suppress rebels in south-west and in central China.

47 Yingcong Dai, *The White Lotus War: Rebellion and Suppression in Late Imperial China* (Seattle, 2019), p. 437.

By design the Qing treasury could never have generated funds for the reform or restructuring of its military, let alone the amassing of an infantry force comparable to many land armies of the nineteenth century. From an early point in their history, the Qing emperors had displayed a marked preference for a small state, sustained by a relatively low tax burden, for promoting economic recovery after the wars of the mid-seventeenth century and appeasing their newly conquered subjects. Amassing revenue on the scale needed for a truly transformative military, educational, or industrial programme was neither compatible with this structure nor with the objectives of the Qing emperors. In the design of their governmental structures and in their strategic imagination, the Qing were characteristic of the great land empires of the early modern period, including the Ottomans and the Mughals: all were based on tiny governments with constricted budgets, intensely agricultural and commercial economies with low taxes, and, after the mid-eighteenth century, a distinctly defensive military posture. Military innovations were hardly unknown, but nor were they were a high priority in dealing with enemies not equipped with the industrialising military and shipping operations that the Europeans would bring in force to the Chinese coast by the middle of the nineteenth century.

20

# The seventh continent: Russian territorial expansion, 1450–1850

DAVID SCHIMMELPENNINCK VAN DER OYE*

The nineteenth-century Russian poet Fedor Tiutchev rightly marvelled at his motherland's remarkable growth: 'Moscow, and Peter's town, and Constantine's city,[1] these are the Russian realm's cherished capitals . . . But where is its limit? Where are its borders? The fates will reveal them in times to come . . .' Over the past 400 years an obscure principality deep in the forests on Europe's eastern edge had expanded to become the largest continental empire on Earth – a domain whose immense territory stretched from the Baltic Sea to the Pacific Ocean and beyond, covering one-sixth of the planet's dry surface. Although somewhat diminished in size after the collapse of the Soviet Union in 1991, Russians today still nickname their country 'the seventh continent'.

Equally impressive had been that growth's steady course. Unlike Alexander the Great and Chinggis Khan's ephemeral conquests, Russia had increased the extent of its borders over the centuries at a relatively steady pace. At the turn of the twentieth century, a Prussian historian calculated the rate of the empire's advance since the time of Peter the Great (r. 1682–1725) at 90 square kilometres a day.[2] Furthermore, reverses had been few and were almost invariably overcome. After an ambitious naval officer claimed some territory on the Pacific Ocean in the name of his sovereign in 1850, Emperor Nicholas I (r. 1825–55) boasted, 'Where once the Russian flag has been raised, never will it be taken down.'[3]

* I am grateful for the advice of Paul Bushkovitch and Carol Stevens, who read and commented on this chapter's draft.

1 In the aspirational geography of many Russians during the nineteenth century, Constantinople would be added to their empire. See p. 503 below.

2 Graf Maximilian Yorck von Wartenburg, *Weltgeschichte in Umrissen* (Berlin, 1901), p. 495.

3 'Amurskaia ekspeditsiia', in *Voennaia entsiklopediia*, ed. V. F. Novitskii (St Petersburg, 1911).

Because of the remorseless tempo of Russia's expansion and the secretive, semi-Oriental character of its rulers, for many in the West this impressive territorial growth acquired a sinister if not occult aspect. Some readily believed that its origins lay in Peter the Great's putative deathbed summons to his heirs to 'keep the nation on a constant war footing' and extend their sceptre over Europe and much of Asia.[4] Although historians have shown it to be nothing more than Napoleonic disinformation,[5] Peter's 'Testament' could still be invoked late in the twentieth century as the motive of Soviet foreign policy.[6]

There are a host of more prosaic explanations. The great nineteenth-century Russian historian Vasilii Kliuchevskii saw Russian expansion as a spontaneous process driven by the quest of peasants for new land and freedom from serfdom. As he famously put it, 'The history of Russia is the history of a country that colonises itself.'[7] Writing in the 1940s, Berkeley University's Robert Kerner attributed it to 'the urge to the sea', a landlocked people's primordial quest for access to warm water ports.[8] More recently John LeDonne at Harvard rehabilitated Halford Mackinder's geopolitical approach by arguing that an instinctive desire to reach the natural frontiers of the Eurasian 'Heartland' motivated Russian foreign policy.[9]

A classic variant of the geographical imperative suggests that, as a vast expanse of land with few mountain ranges or other physical obstacles to block hostile invasion, Russia's rulers had always endeavoured to put more space between themselves and their enemies by expanding their borders. In a somewhat similar vein, to justify tsarist conquests in Central Asia during the 1860s, Foreign Minister Prince Alexander Gorchakov famously declared that the tsar was merely behaving like 'all civilised states that come into contact with half-savage wandering tribes'.[10] And a straightforward Schumpeterian lust for conquest is also seen as the agent of Russia's advance.[11] One tsarist

4 Cyprien Robert (ed.), *Testament de Pierre le Grand* (Paris, 1860), pp. 1–7.

5 Simone Blanc, 'Histoire d'une phobie: Le Testament de Pierre le Grand', *Cahiers du monde russe et soviétique* 9:3–4 (1968), 265–93.

6 'The Crescent of Crisis: Iran and a Region of Rising Instability,' *Time* 113:3 (15 Jan. 1979), 23.

7 V. O. Kliuchevskii, *Sobranie sochineniia v deviati tomakh*, 9 vols. (Moscow, 1987), vol. 1, p. 50.

8 Robert Kerner, *The Urge to the Sea: The Course of Russian History* (Berkeley, 1942).

9 John P. LeDonne, *The Russian Empire and the World 1700–1917* (Oxford, 1997).

10 *Foreign Office Confidential Print*, ed. Kenneth Bourne and D. Cameron Watt (University Press of America, 1983), Part I, Series A, vol. 1, p. 287.

11 David Schimmelpenninck van der Oye, *Toward the Rising Sun: Russian Ideologies of Empire and the Path to War with Japan* (DeKalb, 2001), pp. 40–1.

statesman suggested that there was 'something erotic' in the storming of the Kokanese city of Tashkent in 1865.[12]

Trying to find a single, overarching explanation for Russia's remarkable territorial expansion during the four centuries from 1450 to 1850 is fruitless. For one thing, it is impossible fully to divine what was going on in the minds of its tsars and tsarinas. Nevertheless, despite the fears of their more pessimistic foes, they never had a plan for world conquest. Indeed, from its very beginnings in the fourteenth century, Moscow's growth was often opportunistic and haphazard. If Ivan IV (r. 1547–84) implemented a well-considered strategy when he conquered the khanate of Kazan in 1552, Catherine II (r. 1762–96) acquired large swaths of Poland through diplomacy while much of Siberia was absorbed largely through the independent efforts of fur traders and peasant settlers.

For the military historian, the relevant question is not *why* Russia conquered an empire but *how* it did so. The answers lie both in the way that its rulers fashioned a highly centralised autocratic state that effectively subordinated their subjects to their military needs and in how they adapted the armed forces to meet the tactical and technological challenges of the day. The latter both changed over time and often varied by theatre. After some general remarks about Moscow's metamorphosis into an immense state, this chapter will survey Russian expansion along three axes: eastward, from the Volga river basin across Siberia to the Pacific Ocean; south, over the Steppe and the Caucasus Mountains; and west, to the Baltic Sea and into Central Europe. (See Map 17.)

In 1450 there was little to suggest that Muscovy would evolve into an empire covering most of Northern Eurasia. The principality was only one among several that eastern Slavs had established on the north-eastern frontier of the former Kievan Rus, a loosely governed state that had fallen to Batu Khan's Mongol hordes in the early thirteenth century. First mentioned in the chronicles in 1147, Moscow was a relative latecomer. It only acquired a separate identity in the second half of the thirteenth century as a minor principality roughly the size of modern Belgium under Daniil Aleksandrovich, the grand prince Alexander Nevsky's youngest son.

Despite its modest beginnings, Moscow benefited from its favourable location at the headwaters of four major rivers, including the Volga and the Don, which boosted trade revenues. Meanwhile, the principality's prestige received a tremendous fillip in 1328 when the metropolitan (then the

12 P. A. Valuev, *Dnevnik* (Moscow, 1961), vol. 1, p. 60.

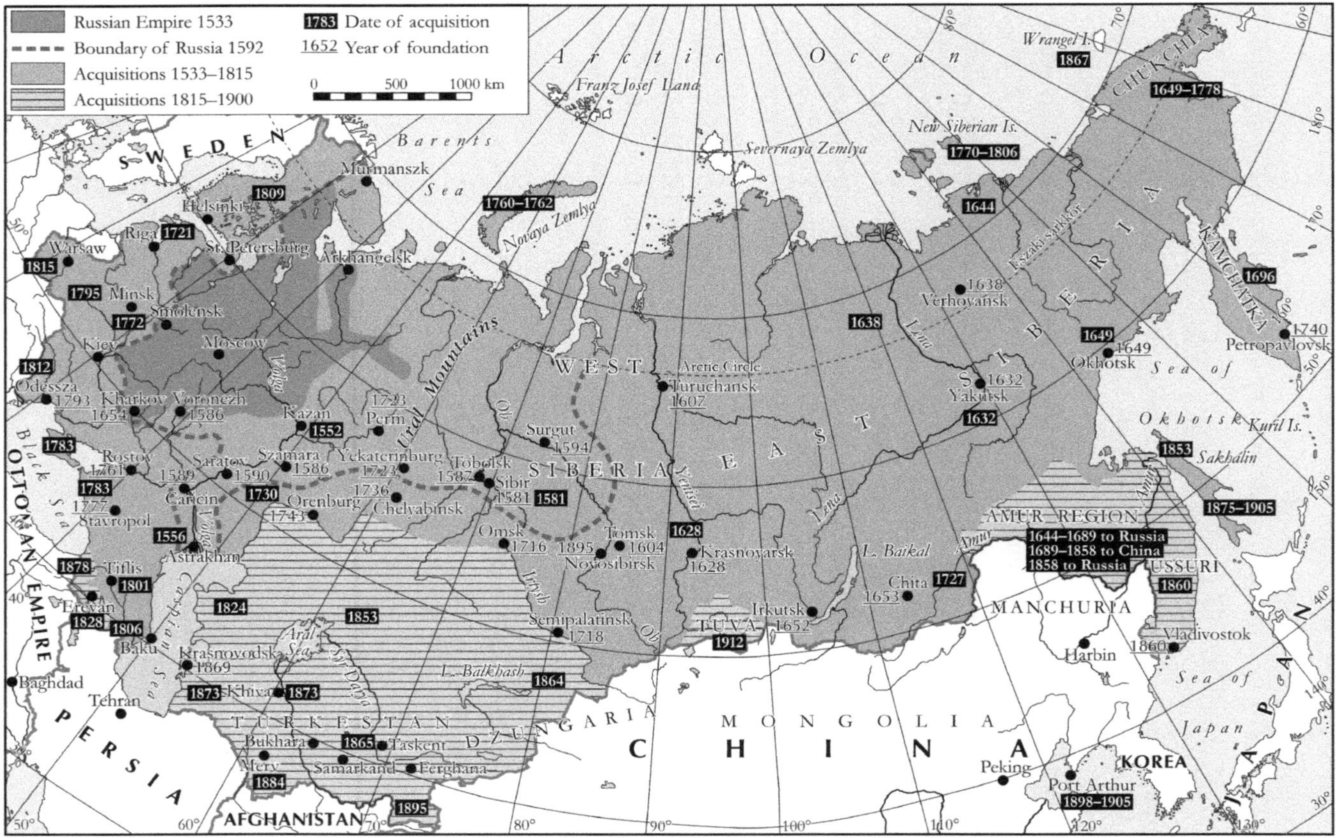


MAP 17. The expansion of Russia in Europe and Asia, 1533–1900. Created by Béla Nagy.

Russian Orthodox Church's senior cleric) permanently established his residence there. Over the next two centuries the Daniilovichi, as Daniil's heirs came to be known, increased the size of their domain eightfold by absorbing neighbouring principalities through annexation or more peaceful means, such as marriage and diplomacy. Moscow's rulers also proved adept at gaining the favour of the Golden Horde, Russia's Mongol overlord at the time, which helped them in their interminable struggles with rival princes.

By the time Ivan III 'the Great' (r. 1462–1502) acceded to the throne, Muscovy dominated central Russia, although it still faced challenges from such principalities as the merchant republic of Novgorod to the north. Powerful states, like the large dual kingdom of Poland-Lithuania in the west as well as the Islamic khanate of Crimea and its eventual Ottoman overlord to the south, also posed serious threats. However, the Golden Horde had been substantially weakened by the secession of Crimea and Kazan two decades earlier, and now the three khanates were often distracted by internecine quarrels. The 'Mongol Yoke' came to a definitive end less than twenty years into Ivan's reign when he stopped paying the annual tribute to the Golden Horde's khan, who proved too weak to enforce his hold on Muscovy

Ivan III would come to be known as 'the gatherer of the Russian lands'. The implication that he entirely reconstituted the erstwhile Kievan Rus is somewhat misleading, since many of its former western domains were now under Lithuanian rule, while his son, Vasilii III (r. 1505–33), would annex Pskov and eastern Riazan. Nevertheless, Ivan did more than his predecessors to complete the process by absorbing Tver, Iaroslavl, and Rostov, among other principalities. His most significant conquest was Novgorod, the immense republic whose northern lands stretched from the Gulf of Finland to beyond the Ural Mountains.

Although Novgorod had largely escaped the ravages of Mongol occupation, because of its fractious politics, it lacked the strong authority of an increasingly autocratic principality like Moscow. In 1471, suspecting that its boyar elite had been dallying with his Lithuanian foe, Ivan invaded the republic, easily overwhelming the latter's much weaker defences. Rather than annexing Novgorod outright, he reduced its autonomy and exacted reparations. Dissatisfied with only partial authority over his neighbour, Ivan launched another campaign in 1478 and fully incorporated it into his own domain. To confirm his hold on the former republic, the grand prince took away the bell that summoned its assembly and expropriated its lands from the local boyars, redistributing them to his own nobles. Not long thereafter, he also absorbed the once formidable rival of Tver.

By the time of his death in 1505, Ivan the Great had expanded Russia's size fourfold. Indeed, as sovereign of an increasingly immense land, he was emboldened to take the imperial title of 'tsar' (Slavonic for Caesar). And, having wed Sophia Palailologina, a niece of Byzantium's last emperor, he also adopted Eastern Rome's double-headed eagle as his state's symbol.[13]

In addition to dramatically increasing the size of his realm, conquering Novgorod also enabled Ivan to establish a new category of military servitors who were entirely beholden to his person. Much as in pre-Mongol Kiev, the Daniilovichi had traditionally commanded relatively small retinues and therefore relied on other princes as well as, on occasion, even friendly khans for major campaigns. Such allies were not always dependable. When Ivan's great-grandfather, Dmitrii Donskoi (r. 1359–89), gathered an army of some 20,000 men to confront the Golden Horde in the celebrated battle of Kulikovo in 1380, a number of other princes did not heed his summons.[14] But as Dmitrii's heirs extended their hold over other principalities, they steadily bound its princes more tightly to their will.

Submission to Moscow could be voluntary. Some nobles saw greater opportunities by serving the increasingly powerful principality. At the same time, a lengthy civil war in the early fourteenth century convinced many others that Muscovite order might be preferable to fratricidal chaos. Moscow also won the loyalty of numerous non-Russian princes, including Lithuanians and Tatars. Like their nomadic foes on the Steppe, the Daniilovichi had no qualms about recruiting former foes into their ranks, without even necessarily insisting that they convert to Orthodox Christianity. Thus, in 1445, Ivan III's father had already incorporated two Tatar princes and their followers into the dependent Kasimov khanate. Right up to the Revolution of 1917, the upper echelons of Russia's army and bureaucracy were strikingly cosmopolitan in their composition.

Having absorbed Novgorod, Ivan expropriated most of its aristocracy and turned their estates over to a new class of gentry who possessed them in exchange for military service. Unlike Muscovy's boyars and other nobles, who inherited their domains, the gentry held them only for as long as they served the prince. By holding land 'in place' (*pomest'e*) at the pleasure of their master, they were entirely obligated to do his bidding. The relationship was different from that between lord and vassal characteristic of Western European

13 Thereby implicitly raising Russia to the same rank as the Holy Roman Empire, which also used the bicephalous raptor as its arms.

14 Iu. G. Alekseev, *Voennaia istoriia dopetrovskoi Rossii* (St Petersburg, 2018), p. 511.

FIGURE 24. Muscovite cavalry, *c.* 1550. Source: Getty Images.

feudalism, since Muscovy's prince owed no reciprocal obligations to his *pomeshchik* (landowner).

To ensure that the gentry had sufficient manpower, Ivan also restricted the ability of peasants to leave their estates – a process that would reach its logical conclusion with the imposition of serfdom 200 years later. As for the higher nobility, Ivan continued to restrict their independence as well. According to John Keep, 'perhaps the main ingredient in Moscow's success was its skilful manipulation of the warrior élite'.[15]

Muscovy's army bore the heavy imprint of steppe warfare. Organised for raiding, its predominant combat arm was light cavalry. (See Figure 24.) Tactics and weapons imitated those of the Tatar khanates on the eastern and southern frontier, who constituted the principal threat until the seventeenth century.

15 John Keep, *Soldiers of the Tsar: Army and Society in Russia, 1462–1874* (Oxford, 1985), p. 13.

The descriptions of Muscovite warfare by the sixteenth-century Habsburg diplomat Baron Sigismund von Herberstein resemble those of the Scythians by Herodotus two millennia earlier:

> Whatever they do, whether they are attacking, or pursuing, or fleeing from the enemy, they do everything suddenly and rapidly, so that neither infantry nor artillery can be of any use to them. They make the first charge on the enemy with great impetuosity; but their valour does not hold out very long, for they seem as if they would give a hint to the enemy, as much as to say, 'if you do not flee, we must.'[16]

Like the Tatars, Muscovite cavalry rode on small but hardy mounts and were armed with composite bows, which, with an effective range of about 300 metres, outshot the bows of medieval Western armies.

While primarily used for siege guns, gunpowder was already slowly making its way to Muscovy during the early fifteenth century. Ivan III used his Byzantine wife's contacts with the West to recruit foreign specialists to improve his artillery. Most prominent among them was the Bolognese architect and military engineer Aristotele Fioravanti, who set up a cannon foundry near the Kremlin. Ivan also made limited use of arquebuses. However, as an effective weapon for the infantry, firearms only came into their own a century later, when Ivan III's grandson, Ivan IV 'the Terrible', founded regiments of musketeers (*strel'tsy*, lit. 'shooters'), who would number 20,000–25,000 men by the turn of the seventeenth century.[17]

Ivan IV's musketeer regiments constituted Muscovy's first permanent, salaried soldiers. This was in contrast to the cavalry, which the tsar only mustered during campaigns. The *strel'tsy* were created in part to man the defensive lines to protect the capital against Crimean Tatar raids from the south, although the first unit formed Ivan's palace guard. Recruited from freemen, they lived in separate neighbourhoods and would come to form a caste that at times took an active part in court politics. Meanwhile, Ivan IV also issued regulations standardising the landed nobility's military obligations.

By Ivan IV's reign, Muscovy's military had been transformed from a heterogeneous and unreliable assembly of princes with their own retinues to a disciplined cavalry answerable only to the tsar as well as a regular force of infantry. With the help of a growing number of foreign specialists – there were some 4,000 of them in Ivan's army[18] – it had also developed the artillery

16 Sigismund von Herberstein, *Notes upon Russia*, trans. R. H. Major (London, 1851), p. 97.
17 Carol B. Stevens, *Russia's Wars of Emergence 1460–1730* (London, 2007), p. 73.
18 Ibid., p. 69.

into an effective combat arm. Yet if the army was slowly adapting to Western ways, until the fall of the Daniilovichi at the end of the sixteenth century the emphasis still remained on the highly mobile warfare of the steppe. The military revolution would have to await a new dynasty.

While Muscovy's expansion was impressive, there were setbacks. The most dramatic began during Ivan IV's reign, which saw a bloody purge of his boyars during the bizarre terror of the *oprichnina*. Combined with the increasingly heavy social and economic strains of almost constant war as well as a dynastic vacuum, when Ivan's successor died without an heir in 1598 these stresses ultimately led to a complete collapse of the state during the fifteen-year 'Time of Troubles'. At its nadir, Polish troops even occupied the Kremlin for some two years until a popular army drove them out in 1612. The following year an assembly elected the teenaged Mikhail Romanov as their tsar, founding a new dynasty.

## To the east

If Ivan III gathered the Russian lands, Ivan IV transformed them into an empire. The principal axis of advance was to the east, where, little over half a century after crossing the Ural Mountains, the Kremlin's reach would extend to the northern Pacific Ocean's waters. More than in the south and the west, much of this expansion was driven by the ambitions of individuals rather than the will of the tsar – although the autocracy's authority invariably followed in their wake.

The first step in Russia's eastward march came relatively early in Ivan IV's reign, with the conquest of Kazan. As one of the Golden Horde's successors, the khanate on the middle Volga had long been a thorn in Muscovy's side. Ivan III had forced Kazan into submission when he conquered it in 1487. However, when his grandson inherited the throne at the tender age of three in 1533, the khanate took advantage of the political uncertainty by joining forces with Crimea, thereby menacing Muscovy from both the east and the south. Within less than a decade, the two khans were launching joint attacks on the young tsar's realm.

When Ivan IV effectively took control of his government in 1547, one of his first acts was to order an assault on Kazan. The logistics of mounting a major campaign with 150,000 troops some 800 kilometres east of Moscow frustrated both this attempt as well as a second one two years later.[19] Building a fortress

19 A. P. Degtiarev and V. P. Demin, *Voennaia istoriia Rossii: Vneshnie i vnutrennie konflikty* (Moscow, 2011), p. 60.

at Sviazhsk, an elevation 40 kilometres upriver from the foe's capital, provided a more effective base for a third campaign in 1552. Now personally commanded by the tsar, Muscovite forces stormed the city after a textbook siege involving trenches, wooden towers, artillery barrages, and mines. Although it would take some time to subdue the surrounding lands, four years later Ivan also absorbed the khanate of Astrakhan downriver.

Because Kazan's incorporation into Muscovy marked the first major conquest of alien nationalities, historians often consider the year 1552 to mark the birth date of the Russian empire. Yet although Ivan IV was the first sovereign to crown himself with the imperial title of tsar, Russia would not formally consider itself an empire until late into the reign of Peter the Great. Nevertheless, imposing Russian rule over the khanate marked the beginning of a host of new challenges as Muscovy evolved into a multiethnic state. Indeed, the first formal census, which was carried out in 1897, revealed that Russians comprised less than half of the empire's population.[20]

The Church portrayed the campaign against Islamic Kazan and Astrakhan as a crusade against the infidel Saracen, and Ivan commissioned the Cathedral of St Basil the Blessed on Red Square to commemorate the occasion. As a result, some scholars subsequently endowed the struggle with deep religious significance.[21] Nevertheless, the tsar's motives were largely geopolitical; namely, to defend Muscovy's eastern frontier as well as to secure access to the Volga river's trade links.

Control over the Volga opened paths to the Caucasus in the south as well as into vast resource-rich eastern lands stretching to the Urals. Among those who quickly took advantage of the latter were the Stroganovs, a prominent merchant family that had made its fortune by distilling salt. In 1558, Grigorii Stroganov convinced the tsar to grant him an enormous concession along the Kama river, a tributary of the Volga that flows from the foothills of the Urals. Together with his brothers, he established saltworks, iron foundries, and farms in the region. Among the Stroganovs' privileges was the right to maintain troops to defend their vast estates. These were necessary since the native population did not always greet the Muscovite interlopers with open arms.

20 Andreas Kappeler, *The Russian Empire: A Multi-ethnic History*, trans. Alfred Clayton (London, 2001), p. 397.

21 Jaroslav Z. Pelenskyj, *Russia and Kazan: Conquest and Imperial Ideology (1438–1560s)* (The Hague, 1974); M. Khudiakov, *Ocherki po istorii kazansogo khanstva* (1923; repr. Kazan, 1990).

The Tatar chief Kuchum was their most formidable adversary. Having seized control of the khanate of Sibir just across the Urals in 1563, he aggressively opposed the Russians by raiding their settlements and terrorising friendly tribes. In addition to jeopardising the various enterprises of the Stroganovs, Kuchum also controlled a region with valuable minerals as well as fur-bearing marten. Ivan was eager to add the khanate to his domains. Indeed, having conquered Kazan and Astrakhan he had already presumptuously added 'master of all Siberian lands' to his growing collection of titles.[22] But because the tsar was already engaged in his long and costly Livonian War on the Baltic Sea, he had few soldiers to spare. In 1574, the tsar therefore gave the Stroganovs a new charter that also allowed them to invade Kuchum's territory.

In circumstances that remain vague, the merchants recruited a Cossack adventurer, Ermak Timofeevich, to march against Kuchum. In early September 1582 Ermak set out for the khanate at the head of some 540 fellow Cossacks. After an epic 1,500 kilometre trek over the Urals, his modest force stormed Kuchum's capital that autumn. Although the defenders had far greater numbers, their archers proved no match for the Cossacks' firearms. Ermak's conquest set the pattern for subsequent sorties that would ultimately bring Siberia's forests entirely under Russian control: small bands of Cossacks armed with muskets making their way ever further into the *taiga*, easily overwhelming indigenous tribes entirely unacquainted with gunpowder.

Much as New France's *coureurs des bois* penetrated deep into North America's wilderness, these men were drawn by the soft gold of fur. Sable pelts, a highly prized commodity in Europe, provided about 10 per cent of the tsar's revenues in the early seventeenth century.[23] Once they had cleared the way, the autocracy invariably sought to establish its hold by building forts and exacting tribute from the local population. Nevertheless, because of the vast distances involved, its grip was considerably looser than in European Russia. Nowhere was the saying 'The Tsar is far away, but God is up high' (*Tsar' daleko, a Bog vysoko*) more apt. As a result, over time Siberia attracted large numbers of runaway serfs, religious sectarians, and other settlers attracted by the relative freedom of the frontier.

The lust for soft gold also resulted in Russia's only attempt to found a colony overseas. A few decades after Vitus Bering, a Danish captain in Peter the Great's service, discovered a maritime route to Alaska in the early

22 Boris Nolde, *La Formation de l'empire russe*, 2 vols. (Paris, 1952), vol. 1, p. 157.
23 Raymond Fisher, *The Russian Fur Trade* (Berkeley, 1943), p. 119.

eighteenth century, small groups of fur traders began to seek their fortunes in North America. Modest settlement eventually followed, and by 1799 Emperor Paul I (r. 1796–1801) chartered the Russian-American Company to develop his North American possessions. However, because of the great distances involved and the rapid exhaustion of fur-bearing animals, the venture was never particularly profitable. Worried that it might eventually fall into British hands, Alexander II (r. 1855–81) sold Alaska to the United States in 1867. Despite the efforts of Peter and his heirs to transform their realm into a naval power, Russia proved neither able nor particularly willing to expand on alien shores. Like Eurasia's other continental empires, it was better at absorbing lands directly contiguous to its borders.

Feeding Siberia's growing population was a major challenge. Because the coniferous taiga's soil is highly acidic and the growing season short, farming often proved impractical. The situation was particularly challenging in the region east of Lake Baikal. When the local governor learned that the local population cultivated grain in the Amur river basin to the south, he organised an expedition in 1643 to investigate. While the mission confirmed that agriculture was indeed feasible there, its leader's harsh ways thoroughly alienated the area's tribes. Subsequent forays by Erofei Khabarov proved even more sanguinary, and in 1650 he occupied the village of Albazin on the Amur.

For a time, Khabarov's superior arms enabled him ruthlessly to exploit the Amur's Daur people. But as tributaries of China's young Qing dynasty, the Daur had powerful patrons. When reports of the Russian incursion reached Beijing, the Kangxi Emperor soon ordered his army to expel the red-haired barbarians. Also armed with cannon and muskets, the Chinese force could easily have carried out its task. However, its commander had been given instructions not to kill the Cossacks and he ultimately withdrew. In subsequent clashes, including a naval battle on the Sungari river in 1658, the Qing tended to prevail and Russian authorities soon ordered a stop to their efforts on the Amur.

There were sound diplomatic reasons for avoiding any further complications with Beijing as well, since Moscow had been endeavouring to establish trade links with its eastern neighbour for some time. Meanwhile, the Qing had more pressing problems in the south, where they were still fighting loyalists of the former Ming dynasty. By 1683 Kangxi had cleared the last remnants of his opponents from their final holdout in Taiwan, and he could now turn his attention northwards. With a proper fortress and a monastery, Albazin now was a Russian city that the tsar had already incorporated into his Siberian possessions. This was intolerable to the Son of Heaven, who considered the Amur to lie well within his own patrimony, and he ordered an assault on the position in June 1685.

With some 3,000 troops and plentiful artillery the Chinese forced the 450-man garrison to surrender within a few days. Yet, having taken Albazin, the siege's commander neglected to secure the position and marched his troops back home. Moscow understood that its hold on the Amur was untenable, and ordered the local military governor to abandon it. Nevertheless, shortly after the Chinese had left, the local governor ordered its former inhabitants to return. With the help of a Prussian military expert, they now built a more robust citadel. The Chinese were back in July 1686. Unable to blast the new stronghold into submission, they resorted to starving its occupants by cutting off all access. Although by December only two dozen of the original 800 defenders were still alive, Kangxi lifted the siege when the tsar's envoys offered to enter into talks.[24]

Where force of arms did not compel the Russians to leave the Amur, diplomacy did the job. With Jesuit interpreters acting as intermediaries, Moscow and Beijing signed a treaty at Nerchinsk on 27 August 1689. In exchange for quitting the Amur, the agreement promised to establish regular trade between the two former adversaries. The Treaty of Nerchinsk was remarkable since it was the first time that China negotiated with another power as equals. It also proved durable, and would secure its border with Russia for the next century and a half.

## To the south

As on its eastern marches, Muscovy's greatest challenge in the south had traditionally been containing the Golden Horde's heirs. In the fifteenth century, the principality's southern frontier abutted on the Pontic Steppe, a fertile, treeless prairie along the Black Sea's northern coast. As the western extension of the Eurasian Steppe, these grasslands had periodically been invaded by Inner Asian nomadic confederations, beginning long before the Scythians of Herodotus' day and continuing up to the Mongols in the thirteenth century. Before the advent of more effective firearms in the eighteenth century, the warriors of the steppe outfought Muscovite cavalry, thereby preventing tsars from fully consolidating their hold over the 'wild field' (*dikoe pole*), a no-man's land.

24 Tonio Andrade, *The Gunpowder Age: China, Military Innovation, and the Rise of the West in World History* (Princeton, 2016), pp. 219–30.

Until its conquest in the late eighteenth century, Crimea was Russia's main rival on the Pontic Steppe. Like Kazan and Astrakhan, the khanate had seceded from the Golden Horde in the first half of the fifteenth century. Its relations with Moscow were cordial at first. Thus Ivan III had concluded an alliance with his southern neighbour to join forces against mutual foes like the Golden Horde. However, when the latter's disintegration around 1500 ended the treaty's *raison d'être*, relations became increasingly hostile. Even before Crimea formally abrogated the pact, its Tatars launched seventeen major raids against Muscovy. In 1521 the khan even besieged the Kremlin, although he did not possess the artillery to break through the walls. During another incursion fifty years later, Khan Devlet Girey torched Moscow, killing most of its inhabitants.[25]

At the time, launching a campaign against Crimea was hardly an option. The logistical challenges were simply too daunting. Meanwhile, the khan had found a formidable patron when he became a vassal of the Ottoman Empire in 1475. Muscovy therefore reverted to a tactic dating from Kievan times by erecting wooden palisades and earthworks in the clearings between forests to obstruct Tatar horsemen. Towards the end of the fifteenth century, a defensive line had already been established along the Oka river, about 100 kilometres south of Moscow. Much like the Great Wall of China, these lines were not so much meant to block Tatar raids as to delay them, giving local military governors time to organise their defences.

Over time Russians settled the lands beyond the Oka, necessitating a new cordon by the 1520s 100 kilometres further south. This gradual process of colonists moving ahead of defensive lines, thereby spurring the construction of newer lines deeper into the steppe, repeated itself several times and drew Russia ever closer to the Black Sea. By the early 1680s the furthest barrier passed through Izium, some 800 kilometres south of the capital.

After the conquest of Astrakhan in 1556, Muscovy also bordered on the Caspian Steppe to the east, which exposed its south-eastern flank to the Nogai, a more loosely organised offshoot of the Golden Horde. Although most had accepted the tsar's rule, groups periodically raided the settlements of Russian colonists. The situation stabilised somewhat in the early seventeenth century with the arrival of the Kalmyks, Mongolian nomads who had been driven from their Inner Asian pastures by the more aggressive Zunghars. As Buddhists faithful to the Dalai Lama's Yellow Hat sect, the

25 Degtiarev and Demin, *Voennaia istoriia Rossii*, p. 63.

Kalmyks shared Muscovy's hostility to the Islamic Nogai and swore allegiance to the tsar.

Like most nomads, the Kalmyks understood fealty very differently from the Russians. Whereas the latter considered such submission to be permanent, Kalmyk chiefs regarded it a temporary arrangement that was binding only during the lifetimes of those who had concluded it.[26] Tensions began to develop in the latter half of the eighteenth century as Empress Catherine II encouraged German settlers to colonise the Caspian steppe's rich farmlands, thereby depriving the nomads of pastures for their flocks. Mindful of their discontent, China's Qianlong Emperor enticed the Kalmyks back to their former homeland, and in 1771 most fled Russian rule in a great exodus.

Further to the east, on the steppes beyond the Caspian Sea, Russians also encountered the Kazakhs. Not to be confused with the Cossacks, the Kazakhs were another Turkic nomadic nation who ranged over much of what is now Kazakhstan. By the eighteenth century they had split into three groups, with the 'Junior Horde' (*Malaia orda* or *Mladshii zhuz*) bordering on Russia's frontier. Like the Kalmyks a century earlier, in 1731 the Junior Horde's khan pledged his submission to Russia's sovereign. Seven years later the Middle Horde's leader followed suit. To ensure obedience, the Romanovs took sons of the khans to St Petersburg as hostages and built a fortress on the Or river. Christened Orenburg, the settlement served as both an observation post over the Kazakh Steppe and an entrepôt for trade with Central Asia.

When Kazakh khans swore fealty to the 'White Tsar', their nominal subjects did not necessarily follow suit. A number of the Little Horde's nobles had already opposed their leader's submission to the Romanovs in 1731, and many Kazakhs joined the Cossack Emelian Pugachov's great rebellion forty years later. The source of their discontent was the basic incompatibility between the nomads' need for extensive pastures to graze their flocks and sedentary society's hunger for farmland.

Much like the white man's encounter with the natives of America's Great Plains, by the nineteenth century it was an unequal contest. In 1822, Siberia's governor-general, Mikhail Speranskii, abolished the title of the Middle Horde's khan and two years later that of the Lesser Horde's khan was similarly abolished. After tsarist troops put down one last major uprising, under Kenisary Qasimov in the 1840s, all three Kazakh hordes were gradually integrated into the Russian Empire.

26 Michael Khodarkovsky, *Where Two Worlds Met: The Russian State and the Kalmyk Nomads, 1600–1771* (Ithaca, 1992), pp. 71–3.

Various smaller successor khanates to the Mongols ruled the deserts south of the Kazakh Steppe. Already during Ivan IV's reign intrepid Englishmen in the Muscovy Company's employ began to explore them as a possible overland trade route to India. Peter the Great had similar thoughts. In 1717, he sent Prince Aleksandr Bekovich-Cherkasskii on an expedition to Khiva in part to establish trade links with India's Mughal emperor. However, by means of guile, Khiva's khan defeated the prince's troops, annihilating his 7,000-man force. Peter's more cautious successors would wait until the Kazakh Steppe was relatively secure before launching their campaign of conquest in the latter half of the nineteenth century.

The steppes on Muscovy's southern frontier were also home to the Cossacks, who would come to play an important military role once the autocracy had tamed them. In the wake of the Golden Horde's collapse these grasslands were the 'wild field'. While they provided pastures for various nomadic tribes as well as a good conduit for Tatar raids, for several centuries no sedentary power possessed them. As serfdom slowly began to tighten its grip on the villages to the north, groups of peasants sought freedom in this frontier where they blended with Tatars, Turks, Poles, and other migrants (the word 'Cossack', or *kazak* in Russian, most likely originates from 'free man' in Turkic).

As people of the steppe, the Cossacks formed excellent cavalry that sold its services to the highest bidder, be it Muscovy, Poland, or Turkey. By the sixteenth century, this varied population began to form hosts, or self-governed communities headed by an elected hetman (*ataman*). The most important groups were the Zaporozhian Cossacks, who were based south of the Dnepr river's rapids in Ukraine, and the Don Cossack Host on the lower Don river. Over time other hosts would establish themselves all along Asian Russia's southern border, which they helped to guard in exchange for special privileges.

Although Crimean raids continued well into the seventeenth century, by now the main obstacle to further southward expansion was the Ottoman Empire. The Turks had already registered their displeasure at Muscovy's progress with an attempt to take Astrakhan in 1569. But they were particularly anxious to keep full control over the Black Sea. The Ottomans well understood that the Russians strove for access to its shore in their long quest for warm water outlets. Thus, as a member of the coalition that waged the Great Turkish War (1683–99), Peter the Great seized the Turkish fortress of Azov on the eponymous waters that join the Black Sea. Less than a century later, another war under Catherine the Great also extended Russian control over Crimea, whose khanate she abolished in 1783.

Catherine's ambitions did not stop on the Black Sea's northern shore. Optimistic that the Ottoman Empire, already seemingly in its dotage, might well disintegrate, she drew up a plan in 1780 for its partition with Austria and other interested European powers.[27] In her enthusiasm she even named her second grandson Konstantin after Constantinople and had him raised by Greek nannies to prepare him to be the potential ruler of a reborn Orthodox empire to be established in Turkey's place. In the event, the Ottomans did not cooperate, and the empire survived for another century.

To Catherine and her heirs, the great prize was the Turkish capital of Constantinople. Not only did 'Tsargrad' have great spiritual value as the former centre of their Orthodox faith, but it also commanded the Bosporus, one of the two straits that link the Black Sea with the Mediterranean. Over the course of the nineteenth century, the Romanovs would wage four more wars with Turkey, but the Straits remained maddeningly elusive.

Across the isthmus that separates the Black Sea from the Caspian lie the Caucasus Mountains, an 800-kilometre-wide wall of rock that traditionally marked Russia's border with Turkey and Persia. Ivan IV's conquest of Astrakhan had already brought his domains up to Kabarda and Dagestan in its north-eastern foothills, which soon drew Muscovy into the region's fractious politics. However, it would take Catherine's subjugation of Crimea two centuries later to secure the range's entire approaches from the north.

While the many nations that inhabit the Caucasus are mostly Muslim, the small Orthodox kingdom of Georgia lay to the south. Highly vulnerable to the two Islamic empires that bordered it, Georgia's King Erkele II sought protection from Catherine II during the uncertainty of her first Turkish war (1768–74). Her support only lasted as long as the conflict. When the two empires were again at peace, she withdrew her troops.

The empress only took notice in 1795, when Persia's new Qajar shah sacked Tiflis, Georgia's capital. While Catherine's death a year later cut her Persian campaign short, in 1801 her grandson, Alexander I, accepted the Georgian king's 'humble request' to become his subject. Alexander did not limit himself to Georgia, and his troops soon marched into Armenia and Azerbaijan as well. It would take two unsuccessful wars for Persia to recognise Russia's hold of Transcaucasia with the Treaty of Turkmanchai in 1828.

27 Edgar Hösch, 'Das sogenannte "griechische Projekt" Katharinas II. Ideologie und Wirklichkeit der russischen Orientpolitik in der zweiten Hälfte des 18. Jahrhunderts', *Jahrbücher für Geschichte Osteuropas* New Series, 12 (1964), 168–206.

The shah's surrender did not bring peace to the range itself, for its tribes strongly opposed the Christian tsar's authority. And their resistance to the occupiers, which had already been raging for some time, only intensified as Russia sought to tighten its grip on the mountains. The tsar's response was equally uncompromising: 'pacifying the highlanders permanently or exterminating the unsubmissive'.[28] Alexander had already raised a 50,000-man Caucasus Corps in 1820, which for the next forty years would conduct a classic counter-insurgency campaign that involved no less than a quarter of a million troops at its height.

To establish a secure line of supplies, the corps' soldiers hacked the Georgian Military Road through the range's middle to Tiflis, effectively separating the theatre into two. The most intense fighting was in Dagestan to the east, where a brilliant and charismatic guerrilla leader, Imam Shamyl, called for the tribes to unite under Islam's green banner and wage *gazawat* (holy war) against the tsarist infidel. It would take some time for the Russian army to adapt its tactics to the unfamiliar terrain, but by 1859 it had managed to capture Shamyl. Three years later it also brought Circassia in the west to heel, driving close to 500,000 highlanders into Ottoman exile.

## To the west

The west required an entirely different approach to war. Whereas the nomads Muscovy encountered on its eastern and southern frontiers were horse-mounted archers who fought with the rapid and flexible tactics of the steppe, its western adversaries adopted the European style that favoured infantry and gunpowder. The challenge here was to adapt the military accordingly. Nevertheless, the transformation was never complete. Chris Bellamy argues that Soviet warfare still bore distinct traces of Muscovy's centuries-long struggle with Inner Asian nomadic warriors.[29]

In the fifteenth century, none of Moscow's immediate foes to the west had already undergone a military revolution. Ivan the Great's main rival here was the Grand Duchy of Lithuania. Beyond the reach of Batu's hordes, various pagan tribes on the Baltic's south-eastern shores had united into an independent kingdom in 1253. During the fourteenth century, its grand dukes took

28 Paraphrased from Firuz Kazemzadeh, 'Russian Penetration of the Caucasus', in Taras Hunczak (ed.), *Russian Imperialism: From Ivan the Great to the Revolution* (New Brunswick, 1974), p. 253.

29 Chris Bellamy, 'Heirs of Genghis Khan: The Influence of the Tatar-Mongols on the Imperial Russian and Soviet Armies', *The RUSI Journal* 128 (1983), 52–60.

advantage of Kievan Rus's destruction to absorb vast eastern Slavic lands in what are now Belarus, Ukraine, and western Russia, thereby becoming Europe's largest country. Although many of his subjects were Orthodox, in 1386 Grand Duke Jogaila adopted Catholicism when he wed the heiress to Poland's throne, Queen Jadwiga. By joining their dynasties, they created a powerful dual monarchy that would fight no less than eleven wars with its eastern neighbour until Catherine II fully absorbed much of it towards the end of the eighteenth century.

In his effort to regather the former lands of Kievan Rus, Ivan began to test Lithuania's borders in the 1480s. Carried out over marshes, scrublands, and forests, his campaigns largely consisted of cavalry raids, much as on his other frontiers. By appealing to their common faith, Ivan convinced a number of local princes to switch their allegiance to Moscow. His main objective was the ancient city of Smolensk, which his heir, Vasilii III, took in 1514. The war transferred a third of the Dual Monarchy's lands to Muscovy as well.

Ivan also had ambitions for the Baltic state of Livonia further north, then loosely ruled by the remnants of the German Teutonic Knights. He did not succeed, and his grandson, Ivan IV, took advantage of the order's growing weakness to make another attempt after conquering the khanates on the Volga. The Livonian War, which would drag on for a quarter of a century from 1558 to 1583, proved disastrous. The campaign began auspiciously as Ivan's forces conquered the port of Narva and the bishopric of Dorpat (Tartu) during the first year. However, the tsar's rapid advance also whetted the appetites of other powers with similar ambitions.

By 1562 the tsar found himself embroiled in a four-way struggle involving Poland, Sweden, and Denmark. The conflict's increasing burden weighed heavily on Russia and may well have contributed to Ivan's growing paranoia, which expressed itself in the bizarre terror of the *oprichnina* and the murder of his heir in a fit of rage. By the end of the war in the early 1580s, Ivan had been deprived of his initial gains while Poland and Sweden divided Livonia between themselves.

Ivan the Terrible's premature death in 1584 put his younger son Feodor (r. 1584–98) on the throne. Sickly and ineffectual, the new tsar begat no male heirs, which brought the Daniilovichi line to an end when he died in 1598. Combined with severe economic and social tensions, the ensuing dynastic crisis soon led to a collapse of the Muscovite state during the fifteen-year 'Time of Troubles'. Russia's travails naturally attracted the attention of its Catholic nemesis, which invaded in 1606. Within four years, Polish troops had occupied Moscow's Kremlin. Meanwhile, Sweden, which had initially allied itself with what was left of Muscovy's government, likewise took advantage of the troubles to seize Novgorod in 1611.

Bereft of an effective ruler, a group of Russians formed a militia in Nizhnyi Novgorod led by a prosperous butcher, Kuzma Minin, and Prince Dmitrii Pozharskii that drove the Poles from the Kremlin in summer 1612. The next year an 'assembly of the land' elected a teenaged boyar's son, Michael Romanov (r. 1613–45), to be the tsar, thereby founding the eponymous dynasty that would rule Russia until 1917. Peace with Sweden and Poland soon followed, by the Treaty of Stolbovo in 1617 and the Truce of Deulino in 1619, respectively.

The new tsar's first military objective was to reconquer Smolensk, which had been ceded to Poland at Deulino. It was clear that Michael could no longer rely on mustering gentry-based cavalry to do the job. Indeed, the Livonian War and the Time of Troubles laid bare the fundamental failure of the *pomest'e* system. For one thing, the cavalry was neither consistently equipped nor trained. At the same time, unable to win new lands to reward their officers, tsars found it increasingly difficult to motivate them.[30]

More importantly, the very nature of European warfare had been undergoing a fundamental change. The new campaign called for infantry arrayed in lines to deliver volleys of lead with muskets according to the drill devised in the West by generals like Prince Maurice of Orange. Hiring mercenaries was too expensive and risky. Therefore, Michael began to recruit foreign officers to lead 'new formation' infantry and cavalry regiments whose ranks were filled by his own subjects. Reform was slow and hesitant. Nevertheless, about half of the 35,000 troops he fielded in his unsuccessful attempt to take Smolensk in a short war from 1632 to 1634 consisted of new formation units.[31]

The Russian army underwent a more thorough transformation under Tsar Alexis (r. 1645–76) during the Thirteen Years' War with Poland. The confrontation's *casus belli* was a revolt among Ukraine's Cossacks, who were then under Polish rule. For some time, the Polish crown had been attempting to tame the Zaporozhians, but such efforts predictably led to rebellion. The most dramatic rising erupted in 1648 under the leadership of Hetman Bohdan Khmelnytsky. The hetman initially turned to the Crimean Tatars for support, but when rebuffed he appealed to the Russian tsar.

30 I am grateful to Carol Stevens for this point.

31 Michael Fredholm von Essen, *Muscovy's Soldiers: The Emergence of the Russian Army* (Warwick, 2018), p. 94.

Although Alexis sympathised with his co-religionists, he was also wary of becoming too closely tied to the unruly Cossacks. Nevertheless, the opportunity to expand his western border at Poland's expense was too enticing to pass up and in 1654 he formally accepted Khmelnytsky's offer of 'eternal fealty' at Pereiaslavl. Like many such oaths on the steppe, the signatories understood allegiance in very different ways. But there was no ambiguity for the Poles, who immediately declared war on Muscovy.

The enemy Alexis confronted was still a formidable military power, but it suffered from fatal political weaknesses. Less than a century earlier, the Union of Lublin in 1569 had merged the dual monarchy of Poland and Lithuania into a single commonwealth, the Rzeczpospolita. According to the union, the commonwealth's elected king ceded much power to the nobility. Further hampering decisive rule was the emergence of the principle of the *liberum veto* from the middle of the seventeenth century, whereby a single vote could block decisions made by the kingdom's legislature, the Sejm. As the *bon mot* had it, the result tended increasingly towards 'anarchy tempered by civil war'. Although there were still a few notable cases when the Polish armies showed themselves redoubtable, these circumstances increasingly hampered the Polish kings' ability to wage lengthy and effective campaigns.

When Poland went to war with Muscovy in 1654, the latter's army of 40,000 men still largely comprised gentry cavalry and *strel'tsy*.[32] As had been the case with his father's march on Smolensk some twenty years earlier, Alexis hired large numbers of foreign officers, who again instructed soldiers in the new military ways of the West. There was some resistance among cavalrymen, because of fears that the reforms would lower their status. However, high rates of casualties in the field hastened the conversion. By 1663, four out of every five soldiers served in new formation regiments. As for the war itself, exhausted by thirteen years of fighting and threatened by Ottoman intervention, in 1667 the combatants signed a truce at Andrusovo that ceded Smolensk and much of Belarus together with Ukraine's left bank to Russia.

When Alexis's son, Peter the Great, took effective control of the realm in 1689, his military was already well on the way towards westernisation. (See Figure 25.) Nevertheless, the dramatic changes that he forced on his subjects were no less important. The greatest deficiency was the lack of a standing army. Whenever the tsar went to war, he had to raise troops anew, only to disband them after the fighting was over. He could call on various military castes, such as the gentry cavalry and the *strel'tsy*, but they also had civilian

32 Von Essen, *Muscovy's Soldiers*, p. 84.

FIGURE 25. Preobrazhenskii Life-Guard Regiment, one of the oldest military corps in Russia, founded by Peter the Great in 1683. From *Historical Description of the Clothing and Armament of the Russian Troops, Compiled by the Highest Command, with Drawings* [in Russian], ed. A. V. Viskovatov (1841–62).

occupations to support themselves in peacetime, which took much of their attention. According to William Fuller, 'suspended in an emulsion of state privileges [these groups] were actually preoccupied with everything *but* the art of war'.[33]

33 William Fuller, *Strategy and Power in Russia 1600–1914* (New York, 1992), p. 33.

The impetus for change was the Great Northern War with Sweden from 1700 to 1721. During the first years of his reign, Peter had been mostly interested in taking on the Ottomans. Indeed, the goal of the celebrated Grand Embassy he undertook to the west in 1697 was to assemble another anti-Turkish alliance. His journey proved unsuccessful. However, on his way back he met with Augustus II, the elector of Saxony and Poland's new king, who suggested he join him as well as the Danish king in a joint attack on Sweden. The Treaty of Stolbovo some eighty years earlier had deprived Muscovy of Ingria, its last possessions on the Baltic, and Peter was eager to regain a foothold on its waters. The time seemed right. Sweden's new king, Charles XII, had just inherited the throne as a fifteen-year-old.

Much to everyone's surprise, the teenaged monarch proved to be a brilliant general. Within a few months his navy forced Denmark out of the war, and he deployed his army to meet Peter's offensive. The clash at Narva, which pitted 9,000 Swedes against 40,000 Russians during an early blizzard in November 1700, thoroughly routed Peter's forces. But, against the advice of his senior commanders, Charles did not pursue the defeated tsar and instead took on Augustus in Poland. It proved to be a fateful choice, for the tsar took full advantage of the respite to build an entirely new army.

To secure a steady supply of officers, Peter cautiously began to make state service mandatory for all noblemen. Upon pain of having his estate confiscated, every adult male born into the gentry was now required to become either an officer or a government administrator. While the principle of such service to a Russian tsar was nothing new, the requirement became more methodical and far-reaching. At the same time, in 1705 Peter also decreed a system of conscription that obligated every twenty peasant households to provide one recruit per year. According to Carol Stevens, Peter's new system put some 250,000–300,000 men in uniform during his reign.[34] As with the gentry, service was for life. The tsar thereby created a permanent and cohesive armed force comprised of his own subjects, a major innovation in an age when many other European armies still heavily relied on mercenaries.

There were a number of other important changes, including systematic training at all levels and new weapons like flintlock muskets and bayonets, as well as considerable advances in artillery, a combat arm much favoured by the Russians. And there was the navy, which was entirely Peter's creation. The tsar also established new mines, foundries, and factories to produce the necessary arms and equipment. Meanwhile, a new poll tax helped raise the

34 Stevens, *Russia's Wars*, p. 231.

revenues to pay for it all. It would take the rest of the century for these reforms to take full effect, but, by the time of Peter's death in 1725, the solid foundations had been laid.

The new army's first test came in 1707, when Charles again turned his attention to Russia. Like Napoleon and Hitler in the following two centuries, the Swedish king foresaw an easy march to Moscow. His intelligence convinced him that Peter's subjects were seething with anger at the many changes to which they had been subjected. He also gained an ally when Ukraine's hetman, Ivan Mazeppa, secretly changed sides the following year, although in the event only some 3,000 Cossacks followed suit.

Charles's optimism proved unfounded. Because of logistical difficulties, in the autumn of 1708 the king veered south to liaise with Mazeppa rather than advancing straight on Moscow. There was bad news when Peter devastated Charles's supply train at Lesnaia in late September. To make matters worse, the harsh Ukrainian winter – the coldest in a century – reduced the effective strength of Sweden's main force by half. Charles's depleted forces finally gave battle with Peter at Poltava on 8 July 1709. Like Narva nine years earlier, with some 50,000 troops the Russians outnumbered the Swedish king's regular troops by about four to one, but the outcome was very different.[35] Thoroughly routed, Charles and Mazeppa fled into Ottoman exile with a little over 1,000 survivors.

Peter soon took Ingria, and the ports of Riga and Reval (Talinn). By 1713 he was sufficiently confident of his hold on the Baltic to declare St Petersburg, the city he had founded at the Neva river's mouth ten years earlier, to be his new capital, thereby signalling a definitive turn to the west. The next year the tsar occupied Finland and he eventually took the war to Sweden itself, where his navy began to ravage the coast. In 1719 Russians even menaced Stockholm. With the threat of more naval raids, Sweden's new king finally sued for peace at Nystadt in August 1721. According to the treaty, he ceded all of his Baltic provinces as well as much of Karelia to the victor. Peter did return Finland, although that would also be incorporated into the Russian Empire during another war in 1807–8. The Peace of Nystadt permanently reduced Sweden to the second tier and made Russia the dominant power in Northern Europe.

A grateful senate humbly requested Peter to the take the title of emperor, thereby formally promoting his realm to the status of an empire. Meanwhile,

35 N. P. Volkovskii (ed.), *Istoriia russkoi armii ot zarozhdeniia Rusi do voiny 1812 g.* (St Petersburg, 2003), p. 181,

the 'long' eighteenth century became the army's golden age, as Romanov emperors and empresses led their armies to victory after victory in Europe. Between the reigns of Peter the Great and Alexander I (r. 1801–25), Russian territory expanded from some 40 million to nearly 50 million square kilometres, much of it in the west.[36] And during the Seven Years' War (1756–63), tsarist troops even advanced on Berlin, after inflicting a series of defeats on Prussia's King Frederick the Great. By the time Catherine II seized power in a palace coup in 1762, Russia was firmly ensconced in the ranks of Europe's great powers.

It would be Catherine's work to destroy her empire's traditional Western nemesis, Poland. By the eighteenth century, the Commonwealth had virtually become a tsarist satellite. During the Seven Years' War, Russian troops crossed Polish territory at will and by 1763 Catherine had arranged to put a former favourite, Stanisław Poniatowski, on the throne. Such interference led to a major rebellion by the Confederation of Bar in 1768, which triggered a Russian invasion. Despite Bourbon and Ottoman support to the insurgents, by the end of the year Warsaw was firmly in the empress's hands. Rather than annexing the kingdom outright, Catherine shared the spoils with Frederick the Great and Maria-Theresa of Austria in the First Partition of Poland. The arrangement, which was negotiated in 1772, deprived the kingdom of 30 per cent of its territory and a third of its population.

In 1793 and 1795, towards the end of her reign, Catherine arranged two more partitions. As a result, Poland entirely disappeared as an independent state, with Russia in possession of two-thirds of its territory. The partitions proved to be a poisoned chalice. Never reconciled with Romanov rule, in the nineteenth century Poles would enthusiastically fight with Napoleon and subsequently rise in two major revolts that blackened Russia's reputation abroad. While the tsar's army put both of them down, the restive nation, which formed a large salient surrounded on three sides by the increasingly hostile Hohenzollern and Habsburg empires, would be a major strategic liability by the turn of the twentieth century.

Russia's military golden age concluded with Alexander I's dramatic defeat of Napoleon's invasion of 1812. Not only did his generals virtually obliterate the Corsican's 600,000-man *Grande Armée*, but they went on to drive its remnants all the way back to Paris over the next two years, thereby entirely quashing his dreams of continental dominion. Historians have traditionally

36 William C. Fuller, 'The Imperial Army', in Dominic Lieven (ed.), *The Cambridge History of Russia*, vol. 2: *Imperial Russia, 1689–1917* (Cambridge, 2006), p. 531.

credited 'Father Winter's' frigid fury for Napoleon's defeat. However, this does not give proper due to the brilliant performance of Alexander's armies in conducting a lengthy campaign with formidable logistical challenges.[37]

The immediate result of the Napoleonic Wars was to make Russia into Europe's dominant land power, a status it would retain until mid-century. Yet while the immense size and reputation of the Romanovs' military inspired respect, if not fear, among the other great powers, the nineteenth century would not prove to be nearly so glorious as the previous one, as Nicholas's debacle in the Crimea would prove in the mid-1850s. For one thing, success breeds complacency. In his discussion of the decades leading up to the Crimean War, one historian described this phenomenon as 'the baleful consequences of History'.[38] More importantly, if the autocratic will of such reformers as Peter the Great had created an army well up to the standards of its most advanced Western adversaries, in the new century his heirs proved unable to keep their lumbering empire up to speed with the increasingly complex requirements of the industrial revolution.

There is no mystery in Moscow's metamorphosis from a peripheral principality into the planet's largest continental empire over the course of four centuries. The explanation partly comes down to luck: Russia's rulers were often successfully able to exploit the decline of rivals that bordered them, whether they were Tatar khans to the east, Turkish sultans in the south, or Polish and Swedish kings on their western marches. They also proved remarkably adept at adapting to the different styles of warfare in their various theatres. The cavalry that faced Asian nomadic warriors learned a great deal from its foes on the steppe, while in Europe the military generally kept pace with the West's military revolution. Another important advantage was the development of a strict autocratic political order. This enabled tsars and tsarinas to create a highly militarised state that effectively marshalled their realm's often modest resources for a powerful and disciplined military that generally did their bidding. Yet, if strict centralisation made Russia strong during the 400 years that separated the reign of Ivan the Great from that of Nicholas I, by the nineteenth century autocratic rule imposed a rigidity that would make it increasingly difficult for the empire to adapt to the industrial and technological revolutions of the modern age.

37 For a convincing discussion of this argument, see Dominic Lieven, *Russia against Napoleon: The Battle for Europe 1807–1814* (London, 2009).
38 Fuller, *Strategy and Power*, p. 177.

21

# The maintenance of Hispanic dominance in Latin America

GABRIEL PAQUETTE

After the conquest of the Caribbean, Mexico, and Peru, Spanish dominance in the Americas was maintained through a combination of "soft" and "hard" power: a mixture of armed coercion and an elaborate legal-administrative apparatus which ensured that tension rarely escalated into full-blown conflict. The sturdiness of Spain's empire may also be attributed to other significant factors, including epidemiology (differential immunity), topography, and the avoidance of certain types of military engagement, all of which tended to intersect with or reinforce the deployment of "soft" and "hard" power. There were at least three broad threats to Spain's dominance: external enemies, particularly rival European states covetous of the economic advantages Spain obtained from its New World dominions; unsubdued Amerindians on the fringes of Spanish settlement, who clung to their autonomy and effectively controlled vast swathes of territory through to the end of the colonial period; and an internal, heterogeneous group from all rungs of the socioeconomic ladder, from wealthy, privileged merchants to mistreated African slaves. At some point or other from 1521 until 1808, an internal challenge to Spanish dominance emerged from every sector of colonial society. Whether by design or felicitous coincidence, external and internal threats to Spanish dominance were rarely coterminous, which may help to explain the empire's resilience and longevity.

Though the sixteenth-century conquest had been achieved initially by force of arms, it was not the work of a professional or standing army. The conquistadors had no steady connection to a permanent, public military organization. Initially they retained personal command of their private armies, which were maintained at their own expense. This arrangement was at first welcomed by the crown as it encouraged military service for the greater glory of the king without requiring an outlay of funds. These conquistadors acquired land grants and titles, becoming *encomenderos*, in possession of Indian tribute and labor. The *encomienda* itself was premised on indigenous systems of authority, specifically the figure of the *cacique* (native leader). The Spanish *encomendero* would inform the local *cacique* of his labor requirements and the *cacique* would activate "pre-Conquest" labor and tribute mechanisms to meet the request. In exchange for the extensive privileges they enjoyed, *encomenderos* were expected to put soldiers into the field as well as to bear arms during emergencies. Until 1582, *encomenderos* were also expected to supply certain forts and garrisons with men whom they paid. The *encomenderos* and their retinues did not constitute, then, a true army. Though the cost-cutting virtue of an inexpensive defense was extolled, high-ranking officials were acutely aware of the potential danger it posed. In 1615, the Peruvian viceroy Montesclaros informed the king that "in any other area, these *soldados* would be regarded as dangerous vagabonds who posed a threat to public safety."[1]

Perhaps due to the paucity of external threats in the mid-sixteenth century, little attention was lavished on Spanish American defense, narrowly defined, in the immediate aftermath of the conquest. Nevertheless, copious coercive legislation, and the corresponding enforcement mechanisms, sought to control the movement, labor, and comportment of Amerindians. Besides the *encomienda*, the *reducción* resettled Indians into towns, ostensibly for the purposes of evangelization, which served to control them. The *mita*, the draft labor system, compelled them to work in silver and mercury mines. Such institutions diminished the likelihood that unrest or discontent would amount to much. In European and creole communities, various formal and informal mechanisms mediated challenges to authority. As John Elliott has argued, "the political culture of the Hispanic World . . . was postulated on the basis of a reciprocal process of negotiation between the monarch and his subjects. Lobbying and petitioning, compromise and counter-compromise, formed the everyday stuff of political life in the Indies."[2]

But military matters were not entirely neglected. Coastal defense was prioritized, with a few well-fortified sites, manned by regular army units composed of troops from the Iberian peninsula, rather than the creole population, ensuring that the entry points to the wealthier, populated interior were guarded. In New Spain (modern Mexico), for example, intrusion by a foreign expeditionary force in all but a handful of places would have been

1 Quoted in Leon Campbell, "The Changing Racial and Administrative Structure of the Peruvian Military under the Later Bourbons," *The Americas* 32 (1975), 117–33, 117.

2 John Elliott, *Empires of the Atlantic World: Britain and Spain in the Americas, 1492–1830* (New Haven and London, 2006), p. 133.

a pointless, foolhardy venture, as mountain ranges, deserts, vast distances, and an absence of roads stood between potential invaders and their desired destinations.[3] But the initial absence of rival European states and the convenience of the conquistador-*encomendero* defense complex, at no direct expense to the crown, necessitated little change. As a result, the number of regular soldiers sent from Spain to the Americas until the early eighteenth century remained minuscule.

Even after 1700, there were only 6,000 regulars in all of Spanish America and the figure never exceeded 14,000 before 1750. The few full-time soldiers were either assigned to garrisons in strategic port cities (Vera Cruz, Havana, and Cartagena) or else sent to *presidios*, frontier military outposts, in northern New Spain, in Southern Chile, and in the unpopulated zones of the Río de la Plata (modern Argentina). In 1723, there were nineteen garrisons on the northern Mexico frontier, with a total of 1,000 men. The Spanish presence on the frontier with Chile's Araucanian Indians was larger, where a string of forts manned by a standing force of up to 2,000 was subsidized by Lima's treasury. Nor was the crown's overall contribution to defense very substantial: between 1535 and 1585, it spent a mere 5 million ducats on American defense whereas 2.5 million ducats were squandered each year in Spain's incessant, ultimately futile war against the Dutch Republic during the same period. Endless European wars meant that Spain had no spare funds to allocate to Spanish American defense. Nor could the crown borrow for this purpose: it was paying 49 percent interest on loans in 1550 and, upon his accession in 1556, Philip II discovered that all revenues had been pledged to repay loans or interest until 1561![4] Spanish America would have to finance its own defense.

If the Spanish Main was not particularly vulnerable to foreign attack in the sixteenth century, the treasure fleet quickly became the object of every corsair's desire and the envy of every European monarch. In fact, the defense of the Indies largely amounted to the protection of the treasure remitted to Spain from Mexico via the Caribbean and Tierra Firme ports. By 1564, an elaborate convoy system had been established to escort merchant ships and other vessels laden with bullion to and from Spanish America. Two fleets per year embarked from Spain for Spanish America, the so-called "Indies Run" (*Carrera de Indias*), laden with goods for sale. In April, the ships would depart from Andalusian ports protected by two heavily armed galleons, the

3 Christon Archer, *The Army in Bourbon Mexico 1760–1810* (Albuquerque, 1977), pp. 2–3.
4 Geoffrey Parker, *The Military Revolution: Military Innovation and the Rise of the West, 1500–1800*, 2nd ed. (Cambridge, 1998), p. 63.

so-called Armada of New Spain. In August, a second fleet of merchant ships embarked from Seville en route to the isthmus of Panama, escorted by a varying number of galleons, collectively called the Armada of Barlovento. Consistent with its policy of forcing the beneficiaries of defense to pay for it, the crown imposed a tax on trade, the *avería*, to underwrite the convoy. It was increased from 1.7 percent in 1585 to 8 percent in 1591, and fluctuated according to the vicissitudes of war.[5] In spite of its obvious inefficiencies, this "fleet system," *Flota de Indias*, long remained the dominant mode of crossing and recrossing the Atlantic Ocean. Until 1650, 85 percent of Spain's transatlantic sailing was undertaken as part of a convoy. In altered form, some version of the *Flota* would survive until the end of the colonial period, mainly because bullion remained Spanish America's chief export.[6]

Though the Atlantic was initially a Spanish lake, the continental littorals and Caribbean islands were soon swarming with predators. At first these consisted mainly of non-state agents, especially corsairs. There were at least 100 piratical coastal raids before 1585, when Sir Francis Drake's massive raid – boasting twenty ships and over 2,500 men – exposed the inadequacy of the Spanish defenses. In the grim aftermath of Drake's raid, Philip II's government began to fortify the Caribbean ports to prevent further debacles. Funding came indirectly, through the establishment of a *situado* system, which diverted silver from the treasury of New Spain to subsidize the fortifications and garrisons in the capital-poor port cities of the circum-Caribbean instead of being remitted directly to the metropole. The Pacific was no less vulnerable: fear regarding the threat posed to Potosí's silver prompted the creation of the *Armada del Mar del Sur*, responsible for policing the Pacific coast as far north as Panama.

Pirates remained a costly nuisance, routinely working in concert with and even at the behest of sovereign states, thus blurring the line between irregular inter-state warfare and piracy, but the greatest external threat to Spanish dominance and dominion in the Americas was posed by the extension of the wars of the Old World to the New World theater. The Dutch, and later the British, contested Iberian hegemony in the Americas. As Protestant powers, they defiantly refused to acknowledge the validity of the papal bulls granting

5 John Lynch, *Spain under the Habsburgs*, 2 vols. (Oxford, 1965), vol. 1, p. 157.

6 In the late Bourbon period, bullion accounted for 77% of the cargo value of Spanish American imports into Spain. Even in places with rapidly diversifying economies, such as Buenos Aires, bullion sent down from Upper Peru to the Río de la Plata accounted for 80% of the value of its exports. David Brading, "Bourbon Spain and Its American Empire," *Historical Journal* 39 (1981), 961–69.

Spain and Portugal exclusive dominion and scoffed at the Treaty of Tordesillas by which the Iberian powers divided the extra-European world between themselves. In revolt against Spain in Europe from 1568, the victorious Dutch soon contested their former overlord's oceanic supremacy as well. This threat necessitated Spanish expeditions to the Philippines as well as to the Atlantic, which were costly and paid for largely from New Spain's treasury. Spain's conflict with England in Europe also resonated in the New World. Spain lost almost fifty ships from its vaunted Armada in its disastrous 1588 planned invasion of England: a debacle which not only made possible the English sack of Cádiz in 1596, but also permitted buccaneers to operate with impunity in the Caribbean and paved the way for rampant contraband trade. By the first decades of the seventeenth century, the Dutch, French, and English had established territorial footholds in the Guianas, the northern coast of South America, from which they would never be dislodged.

But it was during, and immediately after, the Thirty Years' War (1618–48) that Spain's American empire was most directly challenged by an external power. The incorporation of Portugal and its colony of Brazil into the Spanish crown from 1580 brought Spanish territorial claims to their greatest point. But never had Spain faced so many rivals on multiple fronts, experienced comparable levels of internal dissent, or been plagued by worse economic mismanagement than in the middle decades of the seventeenth century. The formation of a nimble, aggressive Dutch West India Company (WIC) in the first decades of the seventeenth century presaged further trouble. In 1624, the Dutch captured Salvador da Bahia, Brazil's most important city, forcing Spain to outfit an armada of fifty-six ships carrying over 12,000 men to expel the Dutch, retake the city, and save face. But the Dutch threat did not subside. In 1628, the treasure fleet returning to Spain was captured in Matanzas Bay by the WIC, resulting in the seizure of 6 million pesos (roughly 170,000 kg of silver), which the Dutch reallocated to fund their twenty-five-year occupation of Pernambuco (1630–54), in northeastern Brazil, as well as the conquest of Curaçăo (1634). The impact on Spain's European war effort, reliant on regular infusions of American silver, was dire.

The English too began to pick off and occupy strategic island outposts and enclaves in the Caribbean: St. Kitts (1624), Barbados (1627), the Mosquito Coast of Nicaragua (sporadically during the 1630s), and modern Belize (1642). By the 1640s, as fully fledged revolts against Spanish rule raged in Portugal, Catalonia, and Naples, the New Spain treasury was spending a third of its total revenue on defense, not counting its *situado* subsidies to the bullion-poor Caribbean colonies. Mounting naval losses both in the European and

Caribbean theater permitted renewed challenges to Spain's sovereignty in the Americas. Though an audacious effort to attack Hispaniola failed, Oliver Cromwell's "Western Design" resulted in the capture and settlement of Jamaica (1655). France insinuated itself into the fertile western fringe of the island of Hispaniola, calling it Saint-Domingue. Buccaneers, too, continued to harass and plunder. In 1668, Captain Henry Morgan led a buccaneer attack on Portobello, whose capture precipitated a two-week orgy of looting. Spain's dominance in the Americas, then, was clearly threatened, not least because its control over vital sea lanes was severely compromised.

Eventually, with its treasury hemorrhaging, Spain sought refuge at the diplomatic bargaining table. Multiple bankruptcies had left the treasury exhausted and its currency utterly discredited by the time of Carlos II's accession in 1665. While not abandoning its long-asserted claims to monopoly over American trade, the crown was compelled to concede territory that had been seized by its European rivals during war. By the Treaty of Münster (1648), Spain accepted Dutch control of all territory that the Republic actually occupied in America in exchange for a pledge to desist from trading with Spanish America. By the Treaty of Madrid (1670), Spain acknowledged Britain's territorial claims in exchange for a similar unfulfilled promise. Spain's dismal state, however, made it incapable of enforcing even these unfavorable terms. In 1697, Spain was forced to recognize French sovereignty over Saint-Domingue. There were, of course, gains as well as concessions: the European powers did strive to exterminate piracy from the Caribbean, thus extending the Westphalian model of inter-state relations to the colonies. Yet concessions and an uneasy peace did not result in demobilization. There was no great reduction in Spain's military expenditure because renewed war remained an ever-present threat. Military expenditure to protect and preserve the Americas thus exceeded the remission of bullion to the peninsula throughout the final decades of the seventeenth century.

Spain suffered further reverses during the War of the Spanish Succession, including a naval defeat off Vigo (1702) and England's conspicuously provocative occupation of Gibraltar (1704). The Treaty of Utrecht (1713), which brought the war to a close and sanctioned the change from the Habsburg to the Bourbon dynasty, resulted in further concessions to Spain's rivals. By the treaty's terms, the English South Sea Company received the lucrative *asiento* contract, the exclusive right to import slaves into Spanish America. The *asiento* also served as cover for contraband trade. As the likelihood of outright conquest of Spanish America dimmed and the wealth and population of its cities grew apace, contraband trade became widespread. Burgeoning demand, combined with the breakdown of the *Flota* convoy system, ensured that a monopoly of colonial trade in the hands of Spanish merchants was little more than a legal fiction for much of the eighteenth century. Nevertheless, despite innumerable concessions and high operational costs, Spanish America remained largely intact. (See Map 18.) Small enclaves and some entrepôts had been surrendered irrevocably, but Spanish control over both its less developed peripheries as well as its core viceroyalties was undiminished.

The advent of the Bourbon dynasty was accompanied by a brief respite from war. Spain took advantage, embarking on a naval reconstruction program. In 1722, there were a meagre twenty-two ships of the line whereas by 1760 there were fifty-two, the upkeep of which accounted for 18 percent of state expenditure.[7] In spite of increased funding, Spain's navy lagged behind Britain's, which necessitated a continued reliance on fortified ports. Although Spain would implement an elaborate coast guard (*guarda costas*) system in the eighteenth century in a futile, belated effort to keep out interloping contraband traders, it studiously avoided naval battles. Spain fought a single naval battle with Britain between 1713 and 1762. The first half of the eighteenth century also witnessed the gradual reorganization of the garrisons of the fortified ports into fixed (*fijo*) or permanent battalions: the first was established at Havana (1719), followed by Cartagena (1736), Santo Domingo (1738), Vera Cruz (1740), Panama (1741), and San Juan (1741).[8] But maintaining effective fortifications, troops, and coast guards was expensive. By the time of the War of Jenkins' Ear (1739–42), 57 percent of Lima's treasury revenue was earmarked for defense. Havana offers another example of the high costs necessitated by military preparedness. Besides a garrison of 1,500 peninsular soldiers, whose upkeep in the Americas was three times as costly as it would have been in Europe, a supplemental militia raised the total force to just over 6,300 troops.[9] It is remarkable that so few men in a mere handful of fortresses managed to stave off superior forces for so long. Part of the explanation lies in differential immunity to Yellow Fever, which was endemic to the Caribbean from the late seventeenth century. As John McNeill has observed, "Spain's hold on the region was inexpensively buttressed by mosquitos and microbes." The toll of disease on invaders was ghastly. When Admiral Vernon led 29,000 soldiers and sailors in amphibious invasions of Santiago de Cuba and Cartagena (1741), 22,000 of his men were dead within a year, only 1,000 of whom

7 John McNeill, *Atlantic Empires of Spain and France: Louisbourg and Havana, 1700–1763* (Chapel Hill and London, 1985), p. 68.
8 Allan Kuethe, *Cuba, 1753–1815: Crown, Military and Society* (Knoxville, 1986), pp. 4–5.
9 McNeill, *Atlantic Empires*, pp. 101–2; figures for upkeep from 1750; militia figures from 1760.

MAP 18. Spanish Central and South America

had been lost on the battlefield. Yellow Fever proved a crucial, and reliable, ally in maintaining Spanish dominance in the Americas.[10]

Spain's bungled entry into the Seven Years' War severely jeopardized its dominance in the Americas. The British capture of Havana and Manila in 1762

10 John McNeill, *Mosquito Empires: Ecology and War in the Greater Caribbean, 1620–1914* (Cambridge, 2010), pp. 136, 166.

underscored not only long-established British naval supremacy, but also the inadequacy of existing fortifications and the shortcomings of manpower. After the establishment of peace in 1763, accompanied by the return of Havana and Manila, the crown undertook to create a genuine colonial army for the first time. Military reform was an important component in a broader attempt to redesign policy in its colonies, an effort that included the reconfiguration of its territorial and administrative units. The viceroyalty of Río de la Plata was created in 1776 and a half dozen smaller provinces on the northern coast of South America were consolidated into the Intendencia of Venezuela in the same year, with Caracas as its capital. Chile was governed as a Capitanía General from 1778. The 1770s were also a decade of remarkable expansion northward: garrisons were established in California, at San Diego and Monterey, in the early 1770s, and one christened San Francisco was established in 1776. Nor were the core viceroyalties – the vital mitochondria of empire – of New Spain (Mexico) and Peru neglected. There was a flurry of new taxes and imposts, and measures to bring Spain and America closer together, such as the creation of a postal service. As David Brading has written, by 1776, "a small army of officials, clerks and guards, stationed in all of the chief towns of New Spain, administered taxes, new and old, with unparalleled vigour and efficiency."[11]

Though the scope and timing were specific to each viceroyalty, the number of regular forces in the two most important viceroyalties of New Spain and Peru tripled in the last four decades of the eighteenth century. Given its geographical expanse, the number of regulars remained paltry, a mere 6,100 in New Spain and 3,000 in Peru. Nevertheless, the creation of standing armies had important repercussions. The arrival of massive numbers of officers from the peninsula "aggravated the discrimination that creoles already felt in the civil and ecclesiastical hierarchies."[12] The majority of the army had always been composed of peninsular Spaniards. Between 1740 and 1760, the proportion of Spaniards in the regular army never fell below 62 percent. The remainder was made up of white creoles, since legislation dating from the seventeenth century forbade Amerindians, *mestizos*, *mulatos*, and

11 David Brading, *Miners and Merchants in Bourbon Mexico 1763–1810* (Cambridge, 1971), p. 29. On the Bourbon reforms more generally, see Gabriel Paquette, *Enlightenment, Governance and Reform in Spain and Its Empire, 1759–1808* (New York and Basingstoke, 2008).

12 Mark Burkholder and Lyman Johnson, *Colonial Latin America*, 6th ed. (Oxford, 2004), p. 323.

*negros* from serving in the regular army.[13] Only at the very end of the colonial period would the Spanish crown resort to slave soldiers, first in the effort to beat back the British invasion of Buenos Aires in 1806 and later during the first phase of the Wars of Independence (1810–13). While Amerindians were not eligible for military service of any kind, and were excluded from the military census, auxiliary companies of Amerindians were nonetheless utilized and tolerated, particularly on the frontier. By the mid-eighteenth century, half of the *presidio* forces were manned by *castas* or Hispanized Amerindians. One of the incentives to induce service in the regular military, as well as the militia, was the extension of the *fuero*, the judicial right of officers, soldiers, and their dependents to be tried before a military instead of a civil court for certain crimes. One historian has claimed that the *fuero*, "with subsequent amplifications, tended to create a military class exempt from civil responsibility and liability."[14] But even these advantages were an insufficient lure to Spanish Americans. Since military service was marked by "low pay, harsh discipline, and poor prospects for advancement," the paucity of voluntary enlistments among Spanish Americans forced "recruiting teams to scour taverns, gambling dens, and jails" in search of prospective soldiers.[15] The recruitment of such undesirable types deepened negative popular perceptions of the regular military.

Regular forces were supplemented by the colonial militia, which was reorganized and expanded after 1763. As a formal institution the militia dated back to 1540, requiring all able-bodied men to serve if called upon, brandishing weapons they were responsible for acquiring. The reliance on the militia made a virtue out of necessity, for Spain did not possess unlimited resources to lavish on America. As the minister of the Indies, José de Gálvez, noted in 1779, to supply all the troops, military supplies, and fortifications that Spain's colonies required "would be an impossible enterprise, even if the King of Spain might have at his disposal all the treasure, the armies and the storehouses of Europe."[16] The perils of tropical disease for newcomers to the region also encouraged the pragmatic embrace of creoles, whose immunity to Yellow Fever gave them a distinct advantage. The

13 Ben Vinson III, *Bearing Arms for His Majesty: The Free-Colored Militia in Colonial Mexico* (Stanford, 2001), p. 11.

14 Lyle N. McAlister, "The Reorganization of the Army of New Spain, 1763–66," *Hispanic American Historical Review* 33:1 (1953), 1–32, 8.

15 Burkholder and Johnson, *Colonial Latin America*, p. 323.

16 Quoted in David Weber, *Bárbaros: Spaniards and Their Savages in the Age of Enlightenment* (New Haven and London, 2006), p. 162.

militia, however, was not a roving force that could be dispatched anywhere an enemy – external or internal – lurked. Instead, militias were tied to their locality and were a defensive force, unavailable for distant campaigns. Initially, militia service was limited to whites, but later it was extended to *mestizos*, free blacks, and those of mixed racial ancestry. By 1600, freedmen of color were serving in militias throughout the Caribbean and in 1778 the Spanish crown actively promoted free blacks in the militia.

The quality of the militia as a fighting force varied. The valor displayed by the free colored militia of Lima who were confronted with the Dutch threat in the early 1620s, for example, earned its membership exemption from tribute as a reward for their service. But other militias did not fare as well: sixteenth-century pirates were often better armed than the militias against which they fought, a factor which undoubtedly contributed to their success. Although older scholarship on the pre-1763 militias stresses that they were untrained, undisciplined, and infrequently summoned, new scholarship vigorously contests this view. As Ben Vinson III has argued, militia units were "greatly involved in patrol duties, *vigía* watches, and a panoply of ancillary functions requested by municipal and provincial authorities."[17] The growth of the militia was prodigious: by 1800, there were approximately 24,000 militiamen in New Spain and over 18,000 in Peru. Maintaining this somewhat revamped, post-1763 army and militia proved expensive. In late eighteenth-century New Spain, for example, over 60 percent of the treasury's expenditures went to defense, and the total outlay doubled between the early 1760s and 1780s. The economic expansion and efflorescence in late colonial Spanish America, however, meant that revenues continued to pour into Spain's treasury. In 1770, for example, revenue from the Indies accounted for 23 percent of the treasury's overall receipts.[18]

Threats to Spain's dominance in the Americas did not come solely from rival European powers. There were internal challenges as well, primarily in the form of indigenous responses to territorial aggrandizement and economic exploitation. Beyond the frontier were Amerindians whose continued autonomy was deemed a threat to Spanish rule. An obsession with both policing and expanding the frontier animated colonial administration. Even in the late eighteenth century, however, the Spanish presence remained nonexistent in many territories claimed by the crown. At the close of the colonial period, independent Amerindians controlled over half of the landmass of modern

17 Vinson, *Bearing Arms for His Majesty*, p. 223.
18 Elliott, *Empires of the Atlantic World*, p. 355.

Latin America, from Tierra del Fuego to Mexico, and it has been estimated that 2.7 million independent Amerindians lived within the boundaries of the Spanish Empire. As late as 1780, two-thirds of Chile's Amerindians did not recognize Spanish sovereignty. Moreover, Comanche country, the *Comanchería*, exceeded the size of Central America and served as a barrier to Spanish expansion. The lack of soldiers and funds, combined with the geographical size of the territories involved, forced Spain to abandon plans for further offensive wars of conquest and instead recognize containment and accommodation as strategies better suited to reality.[19]

Colonial Spanish America was a diversified economy, but a key component of its overall economy – from sugarcane plantations in Cuba to gold mines in New Granada (Colombia) – was dependence on African slave labor. Seeking respite from such a dehumanizing labor regime, some slaves sought to flee. Runaway (*cimarrón*, or maroon) slaves formed communities beyond the reach of Spanish law called *palenques*. As early as the mid-sixteenth century, there were between 2,000 and 7,000 maroon slaves in various *palenques* in Hispaniola. Drake claimed to have encountered 3,000 maroons in Panama in the 1570s, and by 1720 there were 20,000 maroons in Venezuela alone.[20] These *palenques* did not threaten Spanish dominance except insofar as they stood as symbols of defiance of Spanish law, serving as a beacon for those slaves brutalized by the plantation system. It was only later, after the beginning of the Haitian revolution in the 1790s, that maroon communities came to be perceived as a *bona fide* threat to the peace, stability, and prosperity of Spanish America.

Though external threats to Spanish dominance had been the impetus behind the increased size of the standing army and militia, military reform also proved its value in the pacification of the various internal threats facing the Spanish monarchy in the late eighteenth century. The increasingly strident fiscal demands made by the Spanish crown to finance its post-1763 reform program sparked numerous riots and rebellions among creole elites and plebeians in urban areas as well, of which the Quito Insurrection (1765) and the Comunero Rebellion (1781–82) were the most prominent.[21] Though

19 Weber, *Bárbaros*, pp. 12, 61, 72, 164–65.

20 Herbert Klein and Ben Vinson III, *African Slavery in Latin America and the Caribbean*, 2nd ed. (Oxford, 2007), pp. 179–83.

21 Kenneth Andrien, "Economic Crisis, Taxes and the Quito Insurrection of 1765," *Past & Present* 129 (1990), 104–31; John Phelan, *The People and the King: The Comunero Revolution in Colombia, 1781* (Madison, 1978); Anthony McFarlane, "Rebellion in Late Colonial Spanish America: A Comparative Perspective," *Bulletin of Latin American Research* 14 (1995), 313–38.

ubiquitous, the rising tide of discontent produced only paroxysms of protest, not translocal, empire-wide revolt. The same period was a veritable "age of insurrection" in Amerindian communities in the Andes, with at least eighty-two local riots and revolts occurring between 1750 and 1780, culminating in the major rebellion (1781–83) led by José Condorcanqui, better known by his grandiose and neo-Incan title Túpac Amaru II. While ultimately suppressed, the Túpac Amaru rebellion cost 100,000 lives.[22] Here collaboration was the key to the preservation of order. The Túpac Amaru rebellion could never have been suppressed without the native nobility's firm rejection of the demands of the rebel leadership. This nobility enjoyed special privileges and high status under Spanish law, both of which were unavailable to non-elite Amerindians. Some of these privileges or exemptions – such as a license to ride a horse or carry a sword or dagger – might appear trivial to contemporary eyes, but they were an integral part of situating oneself visibly in the legal and social order.[23]

Nevertheless, emphasis on the limits of Spanish military power in the Americas should not diminish its considerable achievements in the late eighteenth century. An enlarged army and revivified navy permitted Spain to join France in aiding the revolutionaries of British America as they threw off the colonial yoke (1779–83), a triumph which in turn made possible Spain's reconquest of West Florida and the Gulf Coast. Distilled to its essence, France pledged in the Treaty of Aranjuez (1779) not to make peace without Spain's consent and to continue the war until Gibraltar and Minorca (and other sundry territories) were recaptured and restored to Spain. As Richard Morris observed, "France had in effect modified her alliance with America, and changed and enlarged the purposes of the war without America's consent and even without her knowledge."[24] The Treaty of Aranjuez was the closest Spain came to an alliance with the United States itself. In effect, Spain committed itself to a treaty with France in support of its alliance with the US rebels against England, but not to an alliance with the rebels themselves.

Yet, even before the formal treaty was concluded, Spain had officially proclaimed its neutrality whilst simultaneously doing everything that it could

22 Peter Bakewell, *A History of Latin America: Empires and Sequels 1450–1930* (Oxford, 1997), p. 277.

23 Kevin Terraciano, "Indigenous Peoples in Colonial Spanish American Society," in Thomas Holloway (ed.), *A Companion to Latin American History* (Oxford, 2008), pp. 126, 135; David Garrett, "'His Majesty's Most Loyal Vassals': The Indian Nobility and Túpac Amaru," *Hispanic American Historical Review* 84:4 (2004), 575–617, 580, 617.

24 Richard Morris, *Peacemakers: The Great Powers and American Independence* (New York, 1965), pp. 15–16.

to help the rebels' cause; this support was mainly aimed at weakening the British rather than strengthening the Americans. The initial aid offered by Spain was 2 million French *livres* sent to the rebels in late 1775, which was followed by shipments of arms and matériel through Spanish Louisiana. Through this hospitality policy, Spain would grant asylum to those who fled to its territories from the war in the Thirteen Colonies, but also used it to carry out all sorts of covert actions in support of the Anglo-American rebels.

The official entry of Spain into the war not only tipped the balance of the conflict, giving France and Spain numerical superiority both at land and sea, but also profoundly changed the general strategy of the war. The combined Bourbon navy had 121 ships of the line compared with 90 for Britain in 1779, 117 compared with 95 in 1780, and 124 compared with 94 in 1781.[25] This clear superiority opened up new theaters in the global war, spreading British resources thin. It compelled Britain to fight on multiple fronts. Britain would be forced to abandon a purely American perspective on the conflict and to adopt a more global view of the war and to relinquish her freedom to choose when and where to strike and instead to assume a defensive posture, which prevented concentration of its forces against the North American rebels.[26]

All of this raised the overall cost of the war for Spain. Though there is no separate record of the amounts effectively spent by Spain on its war with Britain, by comparing the amount expended on defending Spanish territories before and after the war with outlays during the war the cost of the war can be reckoned. The estimate of the total cost for Spain of its war with Britain was approximately 431 million *reales de vellón*, a significant sum, equivalent to the total annual revenue of the Spanish Royal Treasury during this period.[27] Tallying victories and defeats, and aggregating expenditure, however, may shroud the significant benefits Spain derived from its involvement in the American Revolution. Viewed from Madrid, once the conflict had commenced, Spain clearly profited from its continuation. The stalemate in North America helped its Mediterranean policy, for example. In August 1780, a Franco-Spanish fleet came upon a convoy of English merchant

25 David French, *The British Way in Warfare, 1688–2000* (London, 1990), p. 76; see also Jonathan Dull, *The French Navy and American Independence: A Study of Arms and Diplomacy* (Princeton, 1975).

26 Hamish Scott, *British Foreign Policy in the Age of the American Revolution* (Oxford, 1990), p. 277.

27 For further details, see Gabriel Paquette and Gonzalo Quintero Saravia, "Introduction: Spain and the American Revolution," in Gabriel Paquette and Gonzalo Quintero Saravia (eds.), *Spain and the American Revolution: New Approaches and Perspectives* (London and New York, 2020), pp. 12–14.

ships bound for the Caribbean and Indian Ocean off Cape St. Vincent and managed to capture fifty-nine of the sixty-three vessels and brought them triumphantly to Cádiz. The loss of cargo was estimated at £1.5 million, plus 1,350 seamen and 1,255 troops.[28]

In mainland North America, too, the inconclusive, protracted nature of the war created a large buffer zone separating New Spain from the nascent United States. Madrid's ambitions were compatible with, and even assisted by, strong allied, non-European nations, whether Anglophone creoles or Chickasaws, Creeks, and Choctaw Amerindians. Spain entertained a vision of the Gulf Coast region filled with French- and English-speaking landowners who remained or became Spanish subjects.[29] Spain also gained the flexibility to enact new commercial policies. Spanish involvement in war actually boosted free trade within the empire, decreed in 1778. In 1779, for example, Carlos III gave Philippines merchants the freedom to bring ships laden with Asian goods to Latin America.[30] Spain was able to realize dreams it had long harbored. Britain was removed from the southeast of the modern United States. Both Floridas – West and East – were recovered. Britain retained its Caribbean colonies, of course, but without mainland ports. Spanish shipping was safer than it had been for well over 100 years and the treasure fleet more secure than it had been for nearly two centuries. A British invasion and occupation of Spanish America was now implausible. If anything, Spain now appeared poised to expand in North America, to make good on its legal claims in the American Southwest.

The final decades of the eighteenth century witnessed further attempts by the Spanish crown beyond North America to consolidate its control over territory that it claimed but in which it had established no settlements. Strenuous efforts were made to encourage emigration and found new towns in the rustic peripheries of Patagonia, the Mosquito Coast, and Florida. Perhaps most notable were the state-led and financed naval expeditions of the 1780s and 1790s, entrusted with the task of mapping and surveying the economic resources of these frontier zones.[31] Spain was especially concerned about Russian penetration of the northern coast of North America, the United States' westward expansion, and British interest in what is modern-day British Columbia. While the scientific expeditions produced a great deal of

28 Samuel Bemis, *The Diplomacy of the American Revolution* (Bloomington, 1957), pp. 85–88.
29 Kathleen DuVal, *Independence Lost: Lives on the Edge of the American Revolution* (New York, 2015), pp. 223, 227, 258.
30 Mariano Bonialian, *El Pacifico Hispanoamericano* (Mexico City, 2012), pp. 427, 436.
31 Weber, *Bárbaros*; Paula de Vos, "Research, Development, and Empire: State Support of Science in the Later Spanish Empire," *Colonial Latin American Review* 15 (2006), 55–79.

"useful knowledge," Spanish schemes for a fully fledged Pacific empire "collapsed as a result of insufficient funding, inertia or instability in their settled colonies . . . there were no fur traders, whalers, private investors or merchants to follow in the wake of the scientific expeditions."[32] Spain could not assert its claims on the distant periphery with its treasury stretched so thinly. In spite of sabre-rattling, Spain acquiesced to British claims in the Pacific northwest after the Nootka Sound dispute (1789–93). The limits of a self-financing empire could not be pushed too far. The British were also forced to abandon the Mosquito Coast in 1786 (though in exchange for a clear title in Belize). Giddy Spanish officials viewed these successes as a prelude to the repossession of the Malvinas/Falklands and Gibraltar.

But these fantasies were soon dashed by the outbreak of the French Revolution. Spain's involvement proved an unmitigated disaster, attenuating the links between Europe and America. In particular, Spain's ill-fated alliance with France incurred the wrath of the British navy. The defeat of the Spanish fleet off Cape St. Vincent in 1797 foreshadowed Britain's three-year blockade of Cádiz, which disrupted irreparably the supply chain between peninsular merchants and Spanish American markets. The British Navy wrought further havoc by seizing Trinidad in 1797. The centuries-old link between Spain and its Americas had been severed, then, well before the *coup de grâce* delivered at Trafalgar in 1805, where most of Spain's warships were lost.

One consequence of Spanish involvement in the French Revolutionary Wars was that it could no longer spare (or safely send) regular troops to the Americas. Spain had taken the decision in 1786 to cease employing Spanish battalions in America to reinforce the fixed (*fijo*) garrisons because the costs entailed by rotation were exorbitant. By the first decade of the nineteenth century, 60 percent of the officer corps was composed of creoles. Creoles jumped at the opportunity to purchase commissions, on sale (and subject to no standards and few procedures) from 1780 as a revenue-generating measure, and the cash-strapped Spanish treasury continued the practice thereafter, putting profit over both principle and morale.[33] The changes in the composition of the officers' ranks meant that "the reformed military proved an uncertain pillar of royal authority."[34]

By cutting off the peninsula from Spanish America after 1796, the British Navy further undermined Spain's increasingly tenuous hold over its American

32 Christon Archer, "Spain and the Defence of the Pacific Ocean Empire, 1750–1810," *Canadian Journal of Latin American and Caribbean Studies* 11:21 (1986), 15–41, 36.

33 Kuethe, *Cuba*, pp. 128, 149–51.

34 Allan Kuethe, *Military Reform and Society in New Granada, 1773–1808* (Gainesville, 1978), p. 186.

empire, forcing colonial economies to become either self-sufficient or reliant on contraband. The problem had been foreseen by many Spanish officials in the late eighteenth century, including the Peruvian viceroy Gil de Taboada, who predicted that "on the day when [Americans] can supply themselves with all that is necessary, their dependence will be voluntary, and neither the military forces that we have there nor the gentleness of government nor the best-administered justice will be sufficient to secure our possession."[35] Following the invasion of the Iberian peninsula by Napoleon's armies, forcing the abdications of Carlos IV and his son Ferdinand VII, the empire unraveled. In response, each part of the empire devised *ad hoc* solutions to the problem of sovereignty. But the end of Spanish dominion came only after almost two decades of warfare. A restored Ferdinand VII sent an army of reconquest to America, but the lion's share of the 16,000 soldiers succumbed to tropical disease. Royalists clung to power in Peru before the 1824 battle of Ayacucho determined that Spanish America would be independent and republican. Shorn of its American empire, Spain held on to the lucrative sugar-producing islands of Cuba and Puerto Rico for the rest of the nineteenth century. But its dominance had come to a definitive end.

35 Quoted in Patricia Marks, "Confronting a Mercantile Elite: Bourbon Reformers and the Merchants of Lima, 1765–1796," *The Americas* 60:4 (2004), 519–58, 525.

22

# The United States from Independence to the war with Mexico, 1776–1848

KEVIN D. MCCRANIE*

Discontent in Britain's Thirteen Colonies had built to open violence by the mid-1770s, much of it occurring in and around Boston. (See Map 19.) A lack of representation and perceptions that British leaders pursued overbearing policies because they were indifferent or even hostile to the plight of the inhabitants pushed ever more colonists towards open rebellion. In response, the tools Britain possessed to confront its colonial troubles were limited by the nature of its government and the few instruments at its disposal. These included the army and navy, but their use at Boston only exacerbated tensions. Fighting flared on 19 April 1775 when British soldiers attempted to seize munitions at Concord, Massachusetts. Along the way, at Lexington, shots were fired and several colonists were killed. Afterwards, colonists sniped at and harried the British on their return to Boston. In the wake of Lexington and Concord, American militia gathered around Boston, surrounding its British garrison. Nearly two months after the outbreak of hostilities, the Americans seized and fortified the strategic Charlestown Peninsula overlooking Boston harbour. In response, the British stormed the position in what became known as the battle of Bunker Hill: the first major battle of the American Revolution. At the end of the day, the British held the field, but at the cost of nearly a quarter of their army in Boston.

The initial engagements at Lexington and Concord and Bunker Hill shed light on the nature of the American War of Independence. The British army was small, particularly in relation to the size of the Thirteen Colonies. Its soldiers were well trained and possessed superb battlefield discipline. Capable

* The views I have expressed in this volume are mine alone. They do not necessarily represent the views, policies, or positions of the US Naval War College, the US Navy, the Department of Defense, or the US Government.

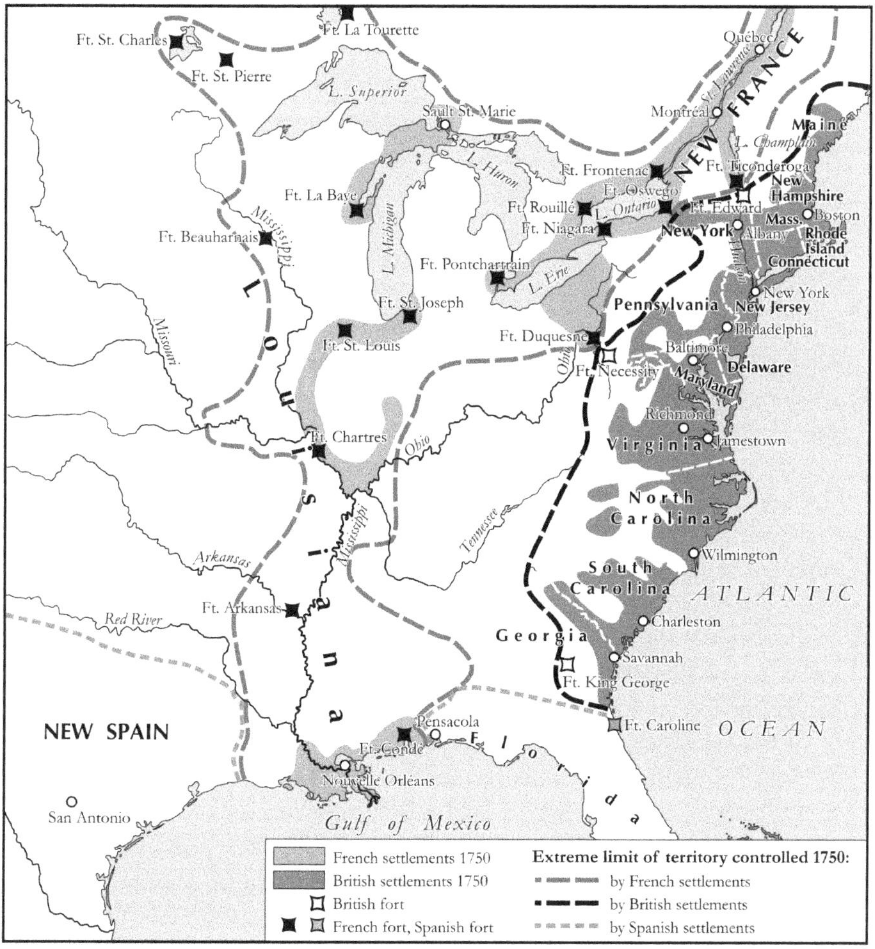


MAP 19. European Colonies in North America, *c.* 1750. Created by Béla Nagy.

of linear tactics, advancing shoulder-to-shoulder under withering fire as at Bunker Hill, they would demonstrate their flexibility on other occasions.

Britain's opponents, labelled Colonists, Rebels, Patriots, or merely Americans, tended to prefer fighting behind the cover of fortifications or trees. Discipline came slowly as their military developed from militia and volunteers. Aspiring national leadership vied for power against local leaders, and significant regional differences existed. Money always proved scarce and America's industrial base weak, making uniforms, provisions, arms, and powder difficult to procure. A lack of ammunition was one reason why the Americans had been defeated at Bunker Hill. The first imperative for the Americans was to develop leadership capable of building, using, and sustaining a fragile military effort, whilst their second imperative was the ability to protract the war. This entailed procuring money and supplies, or enlisting another country willing to part with these items.

Two weeks after Bunker Hill, George Washington took command of the American army outside of Boston. Though most of his troops were from New England, George Washington was a Virginian. This reflected an attempt by the Continental Congress to draw the southern colonies more fully into the war effort. Washington, a man of significant means, status, and military experience, came to dominate the history of the American War of Independence. Though perhaps not a great battlefield commander, he understood the object of the struggle and the means he had available to accomplish that objective. At Boston, Washington managed to keep his army together and maintained the siege throughout the winter. In early 1776, heavy artillery captured at Fort Ticonderoga arrived to bolster the American army. Once this artillery had been positioned on the Dorchester Heights overlooking the harbour, the British were forced to choose between two options: they could either storm the Heights, most likely leading to casualties on the scale of Bunker Hill, or evacuate Boston. They chose the latter.

As the British sailed away, America had never seemed more secure from British rule. Washington, however, understood that the British would return and calculated that their next target would most likely be New York. Yet, without a substantial American naval presence, defending the area with its islands and inland waterways posed significant challenges.

The British arrived off New York in mid-1776. Naval supremacy allowed them to select the time and location of their landings. They outmanoeuvred Washington in a series of encounters beginning at the battle of Long Island on 27 August. A succession of stinging defeats and fortuitous escapes followed as an ever-dwindling number of Americans withdrew. However, the British commanders, the brothers Admiral Richard, Lord Howe and General Sir William Howe, failed to inflict a fatal blow on Washington's army, which managed to survive. Rather, the strategy of the Howe brothers indicated their preoccupation with defeating the Revolutionary forces with a minimum of bloodshed.

As winter approached, the British conducted a pair of operations. First, the Howe brothers sent a force to take Newport, Rhode Island. With its deep-water harbour and moderate climate, this was an excellent location to winter the fleet. Second, British forces advanced into the unspoiled lands of New Jersey, but were hampered by insufficient numbers, despite the fact that the British army had been bolstered by German auxiliaries, generically referred to as 'Hessians'. General Howe hoped that New Jersey would provide food and fodder for the army, and in the process, he sought to kindle popular support for the crown. Yet, foraging led to plundering and other excesses that did little to rebuild loyalty among the population.

By late 1776, Washington's forces had dwindled to a mere skeleton of the army he had led six months before. Even so, he understood the need to strike a blow to restore morale and provide a reason for his men to re-enlist. With this in mind, his army crossed the Delaware, ambushing a Hessian unit bivouacked at Trenton in an attack on the morning of 26 December. Virtually destroying this outpost, Washington secured a second victory eight days later at Princeton: the Americans had regained the initiative. British success around New York in 1776 had yielded them the city and its immediate environs, but they had neither destroyed Washington's army nor broken the American will to fight on.

Not until the summer of 1777 did the British begin their next major series of campaigns. Neither the British commanders in North America nor ministerial leadership in London effectively coordinated efforts. General John Burgoyne advanced from Canada, intending to secure Lake Champlain and the Hudson river valley so as to cut off rebellious New England from the southern colonies, which the British perceived as more loyal. Burgoyne foresaw his forces linking up with other British units advancing north up the Hudson from New York City. The Howe brothers at New York, however, devised their own plan that involved an amphibious assault on Philadelphia: a large, prosperous city and the home of the Continental Congress. A strategically important city, William Howe thought that Philadelphia's occupation would weaken American resolve. Coupled with the occupation of New York, its loss would increase economic hardship in the colonies. The British fleet departed New York on 23 July carrying 16,000 men, eventually making its way up the Chesapeake. In late August, the troops disembarked and advanced on Philadelphia.

With the main British army advancing on Philadelphia, Burgoyne led the second army southward from Canada. He found the terrain more difficult than expected, and the Americans defeated several of his columns. Just south of Saratoga at Bemis Heights, Burgoyne encountered an entrenched American force under the leadership of Horatio Gates and his subordinate Benedict Arnold. Burgoyne attempted to outflank the American position on

19 September in what became known as the battle of Freeman's Farm. The Americans advanced to engage Burgoyne's force. Though the British held the field at the end of the engagement, the Americans managed to withdraw to the Bemis Heights in good order. Over the following weeks, Burgoyne's situation became more tenuous. At least one of his generals advised him to withdraw, but he dithered, still expecting assistance from the British forces that he believed were advancing north from New York City. A force under General Sir Henry Clinton did in fact advance from New York, but the troops were dispatched too late and came nowhere near a rendezvous with Burgoyne's army. Meanwhile, Burgoyne's position was becoming increasingly desperate. Rations were running short, and thousands of reinforcements bolstered Gates's force. After a failed assault on American positions at Bemis Heights on 7 October, Burgoyne held a council of war. All his generals, regimental commanders, and officers as junior as captains received a voice in the discussion. They advocated surrender. Burgoyne agreed: outnumbered more than two-to-one, his command had dwindled to 6,000, with more than a third of these invalids.

Meanwhile, Washington confronted General William Howe's advance on Philadelphia. Though he could not prevent the British from taking the city, he had to avoid a potentially morale-crushing defeat that would result from abandoning this important city without a fight. The expected engagement occurred on 11 September at Brandywine. The British were victorious, but Washington's army once again escaped. Afterwards, Howe occupied Philadelphia. As in the previous year, he dispersed his army, holding a key city and protecting lines of communication. Taking advantage of this, Washington attacked the largest concentration of British troops on 4 October in what became known as the battle of Germantown. While Washington managed to surprise the British, the attack miscarried and the British rallied, forcing the American army to withdraw.

Though Philadelphia had fallen, its loss was offset by several factors. Washington's army remained combat-effective, and the British surrender at Saratoga gave American diplomats in France significant leverage in securing an alliance with a powerful European state already disposed to challenge British colonial power. From early in the war, France had been providing material assistance to the rebellion. In February 1778, that aid was converted into a fully fledged military alliance. France took this decision for several reasons. Saratoga simultaneously engendered optimism about the prospects of the Patriot cause and stoked fears that the British would make concessions to terminate hostilities. France wanted war in the Thirteen

Colonies to fester: it seemed the perfect opportunity for humbling Britain and avenging the French defeat in the Seven Years' War. The French entry into the war changed the nature of the conflict: Britain's entire empire, and even its home islands, now became vulnerable. Faced with such threats, America was relegated to at best a secondary theatre. This shift in priorities only became more pronounced when Spain joined the war in 1779 and the Dutch became involved the following year.

Washington's army spent the winter of 1777/8 just outside of Philadelphia at Valley Forge. Washington used this time to meld his men into a more effective fighting force. An important cog in this process involved the drilling and discipline instilled by Friedrich Wilhelm von Steuben, a soldier of fortune who had served in the Prussian army during the Seven Years' War and had subsequently had a chequered career before being hired by the Revolutionary leadership. The work of Steuben and others turned the Americans into a more competent battlefield opponent. They demonstrated their newfound skill at the battle of Monmouth Courthouse on 28 June 1778, which occurred during the British army's withdrawal from Philadelphia to New York.

The widening scope of the war necessitated the British reducing their commitments in the Thirteen Colonies to free up troops for other theatres. The conflict's expanding scope also provided Washington with direct French military assistance. The effectiveness of the coalition forces was first tested in August at Newport, Rhode Island, which was occupied by 6,700 British soldiers. Washington sought to capture the force and shatter British resolve to continue its war for America. The Americans provided much of the land force while a French fleet with additional soldiers arrived off the coast. Problems occurred. On land, leaders dithered and militia units proved slow to arrive. At sea, Admiral Howe's British squadron intercepted the French naval force and outmanoeuvred it before a storm crippled both squadrons. Yet a draw at sea gave the British enough breathing space to relieve their garrison at Newport while the French fleet eventually sailed for the West Indies.

By 1779, the largest British force left in America was concentrated at New York City. Washington lingered, keeping his distance and avoiding a decisive battle. He followed this course of action to keep his army's strength from dissipating and buy time for the arrival of the French army. Meanwhile, the British government's interest in the American conflict waned as the global war blossomed. Reinforcements from Britain never materialised in force, and General Clinton, now in overall command, decided to abandon Newport. He needed its garrison for operations in Georgia and the Carolinas.

The British decision to shift strategic priorities and to conduct active operations in the southern colonies emerged in fits and starts. In 1776, an attempt to take Charleston, South Carolina, had failed, but Savannah, Georgia, fell to a small British force in December 1778 and an attempt the following year by French and American forces to retake the city miscarried. A powerful joint British army and naval force again attacked Charleston in the early months of 1780. British commanders understood how to apply their strengths against America's weaknesses. Cooperation between the army and navy proved flawless, and the British exploited inland waterways for greater mobility. The Americans, much like at New York, tied themselves to a geographic location, but unlike at New York, there was no line of retreat. The American army became entrapped in the city and the result was the worst American military disaster of the war, with General Benjamin Lincoln surrendering his entire army on 12 May 1780. With Charleston and Savannah in British hands, General Charles, Lord Cornwallis assumed command in the south.

The Americans had lost their southern army at Charleston, but Congress reconstituted it under Horatio Gates. Cornwallis moved quickly, met Gates at Camden on 16 August 1780, and routed his army. For the first time in the war, the British could attempt pacification without facing a nearby American army. Cornwallis empowered Loyalists, but his effort backfired since Georgia and the Carolinas were deeply divided between various groups willing to use violence ostensibly in support of the Loyalist and Patriot causes. The result was a brutal civil war that climaxed at Kings Mountain where irregular rebel militias destroyed a large Loyalist contingent. Support for the British wavered, and Cornwallis found pacification ever more difficult in the face of increasingly emboldened Patriots, who formed irregular militias under the leadership of such figures as Francis Marion and Thomas Sumter.

In late 1780, General Nathanael Greene took command of American forces in the Carolinas. He understood both his objective and his instrument of war. He recognised that his forces were too weak to fight Cornwallis conventionally. Instead, Greene divided his force and operated in conjunction with irregular troops to fight a war of attrition. At Cowpens in early 1781, the strategy paid off when Greene's chief lieutenant, Daniel Morgan, virtually destroyed a portion of Cornwallis's army. Such engagements worked to level the field. Greene declined to fight when pursued by Cornwallis, instead leading the British on an epic pursuit through North Carolina. He only fought at Guilford Courthouse when the British were exhausted. Though Cornwallis won a tactical success by forcing Greene's retreat, the severity of his casualties made it a pyrrhic victory. The British lacked the strength to pursue Greene's defeated army. As the remnants of Cornwallis's command moved towards the Chesapeake, British leaders dithered over the future of operations in the Chesapeake and particularly over what to do with Cornwallis's command.

Meanwhile, other British forces remained scattered in outposts across the Carolinas and Georgia, protecting the population and keeping American irregular forces at bay. After several bloody encounters with Greene's army, the British withdrew into the safety of their port enclaves at Wilmington, Charleston, and Savannah, leaving the Patriots in control of the interior by late summer 1781.

After several unsuccessful initiatives, the Franco-American alliance took a more positive turn in 1780. A French expeditionary land force under General Jean, comte de Rochambeau, supported by a powerful naval squadron, reached Newport, Rhode Island. Even with orders to defer to Washington, Rochambeau was aware of his force's weakness and the limitations of Washington's army. He advocated a cautious policy in ensuing discussions about possible operations, whereas Washington, emboldened by the French presence, sought to go on the offensive. It was Rochambeau who steered the campaign planning towards the vulnerable British command in the Chesapeake.

Deception operations distracted Clinton at New York as French and American units moved south, travelling by land and water. A French naval squadron with siege artillery sailed from Newport. A much larger fleet reached the Chesapeake from the West Indies with the twofold object of assisting the Americans and avoiding the hurricane season. Combined land and naval forces trapped Cornwallis at Yorktown. A naval battle off the Chesapeake Capes drove off a British relief force in early September. On land, siege operations began. In New York, the British prepared an extensive relief operation, but this took too long, and in the interim the allies forced Cornwallis to surrender.

For the French and Americans everything had worked to near perfection, even in the face of slow communications and the challenge of coordinating numerous moving parts of an elaborate multinational, joint operation. For the French and Americans, luck and skill coalesced while Cornwallis had become trapped in a vulnerable position. The outcome of Yorktown was decisive. Cornwallis's surrender led to a change of ministers in Westminster, and these new leaders sought peace. In America, the British continued to hold several enclaves, including New York and Charleston, but active land campaigns had virtually ceased.

The war, however, continued into 1783, with the British fighting the French, Dutch, and Spanish in numerous locales. Most significantly, the Spanish laid siege to the critical British base at Gibraltar while extensive French operations occurred in the East and West Indies. For the United States, the path to independence had been won on the battlefield. The Treaty of Paris acknowledged America's independence and British soldiers withdrew from their last coastal enclaves.

As the United States emerged as an independent state, so too did questions about the future of its military. Central authority and taxation had proved to be important causes of the war, a fact reflected in the United States' first national government, the Articles of Confederation. The government did not have an independent executive, and the legislature lacked the power to tax. Americans worried about the coercive power of a standing army, and many preferred to rely on militias composed of citizen soldiers who served in times of need. Fears of a standing army were at least partially vindicated by the events of the Newburgh Conspiracy, when officers of Washington's army, concerned about their pay, openly questioned the government's intentions at the very end of the War of Independence.

Though Americans worried about a standing army, the War of Independence had demonstrated the unreliability of local militias on numerous occasions. It proved difficult to balance the need for an efficient standing military to meet developing threats with the perception that such a force could itself threaten democracy. This was hardly surprising, given the nature of the central government and the people's history. Even when the Articles of Confederation were replaced by a new national government under the authority of the Constitution, concerns about control of the military persisted. The framers of the Constitution divided military powers: making the president commander-in-chief of the military, but authorising Congress to fund the military, summon the militia, and declare war.

During the Republic's first decades, its most serious security challenge came from Native Americans along the western frontier. Since the arrival of Europeans, a relentless push had driven indigenous inhabitants westward, opening up new land for settlement. Western expansion accelerated following the Revolution. Policing the frontier proved difficult, given the country's size, the distance between the frontier and the population centres, the weakness of the military, the policies of the individual states, and the independent spirit of the people. One approach of the Native Americans to the expansionist white European threat involved accommodation. They tried to blend traditional Native American custom with that of the white Europeans. This policy was especially prominent among the Iroquois and Cherokee in the 1790s.

Other Native American groups sought to reverse white encroachment with violence. In the late 1780s, the Shawnee conducted raids across modern Indiana, Ohio, and western Pennsylvania. The United States government responded with expeditions comprising a mixture of regular and militia forces. Training, discipline, and morale were in short supply, and military operations often proved indecisive or disastrous, as in the case of the defeat of Arthur St Clair's force at the battle of the Wabash in November 1791. As a result, the Federal government increased the size of the army and demanded longer terms of service. Such changes allowed General Anthony Wayne to break the power of the Native Americans at the battle of Fallen Timbers in August 1794. The resulting Treaty of Greenville ended hostilities but required the Native Americans to cede territory in modern Ohio to the United States.

These encounters did not, however, break the will of the Shawnee to resist. Over a decade later, two charismatic brothers, Tecumseh and Tenskwatawa, who was often referred to as 'the Prophet', sought to unite Native American tribes by emphasising their shared culture and the rejection of white European custom. The brothers based their operations at Prophetstown in what is today northern Indiana. They drew support from numerous tribes. This alarmed the governor of the Indiana Territory, William Henry Harrison, who led a military campaign that culminated in 1811 with the battle of Tippecanoe and the destruction of Prophetstown. Harrison's success discredited the Prophet, but Tecumseh was travelling at the time and would again emerge in the War of 1812.

Meanwhile, the US Army continued to evolve. Military leadership improved with the establishment of West Point as a military academy in 1802. Moreover, the military played a critical role in the exploration of the Louisiana Territory after its purchase from France. Most notable work here was done by William Clark, a veteran of the Wayne expedition, and Meriwether Lewis, an army officer and one time secretary to President Thomas Jefferson.

In the late 1790s, naval security threats from the east began to compete with western land-based threats. The former stemmed from the instability resulting from the French Revolutionary and Napoleonic Wars coupled with the rapid growth of American commercial shipping. Merchants tried to translate American neutrality in the European conflict into monetary gain by trading with nearly everyone. Neither the French nor British, however, approved of American commercial activities. As a result, both sides attempted to restrict American trading, going so far as to seize numerous merchant vessels. The defence of American interests necessitated a navy. Though a navy was permitted by the Constitution, it was not until 1794 that Congress voted for one, and years passed before any warships became operational.

The navy's first combat deployments occurred against France in the Quasi War of 1798–1800. An undeclared naval war during the presidency of John Adams, fighting mainly took place in the West Indies as the new United States Navy attempted to prevent French depredations on American commerce. The Convention of 1800 brought hostilities to an end. Though the French did not compensate the Americans for shipping losses, French seizures came to an end, and the peace may have paved the way for the Louisiana Purchase.

Naval operations soon spread to the Mediterranean. Several North African states, collectively referred to as the 'Barbary states', took advantage of the dislocation caused by the Wars of the French Revolution to demand tribute for the safe passage of US merchant vessels. Rather than pay tribute, the Americans sent warships. Though it took several years and greater cost than the tribute itself would have been, the protection of commerce occurred and the US Navy gained valuable experience.

As 1812 approached, American maritime grievances increasingly focused on Britain. Possessing the world's most powerful navy, Britain had the ability to control American commerce like no other state. As the Napoleonic Wars continued, trade restrictions adversely affected American merchant commerce. Moreover, many American citizens were snared into British naval service through impressment. Though this oppressive practice was legally restricted to British subjects, similarities in language and the rudimentary nature of documentation made it difficult to distinguish between Americans and British. It is difficult to ascertain exactly how many Americans were impressed into British service, though it is likely the numbers were not as great as American rhetoric made out. Finally, shots were exchanged between British and American ships; especially notorious, from the American perspective, was the 1807 encounter between the British warship *Leopard* and the US frigate *Chesapeake*.

In this charged environment, the United States and Britain drifted towards war. Maritime concerns upset relations between the countries, but probably would not have led to hostilities. Other issues, including American designs on British Canada, Spanish Florida, and the lands of the Native Americans in the west, loomed. Cumulatively, these factors led the Americans to declare war on the British on 18 June 1812. Few countries have been less prepared to open a war. Though the army had expanded, training and recruitment lagged. Moreover, Congress had failed to fund new naval construction, and several existing warships were not operational.

Even so, the first six months of the naval war went surprisingly well. On three occasions, American frigates defeated their British opponents and a host of other victories added to the US Navy's laurels. Considering that it could put barely 10 warships to sea at once against an opponent with an operational strength of more than 500 warships, the result seemed even more impressive. Victories were important for national morale. But, although the US Navy's initial successes made a statement and American privateers captured many British merchant ships, these losses were not catastrophic, and the Royal Navy never lost command of the sea. Over time, the British navy adapted and its overwhelming strength began to tell. American victories and captures became fewer, while their ships dwindled in number. Though the British never eliminated the American threat on the high seas and the naval war lasted longer and came at a higher price than they expected, they were eventually able to marginalise the American naval threat.

On land, failure greeted the Americans at every turn in 1812. At Detroit in mid-August, a British force under General Isaac Brock, supported by Native Americans, forced an entire American army to surrender. Their victory gave the British control over the Michigan Territory, encouraged Canadian resistance, and brought the Native Americans more fully into the British orbit, while simultaneously deepening political divisions in America. The Madison administration looked to avenge the Detroit debacle. In October, forces crossed the Niagara river intent on seizing Queenston Heights. Though General Brock was killed in the fighting, the American attack was repulsed. Retreat turned into a rout. Trapped against the river, the Americans surrendered in large numbers. Land operations had thus proved a dismal failure. Trained personnel were few and leaders often ineffective. Operating on the Canadian border proved a logistical nightmare, and, even early in the war, funds were in short supply. British Canada would prove anything but the easy conquest that the Americans had expected.

During the winter of 1812/13, American reinforcements filtered westward in an attempt to retake Detroit. A mix of British regulars, Canadian militia, and Native Americans surprised and nearly destroyed an American force on 22 January at the battle of Frenchtown on the river Raisin. Considering the vast area of operations, and the small size of the opposing armies, operations relying on surprise and encirclement often yielded considerable effects. Native Americans proved a double-edged sword for the British. They added manpower, but nothing galvanised American support like the fear of atrocities by Native Americans, such as the massacre of wounded American prisoners that occurred the day after the battle of Frenchtown. Atrocities such as these encouraged American reprisals, with villages burnt and forts erected to control communications along the western frontier.

The further west the operation, the smaller was its scale. This mainly reflected the difficulty of projecting power from distant American population centres. Larger military forces operated around Lakes Erie and Ontario. On 27 April, the Americans seized York, the capital of Upper Canada. Though a magazine explosion caused many casualties, the Americans burned the public buildings as well as great quantities of naval stores in the hope of retarding British naval construction on the Great Lakes. The Americans followed their success with the capture of Fort George at the mouth of the Niagara river. Though the Americans had brought significantly larger land and naval forces to bear than the British could muster, their outnumbered opponents managed to slip away. The Americans continued their advance, but, at Stoney Creek on 6 June, the British launched a surprise pre-dawn, morale-shattering attack that captured the American commanders. The Americans, however, retained Fort George until winter, when British operations regained this valuable position as well as Fort Niagara, located on the US side of the river.

Control of the Great Lakes was an integral aspect of the war on the border with British Canada. The road network, and the distances involved, meant that water-borne transport became a necessary supplement to any land operation. On Lake Erie, the Americans won a squadron-sized encounter in September 1813. This effectively destroyed British naval power on that lake and severed communications with British outposts further to the west, making the British hold on Detroit tenuous. Indeed, General William Henry Harrison took advantage of the victory to retake Detroit and pursue his opponents into Canada. At the battle of the Thames on 5 October, an outnumbered force of British soldiers and Native Americans broke. Though Native Americans put up stiff resistance, the death of Tecumseh signalled their defeat and his loss had significant repercussions on the ability of the British to garner Native American support. Moreover, US leaders had begun to better appreciate their opponents and developed an enhanced understanding of how to employ their better-trained regular and militia units.

The largest US operation on the Canadian front in 1813 involved a two-pronged attack aimed at Montreal. Though the Americans enjoyed significant numerical superiority, this was negated by other factors. One axis of the offensive developed under General Wade Hampton, who advanced from the area around Lake Champlain towards Montreal. The attack miscarried for various logistical and leadership reasons. Moreover, morale collapsed when the men learned they would winter on the US side of the border, meaning that any advance on Montreal would be followed by a retreat. The second prong under General James Wilkinson advanced across Lake Ontario and then moved along the St Lawrence river. Though Wilkinson's forces sustained a stinging defeat at Crysler's Farm, they pressed onward. Only news of the first column's retreat, coupled with Wilkinson's poor health, caused their withdrawal. By the end of 1813, a stalemate existed along the front. Smaller British forces had thwarted the American advance. Logistical difficulties had much to do with this. Still, there were signs of an upturn in American fortunes, while US forces had significantly worn down Native American strength.

Operations on the Canadian frontier did not resume in earnest until July 1814 when US forces crossed the Niagara river and seized the weakly defended British position at Fort Erie. They then advanced towards the shores of Lake Ontario but ran into a much smaller British force on the Chippawa river. Winfield Scott's actions here were critical. Scott had already served with distinction in several actions, including Queenston Heights in 1812 and Fort George in 1813, and had been promoted to brigadier-general before he was thirty. Scott's brigade of well-trained US regulars made a substantial difference. The Americans were resolute and demonstrated increasing tactical skill, inflicting heavy losses and forcing the British to retreat. However, they failed to take advantage of the victory by capturing Fort George. Rather than remain in the area, the Americans began to withdraw. As they fell back, Scott's brigade encountered a large British force at Lundy's Lane on 25 July. Both sides fed more troops into the battle. Better training meant that the Americans exchanged volleys and stood up to artillery fire, but this also made Lundy's Lane the bloodiest battle fought on the Canadian frontier. It was an operational success for the British because the American forces continued to withdraw, falling back to Fort Erie. The

British invested the fort, but failed to take it, and it was only in November that the Americans abandoned Fort Erie.

The end of the Napoleonic Wars in spring 1814 transformed Britain's geostrategic position and allowed the British to devote larger forces to the American war. However, when British reinforcements reached Canada, logistical issues constrained where these newly arrived units could operate. Supplies and a developed infrastructure in particular were wanting. The Champlain front seemed to offer both, and it was along this front that the largest British army of the war invaded America in September. Several days later, they arrived at Plattsburgh, New York. Before the British could continue, they needed to secure their line of supply. For this, their small naval squadron on Lake Champlain had to wrest command of the lake from the Americans, but in battle on 11 September, the British suffered defeat. Rather than press on with unsecured supply lines, the British withdrew.

Whereas the two sides possessed relative naval parity on the isolated inland lakes between the United States and Canada, the oceans proved quite different. There, the British had overwhelming superiority. In 1813, they used this advantage, along with a small number of ground troops, to conduct predatory amphibious operations in the Chesapeake. The British hoped to divert American forces from the Canadian front and degrade this area's high level of support for the war. Moreover, the web of bays and rivers were perfect for raiding operations. It was not, however, until 1814, after the Napoleonic Wars came to an end, that the British had sufficient land and naval forces to make such operations effective. In September, the British quickly secured the vulnerable and weakly defended Penobscot region of Maine. The focal point of British amphibious operations, however, remained the Chesapeake. The Americans lacked the naval power to stop British raids, and with most of their regular units on the Canadian border, its coastal defences rested largely upon militia units. In August 1814, the British landed at Benedict, Maryland, marched overland, defeated an American force at Bladensburg, and then occupied Washington, DC, burning its public buildings. However, a second operation, when British forces landed near Baltimore in September, produced very different results. Though the British defeated a unit of Maryland militia, they suffered heavy casualties in the process and lost their commander. The British then arrived outside Baltimore; however, unlike Washington, DC, this city was well fortified. A bombardment of the town's defences, including Fort McHenry, proved inconclusive and the British chose to withdraw, rather than risk heavy losses in storming the city. Naval superiority gave British forces mobility and ensured that the initiative was with them, but their operations proved indecisive. Conversely, the Americans lacked sufficient naval and ground forces to curtail these attacks. In the end, these operations yielded few effects on the ultimate outcome of the war.

The British, however, sought another target: New Orleans. British strategists had been keenly aware even before the war of the strategic advantages that might follow from occupying this port, but sufficient forces to attempt this only became available after the Napoleonic Wars. By this point, though, the Americans had solidified their position in the south. Mobile had been obtained from the Spanish in 1813. Half of the Creek lands had been taken, breaking Native American power in the region and resulting in Andrew Jackson's rise to prominence.

Jackson's moment of glory, however, occurred at New Orleans. Upon learning that the British were intent upon taking the city, he marched to its defence, arriving in early December 1814. Britain's invasion fleet appeared off the coast shortly thereafter. The Americans held Fort St Philip on the main channel of the Mississippi, but the British had more success on Lake Borgne, where they destroyed an American gunboat flotilla. However, the British found movement through the bayous towards the city slow, cold, and difficult. During this time, Jackson received reinforcements and prepared to defend the city. The main attack came on 8 January 1815. The British stormed prepared positions and lost some 2,000 men. The engagement was reminiscent of Bunker Hill except this time the Americans held the field and the British withdrew. Further east at Mobile, the British had greater success, seizing Fort Bowyer in February.

The battle of New Orleans and the taking of Fort Bowyer occurred after the signing of the Treaty of Ghent on 24 December 1814. Ghent, however, was in modern Belgium, and it took time for news of the treaty to travel across the Atlantic. The Senate then had to ratify it and the president sign it, which took until February 1815. After two decades of war in Europe, the British wanted a return to peace and were willing to offer lenient terms in the hope of discouraging a war of revenge that would again threaten Canada. For the Americans, victory at New Orleans obscured how disastrous the war had been. Their seaboard had suffered repeated attack; merchant commerce had been destroyed; part of Maine occupied; and the national government bankrupted, while Canada remained in British hands. As bad as the outcome was for the United States, the Native Americans fared worse. They were no longer able to play Britain off against the United States; Creek power was broken; and no powerful leader remained to unite the tribes.

The War of 1812 had opened American eyes. Militias had once again proved unreliable and contingent upon state government support, whilst regular forces could respond more effectively to threats. The army's peacetime strength was now set at 10,000. Leaders of considerable talent, like Scott and Jackson, had emerged. Renewed emphasis was placed on West Point for the education of future military leaders. Money went to fortifications to help defend coastal regions against raids. These steps created a larger, better-funded, and more professional army, but these decisions had to compete against short memories and limited government revenue. Within a few years, calls for military economies had re-emerged.

Conflict on the border between Spanish Florida and Georgia halted this retrenchment. Weak Spanish colonial rule allowed various groups significant freedom of action. These included the Seminole people, a mixture of various Native American groups, including the previously defeated Creeks, as well as runaway slaves. The First Seminole War erupted in late 1817, and the US government sent Jackson into Spanish Florida. Rather than escalating the conflict into outright war, the Spanish government decided to negotiate, recognising that their sparsely populated colony was at the mercy of its powerful northern neighbour. This resulted in Spain ceding Florida to the United States in the Adams–Onís Treaty of 1819.

Several years later, the army was reduced by half. However, an emphasis on improving the quality of the remaining troops offset some of the detrimental effects of this downsizing. Training improved, and the curriculum at West Point was overhauled: engineering became the school's forte. Though the army became increasingly professional, it lacked the strength to meet all its missions. Its emphasis remained on perimeter defence, be it along the coasts, the Canadian border, or its poorly defined western frontier.

Though names, places, and Native American groups changed, the situation on the frontier during the years between the American Revolution and the Mexican–American War was thematically constant. Expansion raised questions of land ownership. This either resulted in treaties disadvantageous to Native Americans or led to outright hostility. In the years after the War of 1812, the US Army took a greater role on the frontier, establishing forts around which permanent settlements often emerged.

A particularly ugly case of violence occurred in the early 1830s against elements of several tribes under the Sauk leader Black Hawk. Angered by American occupation of Sauk lands, Black Hawk led several expeditions into modern Illinois. Regulars and militia (including a young Abraham Lincoln) undertook a punitive campaign. At Bad Axe river in 1832, they crushed the

Native Americans in a very lopsided fight. Only after the conclusion of hostilities did a large force of regulars arrive from the east coast under the leadership of General Scott. Ravaged by cholera, they not only appeared late but were also combat-ineffective. Bad Axe river demonstrated the problem with American military operations: as the size of the United States grew, the small military found it ever more difficult to confront challenges in a rapidly evolving environment.

Failure to deal quickly with another problem had more serious consequences. The US government had ratified a treaty with the Seminole people in 1834 specifying their removal from Florida to west of the Mississippi. Many Seminoles did not accept the treaty and, on 28 December 1835, hostilities erupted, resulting in the killing of several white Americans, including an agent sent to oversee their removal. On the same day, another Seminole group destroyed a military column in an event that became enshrined as the 'Dade Massacre'. The Seminoles proved a tough opponent, skilled in irregular warfare, and aided by an inhospitable environment. The turning point did not occur until 1841, when Colonel William Worth assumed command. Previous leaders had suspended operations in the summer to avoid sickness, but this allowed the Seminoles a chance to recover and raise crops. Operating in the summer allowed US forces to break Seminole power, bringing the war to an end in 1842. The war proved one of the longest, bloodiest, and costliest Native American wars and resulted in an almost doubling of the US Army.

During the Second Seminole War, problems erupted in Texas. Though an international issue, it had significant repercussions on US military history. Texas was a sparsely populated province of the newly independent state of Mexico. To increase its population, the Mexican government offered generous land grants to US citizens. However, this policy proved more successful than the Mexican ministers could ever have envisioned: US settlers soon became the majority in Texas.

Attempts to bring Texas more fully into the Mexican state led to tension and then war in 1835. The Texans quickly expelled the central government's presence, but the next year troops personally commanded by the political leader of Mexico, Santa Anna, attempted to re-establish control. At San Antonio, his army took the Alamo after a bloody thirteen-day siege before marching north. At San Jacinto, however, the Texans won a significant victory, and Mexican forces were expelled from Texas. Many Texans desired a union with the United States, but leaders in Washington refused. Questions about the future of slavery in the region alienated northern support; moreover, such a move could lead to war with Mexico. The US Army was particularly unprepared for a major war given its ongoing missions against the Native Americans, notably in the Second Seminole War.

However, tensions between Texas and Mexico remained high, and appetite for annexation built in the United States. When this finally occurred in 1845, discontent escalated and within months an American army under General Zachary Taylor had moved towards the disputed Texas–Mexico border. Hostilities erupted in April 1846. The United States considered objectives beyond compelling the Mexican government to accept the annexation of Texas. Many Americans also wanted to take control of what would become the south-western United States including California, but this would require sweeping territorial concessions that comprised approximately half of Mexico's territory.

In May 1846, a Mexican army advanced into contested lands north of the Rio Grande. An engagement with Taylor's outnumbered army followed at Palo Alto. The Mexican force had superior cavalry, but it was well-handled American artillery that inflicted heavy losses and forced the Mexican army to withdraw. The next day, Taylor's army engaged the Mexicans at Resaca de la Palma. This was mainly an infantry clash on broken ground between advancing Americans and their opponents, but again, the Mexicans withdrew.

In the aftermath of Palo Alto and Resaca de la Palma, President James K. Polk and General Winfield Scott devised a multi-prong strategy to defeat Mexico and secure territory north of the Rio Grande for the United States. American columns advanced into modern New Mexico and California. Though small in number, these forces secured vast expanses of north-western Mexico for the United States. The main American thrust, however, involved Taylor's army. It crossed the Rio Grande and moved on Monterrey. Over several days in late September fierce fighting occurred as the two sides battled for the city. With both sides exhausted, Taylor negotiated an armistice which allowed for the withdrawal of the Mexican army. This infuriated Polk, who refused to accept the terms. American leaders continued to demand far more than the Mexicans were willing to concede.

Meanwhile, Santa Anna reassumed command of the Mexican army. He quickly took the offensive against Taylor's forces. The time seemed right, particularly after obtaining intelligence that the American force had been reduced to approximately 5,000 men. Santa Anna led 15,000 soldiers across 200 miles of desert to attack Taylor at Buena Vista on 22–3 February 1847. Though outnumbered nearly three to one, the Americans held a strong position along hills and ravines. Their artillery again provided the critical advantage. Still, it was a near-run fight. Santa Anna withdrew after sustaining approximately 2,000 casualties to 750 for the United States.

Taylor's command had been weakened to prepare for the main operation of 1847. Rather than push overland towards Mexico City, the Americans decided to outflank the Mexicans using an amphibious assault with approximately 12,000 men under General Winfield Scott. This force landed at Veracruz on 9 March. Scott's men forced the Mexican garrison at Veracruz to surrender after a siege. The Americans faced a dilemma: yellow fever was already ravaging the troops. To minimise exposure to the disease, and maximise the strength of his command, Scott advanced out of the disease-ridden coastal plain towards Mexico City. In the process, he severed his lines of communication with the coast. It was a bold move, that appeared to defy accepted military strategy, for if the Americans were defeated, they would have nowhere to retreat to. The operation therefore required fine-tuned logistics, skill, and professionalism. Judging by the history of American military operations dating back before the War of 1812, the prospects did not appear good.

About 40 miles from Veracruz at Cerro Gordo, the Americans encountered Santa Anna's army in fortified positions. Rather than make a frontal attack on what seemed an impregnable position, the Americans flanked and sent the Mexicans fleeing. Several days later, at Jalapa, Scott had to send seven regiments of volunteers back to the coast. Their term of enlistment had nearly expired and most refused to re-enlist, so, in the middle of the most important campaign of the war, Scott now commanded only half the force he landed with at Veracruz.

Undeterred, he continued to advance. Over the next months, reinforcements arrived to strengthen the army as it slowly approached Mexico City. Attempts to negotiate an end to the war failed. In mid-August, Scott had arrived outside the Mexican capital. As had previously occurred at Cerro Gordo, he sought to avoid frontal attacks against strongly defended Mexican positions by resorting to flanking manoeuvres. As the Americans got closer to the city, flanking options became more difficult, and the result was more hard-fought battles such as at Churubusco on 20 August. The Mexicans held strongly fortified positions in the town, including a massive stone convent. The fighting was the most intense yet, but the Mexicans broke when ammunition ran low.

For two weeks after Churubusco, no fighting occurred as both sides regrouped and negotiations began. However, when these went nowhere, Scott reopened hostilities in early September. He assaulted El Molino del Rey in a bloody frontal attack. After taking this position, Scott's men stormed the Castle of Chapultepec on 13 September and then pressed on, assaulting the Belén and San Cosmé Gates. Rather than face the prospect of house-to-house fighting, Mexico City surrendered.

While the Americans held the city, the Mexicans refused to end the war. As a result, for almost a year, US forces occupied central Mexico, balancing governmental duties against actions needed to curtail significant insurgent opposition. The long occupation stemmed from the need to form a new Mexican government that possessed enough legitimacy to undertake serious negotiations. Terms from these discussions did not emerge until February 1848, when both sides agreed to the Treaty of Guadalupe-Hidalgo. Opposition in Mexico to the severity of the terms delayed exchanging the ratified treaty until late May. The United States received about half of Mexico's territory including the region from Texas in the east to California in the west. The Americans provided 15 million dollars in compensation in return.

Between the start of the American War for Independence and the end of the Mexican–American War, warfare had first forged an independent United States and then driven the state's expansion. There were both defensive and offensive incentives to wage war. On the one hand, military action countered outside threats, most notably Britain. But there was also widespread popular support for territorial aggrandisement. Following independence, money and diplomacy provided alternative means to resolve disputes, but often these worked as adjuncts to offensive military operations. By the end of the Mexican–American War, the boundaries of the continental United States approached a form recognisable today.

Land forces were an essential instrument for securing American independence and later expansion. These ranged from militias to the Continental Army and eventually the US Army. Comparing the Patriots at Bunker Hill with the force that Scott commanded in his conquest of Mexico City nearly seventy-five years later, the difference is overwhelming. In fact, the professionalism and training of Scott's forces resembled the British at Bunker Hill more than their Patriot opponents. The development of a professional army had been slow, and often painful. Fear of arbitrary, despotic rule had driven many to arms in the American Revolution, but such worries worked against the security concerns of the young nation. Only after disasters such as New York in 1776, Detroit in 1812, and Washington in 1814 did Americans come to accept the need for a larger, more professional military. By the Seminole Wars, the quality of officers had begun to improve; however, it was only in the course of the Mexican–American War that American leadership proved capable of utilising forces to maximum advantage while developing a high level of training among the troops. Yet, the American military establishment remained small. This would change more than a decade later during the Civil War, as America faced its most cataclysmic conflict to reunify what had become a fragmented country.

23

# The wars of the French Republic and the Napoleonic Wars, 1789–1815

ALAN FORREST

The years of the French Revolution and First Empire are remembered as much for war and imperial expansion as for the great political and social reforms they introduced. The Revolutionaries saw themselves as sons of the Enlightenment, devoted to ideals of freedom and the betterment of humanity. Yet they unleashed a long period of almost continuous warfare, fought across the European continent and beyond, in North Africa and the Near East, in North America, Asia, and the Caribbean. In Europe, France faced a succession of coalitions of other European powers, from the First Coalition of 1792–7 – an international alliance that included Austria, Prussia, Britain, Spain, Piedmont, Naples, and Sardinia – through to the final coalition, the Seventh, which wearily regrouped to defeat Napoleon after his ill-judged return to France in 1815. The other governments of Europe feared France's political ambitions as much as its military might, and they invariably saw themselves as the victims of French aggression, forced to make war to protect their territory from attack. Britain also feared the challenge to its naval and colonial supremacy which a revitalised France would pose; for London the war was as much about Jamaica and India as the balance of power in Continental Europe, about global competition for resources as much as the ideas of the Revolution in France.[1]

## A global conflict

The Revolutionary and Napoleonic Wars consumed vast resources and demanded huge logistical support. The numbers of soldiers deployed were unprecedented in modern warfare, as first France and then the countries ranged against it raised mass armies of young, relatively inexperienced troops. The French *levée en masse* of 1793, for instance, aimed to create an

1 C. A. Bayly, *The Birth of the Modern World, 1780–1914* (Oxford, 2004), pp. 93–6.

army a million strong, and though real recruitment fell substantially short of that figure, the Jacobin republic could still boast three-quarters of a million men under arms. Not all, of course, were sent into battle, and the size of the armies in the field during the 1790s did not rise proportionately. But under the Empire that changed. The *Grande Armée* that set off for the Russian Campaign totalled over 650,000 men; and even after it was destroyed by Cossack sniping and the cold and hunger of a Russian winter, Napoleon was able to raise another of similar size for the 1813 campaigning season. Most of Europe turned to some form of conscription or compulsion to fill their regiments; and even in Britain, which continued to resist conscription as a foreign, Continental practice, the military more than doubled in size between the country's entry into the war in 1793 and ultimate victory at Waterloo. Recruitment drained the European economy of men and resources, not just in France but across the entire continent. It created shortages of labour and left women to till the land and bring up their families. And while war offered career opportunities to army officers and rich pickings for military contractors, these came at a terrible human cost. Between 1792 and 1815 between 3.5 and 5 million Europeans died, according to varying estimates.[2] Theirs was a lost generation, and the young men who marched off to war often returned – those who did return – with broken bodies and damaged minds, condemned to eke out a marginal existence that would be dominated by pain and poverty.[3] Their war did not end with the Congress of Vienna. Indeed, in the view of historians like David Bell and Jean-Yves Guiomar, this should be seen as a 'total war', the first of the modern era.[4]

For most of the land powers of Europe the principal focus of the conflict was Europe itself, the defence of their interests and frontiers, and the extension of their political control. It was in Europe that France saw the principal threat to the social and political gains of the Revolution, and in Europe that first the Directory, then the Consulate and the Empire, sought to impose its rule. Napoleon had a peculiarly Eurocentric view of the world as he sought to recreate his empire in the footsteps of Charlemagne, with the Rhine and northern Italy at its core, the modern counterpart of Carolingian Lotharingia.

2 Ute Planert, 'Napoleon and Beyond: Reshaping Power in Europe and the World', in Ute Planert (ed.), *Napoleon's Empire: European Politics in Global Perspective* (Basingstoke, 2016), p. 1.

3 'Introduction: Nations in Arms, People at War', in Alan Forrest, Karen Hagemann, and Jane Rendall (eds.), *Soldiers, Citizens and Civilians: Experiences and Perceptions of the Revolutionary and Napoleonic Wars, 1790–1820* (Basingstoke, 2009), pp. 1–12.

4 David Bell, *The First Total War: Napoleon's Europe and the Birth of Warfare as We Know It* (Boston, 2007); Jean-Yves Guiomar, *L'Invention de la guerre totale* (Paris, 2004).

At first his project worked admirably, as he played off the German states against one another and appealed to Russian interests in German Central Europe. By 1807 the French had seen off three successive coalitions and, at Tilsit, Napoleon and Alexander effectively divided the European land mass into two imperial zones, those of France and Russia. Only after 1808, when his armies were sucked into the Iberian peninsula and the miseries of guerrilla warfare, did Napoleon's assured touch begin to wane, his troops overextended in Spain and Russia, command too often left to less able generals. By 1812, with his army mired in the Russian countryside and Moscow in flames, the other European leaders began to smell success, and a new coalition was born. But it is too easy to write the history of these years through the eyes of the French emperor. These were much more than just the 'French Wars' that the British liked to call them. Other states and rulers were involved, too, and some conflicts had nothing to do with France, or Napoleon. Further to the east the decline of the Swedish and Ottoman empires had created a power vacuum. Russia developed plans for an assault on India, while the Tsar also pressed ahead with traditional Russian ambitions in the Balkans. Finland succeeded in gaining independence from Sweden after 200 years, welcoming the chance, in 1809, to become a province of the Russian Empire. And Austria, Prussia, and Russia still had their eyes fixed, covetously, on the pickings they might seize in Poland, which Napoleon had reconstituted as the Grand Duchy of Warsaw. For Germans, Slavs, and Scandinavians, France and its ambitions were often a secondary consideration.[5]

The war was also fought outside Europe. In part, but only in part, this marked the continuation of the Anglo-French colonial rivalries of the eighteenth century, the two countries clashing over Egypt and Syria, Toussaint Louverture playing them off against each other in Haiti, and France harbouring plans to attack the British Empire in India. But each country pursued its own colonial interests. The Dutch, in Java and the Cape, were among the major losers outside Europe. Spain made a desperate bid to capitalise on French weakness in the Caribbean. Russia eyed possible expansion in the Balkans. In some spheres non-European troops played a critical role in the fighting – from the Mameluke cavalry who opposed Napoleon in Egypt to the native troops who fought in a succession of wars in South Asia. Britain also began to use well-disciplined Indian units outside of India: in the assault on the Dutch East Indies in 1811 or the invasion of the Sri Lankan kingdom of Kandy in 1815. Napoleon's attention was too often focused elsewhere to

5 Charles Esdaile, *Napoleon's Wars: An International History* (London, 2007), p. 6.

maintain a consistent colonial policy. Nor did he show any direct interest in the various wars of independence which swept Central and South America after 1810, though French Revolutionary ideas did play a part, along with the success of anti-colonial revolutions in their own hemisphere, in the United States and, closer to hand, in Haiti. Britain, in contrast, unsuccessfully attempted to invade Argentina before capitulating in 1807 at Buenos Aires; and in 1812 Britain became embroiled in a further war against its former colony, the United States.[6]

These conflicts were fought at sea as well as on land, and this would prove to be France's Achilles heel, the weakness which cut lines of communication with its armies and would prove a major contributor to France's final defeat. In particular, in spite of a committed construction programme during the 1780s, *La Royale* was no match for the Royal Navy, on which Britain traditionally lavished a far higher proportion of military spending and for which it could mobilise the manpower resources of its merchant marine. France embarked on the Revolutionary Wars with misplaced confidence, born of the fact that it had briefly enjoyed command of the seas at the end of the American War in 1783, but this had merely driven Britain to invest heavily in fleet renewal. And in truth the French had seldom posed a real challenge to Britain at sea. It was British naval superiority that had given Britain such an overwhelming advantage in the Seven Years' War, in America and Canada, India, the Caribbean, and the Philippines. In North America, it had prevented France from sending reinforcements and supplies to its armies, depriving them of goods to trade with the Indian tribes on whose alliances they were dependent. Naval weakness had been instrumental in undermining military success.[7] In 1793, when the French resumed their naval joust with Britain, their prospects looked even grimmer. Whereas the Royal Navy was noticeably stronger than it had been ten years earlier, the Revolution had sown havoc in the officer ranks of the navy, traditionally even more royalist than the army.[8] By the time war was declared resignations and emigration had decimated naval officers to the point where only two vice-admirals remained out of nine, and three rear-admirals out of eighteen. The first naval encounter between the two nations – Admiral Howe's victory on the Glorious First of June – destroyed or disabled twelve French vessels in the Atlantic; and Nelson's subsequent victories over the

6 Jeremy Black, *European Warfare, 1660–1815* (London, 1994), pp. 199–203.

7 Fred Anderson, *Crucible of War: The Seven Years War and the Fate of Empire in British North America, 1754–1766* (London, 2000), pp. 453–4.

8 William Cormack, *Revolution and Political Conflict in the French Navy, 1789–1794* (Cambridge, 1995).

French at Aboukir Bay and the Nile in Egypt, and at Trafalgar off the coast of Spain, did far more than restore British bragging rights. They established control of the seas, ensuring that Britain could supply its armies on the Continent and maintain trade with its overseas empire.[9] This was to prove vital in determining the final outcome of the war.

## Revolution and war

Yet, if the declarations of its politicians are to be believed, Revolutionary France never intended to embroil itself in war, and certainly not in a war of aggression or conquest. As early as 1790 the French National Assembly recognised that they must be able to defend their territory if it came under attack, but seemed explicitly to rule out wars of a conventional sort, to secure a disputed territory, or claim a dynastic succession, or squabble over colonies. War, they declared, would be reserved for tyrants; the French would never make war on other peoples, and their anger was reserved for the inhabitants of castles, not of humble cottages. Admittedly, the parliamentary debate on the question was more concerned with the balance of domestic power in France than with the diplomatic balance in Europe: the principal aim of the decree that emerged on 22 May 1790 was to restrict the executive power of the king, which, as the Assembly recognised, was enhanced in wartime, when it was the royal prerogative to take control of the armies and make war and peace with other monarchs. But that decree was famously explicit about the terms on which the Revolutionary state could contemplate going to war. If the hostilities were seen by the Assembly to constitute 'a culpable aggression' on the part of ministers or other agents of the executive, then, the decree warned, 'the author of that aggression would be pursued in the criminal courts for the crime of *lèse-nation*', a new, more democratic version of the monarchical *lèse-majesté*. Lest there be any confusion, the Assembly went on to decree that 'the French nation renounces all wars made for the purpose of conquest, and that it will never deploy its forces against the liberty of any people'.[10] That might seem to rule out any return to the foreign policy aims of the eighteenth century, to redefine the very function of war.

It certainly shocked contemporaries – a declaration that flew in the face of all previous diplomatic practice – and if it divided opinion in France itself, every

9 Black, *European Warfare*, 1660–1815, p. 191.
10 *Archives Parlementaires*, première série (1789–1800), vol. 15, pp. 661–2.

foreign ministry in Europe took notice, too. Were the French really abjuring war, which every state in the eighteenth-century world saw as a legitimate and even a necessary extension of diplomacy? The terms of the decree were vague and largely abstract, raising questions about just how they would be translated into the practicalities of foreign policy. But it was a highly dangerous tactic, especially if it meant that Revolutionary France unilaterally cut its military expenditure and reduced the size of its armies. What it certainly did do was to curtail the king's executive authority and confirm the widespread impression abroad that France was heading for a period of diplomatic impotence, an impression strengthened by France's hesitant reply to Britain in the Nootka Sound crisis of 1790.[11] No other European power had gone so far in renouncing the use of war, and it left French interests, in Europe and around the world, dangerously exposed to the covetous eyes of others. It was taken as a sign of weakness which other rulers would be tempted to exploit.

Since the reign of Louis XIV Europe had witnessed a succession of wars in which the great powers had traded alliances and sought to profit from the diplomatic balance of power, at home and in the colonies. Few questioned the right of kings, or of sovereign states, to make war when their interests were at stake; indeed, ever since the Middle Ages war had been seen as a legitimate means of maintaining law and order in international society.[12] Could a government hope to survive without at least the possibility of declaring war on its neighbours? Could it indeed hope to avoid war in a continent that had seen so many? The evidence of the previous century suggested that the other great powers would not hesitate to do battle if they thought that one of their number was mortally weakened. As the great Prussian military thinker and strategist Carl von Clausewitz observed, a ruler had to be prepared for war in circumstances where diplomatic manoeuvres failed; and this was all the more true in an age of empire when European nations did not always control their own destiny. War, he maintained, was 'simply the use of violence by states for the enforcement, the protection, or the extension of their political power'.[13] There were times when it was a necessity; it could not simply be abjured.

The foreign policy of the 1790s was in many respects a continuation of the great powers' ambitions of the previous decades, a period marked by

11 Jeremy Whiteman, *Reform, Revolution and French Global Policy, 1787–1791* (Aldershot, 2003), p. 215.
12 Michael Walzer, *Just and Unjust Wars: A Moral Argument with Historical Illustrations* (London, 1977), pp. 58–9.
13 Carl von Clausewitz, *On War*, ed. Michael Howard and Peter Paret (Princeton, 1976), p. 75; Timothy Blanning, *The Origins of the French Revolutionary Wars* (London, 1986), p. 26.

a succession of limited wars in Europe and wars for empire overseas. The Seven Years' War had started in the Americas, before it spread back to Europe, implicating the major European colonial powers, Britain, France, Spain, and, to a more limited degree, Holland, all of which nurtured global aspirations. France's ill-judged intervention in the War of American Independence was rewarded by a victory that helped to offset the humiliation of the Peace of Paris and constituted what Jeremy Whiteman terms 'the high-water mark' of France's diplomatic prestige under Louis XVI. Indeed, in the final years of the Bourbon monarchy, the foreign minister, Vergennes, made it his goal to limit Britain's global supremacy, while counterbalancing British dominance of the seas by containing the military strength of Austria, Prussia, and Russia on land.[14] He hoped to secure these objectives peacefully by diplomatic manoeuvring. But it meant that France must continue to play its role as one of the leading powers in Continental Europe while maintaining a costly colonial and naval presence overseas. It was a policy that would be impossible to implement without the option of resorting to armed force, since unbroken peace could not be guaranteed. The Revolutionaries seemed to be moving into uncharted territory, which many regarded as foolish and utopian.

In practice, of course, the National Assembly was not renouncing its right to wage war any more than it was disbanding the army it inherited from Louis XVI. Rather it was redefining the purpose of war in an era when the French people, rather than the French monarch, was now sovereign. The evidence of the years that followed would suggest that they did nothing to constrain the exercise of military might. Between 1792 and 1815 France was almost continuously at war, and from 1794 it was difficult to pretend that its purpose was defensive. After some stuttering early engagements, the French succeeded in pushing back the Austrian, Prussian, and Spanish forces that initially threatened their frontiers and took the conflict to foreign territory, where it stayed, resolutely, until the final *Campagne de France* in 1814. Much as they might proclaim the needs of the *patrie en danger*, the Revolutionaries soon turned into aggressors, exporting war and revolution to Belgium, the Rhineland, and Italy and creating a buffer of sister republics.[15] Napoleon lived by war; his authority depended on the fruits of war, whether as a Revolutionary general in Lombardy, as commander of the army in Egypt, as First Consul at Marengo, or as emperor at Austerlitz and Jena. In all he

14 Whiteman, *Reform, Revolution and French Global Policy*, 18–19.
15 See Pierre Serna (ed.), *Républiques sœurs. Le Directoire et la Révolution atlantique* (Rennes, 2009).

fought over sixty battles and emerged victorious in most.[16] The governance and the economy of the Empire depended on the continued pursuit of war, and wars that were declared in the name of justice and freedom were soon turned into conventional wars of conquest, annexation, and often brutal imperialism.

## Declaring war

The Revolutionaries had persisted in claiming that their wars were defensive, fought in the name of liberty between a free people and the instruments of the tyrants who threatened them. But how could they justify this claim? The implication must be that there was an unbridgeable ideological gulf between the two sides, and that Europe was divided into two opposed political systems, unable to coexist in a state of peace. The Revolutionaries threw the responsibility for this inflexibility on their opponents, on the crowned heads of Europe who, supposedly, were preparing to invade in order to restore the Old Regime and punish them for their attacks on the monarchy. They were keen to insist that the war was ideological, and that they were defending the French people, the Revolution, and the legal gains they had acquired against those who felt threatened by them. This provided useful propaganda, of course, both for their own people and for the rest of Europe; it served to blacken the reputation of their enemies and to conjure up terrifying images of the fate that awaited innocent Revolutionaries should their country be invaded. The rumour spread, for instance, in the popular neighbourhoods of Paris, that the Austrians were preparing to invade, that they had overrun the plains to the east of the capital, and that they would butcher anyone suspected of Revolutionary sympathies. Fear and paranoia spread, too. But there was nothing at all unusual in that: throughout the Revolution Paris was prey to rumour and scaremongering, much of it actively encouraged by the political leadership.[17] The question is rather whether the rumours had any basis in fact that might give substance to the politicians' claims that their enemies were bent on destroying the regime and reimposing Bourbon rule on France. Were the monarchies of Europe themselves planning an ideological war?

There is little evidence to suggest this, although the French had plenty of grounds for being nervous. The army was in a poor state, and foreign agents,

16 David Gates, *The Napoleonic Wars, 1803–1815* (London, 1997), p. 5.

17 David Andress, *Massacre at the Champ de Mars: Popular Dissent and Political Culture in the French Revolution* (London, 2000), pp. 85–108.

fuelled by British funds, were believed to be at work instigating discontent and subversion. The king's flight to Varennes had had an especially damaging effect on officers' loyalties, and by 1792 over one-third of French army officers had abandoned their posts.[18] Some had sworn their eternal enmity against the Revolution, and one could easily be persuaded that their intention was to offer their swords to the enemy and return with a royalist army at their backs. It was not only the radicals who believed this; the whole country felt vulnerable and exposed. The Austrians and Prussians helped feed these rumours by seeming to endorse the idea of an ideological war against France. In July 1791 the Austrian emperor issued his Padua Circular to the other monarchs of Europe, appealing for joint action to restore the French king's liberty; and, in August, Frederick William of Prussia and Leopold of Austria signed the Declaration of Pillnitz, seemingly promising a war to restore the Bourbons to the throne. The timing was ominous, and the language used at Pillnitz was unnecessarily aggressive, raising the spectre of an imminent invasion and the imposition of regime change. But the degree of international cooperation which such intervention would have required was never likely to be achieved.[19]

Nevertheless, the sabre-rattling language was enough to persuade deputies in the National Assembly that compromise with Leopold was impossible. The very existence of the Revolution and the liberties granted by the Constitution of 1791 were sufficient, they believed, to present a challenge to monarchs everywhere, turning the crowned heads of Europe into France's natural and necessary enemies. On 20 April, after a heated debate, France declared war, not on the Austro-Hungarian Empire and certainly not on the Austrian people, but specifically, and seemingly personally, on Leopold, the 'King of Hungary and Bohemia', because, they claimed, by his machinations he had broken the peace treaty that bound him to France, a treaty that went back to 1756. It was through this treaty, speakers claimed, that France had been drawn into wars in the interests of Austria and had spilt much blood in respecting its treaty obligations. But now Austria was reneging on the spirit of the treaty: in the words of the French general Charles-François Dumouriez, the Austrian royal house had lost all interest in its obligations 'from the moment when it saw that, by our constitution, France could no longer be the servile instrument of its ambition'; and hence, 'no longer being

18 Rafe Blaufarb, *The French Army, 1750–1820: Careers, Talent, Merit* (Manchester, 2002), pp. 85–91.

19 Blanning, *The Origins of the French Revolutionary Wars*, pp. 86–7.

able to dominate the French nation it became its implacable foe'.[20] The leading Girondin deputy Pierre-Victurnien Vergniaud maintained that the situation had become untenable, since Austria could break its treaty obligations towards France at any time, and seemingly with impunity. 'It is easy to see', he declared to warm applause from the deputies, 'that breaking this treaty is as necessary a revolution for Europe as the demolition of the Bastille was for France.'[21]

All that was missing was any evidence of Austria's hostile intent, which two other prominent Girondins, Jacques-Pierre Brissot and Armand Gensonné, appeared to provide in the excited speeches they delivered in May 1792, a month after France declared war on Austria. They described a plot that had been hatched to overthrow the Convention and reverse the gains of the Revolution, a plot conceived by the Austrian minister Kaunitz and coordinated in France by a supposed 'Austrian Committee' of the king's closest advisers. It was orchestrated, some alleged, by Marie-Antoinette herself. There was little evidence to support this allegation, of course, but that, Brissot reminded his audience, was the nature of conspiracy. Agents were at work; deputies were preparing to betray the Revolution; the king's inner circle, and perhaps the king himself, were implicated in a counter-revolutionary conspiracy, plotting an ideological war against France.[22] Many in Paris were eager to be convinced. For others the rumour merely confirmed their view of the Girondins as crude warmongers, prepared to plunge France into a European war if it would protect their political interests. What mattered, however, is that this gave an appearance of legitimacy to the war and undermined the arguments of those who had urged caution and pragmatism. Most notable among these was Maximilien Robespierre, who warned that war constituted a damaging distraction from the threats of counter-revolution and internal unrest. He reminded the deputies that the other states of Europe were unlikely to be inspired by a sense of ideological purpose, and that the drive towards war was coming from inside France. 'I ask myself who is proposing it, how, in what circumstances, and why?'[23]

20 *Archives Parlementaires*, première série (1789–1800), vol. 42, p. 196.

21 Henry Morse Stephens, *The Principal Speeches of the Statesmen and Orators of the French Revolution*, 2 vols. (Oxford, 1892), vol, 1, p. 280.

22 Timothy Tackett, 'Conspiracy Obsession in a Time of Revolution: French Elites and the Origins of the Terror, 1789–1792', *American Historical Review* 105:3 (2000), 691–713, 691–2.

23 Stephens, *The Principal Speeches*, vol. 2, p. 307.

## A new kind of army

European armies had become more professional and disciplined over the course of the eighteenth century as the authority of states had expanded and rulers had established a monopoly on military power. The most effective forces, such as that of Frederick the Great, had been forged into formidable military machines. Enthusiasm for the regime scarcely counted: it was said of Frederick that 'the secret of his troops' discipline in battle was that they feared death and the enemy less than they feared their NCOs and officers'.[24] Much the same could be said of the Austrian and Russian armies. Men fought from a variety of motives, ranging from patriotism to fealty to the officer class, or out of fear, duty, and institutional loyalty; in his recent re-examination of troop motivation across Europe, Ilya Berkovich talks of a 'counter-culture of honour' in eighteenth-century European armies that was present long before the French Revolution.[25] Armies depended heavily on the use of mercenaries, from countries like Sweden and Switzerland, Ireland and Scotland. This did not make them poor armies: Frederick's was regarded with awe by other generals, and its *esprit de corps* was widely admired. For its part, the French army was composed largely of long-term volunteers, supplemented by regiments of foreign mercenaries paid by the king, and by the hated *milice*, a reserve maintained at around 40,000–60,000 men in peacetime, recruited by ballot, often resentful and semi-trained, who were integrated into regular regiments in times of war.[26]

The Revolution posed a critical problem for an army of this kind. Could it be adapted to serve the cause of the people as the Revolution required? Could officers be trusted to transfer their oath from the king to the French people? Could free men be persuaded to submit to a military discipline that denied them much of their freedom? The first months after 1789 did not bode well: soldiers mutinied for better pay and conditions at Nancy, at Perpignan, and in other garrison towns, while officers left their regiments for Turin and Milan. The outbreak of war did not make things any clearer. Soldiers might now be fighting for a patriotic cause, but that did not mean that they were more inclined to obey orders they disliked. In 1792 a retreat before the Austrians near Tournai turned into a rout; rumours of treason then spread before the troops

24 Geoffrey Best, *War and Society in Revolutionary Europe, 1770–1870* (London, 1982), p. 32.
25 Ilya Berkovich, *Motivation in War: The Experience of Common Soldiers in Old Regime Europe* (Cambridge, 2017), p. 165.
26 John A. Lynn, 'The Pattern of Army Growth, 1445–1945', in John A. Lynn (ed.), *Tools of War: Instruments, Ideas and Institutions of Warfare, 1445–1871* (Urbana, 1990), p. 17.

turned on their commander, General Théobold Dillon, and lynched him.[27] Despite occasional victories – most notably at Valmy in November 1792, which promptly entered Revolutionary legend – within a few months the army the Revolutionaries had inherited from the Bourbons was in a pitiable state, distrusted by the government, and losing men at a frightening rate. By February 1793 it had been reduced to around 228,000 soldiers, the vast majority in line units, compared with over 400,000 just two months before.[28] In the other capitals of Europe Revolutionary politics was seen to have neutralised France's military threat.[29] The reforms that followed – the creation of a 'citizen army', the constitution of the 'nation-in-arms' to defend France against invasion – were born less of idealism than of desperation.

They were, however, dramatic enough to shock military establishments across Europe. In an age when men were routinely press-ganged into service, or commandeered from bars, country markets, and poorhouses, the French sought to create an army of genuine volunteers, of dedicated patriots. Their appeal was answered in 1791 and, to an extent, in the following year, but it soon became apparent that the voluntary principle was inadequate to the military needs of the nation. By 1793 a new source of manpower was desperately needed, and the Revolution turned first to a system of local levies, or *levées*, then to the *levée en masse* of August and the principle that the whole of French society was requisitioned for some form of war service. Every male citizen was liable to perform military service, in person, as one of the obligations of citizenship, though in practice the burden fell largely on bachelors and childless widowers between the ages of eighteen and twenty-five. Social groups, especially the urban bourgeoisie, who had never contributed much by way of military service, now found themselves forced to bear arms in the name of equality. And by the end of the decade, with numbers again falling through losses in battle and deaths through disease, the Directory regularised the system of recruitment to create annual rounds of conscription. This system would remain, with only minor changes, throughout the Napoleonic Wars, and for young Frenchmen, as for young men across the Empire, the fateful annual ballot became a fact of life, a rite of passage from adolescence to manhood. Conscription provided Napoleon

27 S. P. MacKenzie, *Revolutionary Armies in the Modern Era: A Revisionist Approach* (London, 1997), p. 41.

28 Samuel Scott, *The Response of the Royal Army to the French Revolution: The Role and Development of the Line Army, 1787–1793* (Oxford, 1978), p. 178.

29 Gunther Rothenberg, *Napoleon's Great Adversary: Archduke Charles and the Austrian Army, 1792–1814* (Staplehurst, 1985), p. 17.

with the mass armies with which he conquered most of the Continent, and by the end of the war others, most notably Prussia, followed France's example.[30] But it cannot be claimed that conscription helped to achieve national unity. Across Napoleonic Europe the incessant demands of the recruiting officer were widely resented by the young and their families, and draft-dodging, desertion, and village insurrection were festering sores which the Napoleonic state never succeeded in rooting out. The conscription laws remained widely contested, leading to a sharp rise in crime and banditry and canalising popular opposition to the state and its agents.[31]

More difficult was the challenge of renewing and remodelling an officer class weakened by emigration and politically suspect in the eyes of the nation. Like most of Europe, admission to the officer ranks in pre-revolutionary France had been reserved for the landed nobility and great store was placed in the natural bravery and instilled military values of the aristocratic families. Except for the artillery, where knowledge of mathematics was vital, the various branches of the army recruited exclusively from the nobility, who were consequently the only stratum of society who had been provided with officer training. After 1792, moreover, losses in battle were gouging holes in the officers' ranks, so that it became imperative to find some other principle on which to select future officers. Early defeats and the suspected treason of some of the existing generals (most shockingly Dumouriez, who surrendered his whole army to the Austrians) were enough to convince the new government that only the most drastic measures could retrieve the situation for the Republic. Patriotism and commitment to the Revolution were now judged alongside military proficiency in the field, as greater weight was placed on merit and ability. Generals were now appointed by the Convention, who took account of their political reliability and loyalty to the regime. Other officers were promoted from within the army, with the soldiers of each rank voting to elect those of their peers who should pass into the rank above, the government calculating that the men engaged in day-to-day fighting were best placed to recognise courage and strategic ability in others. Outstanding merit and long years of service were equally rewarded as the army was opened to career

30 Alan Forrest, 'Conscription as Ideology: Revolutionary France and the Nation-in-Arms', in Lars Mjøset and Stephen van Holde (eds.), *The Comparative Study of Conscription in the Armed Forces* (Oxford, 2002), pp. 99–105.

31 Alan Forrest, *Conscripts and Deserters: The Army and French Society during the Revolution and Empire* (Oxford, 1989), pp. 146–68.

soldiering, producing a revitalised officer corps of men with battle experience who could command the respect and affection of their men. The new processes produced some able young generals, too, like Lazare Hoche and Napoleon Bonaparte, both promoted to command armies while still in their twenties. They took their place alongside aristocratic officers inherited from the line army of the 1780s, who were able to lend their experience and knowledge of military science, but who remained distrusted and subjected to intense political scrutiny. During the Jacobin Republic, the Terror was carried to the armies as well as to civil society: the Armée du Nord alone lost four generals to the guillotine in 1793 and 1794.[32]

We should not, however, exaggerate the degree of continuity here, even if the line continued to play a major military role in the early Revolution. The years of war and political turbulence inevitably led to a rapid turnover of officers, so that by the beginning of 1794 only 38 of 500 officers who had served with Rochambeau in America were still in the French army.[33] And with the *amalgame* of 1793, the line as a separate entity disappeared, the old regiments being replaced by smaller, more manoeuvrable units or *demi-brigades*, which integrated the battle-hardened line troops inherited from Louis XVI with the new revolutionary volunteers. The names of regionally recruited regiments disappeared, as well, in many cases, as the local aristocrats who had commanded them. There was a clear desire to start again afresh. The continuities between the Revolution and the Empire were much stronger, since Napoleon had no interest in ridding himself of able, battle-seasoned officers and had little difficulty in converting them to the imperial cause. Some went on to serve him as marshals, the acme of the French military. Napoleon rewarded them with honours, lands, and titles to secure their loyalty and commitment by giving them a stake in the imperial order. Their social origins did not concern him: Jean Lannes, for instance, was the son of a peasant family from the south-west. But equally he refused to exclude men on grounds of privilege. Where the Revolutionaries had distrusted the sons of the nobility, Napoleon showed that he was prepared to integrate and promote them, but on merit only. He looked, too, to the schools to fashion

32 John A. Lynn, *The Bayonets of the Republic: Motivation and Tactics in the Army of Revolutionary France, 1791–94* (Urbana, 1984), pp. 13–14.

33 Samuel Scott, *From Yorktown to Valmy: The Transformation of the French Army in an Age of Revolution* (Denver, 1998), p. 182.

a more technocratic kind of officer, a man endowed with talent and ideals. By the end of the Empire, military values were firmly ensconced among the elite values of French society.[34]

Within the army itself, the officers and many of the men shared a new sense of military honour, one that combined a love of rewards and distinctions with a patriotic sense of the well-being of the nation. Many also expressed personal faith in the emperor, a faith not unconnected to his ability to deliver the victories that helped ensure their survival.[35] These were important aspects of morale-building, and they go far to explain the armies' successes in the early years of the Empire. But they could not prevent French troops from losing heart in the face of unending marches, damp, cold, and the fevers and gangrene that beset the wounded in military hospitals, or the melancholy of the battlefield, which explains so many broken lives and crushed spirits and the onset of what French medicine defined as 'nostalgia'.[36] Spirits as well as bodies were broken in war.

## The science of war

If the Revolutionary and Napoleonic Wars were notable for the deployment of larger massed armies in the field, it is more difficult to point to any technical or strategic innovations that might explain the success of French arms. Armies fought with the same weapons as in the 1760s and 1770s; and though they may have placed more emphasis on flexibility and speed, the tactics on the battlefield had not been revolutionised. Artillery pieces were lighter and more manoeuvrable than in previous wars, but they owed the innovation to Gribeauval and the Bourbons. Military drill and training remained largely as they were laid down in the infantry manual of 1791, which itself was deeply rooted in pre-revolutionary practice. And if the new army abandoned formal squares and battlefield formations, that was due less to a new philosophy of warfare than to the fact that the raw young soldiers raised by the Revolutionary levies lacked the levels of training needed to fight classic eighteenth-century battles. They were also short on equipment, arms,

34 Jean-Paul Bertaud, 'La Place de l'officier dans le régime napoléonien', in Paul Viallaneix and Jean Ehrard (eds.), *La Bataille, l'armée, la gloire, 1745–1871*, 2 vols. (Clermont-Ferrand, 1985), vol. 1, pp. 221–30.

35 Michael Hughes, *Forging Napoleon's Grande Armée: Motivation, Military Culture and Masculinity in the French Army, 1800–1808* (New York, 2011).

36 Thomas Dodman, *What Nostalgia Was: War, Empire, and the Time of a Deadly Emotion* (Chicago, 2018), pp. 93–123.

boots, and uniforms.[37] It was this, rather than the natural idealism of the citizen-soldier, which explains their preference for mass assaults, their dependence on loose skirmishing, and the Revolutionary cult of the bayonet. In 1793 whole brigades were deployed in skirmishing order, rushing pell-mell at the enemy. With time, of course, they adapted these tactics and became more sophisticated in their deployment, until by 1795 they were using 'a flexible combination of linear formations, attack columns, skirmishing and sniping'.[38] In French military memory, however, it is the Revolutionary soldiers' qualities of spontaneity and courage that have proved most enduring. In lessons to budding officers at St Cyr under the Third Republic, praise would be lavished on the civic conscience and the selflessness of the young soldiers of Year II; little, however, was said about their technique, little, at least, that could serve to train the new generation of officer cadets.[39]

Napoleon was seen quite differently, as a commander of great strategic subtlety whose careful preparation and incisive decision-making could turn a battle and brought the French a series of remarkable victories. Marengo, Ulm, Austerlitz, and Jena were studied and analysed by generations of French officers seeking inspiration for winning tactics on the battlefield, and Napoleon was widely held up as an example of military genius, his every aphorism praised for its innate wisdom. Above all, he was praised for the boldness of his vision, his preparedness to focus all his firepower on a target and attack. He believed in attacking wherever possible, seeking out battle and making full use of the artillery to pummel the enemy lines. Battles also lasted longer than in previous wars. Clausewitz, one of his greatest admirers, saw him as the epitome of a successful general because of this quality of boldness and fearlessness, throwing men forward at every opportunity. France owed its victories to this quality, not to any change in the character of its soldiers, for no people can fundamentally change its nature over a single generation.[40] 'Essentially war is fighting', Napoleon argued, adding that 'fighting is the only effective principle in the many-fold activities generally designated as war'. He believed that a general should 'achieve the maximum concentration of force and strike the enemy with

37 Gunther Rothenberg, *The Art of Warfare in the Age of Napoleon* (Bloomington, 1978), p. 99.
38 Ibid, p. 115.
39 Ernest Lavisse, *L'Armée à travers les âges. Conférences faites en 1898 à l'École Militaire de Saint-Cyr* (Paris, 1899), p. 200.
40 Raymond Aron, *Clausewitz, Philosopher of War* (London, 1983), p. 21; Bruno Colson, *Clausewitz* (Paris, 2016).

the maximum power. Defence ought to be adopted only if one is too weak to attack.'[41] It was Napoleon's boldness in the Italian Campaign and Frederick's in the Seven Years' War which made them the obvious paragons, in Clausewitz's view, of modern battle technique.

Clausewitz repeated this verdict in 1812 when he wrote that: 'We always have the choice between the most audacious and the most careful solution. Some people think that the theory of war always advises the latter. That assumption is false. If the theory does advise anything, it is the nature of war to advise the most decisive, that is, the most audacious.'[42] Napoleon perfectly encapsulated that quality of audacity, in marked contrast to several of the generals against whom he fought. Most notably, the Austrian army, which was Napoleon's most persistent adversary in the field, developed a reputation for its defensive approach to war and tactical conservatism in battle. The contrast was most vividly demonstrated at Ulm in 1805, where the tactical hesitancy of the Austrian general Karl Mack von Leiberich led to public outcry in Vienna and to Mack's subsequent arrest and court-martial.[43] An over-cautious approach could also affect army morale, as Clausewitz again observed. 'The soldier who is advancing', he wrote pertinently, 'becomes insensible even to danger, the one who is standing still loses his presence of mind.'[44]

Napoleon's brilliance in the field owed as much to planning as to audacity, with skilled staff work a key to his armies' success. Long before the Revolution the French had understood the need for a staff organisation that could direct large and often disparate bodies of men, and after the Seven Years' War they had not, as was customary, disbanded it, but had maintained it even in peacetime. Reforms had followed, and continued during the Revolution. Most notably, Pierre Bourcet proposed a streamlining of staff structures in the 1780s; Lazare Carnot instituted an informal *bureau topographique* in 1793 to take charge of logistical issues, and Louis-Alexandre Berthier reorganised the army field staff in 1796 to contain separate divisions for matters of discipline, logistics and engineering, intelligence and operations, and staff administration. But Napoleon, aided by Berthier, who became his chief of staff, went much further, developing his personal staff (or *maison*) into a staff within a staff, thus enabling him to

41 Azar Gat, *A History of Military Thought: From the Enlightenment to the Cold War* (Oxford, 2001), pp. 204–5.

42 Ibid., p. 207. 43 Rothenberg, *Napoleon's Great Adversary*, pp. 118–22.

44 Rory Muir, *Tactics and the Experience of Battle in the Age of Napoleon* (New Haven and London, 2000), p. 212.

collect intelligence on enemy positions and familiarise himself with the lie of the land before he committed his troops.[45] On the field, too, he oversaw a major streamlining of the all-arms divisions which he had inherited from the Revolution. These were now replaced by army corps consisting of infantry divisions with units of artillery and cavalry attached. As Claus Telp has shown, this brought multiple benefits, allowing the corps to be deployed across a broader front, thereby facilitating the outflanking of the enemy and attacks on the rear of the opposing army. In this way, he concludes, 'manoeuvre for the purpose of battle became easier as chances had improved of driving the enemy into a corner, severing his line of retreat and attacking him from several sides'.[46]

## Defeating Napoleon

Whereas these qualities could win Napoleon battles, they were insufficient to win a war fought over so many years and on so many fronts. For a time, it is true, he drove all before him, conquering the Austrians and Prussians and forcing the Russian tsar to seek peace. At Tilsit the two emperors effectively divided Europe between them, into competing French and Russian zones, with little regard for those they had defeated. Prussia, the principal loser after its mauling at Jena, was stripped of much of its territory and reduced to the status of a regional power within Germany. But by 1807 or 1808 Napoleon had reached the zenith of his power, and the other rulers of Europe began to see opportunities for revenge. The French were feeling the strain of such continuous fighting against so many opponents, while others had begun to emulate France's example and introduce their own programmes of reform to improve the quality and motivation of their troops. Across Europe conscription was either introduced – as in Prussia in 1806 – or reformed to meet the threat from France.[47] Napoleon's sureness of touch was beginning to fail, too, both tactically and strategically. His obsessive concern to undermine the British economy in order to destroy the country's appetite for war led him to bully most of Europe into imposing the Continental System and banning British goods, which many saw as running counter to their

45 Rothenberg, *The Art of Warfare*, p. 209.
46 Claus Telp, *The Evolution of Operational Art, 1740–1813: From Frederick the Great to Napoleon* (London, 2005), p. 69.
47 Donald Stoker, Frederick Schneid, and Harold Blanton (eds.), *Conscription in the Napoleonic Era: A Revolution in Military Affairs?* (London, 2009).

economic self-interest. It led to smuggling, banditry, and a widespread resentment of Napoleonic officialdom. And it was surely folly to launch an attack on Russia in 1812 when his armies were still embroiled in the frustrations of the Peninsular campaign: the removal of troops from Spain gave Wellington the opportunity he needed to attack a weakened French army, while the Russian generals showed both patience and tactical awareness by refusing to engage the French as they retreated, content to harry Napoleon's ragged, dishevelled army as it dragged itself westwards. The campaign ended in disaster for the emperor, with the estimated loss of around 570,000 men and 200,000 horses.[48] But the impact was not confined to the army. The scale of his defeat alerted others to Napoleon's vulnerability, producing a diplomatic revolution in Europe that would lead to the Allied victory at Leipzig and so to the end of the Empire itself. (See Map 20.)

There were other, more structural reasons for his change in fortunes. As Emperor, Napoleon was increasingly involved in affairs of state, and he was forced to leave command in the field to his marshals, some of whom proved to be indifferent tacticians. Napoleon had never learned to delegate decision-making, and he tended to promote men for their loyalty as well as their talent.[49] In the Peninsular War, French morale was sapped by brutal ambushes and the extreme cruelties of guerrilla warfare, which led to indiscipline and savagery in a campaign dogged by hatreds and atrocities. In Spain, but also in large parts of Central and Eastern Europe, rulers found they could appeal to patriotic sentiment and buttressed their campaigns with at least a modicum of popular support. Commercial controls and large-scale requisitioning had antagonised large sections of the population, including influential merchant and industrial communities who felt that their interests had been sacrificed to those of France, while those rulers who had rallied to Napoleon – in Bavaria and Saxony, for instance – often felt patronised and treated with disdain. The Allied powers had no intention of negotiating with Bonaparte: they pushed home their advantage, invading the east of France and fighting their way towards Paris. For many the issue was Napoleon himself, whom they saw as a danger to the peace of Europe. At the Treaty of Vienna his abdication and exile were not negotiable, even if France's position in the new European order was.

48 Richard Riehn, *1812: Napoleon's Russian Campaign* (New York, 1991), p. 395.
49 Jonathon Riley, *Napoleon as a General* (London, 2007), pp. 199–200.

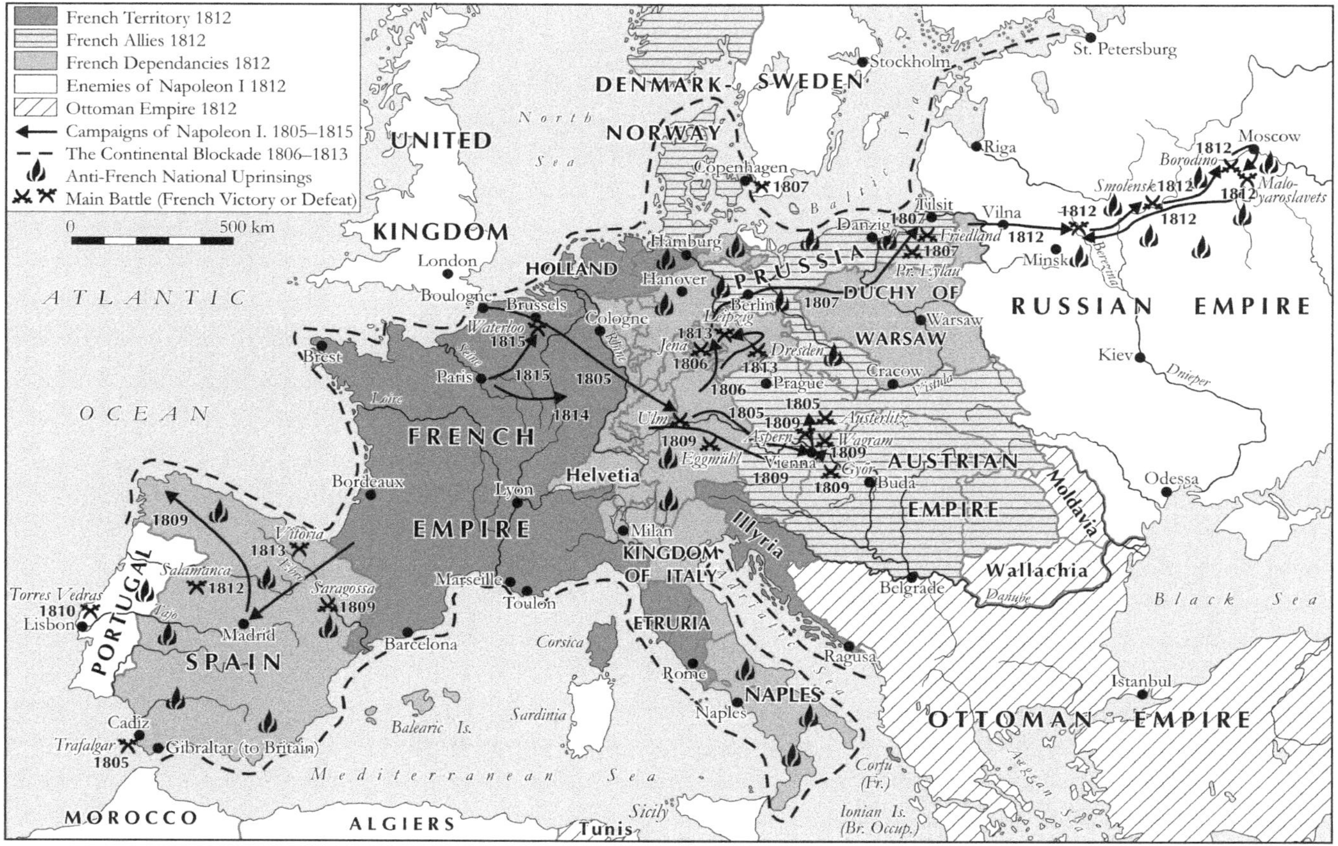


MAP 20. Wars in Europe between 1804 and 1815. Created by Béla Nagy.

For the leaders of the Allied powers Napoleon's responsibility for a quarter-century of bloodshed and destruction was not in doubt, and there could be no thought of compromise, no acceptance of his proposals for an extension of the Empire. The Allied leaders saw no reason to trust him to respect treaties or to maintain the peace. Besides, Napoleon was in no position to negotiate. France had been forced to surrender, and, as war gave way to negotiation, it was for the victorious Allies to dictate the future shape of Europe and the balance of world power.[50]

50 Mark Jarrett, *The Congress of Vienna and Its Legacy: War and Great Power Diplomacy after Napoleon* (London, 2013).

24

# European sea power and the origins of European domination, 1800–1850

ANDREW LAMBERT

Sea power, classically defined as a strategy to control communications, was an essential asset for the creation of European maritime empires, enabling them to secure seaborne trade, open new markets, and acquire territory.[1] It has often been conflated with *seapower*, a concept developed by the ancient Athenians to distinguish states like their own, whose economy, culture, and identity were enmeshed in the maritime sphere, from continental military powers like Sparta and Persia.[2] In this period Britain, the only seapower Great Power, created an extensive maritime empire outside Europe, one combining colonies of settlement, like Australia, with those of occupation, and informal imperialism based on economic power. While many Continental European powers created maritime empires after 1815, they did so in the knowledge that their possessions would be exposed to British naval power in the event of war, and tended to focus on land and security rather than commerce.

Historically seapower empires differed from those of continental states; seapower states were relatively small, and weak in human resources, emphasising the asymmetric advantage of sea control to compensate for those limitations. They tended to focus on port cities as trade hubs and naval stations, rather than territorial expansion, drawing revenues from trade, rather than land. These distinctions remained uneven: British rule in India became increasingly 'continental' after 1800, as the East India Company's business model shifted from trade to regional administration. Between 1800 and 1850 Britain only reluctantly acquired land holdings, preferring to focus on insular or peninsular ports and naval bases with limited terrestrial defence

1 Julian Corbett, *Some Principles of Maritime Strategy* (London, 1911).
2 Andrew Lambert, *Seapower States: Maritime Culture, Continental Empires and the Conflict That Made the Modern World* (New Haven and London, 2018).

needs. Down to 1875 British governments actively opposed the acquisition of new colonial possessions.

The central thesis of this chapter is that the intellectual and technological transformation of naval warfare that occurred between the 1790s and the 1850s, while driven by wars and rivalries between European great powers, proved to be most significant in redrawing the power relationships between those powers and the less developed world. Steam propulsion, artillery firing exploding shells, improved charts, and new medicines opened up new regions to Western exploitation.

The great power conflicts that reshaped the European balance of power between 1792 and 1815 were fought between existing imperial powers, including the United States, and focused on changing the ownership of existing imperial possessions, rather than securing new assets. Campaigns were waged in the Americas, in South-East Asia, and on the coastal margins of Africa. In 1814–15 British oceanic hegemony – a nexus of power combining strategic bases, legal structures, economic control, communications dominance, and overwhelming naval power – enabled Britain to determine the fate of the other extra-European empires. Britain restored Dutch power in Indonesia, along with French colonies in the Caribbean and Indian Ocean, but it retained critical strategic assets, such as Ceylon (Sri Lanka), Mauritius, and Trinidad. It ensured Spain's American empire did not fall into the hands of an imperial rival, namely France, Russia, or the United States, and upheld the legal basis of its power: the right to stop and search neutral shipping on the high seas in time of war.

The British seapower model had evolved across the preceding two centuries as a defensive mechanism to counter the threat posed by hegemonic European powers to English/British insular security. Unable to compete as a military great power, the British evolved a fiscal-military state, largely based on the Dutch experience of seapower imperialism, that harnessed the economic power of capital and commerce to the strategic needs of the state through a constitutionally limited monarchy in which the commercial classes wielded significant political power. After 1688 the Royal Navy became a truly national institution, funded and sustained by Parliament.[3] As the only true seapower state of the era, Britain prioritised the navy and trade, while rival

3 John Brewer, *The Sinews of Power: War, Money and the English State, 1688–1783* (London, 1989); Peter Cain and Anthony Hopkins, *British Imperialism: Innovation and Expansion 1688–1914* (London, 1993); Jan Glete, *Navies and Nations: Warships, Navies and State Building in Europe and America, 1500–1860* (Stockholm, 1993).

powers were obliged to focus on continental borders, and raise revenues from the land. This model was well established by the time the British were drawn into the Wars of the French Revolution in 1793.

## War, technology, and strategy

Britain spent the first five years of the French Revolutionary Wars trying to secure absolute sea control, as the basis of national security and global strategy. On 11 August 1798 Horatio Nelson annihilated the French Mediterranean Fleet in Aboukir Bay, on the coast of Egypt, destroying Napoleon's imperial dream of conquering Asia. This victory enabled British sea power to command the world's oceans for the next century and determined the fate of much of the extra-European world.

Nelson's victory depended on novel ideas, not new technology. His fleet went into action without forming the standard line of battle, leaving his captains to use their initiative within Nelson's permissive doctrinal parameters, which had developed over the preceding months. At the strategic level, Nelson insisted on the complete 'annihilation' of the enemy. He recognised that a new total warfare practised by the armies of the French Revolution rendered the limited methods of eighteenth-century naval warfare irrelevant. If sea power strategy was going to play a role in upholding the European balance, as it had for the past two centuries, it would have to have a much greater impact on land. This required navies to acquire something approaching an absolute command of the sea, so that they could focus on applying power on shore through economic blockade, amphibious landings, and direct assault. At the Nile Nelson took, sank, or burnt eleven of the thirteen French battleships, an unprecedented success that gave Britain control of the Mediterranean. To exploit this newly acquired control of the sea, Nelson split his squadron into three units, each more powerful than the remaining French force. Decisive tactical success had produced truly global strategic power. The fleet blockaded Alexandria, isolating Napoleon's army in Egypt, and escorted its prizes back to Gibraltar, while Nelson took three badly damaged battleships to neutral Naples, which possessed the only accessible dry dock in the Mediterranean. From Naples Nelson could control the eastern Mediterranean, maintain his grip on Egypt, support the blockade of Malta, and secure British trade with Ottoman Turkey. His arrival drew the Neapolitan kingdom into the war against France. Within months, Nelson's new sea power strategy of annihilation helped form a grand alliance against France.

Yet before he left Aboukir Bay Nelson had despatched an officer to India, via Syria and the Persian Gulf, to warn the British regime of the threat posed by an alliance between France and the Sultan of Mysore. Although Nelson's message only arrived in India after the Mysorean capital at Seringapatam had fallen to a British force, it did emphasise British India's absolute dependence on sea power, while his victory at the Nile paved the way for an ever-expanding British imperium on the subcontinent.[4]

Nelson's strategic insight and commitment to annihilation transformed British sea control into a useful counterweight to French military dominion in Europe. By controlling global communications and trade, Britain could sustain its war economy whilst denying France strategic and economic access to the rest of the world. Britain lacked the military power to fight France on the European mainland, but a sustained naval blockade could have devastating economic consequences.

Those economic consequences were felt by neutral states as well as belligerents. In 1800 the Russian-led Armed Neutrality enlisted Denmark, Sweden, and Prussia to back demands that Britain relax its policy of stopping neutral ships trading with the enemy to search their cargoes. The Royal Navy seized, and the High Court of Admiralty condemned, neutral ships that carried cargoes to France either from French colonies or from neutral ports. France desperately needed timber, rope, tar, canvas, and iron to build a new navy, which Baltic States could make a fortune providing them with, but Britain blocked the trade. Without those cargoes, the French navy could not challenge British sea control. Although the Armed Neutrality posed as an even-handed bid for peace, the Continental focus of its demands would have wrecked British strategy. Consequently, Britain negotiated by sending a battle fleet. Nelson's attack on Copenhagen in 1801 was calculated to force Denmark out of the Armed Neutrality, while the assassination of Tsar Paul a few days before the battle demonstrated Russia's dependence on trade with Britain.[5] In 1811 Tsar Alexander risked war rather than uphold the equally disastrous stipulations of Napoleon's Continental System, a counter-blockade that tried to exclude British commerce from Europe.

Nelson's victory at Copenhagen in 1801 secured the legal basis of British sea power, without concession to neutral powers. After 1807 this policy caused problems with the United States, leading to a declaration of war in 1812,

4 Andrew Lambert, *Nelson: Britannia's God of War* (London, 2004), pp. 106–60.
5 Ole Feldbæk, *Denmark and the Armed Neutrality 1800–1801* (Copenhagen, 1980).

although the Americans' primary concern was British imperial expansion. Once again, Britain refused to placate profit-seeking neutrals by disarming the sea power strategy on which it depended to address the existential threat posed by Napoleon.[6] When making peace with the United States at Ghent in December 1814, and helping reshape Europe at the Congress of Vienna, the British simply refused to discuss the legal rights of maritime belligerents, whose suppression had been the key to British use of sea power strategy across the preceding twenty-two years of high-tempo warfare. This position would remain the basis of national strategy for another century.[7]

Having secured the acquiescence of the Western world in their hard-line legal doctrines, the British were free to extend their oceanic dominion to the rest of the world. If France, Russia, and the United States had been obliged to accept British demands, the fate of African kingdoms, South American Republics, and even the Chinese Empire was never in doubt. After 1815 France and the United States constructed different legal regimes to contest British naval strategy, reflecting their shared experience as victims of sea power. This led them to block British attempts to create internationally agreed codes of law to define and facilitate the prosecution of slave ships, lest any further extension of British power damage legitimate commerce. Imperial Russia, while less obviously exposed, nevertheless shared their anxiety to curb the strategic value of sea control. As Britain moved to reduce tariff barriers and increase market access, these legal regimes proved critical.

## Imperial wars

After Nelson's crushing victory at Trafalgar, on 21 October 1805, French military power was confined to the European mainland. Sea power kept vital Baltic arteries for naval stores and grain open, and sustained a British army in Portugal and then Spain, an army fed on grain and beef from North America, and paid with bullion from South America. The British were fighting a global, maritime war. After Trafalgar they seized the imperial possessions of France,

6 Richard Hill, *The Prizes of War: The Naval Prize System in the Napoleonic Wars, 1793–1815* (Gloucestershire, 1998); Henry J. Bourguignon, *Sir William Scott, Lord Stowell* (Cambridge, 1987) for the evolution of British law on belligerent rights. Jan Lemnitzer, *Power, Law and the End of Privateering* (London, 2014); Andrew Lambert, 'Great Britain and Maritime Law from the Declaration of Paris to the Era of Total War', in Rolf Hobson, and Thomas Kristiansen (eds.), *Navies in Northern Waters: 1721–2000* (London, 2004), pp. 11–40.

7 Andrew Lambert, *The Challenge: Britain, America and the Naval War of 1812* (London, 2012); Andrew Lambert, *The Last Sailing Battlefleet: Maintaining Naval Mastery 1815–1850* (London, 1991), pp. 1–16.

Holland, and, before 1808, Spain. The Royal Navy exploited command of the sea to project strategic power in a series of operations: capturing islands, both off the European coast and beyond, securing critical positions to prevent their use by hostile naval forces, the temporary occupation and destruction of hostile naval bases, and punitive measures linked to ongoing diplomatic negotiations. Between 1806 and 1815 the British conquered the French West Indian islands, Mauritius, and the Dutch East Indies; occupied Lisbon; seized the Danish fleet; and destroyed Flushing and then Washington, DC.[8] These operations provided significant strategic advantages for the British, despite their limited military power, and were backed by countless small-scale joint operations and the development of effective methods of purely naval power projection.

The major limit on the application of naval power on the littoral remained the old conundrum that naval firepower was carried in big sailing ships, which could not operate safely near the coast. The only propulsion system available that did not depend on the wind, the oar, was limited to small craft. Existing systems of coastal defence in Europe and beyond – combining massive stone structures that dominated the deep-water passages into harbours, as well as smaller works – proved effective counters to boat attacks. Nelson famously remarked that none but a fool would fight a fort with a ship, but he was frequently obliged to do so. In the absence of serious naval opposition after Trafalgar, the Royal Navy could take more risks in coastal waters, risks that drove the evolution of a new type of naval warfare.

When the United States of America declared war on Britain in 1812, invading British North America (Canada) and attempting to filibuster Spanish Florida, Britain hit back from the sea. America's declaration of war had been propelled by Britain's imperial expansion, fears of slave insurrection, and Native American hostility. American ships challenged Britain on the world's oceans, raiding in the Atlantic, Arctic, Pacific, and Indian Oceans; the China seas; and the Caribbean. Britain responded by imposing the same naval and economic blockade that was undermining Napoleon's empire. Within two years America was bankrupt, and never again challenged British sea power. This small war offered some important pointers for the future.

8 Thomas Munch-Petersen, *Defying Napoleon: How Britain Bombarded Copenhagen and Seized the Danish Fleet in 1807* (Gloucestershire, 2007); Martin Robson, *Britain, Portugal and South America in the Napoleonic Wars: Alliances and Diplomacy in Economic Maritime Conflict* (London, 2010); Gordon Bond, *The Grand Expedition: The British Invasion of Holland in 1809* (Athens, GA, 1979).

Britain's changing strategic outlook after Trafalgar had focused attention on new weapons for coastal operations. Congreve rockets and Fulton torpedoes (in modern terms, mines) were experimented with, and additional mortar vessels were built. These craft mounted two 13-inch mortars, each firing 200-pound exploding shells, the most destructive projectiles of the age. They were the cutting edge of sea power, and it was no accident that the bulk of Britain's bomb vessels were off Baltimore's Fort McHenry in September 1814, supported by a rocket-firing ship.[9] This firepower was used against shore defences, even though it still depended on sail power.

As a global maritime trading power, the British were not greatly interested in land; rather, they wanted secure seas, market access, and increased trade. Seizing the Cape of Good Hope and later Java from the Dutch; Guadeloupe, Martinique, and Mauritius from the French; and cutting Spanish access to South and Central America removed any naval threat to oceanic commerce, turning the world's oceans into a British preserve. These imperial amphibious campaigns were critical for the development of European power projection. They were relatively easy to undertake because the British had already developed a chain of imperial bases and resources to support their warships and troops. The construction of dry docks at Mumbai in the mid-eighteenth century ensured British control of the Indian Ocean and India, as well as access to China.[10] Indian troops, moved on Indian merchant ships, secured Mauritius and Java, ending the privateer threat. The segue from war to peace can be traced in the career of East India Company employee Stamford Raffles, who moved from administering and exploiting the riches of Java to founding the great maritime entrepôt at Singapore after Java was returned to the Dutch.

Following the fiasco that ensued after they attempted to conquer Buenos Aires in 1808, the British rapidly adjusted to the new realities of the Spanish American wars of independence. The nascent republics opened their markets, relying on the British for capital, commerce, and protection. Naval power secured all the benefits of empire without the cost of direct rule. South and Central American bullion fuelled the British war effort against Napoleon after 1807. In the same year, the French invasion of Portugal had given the

9 Fredrick Winter, *The First Golden Age of Rocketry: Congreve and Hale Rockets of the Nineteenth Century* (Washington, DC, 1990), pp. 13–35.

10 Andrew Lambert, 'Strategy, Policy and Ship-Building: The Bombay Dockyard, the Indian Navy and Imperial Security in Eastern Seas, 1784–1869', in Hugh Bowen, Margarette Lincoln, and Nigel Rigby (eds.), *The Worlds of the East India Company* (Woodbridge, 2002), pp. 137–52.

British control of the richer markets of Brazil. British capital and industrial exports secured economic control of the newly independent states, while British naval might both secured them against reconquest and ensured that they paid their debts.[11]

The necessary step-change in the global spread of British naval power had been facilitated by the experience of operating over vast oceanic distances, in little-known waters. This prompted a dramatic increase in scientific and cartographic work. As the British opened new markets, they needed to secure the sea lanes from pirates and marine hazards.[12] By 1814 the British had a unique reservoir of expertise for operations outside Europe, and in the growing armies of British India they possessed the military resources to operate in Asia and East Africa.

## The other side

In wartime, sea power was a unique possession and post-1815 Britain occupied a unique position. Other Western powers found themselves simultaneously vulnerable to sea power and interested in exploiting their own sea power against weaker states, Western or non-Western. However, the threat of a naval or amphibious attack ensured that the ports or commercial harbours of every state that had reason to fear British power were heavily fortified. Throughout the nineteenth century, France, Russia, and the United States committed massive resources to fixed coastal defences. These stone structures provided the most compelling statement of the importance of sea power in global politics; an unwitting testament to the threat posed by sea power. Massive works at Cherbourg, Charleston, and Cronstadt may have marked the limits of naval might, but they also conceded the ocean to the British.[13]

The end of the Napoleonic and Anglo-American wars in 1815 transformed the position of most navies. After years spent under close blockade, impotent and ineffective, they were able to go to sea and promote their nation's

11 Robson, *Britain, Portugal and South America*; John Lynch, *The Spanish-American Revolutions, 1808–1826* (New York, 1973), pp. 337–8; Gerald Graham and Robin Humphreys (eds.), *The Navy and South America, 1807–1823: Correspondence of the Commanders-in-Chief on the South American Station* (London, 1962).

12 The voyage of HMS Beagle may be the best known of such voyages, but the catalogue given in Sir Archibald Day, *The Admiralty Hydrographic Service, 1795–1919* (London, 1967), demonstrates the global reach of the survey effort.

13 Edward Lewis, *Sea Coast Fortifications of the United States* (Annapolis, 1970); Edmond Thin, *Cherbourg. Bastion maritime du Cotentin* (Condé-sur-Noireau, 1990).

interests outside Europe. They emerged to assume a range of tasks that took them into new regions, the suppression of piracy, the conquest of empire, the extension of influence, and exploration. Britain, France, the United States, and Russia all sent out major oceanic expeditions to gather useful knowledge about the sea, astronomy, botany, geology, terrestrial magnetism, trade, and fishing.[14] National interest was engaged, navies gained status, and new regions were examined, as potential colonies or trading links. However, Britain easily maintained a two-power naval standard in battleships – maintaining a larger battlefleet than any two other navies together – ensuring sea control in the event of war.

After 1815 Western navies evolved along broadly similar lines; establishing a common sea power culture of aggressive commercial and imperial expansion, fostering Western values and ideas, from the abolition of slavery to the promotion of Christianity, and supporting the extension of international capital and the evolution of a globalised economy. The marginal spaces of wartime, where corsairs and pirates had operated, were taken under state control, in the interest of improved economic opportunity, reduced insurance premiums, and extended markets. Among those to feel the hard hand of Western sea power were the North African corsair regencies, the Al Quasimi of the Persian Gulf, privateers and pirates operating in the West Indies, Malay sea raiders, and Chinese and Moroccan pirates.[15] The extension of state control, improved warships, and enhanced power drove the threat away from key trade routes. The pirates were left to operate in distant backwaters, where the cost of control outweighed the benefits.[16] Britain upheld the right of states to issue privateer licences, because those states were liable for the actions of their agents.[17]

Western states also began to use sea power to conquer new empires; the French invasion of Algeria in 1830 and Russian use of naval assets to sustain

14 Hermann Viola and Carolyn Margolis, *Magnificent Voyagers: The US Exploring Expeditions, 1838–1842* (Washington, DC, 1965); contrast with Matthew Karp, *This Vast Southern Empire: Slaveholders at the Helm of American Foreign Policy* (Cambridge, MA, 2016).

15 Grace Fox, *British Admirals and Chinese Pirates: 1832–1869* (London, 1939); Charles Pennell, 'Dealing with Pirates: British, French and Moroccans, 1834–56', *The Journal of Imperial and Commonwealth History* 22 (1994), 54–83. Andrew Lambert, 'The Limits of Naval Power: *Three Sisters*, Moroccan Pirates and British Battleships', in Bruce Elleman, Andrew Forbes, and David Rosenberg (eds.), *Piracy and Maritime Crime: Historical and Modern Case Studies* (Newport, RI, 2012), pp. 173–90.

16 Muhammad Al-Qasimi, *The Myth of Arab Piracy in the Gulf* (London, 1986); C. E. Davies, *The Blood Red Flag: Investigation into Qasimi Piracy, 1797–1820* (Cornell, 1997).

17 Matthew McCarthy, *Privateering, Piracy, and British Policy in Spanish America 1810–1830* (Woodbridge, 2013).

their ongoing war in Circassia and Chechnia were typical examples. Navies were also used to increase leverage against weaker states; both France and Russia used warships to coerce Ottoman Turkey.[18] Britain and Holland, close allies in the post-1815 European state system, had very different relations in East Asia. The Dutch East Indies lay astride British trade routes to China, and the British had to remind the Dutch that they could not exclude British trade from the Malacca Straits. Superior sea power enabled Britain to dictate the outcome.[19]

The irruption of new powers into older political structures, one of the key results of European expansion, was not without negative consequences. The decline of imperial authority in China led to a major internal insurrection, the Taiping Rebellion, and the resultant chaos created ample space for a new wave of coastal piracy. When the pirates began to affect Western trade, mostly British and American, the response was swift, and usually highly destructive. From the perspective of non-Western states, these navies were effectively identical. Well-organised states like Japan recognised that resistance was futile. In the early 1850s the only question for Japan was which Western navy would force open the isolationist door, and how much force would have to be used.[20] Warships from the United States, Russia, and Britain arrived to secure treaties within twelve months.

## Steam

Commodore Matthew Perry's 'Black Ships' were led by paddlewheel steamships, a novelty in Japan. The technology was just over forty years old. Robert Fulton's *North River Steamboat* of 1809, the first economically successful steam passenger vessel, prompted a rush into the new technology. The great rivers and low population densities of North America made it ideal for steamshipping. In Britain the *Comet* was at work on the river Clyde in 1812, and by the time peace was restored steamships were ready to cross the English Channel. These early craft were small, mechanically unreliable, and expensive to run, but they promised to improve the speed and above all reliability of communications. Navies were already familiar with the steam engine; they were using steam to

18 John Daly, *Russian Seapower and 'The Eastern Question', 1827–1841* (London, 1991).
19 Gerald Graham, *The Politics of Naval Supremacy* (Cambridge, 1965), pp. 50–6.
20 John Schroeder, *Shaping a Maritime Empire: The Commercial and Diplomatic Role of the American Navy, 1829–1861* (Westport, 1985), pp. 41–2, 88–98.

pump out dry docks, drive rope works, and power bucket dredgers. By 1820 steamships were in regular use towing warships on rivers and harbours.

The first steamship to see war service was the American commercial vessel *Enterprise*, which moved stores and men during the New Orleans campaign of 1814–15. The war prompted Fulton to build the world's first steam warship, the harbour defence battery *Demologos*. Although incomplete when the war ended, it attracted the attention of several states that had felt the hard edge of British sea power, notably Denmark.[21] In his anxiety to counter British sea power, Fulton also deployed improvised explosive devices and underwater weapons against the blockade. Forty years later they would reappear as mines.

Once steamships began to operate in the short seas between Britain and Europe they spread around the European littoral, symbols of modernity and technological prowess that transformed the relationship between states and non-state actors, especially pirates. The superior mobility of the steamship countered the advantages of local actors – pirates and corsairs, with their local knowledge, light-draught vessels, and inshore hiding places. An upsurge of piracy in the Caribbean linked to the Latin American wars of independence threatened American and British economic interests. US Captain David Porter used the hired steamship *Seagull* to support his anti-piracy flotilla in 1822, with impressive results.[22] The steamship enabled major powers to clear the coasts of pirates, or local rivals, and push their control far into the continent. It quickly became the key instrument for imposing nation-state sea control on non-European coastal regions. Despite Porter's success, the United States only acquired first-class steam frigates in 1842 and was slow to build more.[23] The technology was expensive and dominated by British manufacturers.

In the first steam-powered imperial conflict, the British invasion of Burma, 1824–6, three steamships – the *Diana*, *Pluto*, and *Enterprise* – were used as river tugs, transports, and gunboats on the river Irrawaddy. After the annexation of Tenasserim the British East India Company used steamships on the Burmese

21 Donald Canney, *The Old Steam Navy*, vol. 1: *Frigates, Sloops, and Gunboats, 1815–1885* (Annapolis, 1990), pp. 3–7.

22 R. Gardiner and A. Lambert (eds.), *Steam, Steel and Shellfire: The Steam Warship, 1815-1905* (Michigan, 1992), p. 20; Benjamin Armstrong, *Small Boats and Daring Men: Maritime Raiding, Irregular Warfare, and the Early American Navy* (Norman, 2019).

23 Canney, *The Old Steam Navy*, vol. 1, pp. 11–16.

rivers and coasts, *Diana* steaming 500 miles inland after the war.[24] Steam enabled the British to deal with Burmese river craft, and helped alleviate the real problem of the war: disease.

The first Royal Navy steamship to deploy outside Europe was HMS *Lightning*, which accompanied a flotilla of bomb vessels sent to threaten Algiers.[25] While the diplomatic object was achieved by a show of force, the new technology would enable the bomb vessels to be placed in position to fire their weapons without waiting on winds and tides, or the muscle of rowers.

Although histories of technology have tended to impose determinist agendas on technological progress, this period was littered with dead ends and failed programmes. Among the more interesting were the man-powered paddle wheels devised by Royal Navy officers Ryder Burton and Charles Napier. They illustrate a critical point, often overlooked by historians of technology, that navies existed to conduct missions, not pioneer technology. By the 1820s the utility of mechanical power had been established, but the steamship remained a small, vulnerable instrument, quite unsuited to combat. The man-powered paddle wheel connected the new propeller to the muscle-power of the crew in a more efficient way than using large oars, or row boats. As warships needed large crews to man their guns, they had the manpower for tactical propulsion. The technology worked, and Napier used it in ways that reflected his experience with commercial steamships and coastal warfare, including towing battleships in a flat calm.[26] Had the development of the steam engine stalled in the mid-1820s this option might have been pursued further. His auxiliary paddle-wheel frigate offered a combination of fighting power, cruising range, and tactical mobility that would not be matched until the introduction of auxiliary screw propulsion in the late 1840s.

In the event, the development of the marine steam engine, and of improved propulsion systems, fostered by commercial objectives, ensured that, within a decade of Napier's trials, the screw-propeller steamship

24 Basil Greenhill (ed.), *The Advent of Steam: The Merchant Steamship before 1900* (London, 1993), pp. 28–30, 40.

25 Basil Greenhill and Anne Giffard, *Steam, Politics and Patronage: The Transformation of the Royal Navy 1815–54* (London, 1994), pp. 34–5.

26 Charles H. Ward-Jackson, 'Captain Burton's Method of Propelling Ships of War in a Calm', *Royal Society of Arts Journal* 136 (1988), 740–1; Sir William Napier (ed.), *The Navy: Its Past and Present State in a Series of Letters by Rear Admiral Sir Charles Napier KCB* (London, 1851), pp. 50–1.

*Archimedes* had transformed the possibilities of warship design.[27] By 1850 the steam-powered battleship, technically impossible with paddle-wheel propulsion, had become a fact. By retaining the full sailing rig, these ships restored the ability of fleets to conduct sustained blockades and make transoceanic passages, a combination exemplified in Isambard Kingdom Brunel's massive *Great Britain* of 1845. The Royal Navy, already hard at work on screw propulsion, paid close attention to this vessel, and hired Brunel to engineer the first steam screw warship, the sloop HMS *Rattler*. (See Figure 26.) In turn, Brunel used the naval trials data to design the stern lines of his own vessel, ensuring it success.[28] HMS *Rattler* was a classic imperial asset. In a short but global career, the *Rattler* would capture an Atlantic slaver and a Chinese pirate vessel.[29] However, the structural impact of screw propulsion, especially the vibration of the long, unbalanced propeller shaft, called into question the use of wooden hulls.[30] The superior strength and durability of iron hulls made them attractive for screw propulsion.

Once again, Napier had been among the first to see the way ahead. In 1819 he invested his wartime prize money in a project to run an iron steamship service on the river Seine. Although the project was a commercial failure, the 100-ton steamer *Aaron Manby* was the progenitor of every iron-hulled river gunboat from the East India Company's famous *Nemesis* onwards. Iron hulls could be built with a shallower draught than wooden alternatives and had a greater load-bearing capacity. The river gunboat became the workhorse of empire, a mobile artillery platform, troop transport, field hospital, supply depot, and symbol of prestige that secured and sustained nineteenth-century imperialism.[31]

In 1845 the Royal Navy faced a double crisis. The resurgent French navy was building large paddle-wheel frigates, which had to be countered, while the supply of shipbuilding timber and naval slipways on which to build were limited. Rather than build wooden warships in private shipyards, which did not apply the same high standards of timber seasoning, the navy ordered a fleet of iron frigates, both paddle-wheel and screw-propelled. Early

27 Greenhill, *The Advent* of Steam, pp. 7–9.

28 Andrew Lambert, 'Responding to the Nineteenth Century: The Royal Navy and the Introduction of the Screw Propeller', *History of Technology* 21 (1999), 1–28.

29 Andrew Lambert, 'HMS *Rattler*: Brunel's Warship in Service, 1845–1856', in Dennis Griffiths, Andrew Lambert, and Fred Walker (eds.), *Brunel's Ships* (London, 1999), pp. 105–26.

30 Andrew Lambert, *Battleships in Transition: The Creation of the Wooden Steam Battlefleet 1815–1860* (London, 1984).

31 Daniel Headrick, *The Tools of Empire: Technology and European Imperialism in the Nineteenth Century* (Oxford, 1982), pp. 17–57.

FIGURE 26. HM steam sloops *Rattler* and *Alecto* towing stern to stern for the purposes of testing the relative power of the screw propeller and the paddle wheel (1845). National Maritime Museum, Greenwich.

experience with British-built iron vessels suggested that they could resist cannon shot. However, data from conflicts in the warmth of China, Mexico, and Paraguay, was unreliable. Tests in Britain's cooler climate demonstrated that wrought iron hull plates hit by shot would shatter into shrapnel. The iron frigates were reduced to transports. HMS *Birkenhead* was wrecked on the coast of South Africa in 1852, carrying troops for an imperial conflict, with heavy loss of life.[32] However, the technology proved critical to the expansion of European control, since few non-Western states possessed the firepower to damage iron ships.

## Informal empire

In 1815 British statesmen used the Congress of Vienna to shape an empire that combined formal possessions – dominated by small islands and ports, which

32 Gardiner and Lambert (eds.), *Steam, Steel and Shellfire*, pp. 47–51.

enabled them to control access to a far larger economic hinterland – with the 'informal empire' of capital.

European rivals obliged to maintain large armies were unable to compete. While France secured an empire in Algeria, at enormous cost in lives and money, French policy makers recognised the colony would be lost in the event of an Anglo-French conflict, and Britain ensured France did not move into neighbouring Morocco, where Britain had major commercial and strategic interests. Many extra-European states became economic satellites, retaining internal political control, but massively indebted to British banks, and dependent on British purchasing power. Britain had learnt the hard way that informal rule was cheaper, and therefore more profitable, than either direct rule or private administration on the East India Company model.[33] Capital exports to new American republics secured British influence, then and substantial reciprocal trade provided interest on the loans and leverage for sea power. Naval force, applied with varying degrees of discretion, from presence on the horizon to full-scale bombardment, ensured debts were honoured and markets opened. New nations like Argentina, Brazil, and Chile were engaged in an economic system that required improved communications, and occasionally a little naval coercion, to sustain it through internal instability, market fluctuations, and local wars. In the 1840s Britain used naval power to control the process that created Uruguay, levered open the trade of the Rio Paraná, and crushed the Brazilian slaver fleet.[34] This last campaign reached a dénouement on 29 June 1850 when the steam sloop HMS *Cormorant* entered the river at Paranaguá, seizing three slave ships destined for the Congo. When a Brazilian fort opened fire the *Comorant* replied, and then burnt two of the slave ships. The remaining ship was condemned in a British Prize Court. Initially passions ran high, but Brazil 'was virtually powerless to resist the British Navy and . . . any attempt to do so could only aggravate the situation'. With vital coastal and export trades at the mercy of British warships, the government abandoned the trade.[35] Small British warships, could operate with impunity in the territorial waters of foreign powers

33 Thomas Otte, 'A Janus-Like Power: Great Britain and the European Concert, 1814–1853', in W. Pyta and P. Menger (eds.), *Das europäische Mächtekonzert. Friedens- und Sicherheitspolitik vom Wiener Kongreß 1815 bis zum Krimkrieg 1853* (Cologne, 2009), pp. 125–53.

34 Ronald Robinson, John Gallagher, and Alice Denny, *Africa and the Victorians: The Climax of Imperialism in the Dark Continent* (London, 1961), pp. 1–26; S. Roberts, *Charles Hotham: A Biography* (Melbourne. 1985), pp. 30–62.

35 Leslie Bethell, *The Abolition of the Brazilian Slave Trade: Britain, Brazil and the Slave Trade Question, 1807–1869* (Cambridge, 1970), pp. 310–12.

because behind them lay a mighty force. Informal empire worked because the British had the power to destroy local economies, and many capital cities, but preferred to use their power to promote 'free' trade, and the necessary concomitant of good government, with a light touch. To emphasise that these were universal values, Britain often worked with other Western nations, especially France and the United States, because they acted to extend trade, not to restrict it, in British interests. The ultimate symbol of British informal empire was a Royal Navy depot ship, usually an old Napoleonic-era frigate, stocked with food, spare parts, and – increasingly – coal, permanently moored at Rio de Janeiro, Valparaiso, and Callao.[36] This diminutive wooden British territory symbolised an unequal power relationship, one that relied on rapidly improving communications technology.

## Communications

The steamship transformed European empires by improving the speed and reliability of strategic communications with the metropole, communications that imposed control and provided security. They also linked distant markets into an increasingly global market, centred on London, which dominated the linked activities of steamship company formation, capital export, marine engineering, and telecommunications. As a seapower state, Britain had emphasised strategic communications for generations.[37] By 1830 the Royal Navy ran a monthly naval steamship service from Falmouth to Corfu, via Lisbon, Gibraltar, and Malta. By 1840 that network had been privatised and expanded to North America, the Caribbean, and India. British admirals on distant stations quickly adjusted their operational cycles to exploit the new links with London. Admiral Sir George Seymour commanded the Pacific and then North America and West Indies stations between 1845 and 1854, adjusting his cruising patterns to meet the mail steamers, and advising the government to adjust routes to improve their connectivity.[38] By 1850 the United States was developing a similar system, while Austria had a local variation running down the Adriatic and into the Black Sea.[39]

36 The practice dated back to the seventeenth century. David Lyon, *The Sailing Navy List: All the Ships of the Royal Navy – Built, Purchased and Captured 1688–1860* (London, 1993), pp. 344–7.

37 Tony Pawlyn, *The Falmouth Packets: 1689–1851* (Truran, 2001); Lambert, *Seapower States*.

38 Seymour of Ragley MS, Warwickshire Record Office, CR114A.

39 Freda Harcourt, *Flagships of Imperialism: The P&O Company and the Politics of Empire from Its Origins to 1867* (Manchester, 2006); Lawrence Sondhaus, *The Habsburg Empire and the Sea: Austrian Naval Policy 1797–1866* (Indiana, 1989), pp. 86–96, 112–14, 132.

Steamship companies were created to secure government funding by providing regular timetabled communications, which proved highly attractive to commercial interests. The larger steamships needed for oceanic passages provided space for passengers and cargo, both state and private. By the early 1840s, the British had timetabled routes to North America and Asia in operation. They transformed the speed and reliability of communication for diplomacy and war. The ability to move messages and men from London to China in two months transformed the way the British ran their empire; and their relations with extra-European powers. Nineteenth-century technologies rapidly increased the ability of global powers to control actions half a world away from the centre. Although private traffic could sustain key routes in Europe and the Mediterranean, the spread of imperial power often required more state support. Limited local infrastructure obliged Britain to use sailing packets for the South American mail into the 1850s.

The growing importance of British Asian trade, linking India, China, and Britain, and the communication needs of an increasingly imperial regime at Calcutta (Kolkotta) drove the development of steam communications. Despite the failure of early steam voyages, steam power proved successful in the Burma War, as did the development of shallow-draught iron steamships for river transport.[40] Nor was it a coincidence that the world's first all-steam navy was the Bombay Marine. The East India Company recognised that their system depended on fast and reliable communications with London, transforming their navy into a steamship service linking Suez and Bombay (Mumbai), and acquiring an essential coalling station at Aden in January 1839, just ahead of the advancing armies of Mehemet Ali, pasha of Egypt. Behind the Egyptian army lay the more worrying threats of high tariff barriers against British goods, and French influence.[41]

The ships used on the Bombay–Suez route were mostly Indian-built, but fitted with British steam machinery, and run by British officers and engineers.[42] In the 1850s this system was overtaken by the advance of commercial mail steamers, as the Peninsula & Oriental Line steadily expanded across the Indian and Pacific Oceans. This model of private enterprise working in tandem with the state was reproduced in various forms across the Western world. From the outset, these new quasi-national steamships provided critical force multipliers for naval operations. Every state that

40 Clive Dewey, *Steamboats on the Indus: The Limits of Western Technological Superiority in South Asia* (Oxford and New Delhi, 2014).
41 Graham, *The Politics of Naval Supremacy*, pp. 73–5.
42 Gardiner and Lambert (eds.), *Steam, Steel and Shellfire*, pp. 28–9.

subsidised them recognised that large, fast, and capacious steam packets, complete with professional crew, could be armed for use as cruisers or troopships, and mail contracts stipulated that the ships would be available on demand.[43] British mail steamers were used as strategic transports during the Crimean War, and the wars in India and China that followed. Steamships made Britain a global power.

## Operations

Initially steamships were restricted to non-combat roles. Fulton's *Demologos*, a purpose-built warship, had severe limitations, and only appealed to audiences opposed to sea power. The slow integration of steam into combat was driven by experienced naval officers. Whilst Captain John Ross RN was writing a practical treatise on steam warfare that addressed convoy operations and coastal bombardment, Captain Charles Napier was planning the first use of steam in an open-ocean naval battle.[44]

On 5 July 1833 Napier, commanding the Portuguese Constitutional Fleet in the civil war, engaged the larger and more heavily armed Absolutist Fleet off Cape St Vincent. He had sailed from Oporto with a small army, which landed in the Algarve. The enemy fleet responded. To secure command of the sea for the Constitutional party, before moving on to Lisbon, Napier used his extensive experience with steamships to plan a decisive battle. Steamers, hired from a British company, would tow his warships into battle. When the steamship engineers, who were civilians, demanded £2,000 a man before they would risk their lives in battle Napier had to wait for the wind to rise later in the day. Despite the lack of steamships, he still captured the entire enemy fleet, ensuring the Constitutional party won the war.[45] By the mid-1830s, steam had assumed a significant role in naval warfare, and navies began creating permanent corps of engineers, officers, and men, to run the machinery. This development can be linked to the professionalisation of naval gunnery. By 1840 the paddle-wheel steam warship had reached a development plateau in the steam frigate of 1,000 to 1,500 tons, armed with ten to twenty heavy guns, many

43 John Beeler, 'Ploughshares into Swords: The Royal Navy and Merchant Marine Auxiliaries in the Late Nineteenth Century', in G. Kennedy (ed.), *The Merchant Marine in International Affairs, 1850–1950* (London, 2000), pp. 5–30.

44 Napier (ed.), *The Navy*, traces the evolution of his thinking about steam warfare; Sir John Ross, *A Treatise on Navigation by Steam and an Essay towards a System of Naval Tactics Peculiar to Steam Navigation* (London, 1828).

45 Edward Elers-Napier, *Life and Correspondence of Admiral Sir Charles Napier KCB*, 2 vols. (London, 1862), vol. 1, pp. 215–16.

FIGURE 27. Bombardment of St. Jean d'Acre, 1840. Engraved by Charles Frederick de Brocktorff, from a painting by J. K. Wilson, 1841. National Maritime Museum, Greenwich.

firing exploding shells. These ships also carried large boats as the covers over their paddle wheels; these paddle box boats became the standard landing craft of the era, ideal for disembarking horses and artillery. British operations on the Syrian coast and on the coasts and rivers of China between 1840 and 1842 emphasised the utility of steam.[46] On the Syrian coast, steam power facilitated amphibious landings and direct attacks on coastal forts, including at St Jean d'Acre on 3 November 1840, where a shell, probably fired by the steamer HMS *Gorgon*, detonated an Egyptian magazine, bringing the action to a dramatic conclusion. (See Figure 27.)

During the Anglo-Chinese War of 1839–42, the British relied on East India Company steam warships for coastal and riverine operations. While the exploits of the iron paddle-wheel vessel *Nemesis* caught the headlines, it was the more numerous wooden-hulled steamships that enabled the Royal Navy to outflank Chinese coastal defences and tow an amphibious strike force up the Yangtze to Nanjing, where peace was made. (See Figure 28.)

46 Gerald S. Graham, *The China Station: War and Diplomacy 1830–1860* (Oxford, 1978).

FIGURE 28. Signing of the Treaty of Nanking, 29 August 1842, showing the Chinese and British signatories on board HMS *Cornwallis*. Painting by Capt. John Platt, 1846. Anne S. K. Brown Military Collection, Brown University.

Steam allowed the British to strike the Chinese Empire's weakest point, the rice supply from the Yangtze to Beijing. The operation was based on intelligence gathered over the preceding half-century by British travellers and naval officers. In both cases London exercised strategic control, and, in the case of Acre, issued operational orders using the reliable mail delivery provided by commercial steamships, which also conveyed an admiral to China in record time. Steam power provided fast and reliable communications along with tactical advantages that secured impressive strategic effect.

This experience was swiftly digested. The steamship improved reaction times, carrying troops from harbour to harbour, avoiding lengthy embarkation and disembarkation at sea, and improved the speed and reliability of amphibious landings, with paddle-box boats and fire support from shell-firing guns. In the 1840s all four major navies moved to steam, while the smaller fleets of Holland, Naples, and Spain found the steamship to be the ideal instrument for nations that needed to move troops to threatened areas, be they outlying Indonesian islands, potentially Carlist provinces in northern Spain, Imperial Cuba, or the rebellious island of Sicily.[47] Most of the ships, engines, and engineers were British. Steamships were used to great effect in the Crimean War, delivering men, stores, or firepower ashore. They were equally significant in imperial conflicts, where command of the sea was not in doubt.

In contrast to their high-profile role in imperial conflicts, paddle-wheel steamers were never front-line fighting ships, lacking the firepower to engage sailing battleships. They remained auxiliaries in sailing fleets, and an arm of exploitation once sea control had been secured. The full application of steam to naval warfare would require a shift in technology.

## Empires of understanding

While the steamship made it possible for the British to defeat China, the strategy of the conflict was based on a sophisticated, long-term analysis of the enemy. In the process of becoming the first truly global imperial power, Britain emphasised the collection and processing of intelligence – commercial, cartographic, cultural, and strategic. In this it followed the practice of earlier seapower empires, albeit on a larger scale. The Permanent Secretary of the Admiralty for most the period, Sir John Barrow, was also a leading geographer;

47 Carlos Alfaro Zaforteza, 'Sea Power, Strategy and Society in Liberal Spain 1833–1868', PhD diss. (Kings College London, 2011).

he had visited China and South Africa, and created the Royal Geographical Society as a locus for the gathering of useful knowledge. Links between state, navy, and society were necessarily close. Further support came from commercial interests, notably opium smugglers. The creation of the Hydrographic Office at the end of the eighteenth century acknowledged the need for charts of worlds beyond Europe. Under Captain Sir Francis Beaufort it became the centre of cartographic and scientific research that enabled Britain to use the sea ever more effectively for war and commerce. Steamships were in use for surveying by the late 1830s.[48]

## Trade defence and the Cherbourg strategy

In an attempt to counter British naval dominance, France began constructing a large artificial harbour and naval arsenal at Cherbourg in the 1780s. This project had been taken forward by Napoleon and had reached a critical stage in 1829. The French envisaged steam-powered cruisers operating in the English Channel. This threat prompted Britain to evolve an essential countermeasure: steam warships and the bases, engineering resources, and coal depots needed to sustain their operations. The threat of commercial raiding, rather than that of an invasion, drove the development of coastal offensive strategy between 1830 and 1850. The methods in use in 1815 were modernised and steam power improved access to coastal and riverine spaces, while new shell-firing guns gave steamships the firepower to inflict significant damage on shore defences as well as on ships. Henri Paixhans's French shell guns, which had been developed to sink British battleships, found a new role as the main armament of steam warships, for coastal power projection.

After the Syrian campaign of 1840, when it had been unable to protect its protégé Mehemet Ali, France tried to develop the strategic leverage to restrain Britain. This resulted in a major programme of paddle-wheel frigates, linked to the steam arsenal at Cherbourg, the only deep-water naval base on the Channel coast. The British response was systematic and comprehensive: a significantly larger fleet of steam cruisers, a coalling base at Portland to fuel a steam blockade, an advanced base at Alderney, and steam engineering workshops at Portsmouth and Plymouth. By 1850 the Royal Navy was confident it had the steam power to deal with any French challenge. In the process the British developed the screw propeller, which revolutionised blockade operations by enabling warships to combine the endurance of sail

48 Archibald Day, *The Admiralty Hydrographic Service 1795–1919* (London, 1967).

with the tactical mobility of steam, while small screw-propeller gunboats using heavy guns would bombard Cherbourg and other hostile naval bases. In 1845 the British gunboat prototype HMS *Fairy* tested the design, and nine years later it was mass produced to attack Cronstadt and St Petersburg.[49] Later, these gunboats were used in the Second Anglo-Chinese War of 1856–9, and mobilised to deter American aggression in 1861.[50] Ultimately, gunboats designed to attack Cherbourg and built to bombard Cronstadt imposed new trade and diplomatic arrangements on Qing China. They were equally effective in policing areas beyond the formal rule of Western powers. While this use was incidental and did not drive the development of sea power, it was nonetheless startling. The steam gunboat became the emblematic instrument of Western sea power, always backed by overwhelming naval forces.

The sudden emergence of the transatlantic iron-screw steamship, starting with the SS *Great Britain*, transformed the strategic mobility of Western armies. The *Great Britain* was the biggest ship in the world, and the fastest on long ocean passages. By 1850 such ships had transformed the number of troops that could be moved, carrying over 1,000 men, or large numbers of horses. The reliability of the large steamship ensured steady logistical back-up. In contrast to the position before 1815, large armies could now be moved quickly, landed easily, and sustained for long periods by commercial steamships. This development proved critical to the grand strategy of the Crimean War and American Civil War, as well as the two Anglo-Chinese conflicts. These ships provided the key to Allied success in the Crimean campaign. The strategic impact of sea power had never been greater. It would only be challenged in regions of the world where dense rail networks brought similar advantages to armies.

## The screw propeller

Although steam had been a feature of naval operations since the 1820s, the limitations of the paddle wheel restricted its use in extended oceanic operations and denied it any role in fleet battle. Whilst paddlewheel warships were ideal for coastal power projection, riverine operations,

49 Gardiner and Lambert (eds.), *Steam, Steel and Shellfire*, pp. 43–4.

50 Andrew Lambert, 'Winning without Fighting: British Grand Strategy and Its Application to the United States, 1815–1865', in Bradford Lee delee and Karl Walling (eds.), *Strategic Logic and Political Rationality: Essays in Honour of Michael J. Handel* (Newport, RI, 2003).

short sea communications, and towing, the paddle wheels and crankshaft were vulnerable to gunfire and reduced the broadside battery of vessels.[51]

The screw propeller integrated steam power into existing wooden sailing navies armed with full broadsides of cannon. Screw-propeller frigates and battleships restored the capability of dominant navies to blockade inferior forces in harbour. The new technology enabled France to shift the focus of its naval effort to address the vulnerability of its imperial prestige project in Algeria.[52] Algeria tied down much of the professional army, leaving the European frontier exposed to German armies. Unwilling to sacrifice Algeria, French policy makers shifted their naval effort to securing the sea route between Toulon and Algiers. The world's first fast screw-propeller battleship, *Le Napoléon*, was designed to escort troop convoys, rather than act in a regular battlefleet. Within half a decade the rapid increase in the number of screw battleships operated by Britain and France rendered such distinctions moot, but the high sustained speed of the French vessel ensured all new ships of this type were equipped with powerful engines.

Even before the full impact of the industrial revolution had been felt, naval design was already fragmenting into increasingly specialist forms. The United States built very large screw frigates rather than battleships. Many smaller navies shifted to coast-defence types, while the United States, Prussia, and Imperial Russia developed the ultimate response to the steam-powered coastal naval offensive: sea mines both contact-fused and electrically detonated. China created an area access denial system (to use modern terminology) based on forts and fire rafts in the First Anglo-Chinese War, though without success. Tested in 1848, the first serious use of sea mines occurred in the Crimean War, and they became effective ship-sinking weapons in the rivers of the Southern Confederacy.[53]

51 Andrew Lambert, 'Responding to the Nineteenth Century: The Royal Navy and the Introduction of the Screw Propeller', in Graham Hollister-Short (ed.), *History of Technology*, vol. 21 (London 1999), pp. 1–28.

52 On French notions of prestige, see Hugh Brogan, *Alexis de Tocqueville: A Life* (New Haven and London, 2006), esp. pp. 386–40

53 Robert Fulton, *Torpedo War, and Submarine Explosions* (New York, 1810); W. von Siemens, *Inventor and Entrepreneur: Recollections of Werner von Siemens* (London, 1966), pp. 56–70; Philip Lundeberg, *Samuel Colt's Submarine Battery: The Secret and the Enigma* (Washington, DC, 1974).

## New men

Between 1815 and 1850, naval manpower underwent a striking transformation. The British system of impressing seamen from merchant vessels in wartime had led to diplomatic tensions with the United States, was generally unpopular with the seafaring community, and subject to increasing attack by libertarian reformers. After 1830 Britain retained the right to impress, but it was never used after 1814. Similarly old disciplinary systems based on corporal punishment were wound down, the alcohol ration reduced, and good service payments introduced. Jack Tar was transformed into a Victorian hero. The man-of-war sailor became a distinct type, increasingly unrelated to the commercial seafarer. Professional gunnery training, instituted in 1830, introduced the idea of serving in the navy, rather than merely for the commission of a specific ship. By then, many British naval seamen joined the service as boys, and only served in warships. The Cherbourg strategy emphasised the need to man reserve ships quickly, to ensure the French base could be blockaded at the earliest opportunity. This prompted fresh thinking about rating careers. The abolition of the Navigation Acts in 1849, which had ensured that many British seamen served in merchant ships, completed the process. A report of 1852 recommended that future naval ratings enter for ten or twenty years of 'Continuous Service', creating a highly skilled, technically adept man-of-war sailor to serve the increasingly mechanised Victorian navy, and qualify for a pension.[54] The quality gap between these men and the short-service conscripts who crewed most other navies was significant.

## New bases

While steam power, screw propulsion, and iron hulls enhanced the role of naval power as a tool of Western imperialism, they required a whole new arsenal of supporting technologies. Imperial outposts needed steam factories, coal depots, dry-docks, ironworks, and engineering facilities. These were often shared with commercial steamship users, creating a synergy exemplified by the history of imperial dry-dock accommodation. The ability of the British state to exploit obvious commercial needs to support naval activity at the outer reaches of empire enabled it to sustain global naval power on low budgets.[55] While

54 Sarah Palmer, *Politics, Shipping and the Repeal of the Navigation Laws* (Manchester, 1990), pp. 63–5, 137.

55 For dry docks and imperial strategy, see Lambert, 'Strategy, Policy and Ship-Building'; Andrew Lambert, 'Wirtschaftliche Macht, technologischer Vorsprung und imperiale Stärke: Großbritannien als einzigartige globale Macht 1860 bis 1890', in

other states extended their dockyard capacity, none could hope to match the British network and this, when linked to British control of the best steam coal, made the cost of mounting a serious challenge to British sea power unfeasibly large. Britain had control of the world ocean and, in the absence of serious challengers, could run a very small peacetime fleet, mostly focused on applying sea power against continental states, be they advanced Western powers, the great nations of Asia, American Republics, or African chiefdoms. The same instruments and ideas drove the application of sea power at all levels. The only question was the scale and the cost.

## Conclusion

Steam-age sea power provided Western nations with a critical advantage. As European navies developed new ways of fighting each other, or in the British case to combat the coastal defences of other powers, these developments transformed their ability to engage extra-European powers. After centuries in which human resources, isolation, and extent had rendered vast empires like Qing China and Ottoman Turkey invulnerable to Western sea power, new systems of war and communication enabled the West to control regional trade and dominate non-Western coastal and riverine spaces. They acted in pursuit of trade, profit, or prestige. Between 1815 and 1854 Britain fought a series of wars with large and small powers, all of which were driven by the need to open or secure markets for rapidly expanding commerce. They were not driven by territorial expansion, in contrast to the wars and conquests of contemporary continental powers, notably France and Russia, which used naval power to support military campaigns leading to conquest in North Africa and the Caucasus. When Britain defeated Imperial China, it took Hong Kong, a small island, as a base for commercial and naval activity.

Steam power, screw propellers, and iron ships were not military developments, although they were rapidly harnessed to warlike ends. Down to 1850 these maritime developments, rather than any specifically naval or military technologies, were critical to the advance of European imperialism. Yet behind these new technologies the basis of sea power in 1850 was strikingly old-fashioned, resting as it did on fleets of sailing battleships: tools rarely exercised after 1805, but constantly renewed by Britain, France, and Russia.[56]

Michael Epkenhans and Gerhard Groß (eds.), *Das Militär und der Aufbruch in die Moderne 1860–1890* (Munich, 2003).

56 Lambert, *Battleships in Transition*; John Tredrea, and Eduard Sozaev, *Russian Warships in the Age of Sail, 1696–1860* (Barnsley, 2010); Riff Winfield and Stephen Roberts, *French*

Outside Europe the taste for sea power grew. In the 1830s the battleship arms race between Ottoman Turkey and Mehemet Ali's Egypt mirrored the Anglo-French rivalry. Lacking the financial resources to generate and sustain the necessary skilled manpower, however, these proved ephemeral. The oceanic empire of the sultan of Oman recognised the prestige value of the battleship, acquiring the Mumbai-built, seventy-four-gun *Liverpool* in 1826 to sail on an ocean ruled by Britain, and rarely visited by other powers. The ship was given to Britain in 1836, in exchange for a royal yacht.[57] In reality, none of these states aspired to challenge the Western battlefleet powers. Even the United States soon dropped out of the race, laying down its last battleship in 1822.[58] America focused on tools such as large cruising vessels and fixed defences to counter British seapower.

After 1815 Western powers used the enhanced capabilities of sea power in the steam era to invade and conquer new possessions, expand the frontiers of existing possessions or force open the hitherto closed markets of non-Western and other weaker states. Among the major wars, three exemplify the superior strategic utility of sea power. Most were conducted across land frontiers, but the Burma, Caucasian, and Mexican wars involved contiguous land frontiers. In the third case both states were 'Western', but Mexico had fallen behind its northern neighbour in the race to modernity. It ordered new iron steam warships from Britain, but failed to pay for them.[59] When the American overland invasion of Mexico faltered, an improvised amphibious strike from Veracruz secured peace, and a massive transfer of territory. The new frontier left Mexico dangerously exposed to American land power. These wars provided models for the invasion of the Crimea in 1854, the culmination of three decades of development which brought the wooden steamship to a development plateau and produced British naval and commercial steam fleets capable of lifting and sustaining armies on an unprecedented, strategic scale. This new reality was recognised by most observers, notably American army officer George B. McClellan, who would model his own Peninsula Campaign on the Crimea, while France was still planning large-scale amphibious operations in 1870. Both projects were compromised by the spread of rail networks in North America and Western Europe. The

*Warships in the Age of Sail, 1786–1861: Design, Construction, Careers and Fates* (Barnsley, 2015).

57 David Lyon and Riff Winfield, *The Sail and Steam Navy List: All the Ships of the Royal Navy 1815–1889* (London, 2004), p. 101.

58 Donald Canney, *Sailing Warships of the U.S. Navy* (London, 2001), pp. 110–11.

59 Thomas Wells, *Commodore Moore and the Texas Navy* (Austin, 1988).

American response to the devastating impact of British sea power during the War of 1812 was the construction of canals, roads, and increasingly railways to improve inland communications.[60] The costly American 'Third System' of coastal forts and the abandonment of capital ship construction reflected a profound shift as American cultural identity moved from maritime to continental.[61] America and Russia created costly coastal defences, behind which they embarked on decades of continental expansion, decades in which their imperium expanded further and faster than that of any maritime power.

The main limits to Western expansion remained overland distance and the ever-present threat of disease. Disease severely restricted the use of Western military power in sub-Saharan Africa, and compromised campaigns elsewhere. In an age before modern epidemiology, the most effective countermeasures were to avoid areas infested with disease-bearing insects, mostly mosquitoes, and uncertain water supplies. While steamships could limit the risks, leaving dangerous coasts before dusk, and providing distilled water, successive expeditions into Africa, from the *Congo* mission of 1816 to Macgregor Laird's Niger voyage, demonstrated that every step beyond the coastal margin was fraught with danger.[62] As the flawed British invasion of Afghanistan demonstrated all too clearly, the highland fastness of Central Asia, as well as Central Africa, remained beyond the influence of Western power, for war or for trade.

Even high-profile sea power successes like the First Anglo-Chinese War revealed deeper weaknesses. Qing China took a far-sighted view of Western sea power. Unable to compete with Britain on the maritime frontiers, the imperial regime recognised that Britain and other Western seapowers were interested in trade, rather than conquest or rule, and that their aims did not pose an existential threat to the regime. By contrast, Imperial Russia replaced the Mongols as Qing China's principal threat, slowly advancing overland and seizing provinces: it fundamentally imperilled the regime's survival. Consequently, the Chinese concentrated their naval defences in the north, to block access to Beijing, rather than in the south, leaving the trading ports of Canton and Shanghai exposed.[63] The great empires of the nineteenth century would be

60 Robert Angevine, *The Railroad and the State: War, Politics and Technology in Nineteenth Century America* (Redwood, 2004), pp. 6, 13, 42–54, 193–5, 217, 228; John Marszalek, *Commander of All Lincoln's Armies: A Life of General Henry W. Halleck* (Cambridge, MA, 2004), pp. 40–7.

61 Alex Roland, W. Jeffrey Bolster, and Alexander Keyssar, *The Way of the Ship: America's Maritime History Reenvisioned, 1600–2000* (Hoboken, 2008).

62 For the conquest of tropical disease, see Headrick, *The Tools of Empire*, pp. 58–79.

63 John K. Fairbank, 'Introduction: Maritime and Continental in China's History', in John K. Fairbank (ed.), *The Cambridge History of China*, vol. 12: *Republican China 1912–1949* (Cambridge, 1988), pp. 1–27.

continental Russia and America, effectively invulnerable to sea power in the short term: they were the heirs of Rome. Britain's 'Carthaginian' empire followed a very different model. Between these polar opposites lay the European maritime empires of France, Holland, Spain, and Portugal, some of which had been returned to their owners at the Congress of Vienna, and all were held on British sufferance. The British sea power monopoly enabled dominance of global commerce, and sustained the technological and information revolutions that propelled Western imperialism.

# *Select bibliography*

## The legacy of the nomadic empires


Allsen, Thomas T., "The Rise of the Mongolian Empire and Mongolian Rule in North China," in Herbert Franke and Denis Twitchett, eds., *The Cambridge History of China*, vol. 6: *Alien Regimes and Border States, 907–1368*, Cambridge, 1994.

Allsen, Thomas T., "The Circulation of Military Technology in the Mongolian Empire," in Nicola Di Cosmo, ed., *Warfare in Inner Asian History (500–1800)*, Leiden and Boston, 2002.

Avery, Peter, "Nādir Shāh and the Afsharid Legacy," in Peter Avery, Gavin Hambly, and Charles Melville, eds., *Cambridge History of Iran*, vol. 7: *From Nadir Shah to the Islamic Republic*, Cambridge, 1991.

Baipakov, K. M. and B. E. Kumekov, "The Kazakhs," in Chahryar Adle and Irfan Habib, eds., *History of Civilizations of Central Asia*, vol. 5: *Development in Contrast: From the Sixteenth to the Mid-Nineteenth Century*, Paris, 2003.

Biran, Michal, *Chinggis Khan*, Oxford, 2007.

Bregel, Yuri, *An Historical Atlas of Central Asia*, Leiden and Boston, 2003.

Bregel, Yuri, "The New Uzbek States: Bukhara, Khiva and Khoqand: *c.* 1750–1886," in Nicola Di Cosmo, Allen J. Frank, and Peter B. Golden, eds., *The Cambridge History of Inner Asia: The Chinggisid Age*, Cambridge, 2009.

Chase, Kenneth, *Firearms: A Global History to 1700*, Cambridge, 2003.

Cherniavsky, Michael, "Russia," in Orest Ranum, ed., *National Consciousness, History and Political Culture in Early-Modern* Europe, Baltimore and London, 1975.

Collins, L. J. D., "The Military Organization and Tactics of the Crimean Tatars, 16th–17th Centuries," in V. J. Parry and M. E. Yapp, eds., *War, Technology and Society in the Middle East*, London, 1975.

Di Cosmo, Nicola, "Introduction: Inner Asian Ways of Warfare in Historical Perspective," in Nicola Di Cosmo, ed., *Warfare in Inner Asian History (500–1800)*, Leiden and Boston, 2002.

Di Cosmo, Nicola, "Did Guns Matter? Firearms and the Qing Formation," in Lynn A. Struve, ed., *The Qing Formation in World-Historical Time*, Cambridge, MA, 2004.

Di Cosmo, Nicola, "The Qing and Inner Asia: 1636–1800," in Nicola Di Cosmo, Allen J. Frank, and Peter B. Golden, eds., *The Cambridge History of Inner Asia: The Chinggisid Age*, Cambridge, 2009.

Frank, Allen J., "The Qazaqs and Russia," in Nicola Di Cosmo, Allen J. Frank, and Peter B. Golden, eds., *The Cambridge History of Inner Asia: The Chinggisid Age*, Cambridge, 2009.

Kappeler, Andreas, *The Russian Empire*, trans. Alfred Clayton, Edinburgh, 2001.

Khodarkovsky, Michael, *Russia's Steppe Frontier: The Making of a Colonial Empire, 1500–1800*, Bloomington, 2002.

Manz, Beatrice F., *The Rise and Rule of Tamerlane*, Cambridge, 1989.

Manz, Beatrice F., "Nomads and Regional Armies in the Middle East," in Kurt Franz and Wolfgang Holzwarth, eds., *Nomad Military Power: Iran and Adjacent Areas in the Islamic Period*, Wiesbaden, 2015.

May, Timothy, *The Mongol Conquests in World History*, London, 2012.

McChesney, Robert D., "The Chinggisid Restoration in Central Asia: 1500–1785," in Nicola Di Cosmo, Allen J. Frank, and Peter B. Golden, eds., *The Cambridge History of Inner Asia: The Chinggisid Age*, Cambridge, 2009.

Millward, James A., "The Qing Formation, the Mongol Legacy, and the 'End of History' in Early Modern Central Eurasia," in Lynn A. Struve, ed., *The Qing Formation in World-Historical Time*, Cambridge, MA, 2004.

Miyakawi, Junko, "The Legitimacy of Khanship among the Oyirad (Kalmyk) Tribes in Relation to the Chinggisid Principle," in Reuven Amitai-Preiss and David Morgan, eds., *The Mongol Empire and Its Legacy*, Boston and Leiden, 1999.

Morgan, David, "The 'Great *Yasa* of Chinggis Khan' Revisited," in Reuven Amitai and Michael Biran, eds., *Mongols, Turks and Others: Eurasian Nomads and the Sedentary World*, Leiden and Boston, 2005.

Ostrowski, Donald, *Muscovy and the Mongols: Cross-Cultural Influences on the Steppe Frontier, 1304–1589*, Cambridge, 1998.

Perdue, Peter C., *China Marches West: The Qing Conquest of Central Eurasia*, Cambridge, MA, 2005.

Stevens, Carol B., *Russia's Wars of Emergence, 1460–1730*, Harlow, 2007.

Veit, Veronika, "The Eastern Steppe: Mongol Regimes after the Yuan (1368–1636)," in Nicola Di Cosmo, Allen J. Frank, and Peter B. Golden, eds., *The Cambridge History of Inner Asia: The Chinggisid Age*, Cambridge, 2009.

Woods, John E., *The Aqquyunlu: Clan, Confederation, Empire*, Minneapolis and Chicago, 1976.

## Ming conflicts with its neighbors, 1368–1644

Atwell, William, "The T'ai-ch'ang, T'ien-ch'i, and Ch'ung-chen Reigns, 1620–1644," in Frederick Mote and Denis Twitchett, eds., *The Cambridge History of China*, vol. 7 pt. 1: *The Ming Dynasty, 1368–1644*, Cambridge, 1988.

Crossley, Pamela Kyle, *The Manchus*, Oxford, 2002.

Dreyer, Edward, *Early Ming China: A Political History, 1355–1435*, Stanford, 1982.

Dreyer, Edward, *Zheng He: China and the Oceans in the Early Ming Dynasty*, New York, 2005.

Duyvendak, J. J. L., *China's Discovery of Africa*, London, 1949.

Franke, Wolfgang, *An Introduction to the Sources of Ming History*, Kuala Lumpur, 1968.

Hawley, Samuel, *The Imjin War*, Berkeley, 2005.

Huang, Ray, "The Lung-ch'ing and Wan-li reigns, 1567–1620," in Frederick Mote and Denis Twitchett, eds., *The Cambridge History of China*, vol. 7 pt. 1: *The Ming Dynasty, 1368–1644*, Cambridge, 1988.

Kangying, Li, *The Ming Maritime Trade Policy in Transition, 1368 to 1567*, Wiesbaden, 2010.

Kuno, Yoshi, *Japanese Expansion on the Asiatic Continent*, 2 vols., Berkeley, 1937, 1940.

Kwan-wai, So, *Japanese Piracy in Ming China during the Sixteenth Century*, East Lansing, 1975.

Levathes, Louise, *When China Ruled the Seas: The Treasure Fleet of the Dragon Throne, 1405–1433*, New York, 1994.

Liew, Foon Ming, *The Treatises on Military Affairs of the Ming Dynastic History (1368–1644)*, Hamburg, 1998.

Lo, Jung-pang, "The Decline of the Early Ming Navy," *Oriens Extremus* 5 (1958), 149–68.

Lorge, Peter, *The Asian Military Tradition: From Gunpowder to the Bomb*, Cambridge, 2008.

Mote, Frederick, "The T'u-mu Incident of 1449," in Frank A. Kierman and John K. Fairbank, eds., *Chinese Ways in Warfare*, Cambridge, MA, 1974.

Needham, Joseph and Robin D. S. Yates, *Science and Civilisation in China*, vol. 5 pt. 6: *Chemistry and Chemical Technology: Military Technology: Missiles and Sieges*, Cambridge, 1994.

Okada, Hidehiro, "China as a Successor State to the Mongol Empire," in Reuven Amitai-Preiss and David Morgan, eds., *The Mongol Empire and Its Legacy*, Leiden, 1999.

Parsons, James, *The Peasant Rebellions of the Late Ming Dynasty*, Tucson, 1970.

Pokotilov, Dmitrii, *History of the Eastern Mongols during the Ming Dynasty from 1368 to 1644*, trans. Rudolph Löwenthal, Chengtu, 1947–49.

Ricci, Matthew, *China in the Sixteenth Century: The Journals of Matthew Ricci, 1583–1610*, trans. Louis Gallagher, New York, 1953.

Robinson, David, *Bandits, Eunuchs, and the Son of Heaven: Rebellion and the Economy of Violence in Mid-Ming China*, Honolulu, 2001.

Rossabi, Morris, *The Jurchens in the Yuan and Ming*, Ithaca, 1982.

Rossabi, Morris, "Ming Foreign Policy: The Case of Hami," in S. Dabringshaus and R. Ptak, eds., *China and Her Neighbours*, Wiesbaden, 1997.

Rossabi, Morris, "The Ming and Inner Asia," in Denis Twitchett and Frederick Mote, eds., *The Cambridge History of China*, vol. 8 pt. 2: *The Ming Dynasty, 1368–1644*, Cambridge, 1998.

Roth, Gertraude, "The Manchu–Chinese Relationship," in Jonathan Spence and John Wills, eds., *From Ming to Ch'ing: Conquest, Region, and Continuity in Seventeenth-Century China*, New Haven, 1979.

Ryor, Kathleen, "*Wen* and *Wu* in Elite Cultural Practices during the Late Ming," in Nicola Di Cosmo, ed., *Military Culture in Imperial China*, Cambridge, MA, 2009.

Serruys, Henry, *Sino-Mongol Relations during the Ming*, vol. 3: *Trade Relations: The Horse Fairs*, Brussels, 1975.

Shin, Leo, *The Making of the Chinese State: Ethnicity and Expansion in the Ming Borderlands*, Cambridge, 2006.

Sun, E-tu Zen, *Tien-kung k'ai-wu: Chinese Technology in the Seventeenth Century by Sung Ying-hsing*, trans. Sun Shiou-chuan, University Park, PA, 1966.

Swope, Kenneth, "Clearing the Fields and Strengthening the Walls: Defending Small Cities in Late Ming China," in Kenneth Hall, ed., *Secondary Cities and Urban Networks in the Indian Ocean Realm*, Boulder, 2008.

Swope, Kenneth, "Restoring the Double-Edged Sword: Wanli as Supreme Military Commander," in David Robinson, ed., *Culture, Courtiers, and Competition: The Ming Court (1368–1644)*, Cambridge, MA, 2008.
Swope, Kenneth, *A Dragon's Head and a Serpent's Tail: Ming China and the First Great East Asian War, 1592–98*, Norman, 2009.
Swope, Kenneth, *The Military Collapse of China's Ming Dynasty*, Abingdon, 2013.
Taylor, Romeyn, "Yuan Origins of the Wei-so System," in Charles Hucker ed., *Chinese Government in Ming Times: Seven Studies*, New York, 1969.
Tong, James, *Disorder Under Heaven: Collective Violence in the Ming Dynasty*, Stanford, 1992.
Tsai, Shih-san Henry, *The Eunuchs in the Ming Dynasty*, Albany, 1996.
Wakeman, Frederic, *The Great Enterprise: The Manchu Reconstruction of the Imperial Order in Seventeenth-Century China*, 2 vols., Berkeley, 1985.
Waldron, Arthur, *The Great Wall of China: From History to Myth*, Cambridge, 1990.
Yi-t'ung, Wang, *Official Relations between China and Japan, 1368–1549*, Cambridge, MA, 1953.
Yuan, Tsing, "Urban Riots and Disturbances," in Jonathan Spence and John Wills, eds., *From Ming to Ch'ing: Conquest, Region, and Continuity in Seventeenth-Century China*, New Haven, 1979.

## Ottoman wars and military transformation, 1453–1826

Ágoston, Gábor, "Ottoman Warfare in Europe, 1453–1826," in Jeremy Black, ed., *European Warfare, 1453–1815*, New York, 1999.
Ágoston, Gábor, *Guns for the Sultan: Military Power and the Weapons Industry in the Ottoman Empire*, Cambridge, 2005.
Ágoston, Gábor, "Empires and Warfare in East-Central Europe, 1550–1750: The Ottoman–Habsburg Rivalry and Military Transformation," in Frank Tallett and D. J. B. Trim, eds., *European Warfare, 1350–1750*, Cambridge, 2010.
Ágoston, Gábor, "The Ottomans: From Frontier Principality to Empire," in John Andreas Olsen and Colin S. Gray, eds., *The Practice of Strategy: From Alexander the Great to the Present*, Oxford, 2011.
Ágoston, Gábor, *The Last Muslim Conquest: The Ottoman Empire and Its Wars in Europe*, Princeton, 2021.
Aksan, Virginia H., "Whatever Happened to the Janissaries? Mobilization for the 1768–1774 Russo-Ottoman War," in Virginia H. Aksan, ed., *Ottomans and Europeans: Contacts and Conflicts*, Istanbul, 2004.
Aksan, Virginia H., *Ottoman Wars, 1700–1870: An Empire Besieged*, Harlow, 2007.
Brummett, Palmira, *Ottoman Seapower and Levantine Diplomacy in the Age of Discovery*, Albany, 1994.
Faroqhi, Suraiya, ed., *The Cambridge History of Turkey*, vol. 3: *The Latter Ottoman Empire, 1603–1839*, Cambridge, 2006.
Finkel, Caroline, *The Administration of Warfare: The Ottoman Military Campaigns in Hungary, 1593–1606*, Vienna, 1988.

Hess, Andrew C., *The Forgotten Frontier: A History of the Sixteenth Century Ibero-African Frontier*, Chicago, 1978.

Heywood, Colin, *Writing Ottoman History: Documents and Interpretations*, Aldershot, 2002.

Imber, Colin, *The Ottoman Empire, 1300–1650: The Structure of Power*, 2nd ed., Basingstoke, 2009.

Kafadar, Cemal, "Janissaries and Other Riffraff of Ottoman Istanbul: Rebels without a Cause?," in Baki Tezcan and Karl K. Barbir, eds., *Identity and Identity Formation in the Ottoman World: A Volume of Essays in Honor of Norman Itzkowitz*, Madison, 2007.

Khoury, Dina Rizk, *State and Provincial Society in the Ottoman Empire: Mosul, 1540–1834*, Cambridge, 1997.

Kunt, İ. Metin, *The Sultan's Servants: The Transformation of Ottoman Provincial Government, 1550–1650*, New York, 1983.

Murphey, Rhoads, *Ottoman Warfare, 1500–1700*, New Brunswick, 1999.

Ostapchuk, Victor and Svitlana Bilyayeva, "The Ottoman Northern Black Sea Frontier at Akkerman Fortress: The View from a Historical and Archeological Project," in A. C. S. Peacock, ed., *The Frontiers of the Ottoman World*, Oxford, 2009.

Özbaran, Salih, *Ottoman Expansion toward the Indian Ocean in the 16th Century*, Istanbul, 2009.

Pitcher, Donald Edgar, *An Historical Geography of the Ottoman Empire from Earliest Times to the End of the Sixteenth Century*, Leiden, 1973.

Soucek, Svat, *Studies in Ottoman Naval History and Maritime Geography*, Istanbul, 2009.

Stavrides, Theoharis, *The Sultan of Vezirs: The Life and Times of the Ottoman Grand Vezir Mahmud Pasha Angelović (1453–1474)*, Leiden, 2001.

Sugar, Peter, "A Near-Perfect Military Society: The Ottoman Empire," in L. L. Farrar, ed., *War: A Historical, Political and Social Study*, Santa Barbara, 1978.

Tezcan, Baki. *The Second Ottoman Empire*, Cambridge, 2010.

Uyar, Mesut and Edward J. Erickson, *A Military History of the Ottomans from Osman to Atatürk*, Westport, 2009.

## Spain's wars and the rise of the Hispanic world empires, 1516–1828

Casey, James, *The Kingdom of Valencia in the Seventeenth Century*, Cambridge, 1979.

Chavez, Thomas, *Spain and the Independence of the United States: An Intrinsic Gift*, Albuquerque, 2002.

Earle, Rebecca, *Spain and the Independence of Colombia 1810–1825*, Exeter, 2000.

Elliott, John H., *Europe Divided, 1559–98*, 2nd ed., London, 2000.

Elliott, John H., *Imperial Spain, 1469–1716*, 2nd rev. ed., London, 2002.

Fernández-Armesto, Felipe, *The Spanish Armada: The Experience of War in 1588*, Oxford, 1988.

Fernández-Armesto, Felipe, *Millennium*, London, 1995.

Fernández-Armesto, Felipe, *Our America: A Hispanic History of the United States*, New York, 2014.

Fernández-Armesto, Felipe, Manuel Lobo Cabrera, and Antonio Rumeu de Armas, *Canarias e Inglaterra a través de la historia*, Las Palmas, 1994.

Gonzáles de León, Fernando, *The Road to Rocroi: Class, Culture and Command in the Spanish Army of Flanders, 1567–1659*, Leiden, 2009.
Goodman, David, *Spanish Naval Power, 1589–1665: Reconstruction and Defeat*, Cambridge, 2003.
Hämäläinen, Pekka, *The Comanche Empire*, London, 2009.
Hemming, John, *The Conquest of the Incas*, 2nd ed., London, 2004.
Kessell, John L., *Mission of Sorrows: Jesuit Guevavi and the Pimas, 1691–1767*, Tucson, 1970.
Mackay, Ruth, *The Limits of Royal Authority: Resistance and Obedience in Seventeenth-Century Castile*, Cambridge, 1999.
Macleod, Murdo J., "Self-Promotion: The *Relaciones de Méritos y Servicios* and Their Historical and Political Interpretation," *Colonial Latin American Historical Review* 7 (1998), 25–42.
Matthew, Laura E. and Michel R. Oudijk, eds., *Indian Conquistadors: Indigenous Allies in the Conquest of Mesoamerica*, Norman, 2007.
McNeill, John R., *Mosquito Empires: Ecology and War in the Greater Caribbean, 1620–1914*, Cambridge, 2010.
Monroy, David, *Thrown Among Strangers: The Making of Mexican Culture in Frontier California*, Berkeley, 1993.
Parker, Geoffrey, *The Army of Flanders and the Spanish Road: The Logistics of Spanish Victory and Defeat in the Low Countries Wars*, 2nd ed., Cambridge, 2008.
Restall, Matthew, *Maya Conquistador*, Boston, 1998.
Storrs, Christopher, *The Resilience of the Spanish Monarchy, 1665–1700*, Oxford, 2006.
Thomas, Hugh, *Conquest: Montezuma, Cortés and the Fall of Old Mexico*, New York, 1995.
Weber, David J., *The Spanish Frontier in North America*, New Haven and London, 1992.
Williams, Glyn, *Madoc: The Making of a Myth*, London, 1979.

## Wars of the European continent, 1500–1650

Baumgartner, Frederic, *Henry II, King of France*, Durham, NC, 1986.
Bireley, Robert S. J., *The Jesuits and the Thirty Years' War*, Cambridge, 2003.
Bonney, Richard, ed., *Economic Systems and State Finance*, Oxford, 1995.
Bracewell, Wendy, *The Uskoks of Senj: Piracy, Banditry, and Holy War in the Sixteenth Century Adriatic*, Ithaca, 1992
Buisseret, David, *Henri IV: King of France*, London, 1984.
Capponi, Niccolò, *Victory of the West: The Story of the Battle of Lepanto*, London, 2007.
Carroll, Stuart *Martyrs and Murderers: The Guise Family and the Making of Europe*, Oxford, 2009.
Clark, Christopher, *The Iron Kingdom: The Rise and Downfall of Prussia, 1600–1947*, London, 2007.
Crowley, Roger, ed., *Empires of the Sea: The Siege of Malta, the Battle of Lepanto, and the Struggle for the Center of the World*, New York, 2008.
Dávid, Géza and Pál Fodor, eds., *Ottomans, Hungarians, and Habsburgs in Central Europe: The Military Confines in the Era of Ottoman Conquest*, Leiden, 2002.
Fodor, Pál, ed., *The Battle for Central Europe: The Siege of Szigetvár and the Death of Süleyman the Magnificent and Nicholas Zrínyi (1566)*, Leiden and New York, 2019.

Frost, Robert, *The Northern Wars: State and Society in Northern Europe, 1558–1721*, Harlow, 2000.

Guilmartin, John, *Gunpowder and Galleys: Changing Technology and Mediterranean Warfare at Sea in the Sixteenth Century*, Cambridge, 1975.

Heal, Bridget and Ole Peter Grell, eds., *The Impact of the European Reformation*, Burlington, 2008.

Imber, Colin, *The Ottoman Empire: The Structure of Power, 1300–1650*, Basingstoke and New York, 2002.

Ingrao, Charles, *The Habsburg Monarchy, 1618–1815*, Cambridge, 1994.

Kaplan, Benjamin, *Divided by Faith: Religious Conflict and the Practice of Toleration in Early Modern Europe*, Cambridge, MA, 2007.

Knecht, Robert, *Richelieu*, Harlow, 1991.

Knecht, Robert, *Renaissance Warrior and Patron: The Reign of Francis I*, Cambridge, 1994.

Knecht, Robert, *The French Civil Wars, 1562–1598*, Edinburgh, 2000.

Knecht, Robert, *The Rise and Fall of Renaissance France, 1483–1610*, 2nd ed., London, 2002.

Kohler, Alfred, *Ferdinand I, 1503–1564*, Munich, 2003.

Lockhart, Paul Douglas, *Sweden in the 17th Century*, London, 2004.

Lockhart, Paul Douglas, *Denmark, 1513–1660: The Rise and Decline of a Renaissance Monarchy*, Oxford, 2007.

Mallett, Michael and John Hale, *The Military Organization of a Renaissance State: Venice*, Cambridge, 1984.

Moote, Alanson Lloyd, *Louis XIII, the Just*, Berkeley, 1991.

Pálffy, Géza, *The Kingdom of Hungary and the Habsburg Monarchy*, Boulder, 2009.

Pepper, Simon and Nicholas Adams, *Firearms and Fortifications: Military Architecture and Siege Warfare in Sixteenth-Century Siena*, Chicago, 1986.

Perjés, Géza, *The Fall of the Medieval Kingdom of Hungary: Mohács 1526–Buda 1541*, Boulder, 1989.

Pörtner, Regina, *The Counter-Reformation in Central Europe: Styria, 1580–1630*, Oxford, 2001.

Purcell, Brendan C., *The Winter King: Frederick V of the Palatinate and the Coming of the Thirty Years' War*, Aldershot, 2003.

Ragnow, Marguerite and William D. Philips, Jr., eds., *Religious Conflict and Accommodation in the Early Modern World*, Minneapolis, 2011.

Rodriguez-Salgado, Mia J., *The Changing Face of Empire: Charles V, Philip II, and Habsburg Authority, 1551–1555*, Cambridge, 1988.

Sonnino, Paul, *Mazarin's Quest: The Congress of Westphalia and the Coming of the Fronde*, Harvard, 2008.

Stein, Mark L., *Guarding the Frontier: Ottoman Border Fortresses and Garrisons in Europe*, New York and London, 2007.

Tenenti, Alberto, *Piracy and the Decline of Venice*, Berkeley, 1967.

Tracy, James, *Emperor Charles V, Impresario of War: Campaign Strategy, International Finance, and Domestic Politics*, Cambridge, 2002.

Tracy, James, *The Founding of the Dutch Republic: War, Finance, and Politics in Holland, 1572–1588*, Oxford, 2008.

Tracy, James, *Balkan Wars: Habsburg Croatia, Ottoman Bosnia, and Venetian Dalmatia, 1499–1617*, Lanham, 2016.

Van Zanden, Jan Luiten, Marjolein 't Hart, and Joost Jonker, eds., *A Financial History of the Netherlands*, Cambridge, 1997.
Wilson, Peter H., *Europe's Tragedy: The Thirty Years' War*, London, 2009.

## The evolution of naval warfare, 1450–1650

Anderson, Gary M. and Gifford Adam, Jr., "Privateering and the Private Production of Naval Power," *Cato Journal* 11 (1991), 99–122.
Andrews, Kenneth R., *Elizabethan Privateering: English Privateering during the Spanish War, 1585–1603*, Cambridge, 1964.
Beltrame, Carlo and Renato Gianni Ridella, eds., *Ships and Guns: The Sea Ordnance in Venice and Europe between the 15th and 17th Centuries*, London, 2011.
Bruijn, Jaap R. *The Dutch Navy of the Seventeenth and Eighteenth Centuries*, 2nd ed., Liverpool, 2011.
Caruana, Adrian B., *History of English Sea Ordnance*, vol. 1: *The Age of Evolution, 1523–1715*, Rotherfield, 1994.
Cipolla, Carlo M., *Guns, Sails, and Empires: Technological Innovation and the Early Phases of European Expansion, 1400–1700*, New York, 1965.
Engels, Marie-Christine, *Merchants, Interlopers, Seamen and Corsairs: The "Flemish" Community in Livorno and Genoa (1615–1635)*, Hilversum, 1997.
Fodor, Pál, "Between Two Continental Wars: The Ottoman Naval Preparations in 1590–1592," in *In Quest of the Golden Apple: Imperial Ideology, Politics, and Military Administration in the Ottoman Empire*, Istanbul, 2000.
Gardiner, Robert and Richard W. Unger, eds., *Cogs, Caravels and Galleons: The Sailing Ship 1000–1650*, London, 1994.
Gardiner, Robert and John Morrison, eds., *The Age of the Galley: Mediterranean Oared Vessels since Pre-Classical Times*, London, 1995.
Glete, Jan. *Warfare at Sea, 1500–1650: Maritime Conflicts and the Transformation of Europe*, 2 vols., London, 2000.
Guilmartin, John, "The Logistics of Warfare at Sea in the Sixteenth Century: The Spanish Perspective," in John A. Lynn, ed., *Feeding Mars: Logistics in Western Warfare from the Middle Ages to the Present*, Boulder, 1993.
Hammer, Paul Edward J., *Elizabeth's Wars: War, Government and Society in Tudor England, 1544–1604*, Basingstoke, 2003.
Hough, Jerry F. and Robin Grier, *A Long Process of Development: Building Markets and States in Pre-Industrial England, Spain and Their Colonies*, New York, 2015.
Imber, Colin, "The Reconstruction of the Ottoman Fleet after the Battle of Lepanto," in *Studies in Ottoman History and Law*, Istanbul, 1996.
Jones, James Rees, *The Anglo-Dutch Wars of the Seventeenth Century*, Abingdon, 2015.
Kirsch, Peter, *Fireship: The Terror Weapon of the Age of Sail*, trans. J. Harland, Barnsley, 2009.
Lane, Frederic C., "Technology and Productivity in Seaborne Transportation," in Anna Vannini Marx, ed., *Trasporti e sviluppo economico. Secoli XIII–XVIII*, Florence, 1986.
Lane, Frederic C., *Venetian Ships and Shipbuilders of the Renaissance*, Baltimore, 1992.
Lavery, Brian, *The Ship of the Line*, vol. 1: *The Development of the Battlefleet, 1650–1850*, London, 1983.

Loades, David, *The Tudor Navy: An Administrative, Political and Military History*, Aldershot, 1992.

Mahan, Alfred T., *The Influence of Sea Power upon History, 1660–1783*, 12th ed., Boston, 1918.

Martin, Colin, "The Ships of the Spanish Armada," in P. Gallagher and D. W. Cruickshank, eds., *God's Obvious Design: Spanish Armada Symposium, Sligo, 1988*, London, 1990.

Parker, Geoffrey and Colin Martin, *The Spanish Armada*, rev. ed., Manchester, 1999.

Phillips, Carla Rahn, *Six Galleons for the King of Spain: Imperial Defense in the Early Seventeenth Century*, Baltimore, 1986.

Pryor, John, *Geography, Technology, and War: Studies in the Maritime History of the Mediterranean, 649–1571*, Cambridge, 1988.

Rodger, Nicholas, *The Command of the Ocean: A Naval History of Britain, 1649–1815*, London, 2004.

Rodger, Nicholas, *The Safeguard of the Sea: A Naval History of Britain, 660–1649*, London, 2004.

Tunstall, Brian, *Naval Warfare in the Age of Sail: The Evolution of Fighting Tactic, 1650–1815*, London, 1990.

Wilson, Charles, *Profit and Power: A Study of England and the Dutch Wars*, The Hague, Boston and London, 1978.

Wing, John T., *Roots of Empire: Forests and State Power in Early Modern Spain, c. 1500–1750*, Leiden, 2015.

## Military finance, organisation, and state development, 1450–1650

Ailes, Mary Elizabeth, *Courage and Grief: Women and Sweden's Thirty Years War*, Lincoln, NE, 2018.

Black, Jeremy, *European Warfare 1494–1660*, London, 2002.

Bonney, Richard, *Political Change in France under Richelieu and Mazarin 1624–1661*, Oxford, 1978.

Bonney, Richard, ed., *The Rise of the Fiscal State in Europe c. 1200–1815*, Oxford, 1999.

Braddick, Michael, *The Nerves of State: Taxation and the Financing of the English State, 1558–1714*, Manchester, 1996.

Brandon, Pepijn, *War, Capital and the Dutch State (1588–1795)*, Leiden, 2015.

Brewer, John, *The Sinews of Power: War, Money and the English State, 1688–1783*, New York, 1989.

Downing, Brian, *The Military Revolution and Political Change*, Princeton, 1992.

Dreilichman, Mauricio and Hans-Joachim Voth, *Lending to the Borrower from Hell: Debt, Taxes and Default in the Age of Philip II*, Princeton, 2014.

Ertman, Thomas, *Birth of the Leviathan*, Cambridge, 1997.

Frost, Robert, *The Northern Wars 1558–1721*, Harlow, 2000.

Glete, Jan, *War and State in Early Modern Europe: Spain, the Dutch Republic and Sweden as Fiscal-Military States, 1500–1660*, London, 2002.

Godsey, William, *The Sinews of Habsburg Power: Lower Austria as a Fiscal-Military State*, Oxford, 2018.

Gonzáles de León, Fernando, *The Road to Rocroi: Class, Culture and Command in the Spanish Army of Flanders, 1567–1659*, Leiden, 2009.

Göse, Frank, *Otto Christoph Freiherr von Sparr 1605–1668*, Berlin, 2006.

Haberer, Stephanie, *Ott Heinrich Fugger (1592–1644)*, Augsburg, 2004.

Hochedlinger, Michael, *Austria's Wars of Emergence*, Harlow, 2003.
Hoffman, Phillip T. and Kathryn Norberg, eds., *Fiscal Crises, Liberty and Representative Government 1450–1789*, Stanford, 1994.
Howard, Michael, *War in European History*, Oxford, 1976.
Jörgensen, Christer, Michael F. Pavkovic, Rob S. Rice, Frederick C. Schneid, and Chris L. Scott, *Fighting Techniques of the Early Modern World* AD *1500*–AD *1763*, London, 2005.
Knox, MacGregor and Murray, Williamson, eds., *The Dynamics of Military Revolution 1300–2050* (Cambridge, 2001).
Lynn, John A., *Women, Armies and Warfare in Early Modern Europe*, Cambridge, 2008.
Manning, Roger B., *An Apprenticeship in Arms: The Origins of the British Army 1585–1702*, Oxford, 2006.
Mortimer, Geoffrey, *Wallenstein: The Enigma of the Thirty Years War*, Basingstoke, 2010.
Oestreich, Gerhard, *Neostoicism and the Early Modern State*, Cambridge, 1982.
Parrott, David, *Richelieu's Army: War, Government and Society in France, 1624–1642*, Cambridge, 2001.
Parrott, David, *The Business of War*, Cambridge, 2012.
Porter, Bruce, *War and the Rise of the State*, New York, 1994.
Rebitsch, Robert, *Matthias Gallas (1588–1647)*, Münster, 2006.
Redlich, Fritz, *The German Military Enterpriser and His Workforce*, 2 vols., Wiesbaden, 1964–65.
Rogers, Clifford, ed., *The Military Revolution Debate: Readings on the Military Transformation of Early Modern Europe*, Boulder, 1995.
Singer, Peter Warren, *Corporate Warriors: The Rise of the Privatized Military Industry*, 2nd ed., Ithaca, 2008.
Stein, Stanley J. and Barbara H. Stein, *Silver, Trade and War: Spain and America in the Making of Early Modern Europe*, Baltimore, 2000.
't Hart, Marjolein, *The Making of a Bourgeois State: War, Politics and Finance during the Dutch Revolt*, Manchester, 1993.
Tilly, Charles, "War Making and State Making as Organised Crime," in Peter B. Evans, Dietrich Rueschemeyer, and Theda Skocpol, eds., *Bringing the State Back In*, Cambridge, 1985, pp. 169–91.
Tilly, Charles, *Coercion, Capital and European States*, AD *990–1992*, Oxford, 1992.
Tracy, James, *Emperor Charles V, Impresario of War*, Cambridge, 2002.
Upton, A. E., *Charles XI and Swedish Absolutism*, Cambridge, 1998.
Wheeler, James Scott, *The Making of a World Power: War and the Military Revolution in Seventeenth-Century England*, Stroud, 1999.
Wilson, Peter H., *Europe's Tragedy: A History of the Thirty Years War*, London, 2009.
Wilson, Peter H., "War Finance, Policy and Strategy in the Thirty Years War," in Michael Rohrschneider and Anushka Tischer, eds., *Dynamik durch Gewalt? Der Dreißigjährige Krieg (1618–1648) als Faktor der Wandlungsprozesse des 17. Jahrhunderts*, Münster, 2018, pp. 229–50.
Wood, James, *The King's Army: Warfare, Soldiers, and Society during the Wars of Religion in France, 1562–1576*, Cambridge, 1996.
Wunder, H., "Finance in the Economy of Old Europe," in Peter-Christian Witt, ed., *Wealth and Taxation in Central Europe*, Leamington Spa, 1987.
Yun-Casalilla, Bartolomé and Patrick K. O'Brien, eds., *The Rise of Fiscal States: A Global History, 1500–1914*, Cambridge, 2012.

## Fortification and siege warfare

Alder, Ken, *Engineering the Revolution: Arms and Enlightenment in France, 1763–1815*, Princeton, 1997.

Arfaioli, Maurizio, *The Black Bands of Giovanni: Infantry and Diplomacy during the Italian Wars*, Pisa, 2005.

Arnold, Thomas, "Fortifications and the Military Revolution: The Gonzaga Experience, 1530–1630," in Clifford Rogers, ed., *The Military Revolution Debate: Readings on the Military Transformation of Early Modern Europe*, Boulder, 1995, pp. 201–26.

Buchanan, Brenda J., "The Art and Mystery of Making Gunpowder: The English Experience in the Seventeenth and Eighteenth Centuries," in Brett D. Steele and Tamara Dorland, eds., *The Heirs of Archimedes: Science and the Art of War through the Age of Enlightenment*, Cambridge, MA, 2005, pp. 233–74.

Bull, Stephen, *The Furie of the Ordnance: Artillery in the English Civil Wars*, Woodbridge, 2008.

Cipolla, Carlo, *Guns and Sails in the Early Phase of European Expansion 1400–1700*, London, 1965.

DeVries, Kelly. "The Impact of Gunpowder Weaponry on Siege Warfare in the Hundred Years War," in Ivy D. Corfis and Michael Wolfe, eds., *The Medieval City under Siege*, Woodbridge, 1995/99, pp. 227–44.

DeVries, Kelly, "Facing the New Technology: Gunpowder Defenses in Military Architecture before the Trace Italienne, 1350–1500," in Brett D. Steele and Tamara Dorland, eds., *The Heirs of Archimedes: Science and the Art of War through the Age of Enlightenment*, Cambridge, MA, 2005, pp. 37–71.

Duffy, Christopher, *Siege Warfare: The Fortress in the Early Modern World 1494–1660*, London, 1979.

Frazer, Augustus Simon, *Letters of Colonel Sir Augustus Simon Frazer KCB, Commanding the Royal Horse Artillery in the Army under the Duke of Wellington, Written during the Peninsular and Waterloo Campaigns*, ed. Major-General Edward Sabine, London, 1859.

Guicciardini, Francesco, *The History of Italy*, trans. Austin P. Goddard, London, 1754.

Guilmartin, John Francis, *Gunpowder and Galleys: Changing Technology and Mediterranean Warfare at Sea in the Sixteenth Century*, Cambridge, 1974.

Guilmartin, John Francis, "The Siege of Malta, 1565," in David Trim and Mark Charles Fissel, eds., *Amphibious Warfare 1000–1700: Commerce, State Formation and European Expansion*, Leiden, 2006, pp. 149–80.

Hale, John, "The Early Development of the Bastion: An Italian Chronology *c.* 1450–*c.* 1534," in John Hale, Roger Highfield, and Beryl Smalley, eds., *Europe in the Late Middle Ages*, London, 1965, pp. 466–94.

Hall, Christopher D., *Wellington's Navy: Sea Power and the Peninsular War, 1807–1814*, London, 2004.

Hanlon, Gregory, *The Twilight of a Military Tradition: Italian Aristocrats and European Conflicts, 1560–1800*, London, 1998.

Holmes, Richard, *Marlborough: England's Fragile Genius*, London and New York, 2008.

Horward, Donald D., *Napoleon and Iberia – The Twin Sieges of Cuidad Rodrigo and Almeida, 1810*, Tallahassee, 1984.

Hughes, Quentin, *Military Architecture*, London, 1974.

Israel, Jonathan, *The Dutch Republic: Its Rise, Greatness and Fall 1477–1806*, Oxford, 1995.

Jones, John Thomas, *Journals of Sieges Carried on by the Army under the Duke of Wellington in Spain during the Years 1811 to 1814*, 3 vols., London, 1846.

Langins, Janis, *Conserving the Enlightenment: French Military Engineering from Vauban to the Revolution*, Cambridge, MA and London, 2004.

Lipscombe, Nick, *The Peninsular War Atlas*, Oxford, 2010.

Lipscombe, Nick, *Wellington's Guns*, Oxford, 2013.

Lloyd, E. M., *Vauban, Montalembert, Carnot: Engineering Studies*, London, 1887.

Lynn, John A., "The Trace Italienne and the Growth of Armies: The French Case," in Clifford Rogers (ed.), *The Military Revolution Debate: Readings on the Military Transformation of Early Modern Europe*, Boulder, 1995, pp. 169–99.

Mallory, Keith and Arvid Ottar, *Architecture of Aggression: A History of Military Architecture in North West Europe 1900–1945*, London, 1973.

McGuffie, Tom Henderson, *The Siege of Gibraltar*, London, 1965.

Mintzker, Yair, *The Defortification of the German City, 1689–1866*, Cambridge, 2012.

Ostwald, Jamel, *Vauban Under Siege: Engineering Efficiency and Martial Vigor in the War of the Spanish Succession*, Leiden and Boston, 2007.

Parker, Geoffrey, *The Army of Flanders and the Spanish Road, 1567–1659*, Cambridge, 1972 and 2004.

Parker, Geoffrey, *The Military Revolution: Military Innovation and the Rise of the West, 1500–1800*, Cambridge, 1988 and 1996.

Parker, Geoffrey, "The Etiquette of Atrocity: The Laws of War in Early Modern Europe," in *Empire, War, and Faith in Early Modern Europe*, New York, 2002, pp. 144–68.

Pepper, Simon, "Castles and Cannon in the Naples Campaign of 1494–95," in David Abulafia, ed., *The French Descent into Renaissance Italy, 1494–95: Antecedents and Effects*, London, 1995, pp. 263–93.

Pepper, Simon, "The Face of the Siege: Fortification, Tactics and Strategy in the Early Italian Wars," in Christine Shaw, ed., *Italy and the European Powers: The Impact of War, 1500–1530*, Leiden and Boston, 2006, pp. 33–56.

Pepper, Simon and Nicholas Adams, *Firearms and Fortifications: Military Architecture and Siege Warfare in Sixteenth Century Siena*, Chicago, 1986.

Puype, Jan Piet and Marco van der Hoeven, eds., *The Arsenal of the World: The Dutch Arms Trade in the Seventeenth Century*, Amsterdam, 1996.

Saunders, Andrew, *Fortress Britain: Artillery Fortification in the British Isles and Ireland*, Liphook, 1989.

Setton, Kenneth M., *The Papacy and the Levant, 1204–1571*, 4 vols., Philadelphia, 1976).

Smith, Robert Douglas and Kelly DeVries, *The Artillery of the Dukes of Burgundy, 1363–1477*, Woodbridge, 2005.

Spiteri, Stephen C., *The Great Siege: Knights vs Turks MDLXV*, Tarxien, 2005.

Steele, Brett D. and Tamara Dorland, eds., *The Heirs of Archimedes: Science and the Art of War through the Age of Enlightenment*, Cambridge, MA, 2005.

Taylor, Frederick, *The Art of War in Italy, 1494–1529*, Cambridge, 1921.

*A True Historie of the Memorable Siege of Ostend . . . Translated out of French into English by Edward Grimestone*, London, 1604.

Van Nimwegen, Olaf, "Maurits van Nassau and Siege Warfare (1590–1597)," in Marco van der Hoeven, ed., *Exercise of Arms: Warfare in the Netherlands, 1568–1648*, Leiden, 1997, pp. 113–31.

Vauban, Sébastien Le Prestre de, *A Manual on Siegecraft and Fortification*, trans. George A. Rothrock, Ann Arbor, 1968.

Wood, James B., *The King's Army: Warfare, Soldiers, and Society during the Wars of Religion in France, 1562–1576*, Cambridge, 2002.

## European expansion in the Indian Ocean and Pacific, 1450–1850

Asher, Catherine Blanchard and Cynthia Talbot, *India before Europe*, Cambridge, 2006.

Barendse, Rene J., *The Arabian Seas: The Indian Ocean World of the Seventeenth Century*, London, 2002.

Black, Jeremy, *War in the World: A Comparative History 1450–1600*, Basingstoke, 2011.

Blussé, Leonard, *Visible Cities: Canton, Nagasaki and Batavia and the Coming of the Americans*, Cambridge, MA and London, 2008.

Boulle, Pierre H., "French Mercantilism, Commercial Companies and Colonial Profitability," in L. Blussé and F. Gaastra, eds., *Companies and Trade: Essays on Overseas Trading Companies during the Ancien Régime*, Leiden, 1981.

Casale, Giancarlo, *The Ottoman Age of Exploration*, Oxford, 2010.

Chaudhuri, Kirti N., *The English East India Company: The Study of an Early Joint-Stock Company 1600–1640*, London, 1965.

Chaudhuri, Kirti N., "Reflections on the Organizing Principle of Premodern Trade," in James D. Tracy, ed., *The Political Economy of Merchant Empires: State Power and World Trade, 1350–1750*, Cambridge, 1991.

Chaudhuri, Kirti N., "The English East India Company's Shipping (*c.* 1660–1760)," in Femme S. Gaastra and Jaap R. Bruijn, eds., *Ships, Sailors and Spices: East India Companies and Their Shipping in the 16th, 17th and 18th Centuries*, Amsterdam, 1993.

Darwin, John, *After Tamerlane: The Rise and Fall of Global Empires, 1400–2000*, London, 2008.

Fernández-Armesto, Felipe, *Pathfinders: A Global History of Exploration*, New York and London, 2007.

Furber, Holden, *Rival Empires of Trade in the Orient, 1600–1800*, Minneapolis, 1981.

Gaastra, Femme S. and Jaap R. Bruijn, "The Dutch East India Company's Shipping, 1602–1795, in a Comparative Perspective," in Femme S. Gaastra and Jaap R. Bruijn, eds., *Ships, Sailors and Spices: East India Companies and Their Shipping in the 16th, 17th and 18th Centuries*, Amsterdam, 1993.

Gaastra, Femme S., "Dutch, French and English Rivalry in Asia, 1688–1714," in *Guerres maritimes (1688–1713). IVes Journées franco-britanniques d'histoire de la marine* (Vincennes, 1996), pp. 227–43.

Glete, Jan, *Warfare at Sea, 1500–1650: Maritime Conflicts and the Transformation of Europe*, London, 2000.

Gupta, Ashin, *Indian Merchants and the Decline of Surat, c. 1700–1750*, Wiesbaden, 1979.

Kearney, Milo, *The Indian Ocean in World History*, London, 2004.

Knaap, Gerrit, Henk den Heijer, and Michiel de Jong, *Oorlogen overzee: Militair optreden door compagnie en staat buiten Europa, 1595–1814*, Amsterdam, 2015.

Koninckx, Christian, "The Swedish East India Company," in Femme S. Gaastra and Jaap R. Bruijn, eds., *Ships, Sailors and Spices: East India Companies and Their Shipping in the 16th, 17th and 18th Centuries*, Amsterdam, 1993.

Kruijtzer, Gijs, *Xenophobia in Seventeenth-Century India*, Leiden, 2009.

Levathes, Louise, *When China Ruled the Seas: The Treasure Fleet of the Dragon Throne, 1405–1433*, New York and Oxford, 1997.

McAleer, John, *Britain's Maritime Empire: Southern Africa, the South Atlantic and the Indian Ocean, 1763–1820*, Cambridge, 2017.

Özbaran, Salih, "Ottoman Naval Policy in the South," in Metin Kunt and Christine Woodhead, eds., *Süleyman the Magnificent and His Age: The Ottoman Empire in the Early Modern World*, London and New York, 1995.

Parker, Geoffrey, "Europe and the Wider World, 1500–1700: The Military Balance," in James Tracy, ed., *The Political Economy of Merchant Empires: State Power and World Trade, 1350–1750*, Cambridge, 1991.

Pearson, Michael, *The Indian Ocean*, London, 2003.

Pendse, Sachin, "Maratha Sea Power," in Christian Buchet and Gérard Le Bouëdec, eds., *The Sea in History: The Early Modern World*, Woodbridge, 2017.

Phillips, J. R. S., *The Medieval Expansion of Europe*, 2nd ed., New York and Oxford, 1998.

Scammel, Geoffrey V., *The First Imperial Age: European Overseas Expansion c. 1400–1715*, London, 1989.

Sicking, Louis, "Funduq, Fondaco, Feitoria. The Portuguese Contribution to the Globalisation of an Institution of Overseas Trade," in Giampiero Nigro, ed., *Reti marittime come fattori dell'integrazione europea/Maritime Networks as a Factor in European Integration* (Florence, 2019), pp. 198–207.

Subrahmanyam, Sanjay, *The Political Economy of Commerce: Southern India 1500–1650*, Cambridge, 1990.

Subrahmanyam, Sanjay, *The Portuguese Empire in Asia 1500–1700*, Oxford, 2012.

Webster, Anthony, *The Twilight of the East India Company: The Evolution of Anglo-Asian Commerce and Politics, 1790–1860*, Woodbridge, 2009.

Wolters, O. W., *The Fall of Srivijaya in Malay History*, London, 1970.

## Conflict in Northeastern Europe and the emergence of Russia to 1721

Ålberg, Alf, "The Swedish Army from Lützen to Narva," in Michael Roberts, ed., *Sweden's Age of Greatness 1632–1718*, New York, 1973.

Black, Jeremy, *War and the World: Military Power and the Fate of Continents, 1450–2000*, New Haven and London, 1998.

Bushkovitch, Paul, *Peter the Great: The Struggle for Power, 1671–1725*, Cambridge, 2001.

Davies, Brian L., *Warfare, State and Society on the Black Sea Steppe*, London, 2007.

Filipczak-Kocur, Anna, "Poland-Lithuania before Partition," in Richard Bonney, ed., *The Rise of the Fiscal State in Europe, c. 1200–1815*, Oxford, 1999.

Friedrich, Karin, *Brandenburg-Prussia, 1466–1806: The Rise of a Composite State*, London, 2012.

Frost, Robert, *The Northern Wars, 1558–1721*, London, 2000.

Frost, Robert, "The Death of Military Culture? The Citizen Army and the Military Failure of the Polish Lithuanian Commonwealth, 1648–1717," in Winfried Speitkamp, ed., *Gewaltgemeinschaften. Von der Spätantike bis ins 20. Jahrhundert*, Göttingen, 2013.

Glete, Jan, *War and the State in Early Modern Europe, 1500–1660*, London, 2002.

Glete, Jan, "The Swedish Fiscal-Military State in Transition and Decline, 1650–1815," www2.historia.su.se/personal/jan_glete/Glete-Swedish_Fiscal_Mil_State_Trans.pdf (2006).

Hellie, Richard, *Enserfment and Military Change in Muscovy*, Chicago, 1971.

Hennings, Jan, *Russia and Courtly Europe: Ritual and the Culture of Diplomacy, 1648–1725*, Cambridge, 2016.

Kollmann, Nancy Shields, *The Russian Empire, 1450–1801*, Oxford, 2017.

Kupisz, Dariusz, "The Polish-Lithuanian Army in the reign of King Stefan Báthory," in Brian L. Davies, ed., *Warfare in Eastern Europe, 1500–1800*, Leiden, 2012.

Lockhart, Paul D., *Sweden in the Seventeenth Century*, London, 2004.

Lockhart, Paul D., *Denmark, 1513–1660*, Oxford, 2007.

Martin, Janet, "Tatars in the Muscovite Army during the Livonian War," in Eric Lohr and Marshall Poe, eds., *Military and Society in Russia, 1450–1917*, Leiden, 2002.

Martin, Janet, "The Economic Effectiveness of the Muscovite Pomest'e System," in Brian L. Davies, ed., *Warfare in Eastern Europe, 1500–1800*, Leiden, 2012.

Murphey, Rhoads, "Ottoman Military Organization in South-Eastern Europe, *c.* 1420–1720," in Frank Tallett and David Trim, eds., *European Warfare, 1350–1750*, Cambridge, 2010.

Plakans, Andreas, *A Concise History of the Baltic States*, Cambridge, 2011.

Plokhy, Serhii, ed., *Poltava 1709: The Battle and the Myth*, Cambridge, MA, 2012.

Renner, Johannes, *Livonian History, 1556–1561*, ed. and trans. J. C. Smith, W. Urban, and J. W. Jones, Lewiston, 1997.

Scott, Hamish, "The Great Northern War (1700–21) and the Integration of the European States System," in John B. Hattendorf, Åsa Karlsson, Margriet Lacy-Bruijn, Augustus J. Veenendaal, Jr., and Rolof van Hövel tot Westerflier, eds., *Charles XII: Warrior King*, Oxford, 2019.

Stevens, Carol B., *Soldiers on the Steppe*, Dekalb, 1996.

Stevens, Carol B., *Russia's Wars of Emergence, 1460–1730*, London, 2007.

Stevens, Carol B., "Land Warfare," in Hamish Scott, ed., *The Oxford Handbook of Early Modern European History, 1350–1750*, Oxford, 2015.

Stone, Daniel, *The Polish-Lithuanian State, 1386–1795*, Washington, 2001.

Tallett, Frank, and David Trim, eds., *European Warfare, 1350–1750*, Cambridge, 2010.

Upton, Anthony, *Charles XI and Swedish Absolutism, 1660–1697*, Cambridge, 1998.

Wilson, Peter H., *German Armies: War and German Society 1648–1806*, London, 1998.

## The military in Safavid Iran, 1501–1736

Durand-Geudy, David. "Goodbye to the Türkmen? The Military Role of Nomads in Iran after the Saljuq Conquest," in Kurt Franz and Wolfgang Golzwarth, eds., *Nomad Military Power in Iran and Adjacent Areas in the Islamic Period*, Wiesbaden, 2015.

Floor, Willem, *Safavid Government Institutions*, Costa Mesa, 2001.

Matthee, Rudi, "Unwalled Cities and Restless Nomads: Gunpowder and Artillery in Safavid Iran," in Charles Melville, ed., *Safavid Persia: The History and Politics of an Islamic Society*, London, 1996.
Matthee, Rudi, "The History of Firearms in Persia," *Encyclopaedia Iranica* 9 (1999), 619–28.
Mohebbi, Parviz, "Firearms: Production of Cannon and Muskets," *Encyclopaedia Iranica* 9 (1999), 628–31.
Rota, Giorgio, "The Mask of Shah ʿAbbas I," in Maria Szuppe, Anna Krasnolwolska, and Claus V. Pedersen, eds., *Medieval and Modern Iranian Studies, Proceedings of the Sixth European Conference of Iranian Studies*, Paris, 2011.
Savory, Roger M., "The Office of Sipahsālālar (Commander in Chief) in the Safavid State," in Bert G. Fragner, Christa Frager, Gherardo Gnoli, Roxane Haag-Higuchi, Mauro Maggi, and Paula Orsatti, eds., *Proceedings of the Second European Conference of Iranian Studies*, Rome, 1995.

## Wars of dynasties, wars of empires: The nature of European conflicts, 1648–1792

Baugh, Daniel, *The Global Seven Years War, 1754–63: Britain and France in a Great Power Contest*, Harlow, 2011.
Black, Jeremy, *European International Relations 1648–1815*, London, 2002.
Burckhardt, Johannes, "Religious War or Imperial War?: Views of the Seven Years' War from Germany and Rome," in Mark H. Danley and Patrick J. Speelman, eds., *The Seven Years' War: Global Views*, Leiden, 2012.
Croxton, Derek, *Westphalia: The Last Christian Peace*, Basingstoke, 2013.
Dover, Paul M., ed., *Secretaries and Statecraft in the Early Modern World*, Edinburgh, 2016.
Duindam, Jeroen, *Dynasty*, Cambridge, 2015.
Gilbert, Felix, "The 'New Diplomacy' of the Eighteenth Century," in Felix Gilbert, ed., *History: Choice and Commitment*, Cambridge, MA, 1977.
Hatton, Ragnhild M., "Louis XIV and His Fellow Monarchs," in John C. Rule, ed., *Louis XIV and the Craft of Kingship*, Columbus, 1969.
Hennings, Jan, *Russia and Courtly Europe: Ritual and the Culture of Diplomacy, 1648–1725*, Cambridge, 2016.
Luard, Evan, *The Balance of Power: The System of International Relations 1648–1815*, Basingstoke, 1992.
Marshall, P. J., ed., *The Oxford History of the British Empire*, vol. 2: *The Eighteenth Century*, Oxford, 1998.
McKay, Derek and Hamish M. Scott, *The Rise of the Great Powers, 1648–1815*, London, 1983.
Onnekink, David, ed., *War and Religion after Westphalia, 1648–1713*, Farnham, 2009.
Onnekink David and Gijs Rommelse, eds., *Ideology and Foreign Policy in Early Modern Europe (1650–1750)*, Farnham, 2011.
Oresko, Robert, Graham Gibbs, and Hamish Scott, eds., *Royal and Republican Sovereignty in Early Modern Europe: Essays in Memory of Ragnhild Hatton*, Cambridge, 1997.
Parrott, David, "War and International Relations," in Joseph Bergin, ed., *The Seventeenth Century*, Oxford, 2001.
Riches, Daniel, *Protestant Cosmopolitanism and Diplomatic Culture: Brandenberg–Swedish Relations in the Seventeenth Century*, Leiden, 2013.

Rommelse, Gijs, *The Second Anglo-Dutch War (1665–1667): Raison d'état, Mercantilism and Maritime Strife*, Hilversum, 2006.
Rowen, Herbert H., *The King's State: Proprietary Dynasticism in Early Modern France*, New Brunswick, 1980.
Rule, John C., "Colbert de Torcy, an Emergent Bureaucracy, and the Formulation of French Foreign Policy, 1698–1715," in Ragnhild Hatton, ed., *Louis XIV and Europe*, Basingstoke, 1976.
Rule, John C. and Ben S. Trotter, *A World of Paper: Louis XIV, Colbert de Torcy and the Rise of the Information State*, Montreal, 2014.
Schroeder, Paul W., *The Transformation of European Politics, 1763–1848*, Oxford, 1994.
Schroeder, Paul W., "Not Even for the Seventeenth and Eighteenth Centuries: Power and Order in the Early Modern Era," in Ernest R. May, Richard Rosencrance, and Zara Steiner, eds., *History and Neorealism*, Cambridge, 2010.
Scott, Hamish M., *The Emergence of the Eastern Powers, 1756–1775*, Cambridge, 2001.
Scott, Hamish M., *The Birth of a Great Power System, 1740–1815*, Harlow, 2006.
Sonnino, Paul, *Louis XIV and the Origins of the Dutch War*, Cambridge, 1988.
Sorel, Albert, *Europe and the French Revolution: The Political Traditions of the Old Régime*, London, 1969.
Storrs, Christopher, *The Spanish Resurgence 1713–1748*, New Haven, 2016.
Thompson, Andrew C., *Britain, Hanover and the Protestant Interest, 1688–1756*, Woodbridge, 2006.
Whaley, Joachim, *Germany and the Holy Roman Empire*, vol. 2: *The Peace of Westphalia to the Dissolution of the Reich, 1648–1806*, Oxford, 2012.
Wright, Anthony, *The Early Modern Papacy: From the Council of Trent to the French Revolution 1564–1789*, Harlow, 2000.

## The standing army: France and Brandenburg Prussia – a comparison

Anderson, Matthew, *War and Society in Europe of the Old Regime*, Leicester, 1988.
Bannister, Julia, *Masculinity, Militarism and Eighteenth-Century Culture 1689–1815*, Cambridge, 2018.
Behrens, C. B. A., *Society, Government and the Enlightenment: The Experiences of Eighteenth-Century France and Prussia*, London, 1985.
Black, Jeremy, *War in the Eighteenth-Century World*, New York, 2013.
Blanning, Tim, *Frederick the Great: King of Prussia*, London, 2015.
Blaufarb, Rafe, *The French Army 1750–1820: Careers, Talent, Merit*, Manchester, 2002.
Büsch, Otto, *Military System and Social Life in Old Regime Prussia 1713–1807: The Beginnings of the Social Militarization of Prusso-German Society*, Atlantic Highlands, 1997.
Clark, Christopher, *Iron Kingdom: The Rise and Downfall of Prussia 1600–1947*, London, 2006.
Duffy, Christopher, *The Army of Frederick the Great*, 2nd ed., London, 1996.
Finer, Samuel, "State- and Nation-Building in Europe: The Role of the Military," in Charles Tilly, ed., *The Formation of National States in Western Europe*, Princeton, 1975.
Gunn, Steven, David Grummitt, and Hans Cools, "War and the State in Early Modern Europe: Widening the Debate," *War in History* 15 (2008), 371–88.
Gutman, Myron, *War and Rural Life in Early Modern Low Countries*, Assen, 1980.

Hagen, William, *Ordinary Prussians: Brandenburg Junkers and Villagers 1500–1840*, Cambridge, 2002.
Kennett, Lee, *The French Armies in the Seven Years' War: A Study in Military Organization and Administration*, Durham, NC, 1968.
Lynn, John A., *Giant of the Grand Siècle: The French Army 1610–1715*, Cambridge, 1997.
McNeill, William H., *The Pursuit of Power: Technology, Armed Forces and Society since* A.D. *1000*, Chicago, 1982.
Parrott, David, *Richelieu's Army: War, Government and Society in France, 1624–1642*, Cambridge, 2001.
Rowlands, Guy, *The Dynastic State and the Army under Louis XIV: Royal Service and Private Interest, 1661–1701*, Cambridge, 2002.
Scott, Hamish, ed., *The European Nobilities in the Seventeenth and Eighteenth Centuries*, 2 vols., London, 1995.
Scott, Samuel F., *The Response of the Royal Army to the French Revolution: The Development of the Line Army 1787–1793*, Oxford, 1978.
Wilson, Peter H., *German Armies: War and German Politics 1648–1806*, London, 1998.

## Navies, technology, and organisation, 1648–1815

Baugh, Daniel A., *British Naval Administration in the Age of Walpole*, Princeton, 1965.
Black, Jeremy and Philip Woodfine, eds., *The British Navy and the Use of Naval Power in the Eighteenth Century*, Leicester, 1988.
Brandon, Pepijn, *War, Capital and the Dutch State (1588–1795)*, Leiden, 2011.
Bruijn, Jaap R., *The Dutch Navy of the Seventeenth and Eighteenth Centuries*, Liverpool, 1993.
Capp, Bernard S., *Cromwell's Navy: The Fleet and the English Revolution, 1648–1660*, Oxford, 1989.
Cormack, William S., *Revolution and Political Conflict in the French Navy, 1789–1794*, Cambridge, 1995.
Dull, Jonathan R., *The French Navy and the Seven Years War*, Lincoln, NE, 2005.
Dull, Jonathan R., *The Age of the Ship of the Line: The British and French Navies, 1650–1815*, Lincoln, NE, 2009.
Glete, Jan, *Navies and Nations: Warships, Navies and State Building in Europe and America, 1500–1860*, 2 vols., Stockholm, 1993.
Hall, Christopher D., *Wellington's Navy: Sea Power and the Peninsular War, 1807–14*, Chatham, 2004.
Harbron, John D., *Trafalgar and the Spanish Navy: The Spanish Experience of Sea Power*, London, 1988.
Harland, John, *Seamanship in the Age of Sail*, London, 1984.
Harris, J. R., *Industrial Espionage and Technology Transfer: Britain and France in the Eighteenth Century*, Aldershot, 1998.
Haycock, David Boyd and Sally Archer, eds., *Health and Medicine at Sea, 1700–1900*, Basingstoke, 2009.
Hill, John Richard, *The Prizes of War: The Naval Prize System in the Napoleonic Wars, 1793–1815*, London, 1998.

Knight, Roger J. B., *Britain against Napoleon: The Organisation of Victory, 1793–1815*, London, 2013.

Knight, Roger J. B. and Martin H. Wilcox, *Sustaining the Fleet, 1793–1815: War, the British Navy, and the Contractor State*, Basingstoke, 2010.

Lavery, Brian, *Nelson's Navy: The Ships, Men and Organisation, 1793–1815*, London, 1989.

Morriss, Roger, *The Foundations of British Maritime Ascendancy: Resources, Logistics and the State, 1755–1815*, Cambridge, 2011.

Pritchard, James, *Louis XV's Navy, 1748–1762*, Kingston, ON, 1987.

Rodger, Nicholas A. M., *The Wooden World: An Anatomy of the Georgian Navy*, Glasgow, 1986.

Rodger, Nicholas A. M., *The Insatiable Earl: A Life of John Montagu, 4th Earl of Sandwich, 1718–1792*, London, 1993.

Symcox, Geoffrey, *The Crisis of French Sea Power, 1688–1697: From the Guerre d'Escadre to the Guerre de Course*, The Hague, 1974.

Van Der Merwe, Pietre, ed., *Science and the French and British Navies, 1700–1850*, London, 2003.

Willis, Sam, *Fighting at Sea in the Eighteenth Century*, Woodbridge, 2008.

## Warfare in early modern Africa, *c.* 1450–*c.* 1850

Abir, Mordechai, *Ethiopia: The Era of the Princes*, London, 1968.

Ajayi, J. F. Ade, "Professional Warriors in Nineteenth-Century Yoruba Politics," *Tarikh* 1 (1965), 72–81.

Bennett, Norman R., *Mirambo of Tanzania, 1840?–1884*, New York, 1971.

Heywood, L. M., *Njinga of Angola: Africa's Warrior Queen*, Cambridge, MA, 2017.

Lamphear, J., "Brothers in Arms: Military Aspects of East African Age-Class Systems in Historical Perspective," in E. Kurimoto and S. Simonse, eds., *Conflict, Age and Power in North East Africa*, Oxford, 1998.

Law, Robin, *The Oyo Empire: A West African Imperialism in the Era of the Atlantic Slave Trade*, Oxford, 1977.

Law, Robin, *The Slave Coast of West Africa, 1550–1750*, Oxford, 1991.

Reid, Richard, *War in Pre-Colonial Eastern Africa: The Patterns and Meanings of State-Level Conflict in the Nineteenth Century*, Oxford, 2007.

Reid, Richard, *Warfare in African History*, New York, 2012.

Smaldone, Joseph, *Warfare in the Sokoto Caliphate: Historical and Sociological Perspectives*, Cambridge, 1977.

Smith, Robert S., *Warfare and Diplomacy in Pre-Colonial West Africa*, 2nd ed., London, 1989.

Smith, Robert and Ade Aayi, *Yoruba Warfare in the Nineteenth Century*, Cambridge, 1971.

Thornton, John, *Warfare in Atlantic Africa 1500–1800*, London, 1999.

Wright, John, "Turbulent Times: Political Transformations in the North and East, 1760s–1830s," in C. Hamilton, B. Mbenga, and R. Ross, eds., *Cambridge History of South Africa*, vol. 1: *From Early Times to 1885*, Cambridge, 2010.

## Warfare in India, 1450–1850

Axworthy, Michael, *The Sword of Persia: Nader Shah from Tribal Warrior to Conquering Tyrant*, London and New York, 2006.

Barua, Pradeep, *The State at War in South Asia*, Lincoln and London, 2005.

Bayly, Christopher, "The British Military-Fiscal State and Indigenous Resistance: India 1750–1820," in *The C. A. Bayly Omnibus*, New Delhi, 2010.

Beveridge, Annette (trans.), *Babur-Nama*, 2 vols., New Delhi, 1989.

Burton, Reginald George, *The Maratha and Pindari War*, Delhi, 1975.

Cooper, Randolf G. S., *The Anglo-Maratha Campaigns and the Contest for India: The Struggle for Control of the South Asian Military Economy*, Cambridge, 2005.

Digby, Simon, *War Horse and Elephant in the Delhi Sultanate*, Karachi, 1971.

Eaton, R. M., "India's Military Revolution: The View from Early Sixteenth-Century Deccan," in Raizuddin Aquil and Kaushik Roy, eds., *Warfare, Religion, and Society in Indian History*, New Delhi, 2012.

Eaton, Richard M. and Philip B. Wagoner, *Power, Memory, Architecture: Contested Sites on India's Deccan Plateau, 1300–1600*, New Delhi, 2015.

Garza, Andrew de la, *The Mughal Empire at War: Babur, Akbar and the Indian Military Revolution, 1500–1605*, London and New York, 2016.

Gommans, Jos, *Mughal Warfare: Indian Frontiers and High Roads to Empire, 1500–1700*, London and New York, 2002.

Gommans, Jos, "Warhorse and Gunpowder in India, *c.* 1000–1850," in Jeremy Black, ed., *War in the Early Modern World: 1450–1815*, London and New York, 2004.

Gommans, Jos, *The Indian Frontier: Horse and Warband in the Making of Empires*, New Delhi, 2018.

Gommans, Jos and Dirk H. A. Kolff, "Introduction: Warfare and Weaponry in South Asia: 1000–1800 AD," in Jos Gommans and Dirk H. A. Kolff, eds., *Warfare and Weaponry in South Asia: 1000–1800*, New Delhi, 2001.

Gupta, Hari Ram, ed., *Marathas and Panipat*, Chandigarh, 1961.

Irvine, William, *The Army of the Indian Moghuls*, Delhi, 1994.

Jackson, Peter, *The Delhi Sultanate: A Political and Military History*, Cambridge, 1999.

Kantak, M. R., *The First Anglo-Maratha War, 1774–1783: A Military Study of Major Battles*, Bombay, 1993.

Khan, Iqtidar Alam, *Gunpowder and Firearms: Warfare in Medieval India*, New Delhi, 2004.

Kolff, Dirk H. A., "The End of an *Ancien Regime*: Colonial War in India, 1798–1818," in J. A. de Moor and H. L. Wesseling, eds., *Imperialism and War: Essays on Colonial Wars in Asia and Africa*, Leiden, 1989.

Kolff, Dirk H. A., *Naukar, Rajput and Sepoy: The Ethnohistory of the Military Labour Market in Hindustan, 1450–1850*, Cambridge, 1990.

Macdougall, Philip, *Naval Resistance to Britain's Growing Power in India, 1660–1800: The Saffron Banner and the Tiger of Mysore*, Suffolk, 2014.

Malleson, George Bruce, *The Decisive Battles of India*, Jaipur, 1986.

Peers, Douglas M., *Between Mars and Mammon: Colonial Armies and the Garrison State in Early Nineteenth-Century India*, London, 1995.

Rosen, Stephen, *Societies and Military Power: India and Its Armies*, New Delhi, 1996.

Roy, Kaushik, *The Oxford Companion to Modern Warfare in India: From the Eighteenth Century to Present Times*, New Delhi, 2009.

Roy, Kaushik, *War, Culture and Society in Early Modern South Asia: 1740–1849*, London and New York, 2011.

Roy, Kaushik, *War and Society in Afghanistan: From the Mughals to the Americans, 1500–2013*, New Delhi, 2015.

Roy, Kaushik, *Warfare in Pre-British India: 1500 BCE–1740 CE*, London and New York, 2015.

Sarkar, Jadunath, *Military History of India*, Bombay, 1970.

Sarkar, Jadunath, *History of Aurangzeb*, 5 vols., Bombay, 1972.

Sarkar, Jadunath, *A Short History of Aurangzeb*, New Delhi, 1979.

Streusand, Douglas, *The Formation of the Mughal Empire*, Delhi, 1989.

Upadhyaya, Rashmi, "Military Strategy of the Rajputs during the Early Medieval Period," in Shahabuddin Iraqi, ed., *Medieval India 2: Essays in Medieval Indian History and Culture*, New Delhi, 2008.

Wickremesekera, Channa, *"Best Black Troops in the World": British Perceptions and the Making of the Sepoy, 1746–1850*, New Delhi, 2002.

## Japan and Korea, 1450–1850

Andreski, Stanislav, *Military Organization and Society*, 2nd ed., Berkeley, 1968.

Berry, Mary Elizabeth, *Hideyoshi*, Cambridge, MA, 1982.

Berry, Mary Elizabeth, *The Culture of Civil War in Kyoto*, Berkeley, 1994.

Blussé, Leonard, *Visible Cities: Canton, Nagasaki, and Batavia and the Coming of the Americans*, Cambridge, MA, 2008.

Bodart-Bailey, Beatrice M., *The Dog Shogun: The Personality and Policies of Tokugawa Tsunayoshi*, Honolulu, 2006.

Bolitho, Harold, "The Myth of the Samurai," in Alan Rix and Ross Mouer, eds., *Japan's Impact on the World*, Canberra, 1984.

Brownlee, John S., *Japanese Historians and the National Myths, 1600–1945: The Age of the Gods and Emperor Jimmu*, Vancouver, 1997.

Bryant, Anthony J., *Sekigahara: The Final Struggle for Power*, Oxford, 1995.

Ch'oe, Yŏngho, Peter H. Lee, and Wm. Theodore de Bary, eds., *Sources of Korean Tradition*, vol. 2: *From the Sixteenth to the Twentieth Centuries*, New York, 2000.

Chung, Chai-Sik, *A Korean Confucian Encounter with the Modern World: Yi Hang-no and the West*, Berkeley, 1995.

Cooper, Michael, ed., *They Came to Japan*, Berkeley, 1965.

Davis, David L., "*Ikki* in Late Medieval Japan," in John W. Hall and Jeffrey P. Mass, eds., *Medieval Japan: Essays in Institutional History*, Stanford, 1974.

Duus, Peter, *The Japanese Discovery of America: A Brief History with Documents*, Boston, 1997.

Earl, David Margarey, *Emperor and Nation in Japan*, Seattle, 1964.

Elison, George, "The Cross and the Sword: Patterns of Momoyama History," in George Elison and Bardwell L. Smith, eds., *Warlords, Artists, and Commoners: Japan in the Sixteenth Century*, Honolulu, 1981.

Elisonas, Jurgis, "The Inseparable Trinity: Japan's Relations with China and Korea," in John W. Hall, ed., *The Cambridge History of Japan*, vol. 4: *Early Modern Japan*, Cambridge, 1991.

Fairbank, John K., "Introduction: Varieties of the Chinese Military Experience," in Frank A. Kierman, Jr. and John K. Fairbank, eds., *Chinese Ways in Warfare*, Cambridge, MA, 1974, pp. 1–26

Farris, William Wayne, *Heavenly Warriors: The Evolution of Japan's Military, 500–1300*, Cambridge, MA, 1992.

Friday, Karl F., *Hired Swords: The Rise of Private Warrior Power in Early Japan*, Stanford, 1992.

Goble, Andrew, "Visions of an Emperor," in Jeffrey P. Mass, ed., *The Origins of Japan's Medieval World: Courtiers, Clerics, Warriors and Peasants in the Fourteenth Century*, Stanford, 1997.

Gordon, Andrew A., *A Modern History of Japan: From Tokugawa Times to the Present*, New York, 2003.

Grossberg, Kenneth Alan, *Japan's Renaissance: The Politics of the Muromachi Bakufu*, Ithaca, 2000.

Haboush, JaHyun Kim, *The Great East Asian War and the Birth of the Korean Nation*, New York, 2016.

Haboush, JaHyun Kim and Martina Deuchler, eds., *Culture and the State in Late Chosŏn* Korea, Cambridge, MA, 1999.

Harootunian, H. D., *Things Seen and Unseen: Discourse and Ideology in Tokugawa Nativism*, Chicago, 1988.

Hawley, Samuel, *The Imjin War*, Seoul, 2005.

Hayashi, Tatsusaburō, "Kyoto in the Muromachi Age," in John W. Hall and Toyoda Takeshi, eds., *Japan in the Muromachi Age*, Berkeley, 1997.

Ikegami, Eiko, *The Taming of the Samurai: Honorific Individualism and the Making of Modern Japan*, Cambridge, 1995.

Jansen, Marius B., *China in the Tokugawa World*, Cambridge, MA, 1992.

Jansen, Marius B., *Sakamoto Ryōma and the Meiji Restoration*, New York, 1997.

Jansen, Marius B., *The Making of Modern Japan*, Cambridge, MA, 2000.

Kawashima, Fujiya, *What Is Yangban? A Legacy for Modern Korea*, Seoul, 2002.

Keene, Donald, *Emperor of Japan: Meiji and His world, 1852–1912*, New York, 2002.

Kibaik, Lee, "Korea: The Military Tradition," in Hugh H. W. Kang, ed., *The Traditional Culture and Society of Korea: Thought and Institutions*, Honolulu, 1975.

Kim, Sang H., trans., *Muye Dobo Tongji: The Comprehensive Illustrated Manual of Martial Arts of Korea*, New York, 2000.

Kim, Sun Joo, *Marginality and Subversion in Korea: The Hong Kyŏngnae Rebellion of 1812*, Seattle, 2007.

Lewis, James B., ed., *The East Asian War, 1592–1598: International Relations, Violence, and Memory*, London, 2015.

Lu, David J., *Japan: A Documentary History*, vol. 1: *The Dawn of History to the Late Tokugawa Period*, New York, 1997.

Ooms, Herman, *Tokugawa Ideology: Early Constructs, 1570–1680*, Princeton, 1985.

Palais, James B., *Politics and Policy in Traditional Korea*, Cambridge, MA, 1975.

Palais, James B., *Confucian Statecraft and Korean Institutions: Yu Hyŏngwŏn and the Late Chosŏn Dynasty*, Seattle, 1996.
Park, Eugene Y., *Between Dreams and Reality: The Military Examination in Late Chosŏn Korea, 1600–1894*, Cambridge, MA, 2007.
Robinson, Kenneth R., "Residence and Foreign Relations in the Peninsular Northeast during the Fifteenth and Sixteenth Centuries," in Sun Joo Kim, ed., *The Northern Region of Korea: History, Identity, and Culture*, Seattle, 2010.
Souyri, Pierre François, *The World Turned Upside Down: Medieval Japanese Society*, trans. Käthe Roth, New York, 2001.
Spafford, David, *A Sense of Place: The Political Landscape in Late Medieval Japan*, Cambridge, MA, 2013.
Susser, Bernard, "The Toyotomi Regime and the Daimyo," in Jeffrey P. Mass and William B. Hauser, eds., *The Bakufu in Japanese History*, Stanford, 1985.
Takeuchi Rizō, "The Rise of the Warriors," in Donald H. Shively and William H. McCullough, eds., *The Cambridge History of Japan*, vol. 2: *Heian Japan*, Cambridge, 1999.
Toby, Ronald P., "Carnival of the Aliens: Korean Embassies in Edo-Period Art and Popular Culture," *Monumenta Nipponica* 41 (1986), 415–56.
Toby, Ronald P., *State and Diplomacy in Early Modern Japan: Asia and the Development of the Tokugawa Bakufu*, Stanford, 1991.
Totman, Conrad D., *Japan Before Perry: A Short History*, Berkeley, 1981.
Totman, Conrad D., *Early Modern Japan*, Berkeley, 1993.
Totman, Conrad D., *Pre-Industrial Korea and Japan in Environmental Perspective*, Leiden, 2004.
Turnbull, Stephen, *Samurai Invasion: Japan's Korean War, 1592–98*, London, 2002.
Varley, Paul, *Warriors of Japan, as Portrayed in the War Tales*, Honolulu, 1994.
Vlastos, Stephen, *Peasant Protests and Uprisings in Tokugawa Japan*, Berkeley, 1986.
Wakabayashi, Bob Tadashi, *Anti-Foreignism and Western Learning in Early-Modern Japan: The "New Theses" of 1825*, Cambridge, MA, 1991.
Walthall, Anne, ed. and trans., *Peasant Uprisings in Japan*, Chicago, 1991.
Yi, Tae-Jin, *The Dynamics of Confucianism and Modernization in Korean History*, Ithaca, 2007.
Yonemoto, Marcia, *Mapping Early Modern Japan: Space, Place, and Culture in the Tokugawa Period (1603–1868)*, Berkeley, 2003.

## Warfare in South-East Asia, 1448–1851

Boomgaard, Peter, "Horse Breeding, Long-Distance Horse Trading and Royal Courts in Indonesian History, 1500–1900," in Greg Bankoff and Sandra Swart, eds., *Breeds of Empire: The "Invention" of the Horse in Southeast Asia and Southern Africa 1500–1950*, Copenhagen, 2007.
Charney, Michael W., *Southeast Asian Warfare 1300–1900*, Leiden, 2004.
Dutton, George Edson, *The Tây Sơn Uprising: Society and Rebellion in Eighteenth-Century Vietnam*, Honolulu, 2006.
Hodgkin, Thomas, *Vietnam: The Revolutionary Path*, New York, 1981.

Lieberman, Victor B., *Strange Parallels: Southeast Asia in Global Context, c. 800–1830*, vol. 1: *Integration on the Mainland*, Cambridge, 2003.
Reid, Anthony, *Southeast Asia in the Age of Commerce, 1450–1680*, vol. 2: *Expansion and Crisis*, New Haven and London, 1993.
Ricklefs, Merle C., *A History of Modern Indonesia since c. 1200*, Stanford, 2001.
Sun, Laichen, "Chinese Military Technology in Dai Viet: *c.* 1390–1497," Asia Research Institute working paper series no. 11, Singapore, 2003.
Surakiat, Pamaree. "The Changing Nature of Conflict between Burma and Siam as Seen from the Growth and Development of Burmese States from the 16th to the 19th Centuries," Asia Research Institute working paper series no. 64, Singapore, 2006.

## War in the era of Qing imperial consolidation and expansion, 1587–1804

Adshead, S. A. M., "Horse Administration under the Ch'ing," *Papers on Far Eastern History: Ming and Early Ch'ing China* 17 (1978), 71–79.
Andrade, Tonio, *Lost Colony: The Untold Story of China's First Great Victory over the West*, Princeton, 2011.
Bartlett, Beatrice S., *Monarchs and Ministers: The Grand Council in Mid-Ch'ing China, 1723–1820*, Berkeley, 1992.
Bergholz, Fred W., *The Partition of the Steppe: The Struggle of the Russians, the Manchus and the Zunghar Mongols for Empire in Central Asia, 1619–1758: A Study in Power Politics*, New York, 1993.
Crossley, Pamela Kyle, *Orphan Warriors: Three Manchu Generations and the End of the Qing World*, Princeton, 1990.
Crossley, Pamela Kyle, *A Translucent Mirror: History and Identity in Qing Imperial Ideology*, Berkeley, 1999.
Crossley, Pamela Kyle, "The Conquest Elite of the Ch'ing Empire," in Willard Peterson, ed., *The Cambridge History of China*, vol. 9 pt. 1: *The Ch'ing Empire to 1800*, Cambridge, 2002.
Crossley, Pamela Kyle, "Flank Contact, Social Contexts, and Riding Patterns in Eurasia, 500-1500," in Morris Rossabi, ed., *How Mongolia Matters: War, Law and Society*, Leiden, 2017, pp. 129–46.
Dai, Yingcong, *The Sichuan Frontier and Tibet: Imperial Strategy in Early Qing*, Seattle, 2009.
Dai, Yingcong, *The White Lotus War: Rebellion and Suppression in Late Imperial China*, Seattle, 2019.
Elliott, Mark C., *The Manchu Way: The Eight Banners and Ethnic Identity in Late Imperial China*, Stanford, 2001.
Lary, Diana, ed., *The Chinese State at the Borders*, Vancouver, 2007.
Li, Gertraude Roth, "The Manchu–Chinese Relationship, 1616–36," in Jonathan D. Spence and John E. Wills, Jr., ed., *From Ming to Ch'ing: Conquest, Region and Continuity in Seventeenth-Century China*, New Haven, 1979.
Li, Gertraude Roth, "State Building before 1644," in Willard Peterson, ed., *The Cambridge History of China*, vol. 9 pt. 1: *The Ch'ing Empire to 1800*, Cambridge, 2002.

Liu, Kwang-ching, "The Military Challenge: The North-West and the Coast," in Denis Twitchett and John K. Fairbank, eds., *The Cambridge History of China*, vol. 11: *Late Ch'ing 1800–1911*, Cambridge, 1980.

Lorge, Peter, *War, Politics and Society in Early Modern China, 900–1795*, London, 2005.

Millward, James A., *Beyond the Pass: Economy, Ethnicity and Empire in Qing Central Asia, 1759–1864*, Stanford, 1998.

Perdue, Peter C. *China Marches West: The Qing Conquest of Central Eurasia*, Cambridge, MA, 2005.

Perdue, Peter C., "Embracing Victory, Effacing Defeat: Rewriting the Qing Frontier Campaigns," in Diana Lary, ed., *The Chinese State at the Borders*, Vancouver, 2007, pp. 105–25.

Rawski, Evelyn S., *The Last Emperors: A Social History of Qing Imperial Institutions*, Berkeley, 2001.

Rhoads, Edward J. M., *Manchus and Han: Ethnic Relations and Political Power in Late Qing and Early Republican China, 1861–1928*, Seattle, 2000.

Rowe, William T., *China's Last Empire: The Great Qing*, Cambridge, MA, 2009.

Spence, Jonathan D., *Ts'ao Yin and the K'ang-hsi Emperor: Bondservant and Master*, New Haven, 1966.

Struve, Lynn, *The Southern Ming, 1664–1662*, New Haven, 1984.

Torbert, Preston M., *The Ch'ing Imperial Household Department: A Study of Its Organization and Principal Functions, 1661–1796*, Cambridge, MA, 1977.

Woodside, Alexander, "The Ch'ien-lung Reign," in Willard Peterson, ed., *The Cambridge History of China*, vol. 9 pt. 1: *The Ch'ing Empire to 1800*, Cambridge, 2002.

## The seventh continent: Russian territorial expansion from 1450 to 1850

Baddeley, Joh F., *The Russian Conquest of the Caucasus*, New York, 1908.

Baumann, Robert F., *Russian–Soviet Unconventional Wars in the Caucasus, Central Asia, and Afghanistan*, Fort Leavenworth, 1993.

Davies, Brian L., *Warfare, State and Society on the Black Sea Steppe, 1700–1700*, London, 2007.

Donnely, Alton S., *The Russian Conquest of Bashkiria 1552–1740: A Case Study in Imperialism*, New Haven, 1968.

Duffy, Christopher, *Russia's Military Way to the West: Origins and Nature of Russian Military Power 1700–1800*, London, 1981.

Englund, Peter, *The Battle That Shook Europe: Poltava and the Birth of the Russian Empire*, London, 2003.

Filjushkin, Alexander, *Ivan the Terrible: A Military History*, London, 2008.

Fisher, Alan W., *The Russian Annexation of the Crimea 1772–1783*, Cambridge, 1970.

Frost, Robert I., *The Northern Wars: War, State and Society in Northern Europe, 1558–1721*. London, 2000.

Fuller, William C., *Strategy and Power in Russia 1600–1914*, New York, 1992.

Grey, Ian, *Ivan III and the Unification of Russia*, New York, 1967.

Hellie, Richard, *Enserfment and Military Change in Muscovy*, Chicago, 1971.

Hunczak, Taras, ed., *Russian Imperialism: From Ivan the Great to the Revolution*, New Brunswick, 1974.
Kagan, Frederick W. and Robin Higham, eds., *The Military History of Tsarist Russia*, New York, 2002.
Keep, John L., *Soldiers of the Tsar: Army and Society in Russia 1462–1874*, Oxford, 1985.
Khodarkovsky, Michael, *Where Two Worlds Met: The Russian State and the Kalmyk Nomads, 1600–1771*, Ithaca, 1992.
Khodarkovsky, Michael, *Russia's Steppe Frontier: The Making of a Colonial Empire, 1500–1800*, Bloomington, 2002.
Lantzeff, George V. and Richard A. Pierce, *Eastward to Empire: Exploration and Conquest on the Russian Open Frontier to 1750*, Montreal, 1973.
LeDonne, John P., *The Russian Empire and the World 1700–1917*, Oxford, 1997.
Lieven, Dominic, *Russia against Napoleon: The Battle for Europe 1807–1814*, London, 2009.
Lincoln, Bruce W., *The Conquest of a Continent: Siberia and the Siberians*, New York, 1994.
Lohr, Eric and Marshall Poe, eds., *The Military and Society in Russia 1450–1917*, Leiden, 2002.
Mancall, Mark, *Russia and China: Their Diplomatic Relations to 1728*, Cambridge, MA, 1971.
March, G. Patrick, *Eastern Destiny: Russia in Asia and the North Pacific*, Westport, 1996.
Payne, S. C. M., *Imperial Rivals: China, Russia and Their Disputed Frontier*, Armonk, 1996.
Pelenskyj, Jaroslav Z., *Russia and Kazan: Conquest and Imperial Ideology (1438–1560s)*, The Hague, 1974.
Presnyakov, A. E., *The Formation of the Great Russian State*, trans. A. E. Moorehouse, Chicago, 1970.
Ralston, David B., *Importing the European Army: The Introduction of European Military Techniques and Institutions into the Extra-European World, 1600–1914*, Chicago, 1990.
Rieber, Alfred, *The Struggle for the Eurasian Borderlands: From the Rise of the Early Modern Empires to the End of the First World War*, Cambridge, 2014.
Rywkin, Michael, ed., *Russian Colonial Expansion to 1917*, London, 1988.
Stevens, Carol Belkin, *Russia's Wars of Emergence 1460–1730*, London, 2007.
Sunderland, Willard, *Taming the Wild Field: Colonization and Empire on the Russian Steppe*, Ithaca, 2004.
Vinkovetsky, Ilya, *Russian America: An Overseas Colony of the Russian Empire, 1804–1867*, Oxford, 2011.

## The maintenance of Hispanic dominance in Latin America

Archer, Christon I., *The Army in Bourbon Mexico 1760–1810*, Albuquerque, 1977.
Bakewell, Peter, *A History of Latin America: Empires and Sequels 1450–1930*, Oxford, 1997.
Bemis, Samuel F., *The Diplomacy of the American Revolution*, Bloomington, 1957.
Brading, David A., *Miners and Merchants in Bourbon Mexico 1763–1810*, Cambridge, 1971.
Burkholder, Mark and Lyman Johnson, *Colonial Latin America*, 6th ed., Oxford, 2004.
Dull, Jonathan R., *The French Navy and American Independence: A Study of Arms and Diplomacy*, Princeton, 1975.
DuVal, Kathleen, *Independence Lost: Lives on the Edge of the American Revolution*, New York, 2015.

Elliott, John H., *Empires of the Atlantic World: Britain and Spain in the Americas, 1492–1830*, New Haven and London, 2006.

French, David, *The British Way in Warfare, 1688–2000*, London, 1990.

Klein, Herbert S. and Ben Vinson III, *African Slavery in Latin America and the Caribbean*, 2nd ed., Oxford, 2007.

Kuethe, Allan J., *Military Reform and Society in New Granada, 1773–1808*, Gainesville, 1978.

Kuethe, Allan J., *Cuba, 1753–1815: Crown, Military and Society*, Knoxville, 1986.

Lynch, John, *Spain under the Habsburgs*, 2 vols., Oxford, 1965.

McNeill, John Robert, *Atlantic Empires of Spain and France: Louisbourg and Havana, 1700–1763*, Chapel Hill and London, 1985.

McNeill, John Robert, *Mosquito Empires: Ecology and War in the Greater Caribbean, 1620–1914*, Cambridge, 2010.

Morris, Richard B., *Peacemakers: The Great Powers and American Independence*, New York, 1965.

Paquette, Gabriel, *Enlightenment, Governance and Reform in Spain and Its Empire, 1759–1808*, New York and Basingstoke, 2008.

Paquette, Gabriel and Quintero Gonzalo M. Saravia, "Introduction: Spain and the American Revolution," in Gabriel Paquette and Quintero Gonzalo M. Saravia, eds., *Spain and the American Revolution: New Approaches and Perspectives*, London and New York, 2020.

Parker, Geoffrey, *The Military Revolution: Military Innovation and the Rise of the West, 1500–1800*, 2nd ed., Cambridge, 1998.

Phelan, John Leddy, *The People and the King: The Comunero Revolution in Colombia, 1781*, Madison, 1978.

Scott, Hamish M., *British Foreign Policy in the Age of the American Revolution*, Oxford, 1990.

Terraciano, Kevin, "Indigenous Peoples in Colonial Spanish American Society," in Thomas Holloway, ed., A *Companion to Latin American History*, Oxford, 2008.

Vinson III, Ben, *Bearing Arms for His Majesty: The Free-Colored Militia in Colonial Mexico*, Stanford, 2001.

Weber, David, *Bárbaros: Spaniards and Their Savages in the Age of Enlightenment*, New Haven and London, 2006.

## The United States from Independence to the war with Mexico, 1776–1848

Bird, Harrison, *War for the West, 1790–1813*, New York, 1971.

Black, Jeremy, *Fighting for America: The Struggle for Mastery in North America, 1519–1871*, Bloomington, 2011.

Eisenhower, John, *So Far from God: The U.S. War with Mexico, 1846–1848*, New York, 1990.

Ferling, John, *Almost a Miracle: American Victory in the War for Independence*, New York, 2007.

Fischer, David Hackett, *Washington's Crossing*, New York, 2004.

Guarding, Peter, *The Dead March: A History of the Mexican–American War*, Cambridge, 2017.

Henderson, Timothy J., *A Glorious Defeat: Mexico and Its War with the United States*, New York, 2007.

Hickey, Donald R., *The War of 1812: A Forgotten Conflict*, Urbana, 1989.
Johnson, Timothy D., *A Gallant Little Army: The Mexico City Campaign*, Lawrence, 2007.
Lambert, Andrew, *The Challenge: Britain against America in the Naval War of 1812*, London, 2012.
Latimer, Jon, *1812: War with America*, Cambridge, MA, 2007.
Lee, Wayne E., *Barbarians and Brothers: Anglo-American Warfare, 1500–1865*, NewYork, 2011.
Mackesy, Piers, *The War for America 1775–178*, Cambridge, MA, 1964.
Mahon, John D., *History of the Second Seminole War, 1835–1842*, rev. ed., Gainesville, FL, 1992.
McCranie, Kevin D., *Utmost Gallantry: The U.S. and Royal Navies at Sea in the War of 1812*, Annapolis, 2011.
Middlekauff, Robert, *The Glorious Cause: The American Revolution, 1763–1789*, New York, 1982.
Prucha, Francis P., *The Sword of the Republic: The United States Army on the Frontier, 1783–1846*, New York, 1969.
Skelton, William B., *An American Profession of Arms: The Army Officer Corps, 1784–1861*, Lawrence, 1992.
Taylor, Alan, *The Civil War of 1812: American Citizens, British Subjects, Irish Rebels, and Indian Allies*, New York, 2010.
Winders, Richard B., *Mr. Polk's Army: The American Military Experience in the Mexican War*, College Station, 1997.

## The wars of the French Republic and the Napoleonic Wars, 1789–1815

Anderson, Fred, *Crucible of War: The Seven Years' War and the Fate of Empire in British North America, 1754–1766*, London, 2000.
Andress, David, *Massacre at the Champ de Mars: Popular Dissent and Political Culture in the French Revolution*, London, 2000.
Aron, Raymond, *Clausewitz, Philosopher of War*, London, 1983.
Bayly, Christopher A., *The Birth of the Modern World, 1780–1914*, Oxford, 2004.
Bell, David A., *The First Total War: Napoleon's Europe and the Birth of Warfare as We Know It*, Boston, 2007.
Berkovich, Ilya, *Motivation in War: The Experience of Common Soldiers in Old-Regime Europe*, Cambridge, 2017.
Best, Geoffrey, *War and Society in Revolutionary Europe, 1770–1870*, London, 1982.
Black, Jeremy, *European Warfare, 1660–1815*, London, 1994.
Blanning, Tim C. W., *The Origins of the French Revolutionary Wars*, London, 1986.
Blaufarb, Rafe, *The French Army, 1750–1820: Careers, Talent, Merit*, Manchester, 2002.
Colson, Bruno, *Clausewitz*, Paris, 2016.
Cormack, William S., *Revolution and Political Conflict in the French Navy, 1789–1794*, Cambridge, 1995.
Dodman, Thomas, *What Nostalgia Was: War, Empire and the Time of a Deadly Emotion*, Chicago, 2018.
Esdaile, Charles, *Fighting Napoleon: Guerrillas, Bandits and Adventurers in Spain, 1801–1814*, New Haven, 2004.

Esdaile, Charles, *Napoleon's Wars: An International History*, London, 2007.

Forrest, Alan, *Conscripts and Deserters: The Army and French Society during the Revolution and Empire*, Oxford, 1989.

Forrest, Alan, "Conscription as Ideology: Revolutionary France and the Nation-in-Arms," in Lars Mjøset and Stephen van Holde, eds., *The Comparative Study of Conscription in the Armed Forces* Oxford, 2002.

Forrest, Alan, Karen Hagemann, and Jane Rendall, eds., *Soldiers, Citizens and Civilians: Experiences and Perceptions of the Revolutionary and Napoleonic Wars, 1790–1820*, Basingstoke, 2009.

Gat, Azar, *A History of Military Thought: From the Enlightenment to the Cold War*, Oxford, 2001.

Gates, David, *The Napoleonic Wars, 1803–1815*, London, 1997.

Hagemann, Karen, *Revisiting Prussia's Wars against Napoleon: History, Culture and Memory*, Cambridge, 2015.

Hagemann, Karen, Gisela Mettele, and Jane Rendall, eds., *Gender, War and Politics: Transatlantic Perspectives, 1775–1830*, Basingstoke, 2010.

Hartley, Janet M., Paul Keenan, and Dominic Lieven, eds., *Russia and the Napoleonic Wars*, Basingstoke, 2015.

Hughes, Michael J., *Forging Napoleon's Grande Armée: Motivation, Military Culture and Masculinity in the French Army, 1800–1808*, New York, 2011.

Jarrett, Mark, *The Congress of Vienna and Its Legacy: War and Great Power Diplomacy after Napoleon*, London, 2013.

Lynn, John A., *The Bayonets of the Republic: Motivation and Tactics in the Army of Revolutionary France, 1791–94*, Urbana, 1984.

Lynn, John A., "The Pattern of Army Growth, 1445–1945," in John A. Lynn, ed., *Tools of War: Instruments, Ideas and Institutions of warfare, 1445–1871*, Urbana, 1990.

MacKenzie, S. P., *Revolutionary Armies in the Modern Era: A Revisionist Approach*, London, 1997.

Mikaberidze, Alexander, *The Napoleonic Wars: A Global History*, Oxford, 2020.

Muir, Rory, *Tactics and the Experience of Battle in the Age of Napoleon*, New Haven and London, 2000.

Planert, Ute, ed., *Napoleon's Empire: European Politics in Global Perspective*, Basingstoke, 2016.

Riehn, Richard, *1812: Napoleon's Russian Campaign*, New York, 1991.

Riley, Jonathon, *Napoleon as a General*, London, 2007.

Rothenberg, Gunther E., *The Art of Warfare in the Age of Napoleon*, Bloomington, 1978.

Rothenberg, Gunther E., *Napoleon's Great Adversary: Archduke Charles and the Austrian Army, 1792–1814*, Staplehurst, 1985.

Scott, Samuel F., *The Response of the Royal Army to the French Revolution: The Role and Development of the Line Army, 1787–1793*, Oxford, 1978.

Scott, Samuel F., *From Yorktown to Valmy: The Transformation of the French Army in an Age of Revolution*, Denver, 1998.

Stoker, Donald, Frederick C. Schneid, and Harold D. Blanton, eds., *Conscription in the Napoleonic Era: A Revolution in Military Affairs?*, London, 2009.

Tackett, Timothy, "Conspiracy Obsession in a Time of Revolution: French Elites and the Origins of the Terror, 1789–92," in Jeremy Black, ed., *Revolutions in the Western World, 1775–1825*, London, 2006.

Telp, Claus, *The Evolution of Operational Art, 1740–1813: From Frederick the Great to Napoleon*, London, 2005.
Whiteman, Jeremy J., *Reform, Revolution and French Global Policy, 1787–1791*, Aldershot, 2003.

## European sea power and the origins of European domination, 1800–1850

Angevine, Robert G., *The Railroad and the State: War, Politics and Technology in Nineteenth-Century America*, Redwood, 2004.
Bethell, Leslie, *The Abolition of the Brazilian Slave Trade: Britain, Brazil and the Slave Trade Question 1807–1869*, London, 1970.
Bond, Gordon C., *The Grand Expedition: The British Invasion of Holland in 1809*, Athens, GA, 1979.
Cain, Peter J. and Anthony G. Hopkins, *British Imperialism: Innovation and Expansion 1688–1914*, London, 1993.
Canney, Donald L., *Sailing Warships of the U.S. Navy*, London, 2001.
Daly, John C. K., *Russian Seapower and "The Eastern Question", 1827–1841*, London, 1991.
Dewey, Clive, *Steamboats on the Indus: The Limits of Western Technological Superiority in South Asia*, Oxford and New Delhi, 2014.
Egerton, Judy, *Turner: The Fighting Temeraire*, London 1995.
Feldbæk, Ole, *Denmark and the Armed Neutrality 1800–1801*, Copenhagen, 1980.
Glete, Jan, *Navies and Nations: Warships, Navies and State Building in Europe and America, 1500*–1860, 2 vols., Stockholm, 1993.
Graham, Gerald S., *The Politics of Naval Supremacy*, Cambridge, 1965.
Graham, Gerald S., *The China Station: War and Diplomacy 1830–1860*, Oxford, 1978.
Graham, Gerald S., and Robin S. Humphreys, eds., *The Navy and South America,1807–1823: Correspondence of the Commanders-in-Chief on the South American Station*, London, 1962.
Grindal, Peter, *Opposing the Slavers: The Royal Navy's Campaign against the Atlantic Slave Trade*, London, 2017.
Harcourt, Freda, *Flagships of Imperialism: The P&O Company and the Politics of Empire from Its Origins to 1867*, Manchester, 2006.
Headrick, Daniel R., *The Tools of Empire: Technology and European Imperialism in the Nineteenth Century*, Oxford, 1982.
Kennedy, Greg, ed., *The Merchant Marine in International Affairs, 1850–1950*, London, 2000.
Lambert, Andrew, *Battleships in Transition: The Creation of the Wooden Steam Battlefleet 1815–1860*, London, 1984.
Lambert, Andrew, *The Last Sailing Battlefleet: Maintaining Naval Mastery, 1815–1850*, London, 1991.
Lambert, Andrew, "HMS *Rattler* – Brunel's Warship in Service, 1845–1856," in D. Griffiths, A. Lambert, and F. Walker, eds., *Brunel's Ships*, London, 1999.
Lambert, Andrew, *Responding to the Nineteenth Century: The Royal Navy and the Introduction of the Screw Propeller*, London, 1999.
Lambert, Andrew, "Strategy, Policy and Ship-Building: The Bombay Dockyard, the Indian Navy and Imperial Security in Eastern Seas, 1784–1869," in H. Bowen, M. Lincoln, and N, Rigby, eds., *The Worlds of the East India Company*, Woodbridge, 2002.

Lambert, Andrew, "Winning without Fighting: British Grand Strategy and Its Application to the United States, 1815–1865," in B. Lee and K. Walling, eds., *Strategic Logic and Political Rationality: Essays in Honour of Michael J. Handel*, Newport, RI, 2003.

Lambert, Andrew, *Nelson: Britannia's God of War*, London, 2004.

Lambert, Andrew, "The Limits of Naval Power: *Three Sisters*, Moroccan Pirates and British Battleships," in Bruce A. Elleman, Andrew Forbes, and David Rosenberg, eds., *Piracy and Maritime Crime: Historical and Modern Case Studies*, Newport, RI, 2010.

Lambert, Andrew, *The Challenge: Britain, America and the Naval War of 1812*, London, 2012.

Lambert, Andrew, *Seapower States: Maritime Culture, Continental Empires and the Conflict That Made the Modern World*, New Haven and London, 2018.

Lemnitzer, Jan, *Power, Law and the End of Privateering*, London, 2014.

McCarthy, Matthew, *Privateering, Piracy, and British Policy in Spanish America 1810–1830*, Woodbridge, 2013.

Munch-Petersen, Thomas, *Defying Napoleon: How Britain Bombarded Copenhagen and Seized the Danish Fleet in 1807*, Gloucestershire, 2007.

Reidy, Michael S., *Tides of History: Ocean Science and Her Majesty's Navy*, Chicago, 2008.

Robson, Martin, *Britain, Portugal and South America in the Napoleonic Wars: Alliances and Diplomacy in Economic Maritime Conflict*, London, 2010.

Roland, Alex, W. J. Bolster, and A. Keyssar, *The Way of the Ship; America's Maritime History Re-Envisioned, 1600–2000*, Hoboken, 2008.

Rozwadowski, Helen M., *Vast Expanses: A History of the Oceans*, London, 2019.

Schroeder, John H., *Shaping a Maritime Empire: The Commercial, and Diplomatic Role of the American Navy, 1829–1861*, Westport, 1985.

Sondhaus, Lawrence, *The Habsburg Empire and the Sea: Austrian Naval Policy 1797–1866*, Indiana, 1989.

Tredrea, John and Edouard Sozaev, *Russian Warships in the Age of Sail, 1696–1860*, Barnsley, 2010.

Winfield, Rif, *British Warships in the Age of Sail: 1817–1863: Design, Construction, Careers and Fates*, Barnsley, 2014.

Winfield, Rif and Stephen Roberts, *French Warships in the Age of Sail, 1786–1861: Design, Construction, Careers and Fates*, Barnsley, 2015.

# Index


*Aaron Manby*, steamship, 585
ʿAbbas I, Safavid shah, 289, 290, 293–94, 301–3
Aboukir Bay, battle of (battle of the Nile), 383, 556, 575
absolutism, 26–27, 203, 269, 315, 320, 332, 343
Abu'l Khayr Khan, 57–59, 67
Aceh, 237–38, 454, 456
Adal, 390
Adams–Onís Treaty, 546
Aden, 240, 250, 589
  Ottoman seizure of, 236
  siege of, 235
Afghanistan, 36, 410, 424
  British invasion of, 600
  Mughal–Safavid conflict. *See also* Qandahar
  supply of horses from, 415, 420
  tribalism, 63
Afghans
  canceled Safavid campaign against, 307
  defeat of the Safavids, 308
  in Mughal army, 415–16
African warfare
  East African campaigns, 1500–1700, 396–97
  environmental influences on the development of military systems, 5, 387–88
  European incursions in the nineteenth century, 406–7
  European perceptions on the nature of warfare in, 408
  expansion of European activity and, 399–400
  fighting formations, 392
  influence of insect-borne disease, 7, 388
  jihadist movements, 403–4
  military revolution in the nineteenth century, 401–7
  military support systems, 394
  military systems and size of armies, 393–95, 397–99
  military systems of the East African pastoralists, 399
  political power of the armies, 393
  processes of militarization in Eastern and Southern Africa, 396
  violence and militarization associated with the slave trade, 391–94, 401–2
  weaponry, 390–91, 392–93, 394, 395–96, 401, 403
Afshar, 293
Agnadello, battle of, 208, 214
Agung, sultan of Mataram, 455
Ahmad ibn Ibrahim, 390
Ahmad Shah Durrani, 424
Ahmed ibn Ibrahim, 400
Aisin khanate, 61, 466, 477
Aix-la-Chapelle (Aachen), Treaty of, 325, 334
Akbar, Jalaluddin Muhammad, Mughal emperor, 416–17
*akıncı* (irregular frontier raiders), 108
Akkerman, 105, 106
Akō Incident, 441
Akwamu, 395
Al Quasimi, 581
*ʿalaf-i shamshir* ("sword fodder"), 300
Alamo, battle of the, 547
Alauddin Khalji, Delhi sultan, 413
Alaunghpaya, Burmese king, 460
Albania, 104
Albazin
  Chinese sieges of, 16
  Russian occupation of, 498–99
Albert V, duke of Bavaria, 165
Albuquerque, Alfonso de, 235
Alexander I, emperor of Russia, 503, 511, 554, 569, 576
Alexander II, emperor of Russia, 498
Alexander VII, pope, 311
Alexandria, blockade of, 575

Alexei Petrovich, grand duke of Russia, 279
Alexis, tsar of Russia, 506–7
Algeria, French invasion, 406, 581, 587, 596
Algiers, 150, 314, 584, 596
al-Hajj Umar Tal, 403
Allada, 395
Alqas Mirza, 296
Altan Khan, 60, 61, 84
*amabutho* (Zulu regiments), 402
Amasya, Peace of, 297
Amboina, 239
Ambon, 240
American Civil War, 595
American War of Independence, 140, 250, 337
  British, French, Dutch, and Spanish territorial conflicts, 538
  British defeat, 537–38
  British New York and New Jersey campaign, 532–33
  early engagements, 530–32
  French intervention, 535–36
  naval warfare, 382–83
  Patriot resurgence, 533
  Philadelphia campaign, 533–35
  Spanish intervention, 525–27
  war in the South, 536–37
Amiens, Peace of, 385
Amur river basin, Russian occupation and abandonment of, 498–99
Ancona, fortifications, 212
Andean revolt. *See* Túpac Amaru rebellion
Andrade, Tonio, 417
Andrusovo, truce, 266
Angelović, Mahmud Pasha, 103
Angelović, Michael, joint regent of Serbia, 103
Anglo-Bourbon conflict, 336
Anglo-Burmese Wars, 464–65, 583, 599
Anglo-Chinese Wars, 591–93, 595, 596, 600
Anglo-Dutch Treaty of 1824, 463
Anglo-Dutch Wars, 17, 250, 329, 333–34
  naval engagements, 186–88, 366, 367–68
Anglo-Maratha Wars, 423–26
Anglo-Mysore Wars, 249, 422–23
Anglo-Portuguese alliance (1662), 244
Angrias, 249
Ankara, battle of, 98
Annam, 32
  Ming invasion of, 87–89, 93
  Qing occupation of, 470
Anne, queen of Great Britain, 321
annihilation, battles of, 383, 415, 575–76
Antigua, 377
Antwerp, citadel, 212
Apaches, 6, 138
  relations with the Spanish, 138–39
Aqqoyunlu ("White Sheep") Turcoman confederation, 56, 59, 104
Arabian Sea, 231, 237, 242, 428
Aragon, 122, 123, 126, 132, 155
Arakan, 37, 460, 463
Aranjuez, Treaty of, 525
Araucanian Indians, 515
archers. *See also* horse archers
  African, 392, 403–4
  Korean, 13, 437, 438
  Mughal, 21
*Archimedes*, screw propeller steamship, 585
Arcola, battle of, 225
Arcot, 245
Argentina, 515, 555, 587
Arizona, 139
Armada campaign, 132, 179, 183
Armada of Barlovento, 516
Armed Neutrality (1800), 576
Armed Neutrality of the North (1780), 379
Armée du Nord, 565
Armenia, 292, 297, 503
  Safavid raids, 292
armor, 11
  African, 391, 403
  Qing, 475–76
  *Qizilbash*, 286
Army of Flanders, 124
Arnold, Benedict, 533
Arras, fortifications, 211
artillery, 9, 14–15
  *bans* (rockets), 416
  bombards, 11, 77, 90, 103, 216, 414
  bow chaser, 176
  cannons, 47, 68
    *drake*, 184
    introduction to Southeast Asian warfare, 452
    manufacture in Muscovy, 494
    manufacture in Mysore, 423
    Ming, 77, 90
    of the Bahmani Sultanate, 413
    of the Bidar Sultanate, 414
    Ottoman foundries, 110
    Qing, 482–84
    Safavid, 297–301
    siege guns, 174, 206, 209, 417
  Congreve rockets, 579
  Deccan, 413–14
  French Revolutionary and Napoleonic Wars, 566

horse, 416
in the First Anglo-Sikh War, 427
Maratha, Gribeauval system, 425
Mughal, 415–16, 417–18
Muscovite, 494
naval, 171–73, 175–76, 177, 180
projectiles, 176
stone-throwing, 184
Ottoman, 176–77
Qing, 478
Red Barbarian Gun (*hongyi pao*), 483
siege warfare, 223, 225
French Revolutionary and Napoleonic wars, 226–27
manufacture, 212–13
siege of Malta, 216–17
strength of in sixteenth century, 212–14
swivel guns, 175, 176, 177
camel-mounted, 299, 416, 427
transportation of, 213
artillery corps, Ottoman, 110
artillery defenses, 227
Arughtai, Eastern Mongol leader, 82–83
Arung Palakka, 455
Asante, 393, 394, 395, 406
Assam, 460, 463
Assaye, battle of, 425
assegais, 399, 402
Astarabad, 296, 307
Astrakhan, khanate of, 500, 502
Ivan IV's conquest of, 59–60, 496–97, 503
Ath, siege of, 221
Augsburgh, Peace of, 164, 311, 313
Augustus II, king of Poland-Lithuania and elector of Saxony, 273, 275–76, 279, 329, 509
Augustus III, king of Poland-Lithuania and elector of Saxony, 324
Aurangzeb, Mughal emperor, 247, 305, 417
Austerlitz, battle of, 567
Australia, 250, 420, 573
Austria. *See also* Russo-Austrian–Ottoman War; War of the Austrian Succession
adoption of territorial recruitment, 338
ambitions for Poland, 554
dynastic marriages, 330
French declaration of war on (French Revolutionary Wars), 560–61
land gains, 339
Lutheran rebellions in Upper Austria, 166
Ottoman wars, 99, 313
strategy and tactics in the Napoleonic Wars, 568
Auton, Jean d', 209
Ava (Upper Burma), 449, 454
Ayacucho, battle of, 529
*ayan* (notables and gentry), 112, 117–18
Ayudhya, 37, 449, 456
Burmese invasions, 459–60
conquered by the first Toungoo dynasty, 453
*azab* (peasant militia footmen), 108
Azadlı gunpowder factory, Istanbul, 110
Azerbaijan, 56, 295, 297, 301, 503
firearms industry, 299
Azov, 99, 105, 502
Aztec state, 44
Aztecs, 2, 128, 133

Babur (Zahiruddin Muhammad) Timurid prince and Mughal emperor, 59, 415–16
Bad Axe, battle of, 546
Badajos, siege of, 227
*badlij* or *balmiz* (siege-cannons), 298
Baghdad, 296, 303, 304
Bahmani Sultanate, 413
Bahrain, 307
Bakhiyari, 307
Balaji Vishwanath, 424
Balkans, 102, 154
Habsburg gains, 273
Ottoman campaigns, 23, 104, 121, 294
Ottoman cannon foundries, 110
Ottoman conquest of, 98, 206
Ottoman defeat in, 99
Russian designs on, 112, 554
Baltimore, battle of, 544–45
Baluchis, 298, 307
Banda islands, 240
Bandar Abbas, 240
*banjaras* (mobile grain traders), 421
banners (military and administrative divisions), 61, 65, 67, 95
Eight Banners of the Qing, 20, 32, 40
Three Superior Banners, 471
Barbados, 517
Barbarossa, Hayreddin (Khair-ad-Din), 106, 150–52, 159, 173
Bardi, Gualterotto de', 182
Barfleur, battle of, 370
Barrow, John, 593
*barusura* (Bunyoro regiments), 404
Barwälde, Treaty of, 167, 168
Bassano, battle of, 225
Batavia (Jakarta), 240, 242, 252, 455

Báthory, Stefan, king of Poland, 264
battle axes, 110, 391
battle tactics
  Maratha, 417, 424, 426
  Ming, 78
  Mongol, 82
  Mughal, 415–16
  Muscovite, 493, 500
  Rajput vs. Turkish conflicts, 412–13
  Safavid, 290–91, 296
  siege warfare, 224
  Zulu, 402
Batu Khan, Mongol ruler, 489, 504
Batu Möngke. *See* Dayan Khan
Batur Khong-Taiji, 62
Bavaria, 147, 162, 166
  taxation, 147
  war of succession, 325–26
Bay of Bengal, 231, 238, 242, 449, 454
Bayani, Khanbaba, 283
Bayezid II, Ottoman sultan, 105–6
bayonets, 102, 422, 509, 567
Beachy Head, battle of, 370
Beaufort, Francis, 594
Becher, Johann Joachim, 337
Beijing, 84, 482, 592, 600
  establishment of the Qing at, 467, 473
  Hideyoshi's ambitions for, 434
  Mongol and Turkic trade and tribute missions to, 87
  Russian relations, 498
  violence and corruption under the Ming, 78
Bekovich-Cherkasskii, Prince Aleksandr, 502
Belgrade
  Ottoman capture of, 149
  Ottoman recapture of, 99, 120
  Treaty of, 99
Belize, 141, 517
Bell, David, 47, 553
Bellamy, Chris, 504
Bemba, 403
Bemis Heights (Saratoga), battle of, 533–34
Bena, 403
Bencoolen, 463
Bengal, 244–46, 249, 415, 418, 464
Berar Raja, 426
Berg, 329
Bering, Vitus, 497
Berkovich, Ilya, 562
Bernstorff, Andreas Peter, 322
Bernstorff, Johan Hartwigg Ernst, 322
Berthier, Louis-Alexandre, 568
Bessarabia, 105
Bestuzhev, Alexey Petrovich von, 322
Bethlen, Gábor, prince of Transylvania, 166
Bielfeld, Jakob Friedrich von, 338
Bijapur, 414, 417
Bío-bío, 133
*Birkenhead*, HMS, 586
Biton Kulibali, 394
Black Hawk, 546
Black, Jeremy, 26
Blake, Robert, 186
Blomfield, Reginald, 221
Bodawhpaya, Burmese king, 460, 462
Boers, 400
Bohemian Revolt, 166
Boigne, Benoit de, 425
Bombay, 377
  English trading post at, 244
Bombay Army, 424
Bombay Marine, 249, 421, 589
Bosnia, 104
Boston, 530–32
Bourbons, 41, 326, 329, 519, 560, 563. *See also* War of the Spanish Succession
  alliance with Spain, 331, 379
  artillery, 15, 566
  establishment of Chartered Companies, 333
  imprisoned by Napoleon, 143
Bourcet, Pierre, 568
bows, 391, 403
  bamboo, 411
  composite, 6, 12, 411
    Manchu, 476
    Mongol, 476
    Muscovite, 494
    Qing, 476
  compound
    Mongol, 68
  Ming, 77
Brading, David, 521
Braganzas, 326
Brandenburg-Prussia, 313
  administrative reorganization, 321
  financing of the military, 268
Brandenburg-Prussian army, 355
  captains and colonel-proprietors, 356–60
  recruitment, 360
    cantonal system, 338, 360–62
  the business of the company, 362–64
Brantôme, Pierre de Bourdeille de, 218
Brazil
  British attack on slaving fleet, 587
  British control of the markets of, 580
  Dutch invasions, 517

incorporated into the Spanish crown, 517
Portuguese settlements, 44, 137
Breitenfeld, battle of, 168
Brest, naval base, 375, 376
Brissot, Jacques-Pierre, 561
Britain. *See also* American War of Independence; Anglo-Burmese Wars; Anglo-Chinese Wars; Anglo-Dutch Wars; Anglo-Maratha Wars; First Anglo-Sikh War; War of 1812
attempt to seize Buenos Aires, 143
collection and processing of intelligence, 593–94
colonial rivalries, 336–37
constitutional monarchy, 320–21
domination of overseas trade, 249–51
imperial wars, 577–80
naval power and maritime conflict, 366, 573–75. *See also* East India Company; Royal Navy
activity, 1794–1810, 383–86
enforcement of informal rule in the colonies, 586–88
enrichment opportunities for seamen, 379
finance and administration, 375–76
health and wellbeing of crews, 378–79
improvements in command and administrative structures, 372
late seventeenth- and eighteenth-century European wars, 370–72
naval bases, 376–77
recruitment and quality of crews, 374–75, 597
selection and training of officers, 377–78
Seven Years' War, 379–80
steamships and communication, 588–90
strategy in the French Revolutionary Wars, 575–77
technological advances, 374
technology and strategy, 575–77
trade defense and the Cherbourg strategy, 594–95
shipbuilding, 375, 381–82
succession struggles, 323
Brito, Philip de, 454, 456
Brock, Isaac, 541
Brown Bess muskets, 427, 428
Brühl, Graf Heinrich von, 322
Brunel, Isambard Kingdom, 585
Buddhism, 55, 60, 62, 430, 439
Yellow Hat sect, 85, 500
Buena Vista, battle of, 548
Buenos Aires, 143, 579
British invasion of, 522
Buganda, 396, 397–98, 404
Bugis, 455–56
Bukhara, 64
Bukovina, Austrian annexation of, 339
Bunker Hill, battle of, 530–32, 545, 550
Bunyoro, 396, 404
Burgoyne, John, 24, 533–34
Burgundy, 158, 159
Burma, 43
conquests of the first Toungoo dynasty and incorporation of Portuguese military technology and manpower, 452–53
eighteenth-century expansionist wars, 460
English trading post at Rangoon, 244
Qing invasion of, 468, 475
seizure of Arakan and Assam, 463
siege of Syriam and execution of de Brito, 454–55
use of captives in the military, 458
war with the East India Company, 464, 583
Burton, Ryder, 584
Büsch, Otto, 361
*bushidō* ("the way of the warrior") code, 431, 439, 440–41, 445

Cabral, Pedro Alvares, 234
Cádiz, 527
British blockade of, 528
English sack of, 517
Caffa, 105
Calcutta, 246, 589
Caliano, battle of, 225
Calicut, 234
California, 548, 550
Spanish colonization, 140–41, 521
Calvinists, Calvinism, 165–66, 311, 313, 314
intervention in the War of Devolution, 334
Cambodia, 453, 462
Cambrai, League of. *See* War of the League of Cambrai
Cambrai, Treaty of, 159
camels, 290, 478
as baggage carriers, 421
used as gun mounts, 299, 416, 427
cameralism, 337–38, 339
Canada, 24, 141, 336, 545
British advance from in the American War of Independence, 533
French loss of, 336, 380
operations in the War of 1812, 541–44
Cañari, 133

Canary Islands, 122
Candia, siege of, 208
Candia War, 99, 115, 189, 217, 305
Canete, 239
Canosa, 209
Cape Colony, 400
Cape of Good Hope, 250
  British seizure of, 579
  Dutch colony, 242
Cape Passaro, battle of, 371
Cape St. Vincent, battle of (1780), 527
Cape St. Vincent, battle of (1797), 528
Cape St. Vincent, battle of (1833), 590
Cape Town, 250, 251
Capo d'Orso, battle of, 176
*caponiers* (pill boxes), 209, 214, 228
Carignano, siege of, 225
Carl Gustav, king of Sweden, 266
Carlos II, king of Spain, 136, 327, 518
Carlos III, king of Spain (Carlos I of Sicily and Carlos IV of Naples), 324, 331, 527
Carlos IV, king of Spain, 143
  abdication, 529
Carnatic wars, 245
Carnot, Lazare, 228, 230, 568
Carolinas, 536–37
Cartagena, 519
  invasion of, 519
  Spanish defense of, 141
cartography, 483, 580, 593–94
Caspian Steppe, 500–1
Castelnuovo, 207
Castiglione, battle of, 225
Castile, 123, 126, 155
  finances, 155
  taxation, 131–32, 201
Castries, marquis de, 382
Catalonia, 122, 123, 132
  rebellion against Spain, 163, 517
Cateau-Cambrésis, Peace of, 160
Cathedral of St. Basil the Blessed (Moscow), 496
Catherine II, the Great, empress of Russia, 322, 489, 501, 503–3, 505, 511
Catherine of Braganza, 244
Catholic League, 161, 167
Catholics
  anti-Catholic policies in Vietnam, 465
  conflict with Protestants, 148, 163–69, 313, 314–16
  declining influence of the Papacy, 311–12
  defection to Spanish armies, 127
  German, 162
  persecution of in Korea, 443
  willingness to collaborate with Lutherans, 154
Caucasus, 105, 292, 503
cavalry. *See also* horse archers
  African, 390–91, 394, 395, 397, 403–4
  Dutch East India Company, 455
  Maratha, 418, 425
  Mughal, 12, 415, 416–17
  Muscovite, 13, 493–94
  Mysore, 422
  of the East India Company, 420
  Ottoman, 102, 104, 107–10, 112, 117
  Poland-Lithuania, 148
  Qing, 478–81
  Rajput, 412–13, 416
  Safavid, 30, 286, 292, 306–7
Celali rebels, 30, 115
central institutions, 195–96
Cerignola, 157
Cerro Gordo, battle of, 549
Ceylon, 241, 574
Chaghatai, son of Chinggis Khan, 54
Chakhar state, 466
Chaldiran, battle of, 98, 289, 290, 295–96, 297, 300, 415
Champa, 449
Champlain, Lake, 533, 543, 544
Chandernagore, 247
Chardin, John, 290, 306
Charles I, king of England, 184
Charles II, king of England, 369, 377
  marriage to Catherine of Braganza, 244
Charles IV, duke of Lorraine, 34, 162
Charles V, Holy Roman Emperor, 41, 148, 155, 173, 193, 202, 212
  abdication, 160
  Italian Wars, 156–60
  provision of firearms to the Safavids, 298
  Schmalkaldic War, 164
  unsuccessful attack on Metz, 213
  war with the Ottomans, 149–53
Charles VI, Holy Roman Emperor, 318
  succession, 326, 330
Charles VIII, king of France, 157, 207
Charles XII, king of Sweden, 24, 330, 509, 510
Charles Albert, elector of Bavaria and Holy Roman Emperor, 325
Charles Emmanuel I, duke of Savoy, 161
Charles Louis, elector Palatine, 168
Charles Theodore, elector of Bavaria, 326
Charleston, 580
Charleston, siege of, 536

Checa, 133
Chechnia, 582
Cherbourg, 580
Cherbourg strategy, 594–95, 597
Cherokee, 539
Chesapeake, 533, 537, 544
  battle of, 537
*Chesapeake* (US frigate), 540
Chickasaws, 527
*Chikunda* (slave soldiers), 399
Child, Josiah, 333
Chile, 124, 515, 587
  governance, 521
  indigenous hostility to Spain, 524
  introduction of horses, 6, 138
China. *See also* Ming dynasty; Qing dynasty
  maritime expeditions, 78–79
  trade with, 251
Chinggis Khan, 4, 20, 39, 53, 54–55, 59, 86
  inclusion in the Buddhist tradition, 60
  military structure, 67
Chinggisids, 54–58, 70
Chippawa river, 543
Choctaws, 527
Choiseul, duc de, 322, 330, 349
Chokwe, 401
Cholula, 133
Chosŏn dynasty. *See* Korea
Christian IV, king of Denmark, 166
Christian, prince of Anhalt, 166
Christianity, prohibition of in Japan, 438
Christians
  Japanese persecution of, 434
  migration from Japan to Southeast Asia, 37, 456
  Nobunaga's support for, 433
Churubusco, battle of, 549
Cipolla, Carlo, 207
Circassia, 504, 582
civil service examinations (China), 79
Clark, William, 539
Clausewitz, Carl von, 557, 567–68
Clement VII, Pope, 158
clergy, assigned to Spanish armies, 130–31
Cleves, Philippe de, 172
Clinton, Henry, 534, 535–36, 537
Clive, Robert, 246
Cochi, 240
Cochin, Dutch conquest of, 241, 243
Cocom, 133
Coëhoorn, Menno van, 220, 223
Coen, Jan Pieterszoon, 240
Colbert family, 350
Colbert, Jean Baptiste, marquis de Torcy, 212, 321, 322, 332–34, 368, 376
Coligny, Gaspard de, 160
Colombo, 250
  Dutch fortification, 243
Colonia del Sacramento, 141
colonial militia, 522–23
*Comanchería* (Comanche country), 524
Comanches, 6, 138
  peace with Spain, 139
*Comet*, steamship, 582
communications, transformation and expansion, 192
*Compagnie des Indes*, 246, 247, 253
Comunero Rebellion, 524
Concord, battle of, 530
*Concordia*, Dutch yacht, 242
Condé, prince of, 344, 351
Condorcanqui, José (Túpac Amaru II), 525
Confucianism
  in Annam, 88
  in Japan, 430, 435, 439, 441, 444–45
  in Korea, 436–37, 444–45
Congress of Vienna, 553, 586, 601
Congreve rockets, 421
*conquistadores*, 127, 133, 134
conscription, 27, 48–49, 196–97
  during the Napoleonic Wars, 553, 569
  European systems of, 338
  French ballot system, 353–55, 562, 563
  in the French Revolutionary army, 49, 563–64
  naval, 374–75
    impressment, 597
  Russian system under Peter I, 48, 338, 509
Constantinople, 98, 154, 503
  Ottoman conquest of, 103–5, 313
  siege of, 11
contractor state, 32–38, 50, 267–68
contractors, 23, 32–34, 35, 49–50, 553
  Africa, 37
  Barbary corsairs as, 36
  France, 351–52
  Korea, 437
  military enterprisers, 194–95, 198
  naval, 37, 137, 375
  Qing dynasty, 38
  Spain, 127
contractors-creditors, 267
Contreras, Alfonso de, 131
Convention of 1800, 540
Cook, James, 250

Cooper, Randolf G. S., 425
Copenhagen
  battle of, 576
  Swedish blockade of, 273
Corbett, Julian, *Some Principles of Maritime Strategy*, 170
Corfu, Ottoman attack on, 150
*Cormorant*, HMS, 587
Cornwallis, Charles, 423, 536–38
Corsica, 160, 339
Cortés, Hernán, 127, 128
Cossack–Polish War, 264–66, 506
Cossacks, 35, 38, 102, 263, 498, 502, 510, 553
  attack on Siberia, 497
  involvement in the Great Northern War, 277–78
  involvement in the Thirteen Years' War, 264–65
  raids on Safavid territories, 307
Council of Trent, 165
Coutinho, Tomé de Sousa, 237
Coutras, battle of, 218
Cowpens, battle of, 536
Creeks, 527, 545, 546
Crépy, Treaty of, 159
Cretan War. *See* Candia War
Crete, 99, 102, 105
Crimean khanate, 56–57, 265, 266
  annexed by Russia, 70, 101, 339, 503
  involvement in East European politics, 59
  Ottoman conquest of, 103, 105
  refusal to assist in the Khmelnytsky uprising, 506
  Tatar horsemen, 108
  threat to Muscovy, 491, 494, 495, 500
Crimean War, 512, 599
  sea mines, 596
  use of steamships, 590, 593, 595
Cromwell, Oliver, 368, 518
Cronstadt, 580, 595
crossbows
  multiple-arrows (Ming), 77
  Ottoman, 110
Crysler's Farm, battle of, 543
Cuba, 44, 519, 524, 529, 593
Cuidad Rodrigo, siege of, 225
Cult of Reason, 316
Curaçāo, conquest of, 517
Curzolaris Islands, 153
Cyprus, 102, 105, 107
Cyprus War, 153, 173, 180

D'Ormea, Carlo Francesco, 322
Da Nang, 465
Dade Massacre, 547
Dagestan, 504
Dahlberg, Eric, 220
Dahomeans, 395
Dahomey, 393–95, 404
Dai Viet, 449, 452–53
*Dal Khalsa* (Khalsa army), 426–28
Dalai Lama, 60, 62
Dalmatia, 99, 152
Daniil Aleksandrovich, prince of Moscow, 489
Daniilovichi, 491, 492, 495
Danish East India Company), 247
Daulat Rao Sindia, 426
Daur, 498
Dayan Khan, 60, 84
De'Marchi, Francesco, 227
Deane, Richard, 186
Deccan, 418, 419, 424
Deccan Wars, 417–18
Delhi, conquered by Nadir Shah Afshar, 64
Delhi Sultanate, 31
  military recruitment, 36
  military system, 410–13
Della Valle, Pietro, 290
*Demologos*, steam warship, 583, 590
Denmark, 145, 255, 264, 279, 280, 381, 583
  attempt to force out of the Armed Neutrality, 576
  consultative monarchy, 257
  decline of representative institutions, 203
  enlisted to the Armed Neutrality, 576
  influence in Northern Europe, 147
  intervention in the Thirty Years' War, 166–67
  Livonian War, 505
  military contractor-creditors, 267
  navy, 374
  Nordic Seven Years' War, 260
  overseas trade, 246–47
  Scanian War, 268
  Thirteen Years' War, 266
  withdrawal from the Great Northern War, 273, 509
Derfflinger, Georg von, 344, 356
Deshima, 242
Detroit
  siege of, 541, 550
  US recapture of, 542
Deulino, Truce of, 506
Devlet Girey, Crimean khan, 500
*devshirme* (Ottoman recruitment system), 21

Dhu'l-Qadr, 293
Dhuka, Yahya, 283
*Diana*, steamship, 583
Diarbakr, 295
Dillon, Théobold, 563
diplomacy, Ming, 74–75, 78
Diplomatic Revolution of 1756, 316
disease, 7–8, 88, 134, 371
  as a factor in Spanish dominance in the Americas, 519–20, 522
  casualties in Spain's attempt to reconquer the Americas, 529
  casualties in the First Anglo-Burmese War, 464
  effect on native populations, 45, 139
  First Anglo-Burmese War, 584
  insect-borne, 388, 600
Diu, siege of, 236
divide and rule tactic, 81
Dmitrii Donskoi, 492
Don Cossack Host, 502
*Don Quixote*, 126, 128–29
Doria, Andrea, 150, 159, 217
Doria, Antonio, 173
Doria, Filippino, 176
Doroshenko, Petro, 266
Douai, siege of, 223
Downs, battle of the, 135, 183, 366
Drake, Francis, 181, 182, 516, 524
*Dreadnought* (galleon), 180
du Mans, Raphaël, 300
Dumouriez, Charles-François, 560, 564
Dupleix, Joseph, 246
Duras, duke of, marshal of France, 350
Durrani dynasty, 424
Durrës, 152
Dutch East India Company (*Verenigde Oost-Indische Companie*, VOC), 15, 154, 238–44, 245, 246, 250, 252, 453, 455, 456
Dutch East Indies, 554, 578, 582
Dutch Republic, 147, 320
  expansion and involvement in Southeast Asian warfare, 455
  expansion of overseas trading networks, 238–44
  fiscal policy, 147
  fortifications, 220
  heavy arms industry, 212
  invasions of Brazil, 517
  loss of overseas possessions, 250
  naval power, 366
  overseas expansion, 253
  reliance on credit, 201
  siege warfare, 208
  war against Javanese rebels, 464
Dutch Revolt, 127, 153, 155, 163, 203, 238
  destruction of the citadels of Utrecht and Antwerp, 212
Dutch War, 329, 334
  naval operations, 369
Dutch West India Company (WIC), 517
Dutch–Portuguese War, 240–42
dynasticism, 316–22
  continuing importance of royal marriage and familial political links, 330–31
  international relations and, 322
  prevention of hostilities, 329–30
  role in major European conflicts, 322–27
    influence of intermarriage, 327–28
dysentery, 7

East Asian War. *See* Imjin War
East India Company (EIC), 8, 37, 38, 43, 154, 244–46, 252, 419. *See also* Danish East India Company; Dutch East India Company; Swedish East India Company
  activity in the Indian Ocean region, 249–51
  Anglo-Maratha Wars, 423–26
  Anglo-Mysore Wars, 422–23
  conflict with the Khalsa/Sikh kingdom, 426–28
  displacement of the Dutch in the Straits of Melaka, 463
  hybrid military system, 420–22
  importance of communications, 589
  navy, 249, 421–22
  steamships, 583
  war with Burma, 464
Eaton, Richard, 413
Edict of Restitution, 167, 168
Efendy, Dourry, 308
Egypt, 390
  military expansionism, 406
  Ottoman conquest of, 235
  Portuguese plans for, 235
Eilat, 231
Elba, Isle of, 178
elephants
  as baggage carriers, 421
  as combat animals, 14, 412, 415, 416, 451, 457
Elisabeth, empress of Russia, 322
Elisabeth of Württemberg, 330
Elizabeth I, queen of England, 142, 182, 244
Elliott, John, 514
Emmanuel-Philibert, duke of Savoy, 160

*encomenderos*, 513
England. *See also* Anglo-Dutch Wars; Britain; East India Company
establishment in the New World, 134
expansion of overseas trade routes, 244–46
first campaigns of the Spanish Armada, 132
invaded by William of Orange, 42, 369
investment in privateering, 182
manufacture of cast-iron artillery, 212
naval strategy, 180
naval technology, 184
religious reformation, 124
representative institutions, 203
seizure of Spanish territories in the Caribbean, 517–18
shipbuilding, 175, 188, 369
English Civil War, 163, 212
English Navigation Acts, 335
*Enterprise*, steamship, 583
Erik XIV, king of Sweden, 261
Erkele II, king of Georgia, 503
Ermes, battle of, 259
Érsekújvár, 99
Esen, Oirat leader, 58
Esen Taishi, Oyirad ruler, 83–84, 94
*esham* system (Ottoman fiscal policy), 119
espionage and intelligence, 421
Ethiopia, 5, 399, 400
adoption of firearms, 13, 390
British campaign against, 406
firearms and military professionalism in the seventeenth and eighteenth centuries, 405
horses used in warfare, 137
military strength in the seventeenth and eighteen centuries, 397
warfare in the sixteenth century, 388–90
Eugene, prince of Savoy, 99, 370
Evans, John, 141–43

*Fairy*, HMS, 595
Falconi, Alessandro, 181
Falkland Islands, 336
Fallen Timbers, battle of, 539
Famagusta, siege of, 153, 209
Farnese, Elisabeth, 321, 324, 325
Fars, 295
Fausto, Vettor, 175
Fawkes, Guy, 127
Fehrbellin, battle of, 268
female fighters, 404
Feodor, tsar of Russia, 505
Ferdinand I, Holy Roman Emperor and Archduke of Austria, 150, 157–58, 160, 164–65
Ferdinand II, Holy Roman Emperor, 156, 162, 166, 167, 168
Ferdinand II, king of Aragon, 155, 157
Ferdinand III, Holy Roman Emperor, 168, 193
Ferdinand VI, king of Spain, 331
Ferdinand VII, king of Spain, 529
abdication and restoration, 529
Ferozeshah, battle of, 427
Finland, 510, 554
Fioravanti, Aristotele, 494
fire arrows, 90, 92, 413
firearms, 8–9, 12–14, 68–69. *See also* matchlock firearms; muskets
adoption of in Africa, 403, 404
advancements in Southeast Asian warfare, 457
arquebus, 110, 298, 477, 494
hand cannon, 476, 477
handguns, 9, 10, 12
Ethiopian, 13
Ming, 477, 482
Persian, 419
Safavid, 298, 300
improvements in, 192
Japanese, 91
Ming, 90
Ottoman, 68, 110
Portuguese, introduction to Southeast Asia, 452
Qing, 476–78
rapid-firing small arms, 230
rifles, 295, 405
breech-loading, 14, 48, 230, 406
Russia, 494
Safavid, 297–301
used in African warfare, 392, 394, 395–96
used in fifteenth-century Southeast Asia, 449
First Anglo-Burmese War, 8, 455, 464, 583
First Anglo-Chinese War, 591–93, 596, 600
First Anglo-Dutch War, 242
naval warfare, 186–88
First Anglo-Maratha War, 424–25
First Anglo-Mysore War, 422
First Anglo-Sikh War, 427
First Coalition of 1792–1797, 552
First Partition of Poland, 511
First Seminole War, 546
Flanders, 41

Fleury, cardinal, 322
Floor, Willem, 283
Florence, 159, 215
  fortifications, 206, 212, 214
Florida
  American designs on, 541, 578
  ceded to the United States, 546
  encouragement of emigration to, 527
  removal of the Seminoles from, 547
  Spanish recovery of, 141, 525, 527
  value to Spain, 140
foreign policy advisors, 322
Formosa, 240, 243
Fort Bowyer, battle of, 545
Fort Dansborg, 246
Fort-Dauphin, 247
Fort Erie, 543, 544
Fort George, 542, 543
Fort Jesus (Mombasa), 237
Fort McHenry, battle of, 544, 579
Fort Niagara, 542
Fort Oranje, 243
Fort St. Philip, 545
Fort Ticonderoga, 532
Fort William (Calcutta), 246
Fort Zeelandia, 240
fortifications, 15–16
  advances in the late seventeenth century, 220–23
  artillery forts, 227
  bastions, 208–9, 218, 223
  city defenses, 211–12
  coastal defenses, 580
  dismantling of, 230–30
  ditches, 209, 218
    designed by Carnot, 228
  evolution of, 208–9
  fieldworks, 216
  Mughal, 415
  Muscovite, 68, 500
  Mysore, 423
  Peninsular War, 227–28
  Qing, 482
  role in Dutch and Portuguese expansion, 242–43
  Safavid, 291
  Spanish America, 519
  *trace italienne*, 147, 148, 423
  trenches, 221
Four-Days Battle, 186
Fourth Anglo-Dutch War, 250
Fourth Anglo-Mysore War, 423
France, 147. *See also* French army; French Revolutionary and Napoleonic Wars; French Wars of Religion
  alliance with the Ottomans, 149
  contestation of British sea power, 577
  fiscal policy, 148
  heavy arms industry, 212
  invasion of Italy, 145, 156, 207–8
  involvement in the American War of Independence, 525–26
  loss of possessions in India to the British, 249
  naval power and maritime conflict, 366
    American War of Independence, 382–83
    finance and administration, 376
    health and wellbeing of crews, 378–79
    late seventeenth- and eighteenth-century European wars, 369–72
    recruitment, 374–75
    selection and training of officers, 377–78
    Seven Years' War, 379–80
    steam-power technology, 596
    technological improvements, 372
  overseas trade, 247
  per-capita tax burden, 201
  shipbuilding, 369, 375, 381–82
  subsidization of the Dutch, 202
  succession struggles, 330
  trade defense, 594–95
  wars with the Habsburgs, 155–63
Francis, duke of Guise, 160
Francis I, Holy Roman Emperor (Francis Stephen, duke of Lorraine), 325
  forced to accept Tuscany in exchange for Lorraine, 339
Francis I, king of France, 156, 157–59
Francis II, Holy Roman Emperor, 330
Franco-Dutch War. *See* Dutch War
Frederick I, king of Prussia, 322, 356, 360
Frederick I, king of Sweden, 330
Frederick II, king of Denmark, 257
Frederick II, the Great, king of Prussia, 319, 329, 358, 359–60, 363, 364, 365, 511
  armies of, 562
  contempt for religion, 316
  invasion of Silesia, 325
Frederick III, duke of Holstein-Gottorp, 301
Frederick III, Palatinate elector, 165
Frederick V, Palatinate elector, 166
Frederick William, elector of Brandenburg, 268, 323

Frederick William I, king of Prussia, 280, 321, 322, 350, 356
Frederick William II, king of Prussia, 331, 359, 560
Frederick William III, king of Prussia, 359
Freeman's Farm, battle of, 534
French army, 344–45
  captains as proprietors, 348–50
  colonel-proprietors, 346–48
  generals, 345
  military administration and recruitment, 350–55
French Revolutionary and Napoleonic Wars, 552, 556–59
  British naval defeat at Mauritius, 250
  British occupation of Java, 463
  continuity and adaptation of the revolutionary armies, 562–66
  defeat of Napoleon, 569–72
  effects on US commercial activity, 539
  global nature of, 552–56
  honor and morale, 566
  invasion of Russia, 511–12
  justification, 559–61
  naval warfare, 575–77
  recruitment and size of armies, 48, 552–53
  siege warfare, 225–30
  Spanish involvement, 143–44, 528–29
  weaponry, strategy and tactics, 566–69
French Wars of Religion, 161, 163, 214, 225, 315
French West Indies, 578
Frenchtown, battle of, 542
*frontière de fer*, 221
Fuenterrabía, 157
Fugger family, 202
Fulani, 403
Fuller, William, 508
Fulton, Robert, 579, 582–83, 590
fur trade, 497–98
Futa Jalon, 403
Futa Toro, 403

Gabbard, battle of, 187
Galdan, Zunghar leader, 62, 468, 484
Galle Fort, 243
Gallileo, 220
Gálvez, Bernardo de, 139
Gálvez, José de, 140, 522
Gama, Vasco da, 172, 234–35
Gangut, battle of, 280
*ganimi kava* tactic, 418, 424, 426
Garza, Andrew de la, 417
Gates, Horatio, 533, 536
Genoa, 161, 217, 339
Genoese, 231
Gensonné, Armand, 561
George I, king of Great Britain, 280, 321
George II, king of Great Britain, 321
George III, king of Great Britain, 321
George William, elector of Brandenburg, 167
Georgia, 292, 294, 297, 301, 503
  Safavid raids on, 292
Georgia (US state), 536–37, 546
Georgian troops, 302, 306
Germany, 124, 147, 155
  Protestantism, 150, 160
  Schmalkaldic War, 164
  Thirty Years' War, 162–63
Ghagra, battle of, 416
*ghazi* warfare, 292
*ghazi*-corsairs, 106
Ghent, Treaty of, 545
*ghulaman-i khassa-yi sharifa* (slaves of the royal household), 289
*ghulams* (slave soldiers), 20, 289, 293, 301–3, 304, 413
Ghurids, 412
Gianfigliazzi, Niccolò, 182
Gibraltar, 136, 324, 525
  British naval base, 377
  capture of during the War of the Spanish Succession, 371, 518
  siege of, 227, 538
Giorgio, Francesco di, 209
Giovio, Paolo, 176, 207
Girona, siege of, 225
Girondins, 561
Glorious First of June, 555
Glorious Revolution, 203
Gneisenau, Field-Marshal, 385
Goa, 235, 240
  Dutch blockade of, 241
  Dutch–Portuguese battles, 241
Goddard, Thomas, 424
Golden Horde, 55, 56–57, 58, 60, 491, 492, 500, 502
Goletta, 150
Golkunda, 417
Gommans, Jos, 409, 414
Góngora, Luis de, 131
Gonzague-Nevers, Charles de, 161
Gorchakov, Alexander, prince, 488
*Gorgon*, HMS, 591
Goslar, 167
Gough, Hugh, 427
Gozo, 152

Granada, 122
reconquest of, 206, 229
Grande, Domenego, 178
Gravelines, battle of, 181
*Great Britain*, SS, 585, 595
Great Indian Rebellion, 251
Great Northern War, 24, 247, 255, 272–80, 329, 507–10
Great Turkish War, 339, 502
Green Standard Armies, 32, 474–75, 485
armor, 476
horses, 481
Greene, Nathanael, 536–37
Greenville, Treaty of, 539
Gribeauval system, 15, 425, 566
Grol (Groenlo), siege of, 213
Guadalupe-Hidalgo, Treaty of, 550
Guadeloupe, 579
Guangzhou, 244
Guaraní war, 141
Guelders, duke of, 159
guerrilla warfare, 143
and the spice trade, 240
Annamese, 88–89
battle of Sarhū, 479
Java War, 464
Maratha, 418
Mongol, 81
naval, 424
Peninsular War, 554, 570
Safavid, 290
Guicciardini, Francesco, 207–8
Guiche, Comte de, 186
Guilford Courthouse, battle of, 537
Guilmartin, John F., 170, 214
Guiomar, Jean-Yves, 47, 553
Gujarat, 237
gunboat diplomacy, 447
gunpowder, 9, 10, 230
application in naval warfare, 16
arrival in Muscovy, 494
Chinese knowledge of, 477
improvements in the manufacture of, 192
knowledge of in India, 413
Ottoman production, 110
spread of from China to Eurasia, 68
used by the Ming, 77
gunpowder weapons, 10–12, 13, 15, 43
first recorded use of, 297
introduction of in the Deccan, 413–14
introduction to India, 428
Mughal, 415
rocket arrow launcher, 77
*gûsa* (socio-military organizations), 470
Gustav Vasa, king of Sweden, 258
Gustavus Adolphus, king of Sweden, 147, 167–68, 193, 415

Habsburgs, 326. *See also* War of the Spanish Succession
acquisition of the thrones of Castile and Aragon, 122
anti-Swedish alliance, 266
dynastic marriages, 330
establishment of a court war council (*Hofkriegsrat*), 195
financial crisis, 203
foreign office, 321
fortifications on the Ottoman frontier, 220
gains in the Balkans, 273
Prussian seizure of Silesia from, 339
reconquest of Hungary, 314
relations with the Ottomans, 42, 111, 313. *See also* Ottoman–Habsburg wars
taxation, 313
territories of, 145–47
wars with France, 41, 155–63
Haitian revolution, 524, 554
Hajji Giray, 56, 58, 67
Hale, John, 208
Halifax, 377
Hami, 87
Haneda, Masashi, 283
Hanoverians, 314, 321, 329
Hanseatic League, 257
Harrison, William Henry, 539, 542
Hausa, 403
Havana, 380, 519
British capture of, 520
Hawkins, John, 182
Hehe, 403
Henri II, king of France, 160, 164
Henri IV, king of France, 161, 165
Henry, duke of Guise, 161
Henry VIII, king of England, 175
Herat, 295
Herberstein, Sigismund von, 13, 494
Herzegovina, 103, 104
Heyn, Piet, 131
Hideyoshi. *See* Toyotomi Hideyoshi
Hindus, 413
Hintze, Otto, 193
Hispaniola, 44, 518
runaway slave communities, 524

Hoche, Lazare, 565
Hohenzollerns, 314, 329
Holy League (1538), 150, 174
Holy League (1571), 153, 178
Holy League (1684), 99, 115, 273, 313
Holy Roman Empire, 148
  fiscal policy, 148
  improvements in postal and printing services, 192
  religious wars, 163–69
Hong Kong, 250
*Hongan-ji*, 433
Hongwu Emperor, 81, 87
Hormuz, 232, 235, 240, 244, 301, 303
  siege of, 236
Hörnigk, Philipp Wilhelm von, 333, 337
horse archers, 428, 504
  Eurasian steppe, 68
  Mughal, 415–16
  Qing, 480
  Turkish, 412
horse artillery, 416
horse-drawn guns, 207, 427
horses, horsemanship, 22, 133
  horse casualties in the Peninsular War, 570
  in nomadic societies, 5, 53, 64, 65, 68
  in the Safavid army, 289
  Indian vs. European, 420–21
  introduced by the Spanish to the Americas, 6, 137
  introduction to Africa, 137
  Ming deficiencies in, 83
  Ming horse fairs, 75
  Mughal, 419
  Qing, 478–81
  Rajputs vs. Turkish, 411–12
House of Orange, 320
household subsistence economy, 196–97
Howard, Charles, Lord of Effingham, 181
Howe, Richard, 532–33, 535, 555
Howe, William, 532–34
howitzers, 223, 229, 421
Hradčany Castle, 166
Hsinhpushin, Burmese king, 460
Huari, 133
Huexotzinca, 128, 133
Hughes, Edward, 249
Hugli, 247
Huguenot (Protestant) party, 161
Huguenots, 315, 357
*Huitong guan* (Interpreters' Institute), 75
Hull, arsenal, 212
Hundred Years' War, 145, 206
Hung Taiji, Qing Emperor, 20, 61, 67, 95–96, 466–67, 471
Hungary, 102, 145
  Catholic offensive against Calvinists and Lutherans, 314
  ceded by the Ottomans to the Habsburgs, 99, 111, 313
  Ottoman cannon foundries, 110
  Ottoman territorial gains in, 98–99
  Ottoman wars in, 105, 109, 114–15, 149–50, 153, 154
Hussite rebellion, 166
Hwang Sayŏng, 443
Hyder Ali, sultan and *de facto* ruler of Mysore, 249, 422
Hydrographic Office, 594

Iam Zapolskii, Truce of, 261
Iaroslavl, 491
Ibrahim Lodhi, Delhi sultan, 36, 415–16
Ibrahim Pasha, Ottoman grand vizier, 236
Île de France, 249
Ilgen, Heinrich Rüdiger von, 322
Illinois, 141, 546
Imam Shamyl, 504
Imbangala, 392–93
Imjin War, 90–93, 434–35, 447, 482
  Korean land forces, 436
  Korean military organization, 438
  Ming weaponry, 477
Imperial Cannon Foundry, Istanbul, 110
Inca, 2, 44, 133
India. *See also* Mughals
  British control of, 249
  caste system, 409
  Danish trade in, 246
  Dutch trade networks, 240
  French trade in, 247
  pre-Mughal-era warfare
    introduction of gunpowder weapons, 413–14
    Turkish and Rajput military systems, 410–13
Indian Ocean, 231
  British control of, 579
  British presence in, 244–45
  Chinese expeditions, 232
  Portuguese expansion in, 232–35
indigenous people. *See also* Native Americans
  hospitable nature of, 132–33
  imperialisms in America, 137
  in foreign armies, 37

in Spanish colonial armies, 521–22, 527
influence of the introduction of horses, 137
prosperity and relations with the Spanish, 134
threat to Spanish rule, 523–24
Industrial Revolution, 251, 252
Ingria, 510
Innocent XI, pope, 311, 313
Innsbruck, 160
international currency transfer, 202
Iran, tribalism, 63–64
Iraq, 105, 236
Iraq-i ʿajam (western Iran), 295
Iroquois, 539
Isabella I, Queen of Castile, 155
Iskander Muda, sultan of Aceh, 238, 456
Islam, 504
and African military culture, 403
and Christian rivalry, 313–14
conversion of the western khans to, 55
in the Indian Ocean region, 234
Isma Ismaʿil I, Safavid shah, 295–96, 297–98, 300
Istanbul, Treaty of (1547), 150
Istanbul, Treaty of (1612), 303
Istanbul, Treaty of (1700), 99
Italian Wars, 145, 155–60, 225
French invasion of Italy (1494), 156, 207–8
French invasion of Italy (1499), 156
siege warfare, 213–16
Ivan III, the Great, prince of Moscow, 491–94, 495, 500, 512
ambitions for Livonia, 505
rivalry with Lithuania, 504–5
Ivan IV, the Terrible, prince of Moscow and tsar of Russia, 260, 262, 489, 494–95, 502, 503
destruction and incorporation of Kazan, 60, 495–96
eastward expansion, 495–97
expansion of the army, 259
westward expansion, 505

Jackson, Andrew, 545–46
Jacobin Republic, 48
Jadwiga, queen of Poland, 505
Jagatai khanate, 412
Jägerndorf, 329
*jagirs* (land grants), 416
Jahangir, Mughal emperor, 244
Jalilis, 112
Jam, battle of, 290, 296
Jamaica, 552
British naval base, 377
English capture of, 368, 518
James I, king of England, 166
James VII and II, king of Scotland and England, 323
Jamestown, 134
Jan II Casimir, king of Poland and grand duke of Lithuania, 265
Jani Khan, 304
Janibeg, 58
Janissaries, 101, 107, 114–16, 302
abolition of, 101
casualty rate, 110
increase in number of during the reign of Mehmed II, 104
metamorphosis of, 115
system of recruitment, 21
weapons, 110
Janusz Radziwiłł, hetman of Lithuania, 265
Japan
expulsion of the Portuguese, 242
invasions of Korea, 90–93, 434–35
isolationist policy (*sakoku*), 12, 32, 438
relations with the Ming, 72–74
reunification, 432, 433–34
stability under the Tokugawa shogunate, 435–36
trade with the Dutch, 242
warrior elite and martial virtue, 438–41
wars in the Sengoku period, 432–33
weaponry, 91
*jasaq* (*yasa*, Mongol law code), 54
Jaswant Rao Holkar, 426
Java, 137, 554
British occupation during the Napoleonic Wars, 463, 579
Chinese expeditions, 232
civil war, 459
Dutch defense of, 243
Dutch involvement in local politics, 455
foundation of Batavia, 240
Java War, 464
javelins, 110, 395
Jefferson, Thomas, 539
Jena, battle of, 567, 569
Jesuits, 312
influence on Qing weapons technology, 478, 482, 483–84
Jiddah, 232
jihadist movements, 403–4
Jin dynasty, 93
*Jinyiwei* (Ming Imperial Guard), 86
Joao II, king of Portugal, 234

Jochi, son of Chinggis Khan, 54, 56
Jogaila, grand duke of Lithuania, 505
John Frederick, elector of Saxony, 164
John George, elector of Saxony, 166–68
John William, duke of Cleves-Jülich, 165
Jones, John, 228
Joseph II, Holy Roman Emperor, 319, 330
Jülich, 329
Jurchen state, 58
Jurchens, 61, 72, 93–95, 138, 436, 437, 471, 477, 483
  supply of horses, 478–79
Justi, Johann Heinrich Gottlob von, 338

Kabalega, ruler of Bunyoro, 405
Kaempfer, Engelbert, 301, 306
*kaizhong* system, 80
Kalmar Union, 260
Kalmyks, 62, 500–1
Kama river, 496
Kamieniec, 99
Kanagawa, Treaty of, 446
Kandy, 554
Kanghwado, Treaty of, 447
Kangxi Emperor, 467–68, 483, 498–99
Kanhoji Angre, 249, 423
*kapılı* (*levend* militiamen), 117
*kapu halkı* (private armies), 117
Karaman, 103, 104
Karamojong, 397
Karl XII, king of Sweden, 273, 276–80
Karlowitz, Treaty of, 99, 111, 305
Karnal, battle of, 419
Kasimov, khanate of, 57, 492
Kassa Hailu. *See* Tewodros, emperor of Ethiopia
Kaunitz-Rietberg, Wenzel Anton von, 319, 322, 561
Kazakhs, 58–59, 62–63, 67, 70, 481, 501–2
Kazakhstan, 70, 501
Kazan, khanate of, 56–57, 60
  Russian conquest of, 489, 495–97
  secession from the Golden Horde, 491, 500
Keep, John, 493
Kemal Reis, Turkish corsair, 106
Kennedy, Paul, 417
Kerner, Robert, 488
*keshig* (Mongol imperial bodyguard), 20, 54, 86
*Kesselschlacht* (cauldron battle), 416
Khabarov, Erofei, 498
Khalkha, 468
Khalsa / Sikh kingdom, war with the East India Company, 426–28
Khanwa, battle of, 416
Khiva, 64, 502
Khlesl, Melchior, 165
Khmelnytsky, Bohdan, 265, 266, 506
Khmelnytsky uprising (Cossack–Polish War), 265, 314, 506
Khoisan, 400
Khubilai Khan, 55, 60, 71
Khurasan, 295, 303, 307
Khwarazm, 63
Kievan Rus, 491, 505
Kilia, 105, 106
Kimbu, 402–3
King's Mountain, battle of, 536
Kitara, 396
Kliszow, battle of, 276
Kliuchevskii, Vasilii, 488
Knights of St. John, 152
Kochinga. *See* Zheng Chenggong
Koja Sinan Paşa, Ottoman grand vizier, 189
Kojong, king of Korea, 447
*kokugaku* (philosophical movement), 439
Kolff, Dirk, 409
Kolkata, 244
Kongo, 391, 400
Korea
  cannon, 483
  civilian and military hierarchy, 431, 443
  government-controlled military organization and warfare, 431, 436–38
  Imjin War, 90–93, 434–35
  military examination (*mukwa*), 432, 437, 438, 443, 445
  military organization and the Confucian scholar officials, 442–43
  supply of military aid to the Qing, 442
Krishna Raya, Vijayanagara emperor, 413
Kuchum, khan of Siberia, 497
Kulikovo, battle of, 492
Kurdan, Treaty of, 112
Kurds, 293
Kwŏn Yul, 438
Kyoto, capture of, 432

La Rochelle, sieges of, 217–18
Laird, Macgregor, 600
Lake, Gerard, 426
Lake Borgne, battle of, 545
Lake Champlain, 544
Lake Erie, 542
Lake Malawi, 398

Lake Ontario, 542, 543
lances, 298, 391, 403, 416, 421, 424, 479
Lao, 453
Law of the Sea, 379
Lê dynasty, 89, 453, 462, 470
Lê Loi, rebel leader, 89
Le Tellier, François-Michel. *See* Louvois, marquis de
Le Tellier, Michel, 322
Le Tellier family, 350
League of Cognac, 158, 217
Leal, Juan Valdés, 136
LeDonne, John, 488
Leiden, siege of, 148
*Leopard* (British warship), 540
Leopold I, Holy Roman Emperor, 315, 323, 327
Leopold II, Holy Roman Emperor, 560
Lepanto, battle of, 17, 42, 107, 120, 131, 174, 178, 181, 182
Lesnaia, battle of, 277, 510
Leszczyński, Stanisław, king of Poland-Lithuania, 276, 339
Levant, 182, 240, 243, 406
Lewenhaupt, Adam Ludwig, 277
Lewis, Meriweather, 539
Lexington, battle of, 530
Lhasa, 468
Li family, 91
Li Zicheng, 96
Liaodong, 83, 473, 483, 485
  fortifications, 482
  horse trade, 478
  Qing campaign against, 96, 466–67, 472, 478
Ligdan, Mongol khan, 61, 467
*Lightning*, HMS, 584
Lille, siege of, 223
Lincoln, Abraham, 546
Lincoln, Benjamin, 536
Lines of Torres Vedras, 228
Lionne, Hugues de, 319
Lisbon, 578, 590
Lithuania, Grand Duchy of, 57, 504–5
Little Ice Age, 7
Little Nogay Horde, 106
*Liverpool*, gunboat, 599
Livonian Order, 255, 257, 259
Livonian War (First Northern War), 255, 259–62, 497, 505, 506
Lockhart, Laurence, 283
Lodhi Sultanate, 415–16
logistics, 31
  African warfare, 394, 398, 407
  challenges for the Russians, 495, 500, 512
  challenges in the War of 1812, 541, 543, 544
  corruption in, 352
  East India Company (EIC), 421
  French Revolutionary and Napoleonic wars, 552, 568
  Imjin War, 93
  in siege warfare, 206, 217, 226
  Mughal, 418
  Ottoman, 11, 22–23, 105, 111, 113, 291
  private entrepreneurship, 343
  problems for the Safavids, 292, 299
  role of galleys in, 174–75, 189
  steamships and, 595
Lombardy, 157, 159, 161
Lonato, battle of, 225
Long Island, battle of, 532
Long Turkish War (1593–1606), 30, 114–15, 154, 313
Lorge, duke of, marshal of France, 350
Los Adaes fort, 124
Louis II Jagellio, king of Hungary, 149
Louis XII, king of France, 157, 214
Louis XIII, king of France, 161, 167
Louis XIV, king of France, 21, 27, 29, 221, 223, 225, 230, 315, 324, 332, 348, 350
  abolition of the *colonel général de l'infanterie*, 345
  attack on the Vaudois, 315
  belief in absolute monarchy, 319
  disbandment of troops in peacetime, 347
  dynastic issues, 323, 327
  hostile relations with the Papacy, 311
  interference in the battles of the Nine Years' War, 370
  introduction of French military uniforms, 349
  invasion of Dutch territory, 334, 369
  naval power, 366
  negotiations with the ministers of Queen Anne, 321
  on the War of the Spanish Succession, 335
  relations with Siam, 458
  succession, 330
  War of Devolution, 323
Louis XV, king of France, 316, 321, 324, 325, 330–31, 336
Louis XVI, king of France, 558, 565
  marriage to Marie Antoinette, 331
Louise of Savoy, 158
Louisiana, 137, 141, 143, 336
  assigned by the English to Spain, 140
  ceded by the French to Spain, 336
  Spanish surrender to the US, 143

Louisiana (cont.)
US exploration of, 539
Louisiana Purchase, 540
Louverture, Toussaint, 554
Louvois, Marquis de, 223, 230, 322, 334, 350–51
Lowestoft, battle of, 187
Lübeck, Treaty of, 167
Lucca, bastions of, 212
Luchuan, 89
Lund, battle of, 268
Lund, Peace of, 269
Lundy's Lane, battle of, 543
Luristan, 301
Lusatia, 166
Lütfi Paşa, 171
Lutherans, 154, 164–65, 166–67, 311, 313, 314
Lützen, battle of, 168, 193
Lynn, John A., 224, 346

Maasai, 397, 399, 401
Mac Dang-dung, Vietnamese leader, 89
Mac dynasty, 453
Macao
Dutch blockade of, 240
Portuguese trade from, 238
Machiavelli, Niccolò, 145, 206
Machiavelli, Piero, 171
Mack von Leiberich, Karl, 568
Mackinder, Halford, 488
Madagascar, 247, 251
Madras, 244, 246
Madrid, Treaty of, 159, 518
Magdeburg, siege of, 167
Magnus, brother of the king of Denmark, 260, 261
Mahadji Sindia, 425
Mahan, Alfred Thayer, *The Influence of Sea Power upon History*, 170
Mahmud II, Ottoman sultan, 101
Mahmud of Ghazni, 413
Majapahit, 449
Makassarese, 455–56
Malaga, battle of, 371
malaria, 7, 400
Malawi, 401, 403
Maldives, 232
Mali, 137
*malikane* (life-term tax farms), 119
Malta, 152, 377
blockade of, 575
siege of, 42, 152, 216–17
Maluku, 455
Mamluks, 98, 102, 390, 413
Mámora, Spanish conquest of, 132
Manaji Angre, 424
Manchuria, 58, 61, 468, 485
fortifications, 482
tribal threat to the Ming, 72
Manchus, 20, 61–62, 76, 93–94, 95–96, 138
banners, 472
bows, 476
cannon, 482
horsemen, 479–80
wars with Korea, 436, 438, 442
Mandans, 143
Mangalore, Treaty of, 422
Manghit tribe (Noghay), 57, 60, 63
Manila, 252, 380
British capture of, 520
British occupation, 463
Dutch blockade of, 239, 240
Portuguese trade with, 238
Mansfeld, Ernest, Count of, 166–67
Mantua, 161
siege of, 225
Manuel I, king of Portugal, 234
Mao Yuanyi, *Wubeizhi (On Military Preparedness)*, 76–77
Mapuche, 6, 138, 141
Marathas, 12, 428. *See also* Anglo-Maratha Wars
navy, 37, 244, 249
reform of the military, 425
requested by the Mysore to join the war against the British, 422
war against the Mughals, 417–18
Maravi, 398
Marciano, battle of, 215
Marengo, battle of, 567
Margareta Theresa of Spain, 327
Maria Theresa, Holy Roman Empress, 321, 324–25, 326, 329, 511
Marie Antoinette, queen of France, 331, 561
Marignano, 157
Marion, Francis, 536
Marlborough, duke of, 225, 370
maroons, 524
Marteilhe, Jean, 174
Martinique, 579
*martoloses* (hired forces in the Ottoman forces), 36
Marx, Karl, 128
Mary II, queen of England, 323, 370
Mary of Hungary, 159, 169
Mashhad, 296
manufacture of cannon, 299

Masqat, 307
Masula, Kalonga ruler, 398
Masulipatam, 246
Matanzas Bay, battle of, 517
Mataram, 455, 457
  civil war, 459
matchlock firearms, 14, 395
  arquebus, 110
  Deccani, 414
  Ming, 477
  Mughal, 415–16
  muskets, 110, 400, 401
  Southeast Asia, 457
Matthias, count of Thurn, 166
Matthias, Holy Roman Emperor, 165–66
Maurice, duke of Saxony, 160, 164
Maurice, prince of Orange, 506
Mauritius, 247, 249, 250, 251, 574, 578, 579
Max Joseph of Bavaria, 326
Maximilian I, duke of Bavaria, 162, 166, 168, 214
Maximilian II, Holy Roman Emperor, 153
May, John, 225
Maya, 128, 136
Mazarin, Cardinal, 163, 327
Mazeppa, Ivan, 277, 278
McClellan, George B., 599
McNeill, John, 519
Mecca, 106, 236
Medici, Ferdinando II, de', grand duke of Tuscany, 182
Medicis, 159, 182, 325
Medina, 106, 236
Medina Sidonia, duke of, 180
Mehemet (Mehmed) Ali. *See* Muhammad Ali, pasha of Egypt
Mehmed II, Ottoman sultan, 98
  conquest of Constantinople, 103–4
  conquest of the southern and northern Black Sea shores, 105
  founding of the Imperial Cannon Foundry, 110
  mobilization of naval power in the Caffa campaign, 106
  victory over Trebizond, 105
Meiji Emperor, 447
Meiji Restoration, 447
Melaka, 252, 449, 456
  Aceh attack on, 237
  British administration, 250, 463
  Dutch blockade of, 239, 240
  Dutch conquest of, 241
  first visit of the Portuguese, 451
  Portuguese capture of, 235, 452, 454
Melilla, Spanish conquest of, 122, 132
Menelik, emperor of Ethiopia, 405
Menorca, 136, 141, 324, 377, 525
Menshikov, Alexander Danilovich, 322
mercantilism, 332–33, 334, 337
mercenaries, 33, 35–36, 37–38, 102, 194, 196
  employed by the Safavids, 298
  in African armies, 392, 395, 400, 407
  in European armies, 509, 562
    Northern Wars, 263–64, 267, 268, 276
  in Indian armies, 410
  in Maratha armies, 426
  indigenous, in the Spanish armies, 127–28
  Japanese, 433
  Southeast Asia, 458
    arrival in the fifteenth century, 449
    Japanese Christian migrants, 456
    Ottoman, 454
    Portuguese, 452, 453, 454–55
Mesopotamia, 296, 303, 304
*mestizos* (free blacks), 521, 523
metallurgy, 14, 192
Metz, siege of, 148, 160, 213
Mexican–American War, 548–50
  sea power strategy, 599
Mexico City, battle for, 549–50
*mfecane* (violent upheaval), 402
Michigan, 141, 541
Middleton, Charles, 381
Mikhail Romanov (Michael I), tsar of Russia, 495, 506
Milan, 136, 145, 157–59, 160, 161
  collapse of, 145
military enterprisers, 198
military examinations, 19, 76, 432, 437, 438, 443, 445
military texts, 19
  Ming, 19, 75–79, 93, 97
military-fiscal states, 25–32, 34, 113, 194, 281–82
  Britain, 574
  Mysore, 422–23
  Northern Wars, 268–72
  Ottoman, 118–20
mines, 77, 205, 413, 496, 579
  sea mines, 583, 596
  Turkish, 208
  used to create ditches, 209
Ming dynasty, 63, 471–72
  battle tactics, 78
    clearing the fields and strengthening the walls, 85–86
  border threats, 71–74

Ming dynasty (cont.)
campaigns against the Mongols, 81–86
campaigns in Korea against Japan, 90–93
conflict on the southern border, 87–89
conflict with the Mongols, 93
decision to refrain from maritime expansion, 232
decline and fall, 93–97
decline in sea power, 79
defeat of Portuguese fleet, 238
foreign relations, 74–75
horses, 477–78
internal insurrections, 96
military organization, 79–80, 86
Mongol influence, 86
military texts, 19, 75–79, 93, 97
relations along the northwest frontier, 86–87
war with the Dutch, 243
war with the Qing, 466
weaponry, 76–77
cannon, 482–83
firearms, 477–78
Minin, Kuzma, 506
Minorsky, Vladimir, 283
Mintzker, Yair, 230
Mir Ali Beg, 237
Mirambo, Nyamwezi warlord, 402
Miranda, Luis de, 130
Mirandola, 216
*miri levend* (state-funded militiamen), 117
Mirza Saru Taqi, Safavid grand vizier, 304
Mirza Taqi, Safavid grand vizier, 304
Missy, Jean Rousset de, *Les Intêrets presens des puissances de l'Europe*, 328
Mocha, 240
Modon, 152
Mohács, battle of, 149
Moldavia, 105
Molino del Rey, battle of, 549
Moluccas, 239–40
Mombasa, battle of, 237, 399
Moncada, Don Ugo de, 176
Monck, George, 186–87
Monferrato, 161
Mongolia, 32, 53, 89, 481, 484
horsemanship, 479
Qing conquest of, 39, 466, 468
Mongols, 468–70
attacks against the Delhi Sultanate, 412
banner organizations, 468
battle tactics, 82
conflicts with the Ming, 71–74
imperial legacy, 54–56
incorporation of military strategy by the Muscovites, 57
introduction of gunpowder to India, 413
invasions of Japan, 72
military organization, 20, 61
relations with the Ming, 81–86, 93
source of horses from, 481
trade with the Ming, 75
weaponry, 68
Monmouth Courthouse, battle of, 535
Monson, George, 426
Monson, William, 181
Montalcino, siege of, 215–16
Montalembert, Marquis de, 227, 230
Montecuccoli, Raimondo, 31, 344
Monterrey, battle of, 548
Montesclaros, Peruvian viceroy, 514
Montmorency, Anne de, constable of France, 159
Montreal, 543
Morea, 103, 104
ceded by the Ottomans to Venice, 99
Ottoman reconquest of, 99, 112, 120
Morgan, Daniel, 536
Morgan, Henry, 518
Morocco, 234, 391, 587
Morris, Richard, 525
Moscow, principality of. *See* Muscovy
Mosquito Coast, 517, 527, 528
mosquitos, 7, 144, 368, 519
Mosul, siege of, 112
Mozambique, 240, 249, 401
Mughals, 10, 39, 58, 59, 284, 486
Akbar's *mansabdari* system, 21, 416–17
attack on Russian trade expedition, 502
Babur's revolution in military affairs, 415–16
conflict with the Safavids, 291, 294, 297, 304–6, 418
decline and fall of, 70, 417–19
fiscal system, 31, 416–17
military recruitment, 36
size of army, 29, 417
use of mercenaries, 298
weaponry, 12, 299, 417, 419, 427
Muhammad Ali, pasha of Egypt, 101, 406, 589, 594, 599
Muhammad bin Tughluq, Delhi sultan, 413
Muhammad Khudabanda, Safavid shah, 297
Muhammad Shah, Mughal emperor, 419
Muhammad Shibani Khan, 59, 65
Mühlberg, battle of, 164
Münster, Peace of, 136, 311, 518

Murad II, Ottoman sultan, 110
Murad III, Ottoman sultan, 110, 153
Muscat, Ottoman capture of, 236
Muscovy, 56–57, 60, 258
conquest of Kazan and Astrakhan, 496
evolution into the Russian Tsardom, 58, 496
expansion and conflicts on the southern borders, 499–501
expansion of territories, 489–92
expansion to the west, 504–5
fortress building, 68
transformation of the military, 492–95
weapons, 494
Muscovy Company, 502
musketeers
African, 395
Burmese, 454, 458
Japanese, 12
Korean, 13, 438, 442
Mughal, 12
Muscovite, 262, 271, 494
Portuguese
in African employ, 390, 391, 400
in Burmese employ, 454
Qing, 475
Safavid, 293, 297, 300, 302
muskets, 9, 11, 14, 47, 506, 509
British, 427, 428
Deccani, 414
flintlock, 14, 110, 394, 400, 401, 402, 417–18, 428, 457, 478, 509
Japanese, 12, 433, 435
Korean, 12
Maratha, 424–25
matchlock, 110, 400, 401
Ming, 77, 90, 476–77
Mughal, 417–18, 419
Mysore, 422, 423
Ottoman, 110, 294
Punjabi, 427
Qing, 476–78, 483, 484
smoothbore, 423
Southeast Asia, 457
training for the use of in Korea, 438
used by the Boers, 400
used in African armies, 391, 394, 395, 399, 401, 402, 404
wheellock, 425, 478
Muslims
Portuguese enmity towards, 232–35
rumors of Ming persecution, 86
Mutapa, 396, 398, 400
Mysore, 249, 422–23, 425, 428, 576

Nabeshima Naoshige, 440
Nadir Shah Afshar, 5, 6, 63, 112, 418–19, 424
Nagasaki, 238, 240
Nagashino, battle of, 12, 433
Namur, siege of, 224
Napier, Charles, 584–85, 590
Naples, 122, 145, 152, 324, 331
Habsburg–French rivalry over, 155, 157, 159, 160
inherited by Charles V, 41
navy, 593
Nelson at, 575
revolt against Spanish rule, 517
shipbuilding, 178
siege warfare, 207, 215, 217, 227
Napoleon, 250, 558, 575–77
Cherbourg project, 594
conscription to the armies of, 563
defeat of, 552, 569–72
early promotion of, 565
inconsistent colonial policy, 553–55
integration and promotion of revolutionary officers, 565–66
occupation of Spain, 143
size of army for 1813 campaign, 553
strategic and tactical talents, 567–69
*Napoleon*, battleship, 596
Napoleonic Wars. *See* French Revolutionary and Napoleonic Wars
naptha fire, 413
Narai, Siamese king, 458
Narva
battle of, 273, 509, 510
retaken by the Swedes, 262
siege of, 259, 505
Nassau, Maurice van, 220
Native Americans
approaches to the white European expansionist threat, 538–39
effects of the War of 1812, 545
involvement in the War of 1812, 541–43
wars with the US army, 546–47
naval power and maritime conflict, 3–4, 150–52, 573–75. *See also* ships, shipbuilding
American War of Independence, 382–83, 526, 532
Anglo-Dutch Wars, 186–88, 366, 367–68
Armada campaign, 132, 181–82
blue-water strategy, 368, 385

naval power and maritime conflict (cont.)
boarding and capture, 17, 177–78, 179–80, 181–82, 183–84
British activity 1794–1810, 383–86
British enforcement of informal empire, 586–88
British imperial wars, 577–80
British technology and strategy, 575–77
Buganda, 404
Burmese–Malay battle for Phuket, 462
commercial and imperial expansion after 1815, 580–82
British activity in India, 245–46, 249
Dutch activity in the East Indies, 238–42
Ottoman–Portuguese conflict, 235–37
Portuguese activity in the Indian Ocean, 232–35
Portuguese blockade of Aceh, 237–38
discipline and quality of crews, 597
Dutch naval power, 366
Dutch War, 369
East India Company, 249, 421–22
France, 30, 366
French attack on the Vietnamese navy at Da Nang, 465
French Revolutionary and Napoleonic Wars, 250, 528, 555–56, 575–77
French steam power technology, 596
health and wellbeing of sailors, 373, 378–79
Imjin War, 91, 437
imperial naval facilities, 597–98
influence of steam power on communications, 588–90
involvement of steamships in military operations, 590–93
line-ahead formation, 187–88, 189–90
long-range exchanges, 179–80
Mahan's principles, 170
Maratha, 37, 244, 249, 423–24
mutinies, 378
Mysore, 423
naval administration, 375–76
naval bases, 376–77
Nine Years' War, 370–71
officers, selection and training, 377–78
opportunities for enrichment, 379
Ottoman, 105–7
Ottoman–Habsburg wars, 150–52
Ottoman–Venetian wars, 106, 153–54, 172–79, 189–90
payment of officers and seamen, 378
recruitment, 374–75
Seven Years' War, 379–80
Southeast Asia, decrease in conflicts, 456–57
Spain, 136, 179
naval reconstruction scheme, 519
protection of the treasure fleet, 456–57
technological advances, 16–18, 47–48
late eighteenth century, 374, 381–82
screw propeller, 595–96
shift to the deployment of firepower, 171–72
steam, 582–86
Thirty Years' War, 135
trade defense and the Cherbourg strategy, 594–95
Tuscan navy, 178, 184–86, 188
United States, 539–41
War of 1812, 540–41, 578
War of 1812, 544–45
War of the Austrian Succession, 371–72
War of the Spanish Succession, 174, 371, 518
Navarre, 122, 123
Navarro, Pedro, 206
Navigation Acts, abolition of, 597
Ndebele, 403
Ndongo, 392
Negapatam
Dutch fortification, 243
siege of, 250
Negroponte, 104
Nelson, Horatio, 190, 378, 386, 555, 578
strategy, 575–76
*Nemesis*, ship, 585
Nepal, 470
Nerchinsk, Treaty of, 61, 468, 499
Netherlands, 123, 124, 145, 381
dynastic ambitions for, 323, 325–26, 329
French ambitions for, 163
French subsidies, 161
occupation of Utrecht, 159
Spanish loss of, 135–36
Netherlands War. *See* Dutch Revolt
Nettuno fortress, 212
Neuf-Brisach, 223
Neutrality Convention, 279
New Granada, 524
New Mexico, 128, 133, 138, 548
increase in population, 139
Spanish reoccupation of and withdrawal from, 136
Spanish settlement, 138
New Model Army, 186
New Orleans, battle of, 545, 583
New Spain, 140, 514, 527
administration, 521

creation of standing armies, 521
growth of the militia, 523
treasury, 516
New York, 532–34, 535–36, 537–38, 544, 550
New Zealand, 250
Newburgh Conspiracy, 538
Newport, Rhode Island, 532, 535–36, 537
Ngoni, 399, 402
Nguyen Anh, 462
Nguyen clan, 453, 462
Nguyen dynasty, 462
Nguyen Hoang, 453
Niagara river, 541, 543
Nicaragua, 517
Nice, 159
Nicholas I, emperor of Russia, 487, 512
Nijmegen, Peace of, 311, 315
Nile, battle of the. *See* Aboukir Bay, battle of
Nine Years' War, 230
as a British succession struggle, 323
naval warfare, 369–71
size of French army and navy, 30
Nizam, Mughal governor of Deccan, 419
Njinga Mbande, Ndongo queen, 392
Noghay. *See* Manghit tribe
nomad armies, 64–67
nomadic warfare, 4–7, 39–40. *See also* Safavids
Nootka Sound dispute, 336, 528, 557
Nordic Seven Years' War, 260, 261, 262
Nördlingen, battle of, 136, 162
*North River Steamboat*, 582
Northern Wars, 255–57
combatants, systems of government, 257–59
effects on Russian military and political affairs, 263
fiscal and organizational difficulties, 261
Great Northern War, 255, 329, 507–10
Livonian War (First Northern War), 255, 259–62, 497, 505, 506
Nordic Seven Years' War, 260, 261, 262
Scanian War, 268
Second Northern War, 265–66, 267–68, 270–71
strength of Poland-Lithuania following First Northern War, 263–64
Thirteen Years' War, 255, 265–66, 270, 507
Novarra, battle of, 157
Novgorod, 491–92, 506
Nurhaci, founder of the Qing dynasty, 61, 466, 476, 477
administrative skills and defeat of the Ming, 94–96
creation of banners, 67, 470, 471–73
fortifications, 482
tactics in the battle of Sarhū, 479–80
weapons, 482, 483
Nyamwezi, 402, 403
Nystadt, Peace of, 280, 510
Nyungu-ya-Mawe, war leader, 402

O'Higgins, Ambrosio, 141
Oda Nobunaga, 12, 90, 432–34
Ögedei Khan, 54, 71
Oghuz Khan, 56
Oirat, 58, 61–62, 67. *See also* Kalmyks
Oldenburgs, 326
Olivares, count-duke of, 132
Oman, sultan of, 599
Ōnin War, 432
opium trade, 245
Opium Wars, 251
*oprichnina*, 495
Oquendo, Antonio de, 183–84
Orang Laut, 449
*ordo* (estates, patrimonies), 471
Orenburg, 501
Oromo, 137, 390, 397, 399
Osaka, battle of, 435
Osman, founder of the Ottoman Empire, 98
Ostend
siege of, 218–20
Spanish reconquest of, 155
Ostend Company (*Generale Indische Compagnie*, GIC), 247–48
Ostermann, Andrey Ivanovich, 322
Ostwald, Jamel, 224
Osuna, duke of, 154
Ottoman–Habsburg wars, 30, 98–99, 114–15, 120, 154–55, 313–14
naval warfare, 107, 150–52
Ottoman–Safavid wars, 30, 98, 102, 115, 295–97, 304–5
military strengths compared, 293–95
Ottoman–Venetian wars, 112, 120, 153–54
naval power, strategy and tactics, 106, 172–79
Ottomans, 10, 25, 39, 56, 58, 60, 64, 70, 381, 486. *See also* Ottoman–Habsburg wars; Ottoman–Safavid wars; Ottoman–Venetian wars; Russo-Ottoman wars; Russo-Ottoman–Austrian War
annexation of Polish-Lithuanian territories, 270
anti-Ottoman offensives, 313–14
armies, 103–4

Ottomans (cont.)
Asakir-i Mansure-i Muhammediye (Victorious Soldiers of Muhammad) conscript army, 101
expansion of the salaried army, 114–16
land forces and military strength, 107–10
*Nizam-i Cedid* (New Order) army, 101
problems of command and control, 22
recruitment, 21
rise of the household armies of governors and grandees, 117–18
strength compared with the Safavid army, 293–95
campaigns, 1521–1617, 149–55
civilization awareness, 149
conquest of Constantinople, 11, 103–5
defense of the empire, 99–101
expansion of the empire, 98–99, 103
geopolitics following Treaty of Karlowitz, 111–13
military and fiscal capabilities, 31, 118–20
military contracting, 36
naval power and maritime conflict, 4, 42, 105–7, 173–74
adaptations following battles against the Venetians, 178, 179
artillery and tactics, 176–77
battles with the Portuguese in the Indian Ocean, 235–37
Jeddah, 172
provision of mercenaries and weaponry to Aceh, 454
recruitment, 30
staying power, 120–21
supply and logistics, 11, 22–23, 31, 105, 111, 113, 291, 292
supply of weapons to India, 414, 415
supply of weapons to the Ming, 477
*Wagenburg* tactic, 296
war against the Mamluks, 390
weaponry, 68, 110–11
Ovimbundu, 401
Oxenstierna, Axel, 167–68, 312
Oyirad tribes, 81–83. *See also* Oirat
Oyo, 393–95, 396, 401, 404
Özi, 99

P'yŏngyang, 91
Pacific Ocean, 43, 231
Dutch–Spanish conflict, 239
expansion of American commerce, 251
Padua, siege of, 209, 211, 213–15
Pagan dynasty, 453
*paiks* (infantry), 413
Paixhans guns, 594
*palenques* (communities of runaway slaves), 524
Palestine, 236
Palmanova fortress, 212
Palo Alto, battle of, 548
Panama, 516, 519
maroon community, 524
*panchayats* (administrative system), 421
Panin, Nikita, 322
Panipat, first battle of, 415
Panipat, third battle of, 424
Papacy, 158
declining influence, 311–12
Paris, Peace of, 336, 558
Paris, Treaty of, 249, 538
Parker, Geoffrey, 170, 213, 224, 299
Parma, 324
Pasamonte, Jerónimo de, 131
Passarowitz, peace treaty of, 111
Paul, grand duke and emperor of Russia, 330, 498, 576
Paul IV, pope, 160
Pavia, battle of, 158, 225
Pawnee, 139
Pech, 133
Pegu (Lower Burma), 449, 452, 460
Penang, 250, 463
Peninsula Campaign (American Civil War), 599
Peninsular War, 529, 554, 570
fortifications and defenses, 227–28
French casualties and morale, 570
naval support, 385
sieges, 225–26
Perpignan, 159
Perry, Matthew C., 446, 582
Persian Gulf, 231, 235, 242, 576
Ottoman presence in, 236–37, 454
Ottoman–Portuguese conflict over, 106, 236
Safavid control of, 303
Peru, 127, 239, 513, 529
creation of standing armies, 521
growth of the militia, 523
natives in Spanish armies, 128
Spanish administration, 521
willingness to form alliances with the Spanish, 133
Perugia, fortifications, 212
Peter I, the Great, tsar of Russia, 339, 487, 496, 512

decision to attack the Safavids, 308
defeated by the Ottomans, 99, 120
despatch of expedition to Khiva, 502
dynastic uncertainty following death of, 330
Great Northern War, 272–73, 277–80, 507–11
military reforms, 273–75, 509–10
seizure of the fortress of Azov, 502
system of conscription, 48, 338, 509
Pfalz-Neuburgs, 314
Phaulkon, Constans, 458
Phélypeaux family, 350
Philip, duke of Parma and Piacenza, 325
Philip II, king of Spain, 131, 152, 153, 155, 515
and the final stages of the Italian Wars, 160
attempts to invade England, 180, 181
becomes king of Portugal, 156
despatch of the Armada to England, 132
fortification of Caribbean ports, 516
intervention in the French Wars of Religion, 161
prohibition of colonial trade between Spain and Portugal, 238
Philip III, king of Spain, 131, 155, 165, 166, 323
Philip IV, king of Spain, 156, 163, 323, 327
Philip V, king of Spain, 318, 321, 331
succession, 323–24
Philip of Hesse, 164
Philippine–American War, 464
Philippines, 37, 122, 123, 140, 456, 464, 517, 555
Dutch blockade, 455
Spanish colonization, 238, 335
Phuket Island, 462
pikes, 102, 192
Pillnitz, Declaration of, 560
Pinto, Fernão Mendes, 454
piracy, 515–17, 581–82, 583
Piri Reis, 106
Pius V, pope, 153
Piyale Paşa, Ottoman admiral, 152
Pizarro, Francisco, 127
plague, 7, 124, 159, 279
Plassey, battle of, 246, 249
Plattsburgh, battle of, 544
Playa Honta, battle of, 239
Pliussa, Truce of, 261
*Pluto*, steamship, 583
Podewils, Heinrich von, 329
Poland-Lithuania, 102, 105, 167, 258, 507. *See also* Northern Wars
as part of the anti-Ottoman coalition, 99
cavalry, 148
occupation of Moscow, 505
partitions, 339, 511
political and military strength following First Northern War, 263–64
relations with the Ottomans, 149
threat to Muscovy, 491
*poligars* (Mysore chieftains), 423
Polish–Ottoman War, 256, 264, 266, 270
Polk, James K., 548
Poltava, battle of, 273, 278–80, 510
Pompadour, Madame de, 316
Pondichéry, 247, 249
Poniatowski, Stanisław, 511
Ponta Delgado, battle of, 181
Pontic Steppe, 499
Port Louis, 249
Porta Codalunga (Padua), 211
Porter, David, 583
Portland, battle of, 186–87
Portobello, 518
Portugal
attempted invasions of Mutapa, 400
conflict with Aceh, 237–38
conflict with the Ottomans in the Indian Ocean, 235–37
expansion and maritime violence in the Indian Ocean, 232–35
French invasion of, 579
incorporated into the Spanish crown, 517
influence on Southeast Asian warfare, 451–55
military activity in Africa, 399–400
naval warfare, 172
relations with the Ming, 238
revolt of, 163
supply of gunpowder technology to India, 414
trade with Japan, 238
union with Spain, 156
Posch, Walter, 283
Potemkin, Grigory Aleksandrovich, 322
Potsdam Edict of Toleration, 357
Potsdam, Treaty of, 280
Pozharskii, Dimitrii, prince, 506
Prague, 165–66
*prazeiros* (Afro-Portuguese warlords), 399
Preveza, battle of, 150, 174
Princes' Revolt, 164
Princeton, battle of, 533
Prithviraj Chauhan, Rajput king, 412
Priuli, Girolamo, 214
privateering, 152, 154, 177, 189, 380
investment in by Elizabeth I, 182
Prophetstown, 539
Protestant Reformation, 257, 327

Protestantism
adoption of in Germany, 150
tensions within, 313
Protestants, 161–62, 164–67, 200
Catholic persecution of, 314–16
Provence, 159
Prussia, 167, 266, 325, 380, 554, 558. *See also* Brandenburg-Prussia
conscription, 564
enlisted to the Armed Neutrality, 576
land gains, 339
loss of territories, 569
Prut river campaign, 99, 279
Pskov, 491
Puerto Rico, 529
Pugachov, Emelian, 501
Punjab, 415, 426–27
Pyrenees, Peace of the, 163

Qajars, 284, 293, 503
*qal'a-kub* (fortress-smasher), 298
Qandahar, 291, 307
Mughal–Safavid conflict over, 294, 297, 304–6, 418
Qaraqoyunlu, 56, 59
Qasimov, Kenisary, 501
*qazaqs* (free adventurers), 58
Qeshm, 236
Qi Jiguang, 438, 442
*Jixiao xinshu* (*New Book on Effectful Discipline*), 76
*Lianbing shiji* (*Practical Arrangement of Military Training*), 76
Qianlong Emperor, 468, 501
Qing dynasty, 6, 58
British imposition of trade and diplomatic arrangements, 595
cannons, 482–84
cartography, 483
conflict with Russia, 442, 498
consolidation of the state and expansion, 60–62, 466–70
Eight Banners and the Green Standard armies, 32, 470–75
establishment of, 97
horses and horsemanship, 478–81
military strategy and financial constraints, mid-eighteenth–early nineteenth century, 485–86
recognition of Western sea powers, 600
struggles with and defeat of the Zunghars, 60–62, 468
weaponry, 475–78
Qinghai, 466, 468
Qiu Fu, Ming commander, 82
*Qizilbash*
weaponry, 286
*Qizilbash* ("Red Heads," Turkoman troops), 20, 286–89, 292–93, 296–97, 300, 304
fighting method, 290
marginalization of, 301–3
rebellion, 296, 301
use of cannon, 297
Quasi War, 540
Queenston Heights, battle of, 541, 543
Quiberon Bay, battle, 380
Quirós, Pedro Fernández de, 128
Quito Insurrection, 524
*qullar* corps, 302
*qurchis* ("quiver bearers"), 302
*qurchis* (Safavid royal bodyguard), 288–89
Quzanlu, Jamil, 283

Raffles, Stamford, 579
Raichur Doab, battle of, 414
Rajputs, 410–13, 415, 416
Raleigh, Walter, 181
Ramaji Pant, 424
Rana Sanga, 415
Ranjit Singh, 426–27
*Rattler*, HMS, 585
Ravenna, siege of, 225
*Realpolitik*, 310, 318
recruitment. *See also* conscription
Brandenburg-Prussia, 360, 363
cantonal system, 48, 360–62
costs of in the French Revolutionary and Napoleonic Wars, 552–53
French system, 562, 563
Korean military examination, 437
of Indians into the army of the EIC, 420
of seamen, 374–75, 597
Ottoman *devshirme* system, 21
seasonal nature of, 196
Swedish, 261
to the French Revolutionary and Napoleonic armies, 563–64
Red Sea, 231, 235, 397
Dutch halt to the Muslim spice trade, 243
Egyptian expansion into, 406
Ottoman presence in, 4, 454
Ottoman–Portuguese conflict over, 106, 236–38
revival of the trading route, 251
Redlich, Fritz, 198
Reduction Act of 1821 (US), 546

Reformation, 200
Regensburg, 167
Reid, Anthony, 456
religious tolerance, 312
religious wars, 41–42, 163–69, 309, 310–16
Resaca de la Palma, battle of, 548
resources, wars over, 309, 331–41
Réunion, 249, 250
Reval (Tallinn), 257, 510
*Revenge* (Drake's flagship), 181
Revocation of the Edict of Nantes, 315
Reynald of Châtillon, lord of Transjordan, 231
Rhodes, 105
  siege of, 149, 152, 208, 209
Riazan, 491
Ricci, Matteo, 83
Richelieu, Armand du Plessis de, Cardinal, 161–63, 217, 230, 345
Richelieu, duc de, marshal of France, 352
Riga, 257, 510
  siege of, 273
Rio Paraná, 587
River Plate, 515, 521
Rivoli, battle of, 225
Robespierre, Maximilien, 561
Rochambeau, Jean, comte de, 537, 565
Rochefort, naval base, 375
Rocroi, battle of, 135, 163
Rodney, Admiral, 382
Rogniat, Joseph, 226
Rohan-Soubise, marshal of France, 352
Romanovs, 118, 263, 326, 501, 503, 511–12
Rome, sack of, 158
Ross, John, 590
Rostov, 491
Rota, Giorgio, 283
Royal Geographical Society, 594
Royal Navy, 24, 190, 367, 378, 578
  activity in the French Revolutionary and Napoleonic Wars, 555–56, 576, 578
  compared with the Spanish navy, 137
  financial security, 375
  financing of, 574
  First Anglo-Chinese War, 591
  importance of managerial culture, 23
  increased strength of, 1850, 594
  investment in the health and wellbeing of crews, 373
  modernization in the late eighteenth century, 381
  policy of aggressive close blockade, 384
  prizes awarded for the capture of enemy ships, 177
  recruitment methods, 374
  Seven Years' War, 380
  ships provided to the East India Company, 421
  size, early nineteenth century, 383
  steamships, 584–86, 588
  technological advantage, 374
  training and promotion of officers, 377–78
  War of 1812, 541
  War of the Spanish Succession, 371
*Rozvi* (Southern-Central African army), 398
Rudolf II, Holy Roman Emperor, 154, 165–66
*ruga ruga* (young African fighters), 402
Rumlu, 293
Russia, 309, 381. *See also* Northern Wars; Russo-Ottoman wars; Thirteen Years' War
  anxiety over British sea power, 577
  border conflicts with Korea, 442
  border conflicts with the Qing, 468
  coastal defenses, 600
  defeat of Napoleon's invasion, 511–12
  dynastic uncertainty following death of Peter I, 330
  effects of the Northern Wars on military and political affairs, 263
  expansion of territories, 339, 487–89
    during the rule of the Muscovite princes, 489–95
    to the east, 60, 495–99
    to the south, 62–63, 499–504
    to the west, 504–11
  fortifications, 495
  military and fiscal capabilities, 118
  military transformation
    under Mikhail Romanov and Tsar Alexis, 506–7
    under Peter I, 507–10
    under the Muscovite princes, 492–95
  Polish invasion of Moscow, 505
  threat to the Ottomans, 112
  Time of Troubles, 495, 505–6
  wars in Chechnia and Circassia, 581
Russo-Ottoman wars, 99–101, 111, 117, 120, 256, 266, 272, 273, 279, 309, 314, 502–3
  Russian territorial gains, 339
Russo-Ottoman–Austrian War, 116, 120
Russo-Persian wars, 503
Russo-Polish War. *See* Thirteen Years' War
Russo-Swedish War (1741–43), 339

's-Hertogenbosch, siege of, 213
Saadat Khan, Mughal governor of Awadh, 419

Sachsen-Weimar, Bernhard von, 34
Safavids, 10, 39, 58, 59, 63, 152, 283–84
army camp, 289–90
battle tactics, 290–91
cavalry, 30, 300
conflict with the Mughals over Qandahar, 294, 304–6, 418
decline in the military, seventeenth and eighteenth century, 304–8
Dutch trade, 240
environmental limitations to campaigns and expansion, 292
firearms, 297–301
inferior weaponry, 294
military strength compared with the Ottomans, 293–95
organization and composition of the army, 286–89, 292–93
problems in military structures, 22
reforms of ʿAbbas I, 301–3
reliance on the Qizilbash, 20
siege warfare, 291–92
strategy and tactics, 290–91, 296
trade with the English, 244
treaty with the Ottomans, 112
wars against the Ottomans, 30, 98, 102, 107, 115, 149, 154, 295–97
wars against the Uzbeks, 290, 295, 296
Safi I, Safavid shah, 304
Saint Domingue, 380, 518
slave revolt, 384
Saintes, battle of the, 382
Saladin, 231
Saltillo, battle of, 548
Saluzzo, 161
Salvador da Bahia
Dutch capture of, 517
Samori Ture, 404
samurai, 90, 431, 436
appearance, 430
governance and the influence of *bushidō*, 439–41, 445
San Francisco, 521
San Jacinto, battle of, 547
San Juan, 519
San Juan de Ulúa, battle of, 182
San Sebastian, siege of, 225, 227, 229
Sandwich, earl of, 381
Sangu, 403
*sankin-kōtai* ("alternate attendance") system, 435, 439
Santa Anna, 547, 548–49
Santiago de Cuba, invasion of, 519
Santo Domingo, 519
Sanudo, Marino, 172
Sarab, Peace of, 303
Saratoga, battle of, 533–34
Sardinia, 136, 321, 552
Sarhū, battle of, 479
Sauk, 546
Savannah, battle of, 536
Savory, Roger, 297
Savoy-Piedmont, 160, 315
Saxe, Maurice de, 352
Saxon Wettins, 314
Saxony, 147
Scanian War, 268
Schellenberg, battle of, 224
Schenck, Martin, 127
Scheveningen, 187
Schmalkaldic League, 164
Schmalkaldic War, 164
scientific expeditions, 581
scorched-earth policy, 277, 291, 295
Scott, Winfield, 543, 546, 547, 548–49, 550
scurvy, 373
*Seagull*, steamship, 583
Second Anglo-Burmese War, 465
Second Anglo-Chinese War, 595
Second Anglo-Dutch War, 187
Second Anglo-Maratha War, 424, 425–26
Second Anglo-Mysore War, 249, 422
Second Anglo-Sikh War, 427
Second Hundred Years' War, 336
Second Northern War, 265–66, 267–68, 270–71
Second Seminole War, 547–48
Seignelay, marquis de, 368
Sekigahara, battle of, 12, 432–33, 435
Selim I, Ottoman sultan, 98, 106, 149
Selim II, Ottoman sultan, 153
Selim III, Ottoman sultan, 101
Seminole Wars, 546, 547–48, 550
Sena, 400
sepoys, 420
Serbia, 103, 104
serfdom, 263, 281, 488, 493, 502
Seringapatnam, siege of, 423
Seven Years' War, 101, 140, 249, 316, 322, 325, 349, 352, 367, 378, 384, 568
artillery, 47
British naval superiority, 555
British occupation of Manila, 463
dynastic dimension, 329
French defeat, 336, 535
French officer casualties, 353

naval power and conflicts, 379–80
origins of the Anglo-French war in North America, 336
Prussian gains, 339
Russian territorial advances, 511
Spanish entry into, 520
Spanish territorial gains, 137
spread from the Americas, 558
troop recruitment in Brandenburg-Prussia, 363
Sevendurg, 424
Seymour, George, 588
Shah Jahan, Mughal emperor, 417
Shāhrukh (Timurid ruler), 87
*shahsivam*, 289
Shaka, Zulu king, 402, 403
Shamlu tribe, 293
Shawnee, 539
Shenyang, 466
Sherley, Anthony, 297
shields
African, 396, 398
Qing, 475, 476
Shintoism, 430, 439
ships, shipbuilding
*bertoni*, 182
caravels, 172, 237
carracks, 172–74, 175, 237
deficiencies in late seventeenth-century design, 369
European shipbuilding industry, 375
frigates, 381, 421, 541, 583
iron, 585
paddle-wheel, 585, 594
screw propeller, 596
galleasses, 17, 175, 178–79, 181, 187, 189, 190
galleons, 17, 43, 45, 174, 180–81, 184–86, 189, 237, 238, 244, 399, 515
galleys, 17, 171, 366
advantages and disadvantages, 174–75
*bastarde*, 175, 179, 185
compared with sailing ships, 172–74
conversion to warships, 175
of the English East India Company, 249
ordnance, 176–77
Ottoman, 106–7
*gallivats*, 421, 424
*grabs*, 421
*gurabs*, 424
*jachten* (yachts), 240
ketches, 421
Korean turtle ships, 92
late fifteenth-century improvements, 171
mortar vessels, 579
*nau*, 237
*pals*, 424
*quinquireme*, 175
sailing ships, 172–75, 235, 237, 578
screw propeller, 595–96
steamships, 18, 251, 582–86
communications, 588–90
operations, 590–93
technological advances in the late eighteenth century, 381–82
VOC (Dutch East India Company) fleet, 242
Shirvan, 294, 295, 297, 301
firearms industry, 299
Shivaji, Maratha leader, 417, 423
Siam, 458, 465
avoidance of European rule, 465
mid-eighteenth–early nineteenth-century conflicts, 460–62
Siberia, 489
Russian expansion into, 16, 60, 497–99
Sibir, khanate of, 57, 497
Sicily, 107, 136, 145, 155, 217, 324, 331, 593
Sidis, 244
siege trains, 206–7, 213, 214
siege warfare, 9, 11
artillery, 223, 229
at the siege of Malta, 216–17
manufacture of, 212–13
Peninsular War, 225
strength of, 212–14
Burma, 454–55
casualties, 217–18, 223
defenses, 208–12
Dutch and French advances in the late seventeenth century, 220–24
Italian Wars, 213–16
French invasion of Italy, 207–8
length of sieges, 225
logistics, 206, 217, 226
Napoleonic and Revolutionary era, 225–30
nocturnal nature of, 228
role of small arms, 209
tactics, 224
development of a more scientific approach, 218–20
zig-zag saps, 221
Siena, 160
fortifications, 212
siege of, 215–16, 225
Sikhs, 426–28
Silesia, 167, 325, 329, 339

silver bullion, 245
Singapore, 250, 463
Siraj ud-Daula, *nawab* of Bengal, 246, 249
*sirdan* (arrow), 476
*Siyi guan* (Translators' Institute), 75
Slade, Thomas, 381
slave communities, Spanish America, 524
slave soldiers, 20, 289, 301–3, 394, 399
 in Spanish colonial army, 522
slave trade, 249, 391–94
 Brazilian, British suppression of, 587
 post-abolition, 401–2, 404
slavery, 253
sleeping sickness (trypanosomiasis), 7, 388
small war, 24, 152, 578
Smolensk War, 263, 264, 505, 506
Sokoto Caliphate, 403
Song Yongxing, *Tiangong kaiwu* (*The Creations of Nature and Man*), 77
Songhay, 137, 390–91
Sophia Palailologina, wife of Ivan III, 492
Sotho, 403
Sound Dues, 257
South China Sea, 231, 454
South Sea Company, 518
South Sulawesi, 455–56
Southeast Asian warfare
 fifteenth century, 449–51
 indigenous wars of mid-eighteenth to early nineteenth century, 459–63
 mid-seventeenth- to mid-eighteenth-century transitions, 456–59
 mid-sixteenth to mid-seventeenth centuries, 451–56
 renewed European expansion, 463–65
*Sovereign of the Seas* (warship), 184
*sowars* (Indian cavalry troopers), 421
Spain
 advantages over European rivals, 124–26
 *alcabala* (tax), 199
 armies
  changes in composition after involvement in French Revolutionary Wars, 528
  creation and composition of the colonial army, 520–23
  financing of, 131–32
  pastoral care, 130–31
  recruitment, 127
  foreign recruits, 127–28
  social composition, 128–30
 civilization awareness, 149
 concession of colonies in the Caribbean, 517–18
 decline and recovery of empire, 135–37
 defense of Latin American territories, 45–46
  Dutch invasions of Brazil, 517
  in the sixteenth century, 514–15
  role of Yellow Fever, 519
  size and financing of army, 515, 519
  sources and pacification of internal threats, 523–25
 dependence on slavery in the colonies, 524
 European campaigns, 132
 expansion in North America, 138–41
 expansion in Southeast Asia, 455–56
 introduction of horses to the Americas, 6, 137
 involvement in and consequences of the French Revolutionary and Napoleonic Wars, 143–44, 528–29
 involvement in and effects of the American War of Independence, 525–27
 naval power and maritime conflict
  Armada campaign, 180–81
  battle of the Downs, 183
  relative weakness of the navy, 136
  ships, 179
 prosperous relations with native people, 134
 protection of the treasure fleet, 45, 515–17
 shipbuilding, 375
 successes in the New World, 132–33
 succession struggles. *See* War of the Spanish Succession
 Thirty Years' War, 135
 value of its overseas possessions, 335–36
 viceroyalties, 521
Spanish Inquisition, 126
Spanish–American War, 463
Sparr, Otto Christoph von, 344
spears, 13, 390, 391
Spencer, Herbert, 193
Speranskii, Mikhail, 501
spice trade, 239, 241, 243, 414, 455
Spinola, Ambrogio, 219
St. Clair, Arthur, 539
St. Jean d'Acre, battle of, 591
St. Kitts, 517
St. Petersburg, 595
St. Quentin, battle of, 160
state formation, shift from domain state to tax state, 197–200
steam power, 582–86
 communications and, 588–90
Stettin, Peace of, 262
Steuben, Friedrich Wilhelm von, 535

Stevens, Carol, 509
Stevin, Simon, 220
stirrups, 14, 412
Stockholm, 510
Stolbovo, Treaty of, 506, 509
Stoney Creek, battle of, 542
Straits of Melaka, 452, 462, 582
  displacement of the Dutch by the British, 463
stranger-effect, 133
*strel'tsy* (musketeers), 262, 271, 494
Streusand, Douglas, 415
Stroganov family, 38, 496–97
Stroganov, Grigorii, 496
Strozzi, Filippo, 181
Stuarts, 323, 326
Suchet, Louis Gabriel, 225
Sudan, 406
Suez Canal
  opening of, 251
  Ottoman ambitions for, 236
Suffren, Le Bailli de, 249
Sufiyan, battle of, 303
Sulayman, Safavid shah, 306
Süleyman I, Ottoman sultan, 98, 105, 106, 171
  army of, 107–9
  campaigns of, 149–53
  expansion of the salaried army, 115
Sultan Husayn, Safavid shah, 306, 307
Sulu Zone, 456
Sumatra, 232, 238, 241, 449, 454, 463
Sumter, Thomas, 536
Sungari river, battle of, 498
Surat, 240, 242, 244, 247
Surian, Antonio, 178, 179
Sviazhsk fortress, 496
Swazi, 403
Sweden, 136, 145, 323, 381. *See also* Northern Wars; Russo-Swedish War
  conscription system, 48, 197, 269, 338
  dynastic upheaval following death of Charles XII, 330
  enlisted to the Armed Neutrality, 576
  generation of revenue, 199
  infantry, 148
  Lutheran tensions with Calvinist Brandenburg-Prussia, 313
  navy, 374
  overseas trade, 248
  production of cannon and cannonballs, 213
  religious war, 163
  representative institutions, 203
  seizure of Novgorod, 506
  succession struggles, 330
  Thirty Years' War, 167–68
Swedish East India Company, 248
Switzerland, 163
swords, 13
  ceremonial, 430, 445
  Indian cavalry, 421
  Maratha, 424
  Ottoman, 110
  Qing, 479, 484
  Rajput, 411, 416
  used in African warfare, 390, 395, 403
Syria, 236
Syriam, siege of, 454, 458
Szapolyai, János, 150
Szigetvár, siege of, 153

Tabkin Kwotto, battle of, 403
Taboada, Gil de, 529
Tabriz, 291, 292, 295, 296, 303, 304
Tahmasb, Safavid shah, 295, 296–97, 298, 301
Taiping Rebellion, 475, 582
Taiwan, 43, 466, 467, 498
Takallu, 293
Talikota, battle of, 414
Tamerlane (Timur), 5, 39, 55, 59, 64, 67, 86, 98
  invasion of India, 413
Tana (Azov), 105
Tanzania, 401, 403
Tanzimat reforms, 101
Tapper, Richard, 283
Tarain, first battle of, 412
  second battle of, 412
Tarragona, siege of, 225
Tashkent, 489
*taulqama* tactics, 415
Taylor, Zachary, 548–49
Tayson uprising, 462
tea, 245
Tecumseh, 539, 542
Tej Singh, 427
Telp, Claus, 569
*tenaille trace* fortification, 216
Tenasserim, 583
Tenochtitlan, 133
Tenpō Famine, 440
Tenskwatawa, 539
Ternate, 239, 240, 243
Ternate sultanate, 455
Terror, Reign of, 565
Tete, 400
Tewodros (Kassa Hailu), emperor of Ethiopia, 405

Texas, 124, 139, 550
annexed by the United States, 548
Texas Revolution, 547–48
Thailand, 462. *See also* Ayudhya
Thames, battle of the, 542
Thérouanne, 213
Third Anglo-Burmese War, 464
Third Anglo-Maratha War, 426, 427
Third Anglo-Mysore War, 423
Thirteen Years' War, 255, 265–66, 270, 506–7
Thirty Years' War, 123, 135, 165–69, 265, 310, 312, 357, 360, 517
consequences for the Spanish, 135
damage caused to German cities, 212
demobilization of troops following, 355
economic consequences of, 358
effects on subsequent military systems, 267, 344
French intervention, 161–63
military outsourcing, 34, 194
mortality, 51
size of armies, 30
strategy, 24
Thirty Years' War (Ottoman), 30, 115
Three Feudatories Rebellion, 467, 475, 483
Three Skills Army (*Samsubyŏng*), 13, 438
throwing knives, 391, 395
Tibet
fortifications, 484
Qing conquest of, 39, 466, 468–70
Qing involvement in the politics of, 62
ties with Mongolia, 60
Tidore, 239
Tilly, Jean t'Serclaes, count, 166–68
Tilsit, 569
*timar* (military fiefs), 22, 104, 107, 117, 119
timariot cavalrymen, 104, 107–9, 112, 117, 119
Timofeevich, Ermak, 497
Timur. *See* Tamerlane
Timurid dynasty, 39, 55–56, 59
Tippecanoe, battle of, 539
Tipu Sultan, ruler of Mysore, 422–23, 425, 427
Tiutchev, Fedr, 487
Tlaxcalteca, 128, 133
Töghus Temür, Mongol khan, 81
Tokugawa Hidetada, 435
Tokugawa Iemitsu, 435
Tokugawa Ieyasu, 432, 435, 440
Tokugawa Mitsukuni, *Great History of Japan* (*Dai Nihon shi*), 439
Tokugawa shogunate, 37, 430, 435–36, 445, 448
social hierarchy and isolationism, 438–41
Tolui, son of Chinggis Khan, 54
Tondibi, battle of, 390
Tong Guogang, 484
*tongshi* (interpreters), 75
*tonjon* (slave soldiers, 394
torpedoes, 579
Torrington, admiral, 370
Toul, 160
Toulon, 159, 371, 375, 376, 382, 596
Toungoo dynasty
first Toungoo dynasty, 452, 454
restored Toungoo dynasty, 454
Tournai, 157
Tournai, siege of, 223
Tourville, admiral, 370
Towerson, William, 172, 187
Toyotomi Hideyori, 435
Toyotomi Hideyoshi, 90, 92, 432, 433–35
*trace italienne* fortresses, 147, 148, 423
Trafalgar, battle of, 18, 143, 375, 383, 385, 528, 556, 577
Trān dynasty, 87, 89
Tranquebar, 246
Transoxiana, 59, 65
Transylvania, 99, 150, 160
Trebizond, 105
Trenton, battle of, 533
Treviso, 214
tribalism, 63–64
Trinh family, 453, 462
Trinidad, 574
British seizure of, 528
Tripoli, Spanish expedition against, 152
Tromp, Maarten, 183, 187
Trunajaya, 455
trunnions, 414
tsetse fly, 7, 388
Tsewan Rabdan, 62
*tufang* (Safavid handguns), 298
*tufangchis* (musketeers), 293, 300, 302
Tukolor, 403
Tulaji Angre, 424
*tulwars* (curved swords), 412
Tunis
captured by Charles V, 150, 155
Ottoman recapture of, 107, 153
*tuntian* (colony system), 80
Túpac Amaru II (José Condorcanqui), 525
Túpac Amaru rebellion, 45, 525
Turco-Mongolian culture, 55–56, 58
Turenne, vicomte de, 334, 344, 351
Turgut Reis, 152
Turin, citadel, 212
Türk Empire, 53–54, 56
Turkana, 401
Turks, 410–12

Tuscany
  awarded to Francis Stephen, 325, 339
  development of the navy, 178, 184–86, 188
  Farnese's claim to, 324
Tver, 491
Twelve Years', Truce of 1609–1621, 155
typhus, 7, 382

*ujen cooha* ("heavy troops" or "cherished troops," 473
Ukraine, 256, 270, 277
  Khmelnytsky uprising (Cossack–Polish War), 265, 314, 506
  "the Ruin," 266–67
Ulan Butung, battle of, 484
Ulm, battle of, 567, 568
Ulrika Eleonora, queen of Sweden, 330
Uluç Ali Paşa, 179
Union of Lublin, 507
United States. *See also* War of 1812
  annexation of Texas, 548
  coastal defenses, 599
  concerns over control of the military following American Revolution, 538
  contestation of British sea power, 577
  evolution of the military, 550–51
  expansion of commerce across the Pacific, 250
  improvements in the military following War of 1812, 546
  military training, 539
  Native American encounters along the western frontier, 538–39
  reduction in size of army, 546
  Seminole Wars, 546, 547–48
  war with Mexico, 548–50
*urdu* (Safavid military camp), 289–90
Uruguay, 587
*Uskoks*, 152, 154
Ustajlu, 293
Utes, 139
Uthman dan Fodio, caliph of Sokoto, 403
Utrecht, 159
  citadel, 212
  Treaty of, 324, 518
Uzbeks, 57–59, 63, 298, 307
  *taulqama* tactics, 415
  wars with the Safavids, 290, 295–96
Uzun Hasan, AqqoyunluAq-Quyunlu ruler, 297

Valmy, battle of, 563
Valturio, Roberto, *De Re Militari*, 171
van Diemen, Antonie, 241
Várad, Ottoman conquest of, 99
Vasas, 326
Vasilii III, prince of Moscow, 491, 505
Vauban, Sébastien Le Prestre de, 11, 16, 206, 224, 227
Vaucelles, Truce of, 160
Vaudois, French attack on the, 315
Velázquez, Diego
  "Mars," 135
  "The Surrender of Breda," 128
Venetians. *See also* Ottoman–Venetian wars
  as mercenaries, 298
  establishment of safe trade routes, 154
  naval power, 172–79, 180, 189–90
  Ottoman territories ceded to, 99
  War of the League of Cambrai, 208, 214
Venezuela, 521
  attempted British invasion, 143
  maroon communities, 524
Venier, Sebastiano, 173
Veracruz, 519, 599
  siege of, 549
Verdun, 160
*Verenigde Oost Indische Compagnie* (VOC). *See* Dutch East India Company
Vergennes, comte de, 322, 558
Vergniaud, Pierre-Victurnien, 561
Vermeer, Johannes, 136
Vernon, Edward, 519
Versailles, Treaty of (second, 1757), 325
Vervins, Peace of, 161
Vienna, Congress of, 577
Vienna, siege of, 99, 313
Vienna, Treaty of, 570
Vietnam. *See also* Annam; Champa; Dai Viet
  Chinese expansion attempts, 74
  Chinese withdrawal from, 79
  civil war, 462
  French attacks on, 465
  military strategy, 462
  war with Thailand for control of Cambodia, 462
Vigo Bay, battle of, 371, 518
Vijayanagara Empire, 240, 413–14
Villagrá, Gonzalo, 128
Vinson III, Ben, 523
Volga river, Russian control of, 496
Volynskii, Artemii, 307
*voynuks* (hired forces in the Ottoman military), 36
Vredenberg castle, 212

Wabash, battle of the, 539

*Wagenburg* tactic, 102, 296
Wagoner, Phillip, 413
*Wakō* (*Wokou*, pirates), 72–74, 436–37, 438
Waldensians, 315
Wallachia, 103, 104
Wallenstein, Albrecht von, 34, 167–68, 194
Wang Minghao, *Dengtan bijiu* (*Necessary Investigations at Mounting the Altar*), 76
Wang Zhen, 83–84
war finance, 29
  borrowing, 26, 202–3
  credit, 201–2
  debt consolidation, 203
  direct taxation, 200–1
  indirect taxation, 199–200
  role of representative institutions, 203–4
  royal revenue, 199
  subsidies, 202
War of 1812, 539, 576, 600
  naval warfare, 540–41, 544–45, 578
  operations along the Canadian border, 541–44
  outcome, 545
  preliminaries, 540–41
  war in the South, 545
War of Devolution, 323, 334
War of Jenkins' Ear, 519
War of the Austrian Succession, 136, 324–25, 352
  naval improvements following, 372–73
  naval warfare, 371–72
War of the Bavarian Succession, 325–26
War of the League of Augsburg. *See* Nine Years' War
War of the League of Cambrai, 208, 213–15
War of the League of Cognac, 158, 217
War of the Polish Succession, 324
War of the Spanish Succession, 136, 247, 273, 277, 279, 280, 323, 370
  importance of overseas possessions, 335
  issues of female inheritance and renunciation of claims, 327–28
  naval battles, 174, 371, 518
  recruitment to the Prussian armies, 360
  sieges, 224
War of the Two Iraqs, 294, 296
Warsaw, 265, 511
Warsaw, Grand Duchy of, 554
Washington, DC, British occupation, 544
Washington, George, 49, 532–33, 534–35, 537
Waterloo, battle of, 553
Watson, Charles, 246
Wayne, Anthony, 539
weaponry, 10–18. *See also* artillery; battle axes; bayonets; bows; firearms; gunpowder weapons; javelins; lances; mines; pikes; spears; swords; throwing knives
Weber, Max, 310, 319
Wei Zhongxian, 94, 96
Wellington, duke of, 385, 425, 570
  siege warfare, 226, 228, 229
West Point military academy, 539
Western Design, 368, 518
Westphalia, Peace of, 40, 41, 162, 168, 309, 310–11, 322, 339
Wettin–Habsburg entente, 165
White Lotus rebellion, 95, 485
White Mountain, battle of, 166
Whiteman, Jeremy, 558
Whydah, 395
Wilkinson, James, 543
William III, king of England (William of Orange), 42, 315, 320–21, 323, 369, 370
William IV, duke of Bavaria, 164
William IV, Dutch *stadtholder*, 331
Winius, George, 454
With, Witte de, 187
Witt, Johan de, 334
Witte, Hans de, 167
women
  Dahomey regiment of, 404
  role in European armies, 35, 197
Wood, James, 218
Worth, William, 547
Wrangel, Carl Gustaf, 344
Wu Sangui, 467, 483

Xhosa, 400
Xinjiang, 466, 467–68, 470, 481, 484, 485
Xiu, 133

Yamaga Sokō, 441
Yamamoto Tsunetomo, *Hagakure (Hidden by the Leaves)*, 440
Yandabo, Treaty of, 464
Yang Hao, Ming military commander, 96
*yangban* ("two orders," civil and military officials), 431, 436, 443, 445, 448
Yao, 401
Yellow Fever, 8, 519, 522, 549
Yerevan, 291, 303, 304, 307
Yi Sun-sin, 91, 92, 437
Yohannes, emperor of Ethiopia, 405

Yongle Emperor, 29, 32, 74, 78
  campaigns against the Mongols, 81–83
  invasion of Annam, 87–89
  relations with the Jurchens, 94
Yongzheng Emperor, 481
Yorktown, battle of, 537
Yoruba, 401, 404
Yuan dynasty, 58, 72, 86, 89, 94
  recovery of the seal of, 61
Yucatán, 133
*yusaeng* (Confucian scholar-officials), 437

*zanburak* (camel-mounted swivel gun), 299
Zanjan, 296
Zaporozhian Cossacks, 102, 106, 152, 502, 506
Zaragoza, sieges of, 225, 226
Zeelandia, fortress, 243
Zhang Xianzhong, 96
Zheng Chenggong, 243, 467
Zheng family, 467
Zheng He, 3, 32, 78, 232, 452
Zhentong Emperor, 83–84
Zimba warriors, 237
Žižka, Jan, 415
Zonchio, battle of, 172, 173, 175
Zsitvatoëroëk, Peace of, 154
Zuhab, Treaty of, 112, 304
Zulu, 402–3
Zunghars, 51, 402–3, 468, 478, 484, 500